T0191620

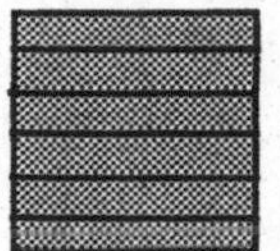

Quantum Field Theory of Point Particles and Strings

Brian Hatfield
Applied Mathematical Physics Research

CRC Press is an imprint of the
Taylor & Francis Group, an informa business

First published 1992 by Westview Press

Published 2018 by CRC Press
Taylor & Francis Group
6000 Broken Sound Parkway NW, Suite 300
Boca Raton, FL 33487-2742

ISBN-13: 978-0-201-36079-0 (pbk)

Library of Congress Cataloging-in-Publication Data
Hatfield, Brian F.
Quantum field theory of point particles and strings.
p. cm. (Frontiers in Physics, vol. 75)
Includes index.
1. Quantum field theory. 2. String models. I. Title.
II. Title: Point particles and strings. III. Series.
QC174.45.H29 1989 530.1'43--dc19 89-130
ISBN 0-201-11982-X
ISBN 0-201-36079-9 (pbk.)

This book was typeset by the author using the TEX typesetting language.

Cover design by Lynne Reed

Frontiers in Physics

David Pines, Editor

Volumes of the Series published from 1961 to 1973 are not officially numbered. The parenthetical numbers shown are designed to aid librarians and bibliographers to check the completeness of their holdings.

Titles published in this series prior to 1987 appear under either the W. A. Benjamin or the Benjamin/Cummings imprint; titles published since 1986 appear under the Addison-Wesley imprint.

1.	N. Bloembergen	Nuclear Magnetic Relaxation: A Reprint Volume, 1961
2.	G. F. Chew	S-Matrix Theory of Strong Interactions: A Lecture Note and Reprint Volume, 1961
3.	R. P. Feynman	Quantum Electrodynamics: A Lecture Note and Reprint Volume
4.	R. P. Feynman	The Theory of Fundamental Processes: A Lecture Note Volume, 1961
5.	L. Van Hove, N. M. Hugenholtz L. P. Howland	Problem in Quantum Theory of Many-Particle Systems: A Lecture Note and Reprint Volume, 1961
6.	D. Pines	The Many-Body Problem: A Lecture Note and Reprint Volume, 1961
7.	H. Frauenfelder	The Mössbauer Effect: A Review—With a Collection of Reprints, 1962
8.	L. P. Kadanoff G. Baym	Quantum Statistical Mechanics: Green's Function Methods in Equilibrium and Nonequilibrium Problems, 1962
9.	G. E. Pake	Paramagnetic Resonance: An Introductory Monograph, 1962 [cr. (42)—2nd edition]G. E. Pake
10.	P. W. Anderson	Concepts in Solids: Lectures on the Theory of Solids, 1963
11.	S. C. Frautschi	Regge Poles and S-Matrix Theory, 1963
12.	R. Hofstadter	Electron Scattering and Nuclear and Nucleon Structure: A Collection of Reprints with an Introduction, 1963
13.	A. M. Lane	Nuclear Theory: Pairing Force Correlations to Collective Motion, 1964
14.	R. Omnès M. Froissart	Mandelstam Theory and Regge Poles: An Introduction for Experimentalists, 1963
15.	E. J. Squires	Complex Angular Momenta and Particle Physics: A Lecture Note and Reprint Volume, 1963
16.	H. L. Frisch J. L. Lebowitz	The Equilibrium Theory of Classical Fluids: A Lecture Note and Reprint Volume, 1964

17. M. Gell-Mann Y. Ne'eman	The Eightfold Way (A Review—With a Collection of Reprints), 1964
18. M. Jacob G. F. Chew	Strong-Interaction Physics: A Lecture Note Volume, 1964
19. P. Nozières	Theory of Interacting Fermi Systems, 1964
20. J. R. Schrieffer	Theory of Superconductivity, 1964 (revised 3rd printing, 1983)
21. N. Bloembergen	Nonlinear Optics: A Lecture Note and Reprint Volume, 1965
22. R. Brout	Phase Transitions, 1965
23. I. M. Khalatnikov	An Introduction to the Theory of Superfluidity, 1965
24. P. G. deGennes	Superconductivity of Metals and Alloys, 1966
25. W. A. Harrison	Pseudopotentials in the Theory of Metals, 1966
26. V. Barger D. Cline	Phenomenological Theories of High Energy Scattering: An Experimental Evaluation, 1967
27. P. Choquàrd	The Anharmonic Crystal, 1967
28. T. Loucks	Augmented Plane Wave Method: A Guide to Performing Electronic Structure Calculations—A Lecture Note and Reprint Volume, 1967
29. Y. Ne'eman	Algebraic Theory of Particle Physics: Hadron Dynamics In Terms of Unitary Spin Current, 1967
30. S. L. Adler R. F. Dashen	Current Algebras and Applications to Particle Physics, 1968
31. A. B. Migdal	Nuclear Theory: The Quasiparticle Method, 1968
32. J. J. J. Kokkedee	The Quark Model, 1969
33. A. B. Migdal	Approximation Methods in Quantum Mechanics, 1969
34. R. Z. Sagdeev	Nonlinear Plasma Theory, 1969
35. J. Schwinger	Quantum Kinematics and Dynamics, 1970
36. R. P. Feynman	Statistical Mechanics: A Set of Lectures, 1972
37. R. P. Feynman	Photon-Hadron Interactions, 1972
38. E. R. Caianiello	Combinatorics and Renormalization in Quantum Field Theory, 1973
39. G. B. Field H. Arp J. N. Bahcall	The Redshift Controversy, 1973
40. D. Horn F. Zachariasen	Hadron Physics at Very High Energies, 1973
41. S. Ichimaru	Basic Principles of Plasma Physics: A Statistical Approach, 1973 (2nd printing, with revisions, 1980)
42. G. E. Pake T. L. Estle	The Physical Principles of Electron Paramagnetic Resonance, 2nd Edition, completely revised, enlarged, and reset, 1973 [cf. (9)—1st edition

Volumes published from 1974 onward are being numbered as an integral part of the bibliography.

43. R. C. Davidson	Theory of Nonneutral Plasmas, 1974
44. S. Doniach E. H. Sondheimer	Green's Functions for Solid State Physicists, 1974

45.	P. H. Frampton	Dual Resonance Models, 1974
46.	S. K. Ma	Modern Theory of Critical Phenomena, 1976
47.	D. Forster	Hydrodynamic Fluctuation, Broken Symmetry, and Correlation Functions, 1975
48.	A. B. Migdal	Qualitative Methods in Quantum Theory, 1977
49.	S. W. Lovesey	Condensed Matter Physics: Dynamic Correlations, 1980
50.	L. D. Faddeev A. A. Slavnov	Gauge Fields: Introduction to Quantum Theory, 1980
51.	P. Ramond	Field Theory: A Modern Primer, 1981 [cf. 74—2nd ed.]
52.	R. A. Broglia A. Winther	Heavy Ion Reactions: Lecture Notes Vol. I, Elastic and Inelastic Reactions, 1981
53.	R. A. Broglia A. Winther	Heavy Ion Reactions: Lecture Notes Vol. II, 1990
54.	H. Georgi	Lie Algebras in Particle Physics: From Isospin to Unified Theories, 1982
55.	P. W. Anderson	Basic Notions of Condensed Matter Physics, 1983
56.	C. Quigg	Gauge Theories of the Strong, Weak, and Electromagnetic Interactions, 1983
57.	S. I. Pekar	Crystal Optics and Additional Light Waves, 1983
58.	S. J. Gates M. T. Grisaru M. Rocek W. Siegel	Superspace *or* One Thousand and One Lessons in Supersymmetry, 1983
59.	R. N. Cahn	Semi-Simple Lie Algebras and Their Representations, 1984
60.	G. G. Ross	Grand Unified Theories, 1984
61.	S. W. Lovesey	Condensed Matter Physics: Dynamic Correlations, 2nd Edition, 1986
62.	P. H. Frampton	Gauge Field Theories, 1986
63.	J. I. Katz	High Energy Astrophysics, 1987
64.	T. J. Ferbel	Experimental Techniques in High Energy Physics, 1987
65.	T. Appelquist A. Chodos P. G. O. Freund	Modern Kaluza-Klein Theories, 1987
66.	G. Parisi	Statistical Field Theory, 1988
67.	R. C. Richardson E. N. Smith	Techniques in Low-Temperature Condensed Matter Physics, 1988
68.	J. W. Negele H. Orland	Quantum Many-Particle Systems, 1987
69.	E. W. Kolb M. S. Turner	The Early Universe, 1990
70.	E. W. Kolb M. S. Turner	The Early Universe: Reprints, 1988
71.	V. Barger R. J. N. Phillips	Collider Physics, 1987
72.	T. Tajima	Computational Plasma Physics, 1989
73.	W. Kruer	The Physics of Laser Plasma Interactions, 1988
74.	P. Ramond	Field Theory: A Modern Primer, 2nd edition, 1989 [cf. 51—1st edition]

75.	B. F. Hatfield	Quantum Field Theory of Point Particles and Strings, 1989
76.	P. Sokolsky	Introduction to Ultrahigh Energy Cosmic Ray Physics, 1989
77.	R. Field	Applications of Perturbative QCD, 1989
80.	J. F. Gunion H. E. Haber G. Kane S. Dawson	The Higgs Hunter's Guide, 1990
81.	R. C. Davidson	Physics of Nonneutral Plasmas, 1990
82.	E. Fradkin	Field Theories of Condensed Matter Systems, 1991
83.	L. D. Faddeev A. A. Slavnov	Gauge Fields, 1990
84.	R. Broglia A. Winther	Heavy Ion Reactions, Parts I and II, 1990
85.	N. Goldenfeld	Lectures on Phase Transitions and the Renormalization Group, 1992
86.	R. D. Hazeltine J. D. Meiss	Plasma Confinement, 1992
87.	S. Ichimaru	Statistical Plasma Physics, Volume I: Basic Principles, 1992
88.	S. Ichimaru	Statistical Plasma Physics, Volume II: Condensed Plasmas, 1994
89.	G. Grüner	Density Waves in Solids, 1994
90.	S. Safran	Statistical Thermodynamics of Surfaces, Interfaces, and Membranes, 1994
91.	B. d'Espagnat	Veiled Reality: An Analysis of Present Day Quantum Mechanical Concepts, 1994
92.	J. Bahcall R. Davis, Jr. P. Parker A. Smirnov R. Ulrich	Solar Neutrinos: The First Thirty Years
93.	R. Feynman F. Morinigo W. Wagner	Feynman Lectures on Gravitation
94.	M. Peskin D. Schoeder	An Introduction to Quantum Field Theory
95.	R. Feynman	Feynman Lectures on Computation
96.	M. Brack R. Bhaduri	Semiclassical Physics
97.	D. Cline	Weak Neutral Currents
98.	T. Tajima K. Shibata	Plasma Astrophysics

Editor's Foreword

The problem of communicating in a coherent fashion recent developments in the most exciting and active fields of physics continues to be with us. The enormous growth in the number of physicists has tended to make the familiar channels of communication considerably less effective. It has become increasingly difficult for experts in a given field to keep up with the current literature; the novice can only be confused. What is needed is both a consistent account of a field and the presentation of a definite "point of view" concerning it. Formal monographs cannot meet such a need in a rapidly developing field, while the review article seems to have fallen into disfavor. Indeed, it would seem that the people most actively engaged in developing a given field are the people least likely to write at length about it.

FRONTIERS IN PHYSICS was conceived in 1961 in an effort to improve the situation in several ways. Leading physicists frequently give a series of lectures, a graduate seminar, or a graduate course in their special fields of interest. Such lectures serve to summarize the present status of a rapidly developing field and may well constitute the only coherent account available at the time. Often, notes on lectures exist (prepared by the lecturer himself, by graduate students, or by postdoctoral fellows) and are distributed in mimeographed form on a limited basis. One of the principal purposes of the FRONTIERS IN PHYSICS series is to make such notes available to a wider audience of physicists.

It should be emphasized that lecture notes are necessarily rough and informal, both in style and content; those in the series will prove no exception. This is as it should be. One point of the series is to offer new, rapid, more informal, and, it is hoped, more effective ways for physicists to teach one another. The point is lost if only elegant notes qualify.

As FRONTIERS IN PHYSICS has evolved, a third category of book, the informal text/monograph, an intermediate step between lecture notes and formal texts or monographs, has played an increasingly important role in the series. In an informal text or monograph an author has reworked his or her lecture notes to the point at which the manuscript represents a coherent summation of a newly developed field, complete with references and problems, suitable for either classroom teaching or individual study.

In the past eight years, string theories have increasingly engaged the attention of theoretical physicists who work on quantum field theories and mathematicians who are interested in solving some of the very difficult mathematical problems posed by their application to particle and gravitational physics. Brian Hatfield's introductory account of string theory is intended to make the recent exciting developments in string physics accessible to a wider audience, on that includes graduate students in physics who are considering entering the field, mature physicists interested in its relation to conventional quantum field theory, and mathematicians who wish to understand the physics behind the mathematical problems that string theories pose. His book is self-contained, in that he does not assume that the reader has a background in quantum field theory. I share his hope that it will serve to introduce strings to physicists and mathematicians alike, and am pleased to welcome him to FRONTIERS IN PHYSICS.

David Pines
Urbana, Illinois
April 1991

Preface

The purpose of this book is to introduce string theory to physicists and mathematicians without assuming any background in quantum field theory. The mathematics used by physicists in quantizing string theories has kindled the interest of many mathematicians in string physics, and in particular, in the physics at the foundation of the application of their mathematics. Thus came a request to me by S.-T. Yau to conduct a seminar on quantum field theory and strings. The seminar spanned two years and was attended by both physicists and mathematicians. This book grew out of an extensive set of notes that I gave out to participants.

Part I of this book follows the development of quantum field theory for point particles, while Part II introduces strings. The choice of topics in Part I on point particles is guided by the requirements for quantizing the string as presented in Part II. All of the tools and concepts that are needed to quantize strings are developed first for point particles. So in addition to presenting the main framework of quantum field theory, the point particle section also provides a structure for a coherent development of the generalization and application of quantum field theory for point particles to the one-dimensional extended objects, strings.

Physicists jump back and forth between representations of quantum field theory, often for convenience, especially for strings. So, one must get used to thinking in these different representations and their varied techniques. All three representations of quantum field theory (operator,

Schrödinger, and path integral) are developed in this book. To emphasize the representation-independent structures of quantum field theory, I also show that one can obtain identical results for the particular field theories considered (φ^4, QED, and Yang-Mills) in each representation.

The first representation of quantum field theory considered is the operator representation (chapters 2-8). This has been the traditional approach to quantum field theory for nearly 40 years. The LSZ reduction formula relates scattering amplitudes to time-ordered products of the field operators in the vacuum. The time-ordered products or Green's functions are computed perturbatively and each term in the series can be represented by a Feynman diagram.

The remaining two representations of quantum field theory considered are functional and require functional calculus. Functional differentiation, integration, and functional differential equations are introduced in chapter 9. The Schrödinger representation is introduced in chapters 10 and 11. In this representation we solve the functional Schrödinger equation to obtain wave functionals that describe the states of the system and are field coordinate representations of state vectors. Scattering amplitudes are simply the overlap between initial and final states, the overlap being a functional integral. There is no need to introduce a reduction formula or to explicitly compute Green's functions or propagators.

The final representation considered, the path integral representation, is introduced first in the context of ordinary quantum mechanics, starting in chapter 12. The path integral naturally represents a generating functional of the time-ordered products of the field operators in the vacuum or Green's functions. The same LSZ reduction formula is used to relate the Green's functions to scattering amplitudes.

In part II, the bosonic string is quantized in several gauges in the various representations to see how the critical space-time dimension (26) and tachyon arise in each case. The treatment is most detailed in the path integral representation where the object of interest, the partition function, is a sum over random surfaces. The relevant mathematics of Riemann surfaces is covered and the reduction of higher genus sums in the critical dimension to integrals over moduli space is carried out. Holomorphic factorization and scattering amplitudes are discussed. The conformal dependence of the functional determinants leading to the conformal anomaly is evaluated explicitly. The difficulties encountered in quantizing the string in noncritical dimensions and ensuing applications are presented. Superstrings are briefly introduced, and the sum over genus 0 supersurfaces is computed. It is hoped that this book will help prepare the reader for more advanced treatments of string theory, such as the 2-volume set by Green, Schwarz, and Witten.

The emphasis in this book is calculational, and most computations are presented in step-by-step *detail*. A road map is provided around the more lengthy calculations. It is hoped that the wealth of computational detail will serve as a reference and will accelerate the transition from quantum mechanics to relativistic field theory and its functional viewpoint.

I would like to thank Shirley Enguehard, John Dawson, Ted Frankel, John Schwarz, Frank Theiss, and Yau for valuable comments and encouragement, and Allan Wylde for his patience. I am indebted to the late Dick Feynman for many valuable conversations we had, often over lunch at Fairway House.

Lexington, Mass.
June 1991

Contents

Preface ix

I. POINT PARTICLES

Chapter 1 **Introduction: Phenomenology Overview** 3

Chapter 2 **First to Second Quantization** 9

2.1 Quantum Mechanics 11
2.2 Second Quantization 20
2.3 Nonlinear Schrödinger Model 30
2.4 Exercises 33

Chapter 3 **Free Scalar Field Theory** 35

3.1 The Klein-Gordon Equation 36
3.2 Relativistic Quantum Field Theory 38
3.3 Free Scalar Field Theory 42
3.4 Charged Scalar Fields 48
3.5 Time-Ordering and Propagators 52
3.6 Exercises 56

Chapter 4 **Free Spinor Field Theory** 59

4.1 The Dirac Equation 59
4.2 Free Spinor Field Theory 69
4.3 The Dirac Propagator 72
4.4 Exercises 75

Chapter 5 **Quantization of the Electromagnetic Field** 77

5.1 Classical Gauge Theory 77
5.2 Quantization in the Coulomb Gauge 81
5.3 Quantization in the Lorentz Gauge 86
5.4 Exercises 91

Chapter 6 **Interacting Field Theories** 93

6.1 Interacting Actions 93
6.2 Discrete Symmetries 98
6.3 Asymptotic States and Perturbation Theory 106
6.4 Exercises 109

Chapter 7 **Self-Interacting Scalar Field Theory** 111

7.1 Canonical Quantization 111
7.2 IN and OUT Fields 112
7.3 Spectral Density 114
7.4 The S-Matrix 118
7.5 Reduction Formula 120
7.6 Perturbation Theory 125
7.7 2- and 4-Point Functions 144
7.8 Cross Sections 150
7.9 Exercises 154

Chapter 8 **Spinor Quantum Electrodynamics** 155

8.1 Quantization in the Coulomb Gauge 155
8.2 IN–OUT Fields and Reduction 158
8.3 Feynman Rules 163
8.4 2- and 3-Point Functions 170
8.5 Electron-Electron Scattering 174
8.6 Exercises 176

Chapter 9 **Functional Calculus** 179

9.1 Functional Derivatives 179
9.2 Functional Differential Equations 182
9.3 Functional Integrals 187
9.4 Exercises 198

Chapter 10 **Free Fields in the Schrödinger Representation** 199

10.1 Free Scalar Field Theory 200
10.2 Photon Field Theory 210
10.3 Spinor Field Theory 217
10.4 Exercises 223

Chapter 11 **Interacting Fields in the Schrödinger Representation** 225

11.1 Self-Interacting Scalar Field Theory 226
11.2 Yang-Mills 243
11.3 Spinor Quantum Electrodynamics 255
11.4 Exercises 270

Chapter 12 **Path Integral Representation of Quantum Mechanics** 271

12.1 Constructive Definition of a Path Integral in Quantum Mechanics 271
12.2 Free Particle Path Integral 281
12.3 Harmonic Oscillator Path Integral 286
12.4 Perturbation Theory 294
12.5 Exercises 300

Chapter 13 **Path Integrals in Free Field Theory** 301

13.1 Free Scalar Field Theory 302
13.2 Free Photon Field Theory 315
13.3 Free Spinor Field Theory 318
13.4 Exercises 327

Chapter 14 **Interacting Fields and Path Integrals** 329

14.1 Self-Interacting Scalar Field Theory 330
14.2 Spinor Quantum Electrodynamics 336
14.3 Exercises 340

Chapter 15 **Yang-Mills and Faddeev-Popov** 341

15.1 A Simple Example 343
15.2 QED 351
15.3 Yang-Mills 353
15.4 Exercises 359

Chapter 16 **Hiding the Infinities** 361

16.1 Overview 361
16.2 Exercises 376

Chapter 17 **Renormalization of QED at 1-Loop** 377

17.1 Preliminaries 377
17.2 Regulators 380
17.3 Renormalization 397
17.4 The Chiral Anomaly 418
17.5 Exercises 428

Chapter 18 **The Effective Action** 431

18.1 The Effective Action 431
18.2 φ^4 at 1-Loop
... from the Operator Representation 445
18.3 ... from the Path Integral Representation 453
18.4 ... from the Schrödinger Representation 456
18.5 The Renormalization Group 463
18.6 The β-Function and Asymptotic Freedom 467
18.7 Exercises 471

II. STRINGS

Chapter 19 **Basic Ideas and Classical Theory** 475

19.1 Overview 475
19.2 Classical String Theory 487
19.3 Exercises 501

Chapter 20 **First Quantization** 503

20.1 Gauge-Fixing and Constraints 503
20.2 Light-Cone Gauge 507
20.3 Covariant Gauge 525
20.4 Exercises 532

Chapter 21 **The Mathematics of Surfaces** 535

21.1 Manifolds, Structures, and Morphisms 537
21.2 Riemann Surfaces 545
21.3 Moduli Spaces 578
21.4 Exercises 601

Chapter 22 **Polyakov's Integral: The Partition Function for Genus 0** 603

22.1 Faddeev-Popov Procedure 604
22.2 Evaluation of Functional Determinants 611
22.3 Conformal Anomaly 616
22.4 Exercises 630

Chapter 23 **Higher-Genus Integrals** 631

23.1 Reduction to Moduli 632
23.2 Genus 1 639
23.3 Holomorphic Factorization 644
23.4 Exercises 652

Chapter 24 **Scattering Amplitudes** 653

24.1 Vertex Operators 653
24.2 Tachyon Tree Amplitudes 658
24.3 1-Loop Tachyon Amplitudes 664
24.4 Exercises 667

Chapter 25 **Noncritical Dimensions** 669

25.1 2-Dimensional Quantum Gravity 670
25.2 QCD and Rigid Strings 677

Chapter 26 **Introduction to Superstrings** 687

26.1 The RNS Action 687
26.2 Supersurface Action 698
26.3 Sum over Genus 0 Supersurfaces 708
26.4 Exercises 720

Bibliography 721

Index 725

To Gaelen,

Deirdre,

and Erik

I. Point Particles

CHAPTER 1

Introduction

Phenomenology Overview

All of the complex structures we see in the world are held together by four fundamental forces. On the microscopic scale, quantum field theory has been successful in describing three of the four forces: the electromagnetic, the weak, and the strong force. In these cases, the fundamental object that carries the force is a zero-dimensional point particle. The exchange of these point particles mediates the force. The fundamental particles that feel the force, that interact with the particles carrying the force, are also point-like objects. This means that the interaction between these particles necessarily occurs at one space-time point. This leads to divergences that occur in each of these theories. The success of quantum field theory rests on the fact that these divergences may be removed in a consistent and physically meaningful way.

The fourth force, gravity, has resisted quantization in this way. This failure is intimately tied to the divergences that appear when quantum field theory is applied to point particles. It now appears that gravity may be consistently quantized if we let the force be carried by a one-dimensional object, a string. All matter would be made up of strings. Just any old string theory won't do the trick. To avoid the problems associated with divergences, the theory should be finite. Mathematical consistency (no anomalies, no tachyons, ...) also restricts the form of the theory. In fact, these restrictions are so severe that only a handful of candidates presently exist. An interesting property of string theory is that the dimension of space-time is not an intrinsic property of the theory. It is possible to define consistent semi-classical expansions in any dimension up to the "critical" space-time dimension of 10. The possible unifying gauge groups that are allowed depend on the choice of space-time dimension.

The known realistic string theories are large enough to contain the other three forces (and lots more). Thus, string theories are presently the only candidates for a unified theory of all forces. So if matter is really made of strings, where are all of these strings? These strings are not very long, only about 10^{-33} cm. At the "low" energies obtainable today in accelerators (about 10^{-16} cm), these strings will look like points. At these energies, string theories can be approximated by point particle theories.

The Forces

Of the four forces, gravity and electromagnetism are the most familiar. They are long-ranged forces. This makes many of their consequences easily detectable at macroscopic scales. Gravity and electromagnetism are mainly responsible for macroscopic structures. On the largest scales gravity dominates and manifests itself through the formation of planets, stars, galaxies, clusters, and so on. On more terrestrial scales, electromagnetism produces the interesting structures such as tables, chairs, cats, etc., through solid-state and chemical forces. Since we see these forces on the macroscopic level, it is not surprising that they were the first to be described successfully by a relativistic classical (nonquantum) field theory. The electromagnetic field was also the first to be quantized and successfully applied.

The strong and the weak force have no classical analogues. They are short-ranged forces, so their presence is not obvious on the macroscopic scale. The strong coupling aspect of the strong force makes the physical classical limit not well defined.

The recognition and acceptance of the weak force had to await the discovery of radioactivity and the muon. A process that characterizes the weak force is "β-decay." For example, a free neutron will decay into a proton, an electron, and a neutrino, with a half-life of about 930 seconds. This decay is unusually slow for the weak force. Typical weak decays are more on the order of nano- to microseconds. For example, the muon decays into an electron and a couple of neutrinos with a lifetime of about 2.2 microseconds. The fact that many weak processes such as the two above can be described by nearly the same coupling strength eventually led to the acceptance of the weak interaction as an independent force.

The notion of the strong force developed only after the discovery of the atomic nucleus. The strong force was originally conceived of as the force that held the nucleons (protons and neutrons) in the nucleus together. We now recognize it as the force that holds a single nucleon together. The large strength of the coupling at low energies allows reactions to proceed more quickly. A typical strong decay has a half-life on the order of 10^{-23} seconds.

The Charges

Associated with each force is a charge. The electric charge is, of course, the most familiar. The charge associated with gravity is mass, or equivalently, energy. Since everything we know about has mass, it appears that gravity couples to everything. The removal of some divergences from quantum field theory requires us to redefine the arbitrary energy scale zero point. We cannot do this if we are trying to quantize gravity since then the energy zero point is no longer arbitrary. This is just one of the difficulties encountered in trying to quantize gravity using point particles.

The charge associated with the weak or strong force is more complicated in that it is a matrix. For the weak hadronic currents, the basis for the matrix representation of the charge is called "flavor." There are five known flavors: up, down, strange, charm, and bottom. We expect to find a sixth flavor, top, in the near future. The elements of the charge matrix are proportional to a "universal" weak coupling that plays a role similar to the electric charge. The coupling is universal in that it appears in both the weak hadronic charge and the weak leptonic charge.

The strong-force charge is called "color." Experimentally, we have never seen any free particle that has a net color. It is possible that color is "confined" and that we will never find a free particle with net color.

The Particles

The fundamental particles can be classified into two groups depending upon their intrinsic angular momentum, called spin. If the particles possess integer spin, then they are called bosons. Particles with half-integral spin are called fermions. In addition, spin 0 bosons are called scalar, spin 1 are called vector, and spin 2, tensor. Spin 1/2 fermions are called spinors, and spin 3/2 fermions are referred to as Rarita- Schwinger.

Fermions and bosons may be subgrouped by whether they feel the strong force or not. Leptons are fermions that do not participate in the strong interaction, while fermions that do are called baryons. The leptons include: the electron, muon, and τ, and their associated neutrinos. Protons and neutrons are examples of baryons. Bosons that feel the strong force are called mesons. The pion is the lightest meson. The mesons and baryons together are called hadrons.

While the leptons are fundamental particles, the hadrons are not. They are composite, bound states of spin 1/2 quarks. Mesons are made up of 2 quarks (a quark and an antiquark) and baryons are made up of three quarks. The quarks do not possess integral electric charge. They have either one-third or two-thirds of the charge of an electron. The quarks come in the three colors and six flavors. Because quarks have net color, they are confined, so we do not expect to find free quarks. Any hadron we do see in nature is a color "neutral" combination of quarks. For example, the neutron is made up of an up quark and 2 down quarks, with one of each color. The neutral pion is made up of an up quark and an up antiquark of say, red and anti-red colors.

The muon, τ lepton, and their associated neutrinos look just like copies of the electron and its neutrino except that they have larger masses. The strange and charmed quarks also look like copies of the up and down quarks except for higher masses. It seems that nature repeats itself in structure at different mass levels and we do not know why. We can combine the electron, its neutrino, the up and (corrected) down quarks, and their associated antiparticles (positron, antineutrino, and antiquarks) into a group called a generation. The muon, its neutrino, and the charmed and (corrected) strange quarks form, with their antipartners, a second generation. The τ, bottom, and (when it is found) top will form a third. Any unified theory of the strong, weak, and electromagnetic forces must explain the existence of these generations.

The Theories

The successful point particle quantum field theories describing the electric, weak and strong forces are based on quantum gauge theories. The construction of these gauge theories requires that the particles transmitting the force be spin 1 bosons, collectively called gauge bosons.

The quantum theory of (pure) electromagnetism is called Quantum Electrodynamics, or QED. The gauge boson is the massless, uncharged photon. Since the photon is massless and unconfined, the force has an infinite range. The photons are coupled to spinors that usually represent electrons and positrons.

The quantum theory of the strong force is known as Quantum Chromodynamics, or QCD. As the name suggests, it is a generalization of QED where the charge is now color. The gauge boson is the gluon. Unlike QED, where there is only one gauge boson, the photon, there are eight different gluons of various color combinations. Since the gluons are not colorless, they are supposed to be confined. Even though the gluons are massless, they are confined and cannot be seen as free particles, so the force described by QCD is short-ranged. The gluons are coupled to spinors that represent the confined quarks. Each flavor of quark comes in three colors.

The successful quantum theory of the weak force actually unifies the weak force with the electromagnetic force and is called the Weinberg-Salam model, the electroweak theory, or (sometimes) Quantum Flavordynamics. The gauge bosons for the electroweak theory are the W and Z bosons for the weak force and the photon for the electric force. The weak force is short-ranged, so the W and Z bosons, which are unconfined, are quite massive, around 80 GeV. (In the units we use, 1 GeV is 1.78×10^{-24} grams.) Pure gauge produce massless gauge bosons. To give the gauge bosons a mass and still maintain a gauge symmetry, the gauge symmetry must be "spontaneously" broken (the action is symmetric but the ground state is not). To accomplish this in the electroweak theory, a scalar field is added called the Higgs field. The Higgs is put into the theory "by hand" and has not been found experimentally. The gauge bosons are coupled to spinor fields that represent the leptons and quarks.

Even though we do not have a successful point particle theory of quantum gravity, we still have a name for the particle that carries the force, the graviton. Unlike the other gauge bosons, the graviton is spin 2. At low energy scales, gravity is so weak that we cannot expect to detect an individual graviton.

When all of the fundamental particles are treated as point-like, the interaction between these particles can only occur on contact, that is, at one space-time point. This causes divergences to appear in the quantum theory. In order to make sense of the results, we must subtract away these divergences. This is done through the renormalization program. One result of renormalization is that the couplings in a theory become scale-dependent. That is, whether a force is weak or strong depends on what energies you are looking at. In QCD, for example, the strong force is only strong at low energies. As we look at higher and higher energies, the force gets weaker and quarks begin to act like free particles. This is called asymptotic freedom. The opposite occurs in QED. As we look at higher energies, or equivalently, shorter distances, the electric charge grows in magnitude. In a grand unified theory, all of the forces have equal intrinsic couplings (strength) at some high energy.

What do we want from a unified theory? The standard model described above contains a large number of parameters that we would like to be able to calculate. In addition, the gauge groups used are dictated by experiment and there are no indications at all as to their origin. The Higgs mechanism does not occur naturally but must be put in by hand. The existence of generations of quarks and leptons is a mystery. Naturally, we would want any truly unified theory to incorporate the standard model at low energies and to answer the questions above that the standard model leaves. Gravity should be incorporated and the theory should be finite to prevent the need to introduce continuous parameters. This is no small order. It is remarkable that supersymmetric string theories presently appear to satisfy most if not all of these requirements. As we have mentioned, string theories incorporate gravity. String theories do not contain free continuous parameters (the string coupling is determined dynamically). Mathematical consistency dictates what the unifying gauge group must be. Superstring theories appear to be perturbatively finite. It also looks possible that the breaking of the gauge group at low energies to that of the standard model may be accomplished naturally and geometrically. The number of generations may also be dictated by the way the number of noncompact space-time dimensions is reduced from 26 or 10 to 4. However, nonperturbative effects of string theory are expected to be important. Presently, we have no completely satisfactory nonperturbative formulation of string theory. A complete proof of finiteness, the identification of how and why symmetries are broken, etc., must await our understanding of the nonperturbative aspects of strings.

In the first half of this book we will develop quantum field theory as applied to point particles. As you will soon discover, almost all of what we know about quantum field theory is based on perturbation theory, including renormalization. We will begin by introducing the notion of second quantization in the so-called operator formalism, and develop a particle interpretation of the resulting quantum field theory. We will then quantize free relativistic scalar, spinor, and vector fields. Interactions will be added and a perturbative solution of the interacting quantum fields will be developed based on the exact free field solutions. We will then discover divergences in the first nonclassical terms in the perturbation series.

Just as ordinary quantum mechanics has three representations, we will develop quantum field theory in the three representations: operator (Heisenberg), Schrödinger, and path integral. The latter two representations require functional calculus, which we introduce in chapter 9.

After obtaining basically the same results and divergences in all three representations, starting in chapter 16 we will face the divergences and apply the renormalization process. After this, we will be ready to apply the same ideas to strings.

Finally, a comment on units. We will use units such that the speed of light, c, and Planck's constant, $\hbar$, are equal to 1. This means that time can be measured in units of length, mass in units of energy, and energy in inverse units of length. Thus, larger energies imply smaller lengths.

CHAPTER **2**

First to Second Quantization

The term "second quantization" is a bit confusing at first. How do you quantize something that is already quantized? The answer to this question, which is the topic of this chapter, is that in second quantization you are quantizing a different thing than in first quantization. We are all familiar with first quantization from ordinary nonrelativistic point particle quantum mechanics. If the position of a classical particle is x and its momentum p, we first quantize by making x and p operators on a Hilbert space. The elements of the Hilbert space describe the possible configurations or states of the one-particle system. The coordinate representation of a state is called the wave function for the system in that state. The wave function is just a function. In the process of second quantization, we take the states of the first quantized system and make them operators. The wave "functions" are no longer functions, but operators. x and p are no longer operators, but continuous indices for our new operators. Now we have a new set of operators but no states for them to operate on. So we find a new set of states, a direct sum of Hilbert spaces, and call it a Fock space. The result is a quantum field theory.

Why do we call this system a quantum field theory? Consider again the coordinate representation of a state of the first quantized system. Call this wave function $\psi(x, t)$. It satisfies some Schrödinger equation, a partial differential operator. Now forget, for a moment, the quantum mechanical origin and interpretation of $\psi(x,t)$. $\psi(x,t)$ is just a function defined on space-time, i.e. a field. It satisfies a differential equation, i.e. a field equation. We could just as well think of the field $\psi(x,t)$ and its field equation as a classical field theory. We no longer mention any first quantization. Now when we quantize by making ψ an operator, we naturally think of it as a quantum field theory.

So we have two ways of arriving at a quantum field theory. We can quantize a classical field theory, or we can "second" quantize a first quantized system (for example, the Dirac equation). We can think of the first quantized system as just another classical field theory if we want to. If the field we are quantizing does have a first quantized interpretation, however, we might expect to see the structure of the first quantized system inside the second quantized system. We will see that this is true later in this chapter.

If we already have a first quantized system, what is the value of second quantizing it? We will see one advantage when we look at how the first quantized system emerges from the quantum field theory. If we have a first quantized system of 2 particles, then we will need 4 operators to give the position and momentum of the particles: x_1, x_2, p_1, and p_2. The wave function, $\psi_2(x_1,x_2,t)$, will satisfy some 2-body Schrödinger equation. Now suppose we want to consider adding another particle. We must come up with another 2 operators, x_3 and p_3, and new wave functions, $\psi_3(x_1,x_2,x_3,t)$. On top of that, we must also use a new (3-body) Schrödinger equation. Every time we add or drop a particle we have to produce a completely different set of equations and solutions. This is only a nuisance if the particle number is fixed, but what are we going to do if at some time, t_0, the number and types of particles change? Do we throw away one set of wave functions and Schrödinger equations at t_0 and grab another? The second quantized system has built into it an infinite set of first quantized systems, one for each particle number. Thus, the quantum field theory can conveniently treat, in a unified setting, the situation where the particle number is changing. The field operator, $\psi(x,t)$, and the operator field equation it solves, do not change in form as the particle number changes. The need to switch from first to second quantization becomes more compelling when we incorporate relativity. The hallmark of relativistic quantum theory is particle creation and destruction. The number of particles is no longer conserved. We can no longer afford to consider one N-body sector at a time.

To illustrate the process of second quantization, we will second quantize the ordinary, nonrelativistic Schrödinger equation. After second quantizing, we will develop a particle interpretation for the field theory. It is certainly not obvious ahead of time how one gets particles out of a field, $\psi(x,t)$. Once we have a particle interpretation, we will show how all of the many-body first quantized systems emerge from the field theory. However, we must not forget that fundamental particles are either fermions or bosons, so we will show how to quantize to get fermions or bosons. Finally, we will second quantize a nonlinear field theory, the Nonlinear Schrödinger model. It is integrable (exactly solvable), so we will be able to illustrate what it means to exactly solve a quantum field theory. As you may have guessed by now, in this case it means solving an infinite set of many-body Schrödinger equations, one equation for each particle number.

For completeness and continuity, and before we march on to second quantizing, we will review first quantization, that is, ordinary nonrelativistic quantum mechanics. In this book, we present quantum field theory in three representations, so that means we must review quantum mechanics in the same three representations. We will do 2 representations here: Heisenberg (operator) and Schrödinger. We will put off path integrals until we introduce functional calculus. As an example, we will consider the harmonic oscillator. It is simple and we will use it over and over again in field theory.

2.1 QUANTUM MECHANICS

The object of interest to calculate in any quantum mechanical process is the probability amplitude. It is a complex number. The square of the modulus of the amplitude is proportional to the probability that the process will occur. The "laws" of quantum mechanics are rules on how to calculate the amplitude for any process.

For a given physical system, quantum mechanics postulates that there is a complete set of states that describes all of the possible configurations of the system. A state will be denoted as $|\,\psi\rangle$. It is a vector in some Hilbert space. $c|\,\psi\rangle$, c a constant, represents the same physical state. The inner product of two states, $|\,\psi_1\rangle$ and $|\,\psi_2\rangle$, is written as $\langle\psi_2\,|\,\psi_1\rangle$. The amplitude for the process that takes the system from the state $|\,\psi_1\rangle$ to $|\,\psi_2\rangle$ is just $\langle\psi_2\,|\,\psi_1\rangle$. The probability for this process to occur is $|\langle\psi_2\,|\,\psi_1\rangle|^2$.

Another postulate of quantum mechanics is that with every physical observable, there is associated an operator. The operator acts on the Hilbert space containing the states. If a measurement is made of the observable, the result of the measurement must be one of the eigenvalues of the operator. No other values are possible. Since the eigenvalues are the result of physical measurements, they must be real. This means that the operators must be Hermitian. If the eigenvalues of some operator $\mathbf{Z}$ for a physical quantity z are just ± 1, then if we measure z, the only possible result is either $+1$ or -1 no matter what state the system is in. We cannot get $1/2$, $1/3$, 2π, etc. The expectation value of the physical quantity z in the state $|\,\psi_1\rangle$ is given by $\langle\psi_1\,|\mathbf{Z}|\,\psi_1\rangle$, where the state vectors are normalized to $\langle\psi_1\,|\,\psi_1\rangle = 1$. This expected value may range between -1 and $+1$. It is not the result of an experimental measurement of z. It is only the expected value. If we made the measurement on a large number of identical systems, all in the state $|\psi_1\rangle$, the result of any one measurement would be $+1$ or -1, but the average value of all the measurements is $\langle\psi_1|\mathbf{Z}|\psi_1\rangle$.

The position of the particle is a physical observable. Let the associated operator be **X**. The eigenvectors of **X** satisfy

$$\mathbf{X}|\,x\rangle = x|\,x\rangle. \tag{2.1}$$

We assume they are complete and form a basis for the Hilbert space of states. The eigenvectors are normalized so that $\langle x'\,|\,x\rangle = \delta(x'-x)$, where δ is the Dirac distribution. (Technically, this normalization means that $|\,x\rangle$ is not an element of any Hilbert space. We are not going to worry about this and will just act as if they are in a Hilbert space. To avoid this snag, we could consider the system to be contained in a very large, finite-volume box instead of an infinite volume. However, since the volume does not enter into physical results, the normalization above presents no difficulties.)

The notation for the eigenvectors can at times be rather confusing. We will use boldface for operators whenever possible. **X** is the position operator, while x is the eigenvalue associated with the eigenvector $|\,x\rangle$. We will often label eigenstates of an operator by their eigenvalues. $|\,x'\rangle$ is the eigenvector of operator **X** with eigenvalue x', while $|-1\rangle$ is the eigenvector of **Z** with eigenvalue -1, and so on. If we expand a state vector, $|\,\psi(t)\rangle$, in the position basis, $|x\rangle$, the component of $|\,\psi(t)\rangle$ in the $|\,x\rangle$ direction is $\langle x\,|\,\psi(t)\rangle$, which is just a number, $\psi(x,t)$. $\psi(x,t)$ is the wave function representing the state $|\,\psi(t)\rangle$.

$$|\,\psi(t)\rangle = \int dx\,\langle x\,|\,\psi(t)\rangle\,|\,x\rangle \tag{2.2}$$

From this expansion, we can see that our assumption of completeness of the basis $|\,x\rangle$ means that

$$\int dx\,|\,x\rangle\,\langle x\,| = 1\,. \tag{2.3}$$

We will use equation (2.3) often.

$|\,x\rangle$ is the state of the system where the particle is at position x. $\langle x\,|\,\psi(t)\rangle = \psi(x,t)$ is the amplitude for the system in state $|\,\psi(t)\rangle$ to also be in the state $|\,x\rangle$. Thus, $|\psi(x,t)|^2$ is the probability that the particle will be found at position x when the system is in state $|\,\psi(t)\rangle$. Since the particle must be found somewhere, $\int |\psi(x,t)|^2\,dx = 1$, that is, the states are normalized so that $\langle\psi(t)\,|\,\psi(t)\rangle = 1$.

So far we have no dynamics, no time evolution. Quantum mechanics provides for time evolution by requiring the undisturbed states of the system to satisfy the Schrödinger equation,

$$i\frac{\partial}{\partial t}\,|\,\psi(t)\rangle = \mathbf{H}\,|\,\psi(t)\rangle\,. \tag{2.4}$$

H is called the Hamiltonian operator and generates infinitesimal time translations. The formal solution to the Schrödinger equation is

$$|\,\psi(t)\rangle = e^{-i\mathbf{H}(t-t_0)}\,|\,\psi(t_0)\rangle\,. \tag{2.5}$$

If the system is initially, at time t_0, in the state $|\,\psi(t_0)\rangle$, the time evolution operator, $\mathbf{U}(t, t_0) = \exp(-i\mathbf{H}(t-t_0))$, evolves the undisturbed system in state $|\,\psi(t_0)\rangle$ to time t, where the system will now be in the state $|\,\psi(t)\rangle$.

Where do we get this Hamiltonian operator? If we have a classical Hamiltonian system to quantize, with a classical Hamiltonian, $H(x,p)$, we may obtain the Hamiltonian operator using $H(x,p)$ as a guide. Since both the position and momentum of a particle are observable quantities, upon quantization, they become operators, **X** and **P**. We can turn the classical Hamiltonian into an operator by replacing x with **X** and p with **P**. To complete the quantization, we must specify quantum conditions for the operators **X** and **P**. **X** and **P** must satisfy

$$[\mathbf{X},\mathbf{P}] = i\,, \tag{2.6}$$

where $[\mathbf{X},\mathbf{P}] = \mathbf{XP} - \mathbf{PX}$ is the commutator of **X** and **P**. The operators of the quantum system are defined such that their commutators are equal to i times the corresponding classical Poisson bracket. That is, we can obtain the quantum conditions for two operators from the classical system by multiplying the Poisson bracket by i and replacing the bracket with a commutator.

If the quantum mechanical system we are trying to describe has no classical analog, such as a system with spin, then we must guess what the Hamiltonian operator is and what the quantum conditions are. In the case where we do have a classical system, the quantum conditions tell us that **X** and **P** do not commute at equal times, so potential operator ordering problems arise in constructing the Hamiltonian operator from the classical Hamiltonian. For example, operator products **XP** and **PX** are inequivalent while xp and px are the same in the classical system. In general, we choose a symmetric ordering, $xp \rightarrow (\mathbf{XP}+\mathbf{PX})/2$. **X** and **P** commute with themselves, $[\mathbf{X},\mathbf{X}] = [\mathbf{P},\mathbf{P}] = 0$, so there is no operator ordering problem with x^2 or p^2.

To work with wave functions, we must use the coordinate representation provided by the position operator eigenvectors, $|\, x\rangle$. As we stated earlier, $\langle x \,|\, \psi(t)\rangle = \psi(x,t)$ is the wave function for state $|\,\psi(t)\rangle$. We use equation (2.3) to obtain a representation for the inner product, $\langle \psi(t') \,|\, \psi(t)\rangle$.

$$\begin{aligned}\langle \psi(t') \,|\, \psi(t)\rangle &= \int dx \langle \psi(t') \,|\, x\rangle\langle x \,|\, \psi(t)\rangle \\ &= \int dx\, \psi^*(x,t')\psi(x,t)\end{aligned} \tag{2.7}$$

The Schrödinger equation becomes

$$\begin{aligned}i\frac{\partial}{\partial t}\langle x \,|\, \psi(t)\rangle &= \langle x \,|\mathbf{H}|\, \psi(t)\rangle \\ &= \int dx' \langle x \,|\mathbf{H}|\, x'\rangle\langle x' \,|\, \psi(t)\rangle\,,\end{aligned}$$

or

$$i\frac{\partial}{\partial t}\psi(x,t) = \int dx' \langle x \,|\mathbf{H}|\, x'\rangle \psi(x',t). \tag{2.8}$$

$\langle x \,|\mathbf{H}|\, x'\rangle$ is called the matrix element of the operator $\mathbf{H}$ in the position basis. To compute $\langle x \,|\mathbf{H}|\, x'\rangle$ we must determine the matrix elements of $\mathbf{X}$, $\langle x \,|\mathbf{X}|\, x'\rangle$, and the momentum $\mathbf{P}$, $\langle x \,|\mathbf{P}|\, x'\rangle$. Since $|\, x\rangle$ is an eigenvector of $\mathbf{X}$, $\langle x \,|\mathbf{X}|\, x'\rangle = x\delta(x - x')$. The delta function tells us that the matrix representing $\mathbf{X}$ in this basis is diagonal. For $\langle x \,|\mathbf{P}|\, x'\rangle$, we turn to the quantum conditions, $[\mathbf{X}, \mathbf{P}] = i$. Since $[x, \frac{\partial}{\partial x}] = -1$, $-i\frac{\partial}{\partial x}$ serves as a representation for $\mathbf{P}$, hence

$$\langle x \,|\mathbf{P}|\, x'\rangle = -i\frac{\partial}{\partial x}\delta(x - x')\,. \tag{2.9}$$

If the Hamiltonian is

$$\mathbf{H} = \frac{1}{2}\mathbf{P}^2 + V(\mathbf{X})\,, \tag{2.10}$$

then

$$\langle x \,|\mathbf{H}|\, x'\rangle = (-\frac{1}{2}\frac{\partial^2}{\partial x^2} + V(x))\,\delta(x - x')\,. \tag{2.11}$$

Substituting eq.(2.11) into eq.(2.8), we recover the familiar Schrödinger operator for the wave function.

So far we have been describing the Schrödinger (coordinate) representation or picture. The states of the system are time dependent; the operators are not. In the Heisenberg (operator) picture, the states are time independent. The operators carry the time dependence. The results obtained in either representation are equivalent.

The transformation between the Schrödinger and Heisenberg pictures is done with the time evolution operator. Recall that the solution of the Schrödinger equation is

$$|\,\psi(t)\rangle = e^{-i\mathbf{H}t}\,|\,\psi(0)\rangle\,. \tag{2.12}$$

$|\,\psi(0)\rangle$ is the initial state of the system at a fixed time, $t = 0$, so it has no time dependence and can be used in the Heisenberg representation. $|\,\psi(t)\rangle$ is time dependent so it belongs to the Schrödinger representation. If we apply $\exp(i\mathbf{H}t)$ to both sides, then

$$e^{i\mathbf{H}t}|\,\psi(t)\rangle = |\,\psi(0)\rangle\,. \tag{2.13}$$

The right-hand side is time independent, so the left-hand side must also be independent. Application of the operator $\exp(i\mathbf{H}t)$ has converted a state in the Schrödinger picture to one in the Heisenberg picture.

To find the transformation for operators, we note that we want to obtain the same results in both representations. In particular, the expectation of any operator, $\mathbf{Z}$, should be the same. Starting in the Schrödinger picture, we have

$$\begin{aligned}\langle\psi(t)\,|\mathbf{Z}_s|\,\psi(t)\rangle &= \langle\psi(0)\,|e^{i\mathbf{H}t}\,\mathbf{Z}_s\,e^{-i\mathbf{H}t}|\,\psi(0)\rangle \\ &\equiv \langle\psi(0)\,|\mathbf{Z}_h(t)|\,\psi(0)\rangle\,.\end{aligned} \tag{2.14}$$

Hence, the time dependent operators in the Heisenberg picture are related to the Schrödinger picture operators by

$$\mathbf{Z}_h(t) = e^{i\mathbf{H}t}\,\mathbf{Z}_s e^{-i\mathbf{H}t}\,. \tag{2.15}$$

Since the states are now time independent, they do not have to satisfy any Schrödinger equation. The dynamics is locked into the time dependent operators. They must satisfy an operator equation of motion, which can be obtained by considering small translations in time, from t to $t+\epsilon$, ϵ small, and expanding (2.15) to first order in ϵ.

$$\begin{aligned}\mathbf{Z}(t+\epsilon) &= e^{i\mathbf{H}\epsilon}\,\mathbf{Z}(t)\,e^{-i\mathbf{H}\epsilon} \\ &= (1+i\mathbf{H}\epsilon)\,\mathbf{Z}(t)\,(1-i\mathbf{H}\epsilon) \\ &= \mathbf{Z}(t) + i\epsilon\,[\mathbf{H},\mathbf{Z}] + \ldots\end{aligned}$$

Expanding the left-hand side to the same order, we have

$$\mathbf{Z}(t+\epsilon) = \mathbf{Z}(t) + \epsilon\frac{d\mathbf{Z}}{dt} + \ldots\,.$$

Equating powers of ϵ yields

$$\frac{d\mathbf{Z}}{dt} = i[\mathbf{H}, \mathbf{Z}]. \tag{2.16}$$

If the operator $\mathbf{Z}$ has an explicit time dependence, then we must add $\partial\mathbf{Z}/\partial t$ to the right-hand side of (2.16). This equation of motion may also be obtained from the classical system by replacing i times the Poisson bracket with a commutator, just as in the quantum conditions. Note that if $\mathbf{Z}$ commutes with $\mathbf{H}$, it is a constant of the motion. So to quantize a classical system in the Heisenberg picture, we elevate observables to operators, specify quantum conditions (now since $\mathbf{X}$ and $\mathbf{P}$ are time dependent, we specify the commutator at equal times, i.e. $[\mathbf{X}(t'), \mathbf{P}(t)] = i$ at $t = t'$), choose an operator ordering, and specify a Hilbert space containing the time independent state vectors. To compute the quantum dynamics, we must solve the operator equations of motion.

It is time to do an explicit example to illustrate the two representations. A simple, but very useful example is the harmonic oscillator. The classical Hamiltonian for the harmonic oscillator is

$$H = \frac{1}{2}p^2 + \frac{1}{2}\omega^2 x^2. \tag{2.17}$$

To quantize this, we make x and p the operators $\mathbf{X}$ and $\mathbf{P}$, and require them to satisfy the quantum conditions, $[\mathbf{X}, \mathbf{P}] = i$.

We will solve this system in the Heisenberg picture first. Let's choose the set of states spanned by the eigenvectors of the Hamiltonian operator. We could "sandwich" the operator equations of motion for $\mathbf{X}$ and $\mathbf{P}$ in these states, that is, compute $\langle\psi_1 |[\mathbf{H}, \mathbf{X}]| \psi_2\rangle$ and $\langle\psi_1 |[\mathbf{H}, \mathbf{P}]| \psi_2\rangle$, to determine the spectrum of $\mathbf{H}$ using the quantum conditions. But instead, we can do a simple change of variables that simplifies the analysis. Define the operator $\mathbf{a}$ as

$$\mathbf{a} = \frac{1}{\sqrt{2}}(\sqrt{\omega}\mathbf{X} + \frac{i}{\sqrt{\omega}}\mathbf{P}). \tag{2.18}$$

The Hermitian conjugate is

$$\mathbf{a}^\dagger = \frac{1}{\sqrt{2}}(\sqrt{\omega}\mathbf{X} - \frac{i}{\sqrt{\omega}}\mathbf{P}). \tag{2.19}$$

Since $\mathbf{a} \neq \mathbf{a}^\dagger$, $\mathbf{a}$ is not Hermitian, so $\mathbf{a}$ does not represent an observable. Computing $[\mathbf{a}, \mathbf{a}^\dagger]$ using the quantum conditions, $[\mathbf{X}, \mathbf{P}] = i$ and $[\mathbf{X}, \mathbf{X}] = [\mathbf{P}, \mathbf{P}] = 0$, we find

$$\begin{aligned}[\mathbf{a}, \mathbf{a}] = [\mathbf{a}^\dagger, \mathbf{a}^\dagger] = 0,\\ [\mathbf{a}, \mathbf{a}^\dagger] = 1.\end{aligned} \tag{2.20}$$

If we substitute $\mathbf{X} = \sqrt{2/\omega}(\mathrm{a} + \mathrm{a}^\dagger)$ and $\mathbf{P} = i\sqrt{2\omega}(\mathrm{a}^\dagger - \mathrm{a})$ into the Hamiltonian operator, and use eq.(2.20) once we find

$$\mathbf{H} = \omega(\mathrm{a}^\dagger \mathrm{a} + \frac{1}{2}). \tag{2.21}$$

Now it is easy to see that if a state $|\,\psi\rangle$ is an eigenvector of $\mathbf{H}$, then it must be an eigenvector of the operator $\mathbf{N} = \mathrm{a}^\dagger \mathrm{a}$. This operator is called the number operator. Let us label the eigenvectors of $\mathbf{N}$ by their eigenvalues,

$$\mathbf{N}|\,n\rangle = n\,|\,n\rangle, \tag{2.22}$$

so that

$$\mathbf{H}|\,n\rangle = \omega\,(n + \frac{1}{2})\,|\,n\rangle. \tag{2.23}$$

Next, consider the norm of the state $\mathrm{a}|\,n\rangle$. The norm (length of the vector) must be greater than or equal to zero.

$$||\,\mathrm{a}|\,n\rangle||^2 \geq 0 \tag{2.24}$$

But

$$||\,\mathrm{a}|\,n\rangle||^2 = (\mathrm{a}|\,n\rangle)^\dagger\,(\mathrm{a}|\,n\rangle) = \langle n\,|\mathrm{a}^\dagger\,\mathrm{a}|\,n\rangle = n\langle n\,|\,n\rangle, \tag{2.25}$$

so

$$n\,\langle n\,|\,n\rangle \geq 0. \tag{2.26}$$

$\langle n\,|\,n\rangle$ is again the square of a norm of a vector and must be greater than or equal to zero. But the only vector of norm zero is the null vector (vector of length zero), which we are not interested in. It does not contribute any probability. Thus $n \geq 0$. We will normalize the states $|\,n\rangle$ such that $\langle n\,|\,n\rangle = 1$. The state of lowest energy, the ground state, is $n = 0$. The ground state energy, from equation (2.23), is $\omega/2$. The first excited state has $n = 1$, the second $n = 2$, etc. The spectrum of $\mathbf{H}$ is $E_n = \omega(n + 1/2)$. The energy levels are evenly spaced with separation ω.

Using eq.(2.25) we see that $\mathrm{a}|\,0\rangle$ is a vector of zero norm, so it must be the null vector,

$$\mathrm{a}\,|\,0\rangle = 0. \tag{2.27}$$

To reach the excited states from the ground state, $|\,0\rangle$, we observe that

$$[\mathbf{N}, \mathrm{a}^\dagger] = \mathrm{a}^\dagger, \tag{2.28}$$

where we have made use of the quantum conditions, eq.(2.20), to compute this commutator. Thus,

$$\begin{aligned} \mathbf{N}\,\mathrm{a}^\dagger\,|\,0\rangle &= (\mathrm{a}^\dagger\,\mathbf{N} + [\mathbf{N}, \mathrm{a}^\dagger])|\,0\rangle \\ &= \mathrm{a}^\dagger\,|\,0\rangle. \end{aligned} \tag{2.29}$$

$a^\dagger|0\rangle$ is an eigenvector of N with eigenvalue 1. $a^\dagger|0\rangle = |1\rangle$. What about $a^\dagger|1\rangle$, $a^\dagger|2\rangle$, and so on?

$$\begin{aligned} N\,a^\dagger\,|n\rangle &= (a^\dagger N + [N, a^\dagger])|n\rangle \\ &= (n+1)a^\dagger\,|n\rangle \\ &= \lambda\,|n+1\rangle \end{aligned} \tag{2.30}$$

where λ is a constant that we will determine in a moment. What about a? From $[N, a] = -a$, we find that

$$a\,|n\rangle = \mu\,|n-1\rangle. \tag{2.31}$$

All of the eigenstates of H can be generated by applying $a^\dagger$ consecutively to $|0\rangle$. The constant λ in equation (2.30) is fixed by making sure the states generated are properly normalized. All states should have unit length. We start by assuming $\langle 0\,|\,0\rangle = 1$. From this it follows that

$$\langle 1\,|\,1\rangle = \langle 0\,|a\,a^\dagger|\,0\rangle = 1$$

where, once again, the quantum conditions, eq.(2.20), are used to do the computation. Similarly,

$$\langle 0\,|a\,a\,a^\dagger\,a^\dagger|\,0\rangle = 2,$$

so the properly normalized state of two excitations is

$$|2\rangle = \frac{1}{\sqrt{2}}\,a^\dagger\,a^\dagger\,|0\rangle.$$

In general,

$$|n\rangle = \frac{1}{\sqrt{n!}}(a^\dagger)^n\,|0\rangle. \tag{2.32}$$

The Schrödinger representation for the harmonic oscillator can be found by solving the Schrödinger equation, eqs. (2.8) and (2.11), using the Hamiltonian, eq.(2.17). It is easier, however, to find the coordinate representation of $|0\rangle$ and apply the transformation (2.12). Let $\psi_0(x,0) = \langle x\,|\,0\rangle$. The state $|0\rangle$ is determined by equation (2.27),

$$a\,|0\rangle = 0.$$

In the coordinate representation,

$$\int dx'\,\langle x\,|a|\,x'\rangle\langle x'\,|\,0\rangle = 0,$$

which is

$$\left(\frac{1}{\sqrt{2}}\sqrt{\omega}x - \frac{1}{\sqrt{\omega}}\frac{\partial}{\partial x}\right)\psi_0(x,0) = 0. \tag{2.33}$$

The solution is

$$\psi_0(x,0) = \left(\frac{2}{\omega}\right)^{\frac{1}{4}} e^{-\omega x^2/2} \tag{2.34}$$

where the normalization was determined from $\langle 0\,|\,0\rangle = 1$. All of the excited state wave functions may be obtained by applying the coordinate representation of equation (2.32). Since $\psi_0(x,0)$ is an eigenstate of $\mathbf{H}$, the computation of $\psi_0(x,t)$ is straightforward.

$$\begin{aligned}\psi_0(x,t) &= e^{-i\mathbf{H}t}\,\psi_0(x,0)\\ &= e^{-i\omega t/2}\,\psi_0(x,0)\end{aligned} \tag{2.35}$$

The operators $\mathbf{a}^\dagger$ and $\mathbf{a}$ are sometimes called ladder operators, or raising and lowering operators, because they make states that march up and down the ladder of excitations. As we shall soon see, free quantum field theories reduce to a collection of independent harmonic oscillators, one for each energy-momentum. The raising and lowering operators of the collection of oscillators provide a particle interpretation. To illustrate this now, let's reinterpret the raising and lowering operators above as particle creation and destruction operators in the following way: Redefine the Hamiltonian, eq.(2.21), by subtracting the constant 1/2 from it, and interpret ω as the mass of a particle. The state $|\,0\rangle$ has no energy. It is empty, the vacuum. The state $|\,1\rangle$ has energy $\omega = m$ so it contains 1 particle, say, at rest. The state $|\,2\rangle$ has energy $2m$ so it contains 2 particles at rest, and so on. $\mathbf{a}^\dagger$ acting on the state $|\,1\rangle$ changes it into $|\,2\rangle$. The energy of the state has increased by m. This means that the operator $\mathbf{a}^\dagger$ has created a particle. Similarly, $\mathbf{a}$ acting on $|\,2\rangle$ changes the state to $|\,1\rangle$. The energy has decreased by m. $\mathbf{a}$ destroys a particle. The number operator, $\mathbf{N} = \mathbf{a}^\dagger\mathbf{a}$, now counts the number of particles in each state. If $\mathbf{a}$ acts on the vacuum, $|\,0\rangle$, where there are no particles to destroy, the result should be zero, (2.27). To find a particle interpretation of any quantum field we look for operators similar to $\mathbf{a}$ and $\mathbf{a}^\dagger$ in the quantum field theory.

So far we have worked with one spatial dimension only. To keep things simple in this chapter, we will continue using only one spatial dimension. In the next chapter we will switch to three spatial dimensions.

2.2 SECOND QUANTIZATION

Let's now second quantize the Schrödinger equation,

$$i\frac{\partial\varphi}{\partial t} = -\frac{1}{2}\frac{\partial^2}{\partial x^2}\varphi + V(x)\,\varphi, \tag{2.36}$$

where $0 \leq x \leq L$. We will assume periodic boundary conditions. To second quantize, we want to make the first quantized wave function an operator. The field operator, $\varphi(x,t)$, is time dependent, hence we will be working in the Heisenberg or operator formalism. Equation (2.36) will then become an operator equation of motion. In quantum mechanics, operator equations of motion take the form of equation (2.16),

$$\frac{d}{dt}\varphi = i[\mathbf{H}, \varphi]. \tag{2.37}$$

Therefore, we must find a field Hamiltonian, $\mathbf{H}$, and quantum conditions (equal-time commutators) such that equation (2.37) reproduces equation (2.36).

The commutator is the field equivalent of $[\mathbf{X}(t), \mathbf{P}(t)] = i$ for the first quantized system. In this field theory, $\varphi(x,t)$ plays the role of $\mathbf{X}$. To write down the quantum commutator involving φ, we must find what the field momentum conjugate to φ is. We could guess, or we can treat the first quantized system as a classical field theory, find an action principle that yields the field equations, and use the action as a crutch to find the field momentum and Hamiltonian.

We will follow here the action principle path because it is used quite frequently, especially in this book. Often we will just start by writing down an action and then finding the equations of motion, instead of the reverse that we are doing here. The use of an action also allows us to easily connect symmetries with conserved currents.

The action, $S[\varphi]$, is a functional of the fields. This means that the action, S, maps the function $\varphi(x,t)$ into the real numbers. The action takes a specific function on space-time and turns it into a number. We use square brackets, $[\cdots]$, to denote the functional nature of S. S is also written as

$$S[\varphi] = \int_{t_i}^{t_f} dt \int_0^L dx\, \mathcal{L}[\varphi], \tag{2.38}$$

where $\int dx\, \mathcal{L}[\varphi]$ is the Lagrangian and $\mathcal{L}$ is the Lagrangian density. The field equations are obtained by varying the action with respect to the fields and setting the variation to zero. If we vary the action by varying the field φ, keeping $\varphi(x,t_i)$ and $\varphi(x,t_f)$ fixed, then

$$\delta S = \int_{t_i}^{t_f} dt \int dx\, \frac{\partial\mathcal{L}}{\partial(\partial_\mu\varphi)}\,\delta\,\partial_m\varphi + \frac{\partial\mathcal{L}}{\partial\varphi}\,\delta\varphi, \tag{2.39}$$

where $\partial_\mu = \partial/\partial x^\mu$, ($x^0 = t$). $\delta\, \partial_\mu \varphi = \partial_\mu\, \delta\varphi$. Integrating the first term by parts (the surface term will vanish because φ is fixed at t_i and t_f), we have

$$\delta S = \int_{t_i}^{t_f} dt \int dx \left(-\partial_\mu \left(\frac{\partial \mathcal{L}}{\partial(\partial_\mu \varphi)} \right) + \frac{\partial \mathcal{L}}{\partial \varphi} \right) \delta\varphi. \tag{2.40}$$

The variation of the action, δS, will vanish for arbitrary variations in φ if the integrand vanishes, namely

$$\partial_\mu \left(\frac{\partial \mathcal{L}}{\partial(\partial_\mu \varphi)} \right) - \frac{\partial \mathcal{L}}{\partial \varphi} = 0. \tag{2.41}$$

We are employing the usual convention where any repeated index is summed over.

The field momentum conjugate to φ is

$$\pi(x,t) = \frac{\partial \mathcal{L}}{\partial(\partial_t \varphi(x,t))}. \tag{2.42}$$

The field Hamiltonian is given by the Legendre transform of the Lagrangian,

$$H = \int dx\, (\pi(x,t) \cdot \partial_t \varphi(x,t) - \mathcal{L}). \tag{2.43}$$

If we are given a Lagrangian, we may quickly find the field equations from eq.(2.41), the Hamiltonian from eq.(2.43), and quantize by making $\varphi(x,t)$ and its conjugate momentum, $\pi(x,t)$, operators satisfying equal-time commutation relations,

$$[\varphi(x,t), \pi(x',t)] = i\delta(x - x'). \tag{2.44}$$

In the absence of difficult operator ordering ambiguities, the field Hamiltonian operator and quantum conditions, eq.(2.44), will reproduce the equations of motion, eq.(2.41), as an operator equation of motion, eq.(2.36).

To find the Lagrangian given the equations of motion, we could work backwards from eq.(2.41) to eq.(2.40) and eq.(2.39) by multiplying the field equations by $\delta\varphi$ and integrating. This will lead to some difficulty when we start with the Schrödinger equation, eq.(2.36), because the equation is first-order in the time derivative.

Instead of working backwards, we will motivate our choice of Lagrangian by observing that a conservation equation can be derived from the Schrödinger equation. This conservation equation tells us that there is a conserved current, here the probability current. The existence of conserved currents implies that the Lagrangian must possess some symmetry, that is, that the Lagrangian is invariant under some symmetry transformation. Let us quickly show that symmetries in the Lagrangian lead to conserved currents. This is known as Nöther's theorem.

Suppose that S is invariant under a transformation of φ where $\varphi \to \varphi + \epsilon$. Now we proceed to compute δS with respect to the variation in the field due to the symmetry transformation, $\delta\varphi = \epsilon\varphi$. $\delta S = 0$ because the action is invariant under the transformation. We have

$$\begin{aligned} \delta S = 0 &= \int dt \int dx \, \frac{\partial \mathcal{L}}{\partial(\partial_\mu \varphi)} \partial_\mu(\delta\varphi) + \frac{\partial \mathcal{L}}{\partial \varphi}\delta\varphi \\ &= \int dt \int dx \, \frac{\partial \mathcal{L}}{\partial(\partial_\mu \varphi)} \partial_\mu(\epsilon\varphi) + \frac{\partial \mathcal{L}}{\partial \varphi}\epsilon\varphi. \end{aligned} \tag{2.45}$$

Using the field equations, eq.(2.41), substitute in for $\partial\mathcal{L}/\partial\varphi$,

$$\begin{aligned} 0 &= \int dt \int dx \, \frac{\partial \mathcal{L}}{\partial(\partial_\mu \varphi)} \partial_\mu(\epsilon\varphi) + \partial_\mu \left(\frac{\partial \mathcal{L}}{\partial(\partial_\mu \varphi)} \right) \epsilon\varphi \\ &= \int dt \int dx \, \epsilon \partial_\mu \left(\frac{\partial \mathcal{L}}{\partial(\partial_\mu \varphi)} \varphi \right). \end{aligned} \tag{2.46}$$

Once again, the integrand must vanish for any ϵ, thus the current,

$$j^\mu = \frac{\partial \mathcal{L}}{\partial(\partial_\mu \varphi)} \varphi, \tag{2.47}$$

is conserved, $\partial_\mu j^\mu = 0$. Each symmetry of S leads to a conserved current.

Now, let's derive the conserved current from the Schrödinger equation, eq.(2.36). After making the "obvious" guess as to what the corresponding symmetry transformation is, we can work backwards through eq.(2.47) to obtain the Lagrangian.

To obtain an expression for the current from the Schrödinger equation, we start by multiplying it by $\varphi^*(x,t)$, the complex conjugate of $\varphi(x,t)$.

$$i\varphi^*(x,t)\frac{\partial \varphi}{\partial t}(x,t) = -\frac{1}{2}\varphi^*(x,t)\frac{\partial^2}{\partial x^2}\varphi(x,t) + V(x)\varphi^*(x,t)\varphi(x,t) \tag{2.48}$$

Next, take the complex conjugate of equation (2.48) and subtract it from (2.48). Assuming the $V(x)$ is a real function of x, the result is

$$\begin{aligned} i\left(\varphi^*\frac{\partial \varphi}{\partial t} + \varphi\frac{\partial \varphi^*}{\partial t}\right) &= -\frac{1}{2}\left(\varphi^*\frac{\partial^2}{\partial x^2}\varphi - \varphi\frac{\partial^2}{\partial x^2}\varphi^*\right) \\ &= -\frac{1}{2}\frac{\partial}{\partial x}\left(\varphi^*\frac{\partial}{\partial x}\varphi - \varphi\frac{\partial}{\partial x}\varphi^*\right). \end{aligned} \tag{2.49}$$

The left-hand side above is the time derivative of $\varphi^*\varphi$, the probability density. Thus, equation (2.49) can be put into the form $\partial_\mu j^\mu = 0$ where $(x^0 = t,\ x^1 = x)$,

$$j^0 = \varphi^*\varphi, \tag{2.50}$$

and

$$j^1 = \frac{i}{2}\left(\varphi^*\frac{\partial}{\partial x}\varphi - \varphi\frac{\partial}{\partial x}\varphi^*\right). \tag{2.51}$$

We have a conserved current that involves the field φ and its complex conjugate. This means that the field φ^* will also appear in the Lagrangian.

What symmetry transformation in the Lagrangian gives us the conserved current? If you recall that $\varphi(x,t)$ originally represented a wave function for a first quantized system, then we know that we can multiply $\varphi(x,t)$ by an arbitrary phase, $\exp(i\alpha)$, where α is a constant, and all of the physical results will be unchanged. Also note that j^0 and j^1 above are independent of phase. So, it is reasonable to try to find a Lagrangian that is invariant under the transformation $\varphi \to \exp(i\alpha)\,\varphi$, $\varphi^* \to \exp(-i\alpha)\,\varphi^*$ and generates the conserved current above, eqs.(2.50)-(2.51), as a result of the symmetry.

Now that we know that the Lagrangian will depend on φ and φ^*, we can rederive the dependence of j^μ on $\mathcal{L}$ in terms of φ and φ^*, following the steps that led to eq.(2.47). Given the expression for j^μ, we can work backwards to find the Lagrangian. The resulting Lagrangian density is

$$\mathcal{L} = \frac{i}{2}(\varphi^*\partial_t\varphi - \varphi\partial_t\varphi^*) - \frac{1}{2}\partial_x\varphi^*\partial_x\varphi + V(x)\varphi^*\varphi. \tag{2.52}$$

You can verify by direct computation that $\delta S/\delta\varphi^* = 0$ results in the Schrödinger equation, (2.36), and that $\delta S/\delta\varphi = 0$ gives the conjugate equation.

$\mathcal{L}$ also has the global phase invariance we wanted. Let's quickly verify that we get the correct conserved current from $\mathcal{L}$. Instead of plugging into an equation like (2.47), we can also find the current by transforming φ and φ^* by a local phase transformation ("local" means α is no longer constant, $\alpha = \alpha(x,t)$). The quantity proportional to α in δS will be $\partial_\mu j^\mu$. So, if

$$\varphi \to e^{i\alpha(x,t)}\varphi,$$

then,

$$\partial_\mu\varphi \to i\partial_\mu\alpha\, e^{i\alpha}\varphi + e^{i\alpha}\partial_\mu\varphi,$$

and so to first order in α,

$$\delta S = \int dt \int dx\ -\varphi^*\varphi\partial_t\alpha - \frac{1}{2}(-i\varphi^*\partial_x\alpha\partial_x\varphi + i\varphi\partial_x\alpha\partial_x\varphi^*).$$

After integrating by parts,

$$\delta S = 0 = \int dt \int dx\, \alpha(\partial_t(\varphi^*\varphi) - \frac{i}{2}\partial_x(\varphi^*\partial_x\varphi - \varphi\partial_x\varphi^*)),$$

and, indeed, we do recover j^0 and j^1 in equations (2.50)-(2.51).

From $\mathcal{L}$, eq.(2.52), we find that the field momentum is

$$\pi(x,t) = \frac{\partial\mathcal{L}}{\partial(\partial_t\varphi)} = i\varphi^*(x,t). \tag{2.53}$$

Thus, the quantum conditions we want to impose on the operators φ and φ^* are

$$\begin{gathered}[\varphi(x,t), \varphi^*(x',t)] = \delta(x-x') \\ [\varphi(x,t), \varphi(x',t)] = [\varphi^*(x,t), \varphi^*(x',t)] = 0.\end{gathered} \tag{2.54}$$

The field Hamiltonian operator is easily computed to be

$$\begin{aligned}\mathbf{H} &= \int dx\, \frac{1}{2}\partial_x\varphi^*\partial_x\varphi + V(x)\varphi^*\varphi \\ &= \int dx\, \varphi^*(x,t)\,(-\frac{1}{2}\partial_x^2 + V(x))\varphi(x,t).\end{aligned} \tag{2.55}$$

We have picked a particular ordering for the operators, called normal ordering, that is not symmetric between φ and φ^*. This ordering was chosen to avoid a divergence, the vacuum energy. We will look at normal ordering more closely in the next chapter.

After all the work we went through to find $\mathcal{L}$, we see that $\mathbf{H}$ and the quantum commutators, eq.(2.54), have an amazingly simple form. All we have done in constructing the field Hamiltonian is sandwich the first quantized Hamiltonian in between the field operators $\varphi^*(x,t)$ and $\varphi(x,t)$. Of course, it was not obvious at first to choose φ^* as the conjugate momentum to φ (instead of $\dot{\varphi}$), but as we shall see again with the Dirac equation, this is what you do when the field equation is first order in the time derivative.

Upon quantization, we are treating φ and φ^* as independent operators. You can easily check that the field Hamiltonian, eq.(2.55), along with the quantum conditions, eq.(2.54), reproduces the Schrödinger equation, eq.(2.36), from eq.(2.37). These equations define a quantum field theory except that we are missing a set of state vectors. Also, where are all of the particles? We need a particle interpretation for the fields φ and φ^*.

The first quantized 1-body Hamiltonian appears in the field Hamiltonian. Let's go back, for a moment, and reconsider the first quantized system. The normalized eigenfunctions $\varphi_n(x)$, eigenvalue e_n, of the first quantized Hamiltonian,

$$\mathbf{h} = -\frac{1}{2}\frac{\partial^2}{\partial x^2} + V(x), \tag{2.56}$$

are assumed to form a complete set, thus any solution φ to the Schrödinger equation can be expanded in terms of the φ_n's :

$$\varphi(x,t) = \sum_n a_n(t)\varphi_n(x)\,. \tag{2.57}$$

In the first quantized system, $\varphi(x,t)$ and $\varphi_n(x)$ are wave functions and $a_n(t)$ is just a number times $\exp(-ie_n t)$. After second quantizing, $\varphi(x,t)$ becomes an operator, so either $a_n(t)$ or $\varphi_n(x)$ must become an operator. We will make $a_n(t)$ an operator and leave $\varphi_n(x)$ a function. Similarly,

$$\varphi^*(x,t) = \sum_n a_n^\dagger(t)\varphi_n^*(x)\,. \tag{2.58}$$

Substitute the expansions (2.57) and (2.58) into the quantum equal-time commutators, eq.(2.54).

$$\begin{aligned}[\varphi(x,t),\varphi^*(x',t)] &= \left[\sum_n \mathbf{a}_n(t)\varphi_n(x)\,,\,\sum_m \mathbf{a}_m^\dagger(t)\varphi_m^*(x')\right] \\ &= \sum_n\sum_m [\mathbf{a}_n(t)\,,\,\mathbf{a}_m^\dagger(t)]\varphi_n(x)\varphi_m^*(x') \\ &= \delta(x-x') \end{aligned} \tag{2.59}$$

Since the first quantized energy eigenfunctions are assumed to be complete,

$$\sum_n \varphi_n^*(x')\varphi_n(x) = \delta(x-x'). \tag{2.60}$$

The last step in equation (2.59) will hold only if

$$[\mathbf{a}_n(t)\,,\,\mathbf{a}_m^\dagger(t)] = \delta_{nm}\,. \tag{2.61}$$

Similarly,

$$[\varphi(x,t)\,,\,\varphi(x',t)] = [\varphi^*(x,t)\,,\,\varphi^*(x',t)] = 0$$

implies

$$[\mathbf{a}_n(t)\,,\,\mathbf{a}_m(t)] = [\mathbf{a}_n^\dagger(t)\,,\,\mathbf{a}_m^\dagger(t)] = 0.$$

Next, substitute the expansion for φ and φ^* into the Hamiltonian operator. The first quantized Hamiltonian, $\mathbf{h}$, lurking inside the field Hamiltonian, will operate on the wave functions, $\varphi_n(x)$, in the expansion of φ. The wave functions, $\varphi_n(x)$, are orthonormal, so they satisfy

$$\int dx\, \varphi_n^*(x)\, \varphi_m(x) = \delta_{nm}\,. \tag{2.62}$$

The orthonormality simplifies the expression of $\mathbf{H}$ into

$$\mathbf{H} = \sum_n e_n\, \mathbf{a}_n^\dagger \mathbf{a}_n. \tag{2.63}$$

For fixed n, we note that $\mathbf{a}_n$ and $\mathbf{a}_n^\dagger$ look identical to the raising and lowering operators of the harmonic oscillator. The field Hamiltonian is just an infinite sum of harmonic oscillator Hamiltonians. The expansions, (2.57) and (2.58), have reduced the quantum field theory to an infinite set of harmonic oscillators. Following the discussion given on the harmonic oscillator, we can now develop a particle interpretation.

The lowest energy state of $\mathbf{H}$, the ground state or bare vacuum, is the one that is empty,

$$\mathbf{a}_n|\,0\rangle = 0. \tag{2.64}$$

The destruction operator for any n, $\mathbf{a}_n$, finds no excitations (particles) to annihilate in the empty vacuum, $|\,0\rangle$, so the result is the null vector. $\mathbf{a}_n^\dagger|\,0\rangle$ is a state of energy e_n. It contains 1 particle of energy e_n, created by $\mathbf{a}_n^\dagger$, the creation operator for mode n. $\mathbf{a}_m^\dagger|\,0\rangle$ is also a 1-particle state except that the energy of the particle is e_m. We have only one field in this theory, so there is only one type of particle. The difference in energy between the two 1-particle states must be due to a difference in momentum between the two. $\mathbf{a}_n^\dagger \mathbf{a}_m^\dagger|\,0\rangle$ is a 2-particle state with energy $e_n + e_m$. The collection of all of the states spanned by the states formed by operating on $|\,0\rangle$ with any number of creation operators for any mode n is called a Fock space.

Now we have our states and our particle interpretation. $\mathbf{a}_n^\dagger(t)|\,0\rangle$ is a state at time t with 1 particle of energy e_n. But where is this particle located? $\varphi(x,t)$ is expanded in terms of $\mathbf{a}_n$ only, while $\varphi^*(x,t)$ is expanded in terms of $\mathbf{a}_n^\dagger$ only. Thus, $\varphi(x,t)$ is a destruction operator and $\varphi^*(x,t)$ is a creation operator. So,

$$\varphi^*(x,t)|\,0\rangle \tag{2.65}$$

is a 1-particle state where the particle is located at position x at time t. In this form, we know where the particle is. But now, what is its energy?

To simplify things just for a moment, let's consider the case where $V(x) = 0$. The normalized eigenfunctions, $\varphi_n(x)$, are plane waves, $L^{-1/2}$

$\exp(i2\pi nx/L)$, that is, momentum eigenfunctions. The expansions (2.57) and (2.58) turn into Fourier series. Inverting eqs.(2.57) and (2.58), we have

$$\mathbf{a}_n(t) = \frac{1}{\sqrt{L}}\int dx'\, e^{-i2\pi nx'/L}\,\varphi(x',t), \tag{2.66}$$

$$\mathbf{a}_n^\dagger(t) = \frac{1}{\sqrt{L}}\int dx'\, e^{i2\pi nx'/L}\,\varphi^*(x',t). \tag{2.67}$$

The state $\mathbf{a}_n^\dagger(t)|\,0\rangle$ is

$$\frac{1}{\sqrt{L}}\int dx'\, e^{i2\pi nx'/L}\,\varphi^*(x',t)|0\rangle. \tag{2.68}$$

Since we are integrating over x', this states creates a particle via $\varphi^*(x',t)$ at every point x' with amplitude $L^{-1/2}\exp(i2\pi nx'/L)$. In other words, the state (2.68) above is a 1-particle state with the particle spread all over space-time, the position probability amplitude at each point given by $L^{-1/2}\exp(i2\pi nx'/L)$. That is, the state (2.68) creates a particle with wave function $L^{-1/2}\exp(i2\pi nx'/L)$. The square of this wave function is independent of x' so $\mathbf{a}_n^\dagger|\,0\rangle$ creates a 1-particle state and we do not know where the particle is. It has equal probability of being anywhere. The particle has a definite momentum n, so by the uncertainty principle, we have no idea what the position is. Likewise, $\varphi^*(x,t)|\,0\rangle$ creates a particle at x, a definite position. Since we know its position, we have no idea what its momentum and energy are. To create a particle definitely located at x, we need to use $\mathbf{a}_n^\dagger$ for every n. The particle has equal probability of having any momentum n.

Now let's return to the case where $V(x) \neq 0$. Suppose we want to make a state containing 1 particle described by wave function $f(x,t)$. To do this, we create a particle at x with amplitude $f(x,t)$ and sum over x:

$$\int dx\, f(x,t)\varphi^*(x,t)|\,0\rangle \tag{2.69}$$

is such a state. Suppose that we want this state to have a definite energy. This limits our choice of f. If the state has a definite energy, then it must be an eigenstate of $\mathbf{H}$, eq.(2.55).

$$\mathbf{H}\int dx f(x,t)\varphi^*(x,t)|\,0\rangle = E\int dx f(x,t)\varphi^*(x,t)|\,0\rangle \tag{2.70}$$

Using the commutators, eq.(2.54), and the fact that $\varphi(x,t)|\,0\rangle = 0$, we find that the state (2.69) will be a solution of eq.(2.70) if $f(x,t)$ satisfies

$$\left(-\frac{1}{2}\frac{\partial^2}{\partial x^2} + V(x)\right) f(x,t) = Ef(x,t). \tag{2.71}$$

In other words, f must be an eigenstate of the first quantized Hamiltonian. $f(x,t) = \varphi_n(x)\exp(-ie_nt)$. So now we see how the first quantized 1-particle system emerges from the quantum field theory.

What about a 2-particle system? A 2-particle state with 1 particle at x_1 and another at x_2 is $\varphi^*(x_1,t)\varphi^*(x_2,t)|0\rangle$. If the 2 particles are spread out with wave function $f_2(x_1,x_2,t)$, then the state is

$$\int dx_1dx_2\, f_2(x_1,x_2,t)\,\varphi^*(x_1,t)\varphi^*(x_2,t)|0\rangle. \tag{2.72}$$

By applying **H** to this state, we can show that the state will be an eigenstate of **H** if f_2 satisfies

$$\left(-\frac{1}{2}\left(\frac{\partial^2}{\partial x_1^2} + \frac{\partial^2}{\partial x_2^2}\right) + V(x_1) + V(x_2)\right) f_2(x_1,x_2,t) = E_2 f_2(x_1,x_2,t).$$

This is a first quantized 2-body Schrödinger equation. Observe that the 2-body potential separates and that the two particles do not directly interact, hence $f_2(x_1,x_2,t) = f(x_1,t)f(x_2,t)$. The lack of particle interaction is due to the fact that **H**, eq.(2.55), is quadratic in φ. If we add an additional term to **H** that is quartic in φ of the form

$$\int dx\, \mathcal{V}(x)\varphi^*(x,t)\varphi^*(x,t)\varphi(x,t)\varphi(x,t), \tag{2.73}$$

we can see, by applying the new **H** to the 2-particle state, eq.(2.72), that the 2-body potential is 2 $\mathcal{V}(x_1)\delta(x_1 - x_2)$. The δ-function arises from use of the commutator, eq.(2.54). The interaction potential is nonzero only if $x_1 = x_2$, a contact interaction. Since we are treating the fundamental particles as points, they can only interact on contact. If we instead add a term to eq.(2.55) that contains $(\varphi^*)^3\varphi^3$, then the 2-body interaction vanishes. However, the 3-body potential is nonzero and is proportional to $\delta(x_1 - x_2)\delta(x_2 - x_3)$. Again the particles only interact on contact, but three must meet to interact. When the first quantized n-body potential is a contact interaction proportional to a δ-function, the quantum field possesses the property called locality. Nonlocal quantum field theories are very difficult to construct and understand, especially relativistic field theories. We will not consider them here. For the same reason, consistent relativistic quantum theories of extended objects are difficult to produce. This is the topic of the second half of this book.

If we apply the interacting field Hamiltonian operator to an n-particle state,

$$\int dx_1 \cdots dx_n\, f_n(x_1, \ldots, x_n)\varphi^*(x_n, t) \cdots \varphi^*(x_1, t)|0\rangle, \tag{2.74}$$

we find that the n-body wave function must satisfy

$$\Big(-\frac{1}{2}\sum_{i=1}^{n} \frac{\partial^2}{\partial x_i^2} + \sum_{i=1}^{n} V(x_i) + \sum_{i \neq j} \mathcal{V}(x_i)\delta(x_i - x_j)\Big) f_n(x_1, \ldots, x_n, t) = E_n f_n(x_1, \ldots, x_n, t). \tag{2.75}$$

In this way, all of the n-body first quantized systems are contained in the corresponding quantum field theory. The operators φ and φ^* change the number of particles present so we can treat physical processes where the particle number is changing in a unified manner.

Before we move on, we should ask about what kind of particles we ended up with, fermions or bosons? The quantum conditions we have chosen, eq.(2.54), specify that $\varphi^*(x, t)$ commutes with itself. Reconsider the 2-particle state, eq.(2.72). Since $\varphi^*(x_1, t)$ and $\varphi^*(x_2, t)$ commute and create identical particles, we can interchange x_1 and x_2 and still have the same physical state. Thus, $f_2(x_1, x_2, t)$ should be symmetric under the exchange of x_1 and x_2.

$$f_2(x_1, x_2, t) = f_2(x_2, x_1, t) \tag{2.76}$$

By the Pauli principle, this exchange symmetry implies that we are dealing with bosons. The symmetric exchange is due to the use of commutators in the quantum conditions, eq.(2.54).

What if we wanted fermions instead? No two identical fermions can be at the same point, so if φ^* creates a fermion, then

$$\varphi^*(x, t)\varphi^*(x, t)|0\rangle = 0. \tag{2.77}$$

Since the state $\varphi^*(x, t)|0\rangle$ is not the null vector, then $(\varphi^*(x, t))^2 = 0$, as an operator equation. If we use commutators in the quantum conditions, then we must conclude that $(\varphi^*(x, t))^2 = 0$ implies that $\varphi^*(x, t) = 0$ as an operator equation. Something is wrong.

The wave function for two identical fermions must be antisymmetric under the exchange of coordinates.

$$\psi_2(x_2, x_1, t) = -\psi_2(x_1, x_2, t) \tag{2.78}$$

Consider a 2-fermion state,

$$|2f\rangle \equiv \int dx_1 dx_2\, \psi_2(x_1, x_2, t)\varphi^*(x_2, t)\varphi^*(x_1, t)|0\rangle. \tag{2.79}$$

Since ψ_2 is antisymmetric, then

$$|2f\rangle = -\int dx_1 dx_2 \psi_2(x_2, x_1, t)\varphi^*(x_2, t)\varphi^*(x_1, t)|0\rangle. \tag{2.80}$$

Since we are integrating over x_1 and x_2, they act as dummy indices. We may exchange them without changing the state.

$$|2f\rangle = -\int dx_1\, dx_2\, \psi_2(x_1, x_2, t)\varphi^*(x_1, t)\varphi^*(x_2, t)|0\rangle \tag{2.81}$$

If $\varphi^*(x_1, t)$ and $\varphi^*(x_2, t)$ commute, then the left-hand side of eq.(2.81) is the original state, so we would have $|2f\rangle = -|2f\rangle$, that is, $|2f\rangle = 0$, the null state. To avoid this, equate (2.79) and (2.81) and note that the only differences are the overall sign and the operator ordering. The antisymmetric nature of ψ_2 needed to describe fermions forces us to conclude that

$$\varphi^*(x_2, t)\varphi^*(x_1, t) = -\varphi^*(x_1, t)\varphi^*(x_2, t). \tag{2.82}$$

In other words, $\varphi^*(x, t)$ must *anti*commute with itself and with φ^* at any other point. We will use braces, $\{\}$, to denote an anticommutator. If we use anticommutators instead of commutators in the quantum conditions, then for $x_1 = x_2$, we have $(\varphi^*(x_1, t))^2 = 0$. However, $\varphi^*(x_1, t) \neq 0$, which is just what we needed. Therefore, to quantize the field φ satisfying eq.(2.36) and obtain fermions, we must use anticommutators for the quantum conditions,

$$\{\varphi(x, t), \varphi^*(x', t)\} = \delta(x - x')$$
$$\{\varphi(x, t), \varphi(x', t)\} = \{\varphi^*(x, t), \varphi^*(x', t)\} = 0. \tag{2.83}$$

2.3 NONLINEAR SCHRÖDINGER MODEL

The Nonlinear Schrödinger model (NLS) is a (1+1)-dimensional quantum field theory describing a nonrelativistic Bose gas. It is based on second quantizing the cubic Schrödinger equation,

$$i\frac{\partial\varphi}{\partial t} = -\frac{\partial^2\varphi}{\partial x^2} + 2c|\varphi|^2\varphi. \tag{2.84}$$

Assume $c > 0$. The field Hamiltonian,

$$\mathbf{H} = \int_0^L dx\, \varphi^*(x, t)(-\partial_x^2)\varphi(x, t) + c\,\varphi^*\varphi^*\varphi\varphi, \tag{2.85}$$

and quantum conditions,

$$[\varphi(x,t),\varphi^*(x',t)] = \delta(x-x')$$
$$[\varphi(x,t),\varphi(x',t)] = [\varphi^*(x,t),\varphi^*(x',t)] = 0, \tag{2.86}$$

reproduce eq.(2.84) as an operator equation of motion. The gas is confined to a box of length L, $0 \le x \le L$. We will impose periodic boundary conditions.

The equivalent first quantized system may be obtained by requiring the N-body state,

$$\int dx_1 \cdots dx_N f_N(x_1,\ldots,x_N,t)\varphi^*(x_N,t)\cdots\varphi^*(x_1,t)|\,0\rangle,$$

to be an eigenstate of the Hamiltonian, eq.(2.85). This will be true if f_N satisfies

$$\left(-\sum_{i=1}^{N}\frac{\partial^2}{\partial x_i^2} + c\sum_{i\neq j}\delta(x_i-x_j)\right) f_N = E_N f_N. \tag{2.87}$$

Periodic boundary conditions mean that

$$f_N(\cdots,x_i=0,\cdots,t) = f_N(\cdots,x_i=L,\cdots,t). \tag{2.88}$$

The particle number operator,

$$\mathbf{N} = \int dx\, \varphi^*(x,t)\varphi(x,t), \tag{2.89}$$

is one of an infinite set of operators that commute with $\mathbf{H}$, eq.(2.85). This means that the particle number is conserved. Since there are an infinite number of operators that commute with $\mathbf{H}$, there are an infinite number of constants of the motion. A quantum field theory that possesses an infinite set of conserved quantities is called integrable. Integrable theories should be exactly solvable.

The general N-body Schrödinger equation, eq.(2.87), was solved by E. Lieb and W. Liniger. The general solution is

$$f_N(x_1,\cdots,x_N)$$
$$= \left(\prod_{i<j}(\theta(x_i-x_j) + e^{i\Delta(k_j-k_i)}\theta(x_j-x_i))\right)\exp\left(i\sum_{j=1}^{N}k_jx_j\right), \tag{2.90}$$

where θ is the Heaviside step function, and

$$e^{i\Delta(q)} = \frac{q-ic}{q+ic}. \tag{2.91}$$

For f_N to be periodic, the k's must satisfy

$$e^{ik_iL} = \prod_{j \neq i} e^{i\Delta(k_j - k_i)}. \tag{2.92}$$

The energy eigenstate of this state is $E_N = \sum_{i=1}^{N} k_i^2$.

This model becomes more interesting physically in the "thermodynamic" limit of the finite density sector. In this sector, we require the states to satisfy the constraint

$$\langle f_N | \varphi^*(x)\varphi(x) | f_N \rangle = \frac{N}{L}. \tag{2.93}$$

The thermodynamic limit is $N \to \infty$, $L \to \infty$, N/L fixed.

Finally, let us examine the 2-point time-independent field-field correlation function, $g(x-y) = \langle 0 | \varphi^*(x)\varphi(y) | 0 \rangle$, in the finite density sector ($\langle 0 | \varphi^*(x)\varphi(x) | 0 \rangle = N/L$), to illustrate the operator product expansion.

Consider the Taylor expansion of $g(x-y)$ about $x = y$.

$$\begin{aligned} g(x-y) & \\ &= g(0) + \partial_x g(x-y)\Big|_{x=y}(x-y) + \frac{1}{2}\partial_x^2 g(x-y)\Big|_{x=y}(x-y)^2 + \cdots \\ &= \langle 0 | \varphi^*(y)\varphi(y) | 0 \rangle + \langle 0 | \partial_x \varphi^*(x)\varphi(y) | 0 \rangle_{x=y}(x-y) \\ &\quad + \frac{1}{2}\langle 0 | \partial_x^2 \varphi^*(x)\varphi(y) | 0 \rangle_{x=y}(x-y)^2 + \cdots \end{aligned} \tag{2.94}$$

The first term is simply given by eq.(2.93). It also follows from the fact that the number operator, eq.(2.89), is conserved. The second term in eq.(2.94) turns out to be related to the second conserved quantity,

$$M_1 = i \int dx\, (\partial_x \varphi^*(x))\varphi(x). \tag{2.95}$$

That is,

$$\langle 0 | \partial_x \varphi^*(x)\varphi(y) | 0 \rangle_{x=y} = -i\frac{M_1}{L}. \tag{2.96}$$

The third term in eq.(2.94) is, in turn, related to the third conserved charge, the Hamiltonian.

$$M_2 = H = \int dx\, ((-\partial_x^2 \varphi^*(x))\varphi(x) + c\varphi^*(x)\varphi^*(x)\varphi(x)\varphi(x)) \tag{2.97}$$

Thus,

$$\langle 0|\partial_x^2\varphi^*(x)\varphi(y)|0\rangle_{x=y} = -\frac{E_0}{L} + \frac{c}{L}\langle 0|\varphi^*(x)\varphi^*(x)\varphi(x)\varphi(x)|0\rangle. \quad (2.98)$$

At this order, the 4-point function evaluated at equal points has appeared. The same thing happens for all higher-order terms in eq.(2.94). In each case the coefficient is related to a conserved charge and higher-point correlation functions evaluated at equal points. Combining eqs.(2.96) and (2.98), we have

$$\begin{aligned} g(x-y) &= \frac{N}{L} - i\frac{M_1}{L}(x-y) \\ &\quad + \frac{1}{2}\Big(-\frac{E_0}{L} + \frac{c}{L}\langle 0|\varphi^*(x)\varphi^*(x)\varphi(x)\varphi(x)|0\rangle\Big)(x-y)^2 + \cdots \quad (2.99) \end{aligned}$$

Written as an operator product expansion, this is

$$\begin{aligned} \varphi^*(x)\varphi(y) &\stackrel{x\to y}{\longrightarrow} \Big(1 - i\frac{M_1}{N}(x-y) - \frac{1}{2}\frac{E_0}{N}(x-y)^2 + \cdots\Big)\varphi^*(y)\varphi(y) \\ &\quad + \frac{1}{2}\frac{c}{L}(x-y)^2\varphi^*(y)\varphi^*(y)\varphi(y)\varphi(y) + \cdots \quad (2.100) \end{aligned}$$

This series truncates at the $N + 1^{\text{st}}$ term for the finite N-body sector. The computation of $g(x-y)$ in closed form in the thermodynamic limit is still an open problem.

2.4 EXERCISES

1. Verify that the n-particle state, eq.(2.74), will be an eigenstate of the interacting field theory Hamiltonian, eqs.(2.55) and (2.73), if the n-body wave function, $f_n(x_1, \cdots, x_n, t)$, satisfies the Schrödinger equation (2.75).
2. Verify that the Hamiltonian, eq.(2.85), and the quantum conditions, eq.(2.86), reproduce the cubic Schrödinger equation (2.84) as an operator equation of motion. Then verify that the N-body state,

$$|N\rangle = \int dx_1 \cdots dx_N \, f_N(x_1, \ldots, x_N, t)\varphi^*(x_N, t)\cdots\varphi^*(x_1, t)|0\rangle,$$

will be an eigenstate of the Hamiltonian, eq.(2.81), if the N-body wave function satisfies eq.(2.87).

3. Show that the particle number is a constant of the motion in the Nonlinear Schrödinger model. Construct the momentum operator and show that it is also conserved. Verify that the quantity

$$M_3 = i^3 \int dx \; \partial_x^3 \varphi^* \; \varphi - 3c\partial_x \varphi^* \; \varphi^* \varphi^2$$

is also a constant of the motion. Compute the next term in the operator product expansion of $\varphi^*(x)\varphi(y)$, eq.(2.100).

4. Show that the N-body wave function, eq.(2.90), satisfies eq.(2.87), with the eigenvalue $E_N = \sum_{i=1}^N k_i^2$. What is the momentum of the N-body state?

5. 2-body sector of the Nonlinear Schrödinger model:

   a. Let $|2\rangle = \int dx_1 \, dx_2 \; \psi(x_1, x_2) \varphi^*(x_2, t) \varphi^*(x_1, t) |0\rangle$ be a 2-body eigenstate of the Hamiltonian, eq.(2.85). What Schrödinger equation does $\psi(x_1, x_2)$ satisfy?

   b. Find the wave function, $\psi_0(x_1, x_2)$, of the vacuum state in the 2-body sector. This wave function will be the solution of the Schrödinger equation of part (a) with the smallest eigenvalue. The vacuum must be translationally invariant, thus the total momentum of this state will vanish. Be sure to normalize the state on the interval $0 \le x \le L$.

   c. Express the equal-time field-field correlation function,

   $$g(x-y) = \langle 0_2 \, |\varphi^*(x)\varphi(y)| \, 0_2 \rangle,$$

   where $|0_2\rangle$ is the 2-body vacuum state, in terms of the vacuum wave function, $\psi_0(x_1, x_2)$. Compute $g(x-y)$ using the result of part (b) for $\psi_0(x_1, x_2)$. Verify that your result is normalized to $g(0) = 2/L$. Why should $g(0) = 2/L$?

   d. Derive the differential equation that $g(x-y)$ satisfies starting from the 2-body Schrödinger equation from part (a). From the resulting equation, interpret $g(x-y)$ as a Green's function.

   e. Verify that your result from part (c) satisfies the operator product expansion.

CHAPTER 3

Free Scalar Field Theory

In chapter 2, we second quantized a nonrelativistic Schrödinger equation to obtain a quantum field theory. This field theory is nonrelativistic, or Galiean invariant, but not Lorentz invariant. The requirement of Lorentz invariance has a profound influence on field theory, right down to the notion of the field description itself. It also leads to the existence of antiparticles, and with it particle creation and destruction, and to the connection of spin and statistics.

To find an interesting relativistic quantum field theory along the same path as in the previous chapter, we need a relativistic Schrödinger equation to second quantize. With little else to go by other than the nonrelativistic Schrödinger equation, we can only guess at what an interesting relativistic equation is. This is what happened historically. One natural relativistic generalization leads to the Klein-Gordon equation, the central topic of this chapter. Once we have the equation, we must discover what it describes. Lorentz covariance requires that the field transform as a scalar under the Lorentz group. Since representations of the Lorentz group can be built from the group $SU(2)$ ("spin"), the scalar field must describe spin 0 particles. The requirement of Lorentz covariance means that we must find completely different equations to describe spin 1/2 fermions or spin 1 bosons. We will take this up in the following two chapters.

3.1 THE KLEIN-GORDON EQUATION

The Schrödinger operator for a free particle may be heuristically obtained via the correspondence principle from the expression for the kinetic energy of a particle,

$$E = \frac{p^2}{2m}, \tag{3.1}$$

by substituting $E \to i\partial/\partial t$, and $p_j \to -i\partial/\partial x^j$. The energy-momentum relation above is nonrelativistic. The relativistic generalization is

$$E^2 = p^2 + m^2 . \tag{3.2}$$

The same substitutions for E and p_j above lead to the Klein-Gordon operator. This operator, acting on the scalar wave function, $\varphi(\vec{x}, t)$, gives the Klein-Gordon equation,

$$\left(\frac{\partial^2}{\partial t^2} - \nabla^2\right)\varphi(\vec{x}, t) + m^2\varphi(\vec{x}, t) = 0. \tag{3.3}$$

Eq.(3.3) may be obtained from the action

$$S[\varphi] = \frac{1}{2}\int d^4x(\partial^\mu\varphi(x)\partial_\mu\varphi(x) - m^2\varphi^2(x)), \tag{3.4}$$

where $x^0 = t$ and $x = (\vec{x}, t)$. $S[\varphi]$ will be Lorentz invariant if φ transforms as a Lorentz scalar. Equivalently, the Klein-Gordon equation, eq.(3.3), will be Lorentz covariant if φ transforms as a scalar.

Historically, eq.(3.3) was proposed as a relativistic Schrödinger wave equation by Schrödinger, Gordon, and Klein in 1926-1927, but was abandoned (until 1934) due to difficulties in constructing a 1-particle theory based on it. The first difficulty is that the Klein-Gordon equation admits negative energy solutions, basically because E^2 and not E appears in eq.(3.2). For example, the plane wave,

$$e^{i(\vec{k}\cdot\vec{x}+\omega t)}, \tag{3.5}$$

is a solution of eq.(3.3) with energy, $E = -\omega = -(\vec{k}^2 + m^2)^{1/2}$. Not only must we find an interpretation of what a negative energy particle is, but, in addition, the energy spectrum is not bounded from below. This means that we could, in principle, extract an arbitrarily large amount of energy from a single-particle system. For example, we could start with one positive energy particle and apply some external influence that allows the particle to jump to a negative energy state. The positive difference in energy between these two states could be used to do work. We can repeat the procedure on the negative energy particle, lowering its energy, and extract more energy. Since the spectrum is not bounded from below, we could do this forever.

We could try to avoid negative energy solutions by taking the positive square root of eq.(3.2). But, this would mean that we would have to define the square root of a differential operator. If we did so using a series expansion, that operator would be nonlocal and extremely difficult to work with.

The second difficulty that arises with the 1-particle theory based on the Klein-Gordon equation is the interpretation of $\varphi(\vec{x},t)$ as a wave function, that is, as a probability amplitude. To do so we must find (define) some quantity based on φ that has a nonnegative norm that we can interpret as a probability density. This probability density must be the time component of a conserved probability current so that the total probability is conserved with time. For the nonrelativistic case in the previous chapter, $\varphi^*(\vec{x},t)\varphi(\vec{x},t)$ fulfilled these requirements. The components of the conserved current in the nonrelativistic case are given by eqs. (2.50) and (2.51).

The Klein-Gordon equation does define a conserved current. To find it, we follow the same procedure as in the previous chapter that led to the current given by eqs. (2.50)-(2.51). The result is that we find the same current for the spatial components, namely,

$$j_i = \frac{1}{2im}(\varphi^*\partial_i\varphi - (\partial_i\varphi^*)\varphi). \tag{3.6}$$

The time component, the probability density, however, is different.

$$j_0 = \rho = \frac{i}{2m}(\varphi^*\frac{\partial\varphi}{\partial t} - \frac{\partial\varphi^*}{\partial t}\varphi) \tag{3.7}$$

This probability density is not positive definite. For example, the probability density associated with the plane wave, eq.(3.5), is

$$\rho = -\frac{\omega}{m} = -\frac{(\vec{k}^2+m^2)^{1/2}}{m}. \tag{3.8}$$

We have no way of interpreting a mixture of positive and negative probabilities. Thus, the use of the Klein-Gordon equation appears to exclude the possibility of a probability interpretation.

These difficulties arise because we are trying to construct a 1-particle theory from eq.(3.3). Our inability to do so is a consequence of the imposition of Lorentz covariance on the quantum mechanical wave equation. Next, let us turn to a discussion of some of the implications of imposing Lorentz invariance on quantum mechanics and field theory.

3.2 RELATIVISTIC QUANTUM FIELD THEORY

We introduced second quantization in the previous chapter in the context of nonrelativistic quantum mechanics. Along the way, we made several assumptions, one of which is that the notion of a field is a local concept. By this we mean that the field is determined by a purely differential equation in time and space. This implies that the interaction potential is also local, occuring essentially only on contact at one space-time point. The equivalent first quantized systems have potentials proportional to δ-functions in space-time. Since the Hamiltonian generates infinitesimal time translations, it is natural in quantum mechanics to assume that the field can be described locally in time. In the nonrelativistic case, there is nothing compelling to make us assume that the same is true for spatial dimensions. However, when we introduce Lorentz symmetry, the spatial coordinates must be treated on the same footing as time. If we base the quantization on the Hamiltonian and equal-time commutators, then we should use fields that are local in space as well as time. Equivalently, the notion of an interaction occuring at a single space-time point is Lorentz invariant. A nonlocal interaction can easily be frame-dependent.

Besides restricting us to a local field description for point particles, the introduction of special relativity into quantum mechanics also introduces the notion of causality. The fact that no measurable signal may propagate faster than the speed of light has far-reaching consequences in quantum mechanics. Two local operators that represent physically measurable quantities must commute at space-like separations. In order to accomplish this, we *must* include negative energy states. This in turn introduces particle creation and destruction, and leads to the existence of antiparticles. The vanishing of the commutator of physical operators at space-like separations also provides for the connection between spin and statistics. In nonrelativistic field theory, we must first assume this connection, and then show that this implies that bosons require commutators, while fermions require anticommutators. The introduction of Lorentz invariance removes this assumption.

In order to illustrate the consequences of the introduction of causality, let us consider a simple example of a particle propagating from space-time point x to y. This process is represented in the space-time diagram in figure 3.1(a). If $\varphi^*(x)$ creates a particle at x, and $\varphi(y)$ destroys a particle at y, then the amplitude for this process is

$$\langle 0\,|\varphi(y)\varphi^*(x)|\,0\rangle.$$

Since $\varphi(y)|\,0\rangle = 0$, this amplitude will vanish at space-like separations provided that $[\varphi(y), \varphi^*(x)] = 0$ at space-like separations, $(x-y)^2 < 0$. If we assume that φ and φ^* are nonrelativistic free fields, then the plane wave expansion of the commutator involves a sum over plane waves of

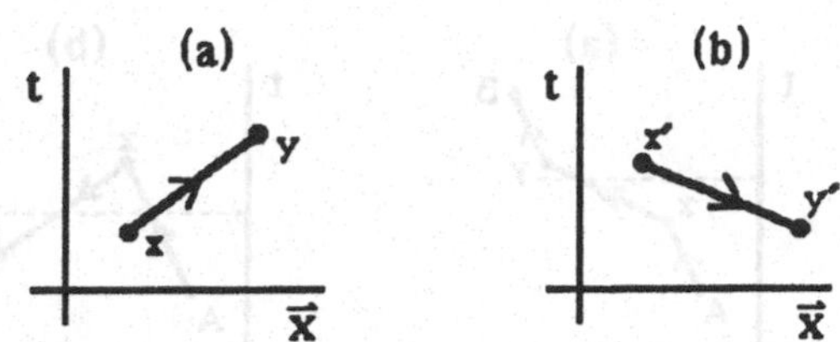

Figure 3.1 (a) Propagation of a positive energy particle. (b) Allowed propagation of a negative energy particle backwards in time. This is entirely equivalent to propagation of an antiparticle forward in time.

positive frequency only. We have only positive frequencies, because the energy for a free nonrelativistic particle (virtual or not) is always positive. Since we have only positive frequencies, it is not mathematically possible to adjust the (operator) coefficients of the plane wave expansions of φ and φ^* such that they commute at space-like separations unless they commute everywhere. In order to accomplish commutation at space-like separations but not everywhere, which must happen if φ and φ^* are to be relativistic, we must introduce negative frequency plane waves into the sum. By the correspondence principle, this means that we must allow the negative energy solutions. If we try to ignore them, we will violate causality.

Since we must include the negative energy states, we must find a way to treat them. The main problem is that they can lead to an energy spectrum that is not bounded from below. Consider the negative energy plane wave, eq.(3.5). Observe that if we reverse time, $t \to -t$, the negative energy states look like positive energy states. That is, negative energy particles traveling backwards in time look like ordinary positive energy particles. The way out of the negative energy problem is to let negative energy states only propagate backwards in time.

Certain attributes of the particle will appear to change when it propagates backwards in time. For example, a negatively charged particle moving backwards in time is equivalent to a positive charge moving forward in time. Thus, negative energy particles moving backwards can be reinterpreted as positive energy particles of the same mass, but opposite charge, moving forward. Such positive energy particles of identical mass but opposite charge are called antiparticles. The negative energy states mathematically describe antiparticles.

In the present situation, particles described by eq.(3.3) do not possess charge or any attribute that easily distinguishes particles from antiparticles. Later in this chapter, we will consider charged scalar fields so that the distinction can be made clear in order to introduce the propagator.

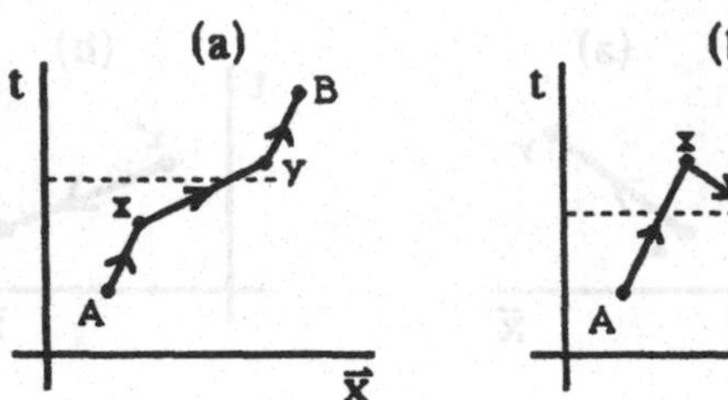

Figure 3.2 Backwards propagation leads to particle creation and destruction. (a) Intermediate state between x and y involves a positive energy particle. At any instant, there is only 1 particle. (b) Intermediate state between x and y involves a negative energy particle. The intermediate state contains 3 particles while initial and final states contain only a single particle. Pair creation occurs at y while pair annihilation occurs at x.

The introduction of charge also allows for an easy reinterpretation of the current density as a charge density. It is certainly acceptable for a charge density not to be positive definite.

Once we accept the negative energy states, as we must do in order to satisfy causality, then we must allow for backwards propagation in time, and for the existence of antiparticles. This in turn implies that physical processes must contain particle creation and destruction. In other words, the number of particles is no longer fixed. We advertised second quantization in the previous chapter on the grounds that it could handle different numbers of particles in a unified setting. The nonrelativistic interacting example, however, still conserved particle number. This is no longer true for relativistic interacting theories. The advantages of second quantization become apparent.

To see that negative energy states, backwards propagation in time, and antiparticles force particle creation and destruction upon us, consider once more the propagation of a real particle between points A and B as illustrated in figure 3.2. In this particular process, the particle interacts twice on the way from A to B. Since positive energy particles only propagate forward in time, and negative energy particles only backwards in time, the intermediate state between x and y in figure 3.2(a) is a positive energy particle, while that in 3.2(b) is a negative energy particle. At any instant in time, we see a state with only one particle in it in figure 3.2(a). However, in figure 3.2(b), we start with a state of one particle. After time y^0 but before x^0, we have a state containing 3 particles. After time x^0, we are once again back to one particle. Since negative energy particles moving backwards in time are really positive energy antiparticles moving forward in time, the physical interpretation of figure 3.2(b) is that pair creation occurs at point y and pair annihilation at point x. In between x and y we have two particles and one antiparticle. In order to get the total amplitude associated with propagation from A to B we must sum over all intermediate states, including the negative energy states.

Since relativity no longer allows us to fix the number of particles, it no longer makes sense to try to build 1-particle theories out of relativistic wave equations for use at all energies. For example, let's consider one scalar point particle whose wavefunction satisfies eq.(3.3). Suppose the particle is at rest in a large box, that is, in a state of definite momentum and energy. By the uncertainty principle, we do not know its position. Now suppose that we try to locate the particle by squeezing the walls of the box. Again, by the uncertainty principle, the energy and momentum of the particle will rise. When the walls of the box are on the order of $1/m$ apart, the Compton wavelength of the particle (its quantum mechanical size), the energy of the particle will exceed twice its rest mass. Any external influence or quantum fluctuations can cause the creation of another particle. Thus, we start with one particle, but when we try to find it by confining it, we end up by creating more particles. However, if the states with energy that is greater than twice the rest mass are unimportant, then the 1-particle interpretation will be valid.

The introduction of causality also provides the connection between spin and statistics. As we shall see in chapter 4, if we try to quantize the Dirac equation using commutators, we will be unable to prevent the noncommutation of physical observables outside of the light-cone. We can only accomplish this, and thereby satisfy causality, if we quantize the Dirac field using anticommutators.

In summary, the introduction of Lorentz symmetry restricts us to a local description of fields and field theory. It also introduces negative energy solutions, and the notion of causality. Causality prevents us from just ignoring the negative energy states, and thereby introduces the possibility of an energy spectrum that is unbounded from below. However, we may still obtain a positive energy spectrum if we allow the positive energy states to only propagate forward in time and the negative energy states only backwards in time. By allowing backwards propagation in time, we no longer have a fixed number of particles. Single-particle relativistic theories no longer make sense unless states with energy greater than twice the rest mass can be ignored. In addition, the negative energy states can be identified with antiparticles, thus the existence of antiparticles is ultimately due to the introduction of relativity into quantum mechanics. Finally, causality provides the connection between spin and statistics.

One last comment on backwards propagation in time. We are forced to propagate negative energy particles backwards in time as a mathematical device to produce an energy spectrum with a lower bound. However, the physics resides in the interpretation of the mathematics. Consider the situation represented in figure 3.1(b), where, mathematically, a negative energy particle propagates backwards in time. The reason that we think of this as backwards propagation is that we associate particle creation with point x' where $\varphi(x')$ creates a negative energy particle, and annihilation with point y' where $\varphi(y')$ absorbs it. Creation must precede absorption, thus the particle travels backwards in time. However, if we reinterpret what creation means (i.e. creation of what?), then our viewpoint of the direction of propagation changes. The creation of a negative energy particle lowers the total energy. However, the annihilation of a positive energy particle also lowers the total energy, etc. Backwards propagation of negative energy particles is entirely equivalent to forward propagation of positive energy antiparticles. Backwards propagation of negative energy particles is just a (useful) mathematical description of antiparticles.

Let us move on now to second quantize the Klein-Gordon equation, eq.(3.3).

3.3 FREE SCALAR FIELD THEORY

We want to canonically quantize the field $\varphi(x)$. This means that $\varphi(x)$ will become an operator. We must find a Hamiltonian and quantum conditions (commutators) such that the Klein-Gordon equation, eq.(3.3), becomes an operator equation of motion. This is accomplished with the aid of the classical action,

$$S[\varphi] = \int d^4x\, \mathcal{L}(x) = \frac{1}{2}\int d^4x\, (\partial^\mu\varphi(x)\partial_\mu\varphi(x) - m^2\varphi^2(x)). \tag{3.4}$$

From the action, we find that the conjugate field momentum, $\pi(x)$, is

$$\pi(x) = \frac{\partial\mathcal{L}}{\partial(\partial_t\varphi)} = \partial_t\varphi \equiv \dot{\varphi}. \tag{3.9}$$

The Hamiltonian quickly follows,

$$H = \frac{1}{2}\int d^3x\, (\pi^2(x) + |\nabla\varphi(x)|^2 + m^2\varphi^2(x)). \tag{3.10}$$

In analogy with ordinary quantum mechanics,($[x,p]=i$), we take

$$\begin{aligned} [\varphi(\vec{x},t), \pi(\vec{y},t)] &= i\delta(\vec{x}-\vec{y}) \\ [\varphi(\vec{x},t), \varphi(\vec{y},t)] &= [\pi(\vec{x},t), \pi(\vec{y},t)] = 0 \end{aligned} \tag{3.11}$$

as our quantum conditions. Using the Hamiltonian and commutators, we find that the operator equation of motion, $\dot{\varphi} = i[H,\varphi]$, reproduces eq.(3.9) as an operator equation, while the operator equation $\dot{\pi} = i[H,\pi]$ reproduces the Klein-Gordon equation as desired.

We now have a quantum field theory, but lack a particle interpretation. To remedy this, we will follow what we did in chapter 2 to arrive at eqs. (2.57)-(2.58). Namely, we expand $\varphi(\vec{x},t)$ in terms of the classical solutions of the Klein-Gordon equation.

$$\begin{aligned}\varphi(\vec{x},t) &\approx \sum_{\vec{k}} a(k)\varphi_k^{(+)}(x) + b(k)\varphi_k^{(-)}(x) \\ &= \int \frac{d^3k}{(2\pi)^3}\frac{1}{2\omega_k}\left(a(k)e^{-ik\cdot x} + b(k)e^{ik\cdot x}\right),\end{aligned} \tag{3.12}$$

where $\omega_k = (\vec{k}^2+m^2)^{1/2}$, and where $\varphi_k^{(+)}(x)$ is a classical positive energy plane wave solution of eq.(3.3) and $\varphi_k^{(-)}(x)$ is a negative energy solution. When $\varphi(\vec{x},t)$ becomes an operator, so do $a(k)$ and $b(k)$. The use of $\varphi_k^{(+)}$ and $\varphi_k^{(-)}$ guarantee that $\varphi(\vec{x},t)$ satisfies the Klein-Gordon equation. $\varphi(\vec{x},t)$ is classically a real field, so it is a Hermitian operator upon quantization. Imposing this on the expansion above, we find $b(k) = a^\dagger(k)$.

The normalization, $1/2\omega_k$, in the volume element above was chosen so as to be Lorentz invariant. This is not obvious in its present form. However, we can rewrite it as

$$\int \frac{d^3k}{(2\pi)^3}\frac{1}{2\omega_k} = \int \frac{d^4k}{(2\pi)^4} 2\pi\delta^4(k^2-m^2)\theta(k^0), \tag{3.13}$$

which is manifestly Lorentz invariant. The identity above follows from the properties of the δ-function,

$$\delta(f(x)) = \sum_i \frac{1}{|\frac{df}{dx}(x_i)|}\delta(x-x_i), \tag{3.14}$$

where x_i is a simple zero of $f(x)$. Using eq.(3.14) we have

$$\delta(k^2-m^2) = \delta(k_0^2-\omega_k^2) = \frac{1}{2\omega_k}(\delta(k_0-\omega_k)+\delta(k_0+\omega_k)). \tag{3.15}$$

$\theta(k_0)$ picks out $\delta(k_0-\omega_k)$, and eq.(3.13) follows upon integration over k_0.

The expansion for $\pi(\vec{x},t)$ follows from eqs.(3.12) and (3.9).

$$\pi(\vec{x},t) = \int \frac{d^3k}{(2\pi)^3}\frac{1}{2\omega_k}\left(-i\omega_k a(k)e^{-ik\cdot x} + i\omega_k a^\dagger(k)e^{ik\cdot x}\right) \tag{3.16}$$

Using the expansion for $\varphi(\vec{x},t)$ and $\pi(\vec{x},t)$, we can solve for $a(k)$.

$$\int d^3x\, e^{-i\vec{k}'\cdot\vec{x}}(\omega_{k'}\varphi(\vec{x},t) + i\pi(\vec{x},t))$$

$$= \int d^3x\, e^{-i\vec{k}'\cdot\vec{x}} \int \frac{d^3k}{(2\pi)^3} \frac{1}{2\omega_k}((\omega_{k'} + \omega_k)a(k)e^{-ik\cdot x} + (\omega_{k'} - \omega_k)a^\dagger(k)e^{ik\cdot x})$$

$$= \int \frac{d^3k}{2\omega_k}(\delta(\vec{k} - \vec{k}')(\omega_{k'} + \omega_k)a(k)e^{i\omega_k t} + \delta(\vec{k} + \vec{k}')(\omega_{k'} - \omega_k)a^\dagger(k)e^{-i\omega_k t})$$

$$= a(k') \exp(i\omega_{k'}t).$$

Thus,

$$a(k) = \int d^3x\, e^{ik\cdot x}(\omega_k\varphi(\vec{x}, t) + i\pi(\vec{x}, t)). \tag{3.17}$$

The quantum commutators for φ and π, eq.(3.11), determine the algebra of $a(k)$ and $a^\dagger(k)$.

$$\begin{aligned}
&[a(k), a^\dagger(k')] \\
&= \int d^3x \int d^3y e^{ik\cdot x}e^{-ik'\cdot y}[\omega_k\varphi(\vec{x}, t) + i\pi(\vec{x}, t), \omega_{k'}\varphi(\vec{y}, t) - i\pi(\vec{y}, t)] \\
&= \int d^3x \int d^3y e^{ik\cdot x}e^{-ik'\cdot y}(\omega_k + \omega_{k'})\,\delta(\vec{x} - \vec{y}) \\
&= (2\pi)^3 2\omega_k \delta^3(k - k').
\end{aligned} \tag{3.18}$$

Similarly, $[a(k), a(k')] = 0 = [a^\dagger(k), a^\dagger(k')]$.

With the commutators for $a(k)$ and $a^\dagger(k)$ at hand, we can now transform the Hamiltonian, eq.(3.10).

$$H = \frac{1}{2}\int \frac{d^3k}{(2\pi)^3}\frac{1}{2\omega_k}\omega_k(a^\dagger(k)a(k) + a(k)a^\dagger(k)) \tag{3.19}$$

As in the field theory in the previous chapter, the Hamiltonian is the continuous sum of harmonic oscillator Hamiltonians, one for each $\vec{k}$. By comparing the commutators, eq.(3.18), with the harmonic oscillator commutators, eq.(2.20), we see that $a^\dagger(k)$ is a creation (raising) operator, while $a(k)$ is a destruction (lowering) operator. The particle interpretation results from considering $a^\dagger(k)$ as an operator that creates a particle of energy ω_k and momentum $\vec{k}$, while $a(k)$ destroys such a particle.

The ground state or bare vacuum is the state in Fock space that satisfies

$$a(k)|\,0\rangle = 0, \tag{3.20}$$

and is normalized so that $\langle 0\,|\,0\rangle = 1$. The state $a^\dagger(k)|\,0\rangle$ is a state containing one particle of energy ω_k, momentum $\vec{k}$; $(a^\dagger(k))^2|\,0\rangle$ contains two such particles, and so on.

The usual probability interpretation requires that all physical states be normalizable, $\langle\,|\,\rangle < \infty$. All along we have been calling $a(k)$ and $a^\dagger(k)$ operators. If we consider the state $a^\dagger(k)|0\rangle$, however, we find that it is not normalizable,

$$\langle 0\,|a(k)a^\dagger(k)|\,0\rangle = \delta(0). \tag{3.21}$$

This is not surprising, because $a^\dagger(k)$ creates a particle of definite energy and momentum. By the uncertainty principle, we have no idea where the particle is located. Thus, its wave function is a plane wave and we know from ordinary quantum mechanics that such states are nonnormalizable if the volume containing the system is infinite (which is what $\delta(0)$ represents). To obtain a normalizable state, we build a wave packet by superposition. That is, we smear out $a^\dagger(k)$. The state,

$$\int d^3k\, f(\vec{k})a^\dagger(k)|\,0\rangle,$$

will be normalizable provided that

$$\int d^3k|f(\vec{k})|^2 < \infty.$$

$a^\dagger(k)$ really only makes sense in an integral with a test function, $f(\vec{k})$. Thus, $a^\dagger(k)$ is a distribution; in fact, an operator-valued distribution. The result of doing the integral containing $a^\dagger(k)$ is an operator (as opposed to the distribution, $\delta(x)$, where the result is just a number). We will continue to be sloppy, however, and just call $a^\dagger(k)$, and similar animals, operators.

Returning to the bare vacuum, $|\,0\rangle$, we find that it is an eigenstate of the Hamiltonian, eq.(3.19), but that its energy, the sum of zero point energies of all of the oscillators, is divergent.

$$\langle 0\,|H|\,0\rangle = \frac{1}{2}\int d^3k\,\omega_k\,\delta^3(0), \tag{3.22}$$

where $(2\pi)^3\delta^3(0) = \int d^3x$. This is the first of many divergences that we will encounter. We will "dispose" of this one by observing that the absolute energy cannot be measured (not so for gravity!). We will merely redefine the zero point of the energy scale. The Hamiltonian, eq.(3.19), is redefined by subtracting this infinite constant, $H \rightarrow H - \langle 0\,|H|\,0\rangle$. This is formally accomplished by "normal ordering" the operators that appear in H. Normal ordering means that all creation operators are to appear to the left of destruction operators. Normal ordering is traditionally denoted by placing colons on both sides of the operator product.

$$: a(k)a^\dagger(k) : \;=\; : a^\dagger(k)a(k) : \;=\; a^\dagger(k)a(k) \tag{3.23}$$

The normal ordered Hamiltonian,

$$: H : = \int \frac{d^3k}{(2\pi)^3} \frac{1}{2\omega_k} \omega_k a^\dagger(k) a(k), \tag{3.24}$$

has vanishing vacuum expectation, $\langle 0 | : H : | 0 \rangle = 0$. From now on, we will assume that the Hamiltonian is normal ordered.

We started with a classical theory that is relativistic. We do not want to destroy this by quantization. In order to canonically quantize, we had to specify equal-time commutators. The choice made, eq.(3.11), is not Lorentz covariant. Thus, to quantize, we must choose a specific Lorentz frame. We want to verify that we get the same quantum theory no matter which frame we choose. One way to do this is to verify that the quantum operator forms of the generators of the Lorentz algebra still satisfy the proper algebra after quantization. The specific computation appears as an exercise at the end of the chapter.

In addition, no signal can travel faster than the speed of light. Two events separated by a space-like distance cannot affect each other. The creation or destruction of a particle is such an event, so if our field theory is to be relativistic, then the commutator $[\varphi(x), \varphi(y)]$ should vanish when the separation of the points x and y is space-like, $(x - y)^2 < 0$. With the aid of the expansion of φ in terms of the a's and $a^\dagger$'s, eq.(3.12), and their commutators, eq.(3.18), the computation of $[\varphi(x), \varphi(y)]$ at different times is straightforward.

$$\begin{aligned}
[\varphi(x), \varphi(y)] &= \int \frac{d^3k}{(2\pi)^3} \frac{d^3k'}{(2\pi)^3} \frac{1}{4\omega_k \omega_{k'}} ([a(k), a^\dagger(k')] e^{-ik\cdot x} e^{ik'\cdot y} \\
&\qquad + [a^\dagger(k), a(k')] e^{ik\cdot x} e^{-ik'\cdot y}) \\
&= \int \frac{d^3k}{(2\pi)^3} \frac{1}{2\omega_k} (e^{-ik\cdot(x-y)} - e^{ik\cdot(x-y)}) \\
&\equiv i\Delta(x-y)
\end{aligned} \tag{3.25}$$

$\Delta(x - y)$ is Lorentz invariant because of the invariant measure and the appearance of only vector dot products. It is important that this commutator be Lorentz invariant. We know from the equal-time commutation relations, eq.(3.11), that $\Delta(\vec{x} - \vec{y})$ vanishes at equal times. Since Δ is Lorentz invariant, it then follows that Δ vanishes for space-like separations as desired. Note that this is true only because eq.(3.15) involves a sum over positive and negative frequency (energy) plane waves. If we had ignored the negative energy states, Δ would no longer vanish at space-like separations, and signals could propagate faster than the speed of light.

Several other interesting properties of Δ are that it is odd, $\Delta(x-y) = -\Delta(y-x)$, that

$$\partial_0 \Delta(x-y)|_{x_0=y_0} = -\delta^3(x-y), \tag{3.26}$$

which follows from the equal-time commutators, eq.(3.11), and that $\Delta(x)$ satisfies the Klein-Gordon equation, $(\partial^\mu \partial_\mu + m^2)\Delta(x) = 0$.

The vanishing of the commutator of two fields at space-like separations is called microscopic causality. The commutator should vanish no matter how close the two points are as long as their separation is space-like. One implication of this is that since the commutator vanishes, we can make precise measurements of the fields at the two points simultaneously without one measurement disturbing the other. But, as we are about to see, we cannot measure the square of a field operator at one point. Thus, the vanishing of the vacuum expectation of the commutator of two fields at space-like separations as the two points approach each other is accomplished by subtracting two infinite quantities!

Consider the vacuum expectation of two fields at two different points.

$$\begin{aligned}\Delta_+(x-y) &\equiv \langle 0\,|\varphi(x)\varphi(y)|\,0\rangle \\ &= \int \frac{d^3k}{(2\pi)^3}\frac{1}{2\omega_k}\frac{d^3k'}{(2\pi)^3}\frac{1}{2\omega_{k'}} e^{-ik\cdot x}e^{ik'\cdot y}\langle 0\,|a(k)a^\dagger(k')|\,0\rangle \\ &= \int \frac{d^3k}{(2\pi)^3}\frac{1}{2\omega_k}e^{-ik\cdot(x-y)},\end{aligned} \tag{3.27}$$

where we have made use of the commutators, eq.(3.18). As $y \to x$,

$$\Delta_+(0) = \langle 0\,|\varphi^2(x)|\,0\rangle = \int \frac{d^3k}{(2\pi)^3}\frac{1}{2\omega_k}, \tag{3.28}$$

which is divergent. As with the energy of the vacuum, the divergence is related to the infinite volume of the system. The average of $\varphi(x)$ in the vacuum vanishes, $\langle 0\,|\varphi(x)|\,0\rangle = 0$. This means that the vacuum fluctuations of $\varphi(\vec{x},t)$, $\Delta\varphi = \sqrt{\langle 0\,|\varphi^2(x)|\,0\rangle - \langle 0\,|\varphi(x)|\,0\rangle^2}$, are infinite. Microcausality states that the two fields commute at space-like separations. In addition, there are many important operators, such as the Hamiltonian, that involve the products of operators at a point. But the quantum mechanical vacuum fluctuations of φ make it impossible to measure the square of the field at one point. Why not just subtract this one away? No simple subtraction works completely. In addition, there are physical effects due to vacuum fluctuations, so we can't eliminate them. Instead, we will just hand-wave it away by noting that if we really do try to measure the square of the field at one point, then the wavelength of the probe must be zero in order to resolve the point. But this requires infinite energies. If we try to make the measurement at any smaller energy scale, then we are actually averaging $\langle 0\,|\varphi^2(x)|\,0\rangle$ over a finite volume and the result will make sense. Finally, the fact that $\langle 0\,|\varphi^2(x)|\,0\rangle$ is divergent never interferes with practical calculations.

In addition to $\Delta(x-y)$ and $\Delta_+(x-y)$, we can define

$$\Delta_-(x-y) = \int \frac{d^3k}{(2\pi)^3}\frac{1}{2\omega_k}\, e^{ik\cdot(x-y)}, \tag{3.29}$$

so that $i\Delta(x-y) = \Delta_+(x-y) - \Delta_-(x-y)$. The $+$ on Δ_+ refers to the positive frequency part of Δ while the $-$ on Δ_- refers to the negative frequency part. The sum of Δ_+ and Δ_- is called $\Delta_1(x-y)$ and is equal to the vacuum expectation of the anticommutator of $\varphi(x)$ and $\varphi(y)$, $\Delta_1(x-y) = \langle 0\,|\{\varphi(x), \varphi(y)\}|\,0\rangle$. Please note that $\Delta_1(x-y)$ does not vanish for space-like separations. Had we quantized $\varphi(x)$ using anticommutators, we would violate causality. Lorentz invariance ultimately provides the connection between spin and statistics.

3.4 CHARGED SCALAR FIELDS

The field theory we have considered so far based on the Hermitian field $\varphi(x)$ does not allow for any distinction between particles and antiparticles. To distinguish between the two, the particles must carry some charge quantum number. The antiparticles will carry the opposite charge. Charge does not necessarily mean electric charge. For example, the $\bar{K}^0$ meson is the antiparticle of the K^0 meson. Both are electrically neutral. In this case, the charge that distinguishes the two is the flavor strangeness.

In order to incorporate a conserved charge, we need a conserved current. Recall that the Klein-Gordon equation does provide a conserved current, eqs. (3.6)-(3.7). The charge density is

$$j_0 = \rho = \frac{i}{2m}(\varphi^* \frac{\partial \varphi}{\partial t} - \frac{\partial \varphi^*}{\partial t}\varphi). \tag{3.7}$$

Notice that the charge density depends explicitly on φ^*, the Hermitian conjugate of φ. If φ is Hermitian, then j_0 vanishes. Thus, to incorporate a charge, we must work with a nonhermitian field, that is, a complex classical field, $\varphi^* \neq \varphi$.

To make φ complex, we introduce 2 Hermitian fields, φ_1 and φ_2, that satisfy the Klein-Gordon equation. The action for the two fields is just a double copy of eq.(3.4).

$$S[\varphi_1, \varphi_2] = \frac{1}{2}\int d^4x\, :\partial_\mu\varphi_1\partial^\mu\varphi_1 + \partial_\mu\varphi_2\partial^\mu\varphi_2 - m^2\varphi_1^2 - m^2\varphi_2^2: \tag{3.30}$$

Following the same development as in the previous section, we have

$$\pi_1(x) = \dot{\varphi}_1(x), \qquad \pi_2(x) = \dot{\varphi}_2(x),$$

and, at equal times,

$$[\varphi_i(\vec{x}), \pi_j(\vec{y})] = i\delta_{ij}\delta^3(x-y).$$

A particle interpretation is introduced by Fourier transforming φ_i.

$$\begin{aligned}\varphi_1(\vec{x},t) &= \int \frac{d^3k}{(2\pi)^3}\frac{1}{2\omega_k}(a_1(k)e^{-ik\cdot x} + a_1^\dagger(k)e^{ik\cdot x})\\ \varphi_2(\vec{x},t) &= \int \frac{d^3k}{(2\pi)^3}\frac{1}{2\omega_k}(a_2(k)e^{ik\cdot x} + a_2^\dagger(k)e^{-ik\cdot x})\end{aligned} \tag{3.31}$$

By introducing 2 independent fields, we have merely doubled the kinds of particles (which we must do to get distinct particles and antiparticles). The ground state or bare vacuum satisfies

$$a_i(k)|\,0\rangle = 0 \qquad \text{for } i = 1,2. \tag{3.32}$$

The operator,

$$N_1 = \int \frac{d^3k}{(2\pi)^3}\frac{1}{2\omega_k}\, a_1^\dagger(k)a_1(k),$$

counts the number of type 1 particles, while

$$N_2 = \int \frac{d^3k}{(2\pi)^3}\frac{1}{2\omega_k}\, a_2^\dagger(k)a_2(k)$$

does the same for type 2.

The complex field, $\varphi(\vec{x},t)$, is introduced by a change of coordinates,

$$\varphi(x) = \frac{1}{\sqrt{2}}(\varphi_1(x) + i\,\varphi_2(x)). \tag{3.33}$$

The action written in terms of the new variables is

$$S[\varphi,\varphi^*] = \frac{1}{2}\int d^4x : \partial^\mu\varphi^*\partial_\mu\varphi - m^2\varphi^*\varphi : . \tag{3.34}$$

From the action we find that

$$\pi(x) = \dot{\varphi}^*(x) \qquad \text{and} \qquad \pi^*(x) = \dot{\varphi}(x), \tag{3.35}$$

so the Hamiltonian is

$$H = \frac{1}{2}\int d^3x \; :\pi^*\pi + \vec{\nabla}\varphi^* \cdot \vec{\nabla}\varphi + m^2\varphi^*\varphi: . \tag{3.36}$$

Both φ and φ^* will satisfy the Klein-Gordon equation if, at equal times,

$$[\varphi(x), \pi(y)] = [\varphi^*(x), \pi^*(y)] = i\delta(\vec{x} - \vec{y}), \tag{3.37}$$

with all other equal-time commutators vanishing. Eq.(3.37) follows from the commutation relations for φ_1 and φ_2.

As usual, we expand φ and φ^* in plane waves to regain a particle interpretation.

$$\begin{aligned} \varphi(x) &= \int \frac{d^3k}{(2\pi)^3}\frac{1}{2\omega_k}\left(a(k)e^{-ik\cdot x} + b^\dagger(k)e^{ik\cdot x}\right) \\ \varphi^*(x) &= \int \frac{d^3k}{(2\pi)^3}\frac{1}{2\omega_k}\left(b(k)e^{-ik\cdot x} + a^\dagger(k)e^{ik\cdot x}\right) \end{aligned} \tag{3.38}$$

In terms of the creation and destruction operators of type 1 and 2 Hermitian scalar fields, we have

$$\begin{aligned} a(k) &= \frac{a_1(k) + ia_2(k)}{\sqrt{2}} \\ b(k) &= \frac{a_1(k) - ia_2(k)}{\sqrt{2}} , \end{aligned} \tag{3.39}$$

etc., and so it follows from

$$[a_i(k), a_j^\dagger(k')] = (2\pi)^3\, 2\omega_k\, \delta_{ij}\, \delta^3(k - k')$$

that

$$[a(k), a^\dagger(k')] = [b(k), b^\dagger(k')] = (2\pi)^3\, 2\omega_k\, \delta^3(k - k'), \tag{3.40}$$

with all others vanishing.

The (identical) bare vacuum, $|0\rangle$, satisfies

$$a(k)|0\rangle = b(k)|0\rangle = 0, \tag{3.41}$$

while

$$N_a = \int \frac{d^3k}{(2\pi)^3}\frac{1}{2\omega_k}a^\dagger(k)a(k)$$

is the number operator for type "a" particles and

$$N_b = \int \frac{d^3k}{(2\pi)^3} \frac{1}{2\omega_k} b^\dagger(k) b(k)$$

for type "b".

N_1, N_2, N_a, and N_b all commute with the Hamiltonian so they are conserved. We may choose our Fock space to be eigenstates of N_a and N_b, or N_1 and N_2. The former choice is of interest here. In any case, $N_1 + N_2 = N_a + N_b$.

The action, $S[\varphi, \varphi^*]$, eq.(3.34), possesses a global $U(1)$ invariance, that is, it is invariant under rotation in (φ_1, φ_2) space. If we take

$$\begin{aligned} \varphi &\to e^{i\theta} \varphi \\ \varphi^* &\to e^{-i\theta} \varphi^*, \end{aligned} \tag{3.42}$$

where θ is a constant (the phase $e^{i\theta}$ is an element of the group $U(1)$), the action is unchanged. The action, $S[\varphi_1, \varphi_2]$, eq.(3.30), must possess the same symmetry, but the action for a single Hermitian field, eq.(3.4), does not. In addition, the symmetry will only exist if the mass of type 1 particles is equal to the mass of type 2, $m_1 = m_2$, or equivalently, $m_a = m_b$.

Since the action, eq.(3.34), possesses a symmetry, there must be a conserved current. Following the development in the previous chapter immediately following eq.(2.52), we find that the conserved current is precisely the one we found directly from the Klein-Gordon equation, eqs. (3.6)-(3.7). The conserved charge is

$$Q = \int d^3x : j_0 := \int d^3x : (\varphi^* \dot{\varphi} - \dot{\varphi}^* \varphi) : . \tag{3.43}$$

If we drop in the expansions for φ and φ^*, eq.(3.38), Q can be written as

$$\begin{aligned} Q &= \int \frac{d^3k}{(2\pi)^3} \frac{1}{2\omega_k} (a^\dagger(k) a(k) - b^\dagger(k) b(k)) \\ &= N_a - N_b . \end{aligned} \tag{3.44}$$

Therefore, the type "a" particles, which are created by φ^* and destroyed by φ, have charge $+1$, while the type "b" particles, which are created by φ and destroyed by φ^*, carry charge -1. The net effect of operating with φ is to lower the charge by one unit, either by destroying a $+1$ charge or by creating a -1 charge, while operating with φ^* raises the total charge by one unit.

In any state $|\, n_a, n_b\rangle$ where $n_b > n_a$ (there are more b particles than a particles), the total charge is less than zero. This is perfectly acceptable. So we see that by reinterpreting the probability density as a charge density, we sidestep the problem of negative probabilities. What we must really watch out for are Fock space states of negative norm. As we shall see in future chapters, the Fock space states of negative norm will destroy the probability interpretation. Following the steps leading to eq.(2.24), we can show that $\langle n_a, n_b \,|\, n_a, n_b\rangle \geq 0$, so there is no problem here.

As we stated earlier, antiparticles are distinguished from particles by carrying an opposite charge. Here, type b particles are antiparticles of type a. Which particle is taken as the "antiparticle" is a matter of convention. As we can see here, the particles and antiparticles enter the theory symmetrically and we could say that the charge symmetry of the action, the global $U(1)$ phase invariance, is equivalent to saying that there is a symmetry under the exchange of particles with antiparticles. The charge symmetry requires that the masses of the two particles be identical, hence particles and antiparticles always have the same mass. To physically measure the charge, we must couple these particles to an external field (e.g. the electromagnetic field).

3.5 TIME-ORDERING AND PROPAGATORS

Let's consider the quantum mechanical propagation of a particle at $x = (\vec{x}, t)$ to $x' = (\vec{x}', t')$. Let the state describing the particle located at x be $|\,\psi(\vec{x}, t)\rangle$ and the state corresponding to the particle located at x' be $|\,\psi(\vec{x}', t')\rangle$. The quantum mechanical amplitude for the particle to start at x and propagate to x' is the overlap between the two states, $\langle\psi(\vec{x}', t')\,|\,\psi(\vec{x}, t)\rangle$. The square of this amplitude is the probability that this process will occur.

Now consider the propagation of charge in the charge scalar field theory. The state corresponding to a particle of charge $+1$ at x is $\varphi^*(x)|\,0\rangle$, while the state corresponding to the particle at x' is $\varphi^*(x')|\,0\rangle$. Thus, the quantum mechanical amplitude to transport the charge from x to x' is

$$\langle 0\,|\varphi(x')\varphi^*(x)|\,0\rangle.$$

Apparently, we can interpret the propagation of charge as the creation of a particle of $+1$ charge, an a particle, out of the vacuum at x, the transport from x to x', and the reabsorption of the a particle into the vacuum at x'. Since we can't absorb the particle before it is created, this process only makes sense if $t' > t$.

This is not the total amplitude for propagation of $+1$ charge. The total amplitude is the sum of all of the amplitudes of different processes that give equivalent physical results. In the process above, the charge at x

was increased by one unit, while the charge at x' was lowered by one unit. We can accomplish the same thing by creating a particle of -1 charge, a b particle, at x' and transporting it to x, then destroying it. Since φ creates b particles and φ^* destroys them, the amplitude for this process is

$$\langle 0\,|\varphi^*(x)\varphi(x')|\,0\rangle.$$

As before, we can't destroy a particle before it is created, so this process only makes sense if $t > t'$.

The total amplitude, $G(x', x)$, for propagation is the sum of the two amplitudes,

$$G(x',x) = \theta(t'-t)\langle 0\,|\varphi(x')\varphi^*(x)|\,0\rangle + \theta(t-t')\langle 0\,|\varphi^*(x)\varphi(x')|\,0\rangle. \tag{3.45}$$

We can write this in a more compact form by introducing the Dyson time-ordering operator, T.

$$T\varphi(x)\varphi^*(x') = \begin{cases}\varphi(x)\varphi^*(x') & \text{if } t' < t\\ \varphi^*(x')\varphi(x) & \text{if } t < t'\end{cases} \tag{3.46}$$

The operator T orders operators by their time-ordering. Operators that occur at later times appear to the left of operators that occur at earlier times. Using T, the propagator is

$$G(x',x) = \langle 0\,|T(\varphi(x)\varphi^*(x'))|\,0\rangle. \tag{3.47}$$

In classical field theory, propagators are Green's functions. The same is true for the quantum case, which we can see by operating on $G(x', x)$ with the Klein-Gordon operator. The boundary conditions implied by the step functions, or time-ordering, in eq.(3.47) are quite different from those arising in classical contexts. To make this calculation easier, we will compute $G(x', x)$ by replacing φ and φ^* with their plane wave expansions, eq. (3.38). Using eq.(3.41) and the a and b commutators, we find

$$G(x',x) = \int \frac{d^3k}{(2\pi)^3}\frac{1}{2\omega_k}\left(\theta(t'-t)e^{-ik\cdot(x'-x)} + \theta(t-t')e^{ik\cdot(x'-x)}\right). \tag{3.48}$$

The boundary conditions are implemented by the step functions. We can cast eq.(3.48) into a covariant form by using an integral representation for the Heaviside step function,

$$\theta(t) = \lim_{\epsilon\to 0^+}\int \frac{d\omega}{2\pi i}\frac{e^{i\omega t}}{\omega - i\epsilon}. \tag{3.49}$$

Subtracting $i\epsilon$ displaces the pole above the $Re(\omega)$ axis. If $t > 0$, we close the contour above the $Re(\omega)$ axis enclosing the pole. If $t < 0$, we close

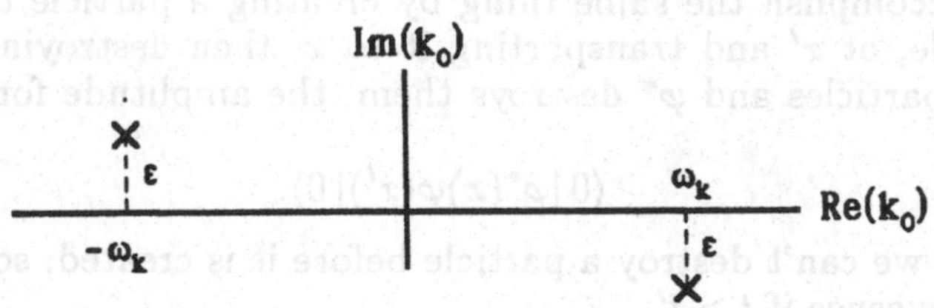

Figure 3.3 Displacement of the poles in the propagator. Positive energies are propagated forward in time, negative energies backwards in time.

the contour in the bottom half-plane, missing the pole, so the integral vanishes.

Inserting the integral into eq.(3.48), we have

$$G(x',x) = \lim_{\epsilon\to 0^+} -i\int \frac{d^3k}{(2\pi)^3}\frac{1}{2\omega_k}\frac{1}{\omega - i\epsilon}\Big(e^{i(\omega-\omega_k)(t'-t)}e^{i\vec{k}\cdot(\vec{x}'-\vec{x})} + e^{-i(\omega-\omega_k)(t'-t)}e^{-i\vec{k}\cdot(\vec{x}'-\vec{x})}\Big). \quad (3.50)$$

In the first integral we change variables from ω to $k_0 = \omega_k - \omega$ and in the second from ω to $k_0 = \omega - \omega_k$ and $\vec{k}$ to $-\vec{k}$. The factors of $\exp(\pm i\omega_k(t'-t))$ cancel out leaving

$$G(x',x) = \lim_{\epsilon\to 0^+} -i\int \frac{d^4k}{(2\pi)^4}\frac{e^{-ik\cdot(x'-x)}}{2\omega_k}\Big(\frac{1}{\omega_k - k_0 - i\epsilon} + \frac{1}{k_0 + \omega_k - i\epsilon}\Big). \quad (3.51)$$

Since ϵ is small, we can drop terms of order ϵ^2 and write the results as

$$\begin{aligned} G(x',x) &= \lim_{\epsilon\to 0^+} i\int \frac{d^4k}{(2\pi)^4}\frac{e^{-ik\cdot(x'-x)}}{k^2 - m^2 + i\epsilon} \quad (3.52)\\ &\equiv i\Delta_F(x'-x). \end{aligned}$$

In this form, it is clear that $\Delta_F(x'-x)$ satisfies

$$(\partial_{\mu'}\partial^{\mu'} + m^2)\Delta_F(x'-x) = -\delta^4(x'-x), \quad (3.53)$$

so $G(x',x)$ is a Green's function for the Klein-Gordon equation.

The "$i\epsilon$" prescription in eq.(3.52) implements the boundary conditions, namely, that Δ_F describes the propagation of a particle from x

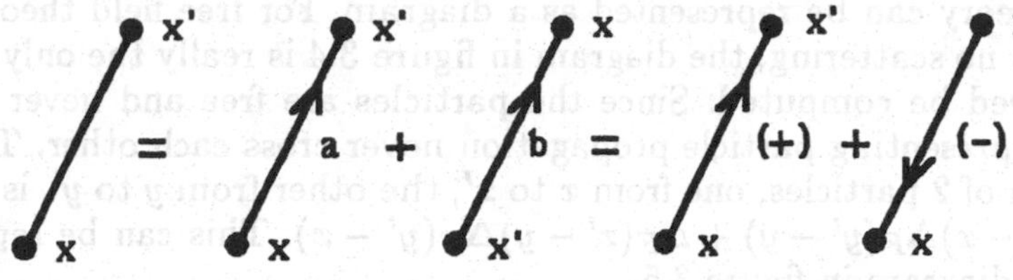

Figure 3.4 Diagram representing the propagator for scalar particles.

to x' when $t' > t$, and the propagation of an antiparticle from x' to x when $t > t'$. $i\epsilon$ accomplishes this by displacing the positive energy pole ($k_0 = \omega_k$) below the k_0 axis and the negative energy pole ($k_0 = -\omega_k$) above the axis (see figure 3.3). For $t' > t$, we must close the contour in the lower half-plane; thus only the positive energy pole will be included. For $t' < t$, the opposite is true. If we fix $t' > t$, then Δ_F will only propagate positive energy states forward in time and negative energy states backwards in time. From this observation, we can see that antiparticles of positive energy propagating forward in time can be interpreted as negative energy particles propagating backwards in time. This propagation is represented by the diagrams in figure 3.4. The single line on the left in figure 3.4 represents both means of propagation (a, b or positive energy ($+$) forward, negative energy ($-$) backwards), that is, Δ_F.

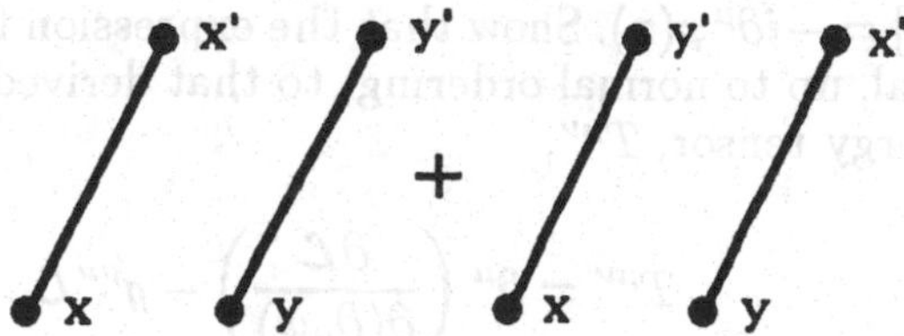

Figure 3.5 Diagram representing the propagation of 2 noninteracting identical scalar particles.

The diagram in figure 3.4 is one of the simplest of a set called Feynman diagrams, after their inventor, R. P. Feynman. As we shall see in future chapters, each term in a perturbative solution to an interacting quantum field theory can be represented as a diagram. For free field theory where there is no scattering, the diagram in figure 3.4 is really the only diagram that need be computed. Since the particles are free and never interact, lines representing particle propagation never cross each other. The propagation of 2 particles, one from x to x', the other from y to y', is given by $\Delta_F(x'-x)\Delta_F(y'-y)+\Delta_F(x'-y)\Delta_F(y'-x)$. This can be represented by the diagram in figure 3.5.

3.6 EXERCISES

1. Starting from the number operator, N, of a real Hermitian field, $\varphi(x)$,

$$N=\int\frac{d^3k}{(2\pi)^3}\frac{1}{2\omega_k}\,a^\dagger(k)a(k),$$

find the corresponding expression for N in terms of $\varphi(x)$ and $\pi(x)$. Compare the result with j^0, eq.(3.47), for a complex field. It is obvious from the form of N above that it commutes with the Hamiltonian, eq.(3.24). Verify that this is true for N and H expressed in terms of $\varphi(x)$ and $\pi(x)$.

2. The total 4-momentum operator is

$$P^\mu=\int\frac{d^3k}{(2\pi)^3}\frac{1}{2\omega_k}\,k^\mu\,a^\dagger(k)a(k).$$

Express P^μ in terms of $\varphi(x)$ and $\pi(x)$ and show that P^μ satisfies $[P^\mu,\varphi(x)]=-i\partial^\mu\varphi(x)$. Show that the expression for $P^\mu=\int d^3x\,T^{0\mu}$ is identical, up to normal ordering, to that derived from the classical stress-energy tensor, $T^{\mu\nu}$,

$$T^{\mu\nu}=\partial^\mu\left(\frac{\partial\mathcal{L}}{\partial(\partial_\nu\varphi)}\right)-g^{\mu\nu}\mathcal{L}.$$

Note that $\partial_\mu P^\mu=0$. For what symmetry in the action does this conserved current arise via eq.(2.45)?

3. The total angular momentum operator is

$$M^{\mu\nu}=\int d^3x\,(x^\mu P^\nu-x^\nu P^\mu).$$

What is the angular momentum of the vacuum? Compute the angular momentum of a classical particle of rest mass m with 4-momentum k^μ_{cl}. Construct a 1-particle state of momentum k^μ_{cl}, and apply $M^{\mu\nu}$ to it. Since the two results are identical, we can conclude that the field φ creates particles of zero intrinsic spin.

4. Lorentz invariance: In order to canonically quantize, we had to choose a specific Lorentz frame. To verify that the resulting quantum field theory is Lorentz invariant, we can check to see that the operators that generate infinitesimal Lorentz transformations still satisfy the Lorentz algebra. The Lorentz algebra is

$$[M^{\mu\nu}, M^{\lambda\sigma}] = i(\eta^{\mu\lambda}M^{\nu\sigma} - \eta^{\nu\lambda}M^{\mu\sigma} - \eta^{\mu\sigma}M^{\nu\lambda} + \eta^{\nu\sigma}M^{\mu\lambda}),$$

where the generators $M^{\mu\nu}$ are given in exercise 3, and where $\eta^{\mu\nu}$ is the Minkowski metric, $\eta^{\mu\nu} = \text{diag}(1, -1, -1, -1)$. Rewrite $M^{\mu\nu}$ in terms of the operators $a(k)$ and $a^\dagger(k)$, and show that the algebra above still holds after quantization.

5. Express $\Delta_F(x-y)$, eq.(3.52), in terms of $\Delta(x-y)$, eq.(3.25), and $\Delta_1(x-y)$ (definition follows eq.(3.29)). Δ vanishes outside of the light-cone, while Δ_1 does not. Thus, the time-ordered product or propagator, Δ_F, does not vanish outside of the light-cone. Explain why this does not violate causality.

CHAPTER **4**

Free Spinor Field Theory

4.1 THE DIRAC EQUATION

In the previous chapter, we discussed the difficulties encountered in constructing a single-particle theory from the Klein-Gordon equation. The conserved current associated with the Klein-Gordon equation does not allow a probability interpretation and the Klein-Gordon equation possesses negative energy solutions. We also discussed how to resolve these difficulties by abandoning the single-particle theory and building a quantum field theory instead. In addition, the existence of antiparticles allowed us to reinterpret the conserved current as a charge current. By allowing the negative energy solutions to only propagate backwards in time, so that they look like positive energy particles, we could identify these solutions with the antiparticles. But, in 1928 when Dirac produced his famous equation, antiparticles had not been discovered yet (not until five years later) and quantum field theory hadn't been "invented." Dirac approached the difficulties of the Klein-Gordon equation by abandoning it and seeking a new equation. He realized that the nonpositive current density was related to the appearance of the second time derivative in the Klein-Gordon equation. So, he looked for a first-order equation. To be satisfactory, the equation had to be relativistic, that is, Lorentz covariant, and its solutions also had to satisfy the Klein-Gordon equation. He succeeded in finding

such an equation, but it still admitted negative energy solutions. In hindsight, this is to be expected since these solutions go with the antiparticles and are a manifestation of the Lorentz symmetry. Faced with the negative energy solutions, Dirac proposed a clever resolution by completely filling the negative energy "sea" and interpreting the "holes" in this sea as positive energy particles. Since these "holes" carried a positive charge, Dirac and others originally hoped to identify them with protons, the only positively charged particles known at that time. This identification met several difficulties. For example, Dirac, Oppenheimer, and Tamm calculated that when an electron met a hole, the electron *quickly* jumped to fill the negative energy vacancy, resulting physically in the annihilation of the electron and the hole. Electrons and protons clearly did not annihilate each other. The interpretation of holes as antiparticles followed the discovery in 1933 of the positron. The annoying negative energy solutions, which had appeared as a shortcoming, became a blessing.

Following Dirac, we look for a relativistic equation that is first-order in the time derivative so that it may yield a nonnegative probability density. Since we want the equation to be Lorentz covariant, it seems reasonable that the equation should be first-order in the spatial derivatives as well so that time and space are treated equally.

$$i\frac{\partial\psi}{\partial t} = -i\,\vec{\alpha}\cdot\frac{\partial\psi}{\partial\vec{x}} + \beta m\psi \tag{4.1}$$

There is no hope for Lorentz covariance if the α_i's and β are just ordinary numbers. We also know from the nonrelativistic case that ψ should have 2 components, one for spin up, the other for spin down. Therefore, we should treat eq.(4.1) as a matrix equation. The α_i's and β are matrices with constant entries. ψ is a column vector.

We want the components of ψ to satisfy the Klein-Gordon equation in the interest of relativistic covariance, and so that the correct relativistic energy-momentum relation for a free particle, eq.(3.2), is satisfied. From the correspondence principle, $E = i\partial/\partial t$, $p_j = -i\partial/\partial x_j$, this means that eq.(4.1) must be a kind of square root of the Klein-Gordon equation. If we apply the operators on the left- and right-hand sides of eq. (4.1) twice, then the Klein-Gordon equation should appear. This will be true if the matrices, α_i and β, satisfy

$$\begin{aligned} \{\alpha_i, \alpha_j\} &= 2\delta_{ij} \\ \{\alpha_i, \beta\} &= 0 \\ \beta^2 &= 1\,. \end{aligned} \tag{4.2}$$

The first anticommutator above implies that $\alpha_i^2 = 1$. Therefore, the eigenvalues of α_i and β are ± 1. Also, the α_i and β must be traceless,

$$\operatorname{tr}\alpha_i = \operatorname{tr}\alpha_i\beta^2 = \operatorname{tr}\beta\alpha_i\beta = -\operatorname{tr}\beta^2\alpha_i = -\operatorname{tr}\alpha_i = 0, \tag{4.3}$$

where we have used the cyclic properties of the trace in going from the second to the third step, and the anticommutators, eq.(4.2), in going from the third to the fourth step. Since the eigenvalues are ± 1 and the matrices are traceless, they must be even-dimensional. The smallest dimension, 2, will not work because there are only three anticommuting matrices, the Pauli spin matrices,

$$\sigma_1 = \begin{pmatrix} 0 & 1 \\ 1 & 0 \end{pmatrix} \quad \sigma_2 = \begin{pmatrix} 0 & -i \\ i & 0 \end{pmatrix} \quad \sigma_3 = \begin{pmatrix} 1 & 0 \\ 0 & -1 \end{pmatrix} . \tag{4.4}$$

If the particles were massless, $m = 0$, then we would only need three matrices, so the Pauli matrices would suffice. But for massive particles, we need four matrices, so they must be at least 4×4. In fact, 4 dimensions are sufficient. One representation that realizes the algebra, eq.(4.2), is the so-called Dirac representation,

$$\alpha_i = \begin{pmatrix} 0 & \sigma_i \\ \sigma_i & 0 \end{pmatrix} \qquad \beta = \begin{pmatrix} 1 & 0 \\ 0 & -1 \end{pmatrix} , \tag{4.5}$$

where we have written the 4x4 matrices in block matrix form. The σ_i's are the Pauli matrices, eq.(4.4). Actually, it is more convenient to work with the "gamma" matrices, defined as $\gamma^0 = \beta$ and $\gamma^i = \beta\alpha_i$. The γ's satisfy

$$\{\gamma^\mu, \gamma^\nu\} = 2g^{\mu\nu} . \tag{4.6}$$

The γ^i's are antihermitian ($\gamma^{i\dagger} = -\gamma^i$) and square to -1. γ^0 is, of course, still Hermitian. In terms of the γ's, the Dirac equation, eq.(4.1), becomes

$$(i\partial\!\!\!/ - m)\psi = 0, \tag{4.7}$$

where we have employed Feynman's slash notation, $\partial\!\!\!/ \equiv \gamma^\mu \partial/\partial x^\mu$. The reason for the preference of using the γ^μ's is that the algebra is manifestly Lorentz invariant. This simplifies the demonstration of the Lorentz covariance of eq.(4.7). We will not explicitly do this here. There are many excellent references that do so (see the Bibliography). The proof of covariance also establishes that the Dirac equation describes spin 1/2 particles.

Now that we have an equation we can see what conserved current it produces. To do so we follow the same procedure as in the previous two chapters. This requires that we know the conjugate equation, the derivation of which is slightly more involved because of the matrices. Starting from eq.(4.1) and taking the Hermitian conjugate, we have

$$-i\frac{\partial \psi^\dagger}{\partial t} = i\frac{\partial \psi^\dagger}{\partial \vec{x}} \cdot \vec{\alpha} + \psi^\dagger \beta m, \tag{4.8}$$

because α_i and β are Hermitian. To recover the form corresponding to eq.(4.7) we multiply on the right by $1 = \beta^2$. $\psi^\dagger$ appears everywhere multiplied by β on the right so we define

$$\bar{\psi} \equiv \psi^\dagger \beta = \psi^\dagger \gamma^0. \tag{4.9}$$

The resulting equation is

$$-i\frac{\partial \bar{\psi}}{\partial t}\gamma^0 = i\frac{\partial \bar{\psi}}{\partial \vec{x}} \cdot \vec{\gamma} + \bar{\psi} m\,,$$

or

$$\bar{\psi}(i\overleftarrow{\not{\partial}} + m) = 0\,, \tag{4.10}$$

where the arrow above the derivative indicates that it acts to the left.

To find the current, we multiply eq.(4.7) on the left by $\bar{\psi}$, multiply eq.(4.10) on the right by ψ, and add the two. The result is

$$\partial_\mu \bar{\psi} \gamma^\mu \psi + \bar{\psi} \gamma^\mu \partial_\mu \psi = 0 = \partial_\mu(\bar{\psi}\gamma^\mu \psi), \tag{4.11}$$

so the conserved current is

$$j^\mu = \bar{\psi}\gamma^\mu \psi. \tag{4.12}$$

The density, $j^0 = \psi^\dagger \psi$, is indeed nonnegative.

Now let's consider the free particle plane wave solutions of eq.(4.7). We will need these to gain a particle interpretation when we second quantize the Dirac equation. ψ is a 4-component column vector called a spinor. In the scalar field case in the previous chapter, we had one component for each particle. For example, the two Hermitian fields, φ_1 and φ_2, that went into the charged scalar field produced two particles, one with a $+$ charge, the other with a $-$ charge. Here, we have 4 components, so we may suspect that, upon quantization, we will have 4 different particles. We started out to describe the electron, with its two spin states, relativistically. But, by requiring the components of ψ to satisfy the Klein-Gordon equation, the matrices α_i and β had to be at least 4×4, so we were forced to introduce 2 more components than we originally thought we should need. By requiring the components to satisfy the Klein-Gordon equation, we can expect that we will introduce negative energy solutions, one with spin up, the other with spin down, and this is what has happened. We know from the previous chapter that these solutions go with the positron, the anti-electron. Once again, we can see that by insisting on a relativistic equation, we must introduce antiparticles.

We will denote the positive energy solutions by

$$\psi_+ = e^{-ip\cdot x} u(p), \tag{4.13}$$

where $u(p)$ is a 4-component spinor. The negative energy solutions are

$$\psi_- = e^{ip\cdot x} v(p). \tag{4.14}$$

If we plug ψ_+ into the Dirac equation, eq.(4.7), we find that the spinor $u(p)$ satisfies

$$(\not p - m)u(p) = 0, \tag{4.15}$$

and, similarly, for negative energy solutions,

$$(\not p + m)v(p) = 0. \tag{4.16}$$

For particles at rest, $\vec{p} = 0$, so $p^0 = E = m$ and eq.(4.15) reduces to

$$(\beta - 1)u(0) = 0, \tag{4.17}$$

which has two independent solutions,

$$u^1(0) = \begin{pmatrix} 1 \\ 0 \\ 0 \\ 0 \end{pmatrix} \quad \text{and} \quad u^2(0) = \begin{pmatrix} 0 \\ 1 \\ 0 \\ 0 \end{pmatrix}. \tag{4.18}$$

u^1 describes a particle with spin up (spin along the $+z$ direction) while u^2 describes a particle with spin down. Eq.(4.16) reduces to

$$(\beta + 1)v(0) = 0, \tag{4.19}$$

which, again, has two independent solutions,

$$v^1(0) = \begin{pmatrix} 0 \\ 0 \\ 1 \\ 0 \end{pmatrix} \quad \text{and} \quad v^2(0) = \begin{pmatrix} 0 \\ 0 \\ 0 \\ 1 \end{pmatrix}. \tag{4.20}$$

v^1 describes a negative energy particle with spin up; v^2 describes spin down.

For particles not at rest, we can solve eqs.(4.15) and (4.16) by noting that $(\not p - m)(\not p + m) = p^2 - m^2 = 0$ so that $u^i(p) = \eta'(\not p + m)u^i(0)$ will satisfy eq.(4.15). Similarly, $v^i(p) = \eta'(-\not p + m)v^i(0)$ will satisfy eq.(4.16). The result is

$$u^1(p) = \eta \begin{pmatrix} 1 \\ 0 \\ \frac{p_z}{E+m} \\ \frac{p_+}{E+m} \end{pmatrix} \quad u^2(p) = \eta \begin{pmatrix} 0 \\ 1 \\ \frac{p_-}{E+m} \\ \frac{-p_z}{E+m} \end{pmatrix}$$

$$v^1(p) = \eta \begin{pmatrix} \frac{p_z}{E+m} \\ \frac{p_+}{E+m} \\ 1 \\ 0 \end{pmatrix} \quad v^2(p) = \eta \begin{pmatrix} \frac{p_-}{E+m} \\ \frac{-p_z}{E+m} \\ 0 \\ 1 \end{pmatrix}, \tag{4.21}$$

where $p_\pm = p_x \pm ip_y$, $E = (p^2 + m^2)^{1/2} > 0$, and the normalization, η (η') is chosen so

$$\bar{u}^i(p)u^j(p) = \delta_{ij} = -\bar{v}^i(p)v^j(p). \tag{4.22}$$

This normalization is Lorentz invariant, since $\bar{\psi}\psi$ transforms like a Lorentz scalar. The charge density, $\rho = \psi^\dagger\psi$, is not a scalar, but a component in a 4-vector. Using the normalization above,

$$u^{i\dagger}(p)u^j(p) = v^{i\dagger}(p)v^j(p) = \frac{E}{m}\delta_{ij}. \tag{4.23}$$

Since the density is 1 over a volume, the factor, E/m, present above is there to compensate for the Lorentz contraction of a unit volume in the direction of motion, thereby insuring an invariant probability.

In addition to eqs.(4.22)-(4.23), the spinors also satisfy

$$\bar{v}^i(p)u^j(p) = 0, \quad \sum_i u^i_\alpha(p)\bar{u}^i_\beta(p) = \left(\frac{\not{p}+m}{2m}\right)_{\alpha\beta},$$

$$\sum_i v^i_\alpha(p)\bar{v}^i_\beta(p) = \left(\frac{\not{p}-m}{2m}\right)_{\alpha\beta}, \tag{4.24}$$

which makes them orthogonal and complete.

While Dirac succeeded in producing a nonnegative current density, he could not shake the negative energy solutions. Faced with the problem of having an energy spectrum that is unbounded from below, Dirac found a clever solution to prevent one from extracting an infinite amount of energy from a particle (as we described in the previous chapter). Dirac made the supposition that all of the negative energy states, the "negative energy sea," were filled. Since spin 1/2 particles are fermions, no two electrons can occupy the same state. Positive energy electrons are prevented from falling into the negative energy sea because there are no vacancies. With a full sea, the system is stable. However, in making this assumption, Dirac implicitly abandoned making a 1-particle theory.

While this filling of the sea solves the stability problem presented by the negative energy solutions, it introduces another dilemma. The ground state or vacuum, the state with no positive energy electrons, is no longer empty! This is certainly counter to what we expect a vacuum to be. The "bare" vacuum, the truly empty vacuum with no positive energy or negative energy electrons, is, however, unstable.

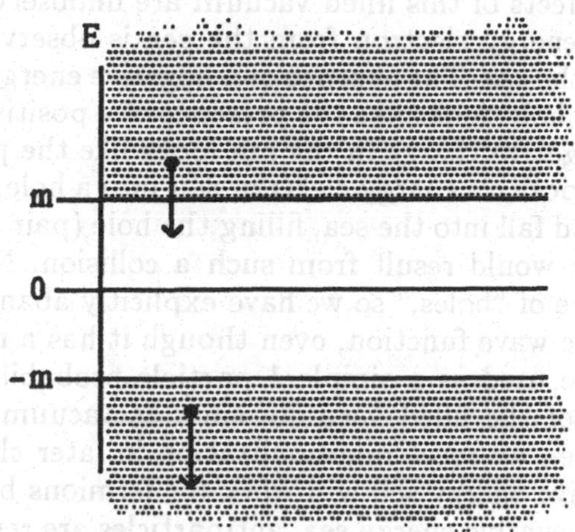

Figure 4.1 Two electrons, produced by a fluctuation, falling to the bottom of the empty negative energy sea.

Using an argument similar to one in the previous chapter, we can see that if we start with the bare vacuum, whose energy is zero, we won't find the bare vacuum again if we look at a later time. There are quantum fluctuations in the bare vacuum, so if we start with an empty vacuum, eventually a fluctuation will create a pair of electrons, one with energy $+E$ $(E \geq m)$, the other with $-E$. The total energy of the new state containing 2 particles is zero, the same as the bare vacuum. But once the 2 particles are created, the system finds that it can lower its energy by letting the 2 particles fall to the "bottom" of the sea as in figure 4.1. (To fall in energy, the electrons would have to radiate the energy, so we would actually need the electrons coupled to something.)

After the fall, we no longer have the bare vacuum and the energy has been lowered. Next, another fluctuation will create another pair, and they too will drop into the sea to lower the energy of the state. Then another fluctuation will create another pair, and so on.

When will this pair production process stop? It will stop when all of the negative energy states are filled. The fluctuations will no longer be able to create a zero-energy pair, because there will be no place to put the negative energy electron of the pair. This state, the physical vacuum, is stable against quantum fluctuations. So, if we start with the bare vacuum, after a while we end up with the negative energy sea filled, the physical vacuum. The physical vacuum has a negative infinite energy, but we can scale this to zero just as we did with the zero-point energy in the previous chapter.

While any effects of this filled vacuum are unobservable, the absence of one negative energy electron from the sea is observable. Dirac called these "holes" in the sea. The absence of a negative energy electron, that is, the presence of a hole, looks like the presence of a positive energy particle. The lack of a negative charge in the sea looks like the presence of a positive charge. If a positive energy electron ran into a hole, then the electron would radiate and fall into the sea, filling the hole (pair annihilation). The physical vacuum would result from such a collision. Now we must consider the presence of "holes," so we have explicitly abandoned a 1-particle theory. Thus, the wave function, even though it has a nonnegative norm, can no longer be used as a simple 1 particle probability density. Other physical effects of the filled vacuum, such as vacuum polarization, will be discussed when we meet renormalization in later chapters. Note that filling the negative energy sea will work for fermions but not for bosons. In terms of the negative energy sea, antiparticles are represented as holes. The representation of antiparticles as negative energy particles propagating backwards in time is a little more general in the sense that it works for bosons and fermions. The "sea"-and-"hole" picture is, nonetheless, conceptually very useful.

As we will see explicitly in the next section, the conserved quantity, $Q = \int d^3x\, j^0$, coming from the conserved current can be interpreted as a charge. The particles (electrons) can be distinguished from their antiparticles (positrons) by their charge. What if we wanted to describe spin 1/2 particles without charge? This is similar to the difference of working with a real field, $\varphi = \varphi^*$, versus a complex, charged field, $\varphi \neq \varphi^*$, in the previous chapter. The real field has half the degrees of freedom of the complex field. We can make ψ real by working with a different representation of the Dirac gamma matrices, γ_μ, the so-called Majorana representation. In this representation we take

$$\gamma^0 = \begin{pmatrix} 0 & \sigma_2 \\ \sigma_2 & 0 \end{pmatrix} \quad \gamma^1 = \begin{pmatrix} i\sigma_3 & 0 \\ 0 & i\sigma_3 \end{pmatrix}$$

$$\gamma^2 = \begin{pmatrix} 0 & -\sigma_2 \\ \sigma_2 & 0 \end{pmatrix} \quad \gamma^3 = \begin{pmatrix} -i\sigma_1 & 0 \\ 0 & -i\sigma_1 \end{pmatrix}. \tag{4.25}$$

One can check that $\{\gamma^\mu, \gamma^\nu\} = 2g^{\mu\nu}$ is satisfied. If we multiply γ^μ by i, we see that all four matrices are real. Thus, the Dirac equation,

$$(i\not{\partial} - m)\psi = (i\gamma^\mu\partial_\mu - m)\psi = 0, \tag{4.7}$$

is real along with its solutions. The number of degrees of freedom in ψ is reduced by a factor of two. With Majorana fermions, we cannot expect to distinguish between particles and antiparticles.

Another interesting case, which we touched upon earlier, is when $m = 0$. A physical realization of this spin 1/2 particle is the neutrino. Any massive particle is also expected to behave like $m = 0$ at very high energies. For $m = 0$, we only need to find three anticommuting matrices instead of four, so the Pauli matrices will suffice and the spinor need only have 2 components. Keeping this in mind, if we consider the 4-component Dirac equation with $m = 0$,

$$i\partial\!\!\!/\psi = 0, \tag{4.26}$$

we can expect, by the proper choice of representation of the γ matrices, to decouple the two components of the spinor, ψ, from the bottom two. Such a choice is called the Weyl or chiral representation.

$$\gamma^0 = \begin{pmatrix} 0 & -1 \\ -1 & 0 \end{pmatrix} \quad \gamma^i = \begin{pmatrix} \sigma_i & 0 \\ 0 & -\sigma_i \end{pmatrix} \tag{4.27}$$

If we restrict ψ to 2 components, we will lose the antiparticles. For the 4-component case, however, there is no charge to distinguish particles from antiparticles. They can be distinguished, though, by their chirality. Since the particles are massless, they move at the speed of light, and there exist no rest frames. Therefore, the projection of spin onto the direction of motion is an invariant. This, of course, is not true if the particle has mass, because we can go to the rest frame of the particle. There we do not know what it means to project the spin onto the direction of motion.

If the projection of spin onto the direction of motion is positive, $\vec{\sigma}\cdot\vec{p} > 0$, the particle is said to have + chirality. If the projection is opposite, the particle has negative chirality. For spin 1 photons, the state of + chirality is traditionally called right circular polarization; hence the motivation for the use of the term chirality.

Particles and antiparticles have opposite chirality. The operator that distinguishes the two is the chirality operator, γ^5.

$$\gamma^5 = \gamma_5 = i\gamma^0\gamma^1\gamma^2\gamma^3 \tag{4.28}$$

γ^5 anticommutes with γ^μ and squares to 1. In the chiral representation, the chirality operator is diagonal,

$$\gamma_5 = \begin{pmatrix} 1 & 0 \\ 0 & -1 \end{pmatrix}. \tag{4.29}$$

Particles of definite chirality are eigenstates of γ_5. From eq.(4.29), we see that the top 2 components of ψ in the chiral representation correspond to a + chiral 2-spinor, and the bottom 2 form a − chirality spinor. If the direction of motion is taken as the z axis, then the positive energy, + chirality spinor is simply

$$u(p) = \begin{pmatrix} 1 \\ 0 \\ 0 \\ 0 \end{pmatrix}, \tag{4.30}$$

while the negative energy, − chirality spinor is

$$v(p) = \begin{pmatrix} 0 \\ 0 \\ 0 \\ 1 \end{pmatrix}. \tag{4.31}$$

A Weyl 4-spinor (a spinor that is an eigenspinor of the chirality operator, γ^5) has the same number of degrees of freedom as a Majorana spinor, that is, half the number of a Dirac spinor. In four dimensions, a spinor cannot be both Weyl and Majorana. In 10 dimensions, however, a spinor can be both Majorana and Weyl. A Dirac spinor in 10 dimensions has 32 degrees of freedom, while a Majorana spinor or a Weyl spinor has 16. A Majorana-Weyl spinor has only 8. This exactly matches the number of transverse modes of a vector in 10 dimensions. This balance is used to construct supersymmetric theories in 10 dimensions where the number of fermionic degrees of freedom must equal the number of bosonic degrees of freedom.

Massless spin 1/2 Dirac particles were first considered by Weyl in 1929. They were ignored because, in 2-component form, there is no way to define the parity operation (space inversion) since parity reverses chirality. Around 1957, however, the discovery was made that the weak interaction violated parity, and that only left-handed (− chirality) neutrinos and right-handed antineutrinos exist in nature, so the parity objection turned to dust.

The conserved current that we derived from the Dirac equation is of the form $j^\mu = \bar{\psi}\gamma^\mu\psi$ and it transformed as a Lorentz 4-vector. $\bar{\psi}\gamma^\mu\psi$ is one of several bilinear covariants (bilinear in ψ and transforming covariantly). Using the four γ matrices, it is possible to construct 16 linearly independent 4×4 matrices. Sandwiched between $\bar{\psi}$ and ψ, these form the covariants. The 16 independent matrices are: the unit matrix (γ_i^2), the gamma matrices themselves, γ^μ, the chirality operator, γ_5, γ_5 multiplying the gamma matrices, $\gamma_5\gamma^\mu$, and $\sigma_{\mu\nu} \equiv i/2[\gamma_\mu, \gamma_\nu]$. The bilinear covariants are:

$\bar{\psi}\psi$, which transforms as a scalar;

$\bar{\psi}\gamma_5\psi$, which transforms as a pseudoscalar;

$\bar{\psi}\gamma_\mu\psi$, which transforms as a vector;

$\bar{\psi}\gamma_5\gamma_\mu\psi$, which transforms as a pseudovector;

$\bar{\psi}\sigma_{\mu\nu}\psi$, which transforms as a second-rank tensor.

The "pseudo" on "scalar" and "vector" above arises because γ_5 anticommutes with parity (space inversion). Therefore, a pseudoscalar transforms into itself under a proper Lorentz transformation and into minus itself under an improper Lorentz transformation.

4.2 FREE SPINOR FIELD THEORY

Treating the Dirac equation as a classical field equation, we now proceed to canonically quantize it. Since we are dealing with fermions, we will use anticommutators instead of commutators.

The Dirac equation,

$$(i\partial\!\!\!/ - m)\psi = 0, \tag{4.7}$$

can be obtained from the action,

$$S[\bar{\psi}, \psi] = \int d^4x\, \mathcal{L} = \int d^4x\, \bar{\psi}(x)(i\partial\!\!\!/ - m)\psi(x), \tag{4.32}$$

by varying $\bar{\psi}$. If we vary ψ, we obtain the conjugate equation. The field momentum conjugate to ψ is

$$\pi = \frac{\partial \mathcal{L}}{\partial \dot{\psi}} = i\psi^\dagger, \tag{4.33}$$

and the Hamiltonian is

$$H = \int d^3x \pi\dot{\psi} - \mathcal{L} = \int d^3x \psi^\dagger(-i\vec{\alpha}\cdot\nabla + \beta m)\psi. \tag{4.34}$$

Quantization is completed by specifying the quantum conditions, the equal-time anticommutators.

$$\{\psi_\alpha(\vec{x},t), \psi_\beta^\dagger(\vec{y},t)\} = \delta_{\alpha\beta}\delta^3(\vec{x}-\vec{y})$$

$$\{\psi_\alpha(\vec{x},t), \psi_\beta(\vec{y},t)\} = \{\psi_\alpha^\dagger(\vec{x},t), \psi_\beta^\dagger(\vec{y},t)\} = 0 \tag{4.35}$$

Once again, the use of noncovariant equal-time anticommutators forces us to choose a specific Lorentz frame to quantize in. We must verify that the Lorentz symmetry remains unbroken after quantization. This is left for the exercises.

To gain a particle interpretation, expand ψ and $\psi^\dagger$ in plane waves,

$$\psi(\vec{x},t) = \int \frac{d^3p}{(2\pi)^3}\frac{m}{E}\sum_{i=1}^{2}(b_i(p)u^i(p)e^{-ip\cdot x} + d_i^\dagger(p)v^i(p)e^{ip\cdot x})$$

$$\psi^\dagger(\vec{x},t) = \int \frac{d^3p}{(2\pi)^3}\frac{m}{E}\sum_{i=1}^{2}(b_i^\dagger(p)\bar{u}^i(p)\gamma^0 e^{ip\cdot x} + d_i(p)\bar{v}^i(p)\gamma^0 e^{-ip\cdot x}). \tag{4.36}$$

Using the properties of the spinors $u^i(p)$ and $v^i(p)$ given in the previous section, the expansion above will satisfy eq.(4.35) if the operators b_i, $b_i^\dagger$, d_i, and $d_i^\dagger$ satisfy

$$\begin{aligned}\{b_\alpha(p), b_\beta^\dagger(p')\} &= (2\pi)^3\frac{E}{m}\delta^3(\vec{p}-\vec{p}')\delta_{\alpha\beta}\\ \{d_\alpha(p), d_\beta^\dagger(p')\} &= (2\pi)^3\frac{E}{m}\delta^3(\vec{p}-\vec{p}')\delta_{\alpha\beta},\end{aligned} \tag{4.37}$$

with the other anticommutators vanishing.

Now for the interpretation. $b_1^\dagger$ creates a positive energy electron with spin up, $b_2^\dagger$ with spin down, while d_1 creates a negative energy electron with spin up, d_2 with spin down. $d_2^\dagger$ destroys a negative energy electron with spin down; that is, it creates a positive energy hole with spin up. The holes are to be interpreted as positrons, hence $d_2^\dagger$ creates a positive energy positron with spin up. $d_2(p)$ destroys a positron of 4-momentum p by filling the hole with a negative energy electron.

Let $|\,0\rangle$ denote the physical vacuum, the state where all the negative states are filled. Let $|\,0\rangle_b$ be the bare, the truly empty, vacuum. Since $d_\alpha(p)$ creates negative energy electrons,

$$|\,0\rangle = \prod_{\alpha=1}^{2}\prod_{p} d_\alpha(p)|\,0\rangle_b\,. \tag{4.38}$$

The d_α's anticommute, so $d_\alpha^2 = 0$, and $d_\alpha(p')|\,0\rangle = 0$. That is, since $d_\alpha(p')$ will try to create a negative energy electron in a state already occupied, the result must be zero. On the other hand, $d_\alpha^\dagger(p')|\,0\rangle$ creates a state with one hole by destroying one negative energy electron in the filled sea. Thus, $d_\alpha^\dagger(p')|\,0\rangle$ is a state with one positive energy positron.

Expressed in terms of the b's and d's, the Hamiltonian is

$$H = \sum_{i=1}^{2}\int\frac{d^3p}{(2\pi)^3}m(b_i^\dagger(p)b_i(p) - d_i(p)d_i^\dagger(p)). \tag{4.39}$$

The expectation of the Hamiltonian in the physical vacuum is negatively infinite. The reason is that while $b_i(p)|0\rangle = 0$, $d_i^\dagger(p)|0\rangle \neq 0$. Once again, we must normal order the operators to remove this infinite constant. This means that positive energy parts stand to the right of negative energy parts, that is, that $d_i^\dagger(p)$ must be on the left of $d_i(p)$. Since they anticommute, when we bring $d_i^\dagger(p)$ to the left in eq.(4.39), that term changes sign leaving

$$:H: = \sum_{i=1}^{2} \int \frac{d^3p}{(2\pi)^3} m(b_i^\dagger(p)b_i(p) + d_i^\dagger(p)d_i(p)). \tag{4.40}$$

Now we can see that the physical vacuum is the state of lowest energy so the system will be stable.

If we rewrite the charge $Q = \int d^3x \psi^\dagger \psi$ in terms of the b's and d's we find that

$$Q = \sum_{i=1}^{2} \int \frac{d^3p}{(2\pi)^3} \left(\frac{m}{E}\right) (b_i^\dagger(p)b_i(p) + d_i(p)d_i^\dagger(p)). \tag{4.41}$$

The charge of the physical vacuum is infinite since $d_i^\dagger$ operates before d_i in the charge operator above. To subtract away this infinite constant we also normal order the charge.

$$\begin{aligned} :Q: = \int d^3x : \psi^\dagger(x)\psi(x) : &= \sum_{i=1}^{2} \int \frac{d^3p}{(2\pi)^3} \left(\frac{m}{E}\right) (b_i^\dagger(p)b_i(p) - d_i^\dagger(p)d_i(p)) \\ &= \sum_{i=1}^{2} \int \frac{d^3p}{(2\pi)^3} \left(\frac{m}{E}\right) (N_i^+(p) - N_i^-(p)), \end{aligned} \tag{4.42}$$

where $N_i^+(p)$ is the number operator for positive energy electrons with 4-momentum p, while $N_i^-(p)$ is the number operator for positive energy positrons. As in the charged scalar field, the particles and antiparticles carry opposite charges and we can interpret $j^0 = :\psi^\dagger\psi:$ as a charge density.

Using the expansions for ψ and $\psi^\dagger$, eq.(4.36), the anticommutators, eq.(4.37), and the properties of the spinors $u^i(p)$ and $v^i(p)$, eqs.(4.22)-(4.24), we can compute the anticommutator of ψ and $\bar{\psi}$ at different times.

$$\{\psi_\alpha(x), \bar{\psi}_\beta(y)\} = (i\partial\!\!\!/_x + m)_{\alpha\beta}\, i\Delta(x-y), \tag{4.43}$$

where $\Delta(x-y)$ is the function encountered in the scalar commutator at different times, eq.(3.25). $\Delta(x-y)$ vanishes outside of the light-cone, so we can conclude, after showing $\{\psi_\alpha(x), \psi_\beta(y)\} = 0 = \{\bar{\psi}_\alpha(x), \bar{\psi}_\beta(y)\}$, that this field theory satisfies micro-causality. If we had mistakenly quantized using commutators instead of anticommutators, not only would we have an unstable system, but the commutator of $\psi_\alpha(x)$ with $\bar{\psi}_\beta(y)$ would be proportional to $\Delta_1(x-y)$, which was defined in the previous chapter. $\Delta_1(x-y)$ does not vanish outside the light-cone. We would lose causality and locality. This is the connection between spin and statistics provided by Lorentz invariance.

4.3 THE DIRAC PROPAGATOR

The amplitude for a positive energy electron to go from x to x' is the overlap of the initial and final states,

$$\langle \Psi(x') \,|\, \Psi(x) \rangle . \tag{4.44}$$

$|\,\Psi(x)\rangle$ is the state containing 1 positive energy electron. From eq.(4.36) we see that the state containing 1 electron at x is $\psi^\dagger(x)|\,0\rangle$ where $|\,0\rangle$ is the physical vacuum. Therefore, the amplitude (4.44) is

$$\langle 0\,|\psi(x')\psi^\dagger(x)|\,0\rangle . \tag{4.45}$$

The quantum mechanical propagation of the electron, like the charged scalar particle, is the process where a positive energy electron is created out of the vacuum at x, transported to x', and reabsorbed into the vacuum at x'. This will only make sense if $t' > t$ since we can't destroy the electron before we create it.

As in the charged scalar case, there is a distinct process that is physically equivalent that we must also consider. If we consider positive energy electrons to carry $-$ charge, then the process above lowers the charge by one unit at x and raises it by one unit at x'. Negative energy electrons running backwards in time, that is, positive energy positrons, carry the opposite charge. If we create a positron at x', move it to x, and destroy it, then we are also raising the charge at x' by one unit and lowering it by the same amount at x. From eq.(4.36) and the particle interpretation, we see that ψ creates positrons so the amplitude for this process is

$$\langle 0\,|\psi^\dagger(x)\psi(x')|\,0\rangle . \tag{4.46}$$

Once again, we can't destroy the positron before we create it, hence this process makes sense only if $t > t'$. The total amplitude or propagator is the *difference* between the amplitudes for the two equivalent processes,

$$iS_F(x', x)\gamma^0 = \langle 0\,|\psi(x')\psi^\dagger(x)|\,0\rangle\theta(t'-t) - \langle 0\,|\psi^\dagger(x)\psi(x')|\,0\rangle\theta(t-t'),$$

or

$$iS_F(x',x) = \langle 0\,|T(\psi(x')\bar{\psi}(x))|\,0\rangle, \tag{4.47}$$

where T is the time-ordering operator, defined in the previous chapter for bosons. For fermions,

$$T\psi(x)\psi(x') = \begin{cases} \psi(x)\psi(x') & t > t' \\ -\psi(x')\psi(x) & t' > t. \end{cases} \tag{4.48}$$

The second term in eq.(4.47) is subtracted because electrons obey Fermi-Dirac statistics. Fermionic amplitudes are antisymmetric under the exchange of $x \leftrightarrow x'$. We may equivalently consider the $-$ sign arising in eqs.(4.47) and (4.48) from the fact that the ψ's obey anticommutation relations.

Since ψ and $\bar{\psi}$ are spinors, S_F is a matrix. Written in matrix notation,

$$iS_F(x',x)_{ij} = \langle 0\,|T(\psi_i(x')\bar{\psi}_j(x))|\,0\rangle \tag{4.49}$$

Using the expansions for ψ and $\bar{\psi}$, eq.(4.36), and the properties of the spinors u_i and v_j, including completeness,

$$\sum_i u^i_\alpha(p)\bar{u}^i_\beta(p) = \left(\frac{\not{p}+m}{2m}\right)_{\alpha\beta}, \qquad \sum_i v^i_\alpha(p)\bar{v}^i_\beta(p) = \left(\frac{\not{p}-m}{2m}\right)_{\alpha\beta}, \tag{4.50}$$

we obtain an explicit expression for S_F,

$$iS_F(x'-x) = \int \frac{d^3p}{(2\pi)^3}\frac{1}{2E}\left(\theta(t'-t)(\not{p}+m)e^{-ip\cdot(x'-x)} - \theta(t-t')(\not{p}-m)e^{ip\cdot(x'-x)}\right). \tag{4.51}$$

The expression above can be put into a covariant form by the same procedure followed in the scalar case in the previous chapter. After inserting the integral representation for θ, eq.(3.49), and changing variables from ω to $p_0 = E-\omega$ in the first term and ω to $p_0 = \omega - E$ in the second, we have

$$S_F(x'-x) = \int \frac{d^4p}{(2\pi)^4}e^{-ip\cdot(x'-x)}\frac{\not{p}+m}{p^2-m^2+i\epsilon}, \tag{4.52}$$

where the limit, $\epsilon \to 0^+$, is understood. The $i\epsilon$ implements the boundary conditions, namely, that positive energy solutions only propagate forward in time, while negative energy solutions only propagate backwards in time.

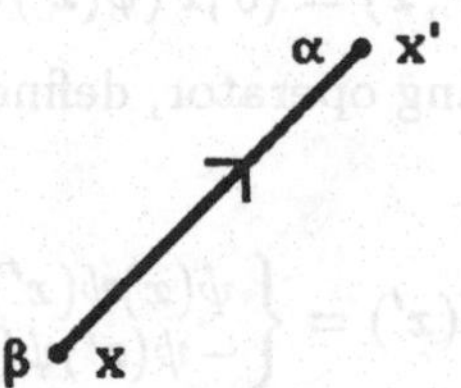

Figure 4.2 Diagram representing the Dirac propagator $\langle 0\,|T(\psi_\alpha(x')\bar{\psi}_\beta(x))|\,0\rangle$.

In $S_F(x'-x)$ above, we recognize the propagator for scalar particles, $\Delta_F(x'-x)$, eq.(3.52). Extracting the numerator, we can write

$$S_F(x'-x) = (i\partial\!\!\!/_{x'} + m)\Delta_F(x'-x). \tag{4.53}$$

We may also write the Fourier transform of $S_F(x'-x)$,

$$S_F(p) = \frac{p\!\!\!/ + m}{p^2 - m^2 + i\epsilon}, \qquad \text{as} \qquad \frac{1}{p\!\!\!/ - m + i\epsilon}. \tag{4.54}$$

In this form, it is fairly easy to see that $S_F(x)$ satisfies

$$(i\partial\!\!\!/ - m)S_F(x) = \delta^4(x). \tag{4.55}$$

The Dirac propagator can be represented by exactly the same diagram as the scalar propagator, figure 3.4, if we identify the "a" quantum in figure 3.4 with the electron and "b" with the positron.

By convention, we associate an orientation to the line representing the propagator. The arrow points in the direction of charge propagation. That is, the arrow points from the point where $\bar{\psi}$ acts to the point where ψ acts. For example, the diagram in figure 4.2 represents the propagator

$$\langle 0\,|T(\psi_\alpha(x')\bar{\psi}_\beta(x))|\,0\rangle = \int \frac{d^4p}{(2\pi)^4} e^{-ip\cdot(x'-x)} \left(\frac{p\!\!\!/ + m}{p^2 - m^2 + i\epsilon}\right)_{\alpha\beta}. \tag{4.56}$$

4.4 EXERCISES

1. Determine the normalization, η, of the spinors, $u^i(p)$ and $v^i(p)$, using eq.(4.22).
2. Using the normalization found in exercise 1 above, verify by direct substitution the orthogonality and completeness relations, eqs.(4.23) and (4.24).
3. Using the positive energy plane wave solutions to the Dirac equation, eq.(4.13), construct the eigenstates of the chirality operator, γ_5, at a given momentum. Do the same for the negative energy solutions, eq.(4.14).
4. Find the transformation, W, that takes the Dirac representation into the Weyl representation. That is, find W such that

$$\gamma^{\mu\prime} = W\gamma^\mu W^{-1},$$

where $\gamma^{\mu\prime}$ are the γ matrices in the Weyl representation, eq.(4.27), and γ^μ are the γ matrices in the Dirac representation. Transform the plane wave solutions, eqs.(4.13), and (4.14), into the Weyl representation and compare with the results of exercise 3.
5. Find the transformation, M, that takes the Dirac representation into the Majorana representation. Transform the plane wave solutions and verify that they satisfy the Dirac equation in the Majorana representation.
6. Prove the following trace identities:

   a. $\mathrm{tr}(\gamma^5\gamma^\mu) = 0$

   b. $\mathrm{tr}(\gamma^\mu\gamma^\nu) = 4g^{\mu\nu}$

   c. $\mathrm{tr}(\sigma^{\mu\nu}) = 0$

   d. $\mathrm{tr}(\gamma^5\gamma^\mu\gamma^\nu) = 0$

   e. $\mathrm{tr}(\gamma^\alpha\gamma^\beta\gamma^\sigma\gamma^\rho) = 4(g^{\alpha\beta}g^{\sigma\rho} - g^{\alpha\sigma}g^{\beta\rho} + g^{\alpha\rho}g^{\beta\sigma})$

   f. $\mathrm{tr}(\text{odd number of } \gamma'\text{s}) = 0$
7. Show that the linear momentum operator, P^j, given by

$$P^j = -i\int d^3x\, \psi^\dagger(x)\partial^j\psi(x),$$

satisfies $[P^j, \psi(x)] = -i\partial^j\psi(x)$. Transform $:P^j:$ into the momentum representation using eq.(4.36) and interpret the results in terms of particles. Show that $\partial_\mu P^\mu = 0$, where $P^0 = H$. What symmetry in the action gives rise to this conserved current?

8. Lorentz invariance: As in the case of scalar particles, in order to canonically quantize, we must choose a specific Lorentz frame. After quantization, we must verify that the resulting quantum field theory is Lorentz invariant.

   a. The generators of infinitesimal Lorentz transformations are given by

$$M^{\mu\nu} = \int d^3x \; x^\mu P^\nu - x^\nu P^\mu + \psi^\dagger(x) \frac{\sigma^{\mu\nu}}{2} \psi(x),$$

   where P^μ is defined in the previous exercise, and where

$$\sigma^{\mu\nu} = \frac{i}{2}[\gamma^\mu, \gamma^\nu].$$

   Examine the spatial components of $M^{\mu\nu}$ (the angular momentum generators) and show that the last term corresponds to an intrinsic angular momentum of spin 1/2.

   b. Use the canonical anticommutators, eq.(4.35), to show that, after quantization, the $M^{\mu\nu}$ still obey the proper algebra (see exercise 4 in chapter 3), or, equivalently, that

$$[M^{\mu\nu}, \psi(x)] = -i(x^\mu \partial^\nu - x^\nu \partial^\mu - \frac{i}{2}\sigma^{\mu\nu})\,\psi(x).$$

   What happens if we use canonical commutators instead of canonical anticommutators for eq.(4.35)?

CHAPTER 5

Quantization of the Electromagnetic Field

5.1 CLASSICAL GAUGE THEORY

In order to produce a quantum field theory describing scalar or spinor particles we had to first find a relativistic Schrödinger equation to second quantize. Historically, these two equations, the Klein-Gordon and Dirac equations, were essentially guessed at. The fields associated with these particles, φ and ψ, are not classically observable. There is no well-defined (measurable) classical limit or classical field theory to guide us through quantization. This is not true for the photon (free, massless, spin 1 boson). The fields associated with the photon, the electric and magnetic fields, are classically observable, and the classical field equations, Maxwell's equations, have been known for a long time. Yet, even though we have an experimentally well-defined classical theory to guide us, the quantization of the electromagnetic field is by far the most delicate.

The trouble begins when we choose to quantize using the vector potential instead of the electric and magnetic fields. We choose the potential over the fields because classically we already know that the potential naturally appears in the coupling of the electromagnetic field to charged currents. In addition, if we use the electric and magnetic fields, the equal-time commutators involve derivatives of the δ-function, and this introduces doubts about preserving the local character of the field theory.

The difficulty with using the vector potential is that it has more components than dynamical degrees of freedom. In the charged scalar case, we have two Hermitian fields and we end up with 2 distinct particles, a particle and an antiparticle. In the spinor case, ψ has 4 components, which give us 4 particles: an electron with spin up, an electron with spin down, a positron with spin up, and a positron with spin down. The vector potential has 4 components, but we know classically that there are only 2 kinds of photons: one with right circular polarization (helicity, chirality), and one with left circular polarization. If the vector potential is to describe photons, then two components should be nondynamical. We should only quantize the dynamical variables.

As is often the case, the presence of nondynamical variables is due to the existence of a symmetry, here a local internal symmetry called a local gauge symmetry. A gauge symmetry expresses the fact that physical results should not depend upon the arbitrary choice of coordinates we make in order to compute the result. When coupled to charged currents, the gauge symmetry of electromagnetism enforces the invariance of physical effects under a change in the choice of phase of the charged matter fields at a point (hence the use of the word "local").

The addition of the gauge symmetry will force us to modify our canonical quantization procedure. If our goal is to quantize only the dynamical degrees of freedom in the vector potential, then we must identify them first.

From the electromagnetic vector potential, $A^\mu(x)$, we can define a second-rank antisymmetric tensor $F^{\mu\nu}$ by

$$F^{\mu\nu} = \partial^\mu A^\nu - \partial^\nu A^\mu. \tag{5.1}$$

$F^{\mu\nu}$ is the field strength tensor. Using the usual definitions of the electric and magnetic fields in terms of the vector potential,

$$\vec{E} = -\nabla A^0 - \frac{\partial \vec{A}}{\partial t}, \quad \vec{B} = \nabla \times \vec{A}, \tag{5.2}$$

we find that

$$F^{\mu\nu} = \begin{pmatrix} 0 & -E^x & -E^y & -E^z \\ E^x & 0 & -B^z & B^y \\ E^y & B^z & 0 & -B^x \\ E^z & -B^y & B^x & 0 \end{pmatrix}. \tag{5.3}$$

The dual antisymmetric tensor to $F^{\mu\nu}$, denoted ${}^*F^{\mu\nu}$, is defined as

$${}^*F^{\mu\nu} = \frac{1}{2}\epsilon^{\mu\nu\rho\sigma} F_{\rho\sigma}, \tag{5.4}$$

where $\epsilon^{\mu\nu\rho\sigma} = -\epsilon_{\mu\nu\rho\sigma}$ is the antisymmetric symbol. $\epsilon^{txyz} \equiv \epsilon^{0123} = 1$ and $\epsilon^{\mu\nu\rho\sigma}$ is 1 if $\mu\nu\rho\sigma$ is an even permutation of 0123, while $\epsilon^{\mu\nu\rho\sigma}$ is -1 if $\mu\nu\rho\sigma$ is an odd permutation. In component form, the dual tensor is

$$ {}^*F^{\mu\nu} = \begin{pmatrix} 0 & -B^x & -B^y & -B^z \\ B^x & 0 & E^z & -E^y \\ B^y & -E^z & 0 & E^x \\ B^z & E^y & -E^x & 0 \end{pmatrix}. \tag{5.5} $$

The homogeneous Maxwell equations are

$$ \partial_\mu {}^*F^{\mu\nu} = 0, \tag{5.6} $$

that is,

$$ \nabla \times \vec{E} = -\dot{\vec{B}}, \quad \nabla \cdot \vec{B} = 0, \tag{5.7} $$

while the inhomogeneous Maxwell equations in the presence of a current, j_μ, are

$$ \partial_\mu F^{\mu\nu} = j^\nu, \tag{5.8} $$

that is,

$$ \nabla \cdot \vec{E} = j^0 = \rho, \quad \nabla \times \vec{B} - \frac{\partial \vec{E}}{\partial t} = \vec{j}. \tag{5.9} $$

Presently, we are considering the electromagnetic fields in vacuum, that is, in the absence of sources and currents, $j^\mu = 0$. The classical action that yields Maxwell's equation by least action is

$$ S = -\frac{1}{4} \int d^4x F^{\mu\nu}(x) F_{\mu\nu}(x). \tag{5.10} $$

The gauge symmetry arises from the equality of mixed partials. Since the field strength is the curl of the vector potential, A^μ, we can add the gradient of an arbitrary function to A^μ without changing $F^{\mu\nu}$.

$$ \begin{aligned} A^\mu(x) &\to A^\mu(x) - \partial^\mu \Lambda(x) \\ F^{\mu\nu}(x) &\to F^{\mu\nu}(x) \end{aligned} \tag{5.11} $$

Transforming A^μ by the addition of a gradient is called a local gauge transformation. Since $F^{\mu\nu}(x)$ is unchanged by a gauge transformation, it is gauge invariant. Similarly, since A^μ does change under a gauge transformation, it is gauge dependent.

We can use the freedom of adding a gradient to A^μ to make A^μ satisfy certain relations. For example, for an arbitrary $A^\mu(x)$, we can take

$$ \Lambda(x) = \int_{-\infty}^{t} A^0(x, y, z, t') dt' \tag{5.12} $$

so that the transformed potential, $A^\mu_\Lambda = A^\mu - \partial^\mu \Lambda$, will satisfy $A^0_\Lambda(x) = 0$. Choosing A^μ to satisfy some condition is called a choice of gauge, or "fixing" the gauge. Actually, the choice $A^0(x) = 0$, called the temporal gauge, does not fully fix the gauge. If we choose a new Λ that is independent of t, then the spatial components of A^μ will change but A^0 will remain zero. Once in the temporal gauge, we can fully fix the gauge by choosing Λ to satisfy $\nabla_i A_i - \nabla^2 \Lambda = 0$. The transformed potential will satisfy

$$\nabla \cdot \vec{A} = 0. \tag{5.13}$$

This is known as the Coulomb gauge or radiation gauge. This gauge choice is noncovariant. A covariant choice is the Lorentz gauge, where A^μ satisfies

$$\partial_\mu A^\mu = 0. \tag{5.14}$$

One other common choice is the axial gauge, where

$$A_z = 0. \tag{5.15}$$

The temporal gauge could equally be called the time-axial gauge. Neither the Lorentz gauge condition nor the axial gauge condition fully "fix" the gauge.

The dynamical variables are those that are independent of the arbitrary function Λ, namely, the gauge invariant variables. Any quantity that depends on how we choose Λ can't be truly physical. We can change its measured value arbitrarily by changing Λ. The time variation of a gauge dependent variable is also arbitrary since Λ may be an arbitrary function of time. Therefore, gauge dependent variables are nondynamical.

We can pick out the dynamical components of A^μ by seeing which ones are gauge invariant. We have already said that $F^{\mu\nu}$ is gauge invariant, hence the combinations of components of A^μ that make $F^{\mu\nu}$ are dynamical. With no charges present, we can set $A^0 = 0$, so A^0 is not dynamical. Since $F^{\mu\nu}$ is the curl of $\vec{A}$, then the longitudinal component of $\vec{A}$ (call it $\vec{A}_L$) is also nondynamical, since by definition, the longitudinal component of $\vec{A}$ satisfies $\nabla \times \vec{A}_L = 0$. This leaves the transverse components, $\vec{A}_T$. The transverse components are the dynamical degrees of freedom. This is a well-known result of the classical field theory.

Since the transverse components satisfy $\nabla \cdot \vec{A}_T = 0$, the Coulomb gauge, $\nabla \cdot \vec{A} = 0$, is the gauge where $\vec{A}_L = 0$. The only nonvanishing components are the dynamical ones.

If we try to quantize using the canonical procedure without fixing the gauge, we will run into trouble. First of all, we will overquantize by quantizing nondynamical, gauge dependent degrees of freedom. In addition, the canonical procedure involves specifying equal-time commutators. But specifying equal-time commutators involving a gauge dependent field variable makes no sense since the gauge dependent variable has arbitrary time dependence. To avoid these troubles, we quantize in a particular gauge. If we quantize in the Coulomb gauge, we will only quantize dynamical degrees of freedom since all the gauge dependent components have been set to zero. The drawback to quantization in the Coulomb gauge is that since this gauge choice is noncovariant, we lose manifest Lorentz covariance. This can make some computations more difficult. To maintain manifest Lorentz covariance, we can quantize in the Lorentz gauge. But, by doing so, we necessarily overquantize so we must modify the canonical quantization procedure to make sure that no nondynamical variables manifest themselves in physical results. We will illustrate both methods by quantizing first in the Coloumb gauge, then in the Lorentz gauge.

5.2 QUANTIZATION IN THE COULOMB GAUGE

We will begin by trying to follow the canonical procedure. From the classical Lagrangian density,

$$\mathcal{L}(x) = -\frac{1}{4}F^{\mu\nu}F_{\mu\nu} = \frac{1}{2}(\vec{E}^2 - \vec{B}^2), \tag{5.16}$$

we find the conjugate momentum to A^μ,

$$\begin{aligned} \pi_0(x) &= \frac{\partial\mathcal{L}}{\partial(\partial_t A^0)} = 0 \\ \pi_i(x) &= \frac{\partial\mathcal{L}}{\partial(\partial_t A^i)} = -E_i, \end{aligned} \tag{5.17}$$

and the Hamiltonian,

$$H = \int d^3x \frac{1}{2}(\vec{E}^2 + \vec{B}^2) + \vec{E}\cdot\nabla A^0. \tag{5.18}$$

If we try to impose the canonical equal-time commutators, we immediately run into trouble because $\pi_0 = 0$. ($F^{\mu\nu}$ is antisymmetric so the diagonal elements vanish.) Dumping the A^0, π^0 commutator, the remaining are:

$$\begin{aligned} &[A_i(\vec{x},t), E_j(\vec{y},t)] = -i\delta_{ij}\delta(\vec{x}-\vec{y}), \\ &[A_\mu(\vec{x},t), A_\nu(\vec{y},t)] = 0, \\ &[E_i(\vec{x},t), E_j(\vec{y},t)] = 0, \\ &[E_i(\vec{x},t), A_0(\vec{y},t)] = 0. \end{aligned} \tag{5.19}$$

By dumping the A^0, π^0 commutator, we sidestep the problem imposed by $\pi_0 = 0$, but we also give up manifest Lorentz covariance. This means that after we are finished quantizing, we must verify that the Lorentz symmetry is still intact.

Since $\pi_0(x) = 0$, A_0 commutes with everything and is not a dynamical variable. It can be taken as a number instead of an operator. Since it is not dynamical, we can eliminate it from the Hamiltonian by solving for it in terms of the dynamical variables. Since $\dot{A}_0 = 0$, then

$$\frac{\partial \mathcal{L}}{\partial A_0} = 0 \quad \text{or} \quad \nabla \cdot \vec{E} = 0, \tag{5.20}$$

which is Gauss's law. The Hamiltonian simplifies to

$$H = \frac{1}{2} \int d^3x \, (\vec{E}^2 + \vec{B}^2) \tag{5.21}$$

with Gauss's law imposed as a constraint.

We must be careful implementing the constraint. If we take $\nabla \cdot \vec{E} = 0$ as an operator equation, we find this incompatible with the commutation relations, eq.(5.19). In particular,

$$[A_i(\vec{x}, t), E_j(\vec{y}, t)] = -i\delta_{ij}\delta(\vec{x} - \vec{y})$$

implies that

$$[A_i(\vec{x}, t), \nabla \cdot \vec{E}(\vec{y}, t)] = -i\partial_i \delta(\vec{x} - \vec{y}). \tag{5.22}$$

The left-hand side of eq.(5.22) vanishes if $\nabla \cdot \vec{E}$ vanishes as an operator. The right-hand side, however, is certainly not zero.

Now we have a choice as to what to do. If we want $\nabla \cdot \vec{E} = 0$ as an operator equation, then we must modify eq.(5.19) to make the two compatible. If we abandon $\nabla \cdot \vec{E} = 0$ as an operator equation and keep eq.(5.19), then we must somehow still impose the constraint. One way to do this would be to impose a "weaker" constraint on the states in the Fock space.

$$\nabla \cdot \vec{E} |\psi\rangle = 0 \tag{5.23}$$

States that satisfy the constraint above are called "physical," while those states that do not are unphysical and are not used. Imposing the constraint this way means that we will only use states that are physical and ignore the rest. The constraint, eq.(5.23), is still incompatible with eq.(5.19) for the same reason.

Alternatively, we can try to impose the constraint using an even weaker condition,

$$\langle \psi | \nabla \cdot \vec{E} | \psi \rangle = 0. \tag{5.24}$$

Then Maxwell's equations that we know from the classical theory only result as averages in the quantum theory.

If we imposed the Gauss's law constraint as eq.(5.23) we would end up in the temporal gauge. For now, however, we are going to maintain $\nabla \cdot \vec{E} = 0$ as an operator equation and modify eq.(5.19) to make the two compatible.

From eq.(5.22) it is clear that we need to replace the δ-function with a function $f(\vec{x} - \vec{y})$ such that $\partial_i f(\vec{x} - \vec{y}) = 0$. We do not want to depart too far from the canonical procedure, so we will choose $f(x' - x)$ to be the transverse δ-function,

$$\delta^{\text{tr}}_{ij}(\vec{x} - \vec{y}) \equiv \int \frac{d^3k}{(2\pi)^3} e^{i\vec{k}\cdot(\vec{x}-\vec{y})} (\delta_{ij} - \frac{k_i k_j}{\vec{k}^2}). \tag{5.25}$$

$\delta^{\text{tr}}_{ij}(\vec{x})$ has the desired property: $\partial_i \delta^{\text{tr}}_{ij}(\vec{x}) = \partial_j \delta^{\text{tr}}_{ij}(\vec{x}) = 0$. We modify the canonical commutation relation to read

$$[A_i(\vec{x}, t), E_j(\vec{y}, t)] = -i\delta^{\text{tr}}_{ij}(\vec{x} - \vec{y}). \tag{5.26}$$

Now, the operator $\nabla \cdot \vec{E}$ commutes with everything so we can set it to zero. However, eq.(5.26) also implies that

$$[\nabla \cdot \vec{A}(\vec{x}, t), E_j(\vec{y}, t)] = 0. \tag{5.27}$$

Thus, $\nabla \cdot \vec{A}$ vanishes as an operator equation so we have arrived at the Coulomb gauge.

One might complain that the new quantum conditions, eq.(5.26), imply that A_i and E_i do not commute at space-like separations so that we have violated causality. But A_i is gauge dependent so we really can't physically measure it. Since A_i isn't physical, there is no harm necessarily done to causality by eq.(5.26). The electric and magnetic fields are gauge invariant and can be measured. Using eq.(5.26) to derive the commutators for E_i and B_i, we find that they commute at space-like separations as expected. By the way, we get the same commutators for the electric and magnetic fields whether we use $\delta^{\text{tr}}_{ij}(x' - x)$ in eq.(5.26) or just plain $\delta(x' - x)$.

Since $\nabla \cdot \vec{A} = 0$ is an operator equation, the operator A is completely transverse so we have only quantized dynamical degrees of freedom. In the Coulomb gauge, the classical equations of motion are $\partial_\mu \partial^\mu \vec{A} = 0$, so once again the classical solutions are plane waves. This means that we should expand the operator $\vec{A}$ in plane waves for a particle interpretation.

$$\vec{A}(\vec{x}, t) = \int \frac{d^3k}{(2\pi)^3} \frac{1}{2\omega} \sum_{\lambda=1}^{2} \vec{\varepsilon}(k, \lambda)(a(k, \lambda)e^{-ik\cdot x} + a^\dagger(k, \lambda)e^{ik\cdot x}), \tag{5.28}$$

where $\omega = |\vec{k}|$. $\vec{\varepsilon}(k,\lambda)$ are two (linear) polarization vectors that are transverse,

$$\vec{k}\cdot\vec{\varepsilon}(k,\lambda) = 0, \tag{5.29}$$

and orthonormal,

$$\vec{\varepsilon}(k,\lambda)\cdot\vec{\varepsilon}(k,\lambda') = \delta_{\lambda\lambda'}. \tag{5.30}$$

By convention we choose $\vec{\varepsilon}(k,1)\times\vec{\varepsilon}(k,2) = \vec{k}/|\vec{k}|$, and $\vec{\varepsilon}(-k,1) = -\vec{\varepsilon}(k,1)$ while $\vec{\varepsilon}(-k,2) = \vec{\varepsilon}(k,2)$. Observe that eq.(5.28) is nearly identical to the scalar case. The "spin" information is carried by $\vec{\varepsilon}$.

The expansion, eq.(5.28), will reproduce the commutators for A_i and E_i if

$$\begin{aligned} &[a(k,\lambda), a^\dagger(k',\lambda')] = (2\pi)^3 2\omega\delta^3(k-k')\delta_{\lambda\lambda'} \\ &[a(k,\lambda), a(k',\lambda')] = [a^\dagger(k,\lambda), a^\dagger(k',\lambda')] = 0. \end{aligned} \tag{5.31}$$

In analogy with the scalar case, we interpret $a^\dagger(k,\lambda)$ as the creation operator of a photon with linear polarization $\vec{\varepsilon}(k,\lambda)$ and with energy $\omega = |\vec{k}|$ and momentum $\vec{k}$. $a(k,\lambda)$ is the destruction operator. The bare vacuum, $|0\rangle$, is the state in Fock space with no photons present.

$$a(k,\lambda)|0\rangle = 0 \tag{5.32}$$

To avoid the infinite zero-point energy, the Hamiltonian must be normal ordered.

$$H = \frac{1}{2}\int d^3x : \vec{E}^2 + \vec{B}^2 := \sum_{\lambda=1}^{2}\int \frac{d^3k}{(2\pi)^3}\frac{1}{2\omega}\omega a^\dagger(k,\lambda)a(k,\lambda) \tag{5.33}$$

The creation operator for right circular polarization is

$$a_R^\dagger(k) = \frac{1}{\sqrt{2}}(a^\dagger(k,1) + ia^\dagger(k,2)), \tag{5.34}$$

while

$$a_L^\dagger(k) = \frac{1}{\sqrt{2}}(a^\dagger(k,1) - ia^\dagger(k,2)) \tag{5.35}$$

creates left circularly polarized photons. Note the similarity between eqs. (5.34) and (5.30) and the charged scalar field theory with the antiparticle interpretation.

In quantizing in the Coulomb gauge we have given up manifest Lorentz covariance. Therefore, we should check that the quantized theory is still Lorentz invariant. To do so we would write the generators of the Lorentz group in terms of the quantized fields A_μ, and, using the commutation relations for A_μ, check that the generators satisfy the proper algebra. As you probably discovered in the previous 2 chapters, these calculations are, in general, laborious. In addition to verifying the Lorentz invariance of the quantization, one also shows that the photon has spin 1. The reader interested in the details should consult one of the references.

Coulomb Gauge Propagator

The photon propagator is

$$iD^{\text{tr}}_{\mu\nu}(x',x) = \langle 0\,|T(A_\mu(x')A_\nu(x))|\,0\rangle. \tag{5.36}$$

The time-ordering insures that positive energies are only propagated forward in time and negative energies only backwards in time. The "tr" notation on $D_{\mu\nu}$ indicates that it is the transverse propagator, the propagator in the Coulomb gauge. Since A_μ is gauge dependent, so is the propagator. By following the computation in the scalar field case, we can use the expansions, eq.(5.28), to express $D^{\text{tr}}_{\mu\nu}$ as

$$D^{\text{tr}}_{\mu\nu}(x',x) = \int \frac{d^4k}{(2\pi)^4}\frac{e^{-ik\cdot(x'-x)}}{k^2+i\epsilon}\sum_{\lambda=1}^{2}\varepsilon_\nu(k,\lambda)\varepsilon_\mu(k,\lambda). \tag{5.37}$$

Once again, the "$i\epsilon$" prescription implements the boundary conditions, the time-ordering, and the spin information is locked up in the polarization vectors ($\varepsilon_0 = 0$). The polarization vectors require a choice of frame to define so $D^{\text{tr}}_{\mu\nu}$ is noncovariant.

We can cast $D^{\text{tr}}_{\mu\nu}$ into a more useful form explicitly separating the covariant and noncovariant parts by introducing the coordinates of an orthonormal frame in which quantization was carried out (that is, the frame in which the polarization vectors are defined). Let η^μ be a time-like unit vector. The polarization vectors are vectors in space, hence they are orthogonal to η^μ, $\eta^\mu\varepsilon_\mu(k,\lambda) = 0$. k^μ is orthogonal to $\varepsilon^\mu(k,\lambda)$ by construction, but k^μ is not orthogonal to η^μ. To complete the orthonormal construction, project out the part of k^μ along η^μ.

$$\hat{k}^\mu = \frac{k^\mu - (k\cdot\eta)\eta^\mu}{\sqrt{(k\cdot\eta)^2 - k^2}} \tag{5.38}$$

$\hat{k}\cdot\eta = 0 = \hat{k}\cdot\varepsilon(k,\lambda)$. With this basis we can express the polarization portion of $D^{\text{tr}}_{\mu\nu}$ as

$$\sum_{\lambda=1}^{2}\varepsilon_\mu(k,\lambda)\varepsilon_\nu(k\lambda) = -g_{\mu\nu} + \eta_\mu\eta_\nu - \hat{k}_\mu\hat{k}_\nu. \tag{5.39}$$

Plugging this into eq.(5.37), we have

$$
\begin{aligned}
&D^{\text{tr}}_{\mu\nu}(x'-x) \\
&= \int \frac{d^4k}{(2\pi)^4}\frac{e^{-ik\cdot(x'-x)}}{k^2+i\epsilon}\left(-g_{\mu\nu} - \frac{k^2\eta_\mu\eta_\nu - (k\cdot\eta)(k_\mu\eta_\nu+\eta_\mu k_\nu)+k_\mu k_\nu}{(k\cdot\eta)^2-k^2}\right).
\end{aligned} \tag{5.40}
$$

The first term above,

$$
g_{\mu\nu}D(x'-x) \equiv -g_{\mu\nu}\Delta_F(x'-x, m=0), \tag{5.41}
$$

where $\Delta_F(x'-x, m=0)$ is the propagator for massless scalar particles, eq.(3.52), is covariant. The remaining ugly mess is why we like to work in gauges that are manifestly covariant if possible. In practical calculations, we use only $D(x'-x)$, eq.(5.41). This is because the photon is coupled to conserved currents where $\partial_\mu j^\mu = 0$ implies $k_\mu \tilde{j}^\mu = 0$. Any part of the propagator containing a free k_ν vanishes in a practical calculation. The $\eta_\mu\eta_\nu$ term can be shown to be equal to the static Coulomb energy when coupled to $j^0 = \rho$. This is canceled by an identical contribution from the Hamiltonian for the charged current-photon system in the Coulomb gauge.

5.3 QUANTIZATION IN THE LORENTZ GAUGE

The form of the Coulomb gauge propagator motivates us to try to quantize in a covariant gauge thereby preserving manifest covariance. To do so, the equal-time commutation relations must be manifestly covariant. Thus,

$$
[A_\mu(\vec{x},t), \pi_\nu(\vec{y},t)] = ig_{\mu\nu}\delta(\vec{x}-\vec{y}). \tag{5.42}
$$

For Minkowski space, $g_{\mu\nu} = \eta_{\mu\nu}$.To maintain covariance, all four components of A_μ (that is, transverse, longitudinal, and temporal) must appear in the commutator, so we are overquantizing. Not only do we end up with the two physical transverse photons, but also an unphysical longitudinal photon and an unphysical temporal photon. Along the path of quantization we will have to make sure that these unphysical photons decouple from the transverse photons and never enter into physically measurable quantities.

Before we can go anywhere, we must face the problem that $\pi_0 = 0$, yet eq.(5.42) implies that it is not zero. To cure this, we modify the Lagrangian. Let

$$
\mathcal{L}(x) = -\frac{1}{4}F^{\mu\nu}F_{\mu\nu} - \frac{\alpha}{2}(\partial\cdot A)^2, \tag{5.43}
$$

from which we derive

$$\pi^\mu = \frac{\partial \mathcal{L}}{\partial(\partial_t A_\mu)} = F^{\mu 0} - \alpha g^{\mu 0}(\partial \cdot A). \tag{5.44}$$

Thus, $\pi^i = -E^i$, the same as the $\alpha = 0$ case, and $\pi^0 = -\alpha(\partial \cdot A)$, which no longer vanishes. Maxwell's equations become

$$\partial_\mu \partial^\mu A_\nu - (1-\alpha)\partial_\nu(\partial \cdot A) = 0. \tag{5.45}$$

The nice choice $\alpha = 1$ simplifies eq.(5.45) and is called the Feynman gauge. From now on we will take $\alpha = 1$. Any physical result had better not depend on the artificial α. We could leave its value arbitrary to verify this, but for simplicity we will stick with $\alpha = 1$.

To recover the original theory, we must impose the constraint $\partial \cdot A = 0$. This will put us in the covariant Lorentz gauge. As with the Coulomb gauge quantization, we must decide how we are going to implement this constraint. We have already decided on the commutators we want to use, eq. (5.42), so we can't adjust them to help as we did in the Coulomb gauge.

We cannot impose $\partial \cdot A = 0$ as an operator equation because $\pi^0 = -(\partial \cdot A)$ would vanish and the point of adding the extra term to the action is to avoid this. Next, consider the weaker condition,

$$(\partial \cdot A)|\,\psi\rangle = 0. \tag{5.46}$$

This, however, is inconsistent with the commutators, eq.(5.42). To see this, consider the expectation of the $\mu = 0 = \nu$ commutator of eq.(5.42) in a state $|\,\psi\rangle$ satisfying eq.(5.46).

$$\langle\psi\,|[A^0(\vec{x},t), \pi^0(\vec{y},t)]|\,\psi\rangle = -i\delta(\vec{x}-\vec{y})\langle\psi\,|\,\psi\rangle \tag{5.47}$$

Since $\pi^0 = -(\partial \cdot A)$, the left-hand side vanishes while the right-hand side does not.

We must enforce the constraint as an expectation,

$$\langle\psi\,|\partial \cdot A|\,\psi\rangle = 0. \tag{5.48}$$

Maxwell's equations are only recovered as averages. To implement the constraint, eq.(5.48), we need only take

$$\partial \cdot A^{(+)}|\,\psi\rangle = 0, \tag{5.49}$$

where $\partial \cdot A^{(+)}$ refers to the positive energy (frequency) part of the operator $\partial \cdot A$. Eq.(5.49) suffices because

$$\begin{aligned}\langle\psi\,|\partial\cdot A|\,\psi\rangle &= \langle\psi\,|\partial\cdot A^{(-)}+\partial\cdot A^{(+)}|\,\psi\rangle = \langle\psi\,|\partial\cdot A^{(-)}|\,\psi\rangle \\ &= (\partial\cdot A^{(+)}|\,\psi\rangle)^{\dagger}|\,\psi\rangle = 0.\end{aligned} \tag{5.50}$$

For $\alpha = 1$, A^{μ} satisfies the wave equation, so once again we expand A^{μ} in plane waves.

$$A_{\mu}(x) = \int \frac{d^3k}{(2\pi)^3}\frac{1}{2\omega}\sum_{\lambda=0}^{3}\left(a(k,\lambda)\varepsilon_{\mu}(k,\lambda)e^{-ik\cdot x} + a^{\dagger}(k,\lambda)\varepsilon^{*}_{\mu}(k,\lambda)e^{ik\cdot x}\right), \tag{5.51}$$

where $\varepsilon_{\mu}(k,\lambda)$ is the polarization vector. For $\lambda = 1, 2$ we can use the transverse polarization vectors that we defined for the Coulomb gauge. Thus, $\varepsilon_0(k,1) = \varepsilon_0(k,2) = 0$ and $\varepsilon_{\mu}(k,1)k^{\mu} = \varepsilon_{\mu}(k,2)k^{\mu} = 0$. $\lambda = 1, 2$ represent the physical degrees of freedom. $\lambda = 0$ is called the scalar or temporal polarization and we will choose $\varepsilon(k,0)$ to point along the time axis in the frame we are in. Thus, $\varepsilon_i(k,0) = 0$, $\varepsilon_0(k,0) = 1$. Finally, we choose $\varepsilon_{\mu}(k,3)$ to point along $\vec{k}$, $\varepsilon_i(k,3) = k_i/|\vec{k}|$, $\varepsilon_0(k,3) = 0$. $\lambda = 3$ is called the longitudinal polarization. For example, if $\vec{k}$ points along the z direction, then

$$\varepsilon(\lambda=0) = \begin{pmatrix}1\\0\\0\\0\end{pmatrix} \quad \varepsilon(1) = \begin{pmatrix}0\\1\\0\\0\end{pmatrix} \quad \varepsilon(2) = \begin{pmatrix}0\\0\\1\\0\end{pmatrix} \quad \varepsilon(3) = \begin{pmatrix}0\\0\\0\\1\end{pmatrix}, \tag{5.52}$$

and, in general,

$$\varepsilon(k,\lambda)\cdot\varepsilon^{*}(k,\lambda') = g^{\lambda\lambda'}. \tag{5.53}$$

Using the expansion eq.(5.51), the covariant commutation relations will be satisfied if

$$[a(k,\lambda), a^{\dagger}(k',\lambda')] = -(2\pi)^3(2\omega)g^{\lambda\lambda'}\delta^3(k-k')$$

$$[a(k,\lambda), a(k',\lambda')] = [a^{\dagger}(k,\lambda), a^{\dagger}(k',\lambda')] = 0. \tag{5.54}$$

$a(k,\lambda)$ is interpreted as a destruction operator, while $a^{\dagger}(k,\lambda)$ is a creation operator. The bare vacuum, $|\,0\rangle$, satisfies

$$a(k,\lambda)|\,0\rangle = 0. \tag{5.55}$$

The zero-point energy is eliminated by normal ordering. The normal ordered Hamiltonian is

$$H = \int \frac{d^3k}{(2\pi)^3} \frac{1}{2\omega}\omega \sum_{\lambda=1}^{3} a^\dagger(k,\lambda)a(k,\lambda) - a^\dagger(k,0)a(k,0). \quad (5.56)$$

Now we have run into a very serious problem. The sign in front of the $a^\dagger(k,0)a(k,0)$ term in the Hamiltonian is the opposite of what we want. Once again, it appears that we have an energy spectrum that is unbounded from below. In addition, by insisting on maintaining manifest covariance, the metric, $g_{\mu\nu}$, had to appear on the right-hand side of the commutators, eqs.(5.42) and (5.54). For the scalar polarization, this means that the wrong sign appears in the commutator, reversing the roles of $a^\dagger(k,0)$ and $a(k,0)$. The consequence of this is that some of the states in the Fock space have a negative norm (called ghost states). Consider a state with one scalar photon,

$$|1_S\rangle \equiv \int \frac{d^3k}{(2\pi)^3} \frac{1}{2\omega} f(k) a^\dagger(k,0)|0\rangle. \quad (5.57)$$

The norm of this state is negative,

$$\begin{aligned} \langle 1_S | 1_S\rangle &= \int \frac{d^3k}{(2\pi)^3} \frac{d^3k'}{(2\pi)^3} \frac{1}{4\omega\omega'} f(k) f(k') \langle 0 | a(k',0) a^\dagger(k,0)|0\rangle \\ &= -\langle 0|0\rangle \int \frac{d^3k}{(2\pi)^3} \frac{1}{2\omega} |f(k)|^2 \end{aligned} \quad (5.58)$$

The origin of the minus sign in eq.(5.58) is the reversed sign in eq.(5.54) forced upon us by the use of $g^{\lambda\lambda'}$. The mixture of positive and negative norm states completely destroys the probability interpretation.

Has this quantization failed? No, because we must still impose the constraint, $\partial \cdot A^{(+)}|\psi\rangle = 0$. What does this constraint translate into? The positive frequency part of A_μ is the part associated with the destruction operators, $a(k,\lambda)$. If we place the positive energy part of the expansion of $A_\mu(x)$, eq.(5.51), into the constraint we get

$$\partial \cdot A^{(+)} = -i \int \frac{d^3k}{(2\pi)^3} \frac{1}{2\omega} \sum_{\lambda=0}^{3} a(k,\lambda)(k \cdot \varepsilon(k,\lambda)) e^{-ik\cdot x}. \quad (5.59)$$

But $k \cdot \varepsilon(k,\lambda) = 0$ for $\lambda = 1,2$ because they are transverse. k_μ is a null vector, hence $k \cdot \varepsilon(k,3) = -k \cdot \varepsilon(k,0)$. Thus, the constraint translates into

$$(a(k,3) - a(k,0))|\psi\rangle = 0. \quad (5.60)$$

The constraint tells us that the only physical states in the Fock space are those that always exactly balance longitudinal photons with scalar photons. The state $|\,1_S\rangle$, eq.(5.57), does not satisfy the constraint because there is not a longitudinal photon to match the temporal one present.

By using states that balance the number of unphysical photons, the bad effects cancel. For example, let's write the physical Fock space states as $|\,\psi\rangle = |\,\psi_P\rangle|\,\psi_U\rangle$. $|\,\psi_P\rangle$ is a state containing only physical, that is, transverse photons, while $|\,\psi_U\rangle$ is a state containing only unphysical photons. We can break up the states in this way because the constraint is linear. The constraint has no effect on $|\,\psi_P\rangle$. For $|\,\psi_U\rangle$, $(a(k,3)-a(k,0)|\,\psi_U\rangle = 0$. The expectation of the Hamiltonian in one of these states is

$$\langle\,\psi|H|\,\psi\rangle = \int \frac{d^3k}{(2\pi)^3}\frac{1}{2\omega}\omega\sum_{\lambda=1}^{2}\langle\psi_P\,|a^\dagger(k,\lambda)a(k,\lambda)|\,\psi_P\rangle, \tag{5.61}$$

where the sum runs only over $\lambda = 1, 2$ because

$$\langle\psi_U\,|a^\dagger(k,3)a(k,3) - a^\dagger(k,0)a(k,0)|\,\psi_U\rangle = 0 \tag{5.62}$$

due to the constraint. Eq.(5.61) is always positive definite. The misplaced minus sign in H has no effect on the constrained states. For exactly the same reasons, the norm of all states obeying the constraint is also positive definite, and we regain our probability interpretation (see exercise 5).

The constraint does not uniquely fix $|\,\psi_U\rangle$. For convenience, we can choose a fiducial state to use in computations. A simple choice for $|\,\psi_U\rangle$ is the state $|\,0_U\rangle$, the state containing no temporal or longitudinal photons. The reason the constraint does not uniquely fix $|\,\psi_U\rangle$ is that the Lorentz gauge condition, $\partial\cdot A = 0$, does not fully fix the gauge. Any gauge transformation, $A_\mu \to A_\mu - \partial_\mu\Lambda$, where $\partial_\mu\partial^\mu\Lambda = 0$, will preserve the Lorentz gauge condition.

To impose the constraint without conflicting with the commutators, we had to decompose the operator $\partial\cdot A$ into positive and negative energy parts. How do we know that we can always do this? The operator equation of motion for $\alpha = 1$ is $\partial_\mu\partial^\mu A_\nu = 0$. If we take the divergence of this equation we have $\partial_\mu\partial^\mu(\partial\cdot A) = 0$. Thus, $\partial\cdot A$ satisfies the Klein-Gordon equation with $m = 0$. $\partial\cdot A$ is just a scalar field. From chapter 3 we know that we can decompose this field operator consistently into positive and negative energy parts.

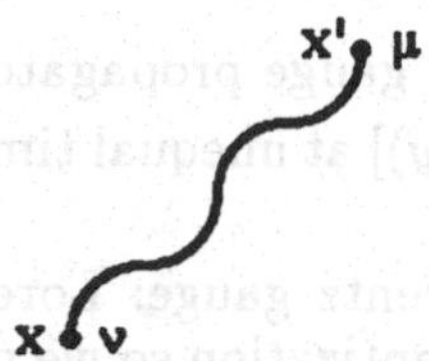

Figure 5.1 Diagram representing the free photon propagator $iD_{\mu\nu}(x'-x)$.

The Lorentz Gauge Propagator

One benefit of working in a covariant gauge will appear now. The photon propagator is

$$iD_{\mu\nu}(x',x) = \langle 0\,|T(A_\mu(x')A_\nu(x))|\,0\rangle, \tag{5.63}$$

where the time-ordering is necessary to handle the negative energy particles. Using the expansion for $A_\mu(x)$ in the Lorentz gauge, we find the Lorentz gauge propagator to be

$$\begin{aligned} D_{\mu\nu}(x'-x) &= -g_{\mu\nu}\int \frac{d^4k}{(2\pi)^4}\,\frac{e^{-ik\cdot(x'-x)}}{k^2+i\epsilon} \\ &= -g_{\mu\nu}\Delta_F(x'-x, m=0) \\ &= g_{\mu\nu}\frac{1}{4i\pi^2}\,\frac{1}{|x'-x|^2+i\epsilon}. \end{aligned} \tag{5.64}$$

This is the covariant term appearing in the transverse, Coulomb gauge propagator, $D^{\rm tr}_{\mu\nu}(x'-x)$.

The photon propagator is traditionally represented in Feynman diagrams as a wavy line.

5.4 EXERCISES

1. Compute the equal-time commutators ($t = t'$) of the electric and magnetic fields,

$$[E^i(\vec{x},t), E^j(\vec{y},t')], \quad [E^i(\vec{x},t), B^j(\vec{y},t')], \quad [B^i(\vec{x},t), B^j(\vec{y},t')],$$

using the commutators, eq.(5.19). Repeat the computation using the modified commutator, eq.(5.26), to demonstrate that the modification does not affect gauge invariant quantities.

2. Calculate the commutators of the electric and magnetic fields at *unequal* times, $t \neq t'$.
3. Compute the Lorentz gauge propagator when $\alpha \neq 1$.
4. Compute $[A_\mu(x), A_\nu(y)]$ at unequal times in the Lorentz gauge. Comment on causality.
5. No ghosts in the Lorentz gauge: Lorentz covariance is manifest in the Lorentz gauge quantization so we need not worry about breaking the Lorentz symmetry. However, there are no free dinners, since to maintain Lorentz covariance we had to introduce states in Fock space with a negative norm. We need to show that the constraint excludes the ghost states from the acceptable physical states.

   a. Show that eq.(5.62) follows from the constraint, eq.(5.60).
   b. Show that eq.(5.62) implies that the norm of $|\psi_U\rangle$ is zero unless $|\psi_U\rangle = |0_U\rangle$, the state with no longitudinal or temporal photons.
   c. Since $|0_U\rangle$ has a positive norm, show that any state $|\psi_U\rangle$ satisfying the constraint also satisfies $\langle\psi_U|\psi_U\rangle \geq 0$. Since $\langle\psi_P|\psi_P\rangle \geq 0$, we may conclude that the constraint eliminates the ghost states and that the probability interpretation remains intact.

CHAPTER 6

Interacting Field Theories

6.1 INTERACTING ACTIONS

So far we have considered quantum field theories that describe free particles. We were able to solve them exactly. Our goal is to describe not only particles, but also the fundamental forces between them. Since the forces couple different kinds of particles together, and the particles are quanta of the fields, then the forces can be described by coupling the different fields together. At the level of the action, this means that we add terms that mix the different fields together to the free field Lagrangians. We do not, however, mix the fields together in any haphazard way. Naturally, we want the entire action to remain Lorentz invariant and local. In addition, we must construct actions that possess symmetries that can generate the conserved currents and conservation laws observed experimentally. Finally, we are only interested in actions that produce quantum theories that we can make sense from. For example, the types of terms that can appear are limited by the requirement that the theory be renormalizable and free from anomalies.

We have already met one interacting field theory in the second chapter, the Nonlinear Schrödinger model. This is a 2-dimensional model describing self-interacting spinless nonrelativistic bosons. The action,

$$S[\varphi^*,\varphi] = \int d^2x\, \frac{i}{2}(\varphi^*\partial_t\varphi - \varphi\partial_t\varphi^*) - \frac{1}{2}(\partial_x\varphi^*\partial_x\varphi - c|\varphi|^4), \qquad (6.1)$$

is the free-field action plus a nonlinear interaction term, $|\varphi|^4$. The first quantized form of the field theory based on eq.(6.1) told us that this interaction, $|\varphi|^4$, translates into a 2-body contact interaction, the contact interaction reflecting the local nature of the field theory.

The 4-dimensional relativistic generalization to this is

$$S[\varphi^*,\varphi] = \int d^4x\, \partial^\mu\varphi^*\partial_\mu\varphi - m^2\varphi^*\varphi - c|\varphi|^4, \qquad (6.2)$$

which is the charged scalar free field action plus the same interaction. Both (6.1) and (6.2) possess a global phase invariance which produces charge conservation. Eq.(6.1) possesses an infinite number of conserved currents, while eq.(6.2) does not, even in 2 dimensions. The Hermitian version of the action above,

$$S[\varphi] = \int d^4x\, \partial^\mu\varphi\partial_\mu\varphi - m^2\varphi^2 - \lambda\varphi^4, \qquad (6.3)$$

is known as "phi to the fourth theory." While not physically realized, this theory provides an excellent example to illustrate the techniques used to perturbatively solve interacting theories. We will study this theory in detail in the next chapter.

We have quantum field theories for free electrons and free photons, so by appropriately coupling the two together we will arrive at quantum electrodynamics, or QED. The appropriate way to couple them together is provided by the well-known classical theory. The classical coupling is $ej^\mu A_\mu$. To translate this into the form for the Dirac action, we use the expression for the conserved current provided by the Dirac equation, namely, $j^\mu = \bar\psi\gamma^\mu\psi$. Thus, the Lagrangian density for QED is

$$\mathcal{L}(x) = -\frac{1}{4}F^{\mu\nu}F_{\mu\nu} + \bar\psi(i\,\partial\!\!\!/ - m)\psi - e\bar\psi\, A\!\!\!/\psi. \qquad (6.4)$$

This approach, using the classical theory as an analogy, does not immediately suggest how to generalize this action to incorporate larger symmetries. One approach that does is the following:

As we have already noted, the Dirac action possesses a global phase invariance that leads to charge conservation. The Dirac action is unchanged under the transformation, $\psi \to \exp(i\theta)\psi$, $\bar\psi \to \exp(i\theta)\bar\psi$, θ a constant.

We use the word "global" because θ does not depend on space-time. But in quantum mechanics, the phase of a wave function is immaterial. We can choose the phase to be whatever we want without affecting the physical results. For convenience we usually take the phase angle to be zero. Nonetheless, it should not matter whether I use some angle and somebody halfway around the world uses some other angle. In other words, the appropriate action should be *locally* phase invariant. Physical results should not be affected by the coordinate system anyone chooses for the phase.

A local phase transformation is $\psi \to \exp(i\theta(x))\psi$, where $\theta(x)$ is now a function of space-time. If we transform the Dirac action by a local phase transformation, we find that the derivative now acts on $\theta(x)$ and the action is not locally phase invariant.

$$\bar{\psi}(i\,\partial\!\!\!/ - m)\psi \to \bar{\psi}(i\,\partial\!\!\!/ - m)\psi - \bar{\psi}\gamma^\mu\psi(\partial_\mu\theta) \tag{6.5}$$

We do not want the action to depend on the choice of phase; therefore we must modify the Dirac action so that the global phase invariance is "elevated" to a local one. We have to cancel the extra term in eq.(6.5). One way to do this is to add a term to the Dirac action of the form

$$-\bar{\psi}\,A\!\!\!/\psi, \tag{6.6}$$

and dictate that A_μ transform under a phase transformation as

$$A_\mu \to A_\mu - \frac{1}{e}\partial_\mu\theta. \tag{6.7}$$

The action

$$\bar{\psi}(i\,D\!\!\!\!/ - m)\psi, \tag{6.8}$$

where $D\!\!\!\!/ = \partial\!\!\!/ + ie\,A\!\!\!/$, is locally phase invariant. We must now find a field A_μ with the property given in eq.(6.7) above. This is easily done, because we recognize eq.(6.7) as a classical gauge transformation. Thus, A_μ is the electromagnetic vector potential. The local phase transformation on ψ is also called a gauge transformation.

The action, eq.(6.8), alone does not allow A_μ to be dynamical. To do so, we add a piece containing time derivatives of A_μ. We want to preserve local phase invariance, that is, gauge invariance.The simplest combination of the A^μ's that is gauge invariant is $F_{\mu\nu}$. To preserve Lorentz invariance, we can only add a piece to the action that is a Lorentz scalar, thus the appropriate piece to add is $-1/4\, F^{\mu\nu}F_{\mu\nu}$. We are back to the QED action, eq.(6.4). $D\!\!\!\!/$ is called the gauge covariant derivative.

QED is called an Abelian or $U(1)$ gauge theory. This is because the gauge transformation on ψ, $\psi \to U\psi$, is done with an element, U, of the "gauge" group $U(1)$. $U(1)$ is the group of unitary 1×1 matrices, i.e.

the unit circle. $U(1)$ is also an Abelian group which means that all of the elements in the group commute with each other under the group multiplication. This method of "gauging" the global phase invariance into a local phase invariance can be generalized to produce more theories by choosing a different group. Using the nonabelian group $SU(2)$ leads to the celebrated Yang-Mills action.

Following the same approach above we generalize the free Dirac equation to possess a global $SU(2)$ invariance. To turn the global $SU(2)$ invariance into a local one, we transform the globally invariant action by a local $SU(2)$ transformation. Extra, noninvariant terms will appear that we must cancel. To do this we add a new field by changing the derivative(s) to a gauge covariant derivative. We then insist that the new field transform under a local $SU(2)$ transformation in just the right way to cancel the extra unwanted terms. In order to make the new field dynamical, we must add one more term to the new action. This term must be gauge and Lorentz invariant, and must contain time derivatives of the new field.

First, we construct the global $SU(2)$ invariance. An element, U, of $SU(2)$ can be written as

$$U = \exp(ig\vec{\theta} \cdot \vec{T}), \tag{6.9}$$

where T^a, $a = 1, 2, 3$, are the generators of the group that satisfy

$$[T^a, T^b] = i\epsilon^{abc} T_c, \tag{6.10}$$

and $\vec{\theta}$ is a constant 3-component vector. Since U is a matrix, ψ can no longer be just one 4-component spinor. We will put ψ in smallest representation of $SU(2)$, the fundamental representation. Then

$$\psi = \begin{pmatrix} \psi^1 \\ \psi^2 \end{pmatrix}, \tag{6.11}$$

where ψ^n, $n = 1, 2$, are each 4-component spinors. A matrix representation for the generators in the fundamental representation is provided by the Pauli matrices, $T^a = \sigma^a/2$. A $SU(2)$ transformation on ψ is $\psi \to U\psi$, $\bar{\psi} \to \bar{\psi}U^\dagger = \bar{\psi}U^{-1}$. The Dirac action,

$$\bar{\psi}(i\,\not{\partial} - m)\psi, \tag{6.12}$$

where ψ has the form eq.(6.11), has the desired global $SU(2)$ invariance. In eq.(6.12), it is understood that $\not{\partial}$ and m multiply a unit 2×2 matrix.

Now we "gauge" the global invariance by allowing $\vec{\theta}(x)$ to depend on space-time. Under a local $SU(2)$ transformation,

$$\bar{\psi}(i\,\not{\partial} - m)\psi \to \bar{\psi}(i\,\not{\partial} - m)\psi + i\bar{\psi}(U^\dagger \gamma^\mu \partial_\mu U)\psi. \tag{6.13}$$

As before, we correct the derivative in eq.(6.12) to remove the undesired second term in eq.(6.13). If we replace the ordinary derivative, $\partial\!\!\!/$, by the gauge covariant derivative, $D\!\!\!\!/ = \partial\!\!\!/ + ig\,A\!\!\!/$, where $A\!\!\!/ = A_\mu\gamma^\mu$, $A_\mu = A_\mu^a T^a$, then this new action,

$$\bar\psi(i\,D\!\!\!\!/ - m)\psi = \bar\psi(i\,\partial\!\!\!/ - g\,A\!\!\!/ - m)\psi, \qquad (6.14)$$

will be locally $SU(2)$ invariant if we insist that A_μ transform as

$$A_\mu \to U\,A_\mu U^\dagger + \frac{i}{g}(\partial_\mu U)U^\dagger. \qquad (6.15)$$

The action, eq.(6.14), is not the whole story. A_μ is not yet dynamical. We need to add the generalization of $F^{\mu\nu}F_{\mu\nu}$. $F_{\mu\nu}$ is the curvature associated to the covariant derivative, D_μ, hence $F_{\mu\nu} = -(i/g)[D_\mu, D_\nu]$. One can check that, in the Abelian case, the commutator produces the usual curl. In the nonabelian case, A_μ and A_ν do not commute (because of the T^a's), hence

$$F_{\mu\nu} = \partial_\mu A_\nu - \partial_\nu A_\mu + ig[A_\mu, A_\nu]. \qquad (6.16)$$

Under the gauge transformation, eq.(6.15), we can see that $F_{\mu\nu}$ is only gauge covariant in the nonabelian case, as opposed to gauge invariant in the Abelian case.

$$F_{\mu\nu} \to U\,F_{\mu\nu}U^\dagger \qquad (6.17)$$

By the cyclic property of the trace, $\mathrm{tr}(F_{\mu\nu}F^{\mu\nu})$ is gauge invariant. The entire Lagrangian is

$$\mathcal{L}(x) = -\frac{1}{2}\mathrm{tr}F^{\mu\nu}F_{\mu\nu} + \bar\psi(i\,D\!\!\!\!/ - m)\psi. \qquad (6.18)$$

The first term involving A_μ only is nonlinear and nontrivial by itself. This term is the Yang-Mills Lagrangian density and is sometimes called a pure gauge theory. If we change the group to $SU(3)$ and include one Dirac term for each quark flavor, then the Lagrangian density in eq.(6.18) is that for the theory of the strong interaction, quantum chromodynamics, or QCD. The particles associated with the A_μ field are called gluons. Since the group $SU(3)$ possesses 8 generators, there are 8 different kinds of gluons. The spin 1/2 particles represented by ψ are quarks. Each 4-component spinor carries 2 indices, ψ_f^a. The "a" index, or color index, runs from 1 to 3 for $SU(3)$ since the fundamental representations of $SU(3)$ are 3-dimensional. The "f" index runs over the quark flavors (up, down, strange, charm, top, and bottom). QCD does not care about quark flavors. The quark flavor index just tags along. The analog of the electric charge is called color. The analog of the photon is the gluon, and the analog of the electron is the quark. The nonabelian nature of the gauge group gives the gluons a charge, unlike the electrically neutral photon.

A different set of models that share some properties with Yang-Mills are called nonlinear sigma-models. These are the quantum theories of maps from 2-dimensional space-time to Riemann surfaces, group manifolds, etc. The action is usually that for free fields. The interactions arise from the constraint that the range of the map lie on the target manifold. One of the simplest versions, the so-called $O(3)$ σ-model, is given by the action

$$S[\vec{\sigma}] = \int d^2x\, \partial_\mu \vec{\sigma} \cdot \partial^\mu \vec{\sigma}, \tag{6.19}$$

where the 3-component vector $\vec{\sigma}$ is constrained to be of unit length, $\vec{\sigma}\cdot\vec{\sigma} = 1$. Thus, $\vec{\sigma}(x,t)$ maps 2-dimensional space-time onto S^2, the sphere. The model possesses an $O(3)$ invariance and also is integrable.

6.2 DISCRETE SYMMETRIES

The symmetries we have discussed so far are continuous symmetries. In addition, there are several common discrete symmetries. These are: charge conjugation or $\mathcal{C}$, parity or $\mathcal{P}$, and time reversal or $\mathcal{T}$. The free fields possess these symmetries separately, but interactions may break them. We may experimentally test whether the fundamental forces break these symmetries. This aids us in constructing models. For example, the weak interactions break parity (space inversion). In nature we only find left-handed neutrinos. Since the right-handed ones are absent, a realistic theory of the weak interactions cannot be invariant under parity. Even though $\mathcal{C}$ or $\mathcal{P}$ may not be a symmetry, experimentally the combination $\mathcal{CP}$ generally is. There is only one known process that violates $\mathcal{CP}$: K^0, $\bar{K}^0$ decay. The violation is tiny and it is not known whether it is due to the the weak interaction or a new force. Even though $\mathcal{CP}$ may not be conserved, the "PCT theorem" tells us that any reasonable interacting theory is symmetric under $\mathcal{PCT}$. Thus, the K^0, $\bar{K}^0$ decay that violates $\mathcal{CP}$ must also violate time reversal so that the combination $\mathcal{PCT}$ is conserved.

Charge Conjugation

Particles and antiparticles are distinguished by charge. The charge conjugation operator, $\mathcal{C}$, changes the sign of the charge. If a theory is symmetric under $\mathcal{C}$, then it is symmetric under the exchange of particles with antiparticles.

Consider again the free charged scalar field. The field φ^* creates particles of charge $+1$ while φ creates particles of charge -1. Since $\mathcal{C}$ changes the charge, then

$$\begin{aligned} \mathcal{C}\varphi(x)\mathcal{C}^{-1} &= \varphi^*(x) \\ \mathcal{C}\varphi^*(x)\mathcal{C}^{-1} &= \varphi(x). \end{aligned} \tag{6.20}$$

In terms of the Hermitian fields, $\varphi_1(x)$ and $\varphi_2(x)$,

$$\begin{aligned} \mathcal{C}\varphi_1(x)\mathcal{C}^{-1} &= \varphi_1(x) \\ \mathcal{C}\varphi_2(x)\mathcal{C}^{-1} &= -\varphi_2(x). \end{aligned} \tag{6.21}$$

If we write eq.(6.20) in terms of the creation and destruction operators we have

$$\begin{aligned} \mathcal{C}a^\dagger(k)\mathcal{C}^{-1} &= b^\dagger(k) \\ \mathcal{C}b^\dagger(k)\mathcal{C}^{-1} &= a^\dagger(k). \end{aligned} \tag{6.22}$$

Let's construct an explicit representation for the operator $\mathcal{C}$. Since $\mathcal{C}$ must be unitary, $\mathcal{C} = \exp(i\mathrm{C})$, where C is Hermitian. Plugging this into eq.(6.22) and expanding the exponential we have

$$\mathcal{C}a^\dagger(k)\mathcal{C}^{-1} = a^\dagger(k) + i[\mathrm{C}, a^\dagger(k)] + \frac{i^2}{2!}[\mathrm{C}[\mathrm{C}, a^\dagger(k)]] + \ldots = b^\dagger(k). \tag{6.23}$$

Since only $a^\dagger(k)$ and $b^\dagger(k)$ show up in eq.(6.23) on the outside of the commutators, then the commutator of C with $a^\dagger(k)$ can only involve $a^\dagger(k)$ and $b^\dagger(k)$. In fact, since no higher powers of $a^\dagger$ and $b^\dagger$ appear outside of the commutators, the C, $a^\dagger$ commutator must be linear in $a^\dagger$ and $b^\dagger$.

$$[\mathrm{C}, a^\dagger(k)] = \alpha a^\dagger(k) + \beta b^\dagger(k). \tag{6.24}$$

The symmetry between $a^\dagger$ and $b^\dagger$ in (6.22) implies that $\beta = \pm\alpha$, thus

$$\begin{aligned} [\mathrm{C}, a^\dagger(k)] &= \alpha(a^\dagger(k) \pm b^\dagger(k)) \\ [\mathrm{C}, b^\dagger(k)] &= \alpha(b^\dagger(k) \pm a^\dagger(k)). \end{aligned} \tag{6.25}$$

Then

$$[\mathrm{C}, [\mathrm{C}, a^\dagger(k)]] = 2\alpha^2(a^\dagger(k) \pm b^\dagger(k)),$$

and so on. Eq.(6.23) becomes

$$\begin{aligned} \mathcal{C}a^\dagger(k)\mathcal{C}^{-1} &= a^\dagger(k) + (i\alpha + \frac{2(i\alpha)^2}{2!} + \frac{2^2(i\alpha)^3}{3!} + \ldots)(a^\dagger(k) \pm b^\dagger(k)) \\ &= a^\dagger(k) + \frac{1}{2}(e^{i2\alpha} - 1)(a^\dagger(k) \pm b^\dagger(k)). \end{aligned} \tag{6.26}$$

To eliminate $a^\dagger(k)$ from the right-hand side, we take $\alpha = \pi/2$. To obtain a positive $b^\dagger(k)$ we take the minus sign in eq.(6.25). Thus,

$$[\mathrm{C}, a^\dagger(k)] = \frac{\pi}{2}(a^\dagger(k) - b^\dagger(k)). \tag{6.27}$$

Using the commutation relations for $a^\dagger$ and $b^\dagger$, eq.(3.40), we find that

$$\begin{aligned}\mathrm{C} &= \frac{\pi}{2}\int \frac{d^3k}{(2\pi)^3}\frac{1}{2\omega_k}(a^\dagger(k) - b^\dagger(k))(a(k) - b(k)) \\ &= -\pi\int \frac{d^3k}{(2\pi)^3}\frac{1}{2\omega_k}a_2^\dagger(k)a_2(k),\end{aligned} \tag{6.28}$$

so

$$\mathcal{C} = \exp\left(-i\pi\int \frac{d^3k}{(2\pi)^3}\frac{1}{2\omega_k}a_2^\dagger(k)a_2(k)\right). \tag{6.29}$$

We could add an arbitrary phase to the definition of $\mathcal{C}$ without changing the physical results. Instead, we set the phase by choosing $\mathcal{C}|\,0\rangle = |\,0\rangle$.

The Hamiltonian for the free charged fields is the sum of the number operators for type 1 and 2 particles. The number operator of type 2 commutes with the number operator of type 1. Since $\mathcal{C}$ only involves the number operator of type 2 particles, it must commute with the Hamiltonian, so it is a symmetry of the system.

What happens if we add interactions? If we consider the action in eq.(6.2), normal ordered, then the only term we must check is $:|\varphi|^4:$. Using eq.(6.20), it is easy to see that this term is symmetric under $\mathcal{C}$ since it contains an equal number of φ^*'s and φ's.

To construct an explicit expression for $\mathcal{C}(t)$ in the interacting theory, we can't just use eq.(6.29), because the plane wave expansion is no longer valid for the interacting operators. But we can translate $\mathcal{C}(t)$ back to $\mathcal{C}(0)$ by

$$\mathcal{C}(0) = e^{-iHt}\mathcal{C}(t)e^{iHt}.$$

At $t = 0$, the interacting fields satisfy the same equal-time commutation relations as the free fields, hence $\mathcal{C}(0) = \mathcal{C}$ of eq.(6.29), where $a_2(k)$ and $a_2^\dagger(k)$ are the expansion coefficients of the interacting field $\varphi_2(\vec{x}, t = 0)$. Since $\mathcal{C}$ is a symmetry, $[\mathcal{C}, H] = 0$, so that $\mathcal{C}(t) = \mathcal{C}(0)$, and we can use eq.(6.29) for all times (keeping in mind that a_2 corresponds to $\varphi_2(\vec{x}, 0)$). If $\mathcal{C}$ is not a symmetry, then $\mathcal{C}(t) = \exp(iHt)\mathcal{C}(0)\exp(-iHt)$.

For free Dirac field theory, we can quickly write down $\mathcal{C}$ using the result for the charged scalar field. $b_i^\dagger(p)$ creates an electron with spin up or down, while $d_i^\dagger(p)$ creates a positron of opposite charge. Therefore, $b_i^\dagger(p)$

corresponds to $a^\dagger(k)$ in the charged scalar theory while $d_i^\dagger(p)$ corresponds to $b^\dagger(k)$. Thus,

$$\mathcal{C} = \exp\left(\frac{i\pi}{2}\int\frac{d^3p}{(2\pi)^3}\frac{1}{2E}\sum_{i=1}^{2}(b_i^\dagger(p) - d_i^\dagger(p))(b_i(p) - d_i(p))\right), \tag{6.30}$$

where we have neglected the inessential unphysical phase that could be included in the definition of $\mathcal{C}$. If we now consider the effects of $\mathcal{C}$ on a spinor $\psi(x)$, we might be tempted to write, in analogy with the scalar case,

$$\mathcal{C}\psi(x)\mathcal{C}^{-1} = \psi^\dagger(x),$$

since $\psi(x)$ and $\psi^\dagger(x)$ create oppositely charged particles. But this is not correct because of the spinor nature of ψ. The left-hand side is a column while the right is a row. To find the correct transformation, we return to the Dirac equation coupled to an external electromagnetic field,

$$(i\,\partial\!\!\!/ - e\,A\!\!\!/ - m)\psi = 0, \tag{6.31}$$

where ψ is the electron wave function. If ψ_c is the positron wave function, then ψ_c satisfies

$$(i\,\partial\!\!\!/ + e\,A\!\!\!/ - m)\psi_c = 0. \tag{6.32}$$

The only change from eq.(6.31) is due to the change in the sign of the electric charge.

Now we want to find the relation between ψ and ψ_c. Taking the complex conjugate of eq.(6.31) we have

$$((i\partial_\mu + eA_\mu)\gamma^{\mu *} + m)\psi^* = 0. \tag{6.33}$$

If we can find a transformation

$$(\mathbf{C}\gamma^0)\gamma^{\mu *}(\mathbf{C}\gamma^0)^{-1} = -\gamma^\mu, \tag{6.34}$$

then we will recover eq.(6.32) by applying $(\mathbf{C}\gamma^0)$ to eq.(6.33) and inserting $1 = (\mathbf{C}\gamma^0)^{-1}(\mathbf{C}\gamma^0)$ between $\gamma^{\mu *}$ and ψ^*. The relation between ψ and ψ_c will be

$$\psi_c = \mathbf{C}\gamma^0\psi^* = \mathbf{C}\bar{\psi}^T, \tag{6.35}$$

where T means "transpose." The inclusion of γ^0 in the definition of the charge conjugation operator on ψ is a convenient convention.

Eq.(6.34) implies $\mathbf{C}\gamma^{\mu T}\mathbf{C}^{-1} = -\gamma^\mu$. The matrix

$$\mathbf{C} = i\gamma^2\gamma^0 \tag{6.36}$$

satisfies the desired properties. Thus, in terms of the spinors,

$$\mathcal{C}\psi(x)\mathcal{C}^{-1} = \psi_c(x) = \mathrm{C}\bar{\psi}^T. \tag{6.37}$$

In the interacting case, the operator defined in eq.(6.30) will serve as $\mathcal{C}(t=0)$ since the equal-time commutators for the interacting fields are identical to the free fields at $t = 0$. If $\mathcal{C}$ is a symmetry, then $\mathcal{C}(t) = \mathcal{C}(0)$. If not, $\mathcal{C}(t) = \exp(iHt)\mathcal{C}(0)\exp(-iHt)$.

Finally, for photons we determine the effect of $\mathcal{C}$ on the vector potential by requiring $\mathcal{C}j_\mu A^\mu \mathcal{C}^{-1} = j_\mu A^\mu$, because the photon is electrically neutral and interacts the same with electrons or positrons. Using the $\mathcal{C}$ for spinors defined above, we find $\mathcal{C}\bar{\psi}\gamma^\mu\psi\mathcal{C}^{-1} = -\bar{\psi}\gamma^\mu\psi$ by construction. To preserve $j_\mu A^\mu$, the vector potential must transform under $\mathcal{C}$ as

$$\mathcal{C}A^\mu(x)\mathcal{C}^{-1} = -A^\mu(x). \tag{6.38}$$

In the Coulomb gauge, this translates into

$$\mathcal{C}a^\dagger(k,\lambda)\mathcal{C}^{-1} = -a^\dagger(k,\lambda). \tag{6.39}$$

Following the same steps as in the scalar case, $\mathcal{C}$ may be represented by

$$\mathcal{C} = \exp\left(i\pi \int \frac{d^3k}{(2\pi)^3}\frac{1}{2\omega}\sum_{\lambda=1}^{2} a^\dagger(k,\lambda)a(k,\lambda)\right). \tag{6.40}$$

The same comments about representing $\mathcal{C}$ for interacting fields apply here as for the scalar and spinor fields.

Parity

Parity is the operation of inverting space, $(\vec{x},t) \to (-\vec{x},t)$. Let $\mathcal{P}$ be the unitary parity operator. If the effect of $\mathcal{P}$ on the Lagrangian density $\mathcal{L}(x)$ is

$$\mathcal{P}\mathcal{L}(\vec{x},t)\mathcal{P}^{-1} = \mathcal{L}(-\vec{x},t), \tag{6.41}$$

so that the equations of motion are unchanged, and if $\mathcal{P}$ leaves the commutation relations invariant, then the system is invariant under parity.

For the free scalar theory,

$$\mathcal{P}\varphi(\vec{x},t)\mathcal{P}^{-1} = \pm\varphi(-\vec{x},t) \tag{6.42}$$

will satisfy eq.(6.41) and leave the commutation relations unchanged. There are two distinct cases corresponding to whether we use the + or − sign in eq.(6.42). The choice is called intrinsic parity. If φ changes sign under $\mathcal{P}$, φ describes a pseudoscalar particle, while if φ is invariant, it is a scalar particle. Compare this terminology with the fermion bilinear covariants of chapter 4.

In terms of the creation and destruction operators,

$$\begin{aligned}\mathcal{P}a(\omega_k,\vec{k})\mathcal{P}^{-1} &= \pm\mathcal{P}a(\omega_k,-\vec{k})\mathcal{P}^{-1}\\ \mathcal{P}a^\dagger(\omega_k,\vec{k})\mathcal{P}^{-1} &= \pm\mathcal{P}a^\dagger(\omega_k,-\vec{k})\mathcal{P}^{-1},\end{aligned} \tag{6.43}$$

where $+$ is for scalar, $-$ for pseudoscalar. To find a representation for $\mathcal{P}$, we follow the same steps as for $\mathcal{C}$. First we write $\mathcal{P}$ as $\exp(i\mathbf{P})$, $\mathbf{P}$ Hermitian, and plug this into eq.(6.43). We can see that $a^\dagger(\omega_k,\vec{k})$ plays the same role, but $b^\dagger(k)$ is replaced here by $a^\dagger(\omega_k,-\vec{k})$. The result is

$$\mathcal{P} = \exp\left(\frac{-i\pi}{2}\int\frac{d^3k}{(2\pi)^3}\frac{1}{2\omega_k}(a^\dagger(k)a(k) \mp a^\dagger(k)a(-k))\right), \tag{6.44}$$

where $-$ is for scalar particles, $+$ for pseudoscalar, and $-k = (\omega_k,-\vec{k})$.

Eq.(6.44) will serve as the parity operator, $\mathcal{P}$, at $t=0$ for interacting scalar fields. If $\mathcal{P}$ is conserved, then it will work for all times. If not, $\mathcal{P}(t) = \exp(iHt)\mathcal{P}(0)\exp(-iHt)$.

For the Dirac field, we can immediately write down $\mathcal{P}$ by comparing with the scalar case and noting that we have 2 types of creation operators, $b_i^\dagger(p)$ and $d_i^\dagger(p)$. Thus,

$$\begin{aligned}\mathcal{P} = \exp\Bigg(\frac{-i\pi}{2}\int\frac{d^3p}{(2\pi)^3}\frac{1}{2E}\sum_{i=1}^{2}&(b_i^\dagger(p)b_i(p) - b_i^\dagger(p)b_i(-p)\\ &+ d_i^\dagger(p)d_i(p) - d_i^\dagger(p)d_i(-p))\Bigg),\end{aligned} \tag{6.45}$$

where $-p = (E,-\vec{p})$. For the corresponding operator on the spinor $\psi(x)$, we go back again to the Dirac equation. The parity operator is an improper Lorentz transformation. Since the Dirac equation is Lorentz covariant, the equation for the parity transformed spinor should be identical to the original equation. If $\psi'(x')$ represents the parity transformed spinor,

$$\psi'(x') = \mathbf{P}\psi(x), \tag{6.46}$$

then $\psi'(x')$ is a solution of

$$(i\,\partial\!\!\!/' - m)\psi'(x') = 0. \tag{6.47}$$

Under parity, $(\vec{x}', t') = (-\vec{x}, t)$, so $\partial_{t'} = \partial_t$ and $\partial_i' = -\partial_i$. In terms of x, eq.(6.46) becomes

$$(i(\gamma^0\partial_0 + \gamma^i\partial_i) - m)\mathbf{P}\psi(x) = 0. \tag{6.48}$$

If we operate on eq.(6.48) with $\mathbf{P}^{-1}$, we will recover the original Dirac equation for $\psi(x)$ if

$$\mathbf{P}^{-1}\gamma^i\mathbf{P} = -\gamma^i, \qquad \mathbf{P}^{-1}\gamma^0\mathbf{P} = \gamma^0. \tag{6.49}$$

Since the γ matrices anticommute, $\mathbf{P} = \gamma^0$ clearly satisfies eq.(6.49). Thus,

$$\mathcal{P}\psi(x)\mathcal{P}^{-1} = \gamma^0\psi(-\vec{x}, t). \tag{6.50}$$

In the Dirac representation, the bottom two diagonal elements of γ^0 are -1, while the top two are 1. Therefore, eq.(6.50) tells us that positive and negative energy particles, that is, particles and antiparticles, have opposite intrinsic parity.

The vector potential in the Coulomb gauge is a spatial vector, so if we invert space, the vector potential will change sign.

$$\mathcal{P}A_i(\vec{x}, t)\mathcal{P}^{-1} = -A_i(-\vec{x}, t) \tag{6.51}$$

A vector that does not change sign under parity is called an axial vector or pseudovector. Using eq.(6.51), we can find a representation for $\mathcal{P}$ for free photons by noting that eq.(6.51) implies

$$\mathcal{P}a(k, \lambda)\mathcal{P}^{-1} = -(-1)^\lambda a(-k, \lambda). \tag{6.52}$$

The factor $(-1)^{\lambda-1}$ arises from the convention chosen for the polarization vectors, $\varepsilon(k, \lambda)$, under $\vec{k} \to -\vec{k}$. Following the same steps as in the scalar case we find $(-k = (\omega, -\vec{k}))$,

$$\mathcal{P} = \exp\left(\frac{-i\pi}{2}\int\frac{d^3k}{(2\pi)^3}\frac{1}{2\omega}\sum_{\lambda=1}^{2} a^\dagger(k, \lambda)a(k, \lambda) - (-1)^\lambda a^\dagger(k, \lambda)a(-k, \lambda)\right). \tag{6.53}$$

The usual comments about parity in interacting theories apply here, too.

Time Reversal

The quantum mechanical equation of motion for the operator A is

$$-i\frac{\partial A(t)}{\partial t} = [H, A(t)]. \tag{6.54}$$

Under time reversal, $t \to t' = -t$, eq.(6.54) becomes

$$i\frac{\partial A(t')}{\partial t'} = [\mathcal{T}H\mathcal{T}^{-1}, A(t')], \tag{6.55}$$

where $\mathcal{T}$ is the time reversal operator. To restore eq.(6.55), we do not want to take $\mathcal{T}H\mathcal{T}^{-1} = -H$, because the spectrum becomes unbounded from below with respect to the vacuum. Instead, we want $\mathcal{T}H\mathcal{T}^{-1} = H$, and still recover eq.(6.54). This can be accomplished if $\mathcal{T}$ is *anti*unitary (unitary + complex conjugation). The effect of $\mathcal{T}$ on the Lagrangian density should be

$$\mathcal{T}\mathcal{L}(\vec{x},t)\mathcal{T}^{-1} = \mathcal{L}(\vec{x},-t) \tag{6.56}$$

if $\mathcal{T}$ is a symmetry operation. As with $\mathcal{C}$ and $\mathcal{P}$, time reversal invariance also requires that $\mathcal{T}$ leave the commutators unchanged.

For the scalar field, these requirements amount to

$$\mathcal{T}\varphi(\vec{x},t)\mathcal{T}^{-1} = \pm\varphi(\vec{x},-t). \tag{6.57}$$

Except for the addition of the complex conjugation, the construction of $\mathcal{T}$ is nearly identical with $\mathcal{P}$.

For the Dirac field,

$$\mathcal{T}\psi(\vec{x},t)\mathcal{T}^{-1} = \mathbf{T}\psi(\vec{x},-t), \tag{6.58}$$

where $\mathbf{T}$ is chosen so that $\mathbf{T}\psi(\vec{x},t)$ satisfies the time reversed Dirac equation. The choice

$$\mathbf{T} = i\gamma^1\gamma^3 \tag{6.59}$$

accomplishes this.

For the electromagnetic field, we want $\mathcal{T}^{-1}j^\mu A_\mu\mathcal{T} = j^\mu A_\mu$. Under time reversal, the current flows in the reverse direction, $\vec{j} \to -\vec{j}$. To preserve $j^\mu A_\mu$,

$$\mathcal{T}A_i(\vec{x},t)\mathcal{T}^{-1} = -A_i(\vec{x},-t). \tag{6.60}$$

The construction of a representation of $\mathcal{T}$ is nearly identical to $\mathcal{P}$.

PCT

Even though a particular theory may not preserve $\mathcal{P}$ or $\mathcal{C}$ or even $\mathcal{CP}$, the PCT theorem states that if the the theory is local, has a reasonable Lagrangian density (Hermitian, Lorentz covariant, etc.), and is quantized with the correct spin-statistics relationship, then the combination $\mathcal{PCT}$ will be a symmetry operation. The proof amounts to showing that under $\mathcal{PCT}$,

$$\mathcal{PCT}\,\mathcal{L}(\vec{x},t)(\mathcal{PCT})^{-1} \to \mathcal{L}(-\vec{x},-t), \tag{6.61}$$

and that the commutation relations are invariant so that the equations of motion are unchanged. Ultimately, the conservation of $\mathcal{PCT}$ is another consequence of Lorentz invariance. See the Bibliography for a rigorous approach. Even though $\mathcal{C}$ may not be conserved, the fact that $\mathcal{PCT}$ is conserved means that stable particles and antiparticles must have the same mass and spin.

6.3 ASYMPTOTIC STATES AND PERTURBATION THEORY

We may canonically quantize the interacting field theories in the same way as the free fields by finding the field momenta, the Hamiltonian, and then by postulating equal-time commutation relations. The difficulty with interacting field theories, though, is that we lose the particle interpretation, partially because we can no longer exactly solve the nonlinear equations of motion.

How do we obtain a particle interpretation for those interacting fields? To answer this, we must decide what it is that we should compute using the interacting field theory to compare with experiment.

In high energy experiments, two particles are accelerated towards each other, they collide and interact, and the particles resulting from the collision fly away from the region of interaction. This is represented schematically in figure 6.1. Before the collision, when the particles are far from each other and the region of interaction, they act like free particles. After the collision is over, the particles leaving the interaction region again act like free particles once they get far away from each other. In an experiment, the cross section for such a process is measured.

Since cross sections are measured, we want to calculate cross sections with our interacting field theories. The cross section for a process is directly related to the probability for the process to occur. In turn, the probability for the process to occur is the square of the amplitude for the process to occur. Therefore, we want to calculate the amplitude for a given process to occur using our field theory.

The amplitude, A, for a process that starts in initial state, $|\,\text{IN}\rangle$, and ends up in a final state, $|\,\text{OUT}\rangle$, is

$$A = \langle \text{OUT}\,|\,\text{IN}\rangle. \tag{6.62}$$

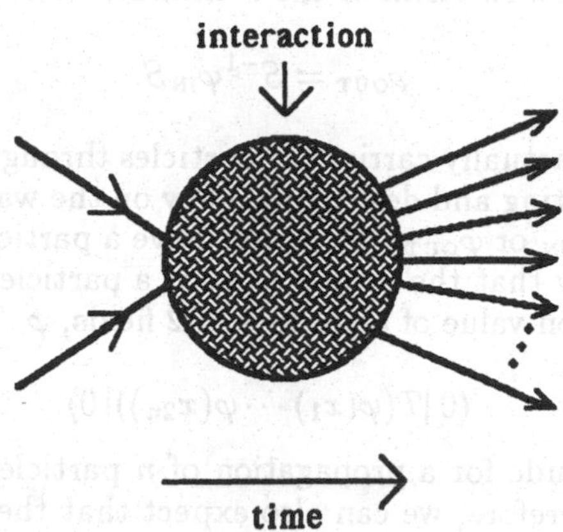

Figure 6.1 Diagram schematically representing a collision of two particles. When the initial-state and final-state particles are far from the region of interaction, they behave as if they were free.

In an accelerator, we prepare the $|\text{IN}\rangle$ state and we measure the kinematics of the $|\text{OUT}\rangle$ state. Since the particles are essentially free in the $|\text{IN}\rangle$ and $|\text{OUT}\rangle$ states, we know how to write these states down using free field theory (because we have a particle interpretation). For physical situations where no bound states arise, the $|\text{IN}\rangle$ states should be isomorphic to (in one-to-one correspondence with) the $|\text{OUT}\rangle$ states. There is really no difference between the $|\text{IN}\rangle$ and $|\text{OUT}\rangle$ states since they belong to the same Fock space corresponding to a free field theory. There will be a mapping that tells us how to map each $|\text{IN}\rangle$ state onto each $|\text{OUT}\rangle$ state. This mapping depends, of course, on the interaction that goes on, that is, on the interacting field theory we are using. This mapping is traditionally denoted as S,

$$S|\text{IN}\rangle = |\text{OUT}\rangle,$$

and is called the S-matrix. In terms of the S-matrix, the amplitude A for a particular process is

$$A = \langle\text{OUT}\,|\,\text{IN}\rangle = \langle\text{IN}\,|S|\,\text{IN}\rangle. \tag{6.63}$$

To calculate a cross section, we need A, so we must relate our interacting field theory to the S-matrix.

Let us say that we are working with a generic field φ. The field describing the $|\text{IN}\rangle$ states is φ_{IN}. φ_{IN} is a free field so it is one of the fields we have quantized in the previous chapters and have exactly solved. φ_{OUT} creates $|\,\text{OUT}\rangle$ states. In terms of the S-matrix,

$$\varphi_{\text{OUT}} = S^{-1}\varphi_{\text{IN}}S. \tag{6.64}$$

The field that actually carries the particles through the interacting region (perhaps creating and destroying a few on the way) is the interacting field, φ, and not φ_{IN} or φ_{OUT}. We do not have a particle interpretation for φ, but we do know that the propagator for a particle is the time-ordered vacuum expectation value of a product of 2 fields, φ. Thus, we can expect that

$$\langle 0\,|T(\varphi(x_1)\cdots\varphi(x_{2n}))|\,0\rangle \tag{6.65}$$

will be the amplitude for a propagation of n particles through the interacting region. Therefore, we can also expect that the S-matrix is related to the propagators, eq.(6.65). The formula that relates the two is called the reduction formula.

To derive a reduction formula we need a particle interpretation. We do not have one for φ, but if we assume that no bound states appear, we can expand φ in terms of a basis provided by φ_{IN}. That is, assume $\varphi = f(\varphi_{\text{IN}})$ and solve the equations of motion for φ perturbatively, by perturbing off the known exact solution for φ_{IN}. By perturbatively solving φ off of φ_{IN}, φ acquires the particle interpretation of φ_{IN}, and we are in business. The time-ordered expectation values of products of φ will be written in terms of a series involving the expectation values of products of φ_{IN}. We know how to compute these exactly. The details of this perturbative program will be given in the next chapter for the self-interacting scalar field theory given by the action eq.(6.3). QED will follow in the chapter after next.

If the coupling in the interaction is strong, then bound states of particles may appear. If bound states live long enough, they will make it out to the asymptotic region where measurements are made. If this happens, our assumption that particles far from the interacting region act like free particles is no longer true. This means that a perturbative solution to the field theory no longer makes physical sense. Thus, if the coupling is strong, we can no longer make use of the particle interpretation provided by free field theory and we are stuck.

QCD, the theory of the strong interactions, is a theory where the asymptotic states are supposed to be bound states of quarks and gluons. In fact, QCD is supposed to confine color, so there are *no* asymptotic states of free quarks and gluons. For this theory, we have no reliable methods to approximately solve it. Finding solutions to strong coupling problems like QCD, even approximate solutions, is perhaps the most challenging problem in theoretical high energy physics today.

6.4 EXERCISES

1. Verify that the gauge transformation for A_μ, eq.(6.15), does indeed make the action eq.(6.14) locally $SU(2)$ invariant.
2. Verify that the commutator,

$$F_{\mu\nu} = -\frac{i}{g}\,[D_\mu, D_\nu],$$

produces eq.(6.16). Using eq.(6.15), show that $F_{\mu\nu}$ is gauge covariant, and that $tr(F^{\mu\nu}F_{\mu\nu})$ is gauge invariant.
3. Construct a representation of the time reversal operator, $\mathcal{T}$, for the free scalar, spinor, and electromagnetic fields.
4. Investigate how each of the bilinear covariants listed at the end of section 4.1 transforms under $\mathcal{P}$, $\mathcal{C}$, $\mathcal{T}$, and the combination $\mathcal{PCT}$.
5. Let $|n\rangle$ be a state of n photons. How does $|n\rangle$ transform under $\mathcal{C}$? The dominant decay mode of the π^0 meson is experimentally seen to be $\pi^0 \to 2\gamma$. Assuming that the strong interactions are invariant under $\mathcal{C}$, how does the π^0 state transform under $\mathcal{C}$? Why is the decay of π^0 into an odd number of photons never seen?
6. Consider a state of 2 photons, $|2\gamma\rangle$, where the total momentum is zero. Show that $|2\gamma\rangle$ is an eigenstate of the parity operator with eigenvalue $+1$ if the two photons have the same polarization, and -1 if their polarizations are perpendicular. Explain how to measure the intrinsic parity of the π^0 and the assumptions that must be made in doing so.
7. Investigate the invariance of the term ${}^*F^{\mu\nu}F_{\mu\nu}$ under $\mathcal{P}$, $\mathcal{C}$, $\mathcal{T}$, $\mathcal{CP}$, and $\mathcal{CPT}$.

CHAPTER 7

Self-Interacting Scalar Field Theory

We will begin exploring interacting quantum field theories and carry out the program discussed in the previous chapter by studying the self-interacting scalar field theory, "$\lambda\,\varphi^4$." We will construct a perturbation series to calculate the amplitude, or S-matrix for any process. But, before we get any distance into the computation of the series, we will meet a wall of divergences. The removal of these divergences will require a renormalization program, which will be taken up later.

7.1 CANONICAL QUANTIZATION

The action for φ^4 theory is the sum of the free scalar field action plus a quartic self-interaction term.

$$S[\varphi] = \int d^4x\,\mathcal{L}(x) = \int d^4x\,\frac{1}{2}(\partial_\mu\varphi\partial^\mu\varphi - m_0^2\varphi^2) - \frac{\lambda_0}{4!}\varphi^4. \tag{7.1}$$

λ_0 is the coupling describing the strength of the interaction. We will consider the interaction to be repulsive, $\lambda_0 > 0$. From the action we find that the field momentum, $\pi(x)$, conjugate to $\varphi(x)$ is

$$\pi(x) = \frac{\partial\mathcal{L}}{\partial(\partial_t\varphi(x))} = \dot{\varphi}(x), \tag{7.2}$$

the same relationship as in the free scalar case. The field Hamiltonian is

$$H = \frac{1}{2} \int d^3x\, \pi^2(x) - |\nabla\varphi|^2 + m_0^2 \varphi^2(x) + \frac{\lambda_0}{4!}\varphi^4(x). \tag{7.3}$$

Canonical quantization implies that the equal-time commutators are

$$[\varphi(\vec{x},t), \pi(\vec{y},t)] = i\delta^3(\vec{x}-\vec{y})$$
$$[\varphi(\vec{x},t), \varphi(\vec{y},t)] = [\pi(\vec{x},t), \pi(\vec{y},t)] = 0. \tag{7.4}$$

The commutators, eq.(7.4), and Hamiltonian, eq.(7.3), reproduce eq.(7.2) and the classical field equation,

$$\Box\, \varphi + m_0^2 \varphi + \frac{\lambda_0}{3!}\varphi^3 = 0, \tag{7.5}$$

as operator equations of motion.

7.2 IN AND OUT FIELDS

The difficulty in computing with nonlinear interacting field theories is that we no longer have a simple particle interpretation. We do not know the classical solutions to eq.(7.5) to expand with. $\varphi(x)$ will no longer just create single particles of a definite position. Because the equal-time commutators, eq.(7.4), are the same as in the free field case, the plane wave expansions are valid at $t = 0$. The time evolution of the operator coefficients in the plane wave expansion is no longer simply $\exp(\pm i\omega_k t)$, but is $\exp(-iHt)a(k)\exp(iHt)$. The interacting Hamiltonian is no longer just a sum of harmonic oscillator Hamiltonians, so the time evolution is difficult to compute. Since we do not have a particle interpretation, we do not even know how to choose the Fock space.

To make any progress, we must make a few assumptions about the states. First of all, we will assume that there is a Lorentz invariant, nondegenerate vacuum, $|\,0\rangle$, that has zero energy (with respect to the interacting theory Hamiltonian) and is translationally invariant (no momentum). We will also assume that there are no bound states. Then we may associate each kind of (stable) particle with a distinct dynamical field. In our present example, we have only one field so we are describing only one type of particle. We will assume for now that the measured mass of the particle is m.

Each one of these assumptions is true for free particles and free field theory. By making these assumptions for the interacting field theory, we are implicitly saying that the interactions are not strong enough to radically change the spectrum of states from the noninteracting case. To avoid bound states and radical changes, the coupling cannot be strong.

The value of making assumptions about the states that are identical to properties of free particle states is that we can borrow the particle interpretation and build a Fock space. As described in the previous chapter, we formally accomplish this by considering isolated particles to be essentially free. In a high energy experiment, isolated particles are accelerated and maneuvered so as to collide. When the two particles collide, the reaction is confined to a small volume. The particles resulting from the interaction fly away from the interaction region and act like isolated free particles.

If the collision occurs at $t = 0$, then the initial state long before the collision, $t \to -\infty$, is a state containing isolated "free" particles. We know how to write down these states. Let $\varphi_{\text{IN}}(x)$ be a free scalar field operator satisfying

$$(\Box + m^2)\varphi_{\text{IN}} = 0. \tag{7.6}$$

Therefore, $\varphi_{\text{IN}}(x)$ creates free particles of physical mass m, that is, the experimentally measured mass of the particle. Note that the term that looks like a mass term in the interacting action, eq.(7.1), contains m_0, not m. Even though the isolated particles are not interacting with other particles, they still may interact with themselves. The influence of such self-interactions is to change the physical mass of a single particle from m_0 to m. m_0 represents the "bare" mass, or the mass of a particle if there were no interaction. In an experiment, we have no way of turning off the self-interaction. Therefore, we can never measure m_0, only m. Once we have added interactions, we can no longer give simple physical interpretations to the parameters appearing in the action.

φ_{IN} creates free particles of mass m. When we think of these particles as isolated parts of an interacting system, then they aren't really free since they self-interact. The use of φ_{IN} is a nice convenient way of treating the self-interaction of a single isolated particle.

Since φ_{IN} is a free field, the usual plane wave expansions are valid and we have a particle interpretation. By assumption, the interacting theory vacuum, $|\,0\rangle$, contains no free isolated particles so

$$a_{\text{IN}}(k)|\,0\rangle = 0. \tag{7.7}$$

An initial n-body state at $t \to -\infty$ is

$$a^\dagger_{\text{IN}}(k_1)\cdots a^\dagger_{\text{IN}}(k_n)|\,0\rangle. \tag{7.8}$$

These free particle states along with $|\,0\rangle$ provide a basis for the Fock space of the interacting theory and are collectively called "IN" states.

Long after the collision, $t \to \infty$, we again have isolated particles propagating at their physical mass, m. Let these particles be created and destroyed by the free scalar field $\varphi_{\text{OUT}}(x)$. Like $\varphi_{\text{IN}}(x)$, $\varphi_{\text{OUT}}(x)$ satisfies

$$(\Box + m^2)\varphi_{\text{OUT}}(x) = 0. \tag{7.9}$$

We can also use the states created with $a^\dagger_{\rm OUT}(k)$ on $|\,0\rangle$ as a basis for the Fock space.

Given this description we might be tempted to write

$$\begin{aligned}\lim_{t\to-\infty}\varphi(x)&=\sqrt{Z}\varphi_{\rm IN}(x)\\ \lim_{t\to\infty}\varphi(x)&=\sqrt{Z}\varphi_{\rm OUT}(x).\end{aligned}\tag{7.10}$$

The factor of $\sqrt{Z}$ is a normalization to indicate that $\varphi(x)$ does more than just create or destroy single particles. If $|\,n\rangle$ is an n-particle IN state such as eq.(7.8), then

$$\langle n\,|\varphi_{\rm IN}(x)|\,0\rangle=0 \qquad \text{unless } n=1, \tag{7.11}$$

whereas

$$\langle n\,|\varphi(x)|\,0\rangle\neq 0 \qquad \text{for } n\neq 1. \tag{7.12}$$

Eq.(7.11) follows from eq.(7.2). Z is determined by

$$\langle 1\,|\varphi(x)|\,0\rangle=\sqrt{Z}\langle 1\,|\varphi_{\rm IN}(x)|\,0\rangle, \tag{7.13}$$

hence we expect Z to be between 0 and 1. $\sqrt{Z}$ is the amplitude for φ to create a 1-particle state out of the vacuum.

Eq.(7.10) *cannot* be implemented as an operator equation. As usual, the reason it cannot is that, as an operator equation, it is in conflict with the equal-time commutation relations. To demonstrate this, and to get a better feeling for what $\varphi(x)$ does, we next consider the commutator of $\varphi(x)$ at unequal times.

7.3 SPECTRAL DENSITY

The IN states are eigenstates of energy and momentum. For example, the state

$$|\,1\rangle=a^\dagger_{\rm IN}(k)|\,0\rangle \tag{7.14}$$

has energy eigenvalue $k^0=\sqrt{\vec{k}^2+m^2}$ and momentum eigenvalue k^i.

$$P^\mu|\,1\rangle=k^\mu|\,1\rangle \tag{7.15}$$

The IN states are also assumed to be complete,

$$1=\sum_n|\,n\rangle\langle n\,|, \tag{7.16}$$

where the label n includes all the quantum numbers necessary to describe the n-body state.

We will now use the properties of the IN states to compute

$$\langle 0 | [\varphi(x), \varphi(y)] | 0 \rangle \equiv i\Delta'(x-y) \tag{7.17}$$

in terms of $\Delta(x-y)$, the free scalar field commutator at unequal times, eq.(3.25). First compute $\langle 0 | \varphi(x)\varphi(y) | 0 \rangle$ by inserting 1 in the form of eq.(7.16) in between $\varphi(x)$ and $\varphi(y)$.

$$\langle 0 | \varphi(x)\varphi(y) | 0 \rangle = \sum_n \langle 0 | \varphi(x) | n \rangle \langle n | \varphi(y) | 0 \rangle \tag{7.18}$$

The momentum operators, P^μ, generate infinitesimal space-time translations. Thus,

$$\varphi(x) = e^{iP\cdot x} \varphi(0) e^{-iP\cdot x}, \tag{7.19}$$

so that

$$\begin{aligned} \langle 0 | \varphi(x) | n \rangle &= \langle 0 | e^{iP\cdot x} \varphi(0) e^{-iP\cdot x} | n \rangle \\ &= e^{-ip_n\cdot x} \langle 0 | \varphi(0) | n \rangle, \end{aligned} \tag{7.20}$$

where p_n^μ is the energy-momentum eigenvalue of the state $| n \rangle$,

$$P^\mu | n \rangle = p_n^\mu | n \rangle, \tag{7.21}$$

and where we have used the fact that the vacuum is translationally invariant. Plugging eq.(7.20) into eq.(7.18) we have

$$\langle 0 | \varphi(x)\varphi(y) | 0 \rangle = \sum_n | \langle 0 | \varphi(0) | n \rangle |^2 e^{-ip_n\cdot(x-y)}, \tag{7.22}$$

and so

$$\Delta'(x-y) = -i \sum_n | \langle 0 | \varphi(0) | n \rangle |^2 (e^{-ip_n\cdot(x-y)} - e^{ip_n\cdot(x-y)}). \tag{7.23}$$

To put $\Delta'(x-y)$ into a more convenient form, insert

$$1 = \int d^4q\, \delta^4(q - p_n) \tag{7.24}$$

into eq.(7.23).

$$\begin{aligned} i\Delta'(x-y) &= \sum_n \int d^4q\, \delta^4(q-p_n) | \langle 0 | \varphi(0) | n \rangle |^2 (e^{-q\cdot(x-y)} - e^{iq\cdot(x-y)}) \\ &\equiv \int \frac{d^4q}{(2\pi)^3} \rho(q) (e^{-iq\cdot(x-y)} - e^{iq\cdot(x-y)}) \end{aligned} \tag{7.25}$$

where

$$\rho(q) = (2\pi)^3 \sum_n \delta^4(q - p_n)|\langle 0\,|\varphi(0)|\,n\rangle|^2 \tag{7.26}$$

is called the spectral density or spectral amplitude. Note that $\rho(q)$ is positive, and that since $p_n^0 > 0$, the presence of the δ-function means that it vanishes outside of the forward light-cone. $\rho(q)$ is also Lorentz invariant because φ transforms as a Lorentz scalar. $\rho(q)$ characterizes what kinds of states $\varphi(x)$ can produce. If $\varphi(x)$ were just a free field, for example $\varphi_{\rm IN}(x)$, only the $n = 1$ term in eq.(7.26) would be nonzero and the spectral density would be $\rho(q) = \int (2\pi)^3\delta^4(q-k)d^4k$. Then eq.(7.25) turns into eq.(3.25).

To express eq.(7.25) in terms of $\Delta(x-y)$, eq.(3.25), we use the properties of $\rho(q)$ to write it as $\rho(q) = \varrho(q^2)\theta(q^0)$ and insert 1 in the form of

$$1 = \int_0^\infty d\sigma^2\delta(q^2 - \sigma^2) \tag{7.27}$$

into eq.(7.25).

$$\begin{aligned} i\Delta'(x-y) &= \int_0^\infty d\sigma^2 \varrho(\sigma^2) \int \frac{d^4q}{(2\pi)^3}\theta(q^0)(e^{-iq\cdot(x-y)} - e^{iq\cdot(x-y)}) \\ &= \int d\sigma^2 \varrho(\sigma^2)\Delta(x-y;\sigma) \end{aligned} \tag{7.28}$$

$\varrho(\sigma^2)$ is a distribution in (mass)^2. By our assumptions about the states, we know that $\varphi(x)$ cannot create a state with mass less than m, the single-particle mass. Thus, $\varrho(\sigma^2) = 0$ for $\sigma^2 < m^2$. If m_1 is the mass where multiparticle production begins, then $\varphi(x)$ cannot create a state with mass above the single-particle state, but below the multiparticle threshold, so $\varrho(\sigma^2) = 0$ for $m^2 < \sigma^2 < m_1^2$. Therefore,

$$\varrho(\sigma^2) \propto \delta(\sigma^2 - m^2) \qquad \text{for } \sigma^2 < m_1^2. \tag{7.29}$$

In our present case with 1 distinct particle of physical mass m, the multiparticle threshold m_1 is $2m$.

By taking the time derivative of $\Delta'(x-y)$, and the limit $y^0 \to x^0$, we find that the equal-time commutators tell us that $(x^0 = t)$,

$$\lim_{y^0\to t} \frac{\partial}{\partial t}\Delta'(x-y) = -i\langle 0\,|[\dot\varphi(\vec{x},t), \varphi(\vec{y},t)]|\,0\rangle = \delta(\vec{x}-\vec{y}). \tag{7.30}$$

Because the equal-time commutators are identical for the free fields, $\varphi_{\rm IN}(\vec{x},t)$, then

$$\lim_{y^0\to t} \frac{\partial}{\partial t}\Delta(x-y;\sigma) = \delta(\vec{x}-\vec{y}). \tag{7.31}$$

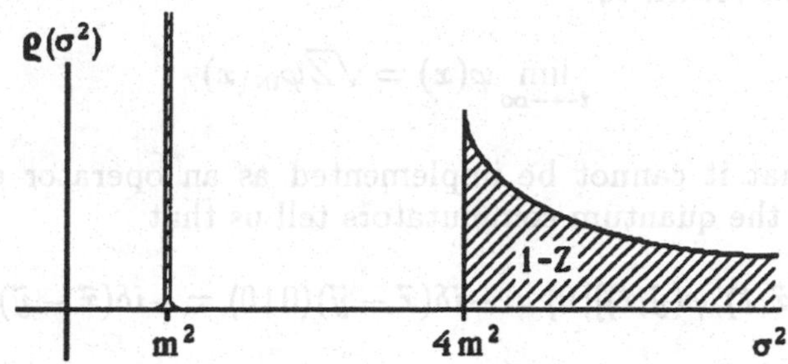

Figure 7.1 Spectral density for an interacting field $\varphi(x)$ describing a single particle of mass m. Multiparticle production begins at $2m$. The total area under $\varrho(\sigma^2)$ must be unity. If $\varphi(x)$ describes a free field, then there is no multiparticle production and $Z = 1$. Likewise, if $Z = 1$, then $\varphi(x)$ is a free field.

If we apply $\lim_{y^0 \to t} \partial/\partial t$ to eq.(7.28) and use eq.(7.30) and eq.(7.31), we find that the equal-time commutators ultimately imply that

$$\int_0^\infty d\sigma^2 \varrho(\sigma^2) = 1. \tag{7.32}$$

For the interacting field $\varphi(x)$, $\varrho(\sigma^2) \neq 0$ for $\sigma^2 \geq 4m^2$, so we must introduce a normalization, Z, to the 1-particle state contribution to $\varrho(\sigma^2)$ so that eq.(7.32) will hold. Therefore,

$$\varrho(\sigma^2) = Z\delta(\sigma^2 - m^2), \tag{7.33}$$

where

$$1 - Z = \int_{4m^2}^\infty \varrho(\sigma^2) d\sigma^2. \tag{7.34}$$

See figure 7.1. Since $\varrho(\sigma^2)$ is positive, then $0 < Z \leq 1$. If $Z = 1$, then $\Delta'(x-y) = \Delta(x-y)$ and $\varphi(x)$ is saturated in 1-particle states so $\varphi(x) = \varphi_{\text{IN}}(x)$.

We carried this calculation through starting with $\langle 0 |\varphi(x)\varphi(y)| 0\rangle$. We can apply the same reasoning to the propagator

$$i\Delta_F'(x-y) = \langle 0 |T(\varphi(x)\varphi(y))| 0\rangle \tag{7.35}$$

and find

$$\Delta_F' = \int_0^\infty d\sigma^2 \varrho(\sigma^2) \Delta_F(x-y), \tag{7.36}$$

where $\Delta_F(x-y)$ is the free scalar propagator, eq.(3.52).

Now let us return to

$$\lim_{t\to-\infty}\varphi(x)=\sqrt{Z}\varphi_{\text{IN}}(x) \tag{7.10}$$

and show that it cannot be implemented as an operator equation. At equal times, the quantum commutators tell us that

$$\langle 0\,|[\dot\varphi(\vec x,t),\varphi(\vec y,t)]|\,0\rangle=-i\delta(\vec x-\vec y)\langle 0\,|\,0\rangle=-i\delta(\vec x-\vec y), \tag{7.37}$$

which holds no matter what t is. Thus,

$$\lim_{t\to-\infty}\langle 0\,|[\dot\varphi(\vec x,t),\varphi(\vec y,t)]|\,0\rangle=-i\delta(\vec x-\vec y). \tag{7.38}$$

If we take eq.(7.10) to be an operator equation, then

$$\lim_{t\to-\infty}\langle 0\,|[\dot\varphi(\vec x,t),\varphi(\vec y,t)]|\,0\rangle=Z\lim_{t\to-\infty}\langle 0\,|[\dot\varphi_{\text{IN}}(\vec x,t),\varphi_{\text{IN}}(\vec y,t)]|\,0\rangle. \tag{7.39}$$

Since the equal-time commutators of φ and φ_{IN} are identical, then eq.(7.39) implies that $Z=1$. But, eq.(7.34) with $Z=1$ implies that $\varrho(\sigma^2)=0$ except at $\sigma^2=m^2$, so $\varphi=\varphi_{\text{IN}}$. Therefore, if we try to implement eq.(7.10) as an operator equation, we end up with $\varphi=\varphi_{\text{IN}}$, a free field theory. In analogy with the free photon quantization, we implement eq.(7.10) in a weaker form. Eq.(7.10) holds for matrix elements of the operators, that is, eq.(7.10) holds when sandwiched between two normalizable states.

7.4 THE S-MATRIX

Our goal is to calculate cross sections from the field theory to compare with experimentally measured cross sections. A typical scattering experiment is shown diagrammatically in figure 7.2. The initial state $|\,\alpha_{\text{IN}}\rangle$ contains two particles that are isolated at times long before the collision. The interaction occurs in the "blob" in the center. The result of the reaction at much later times is the state $|\,\beta_{\text{OUT}}\rangle$. The transition amplitude, T, for such a process to occur is the overlap between the two states,

$$T=\langle\beta_{\text{OUT}}\,|\,\alpha_{\text{IN}}\rangle, \tag{7.40}$$

while the probability that such a process will occur is $|T|^2$. The cross section for the scattering will be proportional to $|T|^2$.

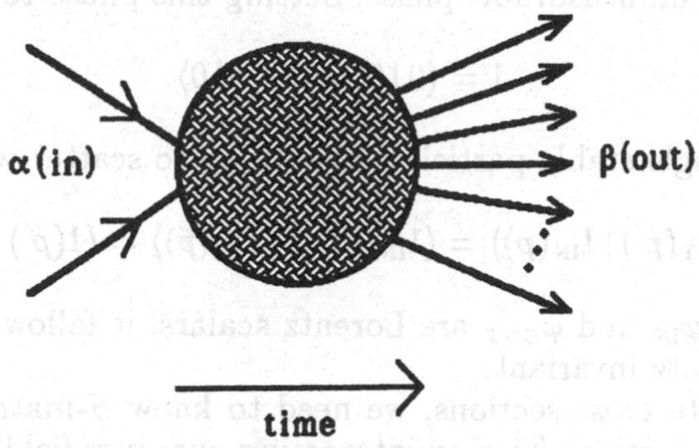

Figure 7.2 A typical scattering experiment.

There is really no difference between the IN and OUT states. Both the IN field theory and the OUT field theory describe the same free scalar particle of mass m. Their spectra are identical. This means that the Fock space built from application of $a_{\text{IN}}(k)$ to $|0\rangle$ should be isomorphic to the OUT Fock space. There should be a mapping, S, that tells us how to relate the OUT states to the IN states.

$$S|\alpha_{\text{OUT}}\rangle = |\alpha_{\text{IN}}\rangle \tag{7.41}$$

In terms of the fields,

$$\varphi_{\text{OUT}}(x) = S^{-1}\varphi_{\text{IN}}(x)S. \tag{7.42}$$

The transformation, S, will, of course, depend on the details of the reactions that go on inside the interacting region, that is, on the interacting field theory.

In terms of our present example,

$$\langle\beta_{\text{OUT}}\,|\,\alpha_{\text{IN}}\rangle = \langle\beta_{\text{IN}}\,|S|\,\alpha_{\text{IN}}\rangle = \langle\beta_{\text{OUT}}\,|S|\,\alpha_{\text{OUT}}\rangle = S_{\beta\alpha}, \tag{7.43}$$

thus the matrix elements of the S operator are the transition amplitudes for scattering processes.

From the definition of S, eq.(7.41), and the IN-OUT states, we see that S is unitary because

$$\delta_{\alpha\beta} = \langle\beta_{\text{IN}}\,|\,\alpha_{\text{IN}}\rangle = \langle\beta_{\text{OUT}}\,|S^{\dagger}S|\,\alpha_{\text{OUT}}\rangle. \tag{7.44}$$

Since the vacuum can't scatter off of itself, then S takes the vacuum into itself up to an unmeasurable phase. Setting this phase to zero, we expect

$$1 = \langle 0 \,|\, 0 \rangle = \langle 0 \,| S |\, 0 \rangle. \tag{7.45}$$

Likewise, a single stable particle has nothing to scatter with so

$$\delta_{p'p} = \langle 1_{\rm OUT}(p') \,|\, 1_{\rm IN}(p) \rangle = \langle 1_{\rm IN}(p') \,| S |\, 1_{\rm IN}(p) \rangle = \langle 1(p') \,|\, 1(p) \rangle. \tag{7.46}$$

Finally, since $\varphi_{\rm IN}$ and $\varphi_{\rm OUT}$ are Lorentz scalars, it follows from eq.(7.42) that S is Lorentz invariant.

To calculate cross sections, we need to know S-matrix elements. To calculate cross sections from an interacting quantum field theory, we need to relate the S-matrix to the fields $\varphi(x)$. This relation is called the reduction formula which we turn to next.

7.5 REDUCTION FORMULA

The reduction formula relates the S-matrix to time-ordered products of the field in the vacuum, the Green's functions of the field theory. Before we construct the formula, we'll introduce some standard notation used in expressing the creation and destruction operators, $a^\dagger_{\rm IN}(k)$ and $a_{\rm IN}(k)$, in terms of $\varphi_{\rm IN}(x)$ and$\pi_{\rm IN}(x)$.

In chapter 3 we expressed $a(k)$ in terms of φ and π in eq.(3.17). For the IN fields,

$$a_{\rm IN}(k) = \int d^3x e^{ik\cdot x}(\omega_k \varphi_{\rm IN}(\vec{x},t) + i\pi_{\rm IN}(\vec{x},t)). \tag{7.47}$$

This integral is independent of time so we may carry it out at a time of our choosing. Now we observe that

$$\omega_k e^{ik\cdot x} = -i\frac{\partial}{\partial t} e^{ik\cdot x} \quad \text{and} \quad \pi_{\rm IN}(\vec{x},t) = \frac{\partial}{\partial t}\varphi_{\rm IN}(\vec{x},t).$$

If we define the operation

$$a\overleftrightarrow{\partial}_t b = -(\partial_t a)b + a\partial_t b, \tag{7.48}$$

then we may rewrite eq.(7.47) more compactly as

$$a_{\rm IN}(k) = \int d^3x e^{ik\cdot x}\overleftrightarrow{\partial}_0 \varphi_{\rm IN}(\vec{x},t). \tag{7.49}$$

To begin constructing the reduction formula, consider an arbitrary element of the S-matrix where the IN state, $|\,\alpha_{\rm IN}\rangle$, contains n particles and the OUT state, $|\,\beta_{\rm IN}\rangle$, contains m particles. Let the momenta of the n IN particles be $k_1^\mu, \ldots, k_n^\mu$, and OUT momenta be $p_1^\mu, \ldots, p_m^\mu$. The states $|\,\alpha_{\rm IN}\rangle$ and $|\,\beta_{\rm OUT}\rangle$ are to be properly normalized, but we will not show the smearing test function explicitly since it is not crucial to the computation. To show the momenta more explicitly, we will write

$$|\,\alpha_{\rm IN}\rangle = |\,k_1 \cdots k_n; {\rm IN}\rangle .$$

Thus,

$$\begin{aligned} S_{\beta\alpha} &= \langle \beta_{\rm OUT}\,|\,\alpha_{\rm IN}\rangle \\ &= \langle p_1 \cdots p_m; {\rm OUT}\,|\,k_1 \cdots k_n; {\rm IN}\rangle . \end{aligned} \tag{7.50}$$

The idea behind deriving the reduction formula is to pull out 1 particle at a time from the IN and OUT states by writing its creation operator explicitly. Then we express the creation operator in terms of the IN or OUT field operator. Since the expression for the creation operators in terms of the IN or OUT fields is independent of time, we will take $t \to \pm\infty$ and use the asymptotic relation between φ and $\varphi_{\rm IN}$ or $\varphi_{\rm OUT}$ to replace the IN or OUT field with the interacting field φ. Once we have "emptied" all of the particles out of the IN and OUT states, we will be left with a vacuum expectation value of a product of interacting fields, $\varphi(x_i)$.

To carry this out we start by removing the particle with momentum k_1^μ from the IN state.

$$\langle p_1 \cdots p_m; {\rm OUT}\,|\,k_1 \cdots k_n; {\rm IN}\rangle = \langle p_1 \cdots p_m; {\rm OUT}\,|a^\dagger_{\rm IN}(k_1)|\,k_2 \cdots k_n; {\rm IN}\rangle .$$

Add and subtract $a^\dagger_{\rm OUT}(k_1)$.

$$\begin{aligned} \langle p_1 \cdots p_m; {\rm OUT}\,|a^\dagger_{\rm IN}(k_1)|\,k_2 \cdots k_n; {\rm IN}\rangle &= \langle p_1 \cdots p_m; {\rm OUT}\,|a^\dagger_{\rm OUT}(k_1)|\,k_2 \cdots k_n; {\rm IN}\rangle \\ &+ \langle p_1 \cdots p_m; {\rm OUT}\,|a^\dagger_{\rm IN}(k_1) - a^\dagger_{\rm OUT}(k_1)|\,k_2 \cdots k_n; {\rm IN}\rangle . \end{aligned} \tag{7.51}$$

The first term in eq.(7.51) will be zero unless one of the p_i is equal to k_1. If one does match, this term represents a "disconnected" process as shown in figure 7.3.

Let's say that $p_1 = k_1$. Then,

$$\begin{aligned} \langle p_1 \cdots p_m; {\rm OUT}\,|a^\dagger_{\rm OUT}(k_1)|\,k_2 \cdots k_n; {\rm IN}\rangle &= (2\pi)^3 2\omega_{k_1} \delta^3(p_1 - k_1) \\ &\times \langle p_2 \cdots p_m; {\rm OUT}\,|\,k_2 \cdots k_n; {\rm IN}\rangle . \end{aligned} \tag{7.52}$$

A process that is not disconnected is called connected. From eq.(7.52) or the diagram in figure 7.3, it's clear that we only need to calculate $S_{\beta\alpha}$

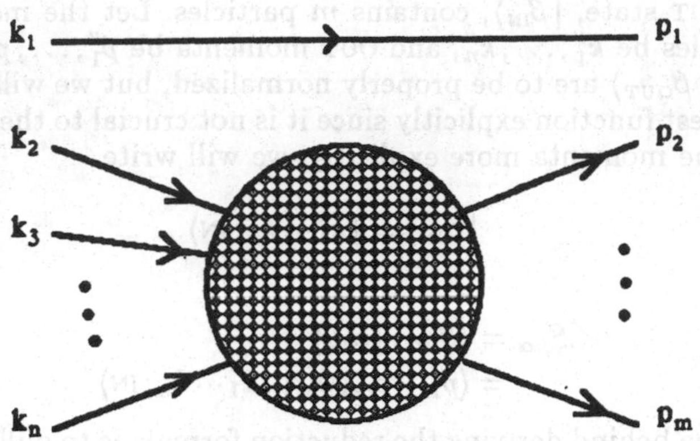

Figure 7.3 An example of a disconnected process.

for connected processes. The disconnected terms can be written as a connected term multiplied by momentum conservation factors as in eq.(7.52) or multiplied by another connected term. Since only the connected terms are nontrivial, we will assume that the original process is connected. This means that the first term in eq.(7.51) vanishes.

Next, express the remaining term in eq.(7.51) in terms of φ_{IN} and φ_{OUT}.

$$S^c_{\beta\alpha} = -i\langle p_1 \cdots p_m ;\text{OUT}| \int d^3x_1 e^{-ik_1\cdot x_1} \overleftrightarrow{\partial}_0 (\varphi_{\text{IN}}(x_1) - \varphi_{\text{OUT}}(x_1))| k_2 \cdots k_n ;\text{IN}\rangle. \quad (7.53)$$

The integral in eq.(7.53) is time independent so we evaluate the term involving $\varphi_{\text{IN}}(x)$ at $t \to -\infty$ and the other term involving $\varphi_{\text{OUT}}(x)$ at $t \to \infty$. Since the operators $\varphi_{\text{IN}}(x)$ and $\varphi_{\text{OUT}}(x)$ are sandwiched between normalizable states, we can use the asymptotic condition, $\varphi_{\text{IN}} \to \varphi/\sqrt{Z}$, as $t \to -\infty$, and $\varphi_{\text{OUT}} \to \varphi/\sqrt{Z}$ as $t \to \infty$. Eq.(7.53) becomes

$$S^c_{\beta\alpha} = \frac{i}{\sqrt{Z}} \Big(\lim_{t_1\to\infty} - \lim_{t_1\to-\infty} \Big) \int d^3x_1 e^{-ik_1\cdot x_1} \overleftrightarrow{\partial}_0 \langle p_1 \cdots p_m ;\text{OUT}|\varphi(x_1)| k_2 \cdots k_n ;\text{IN}\rangle. \quad (7.54)$$

We can rewrite eq.(7.54) in a more convenient form by noting that

$$(\lim_{t\to\infty} - \lim_{t\to-\infty})\int d^3x f(\vec{x},t) = \lim_{\substack{t_f\to\infty\\ t_i\to-\infty}} \int_{t_i}^{t_f} dt \frac{\partial}{\partial t}\int d^3x f(\vec{x},t), \quad (7.55)$$

so that

$$\begin{aligned} S^c_{\beta\alpha} &= \frac{i}{\sqrt{Z}}\int_{-\infty}^{\infty} d^4x_1 \partial_0 \left[e^{-ik_1\cdot x_1}\overleftrightarrow{\partial}_0 \langle p_1 \cdots p_m; \mathrm{OUT}\,|\varphi(x_1)|\, k_2 \cdots k_n; \mathrm{IN}\rangle \right] \\ &= \frac{i}{\sqrt{Z}}\int d^4x_1 (\nabla^2 + m^2) e^{-ik_1\cdot x_1} \langle p_1 \cdots p_m; \mathrm{OUT}\,|\varphi(x_1)|\, k_2 \cdots k_n; \mathrm{IN}\rangle \\ &\quad + \frac{i}{\sqrt{Z}}\int d^4x_1 e^{-ik_1\cdot x_1} \partial_0^2 \langle p_1 \cdots p_m; \mathrm{OUT}\,|\varphi(x_1)|\, k_2 \cdots k_n; \mathrm{IN}\rangle, \end{aligned} \quad (7.56)$$

where we have used the fact that $\exp(-ik_1 \cdot x_1)$ is a solution of the Klein-Gordon equation. To complete the first step in the reduction, we spatially integrate by parts to bring ∇^2 onto $\varphi(x_1)$. We always assume that the interaction is localized so that we can throw away the surface terms. Thus,

$$S^c_{\beta\alpha} = \frac{i}{\sqrt{Z}}\int d^4x_1 e^{-ik_1\cdot x_1} (\Box_{x_1} + m^2) \langle p_1 \cdots p_m; \mathrm{OUT}\,|\varphi(x_1)|\, k_2 \cdots k_n; \mathrm{IN}\rangle. \quad (7.57)$$

To continue, we pull out another particle. For the sake of illustration, we will pull one from the OUT state.

$$\begin{aligned} &\langle p_1 \cdots p_m; \mathrm{OUT}\,|\varphi(x_1)|\, k_2 \cdots k_n; \mathrm{IN}\rangle \\ &\qquad = \langle p_2 \cdots p_m; \mathrm{OUT}\,|a_{\mathrm{OUT}}(p_1)\varphi(x_1)|\, k_2 \cdots k_n; \mathrm{IN}\rangle, \end{aligned}$$

where we have been careful to preserve operator ordering. As before, we add and subtract the matrix elements of $\varphi(x_1)a_{\mathrm{IN}}(p_1)$.

$$\begin{aligned} \langle p_1 \cdots p_m; \mathrm{OUT}\,|\varphi(x_1)|\, k_2 \cdots k_n; \mathrm{IN}\rangle &= \langle p_1 \cdots p_m; \mathrm{OUT}\,|\varphi(x_1)a_{\mathrm{IN}}(p_1)|\, k_2 \cdots k_n; \mathrm{IN}\rangle \\ + \langle p_2 \cdots p_m; \mathrm{OUT}\,|a_{\mathrm{OUT}}(p_1)\varphi(x_1) &- \varphi(x_1)a_{\mathrm{IN}}(p_1)|\, k_2 \cdots k_n; \mathrm{IN}\rangle. \end{aligned} \quad (7.58)$$

Once again, the first term represents a disconnected process so we will ignore it. Expressing the second term in terms of φ_{OUT} and φ_{IN}, we have

$$\begin{aligned} &\langle p_1 \cdots p_m; \mathrm{OUT}\,|\varphi(x_1)|\, k_2 \cdots k_n; \mathrm{IN}\rangle = \\ &\int d^3y_1 e^{ip_1\cdot x_1}\overleftrightarrow{\partial}_0 \langle p_2 \cdots p_m; \mathrm{OUT}\,|\varphi_{\mathrm{OUT}}(y_1)\varphi(x_1) - \varphi(x_1)\varphi_{\mathrm{IN}}(y_1)|\, k_2 \cdots k_n; \mathrm{IN}\rangle. \end{aligned} \quad (7.59)$$

Now we want to apply the asymptotic relation between φ and φ_{IN} or φ_{OUT}. This means that we want to take $y_1^0 \to \infty$ in the first integral in eq.(7.59) and $y_1^0 \to -\infty$ in the second integral. Observe that since $\varphi_{\text{OUT}}(y_1)$ is to the left of $\varphi(x_1)$ and $\varphi_{\text{IN}}(y_1)$ is to the right of $\varphi(x_1)$, the limits we want on y_1^0 naturally correspond to time-ordering. Therefore, the right-hand side of eq.(7.59) becomes

$$\frac{i}{\sqrt{Z}}\Big(\lim_{y_1^0\to\infty} - \lim_{y_1^0\to-\infty}\Big)\int d^3y_1 e^{ip_1\cdot y_1}\overleftrightarrow{\partial}_0$$
$$\langle p_2\cdots p_m;\text{OUT}\,|T(\varphi(x_1)\varphi(y_1))|\,k_2\cdots k_n;\text{IN}\rangle,$$

where T is the time-ordering operator introduced earlier in connection with the propagators. Using eq.(7.55), and following the same steps that led to eq.(7.56), we find that eq.(7.59) can be written as

$$\langle p_1\cdots p_m;\text{OUT}\,|\varphi(x_1)|\,k_2\cdots k_n;\text{IN}\rangle$$
$$=\frac{i}{\sqrt{Z}}\int d^4y_1 e^{ip_1\cdot y_1}(\Box_{y_1}+m^2)\langle p_2\cdots p_m;\text{OUT}\,|T(\varphi(x_1)\varphi(y_1))|\,k_2\cdots k_n;\text{IN}\rangle. \tag{7.60}$$

Combining eq.(7.60) with eq.(7.57) we now have

$$S^c_{\beta\alpha}=\frac{i^2}{Z}\int d^4x_1\,d^4y_1 e^{-ik_1\cdot x_1}e^{ip_1\cdot y_1}(\Box_{x_1}+m^2)(\Box_{y_1}+m^2)$$
$$\times\,\langle p_2\cdots p_m;\text{OUT}\,|T(\varphi(x_1)\varphi(y_1))|\,k_2\cdots k_n;\text{IN}\rangle. \tag{7.61}$$

We continue this process of emptying the IN and OUT states until we are left with the vacuum. The full reduction formula is

$$S^c_{\beta\alpha}=\left(\frac{i}{\sqrt{Z}}\right)^{n+m}\int d^4x_1\cdots d^4x_n\,d^4y_1\cdots d^4y_m e^{-ik_1\cdot x_1}\cdots e^{ip_m\cdot y_m}$$
$$(\Box_{x_1}+m^2)\cdots(\Box_{y_m}+m^2)\langle 0\,|T(\varphi(x_1)\cdots\varphi(x_n)\varphi(y_1)\cdots\varphi(y_m))|\,0\rangle. \tag{7.62}$$

This reduction formula, first obtained by Lehmann, Symanzik, and Zimmermann, is the connection between interacting quantum field theory and observable scattering processes that we need in order to get anything useful out of field theory. The formula has a simple interpretation. Recall the form of the propagator for free particles. The propagator has a pole at $p^2=m^2$, the physical mass of the particle. $\langle 0\,|T(\varphi(x_1)\cdots\varphi(y_m))|\,0\rangle$ is a Green's function for the interacting theory and will have a multipole structure. Each factor of $(\Box_{x_i}+m^2)$ will lower the degree of the pole by one. Thus, the S-matrix is the normalized residue of the multipole of $\langle 0\,|T(\varphi(x_1)\cdots\varphi(y_m))|\,0\rangle$.

To compute cross sections from quantum field theory, we must compute the Green's functions. Our next step in this program is to compute $\langle 0\,|T(\varphi(x_1)\cdots\varphi(y_m))|\,0\rangle$ perturbatively.

7.6 PERTURBATION THEORY

The U-Matrix

φ_{IN} and φ_{OUT} describe identical free particles of mass m, so their corresponding Fock spaces should be isomorphic. The mapping between these spaces is formally written as the S-matrix. Our assumption about the nonexistence of bound states in the interacting spectrum again leads us to expect that φ and φ_{IN} or φ_{OUT} can be related by a unitary operator. Thus, we formally define the U operator,

$$\varphi(\vec{x},t) = U^{\dagger}(t)\varphi_{\text{IN}}(\vec{x},t)U(t). \tag{7.63}$$

In view of the observation that eq.(7.10) cannot be implemented as an operator equation, we know that the same must be true for eq.(7.63). Eq.(7.63) is only meant to make any sense when sandwiched between normalizable states. The main application of U will be to relate the time-ordered products of the interacting field in the vacuum to vacuum expectations of the IN fields. The derivation of the fundamental result, eq.(7.97), is very formal. We may essentially regard eq.(7.97) as a definition.

As is customary, and since the derivation is only formal, we will suppress the factor of $\sqrt{Z}$ that could appear in eq.(7.63). Whether the Z explicitly appears or not depends on whether the theory is written in terms of the so-called bare fields or renormalized fields. The bare field appears in eq.(7.10), and it satisfies canonical equal-time commutators (i.e. commutators identical to free field commutators or normalized to a δ-function). The renormalized fields do not, due to the appearance of Z in the normalization of the commutator. This means that the Hamiltonian written in terms of the renormalized fields will contain some factors of Z so that the operator equations of motion maintain their canonical form, eq.(7.69). The net result of explicitly keeping the Z's will be that they will only appear in $\mathcal{H}_I$, eq.(7.77). However, we can absorb them in a redefinition of the masses and couplings, in which case we get the same result from eq.(7.78) onward. Let us say again that the validity of the analysis here is ultimately complicated by the fact that divergences appear in perturbation theory. In particular, Z diverges. To make sense of the series, the result, eq.(7.97), must be modified by the renormalization process.

To discover what $U(t)$ is, let us find its equation of motion using the known equations of motion for φ and φ_{IN}. Begin by differentiating the relation expressed by eq.(7.63).

$$\frac{\partial}{\partial t}\varphi_{\text{IN}}(\vec{x},t) = \frac{\partial}{\partial t}[U(t)\varphi(\vec{x},t)U^\dagger(t)]$$

$$= \frac{\partial}{\partial t}U(t)\varphi(\vec{x},t)U^\dagger(t) + U(t)\frac{\partial}{\partial t}\varphi(\vec{x},t)U^\dagger(t) + U(t)\varphi(\vec{x},t)\frac{\partial}{\partial t}U^\dagger(t). \quad (7.64)$$

Like the S-matrix, U must be unitary, $U^\dagger U = 1$. This implies that

$$\frac{\partial}{\partial t}(UU^\dagger) = 0 = \frac{\partial U}{\partial t}U^\dagger + U\frac{\partial U^\dagger}{\partial t}. \quad (7.65)$$

Consider the first term on the right-hand side of eq.(7.64) and insert $1 = U^\dagger U$ after $\partial U/\partial t$ but before φ.

$$\frac{\partial U}{\partial t}\varphi U^\dagger = \frac{\partial U}{\partial t}U^\dagger U\varphi U^\dagger = \frac{\partial U}{\partial t}U^\dagger\varphi_{\text{IN}} \quad (7.66)$$

Now insert $1 = U^\dagger U$ after φ in the last term of eq.(7.64).

$$\begin{aligned} U\varphi\frac{\partial U^\dagger}{\partial t} &= U\varphi U^\dagger U\frac{\partial U^\dagger}{\partial t} = \varphi_{\text{IN}}U\frac{\partial U^\dagger}{\partial t} \\ &= -\varphi_{\text{IN}}\frac{\partial U}{\partial t}U^\dagger \end{aligned} \quad (7.67)$$

The last step is accomplished by using eq.(7.65). Eqs.(7.66) and (7.67) combine to form a commutator, hence eq.(7.64) becomes

$$\frac{\partial\varphi_{\text{IN}}}{\partial t} = [\dot{U}U^\dagger, \varphi_{\text{IN}}] + U\dot{\varphi}U^\dagger. \quad (7.68)$$

By using the equation of motion for φ,

$$\dot{\varphi}(\vec{x},t) = i[H(\varphi,\pi),\varphi], \quad (7.69)$$

we can also express the last term in eq.(7.68) as a commutator.

$$U\dot{\varphi}U^\dagger = iU(t)[H(\varphi,\pi),\varphi]U^\dagger(t) \quad (7.70)$$

Even though we do not explicitly show it, we assume that the Hamiltonians for φ and for φ_{IN} are normal ordered. Once again, insert $1 = U^\dagger U$ between H and φ in the commutator above. Eq.(7.70) becomes

$$U\dot{\varphi}U^\dagger = i[U(t)H(\varphi,\pi)U^\dagger(t), \varphi_{\text{IN}}]. \quad (7.71)$$

The interacting Hamiltonian, $H(\varphi, \pi)$, does not depend explicitly on time (time dependence enters H only implicitly through the time dependence of the fields π and φ). Therefore,

$$U(t)H(\varphi,\pi)U^\dagger(t) = H(\varphi_{\rm IN}, \pi_{\rm IN}). \tag{7.72}$$

For example, in the self-interacting scalar theory under present consideration,

$$H = \int d^3x \frac{1}{2}(\pi^2(\vec{x},t) + |\nabla\varphi(\vec{x},t)|^2 + m_0^2\varphi^2) + \frac{\lambda_0}{4!}\varphi^4. \tag{7.73}$$

After sandwiching H between U and $U^\dagger$, insert $1 = UU^\dagger$ between each field operator.

$$\begin{aligned} UHU^\dagger &= \int d^3x \frac{1}{2}(U\pi U^\dagger U\pi u^\dagger + \ldots \\ &= \int d^3x \frac{1}{2}(\pi_{\rm IN}^2(\vec{x},t) + \ldots \\ &= H(\varphi_{\rm IN}, \pi_{\rm IN}) \end{aligned}$$

Inserting eq.(7.72) back into eqs.(7.71) and (7.68), we have

$$\frac{\partial\varphi_{\rm IN}}{\partial t} = [\dot{U}U^\dagger, \varphi_{\rm IN}] + i[H(\varphi_{\rm IN}, \pi_{\rm IN}), \varphi_{\rm IN}]. \tag{7.74}$$

Finally, we use the equation of motion for $\varphi_{\rm IN}$ to replace $\dot{\varphi}_{\rm IN}$ by the commutator, $i[H_{\rm IN}(\varphi_{\rm IN}, \pi_{\rm IN}), \varphi_{\rm IN}]$, and move this to the right-hand side.

$$0 = [\dot{U}U^\dagger, \varphi_{\rm IN}] + i[H(\varphi_{\rm IN}, \pi_{\rm IN}) - H_{\rm IN}(\varphi_{\rm IN}, \pi_{\rm IN}), \varphi_{\rm IN}] \tag{7.75}$$

$\varphi_{\rm IN}$ is a free scalar field describing particles of mass m, so

$$H_{\rm IN}(\varphi_{\rm IN}, \pi_{\rm IN}) = \frac{1}{2}\int d^3x\, \pi_{\rm IN}^2 + |\nabla\varphi_{\rm IN}|^2 + m^2\varphi_{\rm IN}^2. \tag{7.76}$$

$H(\varphi_{\rm IN}, \pi_{\rm IN}) - H_{\rm IN}(\varphi_{\rm IN}, \pi_{\rm IN}) \equiv H_I(\varphi_{\rm IN}, \pi_{\rm IN})$ is just the interacting part of the Hamiltonian.

$$H_I(\varphi_{\rm IN}, \pi_{\rm IN}) = \int d^3x\, \frac{1}{2}(m_0^2 - m^2)\varphi_{\rm IN}^2 + \frac{\lambda_0}{4!}\varphi_{\rm IN}^4 \tag{7.77}$$

Thus, U satisfies

$$[\dot{U}U^\dagger + iH_I(\varphi_{\rm IN}, \pi_{\rm IN}), \varphi_{\rm IN}] = 0. \tag{7.78}$$

If we had started with $\pi_{\text{IN}} = U\pi U^\dagger$ and followed the same steps, we would find that $\dot{U}U^\dagger + iH_I(\varphi_{\text{IN}}, \pi_{\text{IN}})$ also commutes with π_{IN}. Thus, $\dot{U}U^\dagger + iH_I(\varphi_{\text{IN}}, \pi_{\text{IN}})$ must be an ordinary number and not an operator.

$$\dot{U}U^\dagger + iH_I(\varphi_{\text{IN}}, \pi_{\text{IN}}) = f(t) \tag{7.79}$$

$f(t)$ is an arbitrary function that will eventually be absorbed in the normalization. For now, define

$$H_I^f(t) = H_I(t) - f(t) = H_I(\varphi_{\text{IN}}, \pi_{\text{IN}}) - f(t). \tag{7.80}$$

The desired equation of motion for U is obtained by multiplying eq.(7.79) on the right by U.

$$\frac{\partial U}{\partial t} = -iH_I^f(t)U(t) \tag{7.81}$$

The formal solution of eq.(7.81) is

$$U(t) = U(t') - i\int_{t'}^{t} H_I^f(t_1)U(t_1)\,dt_1. \tag{7.82}$$

We may develop a solution to this integral equation in a Neumann series by successive approximation. Let $U^{(k+1)}(t)$ be the $k+1^{\text{st}}$ approximation. $U^{(k+1)}(t)$ satisfies

$$U^{(k+1)}(t) = U(t') - i\int_{t'}^{t} H_I^f(t_1)U^{(k)}(t_1)dt_1. \tag{7.83}$$

To start the series we will take $U^{(0)}(t) = U(t')$. $U(t')$ is constant in time since t' is fixed. Thus,

$$\begin{aligned} U^{(1)}(t) &= U(t') - i\int_{t'}^{t} H_I^f(t_1)U(t_1)dt_1 \\ &= \left(1 - i\int_{t'}^{t} H_I^f(t_1)dt_1\right)U(t'), \end{aligned} \tag{7.84}$$

so

$$\begin{aligned} U^{(2)}(t) &= U(t') - i\int_{t'}^{t} dt_1 H_I^f(t_1)\left(1 - i\int_{t'}^{t} H_I^f(t_2)dt_2\right)U(t') \\ &= \left(1 - i\int_{t'}^{t} H_I^f(t_1)dt_1 + (-i)^2\int_{t'}^{t} dt_1\int_{t'}^{t} dt_2 H_I^f(t_1)H_I^f(t_2)\right)U(t'), \end{aligned} \tag{7.85}$$

and so on. We assume that $\lim_{n\to\infty} U^{(n)}(t) = U(t)$, that is, that the series converges to $U(t)$. Therefore,

$$U(t) = \sum_{n=0}^{\infty}(-i)^n \int_{t'}^{t} dt_1 \int_{t'}^{t} dt_2 \cdots \int_{t'}^{t} dt_n H_I^f(t_1) H_I^f(t_2) \cdots H_I^f(t_n) U(t'). \tag{7.86}$$

From the limits of integration above we see that $t > t_1 > t_2 > \cdots > t_n > t'$. This means that the integrand is naturally time-ordered. If we take $t_2 > t_1$, we will get the same result as above, provided we reorder the operators in the integrand so that they stay time-ordered (that is, move $H_I^f(t_2)$ to the left of $H_I^f(t_1)$). Since there are $n!$ permutations of the ordering of $t_1, \ldots, t_n$, each contributing an identical result to the sum above, we can rewrite eq.(7.86) as

$$\begin{aligned} U(t) &= \sum_{n=0}^{\infty} \frac{(-i)^n}{n!} \int_{t'}^{t} dt_1 \int_{t'}^{t} dt_2 \cdots \int_{t'}^{t} dt_n T(H_I^f(t_1) \cdots H_I^f(t_n)) U(t') \\ &\equiv T \exp\left(-i \int_{t'}^{t} d\tilde{t} \,:H_I^f(\tilde{t}):\right) U(t'), \end{aligned} \tag{7.87}$$

where the time-ordered exponential is a short hand way of writing the sum in eq.(7.87). It is conventional to multiply eq.(7.87) on the right by $U^{-1}(t')$ and define

$$U(t,t') = U(t)U^{-1}(t'). \tag{7.88}$$

From eq.(7.87) we see that

$$U(t,t) = 1 \quad \text{and} \quad U(t,t') = T \exp\left(-i \int_{t'}^{t} d\tilde{t} \,:H_I^f(\tilde{t}):\right). \tag{7.89}$$

In terms of expectation values,

$$\lim_{\substack{t'\to-\infty \\ t\to\infty}} U(t,t') = S,$$

the S-matrix. Therefore, we may formally express S as

$$S = T \exp\left(-i \int_{-\infty}^{\infty} dt \,:H_I^f(t):\right). \tag{7.90}$$

Time-Ordered Products

In order to calculate cross sections, we need to compute the S-matrix. The reduction formula relates the S-matrix to vacuum expectation values of time-ordered products of the fields. We cannot exactly compute these time-ordered products because we do not know the vacuum expectation values of the interacting fields. We do know these values for the IN fields; therefore, we can compute these expectation values for the interacting fields perturbatively. To derive an explicit expression for the vacuum expectation values of interacting fields in terms of the IN fields, we use the U operator defined above.

Consider the "n-point" function,

$$G(x_1,\ldots,x_n) = \langle 0\,|T(\varphi(x_1)\cdots\varphi(x_n))|\,0\rangle, \tag{7.91}$$

and express $\varphi(x_i)$ in terms of U and $\varphi_{\rm IN}(x_i)$.

$$\begin{aligned} G(x_1,\ldots,x_n) &\\ = \langle 0\,|T(U^{-1}(t_1)\varphi_{\rm IN}(x_1)U(t_1)U^{-1}(t_2)\cdots U^{-1}(t_n)\varphi_{\rm IN}(x_n)U(t_n)|\,0\rangle &\\ = \langle 0\,|T(U^{-1}(t_1)\varphi_{\rm IN}(x_1)U(t_1,t_2)\cdots U(t_{n-1},t_n)\varphi_{\rm IN}(x_n)U(t_n)|\,0\rangle . & \end{aligned} \tag{7.92}$$

Insert $1 = U^{-1}(t)U(t) = U^{-1}(-t)U(-t)$ into the time-ordered product of $\varphi_{\rm IN}$'s and U's in eq.(7.92) and take t large enough so that $t > t_i$ and $-t < t_i$. With this large t, we can pull $U^{-1}(t)$ and $U(-t)$ out of the time-ordering and place $U^{-1}(t)$ to the left of every operator and $U(-t)$ to the right.

$$\begin{aligned} G(x_1,\ldots,x_n) &\\ = \langle 0\,|U^{-1}(t)T(\varphi_{\rm IN}(x_1)\cdots\varphi_{\rm IN}(x_n)\exp\left(-i\int_{t'}^{t} H_I^f(t')\,dt'\right))U(-t)|\,0\rangle . & \end{aligned} \tag{7.93}$$

To guarantee that we may pull $U(-t)$ and $U^{-1}(t)$ out from the time-ordering operator, we will only consider the limit as $t\to\infty$ in eq.(7.93) above.

To proceed, we must determine the effect of $U(-t)$ and $U(t)$ on the vacuum as $t\to\infty$. $U(t)$ satisfies eq.(7.86) where the explicit dependence on the initial condition, $U(t')$, can be seen. The choice of time, t', to set the initial condition is arbitrary so we might as well take $t'\to-\infty$. This means that we are essentially free to choose $U(-\infty)$, but we want to make a choice that is consistent with the formalism that we have constructed and the assumptions that we have made so far. Evidently, $U(-\infty)\propto \mathbf{1}$, the unit matrix, is consistent with the matrix representation of eq.(7.10), since U above always appears sandwiched between normalizable states.

With this choice, $U(-\infty)|\,0\rangle = \alpha|\,0\rangle$. We have also assumed that the vacuum is stable, that is, that the vacuum remains the vacuum forever and never decays. Thus, the S-matrix will map the vacuum into itself up to a phase. So

$$S|\,0\rangle = U(\infty)U^{-1}(-\infty)|\,0\rangle = \beta|\,0\rangle. \tag{7.94}$$

Inserting our choice of initial condition, namely $1 = U(-\infty)/\alpha$, into eq.(7.94) to the right of $U^{-1}(-\infty)$, we find that $U(\infty)|\,0\rangle = \beta\alpha|\,0\rangle$, that is, $\langle 0\,|U^{-1}(\infty) = \beta^*\alpha^*\langle 0\,|$. Therefore, the choices of initial condition and phase in eq.(7.93) lead to

$$\begin{aligned} &G(x_1,\ldots,x_n)\\ &\quad= \beta^*\alpha^*\alpha\langle 0\,|T(\varphi_{\rm IN}(x_1)\cdots\varphi_{\rm IN}(x_n)\exp\left(-i\int_{-\infty}^{\infty} H_I^f(t')\,dt'\right))|\,0\rangle. \end{aligned} \tag{7.95}$$

The normalization in eq.(7.95) above may be determined by requiring eq.(7.95) to satisfy the assumptions we have made so far and by making the vacuum unique by setting the unmeasurable phase to zero. Consider removing all of the incoming particles by setting $\varphi_{\rm IN}(x_i) = 1$. By eq.(7.63), this means that we should set $\varphi(x_i) = 1$ on the left-hand side of eq.(7.95) above. Since we are assuming that the vacuum state is normalized and that we are now requiring that the S-matrix map the vacuum into itself with no phase change, then the left-hand side above is equal to 1 when $\varphi_{\rm IN}(x_i) = 1$. Thus,

$$1 = \beta^*\alpha^*\alpha\langle 0\,|T\Big(\exp -i\int_{-\infty}^{\infty} H_I^f(t')\,dt'\Big)|\,0\rangle,$$

that is, the normalization is simply $(\langle 0\,|S|\,0\rangle)^{-1}$. Therefore,

$$G(x_1,\ldots,x_n) = \frac{\langle 0\,|T(\varphi_{\rm IN}(x_1)\cdots\varphi_{\rm IN}(x_n)\exp(-i\int_{-\infty}^{\infty} H_I^f(t')\,dt'))|\,0\rangle}{\langle 0\,|T(\exp(-i\int_{-\infty}^{\infty} H_I^f(t')\,dt'))|\,0\rangle}. \tag{7.96}$$

At this point we may finally cancel the arbitrary function, $f(t)$, in $H_I^f(t)$ that was introduced in eq.(7.97), because it appears identically in the numerator and denominator. Therefore,

$$G(x_1,\ldots,x_n) = \frac{\langle 0\,|T(\varphi_{\text{IN}}(x_1)\cdots\varphi_{\text{IN}}(x_n)\exp(-i\int_{-\infty}^{\infty} H_I(t')\,dt'))|\,0\rangle}{\langle 0\,|T(\exp(-i\int_{-\infty}^{\infty} H_I(t')\,dt'))|\,0\rangle}$$

$$= \frac{\begin{array}{c}\sum_{m=0}^{\infty}\frac{(-i)^m}{m!}\int_{-\infty}^{\infty} d^4y_1\cdots d^4y_m \\ \langle 0\,|T(\varphi_{\text{IN}}(x_1)\cdots\varphi_{\text{IN}}(x_n)\mathcal{H}_I(y_1)\cdots\mathcal{H}_I(y_m))|\,0\rangle\end{array}}{\sum_{m=0}^{\infty}\frac{(-i)^m}{m!}\int_{-\infty}^{\infty} d^4y_1\cdots d^4y_m\langle 0\,|T(\mathcal{H}_I(y_1)\cdots\mathcal{H}_I(y_m))|\,0\rangle}, \tag{7.97}$$

where $\mathcal{H}_I(y) = \mathcal{H}_I(\varphi_{\text{IN}}(y))$ is the normal ordered Hamiltonian density, $H_I(t) = \int d^3x\mathcal{H}_I(x)$.

The vacuum of an interacting theory is a busy and complicated state. The vacuum fluctuations by themselves are unmeasurable. This is why the vacuum appears empty to us. To manifest themselves, the vacuum fluctuations must interact with something we can measure. The 2-point function, eq.(7.97) with $n = 2$, describes the propagation of a particle through this vacuum. As the particle propagates, it interacts with the vacuum fluctuations. In fact, the so-called "self"-interactions of one particle are actually the particle interacting with the vacuum. Fluctuations distinct from the particle, however, do not affect its propagation. These separate vacuum fluctuations should not enter into the calculation of the propagator. The numerator in eq.(7.97) does include these fluctuations, though. To discover the contribution of these fluctuations to the result, we set $\varphi_{\text{IN}}(x_i)$ to 1. To remove these fluctuations, we divide them out, and that is why the denominator in eq.(7.97) is present. The effect of including these superfluous fluctuations is to multiply the final state by a phase, the phase being $\langle 0\,|S|\,0\rangle$. This phase is removed by the denominator.

Feynman Diagrams

Each term in the expansion of the n-point function, eq.(7.97), can be represented by a diagram or graph. In addition, simple rules can be given to reconstruct the term in the expansion from the diagram alone. The correspondence between diagrams and the physical processes they represent is natural. Therefore, in practice, to determine the S-matrix for a given process, instead of formally manipulating eq.(7.97), we draw the associated diagrams and evaluate them using the rules.

In order to compute any term in the expansion (7.97) and to exhibit the diagrammatic representation, we must calculate the vacuum expectation of the time-ordered product of IN fields. Since the vacuum satisfies $a_{\text{IN}}(p)|\,0\rangle = 0$, one way to reduce the computation of the expectation value is to normal order the product.

Consider first normal ordering the time-ordered product of two IN fields. Let us write $\varphi_{\rm IN}(x) = \varphi_{\rm IN}^{(+)}(x) + \varphi_{\rm IN}^{(-)}(x)$, where the positive frequency part, $\varphi_{\rm IN}^{(+)}(x)$, contains the destruction operator, and the negative frequency part, $\varphi_{\rm IN}^{(-)}(x)$, contains the creation operator. Normal ordering means that the positive frequency parts are to the right of the negative frequency parts.

$$
\begin{aligned}
:\varphi_{\rm IN}(x_1)\varphi_{\rm IN}(x_2): = \varphi_{\rm IN}^{(+)}(x_1)\varphi_{\rm IN}^{(+)}(x_2) + \varphi_{\rm IN}^{(-)}(x_1)\varphi_{\rm IN}^{(-)}(x_2) \\
+ \varphi_{\rm IN}^{(-)}(x_1)\varphi_{\rm IN}^{(+)}(x_2) + \varphi_{\rm IN}^{(-)}(x_2)\varphi_{\rm IN}^{(+)}(x_1)
\end{aligned}
\tag{7.98}
$$

By adding and subtracting $\varphi_{\rm IN}^{(+)}(x_1)\varphi_{\rm IN}^{(-)}(x_2)$ in (7.98) above, we have

$$
\varphi_{\rm IN}(x_1)\varphi_{\rm IN}(x_2) =: \varphi_{\rm IN}(x_1)\varphi_{\rm IN}(x_2) : + [\varphi_{\rm IN}^{(+)}(x_1), \varphi_{\rm IN}^{(-)}(x_2)]. \tag{7.99}
$$

Therefore, the time-ordered product of the two IN fields is

$$
\begin{aligned}
T(\varphi_{\rm IN}(x_1)\varphi_{\rm IN}(x_2)) &= \begin{cases} \varphi_{\rm IN}(x_1)\varphi_{\rm IN}(x_2) & t_2 < t_1 \\ \varphi_{\rm IN}(x_2)\varphi_{\rm IN}(x_1) & t_1 < t_2 \end{cases} \\
&= \begin{cases} :\varphi_{\rm IN}(x_1)\varphi_{\rm IN}(x_2): + [\varphi_{\rm IN}^{(+)}(x_1), \varphi_{\rm IN}^{(-)}(x_2)] & t_2 < t_1 \\ :\varphi_{\rm IN}(x_1)\varphi_{\rm IN}(x_2): + [\varphi_{\rm IN}^{(+)}(x_2), \varphi_{\rm IN}^{(-)}(x_1)] & t_1 < t_2 \, . \end{cases}
\end{aligned}
\tag{7.100}
$$

Using the free field plane wave expansions for the IN fields we can explicitly compute the commutators above. From eq.(3.12),

$$
\begin{aligned}
\varphi_{\rm IN}^{(+)}(x) &= \int \frac{d^3k}{(2\pi)^3} \frac{1}{2\omega_k} a_{\rm IN}(k) e^{-ik\cdot x} \\
\varphi_{\rm IN}^{(-)}(x) &= \int \frac{d^3k}{(2\pi)^3} \frac{1}{2\omega_k} a_{\rm IN}^\dagger(k) e^{ik\cdot x} .
\end{aligned}
\tag{7.101}
$$

Thus,

$$
\begin{aligned}
[\varphi_{\rm IN}^{(+)}(x_1), \varphi_{\rm IN}^{(-)}(x_2)] &= \int \frac{d^3k}{(2\pi)^3} \frac{d^3k'}{(2\pi)^3} \frac{1}{4\omega_k\omega_{k'}} [a_{\rm IN}(k), a_{\rm IN}^\dagger(k')] e^{-ik\cdot x_1} e^{ik'\cdot x_2} \\
&= \int \frac{d^3k}{(2\pi)^3} \frac{1}{2\omega_k} e^{-ik\cdot(x_1-x_2)} .
\end{aligned}
\tag{7.102}
$$

Placing this result back into eq.(7.100) we obtain

$$\begin{aligned} T(\varphi_{\rm IN}(x_1)\varphi_{\rm IN}(x_2)) &=: \varphi_{\rm IN}(x_1)\varphi_{\rm IN}(x_2) : \\ &+ \int \frac{d^3k}{(2\pi)^3}\frac{1}{2\omega_k}(e^{-ik\cdot(x_1-x_2)}\theta(t_1-t_2) + e^{ik\cdot(x_1-x_2)}\theta(t_2-t_1)) \\ &=: \varphi_{\rm IN}(x_1)\varphi_{\rm IN}(x_2) : + \langle 0\,|T(\varphi_{\rm IN}(x_1)\varphi_{\rm IN}(x_2))|\,0\rangle \\ &=: \varphi_{\rm IN}(x_1)\varphi_{\rm IN}(x_2) : + i\Delta_F(x_1-x_2), \end{aligned} \tag{7.103}$$

where we recognize the second term as the propagator, eq.(3.48). We could have obtained expression (7.103) without performing the explicit computation by noting that normal ordering differs from time-ordering only by commutators of the scalar field. These commutators are just functions, not operators.

$$T(\varphi_{\rm IN}(x_1)\varphi_{\rm IN}(x_2)) =: \varphi_{\rm IN}(x_1)\varphi_{\rm IN}(x_2) : + f(x_1, x_2) \tag{7.104}$$

To determine the function, $f(x_1, x_2)$, take the vacuum expectation of the entire expression. The vacuum expectation of any normal ordered product is zero (the point of normal ordering), thus

$$f(x_1, x_2) = \langle 0\,|T(\varphi_{\rm IN}(x_1), \varphi_{\rm IN}(x_2))|\,0\rangle, \tag{7.105}$$

which is the free field propagator or 2-point function.

Now let's consider normal ordering the product of three fields. Starting from eq.(7.99) we have

$$\begin{aligned} \varphi_{\rm IN}(x_1)\varphi_{\rm IN}(x_2)\varphi_{\rm IN}(x_3) &=: \varphi_{\rm IN}(x_1)\varphi_{\rm IN}(x_2) : \varphi_{\rm IN}(x_3) \\ &\quad + [\varphi_{\rm IN}^{(+)}(x_1), \varphi_{\rm IN}^{(-)}(x_2)]\varphi_{\rm IN}(x_3) \\ &=: \varphi_{\rm IN}(x_1)\varphi_{\rm IN}(x_2) : (\varphi_{\rm IN}^{(+)}(x_3) + \varphi_{\rm IN}^{(-)}(x_3)) \\ &\quad + [\varphi_{\rm IN}^{(+)}(x_1), \varphi_{\rm IN}^{(-)}(x_2)]\varphi_{\rm IN}(x_3). \end{aligned} \tag{7.106}$$

The commutator is just a function, so the second term above is already normal ordered. Likewise, the first term involving $\varphi_{\rm IN}^{(+)}(x_3)$ is also in normal ordered form. The only part of (7.106) that requires reordering is

$$\left(\varphi_{\rm IN}^{(+)}(x_1)\varphi_{\rm IN}^{(+)}(x_2) + \varphi_{\rm IN}^{(-)}(x_1)\varphi_{\rm IN}^{(+)}(x_2) + \varphi_{\rm IN}^{(-)}(x_2)\varphi_{\rm IN}^{(+)}(x_1)\right)\varphi_{\rm IN}^{(-)}(x_3). \tag{7.107}$$

Adding and subtracting the normal ordered form of (7.107) above to eq.(7.106), we have

$$\begin{aligned} \varphi_{\rm IN}(x_1)\varphi_{\rm IN}(x_2)\varphi_{\rm IN}(x_3) = &: \varphi_{\rm IN}(x_1)\varphi_{\rm IN}(x_2)\varphi_{\rm IN}(x_3) : \\ &+ [\varphi_{\rm IN}^{(+)}(x_1), \varphi_{\rm IN}^{(-)}(x_2)]\varphi_{\rm IN}(x_3) \end{aligned}$$

$$+\varphi_{\rm IN}^{(-)}(x_1)[\varphi_{\rm IN}^{(+)}(x_2),\varphi_{\rm IN}^{(-)}(x_3)]+\varphi_{\rm IN}^{(-)}(x_2)[\varphi_{\rm IN}^{(+)}(x_1),\varphi_{\rm IN}^{(-)}(x_3)]$$
$$+[\varphi_{\rm IN}^{(+)}(x_1)\varphi_{\rm IN}^{(+)}(x_2),\varphi_{\rm IN}^{(-)}(x_3)]. \qquad (7.108)$$

If we add and subtract $\varphi_{\rm IN}^{(+)}(x_1)\varphi_{\rm IN}^{(-)}(x_3)\varphi_{\rm IN}^{(-)}(x_2)$ to eq.(7.108), the last commutator turns into

$$\varphi_{\rm IN}^{(+)}(x_1)[\varphi_{\rm IN}^{(+)}(x_2),\varphi_{\rm IN}^{(-)}(x_3)]+\varphi_{\rm IN}^{(+)}(x_2)[\varphi_{\rm IN}^{(+)}(x_1),\varphi_{\rm IN}^{(-)}(x_3)].$$

Therefore,

$$\varphi_{\rm IN}(x_1)\varphi_{\rm IN}(x_2)\varphi_{\rm IN}(x_3)=:\varphi_{\rm IN}(x_1)\varphi_{\rm IN}(x_2)\varphi_{\rm IN}(x_3):$$
$$+[\varphi_{\rm IN}^{(+)}(x_1),\varphi_{\rm IN}^{(-)}(x_2)]\varphi_{\rm IN}(x_3)$$
$$+[\varphi_{\rm IN}^{(+)}(x_2),\varphi_{\rm IN}^{(-)}(x_3)]\varphi_{\rm IN}(x_1)+[\varphi_{\rm IN}^{(+)}(x_1),\varphi_{\rm IN}^{(-)}(x_3)]\varphi_{\rm IN}(x_2) \qquad (7.109)$$

so

$$T(\varphi_{\rm IN}(x_1)\varphi_{\rm IN}(x_2)\varphi_{\rm IN}(x_3))=:\varphi_{\rm IN}(x_1)\varphi_{\rm IN}(x_2)\varphi_{\rm IN}(x_3):$$
$$+i\Delta_F(x_1-x_2)\varphi_{\rm IN}(x_3)+i\Delta_F(x_2-x_3)\varphi_{\rm IN}(x_1)+i\Delta_F(x_1-x_3)\varphi_{\rm IN}(x_2). \qquad (7.110)$$

From (7.110) we observe that $\langle 0\,|T(\varphi_{\rm IN}(x_1)\varphi_{\rm IN}(x_2)\varphi_{\rm IN}(x_3))|\,0\rangle=0$.

Moving on to the product of four fields, we multiply eq.(7.110) by $\varphi_{\rm IN}(x_4)$. Consider the second term, $i\Delta_F(x_1-x_2)\varphi_{\rm IN}(x_3)\varphi_{\rm IN}(x_4)$. This term can be written as $i\Delta_F(x_1-x_2)(:\varphi_{\rm IN}(x_3)\varphi_{\rm IN}(x_4):+i\Delta_F(x_3-x_4))$ by using eqs.(7.99) and (7.103). The last two terms are similar. Apply the same procedure as before to the first term. We have

$$:\varphi_{\rm IN}(x_1)\varphi_{\rm IN}(x_2)\varphi_{\rm IN}(x_3):\varphi_{\rm IN}(x_4)=:\varphi_{\rm IN}(x_1)\varphi_{\rm IN}(x_2)\varphi_{\rm IN}(x_3)\varphi_{\rm IN}(x_4):$$
$$+:\varphi_{\rm IN}(x_1)\varphi_{\rm IN}(x_2):[\varphi_{\rm IN}^{(+)}(x_3),\varphi_{\rm IN}^{(-)}(x_4)]$$
$$+:\varphi_{\rm IN}(x_1)\varphi_{\rm IN}(x_3):[\varphi_{\rm IN}^{(+)}(x_2),\varphi_{\rm IN}^{(-)}(x_4)]$$
$$+:\varphi_{\rm IN}(x_2)\varphi_{\rm IN}(x_3):[\varphi_{\rm IN}^{(+)}(x_1),\varphi_{\rm IN}^{(-)}(x_4)]. \qquad (7.111)$$

Altogether this makes

$$T(\varphi_{\rm IN}(x_1)\varphi_{\rm IN}(x_2)\varphi_{\rm IN}(x_3)\varphi_{\rm IN}(x_4))=:\varphi_{\rm IN}(x_1)\varphi_{\rm IN}(x_2)\varphi_{\rm IN}(x_3)\varphi_{\rm IN}(x_4):$$
$$+\sum_\sigma:\varphi_{\rm IN}(x_{\sigma(1)})\varphi_{\rm IN}(x_{\sigma(2)}):i\Delta_F(x_{\sigma(3)}-x_{\sigma(4)})$$
$$+(i)^2\Big(\Delta_F(x_1-x_2)\Delta_F(x_3-x_4)+\Delta_F(x_1-x_3)\Delta_F(x_2-x_4)$$
$$+\Delta_F(x_{1-4})\Delta_F(x_2-x_3)\Big), \qquad (7.112)$$

where the sum is carried out over all permutations σ of $\{1,2,3,4\}$ where $\sigma(1) < \sigma(2)$ and $\sigma(3) < \sigma(4)$. If we take the vacuum expectation value of eq.(7.112) only the last term survives.

The result of normal ordering of a product of n fields is known as Wick's theorem.

$$
\begin{aligned}
T(\varphi_{\rm IN}(x_1)\cdots\varphi_{\rm IN}(x_n)) =&: \varphi_{\rm IN}(x_1)\cdots\varphi_{\rm IN}(x_n):\\
&+\sum_\sigma i\Delta_F(x_{\sigma(1)}-x_{\sigma(2)}):\varphi_{\rm IN}(x_{\sigma(3)})\cdots\varphi_{\rm IN}(x_{\sigma(n)}):\\
+\sum_\sigma i\Delta_F(x_{\sigma(1)}-x_{\sigma(2)})&i\Delta_F(x_{\sigma(3)}-x_{\sigma(4)}):\varphi_{\rm IN}(x_{\sigma(5)})\cdots\varphi_{\rm IN}(x_{\sigma(n)}):+\cdots\\
+\begin{cases}\sum_\sigma i\Delta_F(x_{\sigma(1)}-x_{\sigma(2)})\cdots i\Delta_F(x_{\sigma(n-2)}-x_{\sigma(n-1)})\varphi_{\rm IN}(x_{\sigma(n)}) & n \text{ odd}\\ \sum_\sigma i\Delta_F(x_{\sigma(1)}-x_{\sigma(2)})\cdots i\Delta_F(x_{\sigma(n-1)}-x_{\sigma(n)}) & n \text{ even.}\end{cases}
\end{aligned}
\tag{7.113}
$$

where the permutations, σ, are such that $\sigma(1) < \sigma(2)$, $\sigma(3) < \sigma(4)$, ..., $\sigma(n-1) < \sigma(n)$. This general result may be proven by induction.

As one could expect from $n = 2, 3$, and 4, the vacuum expectation value of eq.(7.113) vanishes if n is odd, and is equal to the last line if n is even.

$$
\begin{aligned}
&\langle 0\,|T(\varphi_{\rm IN}(x_1)\cdots\varphi_{\rm IN}(x_n))|\,0\rangle\\
&=\begin{cases}\sum_\sigma i\Delta_F(x_{\sigma(1)}-x_{\sigma(2)})\cdots i\Delta_F(x_{\sigma(n-1)}-x_{\sigma(n)}) & n \text{ even}\\ 0 & n \text{ odd}\end{cases}
\end{aligned}
\tag{7.114}
$$

Thus, the vacuum expectation of a time-ordered product of an even number of IN fields is the product of vacuum expectations of time-ordered pairs of fields, or propagators, summed over all ordered permutations. The appearance of $\langle 0\,|T(\varphi_{\rm IN}(x_i)\varphi_{\rm IN}(x_j))|\,0\rangle$ in the product is called the contraction of $\varphi_{\rm IN}(x_i)$ with $\varphi_{\rm IN}(x_j)$. In the vacuum expectation of the n field product above, each field $\varphi_{\rm IN}(x_i)$ finds itself contracted once with each of the other $n-1$ fields.

In the perturbative expansion of the n-point function, eq.(7.97), vacuum expectations appear of products of IN fields where the IN fields that come from H_I are already normal ordered. When we normal order the entire product, no contractions of fields will appear that already appear together inside one : : symbol. This can easily be seen from the examples we have already done. From eqs.(7.106)-(7.110) we have

$$
\begin{aligned}
T(:\varphi_{\rm IN}(x_1)\varphi_{\rm IN}(x_2):\varphi_{\rm IN}(x_3)) =&:\varphi_{\rm IN}(x_1)\varphi_{\rm IN}(x_2)\varphi_{\rm IN}(x_3):\\
&+ i\Delta_F(x_2-x_3)\varphi_{\rm IN}(x_1) + i\Delta_F(x_1-x_3)\varphi_{\rm IN}(x_2).
\end{aligned}
\tag{7.115}
$$

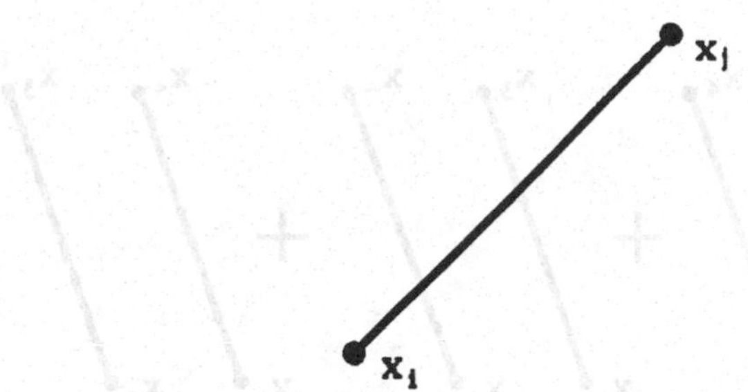

Figure 7.4 Diagrammatic representation of the contraction of $\varphi_{\text{IN}}(x_i)$ with $\varphi_{\text{IN}}(x_j)$, or free field 2-point function, $i\Delta_F(x_i - x_j)$.

The contraction, $\Delta_F(x_1 - x_2)$, does not appear on the right-hand side above because $\varphi_{\text{IN}}(x_1)$ and $\varphi_{\text{IN}}(x_2)$ enter already normal ordered with respect to each other on the left-hand side. Similarly,

$$\langle 0 \,| T(: \varphi_{\text{IN}}(x_1)\varphi_{\text{IN}}(x_2) :: \varphi_{\text{IN}}(x_3)\varphi_{\text{IN}}(x_4) :) |\, 0\rangle$$
$$= (i)^2 \Delta_F(x_1 - x_3)\Delta_F(x_2 - x_4) + (i)^2 \Delta_F(x_1 - x_4)\Delta_F(x_2 - x_3). \tag{7.116}$$

Neither the contraction $\Delta_F(x_1 - x_2)$ nor the contraction $\Delta_F(x_3 - x_4)$ appears.

Each term in the expansion of the n-point function, eq.(7.97), can be reduced via this process of normal ordering to a product of free field 2-point functions or propagators, $\Delta_F(x_i - x_j)$. Which contractions or propagators appear in each term is only limited by the normal ordering initially resident in the factors of H_I present. As we stated in previous chapters, each propagator, $\Delta_F(x_i - x_j)$, is represented by a line as in figure 7.4.

To obtain the diagrammatic representation for each term in the expansion, we merely draw lines for each propagator that appears and identify the common endpoints. For example, the first term in the numerator for the 4-point function $G(x_1, \ldots, x_n)$ is

$$\langle 0 \,| T(\varphi_{\text{IN}}(x_1)\varphi_{\text{IN}}(x_2)\varphi_{\text{IN}}(x_3)\varphi_{\text{IN}}(x_4)) |\, 0\rangle = (i)^2 \Big(\Delta_F(x_1 - x_2)\Delta_F(x_3 - x_4)$$
$$+ \Delta_F(x_1 - x_3)\Delta_F(x_2 - x_4) + \Delta_F(x_1 - x_4)\Delta_F(x_2 - x_3) \Big). \tag{7.117}$$

The diagram representing this term is given in figure 7.5. The interaction Hamiltonian is not present in this term, so this is just the free field theory

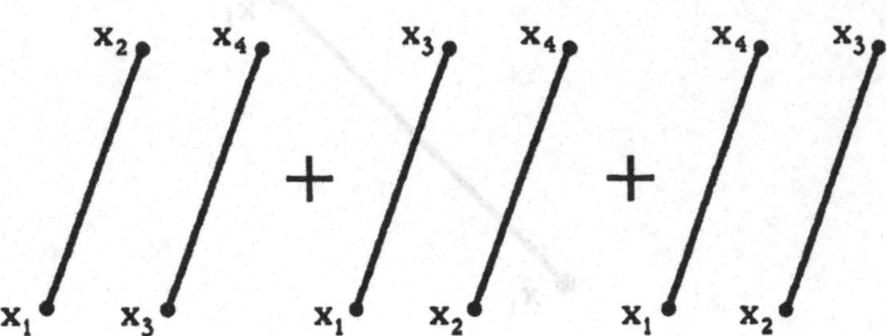

Figure 7.5 Diagrammatic representation of the first term in the series expansion of the 4-point function.

result. None of the propagators cross each other so no interaction takes place.

The next term in the numerator for the 4-point function is

$$-i\int_{-\infty}^{\infty} dy\,\langle 0\,|T(\varphi_{\rm IN}(x_1)\cdots\varphi_{\rm IN}(x_4)\mathcal{H}_I(y))|\,0\rangle, \tag{7.118}$$

where $\mathcal{H}_I = (1/2)\,\delta m^2 : \varphi^2_{\rm IN}(y) : +\lambda_0/4! : \varphi^4_{\rm IN}(y) :$. Consider first the part of $\mathcal{H}_I$ proportional to $\varphi^4_{\rm IN}(y)$.

$$\begin{aligned}\frac{-i\lambda_0}{4!}\int_{-\infty}^{\infty} & dy\langle 0\,|T(\varphi_{\rm IN}(x_1)\varphi_{\rm IN}(x_2)\varphi_{\rm IN}(x_3)\varphi_{\rm IN}(x_4) : \varphi^4_{\rm IN}(y) :)|\,0\rangle \\ &= -i\lambda_0\int_{-\infty}^{\infty} dy\,\Delta_F(x_1-y)\Delta_F(x_2-y)\Delta_F(x_3-y)\Delta_F(x_4-y).\end{aligned} \tag{7.119}$$

This term is represented by the diagram in figure 7.6. From the diagram, we can see that two particles meet, interact at y, and then part ways. The factor of 4! has disappeared from the right-hand side of eq.(7.119) above, because there are 4! equivalent ways of contracting the 4 $\varphi_{\rm IN}(x_i)$'s with the 4 $\varphi_{\rm IN}(y)$'s in $\mathcal{H}_I$.

The point y where the propagators meet is called an elementary vertex. Since the λ_0 part of $\mathcal{H}_I$ is proportional to φ^4, each λ_0 vertex will always have 4 propagators or lines radiating from it. If we had used φ^3, then 3 propagators would intersect at each vertex.

Next, consider the remaining portion of $\mathcal{H}_I$ in eq.(7.118).

$$-i\frac{\delta m^2}{2}\int_{-\infty}^{\infty} d^4y\,\langle 0\,|T(\varphi_{\rm IN}(x_1)\varphi_{\rm IN}(x_2)\varphi_{\rm IN}(x_3)\varphi_{\rm IN}(x_4) : \varphi^2_{\rm IN}(y) :)|\,0\rangle$$

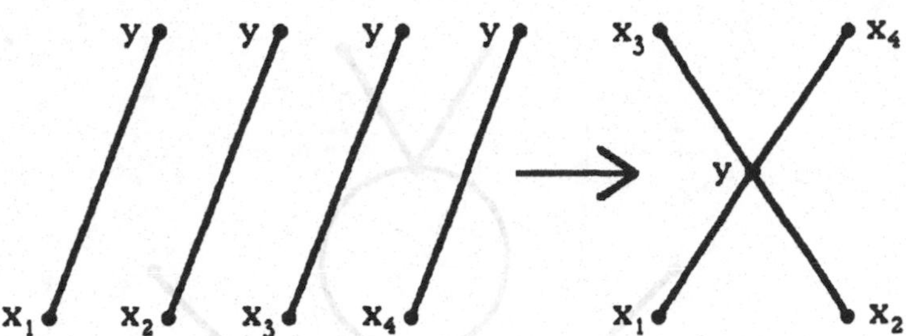

Figure 7.6 Diagrammatic representation of part of the second term in the series expansion of the 4-point function. The 4 contractions present in the term all have a common endpoint, y.

$$= -i\,\delta m^2 \int_{-\infty}^{\infty} d^4y\,(i\Delta_F(x_1 - x_2)\,i\Delta_F(x_3 - y)\,i\Delta_F(x_4 - y) + \text{permut.}) \tag{7.120}$$

The first term on the right-hand side of eq.(7.120) is represented by the diagram in figure 7.7.

This second type of vertex, the δm^2 vertex, is represented by the $\times$ as in figure 7.7. Once again, the factor of 1/2 on the left-hand side of eq.(7.120) is canceled because there are two ways to contract $\varphi_{\rm IN}(x_3)$ with $\varphi^2_{\rm IN}(y)$ that result in the same diagram.

The net result of all of the factors of $\mathcal{H}_I$ that appear in any of the terms in the expansion is to create vertices. Thus, any diagram in the expansion is made up of lines representing propagators connected at points of interaction, the vertices. There will be one vertex for each factor of $\mathcal{H}_I$

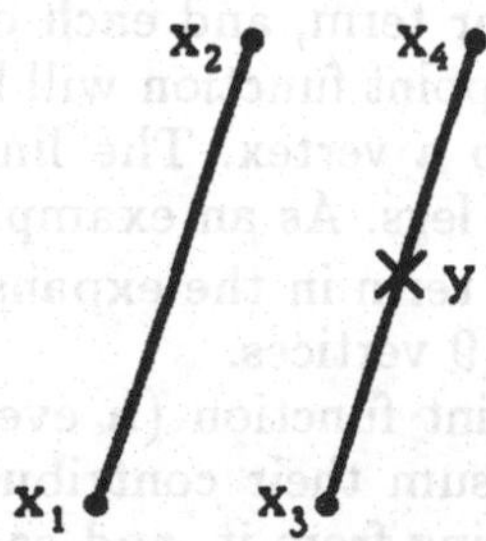

Figure 7.7 Mass shift vertex appearing in a disconnected term of the series for the 4-point function.

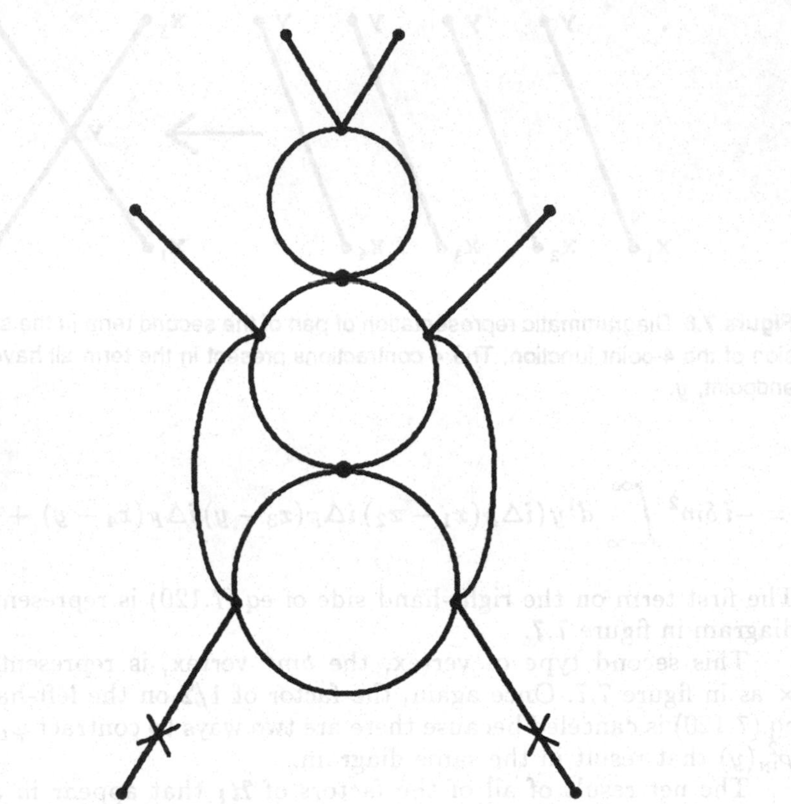

Figure 7.8 Diagrammatic representation of part of the ninth term in the series expansion of the 6-point function.

appearing in the particular term, and each diagram representing a term in the expansion of the n-point function will have n propagators with one free end not connected to a vertex. The lines or propagators with free ends are called (external) legs. As an example, the diagram in figure 7.8 represents part of the 9th term in the expansion of the 6-point function, since there are 6 legs and 9 vertices.

To compute the n-point function (n even), we draw all distinct diagrams with n legs and sum their contributions. Each λ_0 vertex must have 4 propagators emerging from it, and each δm^2 vertex must connect 2 propagators. To compute the actual value of a diagram, we apply a set of rules, called Feynman rules. The origin of these rules can be seen in the diagrams and terms we have already computed.

To compute the value of a given diagram, we associate a propagator $i\Delta_F(z_i - z_j)$ to each line connecting points z_i to z_j. The origin of this rule is the application of Wick's theorem to the vacuum expectation of the product of IN fields. To each λ_0 vertex at y_i, we associate a factor $(-i\lambda_0)\int d^4y_i$. To each δm^2 vertex we associate a factor $-i\delta m^2 \int d^4y_i$. These factors come from H_I. We do not include the factors of 1/4! or 1/2 that appear in H_I since they are usually canceled by combinatorics. The 1/4! factor is not always completely canceled at each vertex, so, finally, we must divide the value of each diagram by a symmetry factor, S, to account for this.

For example, the factor of 4! was completely canceled in the diagram in figure 7.6. However, consider the next order diagram for the 4-point function, figure 7.9.

There are four equivalent ways of contracting $\varphi_{\text{IN}}(x_1)$ with $\varphi_{\text{IN}}(y_1)$ ($\varphi_{\text{IN}}(x_1)$ once with each of the 4 $\varphi_{\text{IN}}(y_1)$'s), and three ways of contracting $\varphi_{\text{IN}}(x_2)$ with the remaining 3 $\varphi_{\text{IN}}(y_1)$'s once the $\varphi_{\text{IN}}(x_1)$ contraction is done. Similarly, there are 4×3 ways of contracting $\varphi_{\text{IN}}(x_3)$ and $\varphi_{\text{IN}}(x_4)$ to $\varphi_{\text{IN}}(y_2)$. Once these contractions are done, there are 2 $\varphi_{\text{IN}}(y_1)$'s and 2 $\varphi_{\text{IN}}(y_2)$'s left to be contracted with each other. There are 2 ways of doing this. Hence, the overall combinatoric factor for this diagram is $4 \times 3 \times 4 \times 3 \times 2 = 4! \times 4!/2!$. The factor of $1/(4! \times 4!)$ originating from the 2 factors of H_I is not completely canceled. The diagram must be divided by a factor of $S = 2$ to account for this.

In summary, the Feynman rules are:

1. Draw all distinct diagrams with n legs.
2. Associate one propagator, $i\Delta_F(z_i - z_j)$, with each line connecting z_i to z_j.
3. Associate a factor $-i\lambda_0 \int d^4y_i$ to each λ_0 vertex and $-i\,\delta m^2 \int d^4y_i$ to each δm^2 vertex.
4. Divide diagram by symmetry factor, S.

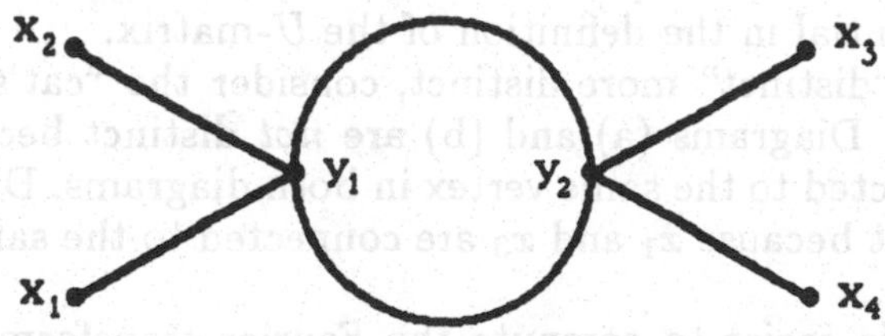

Figure 7.9 An example of a diagram where the symmetry factor, S, is 2.

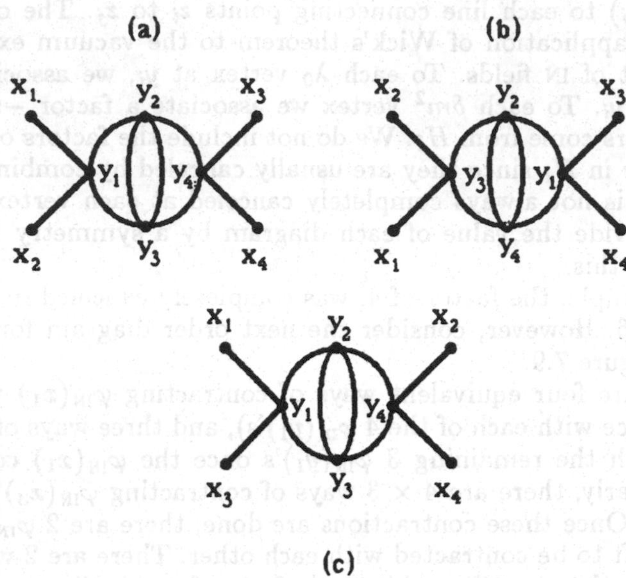

Figure 7.10 "Cat's-eye" diagrams. Diagrams (a) and (b) are not distinct, while diagram (c) is distinct from (a) and (b).

Let us define what we mean by "distinct diagrams." Distinct diagrams are topologically different after labeling the endpoints of the external legs. The labeling of the vertices (with y_i) is unimportant. In fact, since we are integrating over all the y_i's, each permutation of the labels y_i of the vertices contributes an identical result. For a diagram with m vertices, there are $m!$ different permutations. Therefore, we should multiply the result of one diagram with a particular labeling of the vertices by $m!$ to account for the remainder. But, this factor of $m!$ exactly cancels the factor of $1/m!$ present in the expansion of the n-point function, eq.(7.97), due to the exponential in the definition of the U-matrix.

To make "distinct" more distinct, consider the "cat's-eye" diagrams in figure 7.10. Diagrams (a) and (b) are *not* distinct because x_1 and x_2 are still connected to the same vertex in both diagrams. Diagrams (a) and (c) are distinct because x_1 and x_3 are connected to the same vertex in (c) but not in (a).

Often, it is easier to compute the Fourier transform of the n-point function. This is because the free-field propagator, $i\Delta_F(x' - x)$, eq.(3.52),

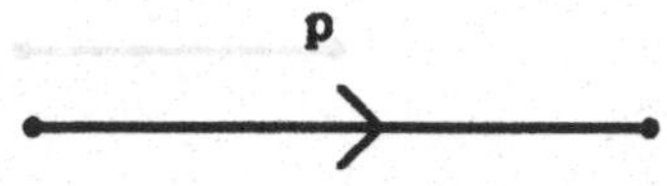

Figure 7.11 Diagrammatic representation of the momentum space propagator, $i\Delta_F(p)$.

is written naturally as a Fourier transform,

$$i\Delta_F(x'-x) = \int \frac{d^4p}{(2\pi)^4} \frac{i}{p^2 - m^2 + i\epsilon} e^{-ip\cdot(x'-x)}.$$

Therefore, the momentum space propagator is

$$i\Delta_F(p) = \frac{i}{p^2 - m^2 + i\epsilon}. \tag{7.121}$$

The diagrammatic representation for $i\Delta_F(p)$ is given in figure 7.11.

Let us state the Feynman rules for computing the Fourier transform, $\tilde{G}(p_1, \ldots, p_n)$, of the n-point function, $G(x_1, \ldots, x_n)$, using diagrams. These rules can be determined by considering the Fourier transform (x_i to p_i and y_i to k_i) of the n-point function computed using the x-space rules. The momentum space rules are:

1. Draw all distinct diagrams with n legs. Once again, a diagram is distinct from another if they are topologically inequivalent after we have labeled the external legs with momenta $p_1, \ldots, p_n$. By convention, all momenta point into the diagram.
2. Label the direction and momentum of all internal propagators. Associate a propagator, $i\Delta_F(p_j)$, to the jth external leg, and the integral $\int d^4k_l/(2\pi)^4 \, i\Delta_F(k_l)$ to the lth internal line.
3. Associate a factor of $-i\lambda_0(2\pi)^4\delta^4(\sum_{i=1}^4 q_i)$ to the λ_0 vertices where q_i are the momenta flowing into the vertex. Associate a factor of $-i\,\delta m^2(2\pi)^4\delta^4(\sum_{i=1}^2 q_i)$ to the δm^2 vertices.
4. Divide the diagram by the symmetry factor S.
5. Multiply the overall result by the factor $(2\pi)^4\delta^4(\sum_{i=1}^n p_i)$.

The origin of the last rule is the translational invariance of the n-point function, $G(x_1, \ldots, x_n)$. The δ-function expresses the momentum conservation of the entire process. The δ-functions in rule 3 associated with the vertices also express momentum conservation at each vertex. The δ-functions arise when Fourier transforming the y_i's after having transformed the x_i's.

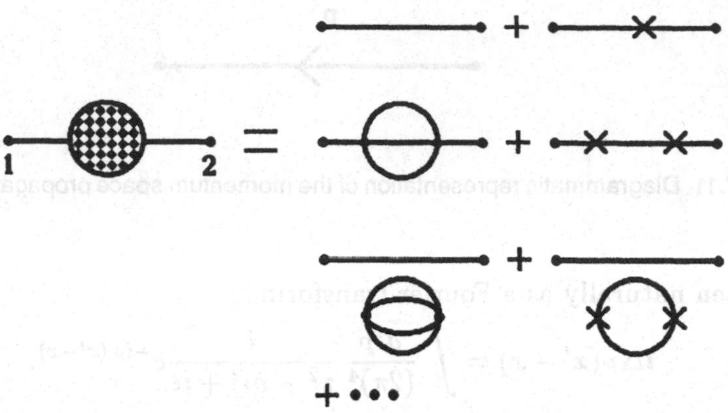

Figure 7.12 Diagrams representing the expansion of the 2-point function to second order.

Let us apply these rules to the computation of the first few terms of the 2-point and 4-point functions in the next section. We will also see, diagrammatically, what to do with the denominator in the expansion of the n-point function, eq.(7.97).

7.7 2- AND 4-POINT FUNCTIONS

We begin by examining the 2-point function, $\tilde{G}(p_1, p_2)$. The diagrams corresponding to the terms in the numerator of eq.(7.97) to second-order have two external legs and at most two vertices, and are in figure 7.12. The denominator in eq.(7.97) is the same for all n-point functions. Since it does not depend on x_i, the diagrams representing these terms will contain no external legs. The diagrams representing these terms to second-order are given in figure 7.13.

Since H_I is normal ordered, $\langle 0 | H_I | 0 \rangle = 0$, there will be no diagrams with just one vertex. If we think of these diagrams in x-space, then the vertices are labeled by y_i. We are integrating over all y_i (rule 3), so all of these diagrams in the denominator are just numbers. They represent disconnected vacuum processes in the interacting vacuum and are called vacuum bubbles.

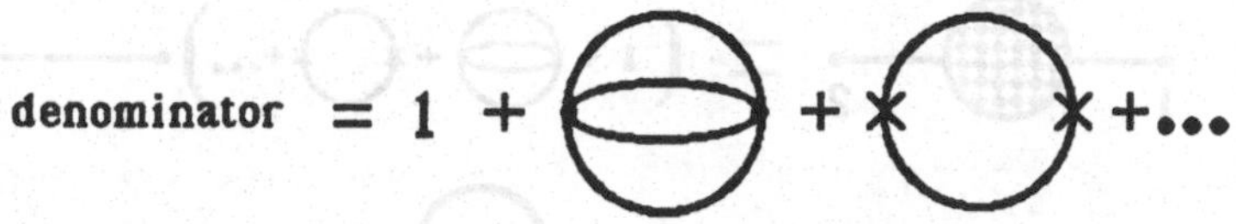

denominator $= 1 + \quad + \quad + \ldots$

Figure 7.13 Diagrammatic representation of the denominator of eq.(7.97).

Now let us return to the numerator. Since the vacuum bubbles just represent numbers, then the diagram in figure 7.14 is just a number times the free field propagator, $i\Delta_F$.

Let us regroup the diagrams in figure 7.12 by factoring out the vacuum bubbles to reflect the fact that they are numbers instead of functions. Now we can see, in figure 7.15, in diagram form, that the denominator exactly cancels the factor in front of the free field propagator. If we wrote out the diagrams to higher order, we would find that the same vacuum bubble factor multiplies each diagram that doesn't contain any vacuum bubbles. Thus, the purpose of the denominator is to remove all diagrams containing disconnected vacuum bubbles. Physically, of course, this is very reasonable since disconnected vacuum processes should have no effect on the propagation of particles. The same vacuum bubbles will show up in diagrams for any n-point function and will be canceled by the denominator in each case. Therefore, we can forget about the denominator in eq.(7.97) as long as we add a rule that says not to calculate diagrams with disconnected vacuum bubbles. The removal of the vacuum bubbles amounts to the removal of the phase generated by the S-matrix acting on the vacuum.

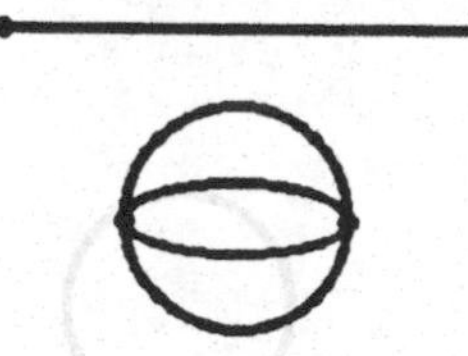

Figure 7.14 The disconnected vacuum bubble in this diagram has no dependence on x_i. It is just a number multiplying the free field 2-point function.

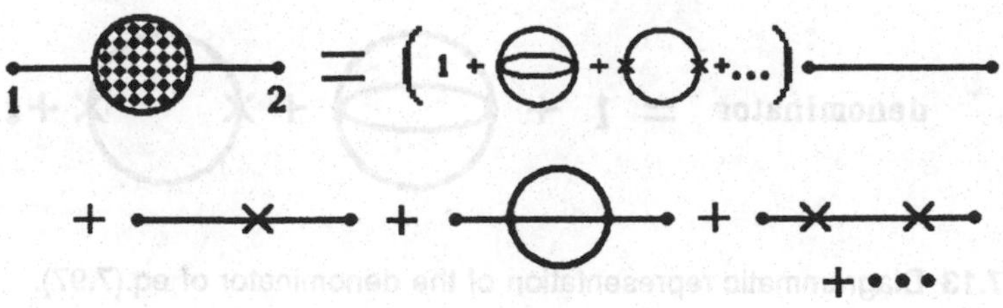

Figure 7.15 Same diagrams as in figure 7.12 except that we have factored out the vacuum bubbles.

Before we begin applying the remaining rules to compute the diagrams, let us mention that the diagram in figure 7.16, a "tadpole" diagram, is absent because we have normal ordered H_I. We could just as well not normal order H_I. Then we would have to include the tadpoles. The tadpoles amount to a mass renormalization, so when all is said and done, we end up with the same results.

The first diagram in the series, figure 7.17(a), is the free field contribution. It is

$$\tilde{G}^{(0)}(p_1, p_2) = \frac{i}{p_1^2 - m^2 + i\epsilon}(2\pi)^4 \delta^4(p_1 + p_2). \tag{7.122}$$

For the 2-point function we will often suppress the overall momentum conservation factor and write for eq.(7.122)

$$G^{(0)}(p, -p) = \frac{i}{p^2 - m^2 + i\epsilon}. \tag{7.123}$$

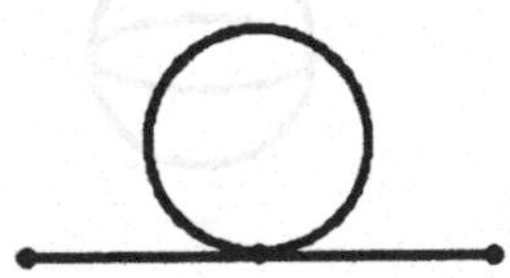

Figure 7.16 A "tadpole" diagram. Such diagrams are not present in the expansion because H_I is normal ordered.

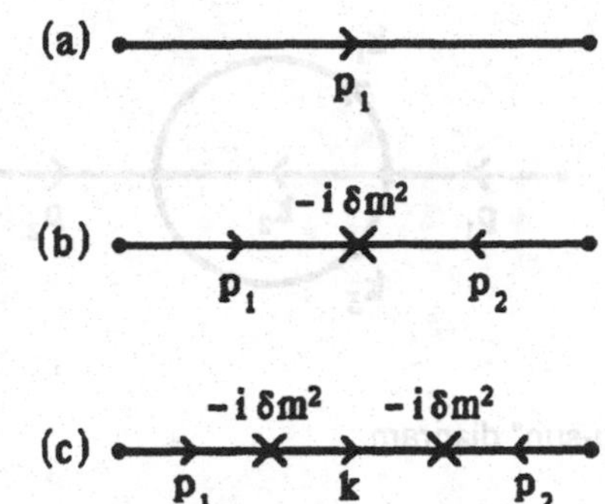

Figure 7.17 First three diagrams in the series for the 2-point function, $G^{(1)}(p_1)$.

The next diagram, figure 7.17(b), involves one δm^2 vertex and is similar to the free field term. We have one propagator for each leg and one factor of $i\delta m^2\,\delta(\sum q_i)$ for the vertex.

$$\tilde{G}^{(1)}(p_1,p_2) = \frac{i}{p_1^2 - m^2 + i\epsilon}\, i\,\delta m^2\, \frac{i}{p_2^2 - m^2 + i\epsilon}(2\pi)^4\delta^4(p_1+p_2)$$

or

$$G^{(1)}(p,-p) = i\,\delta m^2 \left(\frac{i}{p^2 - m^2 + i\epsilon}\right)^2 \tag{7.124}$$

The result for the third diagram, figure 7.17(c), is similar to the one for 7.17(b):

$$\begin{aligned}
\tilde{G}^{(2a)}(p_1,p_2) &= \int \frac{d^4k}{(2\pi)^4}\frac{i}{p_1^2 - m^2 + i\epsilon}\, i\delta m^2\, \frac{i}{k^2 - m^2 + i\epsilon}\delta^4(p_1 - k) \\
&\quad \times i\delta m^2 \frac{i}{p_2 - m^2 + i\epsilon}\delta^4(p_2 + k) \\
&= (i\delta m^2)^2 \left(\frac{i}{p_1^2 - m^2 + i\epsilon}\right)^2 \frac{i}{p_2^2 - m^2 + i\epsilon}(2\pi)^4\delta^4(p_1+p_2),
\end{aligned}$$

or

$$G^{(2a)}(p,-p) = (i\,\delta m^2)^2 \left(\frac{i}{p_1^2 - m^2 + i\epsilon}\right)^3. \tag{7.125}$$

The diagrams involving only δm^2 vertices just guarantee through renormalization that the lone particle propagates through the interacting vacuum at the physical mass m.

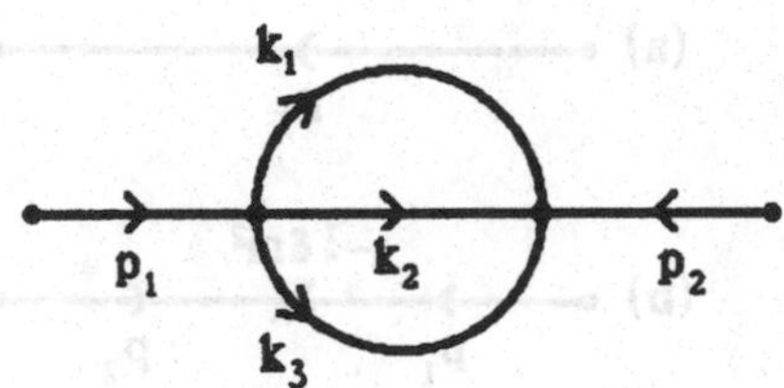

Figure 7.18 The "setting-sun" diagram.

The first nontrivial, interesting diagram is figure 7.18, the "setting-sun" diagram. This diagram is equal to

$$\frac{(i\lambda_0)^2}{3!}(2\pi)^4\delta^4(p_1+p_2)\frac{i}{p_1^2-m^2+i\epsilon}\frac{i}{p_2^2-m^2+i\epsilon}\int\frac{d^4k_1}{(2\pi)^4}\frac{i}{k_1^2-m^2+i\epsilon}$$

$$\int\frac{d^4k_2}{(2\pi)^4}\frac{i}{k_2^2-m^2+i\epsilon}\int\frac{d^4k_3}{(2\pi)^4}\frac{i\delta^4(p_1-k_1-k_2-k_3)}{k_3^2-m^2+i\epsilon}\delta^4(k_1+k_2+k_3+p_2)$$

$$=\frac{(i\lambda_0)^2}{3!}\left(\frac{i}{p_1^2-m^2+i\epsilon}\right)\left(\frac{i}{p_2^2-m^2+i\epsilon}\right)\delta^4(p_1+p_2)$$

$$\int\frac{d^4k_1}{(2\pi)^4}\frac{d^4k_2}{(2\pi)^4}\frac{i}{k_1^2-m^2+i\epsilon}\frac{i}{k_2^2-m^2+i\epsilon}\left(\frac{i}{(p_1-k_1-k_2)^2-m^2+i\epsilon}\right) \tag{7.126}$$

The symmetry factor for this diagram is 3!, because there are three internal lines connecting the two vertices. The remaining integrals in eq.(7.126) are not so easy to do. For large k_1, the k_1 integral is something like $\int d^4k_1/k_1^4$, which diverges. In fact, the integrals above do diverge, and to treat them we must introduce the renormalization program. We will consider this in later chapters.

Moving on to the 4-point function, the diagrams representing the terms to second-order are in figure 7.19. We will ignore for now the diagrams containing the δm^2 vertices. This vertex is used, through renormalization, to absorb (infinite) shifts in the self-energy, so that the particles always propagate at physical mass, m. If we include the δm^2 vertices in the 2-point function, as we have done, and use this 2-point function in place of the free propagator in the diagrams for the higher point functions, then we need not compute the diagrams with δm^2 vertices. The effects of the δm^2 vertices will enter in through the modified 2-point function.

The distinct diagrams corresponding to the first diagram in the series in figure 7.19 are given in figure 7.20(a). They are equal to

$$\tilde{G}^{(0)}(p_1,p_2,p_3,p_4)$$
$$= i^2(2\pi)^8\delta^4(p_1+p_2+p_3+p_4)\left[\frac{\delta^4(p_1+p_2)}{(p_1^2-m^2+i\epsilon)(p_3^2-m^2+i\epsilon)}\right.$$
$$\left.+\frac{\delta^4(p_1+p_3)}{(p_1^2-m^2+i\epsilon)(p_2^2-m^2+i\epsilon)}+\frac{\delta^4(p_1+p_4)}{(p_1^2-m^2+i\epsilon)(p_2^2-m^2+i\epsilon)}\right], \tag{7.127}$$

which is the free field theory result.

The next diagram contains one λ_0 vertex, and is given in figure 7.20(b). Its value is

$$\tilde{G}^{(1)}(p_1,p_2,p_3,p_4) = -i\lambda_0(2\pi)^4\delta^4(p_1+p_2+p_3+p_4)$$
$$\times\left(\frac{1}{(p_1^2-m^2+i\epsilon)(p_2^2-m^2+i\epsilon)(p_3^2-m^2+i\epsilon)(p_4^2-m^2+i\epsilon)}\right). \tag{7.128}$$

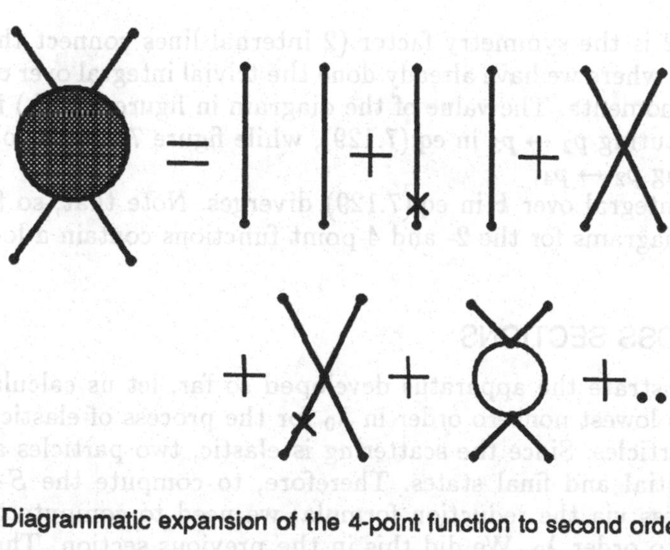

Figure 7.19 Diagrammatic expansion of the 4-point function to second order.

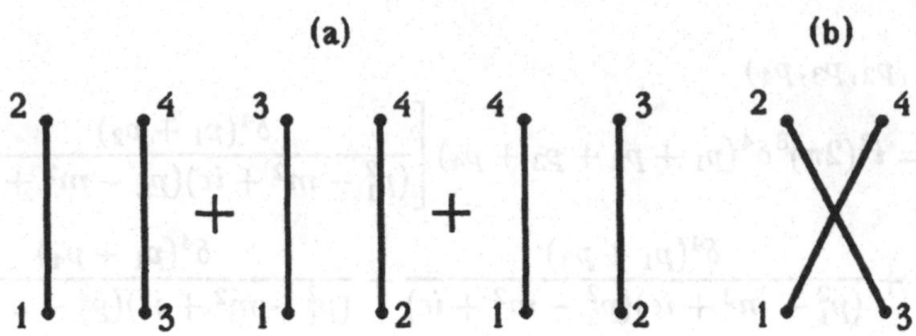

Figure 7.20 Distinct diagrams corresponding to the first two terms in the expansion of the 4-point function.

The distinct diagrams to order λ_0^2 are given in figure 7.21. The first one is equal to

$$\frac{1}{2}(2\pi)^4\delta^4(p_1+p_2+p_3+p_4)\prod_{j=1}^{4}\frac{i}{p_j^2-m^2+i\epsilon}$$
$$\times\int\frac{d^4k}{(2\pi)^4}\frac{i^2}{(k^2-m^2+i\epsilon)((p_1+p_2-k)^2-m^2+i\epsilon)}, \qquad (7.129)$$

where 1/2 is the symmetry factor (2 internal lines connect the two vertices) and where we have already done the trivial integral over one of the 2 internal momenta. The value of the diagram in figure 7.21(b) is obtained by substituting $p_2 \leftrightarrow p_3$ in eq.(7.129), while figure 7.21(c) is obtained by exchanging $p_2 \leftrightarrow p_4$.

The integral over k in eq.(7.129) diverges. Note that, so far, the divergent diagrams for the 2- and 4-point functions contain a loop.

7.8 CROSS SECTIONS

To demonstrate the apparatus developed so far, let us calculate a cross section to lowest nonzero order in λ_0 for the process of elastic scattering of two particles. Since the scattering is elastic, two particles are present in the initial and final states. Therefore, to compute the S-matrix for this process via the reduction formula, we need to compute the 4-point function to order λ_0. We did this in the previous section. The diagrams in figure 7.20(a) represent disconnected processes with no scattering, so they are presently not of interest. The relevant diagram is figure 7.20(b).

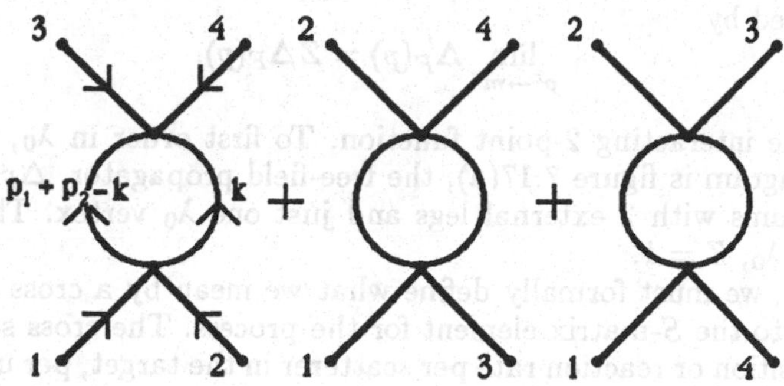

Figure 7.21 Distinct diagrams for the 4-point function to order λ_0^2.

To compute the cross section beyond lowest order in λ_0 would require renormalization to handle the divergent diagrams.

The 4-point function to first order in λ_0, figure 7.20(b), was given in eq.(7.128).

$$\tilde{G}^{(1)}(p_1, p_2, p_3, p_4) = -i\lambda_0(2\pi)^4\delta^4(p_1+p_2+p_3+p_4)\prod_{j=1}^{4}\left(\frac{i}{p_j^2 - m^2 + i\epsilon}\right) \tag{7.128}$$

The S-matrix for this process is obtained from eq.(7.128) through the reduction formula, eq.(7.62). Since we are working in momentum space, we must Fourier transform eq.(7.62). Eq.(7.62) is naturally written as a Fourier transform already, so this is easy to do. Thus, in momentum space, eq.(7.62) becomes

$$S^c_{\beta\alpha} = \left(\frac{-i}{\sqrt{Z}}\right)^{n+m}\prod_{i=1}^{n}\prod_{j=1}^{m}(k_i^2 - m^2)(p_j^2 - m^2)\tilde{G}(k_1, \ldots, k_n, p_1, \ldots, p_m). \tag{7.130}$$

In the present case, $n = 2 = m$, and $\tilde{G}(k_1, k_2, p_1, p_2) = \tilde{G}^{(1)}(p_1, p_2, p_3, p_4)$, eq.(7.128) above. Therefore, the S-matrix for this process to order λ_0 is

$$\begin{aligned} S^c_{fi} &= \lim_{\epsilon\to 0^+}(-i)^4 Z^{-2}(-i\lambda_0)(2\pi)^4\delta^4(p_1+p_2+p_3+p_4) \\ &\quad\cdot\prod_{j=1}^{4}(p_j^2 - m^2)(p_j^2 - m^2 + i\epsilon)^{-1} \qquad (7.131) \\ &= (-i\lambda_0)Z^{-2}(2\pi)^4\delta^4(p_1+p_2+p_3+p_4), \end{aligned}$$

which is the residue at the multipole $p_j = m$ of $\tilde{G}^{(1)}$. Z, the 1-particle amplitude of $\varphi(x)$, is determined by eqs.(7.36) and (7.33). Namely, Z is determined by

$$\lim_{p^2 \to m^2} \Delta'_F(p) = Z\Delta_F(p).$$

Δ'_F is the interacting 2-point function. To first order in λ_0, the only 2-point diagram is figure 7.17(a), the free-field propagator, Δ_F. There are no diagrams with 2 external legs and just one λ_0 vertex. Thus, to first order in λ_0, $Z = 1$.

Next, we must formally define what we mean by a cross section and relate it to the S-matrix element for the process. The cross section, σ, is the transition or reaction rate per scatterer in the target, per unit incident flux. The reaction rate is the transition probability per unit time.

$$\sigma = \text{transition rate} \cdot \frac{1}{\text{\# of scatterers in target}} \cdot \frac{1}{\text{unit incident flux}}$$
$$= \frac{\text{transition prob.}}{\text{unit time}} \cdot \frac{1}{\text{\# of scatterers}} \cdot \frac{1}{\text{unit incident flux}} \tag{7.132}$$

Flux is defined as the number of particles crossing an area, A, in a certain amount of time, T. In a typical scattering experiment, the initial state is carefully prepared so that the incident particles all have nearly the same velocity. We will consider the target particles to be at rest. If the incident particles all carry velocity $\vec{v}_1$, then

$$\text{flux} = \frac{\text{\# of particles}}{AT} \cdot \frac{|\vec{v}_1|}{|\vec{v}_1|} = \frac{\text{\# of particles}\, |\vec{v}_1|}{V},$$

where V is the volume containing the number of particles equal in value to the flux. For our present situation of 1 particle incident on 1 target particle,

$$\text{flux} = \frac{|\vec{v}_1|}{V}. \tag{7.133}$$

Therefore, the cross section is

$$\sigma = \frac{\text{transition probability}}{\text{unit time} \times \text{unit volume}} \cdot 1 \cdot \frac{1}{|\vec{v}_1|}. \tag{7.134}$$

The scattering matrix element, S^c_{fi}, is the probability amplitude for the scattering process to occur ($i \to f : p_1, p_2 \to p_3, p_4$). Thus,

$$\int \frac{|S^c_{fi}|^2}{\langle f \mid f \rangle \langle i \mid i \rangle}\, d^3p_3\, d^3p_4 = \text{transition probability}, \tag{7.135}$$

where we have explicitly included the normalization of the momentum eigenstates since the normalizations are not unity. In fact, from eq.(3.18), we find

$$\langle p_3\, p_4 \,|\, p_3\, p_4\rangle = 4p_3^0 p_4^0 (2\pi)^6. \tag{7.136}$$

All S-matrix elements, S_{fi}, will be proportional to $(2\pi)^4\delta^4(\sum_i p_i)$ because of momentum conservation. Let $\hat{S}^c_{fi}$ be the connected S-matrix element with this factor removed,

$$S^c_{fi} = \hat{S}^c_{fi}(2\pi)^4\delta^4\Big(\sum_i p_i\Big). \tag{7.137}$$

When we square S^c_{fi}, we will be squaring the δ-function. The conventional interpretation of the square of the δ-function is

$$\left|(2\pi)^4\delta^4\Big(\sum_i p_i\Big)\right|^2 = (2\pi)^4\delta^4(0)(2\pi)^4\delta^4\Big(\sum_i p_i\Big). \tag{7.138}$$

By examining the integral definition of the δ-function (the inverse Fourier transform of 1), we can see that $(2\pi)^4\delta^4(0)$ is the volume of space-time. Placing eqs.(7.137) and (7.138) into eq.(7.135), we have

$$(2\pi)^4\delta^4(p_1+p_2-p_3-p_4)\int d^3p_3\, d^3p_4\, \frac{|\hat{S}^c_{fi}|^2}{\langle f\,|\,f\rangle\langle i\,|\,i\rangle} = \frac{\text{transition probability}}{\text{unit time} \times \text{unit volume}}. \tag{7.139}$$

Using eqs.(7.139), (7.136), and (7.134) we can relate the cross section to the S-matrix element.

$$d\sigma = \frac{1}{|\vec{v}_1|}\frac{1}{4p_1^0p_2^0}|\hat{S}^c_{fi}|^2\frac{d^3p_3}{(2\pi)^3 2p_3^0}\frac{d^3p_4}{(2\pi)^3 2p_4^0}(2\pi)^4\delta^4(p_1+p_2-p_3-p_4) \tag{7.140}$$

From eq.(7.131), $\hat{S}^c_{fi}$ for this process is simply $-i\lambda_0$. Finally,

$$d\sigma = \frac{1}{2}\frac{1}{|\vec{v}_1|}\frac{\lambda_0^2}{4p_1^0 m}\frac{d^3p_3}{(2\pi)^3 2p_3^0}\frac{d^3p_4}{(2\pi)^3 2p_4^0}(2\pi)^4\delta^4(p_1+p_2-p_3-p_4), \tag{7.141}$$

where we have set $p_2^0 = m$ since the target is considered to be at rest, and where we have included a factor of 1/2 because the two particles in the final state are identical.

If the target particles are not at rest, then $|\vec{v}_1|$ represents the relative velocity. We may express this in an invariant form.

$$d\sigma = \frac{1}{2}\frac{\lambda_0^2}{4\sqrt{(p_1\cdot p_2)^2 - m^4}}\frac{d^3p_3}{(2\pi)^3 2p_3^0}\frac{d^3p_4}{(2\pi)^3 2p_4^0}(2\pi)^4\delta^4(p_1+p_2-p_3-p_4) \tag{7.142}$$

Now the Lorentz invariance of the expression is manifest.

If we now consider inelastic scattering where the final state contains n identical particles (n even), and the initial state remains the same, we can follow the same analysis to generalize eq.(7.142) to

$$d\sigma = \frac{1}{n!}\frac{1}{4\sqrt{(p_1\cdot p_2)^2 - m^4}}|\hat{S}^c_{fi}|^2(2\pi)^4\delta^4\Big(p_1+p_2-\sum_{j=3}^{n+2}p_j\Big)\prod_{j=3}^{n+2}\frac{d^3p_j}{(2\pi)^3 2p_j^0}, \tag{7.143}$$

where the S-matrix element is determined from the $(n+2)$-point function.

7.9 EXERCISES

1. Draw the diagrams and write down the corresponding expressions for the 2-point and 4-point functions at third order, $O(\lambda_0^3)$.
2. Write down the Feynman rules for the theory,

$$\mathcal{L}(x) = \frac{1}{2}(\partial_\mu\varphi\partial^\mu\varphi - m_0^2\varphi^2) - \frac{\nu_0}{3!}\varphi^3 - \frac{\lambda_0}{4!}\varphi^4.$$

3. Draw the diagrammatic expansion of the 2-point and 3-point functions to $O(\nu_0^3)$ and $O(\lambda_0^0)$, and use the rules from exercise 2 to write down the integrals represented by each diagram.
4. Compute the cross section for elastic 2-body scattering to lowest order, $O(\nu_0^2)$ and $O(\lambda_0)$. What is wrong with a theory in which $\lambda_0 = 0$?

CHAPTER **8**

Spinor Quantum Electrodynamics

The theory of photons and electrons, QED, is perturbatively solved using the ideas and methods presented in the last chapter. Since the concepts involved for spinor and vector fields are identical to those for scalar fields, we will cover only briefly the results leading to the Feynman rules. Using these rules, we will compute the lower order diagrams of the 2-, 3-, and 4-point functions and identify the divergent parts.

8.1 QUANTIZATION IN THE COULOMB GAUGE

Let us first consider the Lagrangian density for QED,

$$\mathcal{L}(x) = -\frac{1}{4}F^{\mu\nu}F_{\mu\nu} + \bar{\psi}(i\not{D} - m_0)\psi, \tag{8.1}$$

as a density defining a classical gauge field theory. $\not{D}$ is the gauge covariant derivative,

$$\not{D} = \not{\partial} + ie_0\not{A}, \tag{8.2}$$

and $F^{\mu\nu}$ was defined in eqs.(5.1) and (5.3). $j^\mu = e_0\bar{\psi}\gamma^\mu\psi$ can then be identified as the classical electric current.

In the Coulomb gauge we have $\nabla\cdot\vec{A}=0$. Let us find the field momenta and Hamiltonian when $\vec{A}$ is transverse. We will then impose equal-time quantum commutators so as to maintain these results as operator equations.

The time component of the vector potential, $A_0(x)$, is still nondynamical; no time derivative of A_0 appears in the density, eq.(8.1). We may eliminate it by solving for it in terms of the dynamical variables. Namely, from the $\mu=0$ field equation, $F^{\mu\nu}{}_{,\nu}=e_0 j^\mu$, we have

$$\nabla\cdot\vec{E}=-\nabla\cdot\dot{\vec{A}}-\nabla^2 A_0=e_0\psi^\dagger\psi, \tag{8.3}$$

which is just Gauss's law. If $\nabla\cdot\vec{A}=0$, then $\nabla\cdot\dot{\vec{A}}=0$, so $\nabla^2 A_0=-e_0\psi^\dagger\psi$. Therefore,

$$A_0=\frac{e_0}{4\pi}\int d^3x'\,\frac{\rho(\vec{x}',t)}{|\vec{x}'-\vec{x}|}, \tag{8.4}$$

where

$$\rho(x)=j^0(x)=\psi^\dagger(x)\psi(x) \tag{8.5}$$

is the static charge density.

The field momenta conjugate to $\vec{A}$ and ψ are

$$\frac{\partial\mathcal{L}}{\partial(\partial_t A_i)}=-E_i=\nabla_i A_0+\dot{A}_i,\quad\text{and}\quad\frac{\partial\mathcal{L}}{\partial\dot{\psi}}=i\psi^\dagger, \tag{8.6}$$

so the Hamiltonian is

$$H=\int d^3x\,\psi^\dagger(-i\vec{\alpha}\cdot\nabla+\beta m_0)\psi+\frac{1}{2}(\vec{E}^2+\vec{B}^2)+\vec{E}\cdot\nabla A_0+e_0\bar{\psi}\gamma_\mu\psi A^\mu. \tag{8.7}$$

If we integrate the third term in the Hamiltonian by parts and drop the surface term, then

$$\begin{aligned}\int\vec{E}\cdot\nabla A_0\,d^3x&=-\int(\nabla\cdot\vec{E})A_0\,d^3x\\&=-e_0\int\rho(x)A_0(x)\,d^3x\end{aligned} \tag{8.8}$$

by Gauss's law, eq.(8.3). This term precisely cancels the $\mu=0$ component of the fourth term in eq.(8.7). Therefore, the Hamiltonian reduces to

$$H=\int d^3x\,\psi^\dagger(-i\vec{\alpha}\cdot\nabla+\beta m_0)\psi+\frac{1}{2}(\vec{E}^2+\vec{B}^2)+e_0\bar{\psi}\gamma_i\psi A^i. \tag{8.9}$$

Since we are working in the Coulomb gauge, the A^i that appears in the last term above is transverse, $\nabla \cdot \vec{A} = 0$. From eq.(8.3) or eq.(8.6), we can see that the electric field appearing in eq.(8.9) is not completely transverse since it is not divergence-free. We want the electric field appearing in the Hamiltonian to be completely transverse so that we may use the same commutators as in the free photon case in chapter 5. This means that we must express the longitudinal part of the electric field, E_L, in terms of the other fields.

To do this, let $\vec{E} = \vec{E}_T + \vec{E}_L$. By definition, $\nabla\cdot\vec{E}_T = 0$, and $\nabla\cdot\vec{E}_L \neq 0$. Thus, from eq.(8.6) we can see that in the Coulomb gauge $\vec{E}_T = -\dot{\vec{A}}$ and $\vec{E}_L = -\nabla A_0$. The E^2 term in eq.(8.9) can be rewritten as

$$\int d^3x\, \vec{E}^2 = \int d^3x\, \vec{E}_T^2 + 2\vec{E}_T\cdot\vec{E}_L + \vec{E}_L^2.$$

The cross term vanishes in the Coulomb gauge because

$$\begin{aligned}\int d^3x\, 2\vec{E}_T\cdot\vec{E}_L &= 2\int d^3x\, \nabla A_0\cdot\dot{\vec{A}} \\ &= -2\int d^3x\, A_0(\nabla\cdot\dot{\vec{A}}) \\ &= 0,\end{aligned}$$

where once again we neglect the surface term arising from the integration by parts. Finally, since $\vec{E}_L = -\nabla A_0$, eq.(8.4) allows us to eliminate $\vec{E}_L$ in favor of ψ and $\psi^\dagger$.

$$\begin{aligned}\int d^3x \vec{E}_L^2 &= \int d^3x \nabla A_0\cdot\nabla A_0 \\ &= -\int d^3x\, A_0\nabla^2 A_0 \\ &= \frac{e_0^2}{4\pi}\int d^3x\, d^3y \frac{\rho(\vec{x},t)\rho(\vec{y},t)}{|\vec{x}-\vec{y}|},\end{aligned} \tag{8.10}$$

which we recognize as the ordinary Coulomb energy due to a static charge distribution. Now, we may regard $\vec{E}$ and $\vec{A}$ as being completely transverse if we write the Hamiltonian as

$$\begin{aligned}H = \int d^3x\, \psi^\dagger(-i\vec{\alpha}\cdot\nabla + \beta m_0)\psi + \frac{1}{2}(\vec{E}^2+\vec{B}^2) + e_0\bar{\psi}A\!\!\!/\psi \\ + \frac{e_0^2}{8\pi}\int d^3x\, d^3y \frac{\rho(\vec{x},t)\rho(\vec{y},t)}{|\vec{x}-\vec{y}|},\end{aligned} \tag{8.11}$$

where $\rho(x) = \psi^\dagger(x)\psi(x)$.

To canonically quantize we must specify equal-time commutation relations. Since $\vec{E}$ and $\vec{A}$ are completely transverse, we may use the transverse δ-function introduced in chapter 5.

$$[A_i(\vec{x},t), E_j(\vec{y},t)] = -i\delta^{\mathrm{tr}}_{ij}(\vec{x}-\vec{y}) \tag{8.12}$$

$$[A_i(\vec{x},t), A_j(\vec{y},t)] = [E_i(\vec{x},t), E_j(\vec{y},t)] = 0$$

$\delta^{\mathrm{tr}}_{ij}(\vec{x}-\vec{y})$ is defined in eq.(5.52). In addition, we may use the usual anticommutation relations for ψ,

$$\{\psi_\alpha(\vec{x},t), \psi^\dagger_\beta(\vec{y},t)\} = \delta_{\alpha\beta}\delta^3(\vec{x}-\vec{y}) \tag{8.13}$$

$$\{\psi_\alpha(\vec{x},t), \psi_\beta(\vec{y},t)\} = \{\psi^\dagger_\alpha(\vec{x},t), \psi^\dagger_\beta(\vec{y},t)\} = 0.$$

All of the commutators mixing A_i and $\dot{A}_i$ with ψ_α and $\psi^\dagger_\alpha$ also vanish. A_0 commutes with the A_i's. However, A_0 must satisfy

$$[A_0(\vec{x},t), \psi_\alpha(\vec{y},t)] = \frac{-e_0}{4\pi|\vec{x}-\vec{y}|}\psi_\alpha(\vec{y},t) \tag{8.14}$$

as a consequence of the operator eq.(8.4).

Although we do not always explicitly note it, we will be normal ordering the action and Hamiltonian.

8.2 IN-OUT FIELDS AND REDUCTION

As soon as we add any realistic interaction to a field theory, we can no longer exactly solve it. Ultimately, this means that we have no particle interpretation for the quantized theory because we no longer know how to expand the interacting operators. The plane wave expansions of the free field are not valid past the initial instant. To proceed, we must assume that the relative couplings are small so that the interactions only mildly perturb the spectrum. We can then assume that the states share many properties with free field states. In particular, we must assume that there are no bound states. This means that we can directly associate particles found in an experiment with quanta of the fields we use in the interacting field theory. In addition, we can treat isolated particles as essentially free so that we can borrow the particle interpretation from free field theory.

For quantum electrodynamics, the relevant coupling is $\alpha = e^2/4\pi$. The experimentally measured value of the fine structure constant is small ($\sim 1/137$) so that we can expect perturbation theory to be valid at laboratory energies. In addition, experimentally, there are no stable bound states. (The closest thing to a bound state is positronium, a system of one electron and one positron. The half-life of the longer 3-photon decay is on the order of 10^{-7} seconds.)

To perturbatively solve QED, we proceed along lines similar to those taken in the previous chapter. Isolated electrons and positrons of physical mass m are created and destroyed with free spinor fields $\psi_{\text{IN}}(x)$ and $\psi_{\text{OUT}}(x)$. Isolated photons of zero rest mass are created and destroyed with the free fields $A^i_{\text{IN}}(x)$ and $A^i_{\text{OUT}}(x)$. Since the kinematic properties of the interacting states are identical to the free field states, we use the free field particle interpretation to build a basis for the Fock space for the interacting theory by operating on the interacting vacuum with $b^\dagger_{\text{IN}}(p,s), d^\dagger_{\text{IN}}(p,s)$, or $a^\dagger_{\text{IN}}(k,\lambda)$ ($s = +$ or $-$ for spin up or down). An identical basis is provided by the OUT fields.

In a scattering experiment, the initial state at $t \to -\infty$ consists of isolated free particles (free up to their self-interactions) and can be written down as one of the IN basis states mentioned above. The final state after the scattering, $t \to \infty$, also consists of free isolated particles and can also be written as one of the OUT basis states. The overlap between the two states is the amplitude for the process to occur.

As in the scalar case, we can define a mapping, the S-matrix, relating the IN fields to the OUT fields. The details of the mapping depend on the interaction so we must relate the S-matrix elements to the interacting theory by a reduction formula. The development of the formula for QED is nearly identical to the scalar case of the previous chapter. We will first treat the electrons and photons separately, and then quickly combine the two.

In analogy to the scalar case, we may try to relate $\psi(x)$ to $\psi_{\text{IN}}(x)$ and $\psi_{\text{OUT}}(x)$ by

$$\lim_{t\to-\infty} \psi(x) = \sqrt{Z_2}\psi_{\text{IN}}(x), \quad \lim_{t\to\infty} \psi(x) = \sqrt{Z_2}\psi_{\text{OUT}}(x). \tag{8.15}$$

Z_2 is defined through the spectral density to be the amplitude for $\psi(x)$ to create a 1-particle state. The spectral density for electrons is a matrix relating the propagator for the interacting electrons, $S'_F(x-y)$, to the propagator for free scalar particles, $\Delta_F(x-y)$. By requiring invariance under Lorentz transformations and parity transformations, the spectral density reduces to 2 functions, $\varrho_1(\sigma^2)$ and $\varrho_2(\sigma^2)$.

$$\begin{aligned}(S'_F(x-y))_{\alpha\beta} &= \int_0^\infty d\sigma^2\,(i\varrho_1(\sigma^2)\not{\partial}_x + \varrho_2(\sigma^2))_{\alpha\beta}\,\Delta_F(x-y;\sigma)\\ &= Z_2 S_F(x-y;m)_{\alpha\beta}\\ &\quad - \int_{4m^2}^\infty d\sigma^2\,(i\varrho_1(\sigma^2)\not{\partial}_x + \varrho_2(\sigma^2))_{\alpha\beta}\Delta_F(x-y;\sigma),\end{aligned} \tag{8.16}$$

where $2m$ is the multiparticle production threshold. If we consider the limit $y^0 \to x^0$ in eq.(8.16) above, the equal-time anticommutators tell us that there must be unit area under $\varrho_1(\sigma^2)$.

$$1 = Z_2 + \int_{4m^2}^{\infty} d\sigma^2 \, \varrho_1(\sigma^2), \tag{8.17}$$

so $0 \leq Z_2 < 1$.

Once again, we will not be able to take eq.(8.15) as an operator statement. If eq.(8.15) is true as an operator equation, then the equal-time anticommutators for ψ, ψ_{IN}, and ψ_{OUT} imply that $Z_2 = 1$. But Z_2 can only be 1 if $\varrho_1(\sigma^2)$ is 0 except at $\sigma^2 = m^2$, that is, only if there is no multiparticle production. Thus, if $Z_2 = 1$, $\psi(x)$ can only create 1-particle states so $\psi = \psi_{\text{IN}}$. To avoid this conflict, we can only use the relation in eq.(8.15) when it appears sandwiched between normalizable states.

The derivation of the reduction formula proceeds along identical lines as in the scalar formula. Let us consider only connected processes. An S-matrix element for a particular process will consist of a product of $b^\dagger_{\text{IN}}(p,s)$, $d^\dagger_{\text{IN}}(p,s)$, $b^\dagger_{\text{OUT}}(p,s)$ and $d^\dagger_{\text{OUT}}(p,s)$ sandwiched between vacuum states. One by one, we replace each $b^\dagger_{\text{IN}}(p,s)$ with

$$b^\dagger_{\text{IN}}(p,s) = \int d^3x \, \bar{\psi}_{\text{IN}}(x)\gamma^0 e^{-ip\cdot x} u(p,s), \tag{8.18}$$

and each $d^\dagger_{\text{IN}}(p,s)$ with

$$d^\dagger_{\text{IN}}(p,s) = \int d^3x \, \bar{v}(p,s) e^{-ip\cdot x}\gamma^0 \psi_{\text{IN}}(x). \tag{8.19}$$

The integrals are independent of time so they can be computed at any desired time, t. Expressions (8.18) and (8.19) are obtained by inverting the plane wave expansions of the free field operators $\psi_{\text{IN}}(x)$ and $\bar{\psi}_{\text{IN}}(x)$ given in eq.(4.36). The OUT operators are replaced by similar expressions involving $\psi_{\text{OUT}}(x)$ and $\bar{\psi}_{\text{OUT}}(x)$.

After this replacement, we take $t \to \pm\infty$ and use eq.(8.15) to replace $\psi_{\text{IN}}(x)$ or $\psi_{\text{OUT}}(x)$ by $\psi(x)$. For example, for the S-matrix element, $S_{\beta\alpha}$, where the initial state contains a positron of momentum p and spin up, the conversion of $d^\dagger_{\text{IN}}(p,+)$ to $\psi(x)$ yields

$$\begin{aligned} S_{\beta\alpha} &= \langle \beta_{\text{OUT}} \,|\, \alpha_{\text{IN}} \rangle \\ &= \frac{i}{\sqrt{Z_2}} \int d^4x \, \bar{v}(p,+) e^{-ip\cdot x} (i\not\partial - m) \langle \beta_{\text{OUT}} \,|\psi(x)|\, \alpha_{\text{IN}} - p \rangle. \end{aligned} \tag{8.20}$$

The state $|\,\alpha_{\text{IN}}-p\rangle$ represents the IN state $|\,\alpha_{\text{IN}}\rangle$ with one particle of momentum p removed. The results for the other 3 possibilities are:

$$\begin{aligned}
\langle\beta_{\text{OUT}}\,|\,\alpha_{\text{IN}}\rangle &= \langle\beta_{\text{OUT}}\,|b^{\dagger}_{\text{IN}}(p,s)|\,\alpha_{\text{IN}}-p\rangle \\
&= \frac{-i}{\sqrt{Z_2}}\int d^4x\,\langle\beta_{\text{OUT}}\,|\bar{\psi}(x)|\,\alpha_{\text{IN}}-p\rangle(i\overleftarrow{\partial\!\!\!/}-m)u(p,s)e^{-ip\cdot x}, \\
\langle\beta_{\text{OUT}}\,|\,\alpha_{\text{IN}}\rangle &= \langle\beta_{\text{OUT}}-q\,|b_{\text{OUT}}(q,s)|\,\alpha_{\text{IN}}\rangle \\
&= \frac{-i}{\sqrt{Z_2}}\int d^4x\,\bar{u}(q,s)e^{iq\cdot x}(i\partial\!\!\!/-m)\langle\beta_{\text{OUT}}-q\,|\psi(x)|\,\alpha_{\text{IN}}\rangle, \\
\langle\beta_{\text{OUT}}\,|\,\alpha_{\text{IN}}\rangle &= \langle\beta_{\text{OUT}}-q\,|d_{\text{OUT}}(q,s)|\,\alpha_{\text{IN}}\rangle \\
&= \frac{i}{\sqrt{Z_2}}\int d^4x\,\langle\beta_{\text{OUT}}-q\,|\bar{\psi}(x)|\,\alpha_{\text{IN}}\rangle(-i\overleftarrow{\partial\!\!\!/}-m)v(q,s)e^{iq\cdot x},
\end{aligned} \tag{8.21}$$

corresponding to removing an electron from the IN-state, an electron from the OUT-state, and a positron from the OUT-state, respectively.

As we convert more of the particle creation or destruction operators extracted from the IN or OUT states, the need to take $t_i\to\pm\infty$ to convert ψ_{IN}, ψ_{OUT} into ψ naturally introduces time-ordering. If the IN state contains n electrons and n' positrons with momenta p_j and p'_j respectively, and if the OUT state contains m electrons and m' positrons with momenta q_j and q'_j, then the final form of the reduction formula is obtained after we have done $n+n'+m+m'$ conversions. The result is

$$\begin{aligned}
S_{\beta\alpha} &= \left(\frac{-i}{\sqrt{Z_2}}\right)^{n+m}\left(\frac{i}{\sqrt{Z_2}}\right)^{n'+m'}\int d^4x_1\cdots d^4x_n\,d^4x'_1\cdots d^4x'_{n'} \\
&\int d^4y_1\cdots d^4y_m\,d^4y_1{}'\cdots d^4y'_{m'}\exp\left(-i\sum p_j\cdot x_j+p'_j\cdot x'_j-q_j\cdot y_j+q'_j\cdot y'_j\right) \\
&\bar{u}(q_1,s_1)(i\partial\!\!\!/_{y_1}-m)\cdots\bar{v}(p'_1,w'_1)(i\partial\!\!\!/_{x'_1}-m) \\
&\langle 0\,|T(\bar{\psi}(y'_1)\cdots\bar{\psi}(y'_{m'})\psi(y_1)\cdots\psi(y_m)\bar{\psi}(x_1)\cdots\bar{\psi}(x_n)\psi(x'_1)\cdots\psi(x'_{n'}))|\,0\rangle \\
&(-i\overleftarrow{\partial\!\!\!/}_{x_1}-m)u(p_1,w_1)\cdots(-i\partial\!\!\!/_{y'_1}-m)v(q'_1,s'_1).
\end{aligned} \tag{8.22}$$

Following the same philosophy as in the scalar and spinor cases, we introduce photon IN and OUT field operators, $\vec{A}_{\text{IN}}(x)$ and $\vec{A}_{\text{OUT}}(x)$, that create and destroy free massless photons from the vacuum. The free field plane wave expansions, eq.(5.28) of chapter 5, are valid for these fields. Inverting the expansions, we have

$$a_{\text{IN}}(k,\lambda) = i\int d^3x\, e^{-ik\cdot x}\vec{\epsilon}(k,\lambda)\cdot \overleftrightarrow{\partial}_0 \vec{A}_{\text{IN}}(x). \tag{8.23}$$

The asymptotic conditions are

$$\begin{aligned} \lim_{t\to-\infty} \vec{A}(\vec{x},t) &= \sqrt{Z_3}\vec{A}_{\text{IN}}(\vec{x},t), \\ \lim_{t\to\infty} \vec{A}(\vec{x},t) &= \sqrt{Z_3}\vec{A}_{\text{OUT}}(\vec{x},t). \end{aligned} \tag{8.24}$$

Z_3 is the amplitude for $\vec{A}(\vec{x},t)$ to create a 1-photon state. By computing the spectral amplitude, we can show that $0 \leq Z_3 < 1$, thus eq.(8.24) must be interpreted as an equation that makes sense only when it appears sandwiched between normalizable states.

The reduction formula is similar to the scalar case. A typical S-matrix element will contain a number of $a^{\dagger}_{\text{IN}}(k,\lambda)$'s and $a_{\text{OUT}}(k',\lambda')$'s sandwiched in the vacuum state. The a_{IN}'s and a_{OUT}'s are converted to $\vec{A}_{\text{IN}}$ and $\vec{A}_{\text{OUT}}$ using eq.(8.23) and an identical one for the OUT states. Since eq.(8.23) is independent of time, we let $t \to \pm\infty$ and use the asymptotic conditions, eq.(8.24) to replace $\vec{A}_{\text{IN}}$ and $\vec{A}_{\text{OUT}}$ with $\vec{A}$. For the removal of one photon from the IN state we have

$$\begin{aligned} S_{\beta\alpha} &= \langle \beta_{\text{OUT}} \,|\, \alpha_{\text{IN}}\rangle \\ &= \frac{i}{\sqrt{Z_3}}\int d^4x\, e^{-ik\cdot x}\epsilon^i(k,\lambda)\Box_x\langle \beta_{\text{OUT}}\,|A^i(x)|\,\alpha_{\text{IN}} - k\rangle \\ &= \frac{-i}{\sqrt{Z_3}}\int d^4x\, e^{-ik\cdot x}\epsilon^{\mu}(k,\lambda)\Box_x\langle \beta_{\text{OUT}}\,|A_{\mu}(x)|\,\alpha_{\text{IN}} - k\rangle, \end{aligned} \tag{8.25}$$

where the minus sign arises in the second expression above because ϵ^{μ} is space-like and we have explicitly introduced the Minkowski metric, $\eta_{\mu\nu}$ by replacing ϵ^i with ϵ^{μ}. If the initial state contains n photons of momentum k_j, polarization λ_j, and the OUT state contains m photons of momentum k'_j, polarization λ'_j, then the full reduction formula is

$$\begin{aligned} S_{\beta\alpha} = &\left(\frac{-i}{\sqrt{Z_3}}\right)^{m+n} \int d^4x_1\cdots d^4x_n\, d^4y_1\cdots d^4y_m \\ &\exp\left(-i\sum_{j=1}^{n} k_j\cdot x_j + i\sum_{j=1}^{m} k'_j\cdot y_j\right) \\ &\epsilon^{\mu}(k_1,\lambda_1)\cdots\epsilon^{\sigma}(k'_m,\lambda'_m)\Box_{x_1}\cdots\Box_{y_m}\langle 0\,|T(A_{\mu}(x_1)\cdots A_{\sigma}(y_m))|\,0\rangle. \end{aligned} \tag{8.26}$$

QED mixes photons with electrons and positrons so a typical scattering process will contain $a^\dagger_{\text{IN}}(k,\lambda), a_{\text{OUT}}(k',\lambda'), b^\dagger_{\text{IN}}(p,s), b_{\text{OUT}}(p',s')$, etc., sandwiched between vacuum states. The reduction formula will contain a mixture of factors from eq.(8.22) and eq.(8.26) that carry the spin information and differential operators to essentially "chop" the external legs off a diagram representing the process. The objects of interest that we must calculate from QED are the Green's functions,

$$\langle 0\,|T(A_\mu(x_1)\cdots A_\sigma(x_n)\bar{\psi}(y_1)\cdots\psi(y_m)|\,0\rangle.$$

These time-ordered products may be computed perturbatively using diagrams. We will give the Feynman rules in the next section.

By the way, once electrons and photons are coupled, Z_2 becomes gauge dependent and loses physical meaning. However, Z_3 remains gauge invariant.

8.3 FEYNMAN RULES

To compute the Green's function $\langle 0\,|T(A_\mu(x_1)\cdots\psi(y_m))|\,0\rangle$ perturbatively, we can define a U operator relating the interacting fields to the IN or OUT fields, and use it to re-express the interacting fields appearing in the time-ordered product as IN fields. The result is a perturbative series expansion of the Green's function that is nearly identical to eq.(7.97).

$$\langle 0\,|T(A_\mu(x_1)\cdots A_\sigma(x_n)\bar{\psi}(y_1)\cdots\psi(y_m))|\,0\rangle$$

$$= \frac{\sum_{j=0}^{\infty}(-i)^j/j!\int d^4z_1\cdots d^4z_j \,\langle 0\,|T(A^\mu_{\text{IN}}(x_1)\cdots\psi_{\text{IN}}(y_m)\mathcal{H}_I(A_{\text{IN}},\psi_{\text{IN}})(z_1)\cdots\mathcal{H}_I(z_j))|\,0\rangle}{\sum_{j=0}^{\infty}(-i)^j/j!\int d^4z_1\cdots d^4z_j\langle 0\,|T(\mathcal{H}_I(z_1)\cdots\mathcal{H}_I(z_j))|\,0\rangle} \tag{8.27}$$

For QED in the Coulomb gauge, the interacting Hamiltonian density, $\mathcal{H}_I$, can be read off from eq.(8.11).

$$\mathcal{H}_I(x) = :\bar{\psi}(x)(\delta m)\psi(x): + e_0 :\bar{\psi}(x)\not{A}(x)\psi(x): + \frac{e_0^2}{8\pi}\int d^3y \frac{:\rho(\vec{x},t)\rho(\vec{y},t):}{|\vec{x}-\vec{y}|} \tag{8.28}$$

After we normal order the time-ordered products in eq.(8.27) and take their vacuum expectation values, we again find a product of contractions or 2-point functions or propagators, either $\langle 0\,|T(\psi_{\text{IN}}(x)\bar{\psi}_{\text{IN}}(y))|\,0\rangle$ or $\langle 0\,|T(A^\mu_{\text{IN}}(x)A^\nu_{\text{IN}}(y))|\,0\rangle$. A diagrammatic representation follows by associating straight lines with the fermion propagators (figure 8.1(a)) and wavy

Figure 8.1 Diagrammatic representation of the contractions or free field 2-point functions for (a) the electron and positron, $iS_F(x-y)_{\alpha\beta}$, and (b) the photon, $D^{\text{tr}}_{\mu\nu}(x-y)$.

lines for the photon propagator (figure 8.1(b)). We have computed the explicit forms of these propagators in chapters 4 and 5.

$$\begin{aligned} iS_F(x-y)_{\alpha\beta} &= i\int \frac{d^4p}{(2\pi)^4} e^{ip\cdot(x-y)} S_F(p)_{\alpha\beta} \\ &= i\int \frac{d^4p}{(2\pi)^4} e^{ip\cdot(x-y)} \left(\frac{\not p + m}{p^2 - m^2 + i\epsilon}\right)_{\alpha\beta} \end{aligned} \tag{8.29}$$

$$\begin{aligned} &D^{\text{tr}}_{\mu\nu}(x-y) \\ &= \int \frac{d^4k}{(2\pi)^4} \frac{e^{-ik\cdot(x-y)}}{k^2+i\epsilon} \left(-g_{\mu\nu} - \frac{k^2\eta_\mu\eta_\nu - (k\cdot\eta)(k_\mu\eta_\nu + \eta_\mu k_\nu) + k_\mu k_\nu}{(k\cdot\eta)^2 - k^2}\right) \end{aligned} \tag{8.30}$$

With practice, the Feynman rules can be read directly off of the action or Hamiltonian. The interacting part of the action or Hamiltonian tells us about the vertices. The mass shift, the first term in eq.(8.28), contains two fermion fields hence it connects two fermion propagators. As in the scalar case, we represent it as an × as in figure 8.2(a). With each vertex of this type, we associate a factor of $-i\delta m$. Similarly, the second term in eq.(8.28) contains two fermion fields and one photon field. Therefore, it connects 2 fermion propagators with one photon propagator as in figure 8.2(b). From eq.(8.28) we see that we should associate a factor of $-ie_0\gamma^\mu$

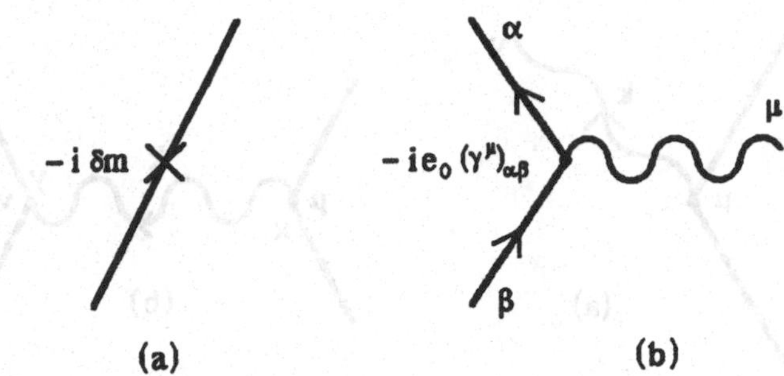

Figure 8.2 Diagrammatic representation of (a) mass shift vertex, and (b) the radiative vertex.

with the radiative vertex. We will ignore the third term in eq.(8.28) for the moment.

All diagrams representing terms in the Green's function in eq.(8.27) must be built from the 2 types of vertices and contain n external photon legs and m external fermion legs. For example, since there is no $A^\mu A^\nu$ term in $\mathcal{H}_I$, we cannot have 2 photon propagators meet at a vertex.

Now that we know the form of the allowed diagrams, let us examine the messy photon propagator, $D^{\text{tr}}_{\mu\nu}(x-y)$, eq.(8.30). The photon propagator can either appear as an external leg (figure 8.3(a)), or as an internal line (figure 8.3(b)). If we are considering the propagator for an external leg, then the photon it represents is a photon that reaches the asymptotic region, that is, a real, detectable photon. Therefore, the momentum vector, k, must be null ($k^2 = 0$). The ν index on the leg indicates that the photon emerges with polarization ϵ_ν. Since it is a real photon, the polarization must be transverse ($k \cdot \epsilon = 0$). Thus, for fixed component ν, the νth component of ϵ will be nonzero if $\nu \neq$ time and if the νth component of k ($= k_\nu$) is zero. In other words, if we choose the z direction to be along the direction of motion, ϵ_x and ϵ_y may not be zero, but $\epsilon_t = \epsilon_z = 0$. To get a nonzero result for figure 8.3(a), ν must be either x or y. However, since k is null, $k_x = k_y = 0$. By definition, $\eta_x = \eta_y = 0$. Therefore, all of the noncovariant terms in eq.(8.30) vanish and we need only use

$$D_{\mu\nu}(x-y) = \int \frac{d^4k}{(2\pi)^4} e^{-ik\cdot(x-y)} D_F(k)_{\mu\nu} = \int \frac{d^4k}{(2\pi)^4} e^{-ik\cdot(x-y)} \frac{-g_{\mu\nu}}{k^2 + i\epsilon}. \tag{8.31}$$

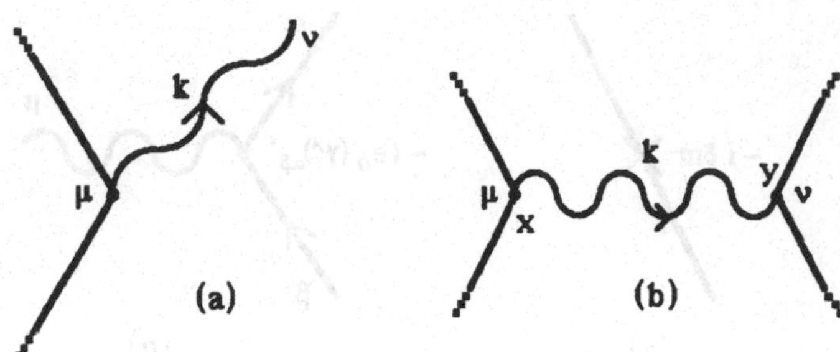

Figure 8.3 (a) Part of a diagram where a real photon appears as an external leg. (b) Part of a diagram where a virtual photon appears as an internal line.

Now consider when the photon propagator is an internal line as in figure 8.3(b). This propagator represents a virtual photon so $k^2 \neq 0$. Figure 8.3(b) will appear as a part of a diagram for any term in eq.(8.27) that contains $\langle 0\,|T(\cdots\bar{\psi}(x)\gamma_\mu\psi(x)A^\mu(x)A^\nu(y)\bar{\psi}(y)\gamma_\nu\psi(y)\cdots)|\,0\rangle$. When $A^\mu(x)$ and $A^\nu(y)$ are contracted, we get a factor of $D^{\rm tr}_{\mu\nu}(x-y)$. However, this factor appears in conjunction with $j_\mu(x) = \bar{\psi}\gamma_\mu\psi(x)$ and $j_\nu(y)$, the conserved currents. Since the currents are conserved, $\partial_\mu j^\mu = 0$, that is, $k^\mu \tilde{j}_\mu = 0$. This means that any term appearing in $D^{\rm tr}_{\mu\nu}(x-y)$ that is proportional to k_μ or k_ν will project out. Thus, there is no need to carry these terms along. This leaves only the covariant part of the propagator and the term proportional to $\eta_\mu\eta_\nu$.

Let's examine the $\eta_\mu\eta_\nu$ term. If we choose frames so that the photon travels along the z direction, then $\eta = (1,0,0,0)$ so $\eta_\mu\eta_\nu = 0$ unless $\mu = \nu =$ time. Thus, $j^\mu\eta_\mu = j^0 = \rho$. If we Fourier transform the $\eta_\mu\eta_\nu$ term in $D^{\rm tr}_{\mu\nu}$ back to x-space we find

$$\int \frac{d^4k}{(2\pi)^4} e^{-ik\cdot x} \frac{k^2\eta_\mu\eta_\nu}{(k\cdot\eta)^2 - k^2} = \frac{\delta(t)\eta_\mu\eta_\nu}{4\pi|\vec{x}|}, \qquad (8.32)$$

so that the $\eta_\mu\eta_\nu$ term in $j^\mu D^{\rm tr}_{\mu\nu} j^\nu$ reproduces the static Coulomb energy, the third term in eq.(8.28), except with the opposite sign. Therefore, the two will cancel each other. So if we ignore the third term in eq.(8.28), we can use the covariant part, eq.(8.31), for the photon propagator on internal lines, too. This is one way to see that quantization in the Coulomb gauge, which breaks manifest Lorentz covariance, still leads to a Lorentz invariant field theory. Alternatively, we can examine the generators of infinitesimal

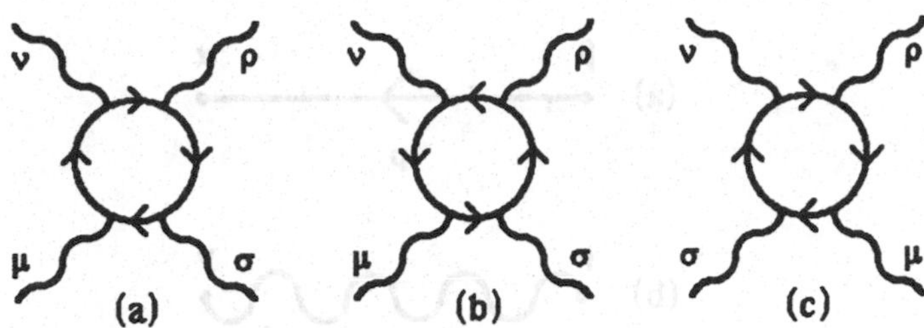

Figure 8.4 Examples of distinct diagrams representing the photon 4-point function or light-light scattering. Diagram (a) is distinct from (b) because of the orientation of the internal fermion loop. Diagram (a) is distinct from (c) because of the labelling of the external photon legs.

Lorentz transformations in this gauge to demonstrate that the invariance remains intact.

We are now ready to state the Feynman rules for QED. For computational convenience, we will work in momentum space.

1. Draw all distinct diagrams for a given process (i.e. fixed number of external oriented fermion and photon legs). Do not include diagrams with disconnected vacuum bubbles.

The denominator in eq.(8.27) will remove diagrams with disconnected vacuum bubbles. Since the fermion lines carry an orientation, "distinct" includes the fermion orientation of the internal fermion lines. Since we are working in momentum space, many of the diagrams that differ only in the orientation of the internal fermion lines will contribute exactly the same result. We define these diagrams as "distinct" in order to get the correct numerical factor in front of the overall result. In x-space, the diagrams with reversed internal fermion orientation correspond to different contractions so they must be counted separately to eliminate the symmetry factor. With a little thought and practice, one can recognize diagrams that give an identical result without having to do the redundant calculation.

As an example, consider the diagrams for light-light scattering in figure 8.4. All three diagrams are to be considered distinct. The second differs from the first by the orientation of the flow of momentum in the internal fermion loop. The third differs from the first by the order of the polarization labels on the external photon lines. Diagram 8.4(b) is equal to diagram 8.4(a). The only difference between the two, the sign of the momentum traveling in the loop, is immaterial since the loop is connected to an even number of photon propagators and we must integrate over the

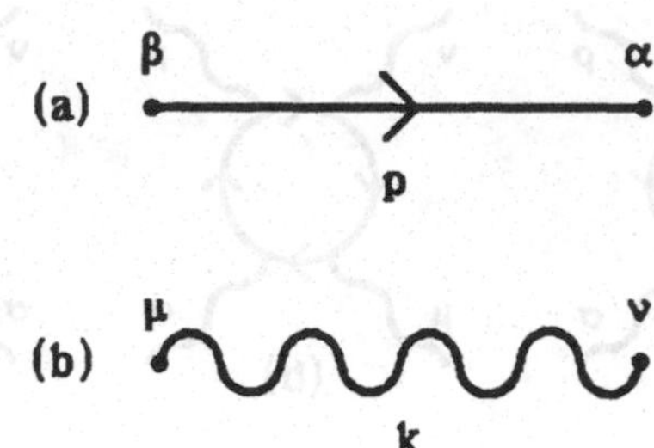

Figure 8.5 Momentum space propagators for (a) the electron or positron, and (b) the photon.

momentum in the loop. Diagram 8.4(c) gives a different result from diagram 8.4(a).

2. For each fermion line, as in figure 8.5(a), assign a fermion propagator, $iS_F(p)_{\alpha\beta}$, eq.(8.29). For each photon line, as in figure 8.5(b), assign a photon propagator, $iD_F(k)_{\mu\nu}$, eq.(8.31)
3. For the mass shift vertex, figure 8.2(a), assign a factor of $-i\,\delta m\,\delta_{\alpha\beta}$. For the radiation vertex, figure 8.2(b), assign a factor of $-ie_0(\gamma^\mu)_{\alpha\beta}$.
4. Conserve momentum at each radiation vertex by including a factor of $(2\pi)^4\,\delta^4(\sum$ momenta leaving − momenta entering vertex).
5. Contract all spinor indices along the fermion lines and vector indices along the photon lines and integrate $\int d^4p_i/(2\pi)^4$ over every internal momentum p_i.
6. Include a factor of (-1) for every closed fermion loop.
7. Assign a global sign to the entire expression depending upon the number of permutations of spinor labels on the external fermion legs.

Let us return again to closed fermion loops. Since the theory is invariant under charge conjugation, any diagram containing a closed fermion loop that has an odd number of photons attached to the loop vanishes. This result is known as Furry's theorem. The individual distinct diagrams are not zero, but diagrams that are identical except for the orientation of the loop cancel.

To show this explicitly, consider a loop attached to an odd number, n, of photons with polarization μ_i as in figure 8.6. For clockwise orientation of the loop, the diagram is proportional to

$$C = \mathrm{tr}(S_F(p_n)\gamma^{\mu_n}\cdots S_F(p_2)\gamma^{\mu_2}S_F(p_1)\gamma^{\mu_1}). \tag{8.33}$$

For counterclockwise orientation, the diagram is proportional to

$$CC = \mathrm{tr}(\gamma^{\mu_1} S_F(p_1)\gamma^{\mu_2} S_F(p_2)\cdots\gamma^{\mu_n} S_F(p_n)). \tag{8.34}$$

Recall from chapter 6 that the charge conjugation operator, $\mathcal{C}$, satisfies

$$\mathcal{C}\gamma^\mu\mathcal{C}^{-1} = -\gamma^{\mu T}, \tag{8.35}$$

thus $\mathcal{C}\not{p}\mathcal{C}^{-1} = -\not{p}^T$, so $\mathcal{C}S_F(p)\mathcal{C}^{-1} = S_F(-p)^T$. Inserting $1 = \mathcal{C}^{-1}\mathcal{C}$ between each of the propagators and γ^{μ_i}'s in eq.(8.34) above, we have

$$\begin{aligned} CC &= \mathrm{tr}(\mathcal{C}^{-1}\mathcal{C}\gamma^{\mu_1}\mathcal{C}^{-1}\mathcal{C}S_F(p_1)\cdots\mathcal{C}^{-1}\mathcal{C}S_F(p_n)) \\ &= (-1)^n\mathrm{tr}({\gamma^{\mu_1}}^T S_F(-p_1)^T\cdots S_F(-p_n)^T) \\ &= (-1)^n\mathrm{tr}(S_F(-p_n)\gamma^{\mu_n}\cdots S_F(-p_1)\gamma^{\mu_1}), \end{aligned} \tag{8.36}$$

where we have used the fact that the trace is unaffected by transposition. All but one of the p_i's can be eliminated by momentum conservation in both eq.(8.33) and eq.(8.36). We must integrate over the remaining momentum, thus eq.(8.36) is equal to eq(8.33) up to the factor $(-1)^n$. For odd n, eq.(8.36) cancels eq.(8.33).

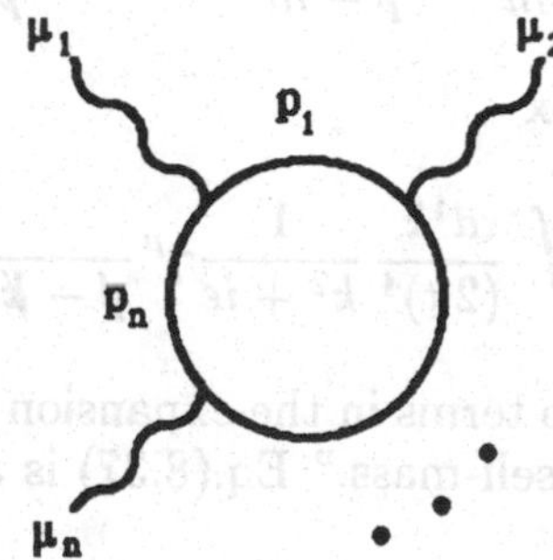

Figure 8.6 An internal fermion loop with n photons attached. When n is odd, the diagram containing this loop vanishes.

8.4 2- AND 3-POINT FUNCTIONS

To illustrate the rules, let us compute the lower order terms of the interacting electron and photon propagators. The diagrammatic representation of the expansion of the electron 2-point function is given in figure 8.7.

The first term in the expansion is just the free field propagator, $iS_F(p)_{\alpha\beta}$, eq.(8.29). The next term contains one "mass counterterm" vertex and is equal to $iS_F(p)_{\alpha\sigma}(-i\,\delta m\delta_{\sigma\rho})iS_F(p)_{\rho\beta} = i\delta m(S_F(p)S_F(p))_{\alpha\beta}$. The third term is the first nontrivial diagram and is called the "self-mass" diagram. Using the momentum labels in figure 8.8 (where we have already accounted for momentum conservation at the vertices), this diagram is equal to

$$(-ie_0)^2\int\frac{d^4k}{(2\pi)^4}ig_{\mu\nu}D_F(k)\,iS_F(p)_{\alpha\omega}\gamma^{\mu}_{\omega\epsilon}iS_F(p-k)_{\epsilon\rho}\gamma^{\nu}_{\rho\sigma}iS_F(p)_{\sigma\beta}$$

$$= -e_0^2\int\frac{d^4k}{(2\pi)^4}\frac{-i}{k^2+i\epsilon}\left(\frac{i}{\not p-m+i\epsilon}\gamma^{\mu}\frac{i}{\not p-\not k-m+i\epsilon}\gamma_{\mu}\frac{i}{\not p-m+i\epsilon}\right)_{\alpha\beta}. \tag{8.37}$$

Ignoring the second diagram in figure 8.7 for the moment, the sum of the first two diagrams is

$$\frac{i}{\not p-m}+\frac{i}{\not p-m}\left(-i\Sigma(p)\right)\frac{i}{\not p-m}, \tag{8.38}$$

where $\Sigma(p)$ is the matrix

$$\Sigma(p) = (-ie_0)^2\int\frac{d^4k}{(2\pi)^4}\frac{1}{k^2+i\epsilon}\gamma^{\mu}\frac{i}{\not p-\not k-m+i\epsilon}\gamma_{\mu}. \tag{8.39}$$

Eq.(8.38) is the first two terms in the expansion of $i/(\not p-m-\Sigma(p))$ which is why it is called the "self-mass." Eq.(8.37) is also divergent.

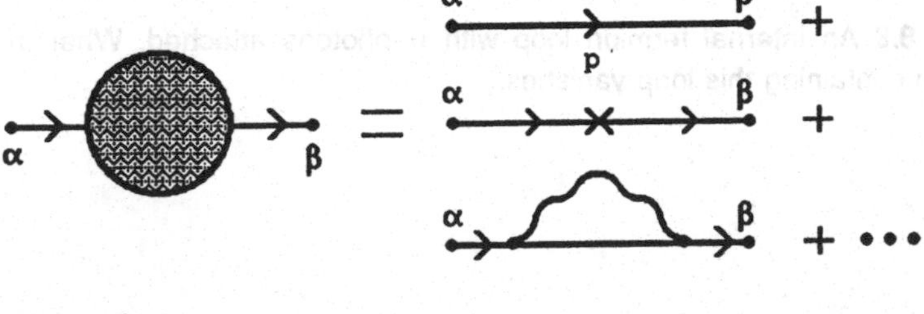

Figure 8.7 Diagrammatic expansion of the interacting electron propagator, $iS'_F(p)_{\alpha\beta}$.

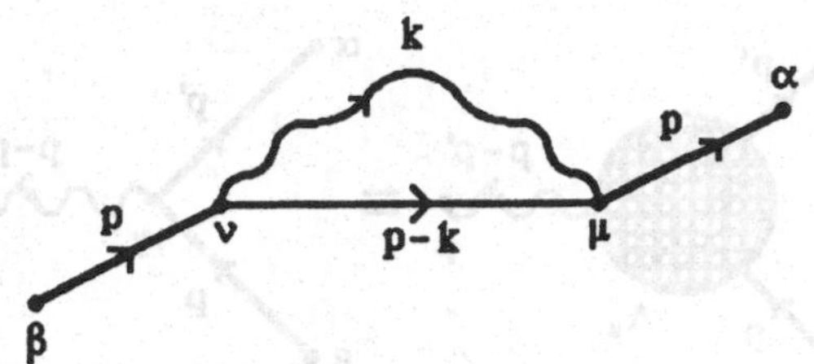

Figure 8.8 Electron "self-mass" diagram.

The expansion of the photon 2-point function is illustrated in figure 8.9. The lowest order term is, again, just the free field result, $ig_{\mu\nu}D_F(k) = -ig_{\mu\nu}/(k^2+i\epsilon)$. The next nontrivial diagram is called the "vacuum polarization" term.

Using the momentum labels in figure 8.9, where we have already incorporated momentum conservation at the vertices, the rules imply that this diagram is equal to

$$iD_F(k)\Pi_{\mu\nu}(k)\,iD_F(k) =$$

$$-\,iD_F(k)(-ie_0)^2\int\frac{d^4p}{(2\pi)^4}\mathrm{tr}\left(\gamma_\mu\frac{i}{\not p-\not k-m+i\epsilon}\gamma_\nu\frac{i}{\not p-m+i\epsilon}\right)iD_F(k), \tag{8.40}$$

where the extra minus sign is present due to the closed fermion loop (rule 6). Once again, the diagram is divergent.

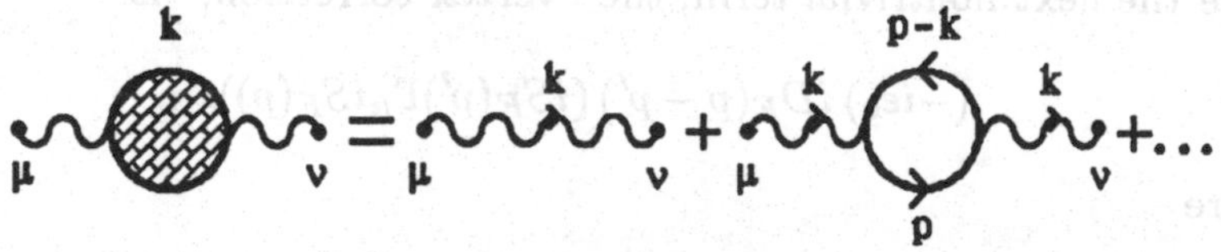

Figure 8.9 Diagrammatic expansion of the interacting photon 2-point function. The first diagram on the right-hand side is the free photon propagator. The second diagram is called the "vacuum polarization" diagram.

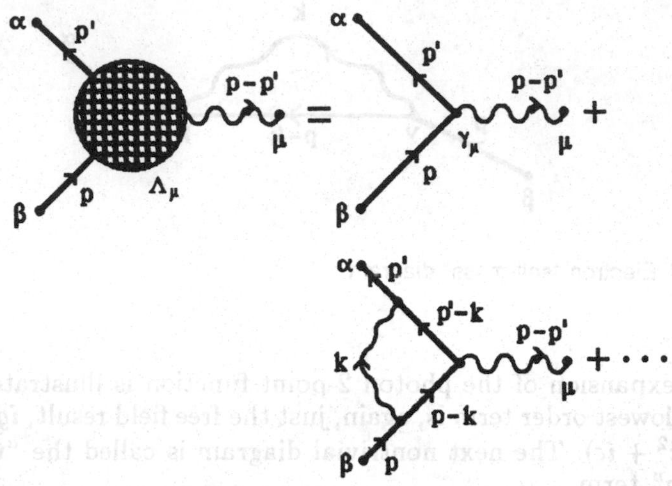

Figure 8.10 Diagrammatic expansion of the full radiative vertex, Λ_μ. The second diagram in the expansion is called the first "vertex correction."

The 3-point function represents the process of *bremsstrahlung*, or the full, interacting photon-electron radiative vertex. The diagrammatic expansion is given in figure 8.10. The lowest order term is

$$\begin{aligned}&(-ie_0)(ig_{\mu\nu}D_F(p-p'))(iS_F(p')\gamma^\nu iS_F(p))_{\alpha\beta}\\&\quad=(-ie_0)\frac{-i}{(p-p')^2+i\epsilon}\left(\frac{i}{\not p'-m+i\epsilon}\gamma_\mu\frac{i}{\not p-m+i\epsilon}\right)_{\alpha\beta}\end{aligned} \tag{8.41}$$

while the next nontrivial term, the "vertex correction," is

$$(-ie_0)\,iD_F(p-p')\,(iS_F(p')\Gamma_\mu iS_F(p))_{\alpha\beta}, \tag{8.42}$$

where

$$\begin{aligned}\Gamma_\mu&=(-ie_0)^2\int\frac{d^4k}{(2\pi)^4}\,iD_F(k)\,(\gamma_\nu iS_F(p'-k)\gamma_\mu iS_F(p-k)\gamma^\nu)\\&=(-ie_0)^2\int\frac{d^4k}{(2\pi)^4}\,\frac{-i}{k^2+i\epsilon}\left(\gamma_\nu\frac{i}{\not p'-\not k-m+i\epsilon}\gamma_\mu\frac{i}{\not p-\not k-m+i\epsilon}\gamma^\nu\right).\end{aligned} \tag{8.43}$$

This diagram is also divergent!

Γ_μ, the vertex correction, is a function of p and p', $\Gamma_\mu = \Gamma_\mu(p,p')$. If we consider the limit $p' \to p$, then the radiated photon carries no momentum and the vertex correction diagram begins to look just like the self-mass diagram. That is, if we remove the radiated photon carrying $p - p'$ from the second term in figure 8.10, we get the self-mass diagram of figure 8.8. This means that eq.(8.43), Γ_μ, is related to $\Sigma(p)$, eq.(8.39), as $p' \to p$. This relation is called a Ward identity, and is formally obtained by realizing that

$$\frac{1}{\not p - m + i\epsilon}\gamma_\mu \frac{1}{\not p - m + i\epsilon} = -\frac{\partial}{\partial p^\mu}\frac{1}{\not p - m + i\epsilon}, \tag{8.44}$$

which implies from eqs.(8.39) and (8.43) that

$$\Gamma_\mu(p,p) = -\frac{\partial}{\partial p^\mu}\Sigma(p). \tag{8.45}$$

In fact, the first two terms in figure 8.10 and the first and third terms of figure 8.7 combine nicely together to give

$$\gamma_\mu + \Gamma_\mu(p,p) = \frac{\partial}{\partial p^\mu}\left(\not p - m - \Sigma(p)\right). \tag{8.46}$$

This identity was generalized by Takahashi to the case where $p' \neq p$. To motivate the form of the generalization, consider the difference between the inverse of the free electron propagator at two different momenta.

$$S_F^{-1}(p') - S_F^{-1}(p) = (\not p' - m + i\epsilon) - (\not p - m + i\epsilon) = \not p' - \not p = (p' - p)^\mu \gamma_\mu \tag{8.47}$$

If we add the results of higher order self-mass diagrams and higher order vertex corrections, then the pole in the propagator shifts from m to $m + \Sigma(p)$, and the vertex shifts from γ_μ to $\Lambda_\mu(p,p')$ (Λ_μ is represented diagrammatically by the "blob" on the left-hand side of figure 8.10). So, from eq.(8.47), we expect that

$$\Sigma(p') - \Sigma(p) = (p' - p)^\mu \Lambda_\mu. \tag{8.48}$$

To convince oneself of the validity of eq.(8.48), notice that all diagrams representing higher order terms of the vertex correction can be obtained from the diagrams representing the higher order terms of the electron 2-point function by inserting 1 more vertex with a photon that carries off momentum $p - p'$. The correspondence is illustrated for one diagram in figure 8.11, where the photon is inserted at point a, b, or c in the self-mass diagram yielding the vertex correction diagrams of figure 8.11 (a), (b), or (c) respectively.

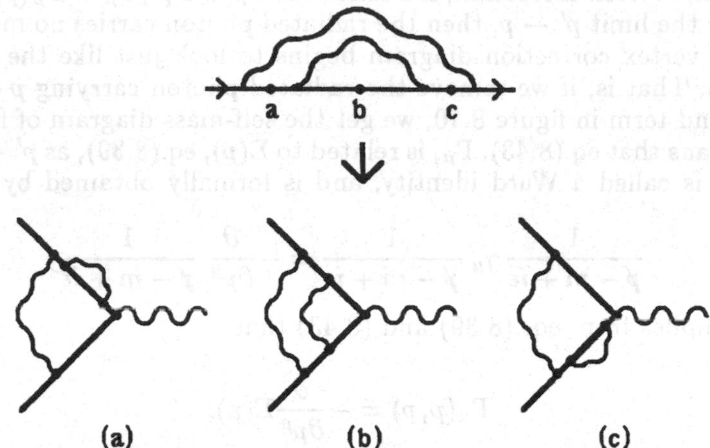

Figure 8.11 Diagrammatic representation of the relation between the mass shift, $\Sigma(p)$, and full radiative vertex, Λ_μ, as expressed in a Ward-Takahashi identity. Insertion of a photon propagator at the point a, b, or c in a second-order self-mass diagram yields second-order vertex correction diagrams.

The value of the Ward-Takahashi identity is that the full interacting electron propagator may be obtained from the full (proper) vertex function. This will be an aid in renormalizing.

8.5 ELECTRON-ELECTRON SCATTERING

As a final illustration of the rules and use of diagrams, let us compute the cross section for electron-electron scattering (to lowest order to avoid the need to renormalize).

Let the momenta of the two electrons before the scattering be p_1 and p_2, and after the scattering, p_1' and p_2'. Momentum conservation requires $p_1 + p_2 = p_1' + p_2'$. There are two equivalent processes that must be computed. The diagrams for these two are given in figure 8.12. The second process, figure 8.12(b), differs from the first by an exchange of electrons in the final state. Therefore, to get the total amplitude for this process to this order we must subtract the result of diagram 8.12(b) from that of 8.12(a).

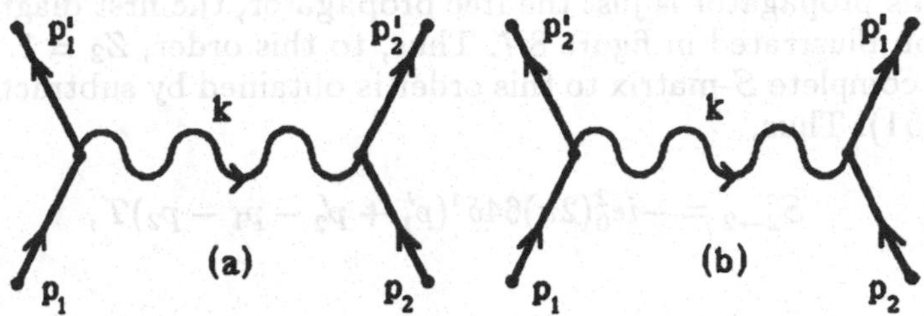

Figure 8.12 Lowest-order diagrams representing elastic electron-electron scattering.

To compute the cross section, we start as usual by computing the 4-point function to lowest nontrivial order using the Feynman rules. For figure 8.12(a) we have

$$(-ie_0)^2 \int \frac{d^4k}{(2\pi)^4} \frac{-i}{k^2 + i\epsilon} \left(\frac{i}{\not{p}_2' - m + i\epsilon} \gamma_\mu \frac{i}{\not{p}_2 - m + i\epsilon} \right)$$
$$\left(\frac{i}{\not{p}_1' - m + i\epsilon} \gamma^\mu \frac{i}{\not{p}_1 - m + i\epsilon} \right) \delta^4(p_1 - p_1' - k). \quad (8.49)$$

Similarly, for figure 8.12(b) we have

$$(-ie_0)^2 \int \frac{d^4k}{(2\pi)^4} \frac{-i}{k^2 + i\epsilon} \left(\frac{i}{\not{p}_1' - m + i\epsilon} \gamma_\mu \frac{i}{\not{p}_2 - m + i\epsilon} \right)$$
$$\left(\frac{i}{\not{p}_2' - m + i\epsilon} \gamma^\mu \frac{i}{\not{p}_1 - m + i\epsilon} \right) \delta^4(p_1 - p_2' - k). \quad (8.50)$$

Next, we must process these two terms through the reduction formula, eq.(8.22). For eq.(8.49),

$$S_{(a)}^c = \frac{(-i)^4}{Z_2^2} (-ie_0)^2 \frac{-i}{(p_1 - p_1')^2} (\bar{u}(p_2', s_2') \gamma_\mu u(p_2, s_2))$$
$$(\bar{u}(p_1', s_1') \gamma^\mu u(p_1, s_1)) (2\pi)^4 \delta^4(p_1' + p_2' - p_1 - p_2), \quad (8.51)$$

while for eq.(8.50),

$$S_{(b)}^c = \frac{(-i)^4}{Z_2^2} (-ie_0)^2 \frac{-i}{(p_1 - p_2')^2} (\bar{u}(p_1', s_1') \gamma_\mu u(p_2, s_2))$$
$$(\bar{u}(p_2', s_2') \gamma^\mu u(p - 1, s_1)) (2\pi)^4 \delta^4(p_1' + p_2' - p_1 - p_2). \quad (8.52)$$

What about Z_2? Z_2 is defined by $\lim_{\not{p} \to m} S_F'(p) = Z_2 S_F(p)$, the 1-particle contribution. $S_F'(p)$ is the full, interacting 2-point function. To lowest order, this propagator is just the free propagator, the first diagram in the expansion illustrated in figure 8.7. Thus, to this order, $Z_2 = 1$.

The complete S-matrix to this order is obtained by subtracting (8.52) from (8.51). Thus,

$$S^c_{2\to 2} = -ie_0^2(2\pi)64\delta^4(p_1' + p_2' - p_1 - p_2)\mathcal{T},$$

where

$$\mathcal{T} = \left[\frac{(\bar{u}(p_2', s_2')\gamma_\mu u(p_2, s_2))(\bar{u}(p_1', s_1')\gamma^\mu u(p_1, s_1))}{(p_1 - p_1')^2} - \frac{(\bar{u}(p_1', s_1')\gamma_\mu u(p_2, s_2))(\bar{u}(p_2', s_2')\gamma^\mu u(p_1, s_1))}{(p_1 - p_2')^2}\right]. \quad (8.53)$$

The definition of the cross section in terms of $|S^c_{2\to 2}|^2$ is identical to that given in the previous chapter. The only difference is that we are dealing with fermions so that the normalization or Lorentz invariant volume element is $\int \frac{d^3p}{(2\pi)^3}\frac{m}{E}$ (fermionic) instead of $\int \frac{d^3k}{(2\pi)^3}\frac{1}{2\omega_k}$ (bosonic). Thus,

$$d\sigma = \frac{m^4 e_0^4}{\sqrt{(p_1\cdot p_2)^2 - m^4)}} \int \frac{d^3p_1'}{(2\pi)^3}\frac{d^3p_2'}{(2\pi)^3}\frac{1}{E_{1'}}\frac{1}{E_{2'}}(2\pi)^4\delta^4(p_1' + p_2' - p_1 - p_2)|\mathcal{T}|^2. \quad (8.54)$$

The square of $|\mathcal{T}|^2$ yields a product of traces involving $u, \bar{u}$, and γ^μ. Generally, in scattering experiments we don't measure the initial and final spins of the electrons; thus we want to average eq.(8.54) over all spin states $s_1', s_2', s_1, s_2 = +$ or $-$. This makes the computation of $|\mathcal{T}|^2$ easier because we may use the completeness of u, eq.(4.50), to replace $u\bar{u}$ by $(\not{p} + m)/2m$. Then, to compute $|\mathcal{T}|^2$, we need only compute the traces of a product of γ matrices. This is a straightforward but sometimes tedious algebraic computation which we will leave for the exercises. The details may be found in the books listed in the Bibliography.

8.6 EXERCISES

1. In preparation for computing the spin-averaged $|\mathcal{T}|^2$ of eq.(8.54), let us consider a simpler case,

$$\sum_{\pm s_1, s_2} |\bar{u}(p_1, s_1)\gamma^0 u(p_2, s_2)|^2,$$

which arises in Mott scattering (of an electron by an external electromagnetic field). First, write the product in the sum above in component form, $\bar{u}_\alpha(p_1, s_1)\gamma^0_{\alpha\beta}u_\beta \cdots$, and insert $1 = \gamma_0^2$ in the appropriate place so that a $u\bar{u}$ sits in the middle of the string. Next, use eq.(4.50) twice. The result is the trace,

$$\mathrm{tr}\left(\gamma_0 \frac{(\not{p}_2 + m)}{2m} \gamma_0 \frac{(\not{p}_1 + m)}{2m}\right).$$

Use the results of exercise 6 in chapter 4 to reduce this to

$$\frac{1}{m^2}(8E_1E_2 - 4p_1 \cdot p_2 + 4m^2).$$

2. Verify that:

   a. $\gamma^\nu \gamma_\nu = 4$

   b. $\gamma^\nu \not{a} \gamma_\nu = -2\not{a}$

   c. $\gamma^\nu \not{a}\not{b} \gamma_\nu = 4a \cdot b$

   d. $\gamma^\nu \not{a}\not{b}\not{c} \gamma_\nu = -2\not{c}\not{b}\not{a}$

   e. $\gamma^\nu \not{a}\not{b}\not{c}\not{d} \gamma_\nu = 2(\not{d}\not{a}\not{b}\not{c} + \not{c}\not{b}\not{a}\not{d})$

3. Follow the same procedure in exercise 1 and apply the results of exercise 2 to compute the spin-averaged cross section of eq.(8.54). First, switch to the center of mass frame where $E_i = E_{i'} = E$ for $i = 1, 2$, and $\vec{p}_2 = -\vec{p}_1$, $\vec{p}_2{}' = -\vec{p}_1{}'$. Let the scattering angle be θ, i.e. the angle between $\vec{p}_1$ and $\vec{p}_1{}'$. Next, integrate over $\vec{p}_2{}'$ in eq.(8.54) to arrive at

$$\frac{d\sigma}{d\Omega} = \frac{m^2 e_0^4}{4E^2(2\pi)^2}|\mathcal{T}|^2.$$

Apply the results of exercises 1 and 2 to $|\mathcal{T}|^2$. Rewrite $p_1 \cdot p_2$, $p_1 \cdot p_1'$, etc., in terms of E, m, and θ to obtain the Møller cross section,

$$\frac{d\sigma}{d\Omega} = \frac{e^4}{16E^2(2\pi)^2} \frac{(2E^2 - m^2)^2}{(E^2 - m^2)^2}\left(\frac{4}{\sin^4\theta} - \frac{3}{\sin^2\theta} + \frac{(E^2 - m^2)^2}{(2E^2 - m^2)^2}\left(1 + \frac{4}{\sin^2\theta}\right)\right).$$

4. Compute the S-matrix element for elastic electron-positron scattering to lowest order, $O(e_0^2)$. Draw the corresponding diagrams.

5. Draw the two lowest order diagrams for Compton scattering and compute the S-matrix element to this order.

6. Draw the lowest order diagrams involving a single virtual photon for $e^+e^- \longrightarrow \mu^+\mu^-$ and $e^+e^- \longrightarrow q\bar{q}$ (quark-antiquark). Write down the S-matrix element and differential cross section in the form eq.(8.54). Argue from the form of the result in the appropriate limit that, if we ignore strong interaction effects,

$$R = \frac{\sigma(e^+e^- \rightarrow \text{hadrons})}{\sigma(e^+e^- \rightarrow \mu^+\mu^-)} = N_c \sum_f Q_f^2,$$

where N_c is the number of quark colors, and Q_f is the electric charge of the quark of flavor f in units of e_0. If E is the center of mass energy of the collision, then the sum extends only over quark flavors such that $2m_f < E$. For $E \sim 1 - 4\text{GeV}$ range, we expect the sum over up, down, and strange flavors ($Q_u = 2/3$, $Q_d = Q_s = 1/3$). The experimental value of R is ~ 2 in this range. How many colors of quarks do we need to agree with this measurement?

7. Write down the action for a charged scalar field coupled to the electromagnetic field. Demonstrate that the action is (gauge) invariant under local phase transformations of the scalar field. Compute the Hamiltonian. Is it gauge invariant?

8. Draw the diagrams for the scalar propagator and the photon propagator to order e_0^4.

9. Draw the lowest order diagrams for scalar-photon Compton scattering and their corresponding expressions. Find the cross section of this process.

10. Write down the Feynman rules for the general Yukawa interaction,

$$\mathcal{L}_I(x) = \bar{\psi}(g - g_5\gamma^5)\psi\varphi.$$

Investigate the properties of this action under $\mathcal{C}$, $\mathcal{P}$, and $\mathcal{T}$. Compute the S-matrix element for the single meson exchange between two different fermions. Draw the diagram.

CHAPTER 9

Functional Calculus

The two remaining representations of quantum field theory, the path integral representation, and the Schrödinger representation, are functional formulations requiring calculus on function spaces, or functional calculus. In this chapter we review the functional calculus that we will need. We begin by defining a functional derivative. We then solve several simple functional differential equations. Finally, we compute the two functional integrals that can be done exactly, once with commuting variables, and once with anticommuting variables.

9.1 FUNCTIONAL DERIVATIVES

Here, a function space is an infinite-dimensional space where each point in the space is a function on space-time. That is, each point in the function space is a mapping of space-time into the real or complex numbers. For our purposes, the function will be a scalar, spinor, or vector on space-time. A particular point in a function space will map points in space-time to scalars, spinors or vectors.

A mapping of the points in the function space to numbers is called a functional. Thus, a functional associates a number with each function on space-time. For example, let a be a point in a scalar function space,

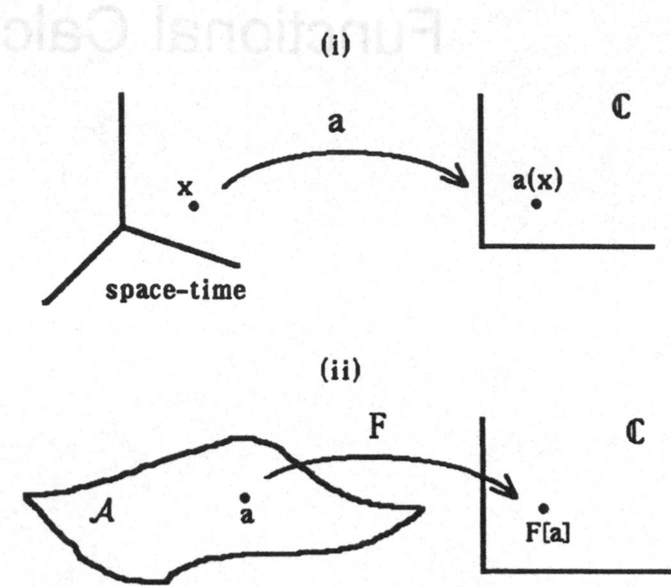

Figure 9.1 (i) a is a function on space-time. It maps points in space-time to real or complex numbers. (ii) a is a point in the function space $\mathcal{A}$. The functional $F[a]$ maps points in the function space $\mathcal{A}$ to real or complex numbers. That is, the functional $F[a]$ turns functions into numbers.

$\mathcal{A}$. a ia a scalar function on space-time, $a = a(x) \in \mathbb{C}$ or $\mathbb{R}$. Let F be a functional on $\mathcal{A}$. F maps points in $\mathcal{A}$ into numbers. We will denote the functional nature of F by using square brackets, $F = F[a] \in \mathbb{C}$ or $\mathbb{R}$. See figure 9.1. A simple example of a functional is $F[a] = \int a(x)\,dx$.

As we move from point to point in the function space, the value of a functional, $F[a]$, will change. The rate of change of F with respect to a change in the function a is expressed as the functional derivative of F, $\delta F/\delta a$. The functional derivative is formally defined as

$$\frac{\delta F[a]}{\delta a} = \lim_{\epsilon \to 0} \frac{F[a + \epsilon\delta] - F[a]}{\epsilon}, \tag{9.1}$$

where δ is the Dirac δ-function distribution. Similarly, we can define a functional directional derivative in the direction of the function $\lambda(x)$ as

$$\frac{\delta_\lambda F[a]}{\delta_\lambda a} = \lim_{\epsilon \to 0} \frac{F[a + \epsilon\lambda] - F[a]}{\epsilon}. \tag{9.2}$$

The ordinary functional derivative, definition (9.1), is a directional derivative in the direction of the δ-function.

We will always explicitly include the dependence of the function on space-time when we write a functional derivative. Thus, by $\delta F[a]/\delta a(x)$ we mean the change in F with respect to a change in the function a at the point x only. Formally,

$$\frac{\delta F[a]}{\delta a(x)} = \lim_{\epsilon \to 0} \frac{F[\hat{a}] - F[a]}{\epsilon}, \tag{9.3}$$

where $\hat{a}$ is the function obtained from a by adding ϵ to a only at the point x. In other words, $\hat{a}$ is equal to a except at the point x where $\hat{a} = a + \epsilon$. If we take $F[a] = \int a(z)\delta(z-y)\,dz = a(y)$, then $F[\hat{a}] = \int (a(z) + \epsilon\delta(x-z))\,\delta(z-y)dz$. From definition (9.3) or (9.1), we can see that

$$\frac{\delta a(y)}{\delta a(x)} = \delta(x-y). \tag{9.4}$$

Evidently, definition (9.2) implies that

$$\frac{\delta_\lambda a(y)}{\delta_\lambda a(x)} = \lambda(x-y). \tag{9.5}$$

The functional derivative formalism, eq.(9.4), may be used to compute the equations of motion from an action without explicitly mentioning the Euler-Lagrange equation. For example, consider the action for φ^4, eq.(7.1).

$$S[\varphi] = \int d^4y \frac{1}{2}(\partial_\mu \varphi \partial^\mu \varphi - m_0^2 \varphi^2) - \frac{\lambda_0}{4!}\varphi^4 \tag{7.1}$$

Applying eq.(9.4) above, we have

$$\begin{aligned} \frac{\delta}{\delta\varphi(x)} \int d^4y \frac{1}{2}(\partial^\mu \varphi \partial_\mu \varphi) &= -\frac{\delta}{\delta\varphi(x)} \int d^4y \frac{1}{2}(\varphi \partial^\mu \partial_\mu \varphi) \\ &= -\int d^4y\, \delta^4(x-y) \partial^\mu \partial_\mu \varphi(y) \\ &= -\partial^\mu \partial_\mu \varphi(x), \end{aligned} \tag{9.6}$$

and

$$\frac{\delta}{\delta\varphi(x)} \int d^4y \varphi^n(y) = n \int d^4y \varphi^{n-1}(y)\delta^4(x-y) = n\varphi^{n-1}(x). \tag{9.7}$$

Thus,

$$\frac{\delta S[\varphi]}{\delta\varphi(x)} = -\left(\partial_\mu \partial^\mu \varphi(x) + m_0^2 \varphi(x) + \frac{\lambda_0}{3!}\varphi^3(x)\right) = 0, \tag{9.8}$$

which is, of course, the same result as obtained from the Euler-Lagrange eq.(2.41).

We may also use the notion of a functional derivative to rewrite the S-matrix operator as a functional differential operator. First, let us define a generating functional for the vacuum expectations of time-ordered products of the fields, $G(x_1, \ldots, x_n) = \langle 0\,|T(\varphi(x_1)\cdots\varphi(x_n))|\,0\rangle$. This generating functional is traditionally denoted as $Z[J]$.

$$\begin{aligned} Z[J] &= \sum_{n=0}^{\infty} \frac{i^n}{n!} \int d^4x_1 \cdots d^4x_n J(x_1) \cdots J(x_n) \langle 0\,|T(\varphi(x_1)\cdots\varphi(x_n))|\,0\rangle \\ &\equiv \langle 0\,|T \exp\left(i \int d^4x J(x)\varphi(x)\right)|\,0\rangle \end{aligned} \tag{9.9}$$

The time-ordered exponential was introduced in chapter 7 in connection with the U-matrix, eq.(7.87). From the expansion in eq.(9.9) we can see that

$$G(x_1, \ldots, x_n) = \frac{\delta}{\delta J(x_1)} \cdots \frac{\delta}{\delta J(x_n)} Z[J]\bigg|_{J=0} . \tag{9.10}$$

Using the reduction formula, eq.(7.62), and the generating functional above, the S-matrix operator can be rewritten as

$$S =: \exp\left(\int d^4y\, \varphi_{\mathrm{IN}}(y)\left((\Box_y + m^2)\frac{1}{\sqrt{Z}}\frac{\delta}{\delta J(y)}\right)\right) : Z[J]\bigg|_{J=0} . \tag{9.11}$$

We will see later that the generating functional $Z[J]$ is the natural thing to compute from the path integral point of view.

9.2 FUNCTIONAL DIFFERENTIAL EQUATIONS

In the Schrödinger representation of field theory, the Schrödinger equation is a functional differential equation, that is, it is an equation involving functional derivatives that the wave functional of the quantum state satisfies. As demonstrated in the previous section, we have already been dealing with functional differential equations when we determine the field equations of motion from an action. For a field $a(x)$, the field equations are

$$\frac{\delta S[a]}{\delta a(x)} = 0, \tag{9.12}$$

where $S[a]$ is the action functional. The functional form of the action is known and eq.(9.12) states that the classical field solutions are the critical points of the functional $S[a]$ in the function space $\mathcal{A}$.

Quantum mechanics requires a different philosophy with regard to what we want from a functional differential equation. In the classical case above, the form of the functional is known and we are looking for isolated points in the function space. In the quantum case, the form of the functional is unknown and is the goal of solving the functional differential equation. We are usually less interested in particular points in the function space, especially in strongly coupled situations.

A functional differential equation can be thought of as an infinite set of coupled partial differential equations or a partial differential equation on an infinite-dimensional (vector) space. As you can imagine, there are very few techniques known to solve these equations. Let us "solve" a few simple equations to demonstrate the complexity and freedom in the choice of solution, and to warm up for the next chapter.

We can borrow one technique from partial differential equations and attempt an infinite-dimensional separation of variables. This will work for linear equations where the solutions are exponential. For example, consider the functional differential equation,

$$\int dy \frac{\delta}{\delta a(y)} F[a] = -\mu F[a]. \tag{9.16}$$

The left-hand side is a sum of uncoupled differential operators, one for each point y of space-time. Therefore, we can write the functional as an infinite product, one factor for each "component" of a. For a fixed space-time point x, let $z = a(x)$. Then each factor will satisfy

$$\frac{d}{dz}\mathcal{F}(z) = -f\mathcal{F}(z).$$

Thus, $\mathcal{F}(z) = \exp(-fz)$, so

$$\begin{aligned} F[a] = \eta \prod_x \exp(-f(x)a(x)) &= \eta \exp\left(-\sum_x f(x)a(x)\right) \\ &= \eta \exp\left(-\int dx\, f(x)a(x)\right). \end{aligned} \tag{9.17}$$

To satisfy eq.(9.16), the arbitrary function $f(x)$ must satisfy

$$\mu = \int dx\, f(x). \tag{9.18}$$

The normalization, η, is determined by the initial or boundary condition we place on F. For example, if $F[\tilde{a}] = 1$ where $\tilde{a}$ is the unit constant function, $\tilde{a}(x) = 1$, then $\eta = \exp(\mu)$.

Similarly,

$$\int dy \frac{\delta}{\delta a(y)} \frac{\delta}{\delta a(y)} F[a] = -\mu F[a] \tag{9.19}$$

is satisfied by the functional

$$F[a] = \eta \exp\left(\pm \int dx\, f(x) a(x)\right), \tag{9.20}$$

where the arbitrary function now satisfies $\mu = \int dx\, f^2(x)$. How arbitrary the function $f(x)$ is depends on the differential equation. For example, consider

$$\int dy \left(\frac{\delta}{\delta a(y)} \frac{\delta}{\delta a(y)} - b(y)\right) F[a] = 0. \tag{9.21}$$

This equation is satisfied by

$$F[a] = \eta \exp\left(\pm \int dx\, \sqrt{b(x)} a(x)\right). \tag{9.22}$$

We can quickly complicate matters and ruin the separation of variables by coupling the function at different points together. Consider adding a "potential" to eq.(9.19).

$$\int dy \frac{\delta}{\delta a(y)} \frac{\delta}{\delta a(y)} F[a] + \int dx\, dy\, f(x,y) a(x) a(y) F[a] = \mu F[a] \tag{9.23}$$

The second term on the left-hand side couples a at one point with a at another. The coupling is given by $f(x, y)$. Assume that $f(x, y)$ is symmetric in x and y. We will meet equations similar to this in field theory. In general, since $F[a]$ appears in each term we may substitute

$$F[a] = \eta \exp(G[a]), \tag{9.24}$$

and solve the resulting equation for $G[a]$.

$$\int dy \frac{\delta}{\delta a(y)} \frac{\delta}{\delta a(y)} G[a] + \left(\frac{\delta}{\delta a(y)} G[a]\right)^2 = \mu - \int dx\, dy\, f(x,y) a(x) a(y) \tag{9.25}$$

To proceed, we note that the first term above, $\delta^2/\delta a(y)^2\, G[a]$, removes 2 factors of $a(y)$ from $G[a]$, while the second removes one factor and squares the result. The right-hand side is the known functional $\mu - \int dx\, dy\, f(x,y) a(x) a(y)$ that we must match with the left side. The simplest guess is to take some functional that is quadratic in a. That way, the second derivative will produce a constant that we can match with μ, while the single derivative squared will still be quadratic in a so as to match with $f(x,y)a(x)a(y)$. Thus, we write $G[a]$ as a general quadratic functional,

$$G[a] = \int dx\, dy\, g(x,y)a(x)a(y). \tag{9.26}$$

$G[a]$ will be a solution provided the symmetric $g(x,y)$ satisfies

$$\mu = \int dy\, g(y,y) \quad \text{and} \quad f(x,y) = \int dz g(x,z)g(y,z). \tag{9.27}$$

The "power counting" technique used above can be used in situations where a number of derivatives of the unknown functional appear on one side of the equation, while a "known" functional that is polynomial in a appears on the other side. As another example, consider the equation

$$\int dy \frac{\delta}{\delta a(y)} \frac{\delta}{\delta a(y)} F[a] = \mu G[a], \quad \text{where} \quad G[a] = \int dx\, g(x)a(x). \tag{9.28}$$

The double derivative on the left-hand side removes 2 factors of a from $F[a]$. The functional on the right contains a single factor of a. One factor of a must remain in F after we remove two, thus $F[a]$ must be cubic in a. We could try

$$F[a] = \int dx\, f(x)a^3(x),$$

but we would find that this leads to divergences when we apply eq.(9.28). In particular, the second derivative with respect to $a(y)$ would give $\delta(0)$. In general, to avoid divergences, we must "spread out" the 3 factors of a and write

$$F[a] = \int dx\, dy\, dz\, f(x,y,z)a(x)a(y)a(z). \tag{9.29}$$

Applying eq.(9.28) we find that $f(x,y,z)$ must satisfy

$$\mu g(x) = 2 \int dy\, f(y,y,x) + f(y,x,y) + f(x,y,y). \tag{9.30}$$

While this functional differential equation seems relatively simple to solve, the work will be in finding a $f(x,y,z)$ that satisfies the integral equation (9.30) for a given $g(x)$.

So far, the arbitrary functions, f, that appear in the functional solutions to the differential equations satisfy integral equations. We can change that by complicating the functional differential equation with the addition of differential operators on space-time. For example, consider a slight variation of eq.(9.28).

$$\int dy \frac{\delta}{\delta a(y)} (\partial_y^2) \frac{\delta}{\delta a(y)} F[a] = \mu \int dx g(x)a(x) \tag{9.31}$$

The same power counting will work here as in eq.(9.28). Thus, eq.(9.29) will be a solution if $f(x, y, z)$ satisfies the integro-differential equation

$$\mu g(z) = \int dy \frac{\partial^2}{\partial y^2} f(y,t,z)\Big|_{t=y} + \int dy \frac{\partial^2}{\partial y^2} f(y,z,t)\Big|_{t=y} + \cdots + \int dy \frac{\partial^2}{\partial y^2} f(t,z,y)\Big|_{t=y}. \tag{9.32}$$

We noted that to avoid divergences, we should not place more than 1 factor of $a(x)$ at the same space-time point in the functional. However, field theory before normal ordering and renormalization is riddled with divergences, so maybe, functionals with divergences and coincident factors of $a(x)$ are not so bad or unavoidable. For example, consider eq.(9.23) again and assume that $f(x, y)$ is diagonal, that is, that $f(x, y) = h(x)\delta(x - y)$. In fact, this situation arises in field theory. To satisfy eq.(9.27), $g(x, y)$ must also be proportional to the δ-function. But this implies that μ, eq.(9.27), must diverge. The divergence is $\delta(0)$ which can be interpreted as the volume of the system. Note that this is the same type of divergence we encountered in free field theory when we computed the vacuum expectation of the Hamiltonian before it was normal ordered.

Having 2 factors of $a(x)$ in a functional doesn't necessarily lead to divergences, however. Suppose that a is now a vector function, $\vec{a}(\vec{x})$. One solution of the equation

$$\vec{a}(\vec{y}) \times \frac{\delta}{\delta \vec{a}(\vec{y})} F[a] = 0 \tag{9.33}$$

is

$$F[a] = \int d^3x \, \vec{a}(\vec{x}) \cdot \vec{a}(\vec{x}). \tag{9.34}$$

As a final example, let us combine the vector nature of $\vec{a}(\vec{x})$ with a differential operator on space-time in a geometric way.

$$\left(\nabla \times \frac{\delta}{\delta \vec{a}(\vec{y})}\right) F[a] = 0 \tag{9.35}$$

The zero on the right-hand side makes the power counting less unique. For example,

$$F[a] = \int d^3x \, \vec{f}(\vec{x}) \cdot \vec{a}(\vec{x}) \tag{9.36}$$

satisfies eq.(9.35) if $\vec{f}$ is irrotational, $\nabla \times \vec{f} = 0$. On the other hand,

$$F[a] = \int d^3x \, d^3y \, f(\vec{x}, \vec{y})(\nabla \cdot \vec{a}(\vec{x}))(\nabla \cdot \vec{a}(\vec{y})) \tag{9.37}$$

also satisfies eq.(9.35). Here, f is arbitrary due to the presence of the divergence operator in the functional.

9.3 FUNCTIONAL INTEGRALS

Bosonic Variables

Functional integrals are integrals of functionals over function spaces and, in general, are not well-defined. There are only 2 types that we can do exactly, so a "CRC" table of functional integrals is not very long.

The first type of functional that we can integrate exactly is the δ-functional, $\delta[a-\xi]$. In fact,

$$\int \mathcal{D}a\, \delta[a-\xi] = 1. \tag{9.38}$$

The fancy $\mathcal{D}$ in the measure indicates that the integral is over the function space containing $a(\vec{x})$. We think of $\mathcal{D}a$ as the infinite product,

$$\mathcal{D}a = \prod_{\vec{x}} da(\vec{x}). \tag{9.39}$$

The δ-functional may also be thought of as an infinite product,

$$\delta[a-\xi] = \prod_{\vec{x}} \delta(a(\vec{x}) - \xi(\vec{x})), \tag{9.40}$$

thus the integral in eq.(9.38) is an infinite product of independent integrals, one for each $\vec{x}$. Using this same reasoning, we can see that if $F[a]$ is an arbitrary functional, then

$$\begin{aligned}\int \mathcal{D}a\, F[a]\, \delta[a-\xi] &\approx \int \prod_{\vec{x}} da(\vec{x})\, \delta(a(\vec{x}) - \xi(\vec{x}))\, F[a] \\ &= \prod_{\vec{x}} \int da(\vec{x})\, \delta(a(\vec{x}) - \xi(\vec{x})) F[a] \\ &= F[\xi].\end{aligned} \tag{9.41}$$

Often it helps to think of the functions, such as $a(x)$, as infinite-dimensional (column) vectors where x plays the role of an index. With this view, a functional integral looks like the natural infinite-dimensional limit of ordinary finite-dimensional integrals.

The second type of functional that we can integrate exactly is the Gaussian functional. Consider

$$\int \mathcal{D}a\, e^{-F[a]}, \quad \text{where} \quad F[a] = \int dx\, a^2(x). \tag{9.42}$$

As in the δ-functional case, we can do this integral exactly because it factorizes into an infinite product of independent Gaussian integrals. To do this integral, we need only recall that

$$\int_{-\infty}^{\infty} dx\, e^{-\alpha x} = \sqrt{\frac{\pi}{\alpha}}. \tag{9.43}$$

We have

$$\begin{aligned}\int \mathcal{D}a\, e^{-\int dx\, a^2(x)} &\to \int \prod_x da(x) e^{-\sum_x a^2(x)} \\ &= \int \prod_x da(x) \prod_x e^{-a^2(x)} \\ &= \prod_x \int da(x)\, e^{-a^2(x)} \\ &= \prod_x \sqrt{\pi},\end{aligned} \tag{9.44}$$

which is divergent. Following the same steps we have

$$\int \mathcal{D}a\, e^{-\int dx\, \alpha\, a^2(x)} = \prod_x \sqrt{\frac{\pi}{\alpha}} \tag{9.45}$$

or, more importantly,

$$\int \mathcal{D}a\, e^{-\int dx\, f(x)\, a^2(x)} = \prod_x \sqrt{\frac{\pi}{f(x)}}. \tag{9.46}$$

If we return to thinking about $a(x)$ as an infinite-dimensional vector, then we may think of $f(x)$ as an infinite-dimensional diagonal matrix,

$$f(x,y) = f(x)\,\delta(x-y).$$

$f(x)$ represents the diagonal elements of the diagonal matrix. They are also the eigenvalues of the matrix. The product of the eigenvalues is the determinant of the matrix, hence

$$\prod_x \sqrt{\frac{\pi}{f(x)}} \to (\sqrt{\pi})^{\infty} \prod_x \frac{1}{\sqrt{f(x)}} = \frac{(\sqrt{\pi})^{\infty}}{\sqrt{\det f}} = \det{}^{-\frac{1}{2}}\left(\frac{f}{\pi}\right),$$

so

$$\int \mathcal{D}a\, e^{-\int dx\, f(x)\, a^2(x)} = \frac{(\sqrt{\pi})^{\infty}}{\sqrt{\det f}} = \det{}^{-\frac{1}{2}}\left(\frac{f}{\pi}\right). \tag{9.47}$$

In a sense, we obtain eq.(9.47) by boldly taking the infinite-dimensional limit of the finite-dimensional result. The potentially bothersome infinite number in the numerator is often absorbed in or canceled by a normalization. Ultimately, we are actually just defining the functional Gaussian integral to be the determinant to the $-1/2$ power of the operator (i.e. matrix) sandwiched between the two factors of $a(x)$.

If the operator is not diagonal, the result is still the same.

$$\int \mathcal{D}a\, e^{-\int dx\, dy\, a(x)g(x,y)a(y)} = \frac{(\sqrt{\pi})^\infty}{\sqrt{\det g}} \tag{9.48}$$

We may justify the result, eq.(9.48), by considering the finite-dimensional case (for example, make space-time a finite lattice) and taking the naive limit. In the finite-dimensional case, a similarity transformation that diagonalizes the matrix g preserves the determinant and turns the integral into the form (9.47). In addition, the basis that diagonalizes g is linearly related to $a(x)$ through the similarity transformation, thus the jacobian of the change of variables is unity.

Another way to justify the result, eq.(9.48), is to Fourier transform the functional. For a translationally invariant g, the Fourier transform diagonalizes the Gaussian. Let $\tilde{a}(k)$ be the Fourier transform of $a(x)$. Again, the two are linearly related so the jacobian of the transformation will be 1. Under the Fourier transform,

$$\int dx\, dy\, a(x)g(x-y)a(y) \to \int dk\, \tilde{g}(k)\tilde{a}^2(k)$$

$$\int \mathcal{D}a \exp\Big(-\int dx\, dy\, a(x)g(x-y)a(y)\Big) \to \int \mathcal{D}\tilde{a} \exp\left(-dk\, \tilde{g}(k)\tilde{a}^2(k)\right),$$

which yields $(\sqrt{\det \tilde{g}})^{-1}$, the Fourier transform of $\det g$ to the $-1/2$ power.

Finally, consider

$$\int \mathcal{D}a\, e^{F[a,J]}, \tag{9.49}$$

where

$$F[a,J] = \int dx\, dy\, a(x)g(x-y)a(y) + \int dx\, J(x)a(x)$$

is the Gaussian functional with the addition of a source term. To do this integral, we complete the squares in the exponential. For the 1-dimensional case,

$$\int_{-\infty}^{\infty} dx\, e^{-\alpha x^2+bx} = \int_{-\infty}^{\infty} dx\, e^{-\alpha(x-b/2)^2}\, e^{1/4(b\frac{1}{\alpha}b)} = e^{1/4(b\frac{1}{\alpha}b)}\int dx\, e^{-\alpha x^2}$$
$$= e^{1/4(b\frac{1}{\alpha}b)}\sqrt{\frac{\pi}{\alpha}}. \tag{9.50}$$

Thus,

$$\int \mathcal{D}a\, e^{-F[a,J]} = \exp\left(\frac{1}{4}\int dx\, dy\, J(x)g^{-1}(x-y)J(y)\right)\frac{(\sqrt{\pi})^{\infty}}{\sqrt{\det g}}, \quad (9.51)$$

where $g^{-1}(x-y)$ is the inverse operator (matrix) of $g(x-y)$,

$$\int dz\, g^{-1}(x-z)g(z-y) = \delta(x-y). \quad (9.52)$$

The source term in the Gaussian allows us to easily compute the functional integral of any moment of the Gaussian. Namely,

$$\int \mathcal{D}a\, a(x_1)\cdots a(x_n)\exp\left(-\int dx\, dy\, a(x)g(x,y)a(y)\right) = \frac{\delta}{\delta J(x_1)}\cdots\frac{\delta}{\delta J(x_n)}\int \mathcal{D}a\, e^{-F[a,J]}\bigg|_{J=0}. \quad (9.53)$$

Using eq.(9.51) we have

$$\int \mathcal{D}a\, a(x_1)\exp\left(-\int dx\, dy\, a(x)g(x,y)a(y)\right) = 0 \quad (9.54)$$

and

$$\int \mathcal{D}a\, a(x_1)a(x_2)\exp\left(-\int dx\, dy\, a(x)g(x,y)a(y)\right) = \frac{1}{4}g^{-1}(x_1-x_2)\frac{(\sqrt{\pi})^{\infty}}{\sqrt{\det g}}. \quad (9.55)$$

In fact, any odd moment vanishes.

Fermionic Variables

So far we have been working with ordinary commuting numbers and functions. Since fermionic operators anticommute, we will have much difficulty using commuting variables in a coordinate representation of them. To represent fermionic integrals, we must introduce anticommuting numbers and functions, which are called "Grassmann variables."

Let η be such an anticommuting variable. This means that η even anticommutes with itself,

$$\{\eta, \eta\} = 0,$$

thus $\eta^2 = 0$. If x is an ordinary, commuting variable, then the derivative operator, d/dx, satisfies

$$\left[\frac{d}{dx}, x\right] = 1.$$

Let us define the derivative operator for an anticommuting variable in an analogous way,

$$\left\{\frac{d}{d\eta}, \eta\right\} = 1. \tag{9.56}$$

For any $f(\eta)$, a function of η, the power series expansion of f must be of the form

$$f(\eta) = a + b\eta, \tag{9.57}$$

since $\eta^2 = 0$. Therefore, $d^2 f/d\eta^2 = 0$, so

$$\left\{\frac{d}{d\eta}, \frac{d}{d\eta}\right\} = 0. \tag{9.58}$$

From the practical (computational) standpoint, $d/d\eta$ may be treated the same as d/dx; we must only be careful of the ordering of anticommuting variables. For example, let a and b in the series expansion of f, eq.(9.57), be ordinary commuting numbers. Then

$$\frac{df}{d\eta} = b. \tag{9.59}$$

However, if a and b are also anticommuting numbers, then

$$\frac{df}{d\eta} = -b. \tag{9.60}$$

The minus sign in the latter case arises because $d/d\eta$ and b anticommute. We will pick up the minus sign when we bring $d/d\eta$ from the left to the right of b.

For anticommuting numbers, we must introduce integration by defining the results of 2 integrals:

$$\begin{aligned} \int d\eta &\equiv 0, \\ \int d\eta\, \eta &\equiv 1. \end{aligned} \tag{9.61}$$

This definition is chosen so that $\int d\eta$ is translationally invariant and linear. It is not the inverse operation of $d/d\eta$; in fact, integration acts just like differentiation. Once again, the anticommuting nature of the variables forces us to keep track of the order of the variables. For example, if η_1 and η_2 are two anticommuting variables, then

$$\begin{aligned}\int d\eta_1\, d\eta_2\, \eta_1\, \eta_2 &= -\int d\eta_1\, d\eta_2 \eta_2\, \eta_1 \\ &= -\int d\eta_2\, d\eta_1\, \eta_1\, \eta_2 \\ &= -1.\end{aligned} \tag{9.62}$$

Let's consider the infinite-dimensional limit. Now, our function space that we want to consider is a space of anticommuting functions, $\eta(x)$. The functional derivative is defined in the same way as in eq.(9.56), namely

$$\left\{\frac{\delta}{\delta\eta(x)}, \eta(y)\right\} = \delta(x-y). \tag{9.63}$$

We can also do the same functional integrals as in the bosonic case. From the definition of integration, eq.(9.61), we see that the δ-functional, $\delta[\eta]$, is

$$\delta[\eta] = \prod_x \eta(x), \tag{9.64}$$

so that

$$\int \mathcal{D}\eta\, \delta[\eta] = 1. \tag{9.65}$$

Once again, by $\mathcal{D}\eta$ we mean $\prod_x d\eta(x)$, but we choose the ordering of the $d\eta(x)$'s to be the opposite of the ordering in the functional so that no extra minus signs are introduced in doing the integral. Similarly,

$$\delta[\eta - \sigma] = \prod_x (\eta(x) - \sigma(x)). \tag{9.66}$$

Then,

$$\begin{aligned}\int \mathcal{D}\eta\, F[\eta]\, \delta[\eta - \sigma] &= \int \prod_x d\eta(x)\, F[\eta] \prod_x (\eta(x) - \sigma(x)) \\ &= F[\sigma].\end{aligned} \tag{9.67}$$

This δ-functional works as follows: If $F[\eta]$ depends on $\eta(x_0)$ at the particular point x_0, then $\eta(x_0)$ will appear in some term in the power series expansion of $F[\eta]$. Denote this generic term by $a_n \eta(x_0) \cdots \eta(x_{n-1})$. Then the integral over $\eta(x_0)$ in the product above yields

$$\int d\eta(x_0)(\eta(x_0) - \sigma(x_0))a_n\,\eta(x_0)\cdots\eta(x_{n-1})$$

$$= \int d\eta(x_0)\,(-\sigma(x_0))a_n\,\eta(x_0)\cdots\eta(x_{n-1})$$

$$= \int d\eta(x_0)\eta(x_0)a_n\sigma(x_0)\eta(x_1)\cdots\eta(x_{n-1})$$

$$= a_n\sigma(x_0)\eta(x_1)\cdots\eta(x_{n-1}).$$

If $\eta(x_0)$ is not in the functional, then the integral over $\eta(x_0)$ is 1 and $\sigma(x_0)$ disappears.

$$\int d\eta(x_0)(\eta(x_0) - \sigma(x_0))a_n\eta(x_1)\cdots\eta(x_n) = a_n\eta(x_1)\cdots\eta(x_n)$$

In this way, each $\eta(x_i)$ in $F[\eta]$ is systematically replaced by $\sigma(x_i)$. Notice that it does not matter which side of $\delta[\eta - \sigma]$ we place $F[\eta]$ on in the integral. To restore the order of the η's and σ's in the functional after each integral is done we must anticommute variables an odd number of times in both cases.

Now consider the Gaussian integral,

$$\int \mathcal{D}\eta\,\exp\left(-\int dx\,dy\,\eta(x)g(x,y)\eta(y)\right). \tag{9.68}$$

$g(x,y)$ must be antisymmetric, $g(y,x) = -g(x,y)$, in order to get a non-vanishing result. To do this integral, consider the finite-dimensional case first. For the 2-dimensional integral we have η_1 and η_2 so (9.68) reduces to

$$\int d\eta_1\,d\eta_2\,e^{-\eta_1\,g_{12}\,\eta_2 - \eta_2\,g_{21}\,\eta_1} = \int d\eta_2 d\eta_1 e^{-2\eta_1\,g_{12}\,\eta_2}, \tag{9.69}$$

where the right-hand side follows because g is antisymmetric, $\eta_2\,g_{21}\,\eta_1 = -\eta_1\,g_{21}\,\eta_2 = \eta_1\,g_{12}\,\eta_2$. Expand the exponential in a power series,

$$\begin{aligned} e^{-2\eta_1\,g_{12}\,\eta_2} &= 1 - 2\eta_1\,g_{12}\,\eta_2 + 2(\eta_1\,g_{12}\,\eta_2)^2 + \ldots \\ &= 1 - 2\eta_1\,g_{12}\,\eta_2. \end{aligned} \tag{9.70}$$

The series terminates because $\eta_1^2 = \eta_2^2 = 0$. Using the definition, (9.61), the integral is easy to do.

$$\int d\eta_1\,d\eta_2\,e^{-2\eta_1\,g_{12}\,\eta_2} = 2g_{12} \tag{9.71}$$

For an antisymmetric 2×2 matrix, $\det g = -g_{12}\, g_{21} = (g_{12})^2$, thus we can rewrite eq.(9.71) in a form that suggests generalization to higher dimensions,

$$\int d\eta_1\, d\eta_2\, e^{-2\eta_1\, g_{12}\, \eta_2} = 2\sqrt{\det g}. \tag{9.72}$$

To see if this is the correct generalization, let's check higher dimensions. For the 3-dimensional case, the integral we want to compute is

$$\int d\eta_1\, d\eta_2\, d\eta_3\, e^{-2(\eta_1\, g_{12}\, \eta_2 + \eta_1\, g_{13}\, \eta_3 + \eta_2\, g_{23}\, \eta_3)}. \tag{9.73}$$

Again, we expand the exponential,

$$e^{-2(\eta_1\, g_{12} \cdots g_{23} \eta_3)} = 1 - 2(\eta_1\, g_{12}\, \eta_2 + \eta_1\, g_{13}\, \eta_3 + \eta_2\, g_{23}\, \eta_3).$$

There is no $\eta_1\, \eta_2\, \eta_3$ term in the expansion so the integral, eq.(9.73), must vanish. In fact, after a little inspection, we can see that all of the odd-dimensional integrals vanish.

Now consider the 4-dimensional case. The expansion of the exponential has 3 terms:

$$\exp(-2(\eta_1 g_{12} \eta_2 + \cdots + \eta_3 g_{34} \eta_4)) =$$

$$1 - 2(\eta_1\, g_{12}\, \eta_2 + \eta_1\, g_{13}\, \eta_3 + \eta_1\, g_{14}\, \eta_4 + \eta_2\, g_{23}\, \eta_3 + \eta_2\, g_{24}\, \eta_4 + \eta_3\, g_{34}\, \eta_4)$$

$$+ \frac{2^2}{2!}(\eta_1\, g_{12}\, \eta_2 + \cdots + \eta_3\, g_{34}\, \eta_4)^2.$$

Only the terms proportional to $\eta_1\, \eta_2\, \eta_3\, \eta_4$ will survive the integration and these arise above only in the last term. The result is

$$\int d\eta_1 \cdots d\eta_4 e^{-2(\eta_1\, g_{12}\, \eta_2 + \cdots + \eta_3\, g_{34}\, \eta_4)}$$

$$= \frac{2^3}{2!}(g_{12}\, g_{34} - g_{13}\, g_{24} + g_{14}\, g_{23}) = 4\sqrt{\det g}. \tag{9.74}$$

From the two cases that we have done, we may expect that the factor in front of $\sqrt{\det g}$ is $(2)^{N/2}$ for the N-dimensional integral (N even). This expectation is born out by calculation or induction.

$$\int d\eta_1 \cdots d\eta_N\, e^{-2(\eta_1\, g_{12}\, \eta_2 + \cdots + \eta_{N-1}\, g_{N-1\, N}\, \eta_N)} = (2)^{N/2}\sqrt{\det g} \tag{9.75}$$

We boldly take the $N \to \infty$ limit and say that

$$\int \mathcal{D}\eta \exp\left(-\int dx\, dy\, \eta(x) g(x,y) \eta(y)\right) = (\sqrt{2})^{\infty} \sqrt{\det g}. \qquad (9.76)$$

We do not always have to take the exact form of the number in front of the square root of the determinant of g seriously. It is a number, and not a function, and in many cases can be normalized away.

It is interesting to compare eq.(9.76) with its bosonic counterpart, eq. (9.48). In summary, bosonic Gaussian integrals lead to determinants of operators to the $-1/2$ power while fermionic Gaussians lead to determinants to the 1/2 power.

Often in field theory we must consider the fields to be complex. The generalization of bosonic Gaussian functional integrals to complex-valued functions is straightforward and yields no surprises. Let $a(x)$ be an ordinary complex function with $a = (a_1(x) + ia_2(x))/\sqrt{2}$ where $a_1(x)$ and $a_2(x)$ are real-valued. If we switch variables from (a^*, a) to (a_1, a_2), the Gaussian integral,

$$\int \mathcal{D}a^* \mathcal{D}a \exp\left(-\int dx\, dy\, a^*(x) f(x,y) a(y)\right),$$

where $f(x, y)$ is symmetric, can be seen to be 2 copies of eq.(9.48). Thus,

$$\int \mathcal{D}a^* \mathcal{D}a \exp\left(-\int dx\, dy\, a^*(x) f(x,y) a(y)\right) = \frac{(i\pi)^{\infty}}{\det f}. \qquad (9.77)$$

The fermionic Gaussian functional integral over complex anticommuting functions requires a little more attention. Let η be a complex anticommuting variable and write $\eta = (\xi_1 + i\xi_2)/\sqrt{2}$ where ξ_1 and ξ_2 are real-valued anticommuting variables. Then, $\eta^* = (\xi_1 - i\xi_2)/\sqrt{2}$. As in the real-valued case, we want to define integration that is linear and translationally invariant. Let

$$\int d\eta\, \eta = 1, \qquad \int d\eta^*\, \eta^* = 1,$$

$$\int d\eta = \int d\eta^* = \int d\eta\, \eta^* = \int d\eta^*\, \eta = 0. \qquad (9.78)$$

This means that $d\eta = (d\xi_1 - id\xi_2)/\sqrt{2}$ and $d\eta^* = (d\xi_1 + id\xi_2)/\sqrt{2}$, and that $d\eta^* d\eta = -id\xi_1 d\xi_2$.

Using the integration rules, eq.(9.78), the 1-dimensional Gaussian integral is simply (m is an ordinary commuting number)

$$\int d\eta^* d\eta \, \exp(-\eta^* m\eta) = \int d\eta^* d\eta \, (1 - \eta^* m\eta) \\ = m. \tag{9.79}$$

Compare this with eq.(9.71) and eq.(9.72). The 2-dimensional Gaussian integral is

$$\begin{aligned} &\int d\eta_1^* d\eta_2^* d\eta_1 d\eta_2 \, \exp\Big(-\sum_{i,j=1}^{2} \eta_i^* m_{ij}\eta_j\Big) \\ &= \frac{1}{2!}\int d\eta_1^* d\eta_2^* d\eta_1 d\eta_2 \, (\eta_1^* m_{11}\eta_1 + \eta_1^* m_{12}\eta_2 + \eta_2^* m_{21}\eta_1 + \eta_2^* m_{22}\eta_2)^2 \\ &= -\int d\eta_1^* d\eta_2^* d\eta_1 d\eta_2 \Big(\eta_2^*\eta_1^*\eta_2\eta_1 (m_{11}m_{22} - m_{12}m_{21})\Big) \\ &= -\text{det} m_{ij}. \end{aligned} \tag{9.80}$$

Compare this to eq.(9.76). Eq.(9.79) can also be written as a determinant (of a 1×1 matrix), so it is easy to generalize, except for the annoying sign difference between eqs.(9.79) and (9.80). Will the 3-dimensional Gaussian be equal to $\det m_{ij}$, $(i,j = 1,2,3)$, or $-\det m_{ij}$?

$$\begin{aligned} &\int d\eta_1^* d\eta_2^* d\eta_3^* d\eta_1 d\eta_2 d\eta_3 \, \exp\Big(-\sum_{i,j=1}^{3} \eta_i^* m_{ij}\eta_j\Big) \\ &\qquad = -\frac{1}{3!}\int d\eta_1^* d\eta_2^* d\eta_3^* d\eta_1 d\eta_2 d\eta_3 \Big(\sum_{i,j=1}^{3} \eta_i^* m_{ij}\eta_j\Big)^3 \\ &\qquad = -\det m_{ij}. \end{aligned} \tag{9.81}$$

Well, maybe the 1-dimensional integral is a special case and all of the higher-dimensional complex fermionic Gaussians yield $-\det m_{ij}$? No, the 4-dimensional integral is $+\det m_{ij}$ and so is the 5-dimensional integral. However, the 6-dimensional integral is back to $-\det m_{ij}$. The general N-dimensional result is

$$\int d\eta_1^* \cdots d\eta_N^* \, d\eta_1 \cdots d\eta_N \, \exp\Big(-\sum_{i,j=1}^{N} \eta_i^* m_{ij}\eta_j\Big) = (-1)^{\frac{N(N+3)}{2}} \det m_{ij}. \tag{9.82}$$

The complex fermionic Gaussian functional integral,

$$\int \mathcal{D}\eta^* \mathcal{D}\eta \exp\left(-\int dx\, dy\, \eta^*(x) g(x,y) \eta(y)\right), \qquad (9.83)$$

is the $N \to \infty$ limit of eq.(9.82). However, eq.(9.82) does not possess a well-defined limit because of the sign ambiguity. We can try to ignore the overall sign problem by throwing it somehow into some normalization (for example, of a wave functional or path integral) and forget about it. However, this will not always work because the constructive definition of the fermionic path integral requires us to compute a general N-dimensional integral. To get the overall sign correct, which is often necessary, we must keep track of all of the factors of (-1).

The solution to the sign ambiguity problem is to realize that the factor of $(-1)^{N(N+3)/2}$ corresponds to a certain ordering of $d\eta_1 \cdots d\eta_N^*$. If a rule can be found for writing down $d\eta_1 \cdots d\eta_N^*$ in the proper order for the general N case, then we can absorb the sign ambiguity by requiring the Gaussian integral to be defined with this ordering.

The origin of the sign ambiguity is that the exponential defines one ordering of $\eta_j^* \eta_i$ while the $d\eta_1^* \cdots d\eta_N^* d\eta_1 \cdots d\eta_N$ defines another. Let's decide that we want the overall sign to be positive. This means that the diagonal term of $\det m_{ij}$, $(m_{11} m_{22} \cdots m_{NN})$, should integrate to a plus sign. One positive ordering of this term is $\eta_N^* \eta_N \cdots \eta_1^* \eta_1$. Therefore, we should order $d\eta_1^* \cdots d\eta_N$ as $d\eta_1^* d\eta_1 d\eta_2^* d\eta_2 \cdots d\eta_N^* d\eta_N$. Then we may say

$$\int \mathcal{D}\eta^* \mathcal{D}\eta \exp\left(-\int dx\, dy\, \eta^*(x) g(x,y) \eta(y)\right) = \det g, \qquad (9.84)$$

where $\mathcal{D}\eta^* \mathcal{D}\eta = \prod_x (d\eta^*(x) d\eta(x))$. Compare this to the bosonic result, eq.(9.77).

The "hyperbolic" fermionic Gaussian functional integral,

$$\int \mathcal{D}\eta^* \mathcal{D}\eta \exp\left(+\int dx\, dy\, \eta^*(x) g(x,y) \eta(y)\right),$$

has no well-defined bosonic analogue. In order to get a well-defined limit, we must reverse the order of $d\eta_i^* d\eta_i$. Thus,

$$\int \mathcal{D}\eta^* \mathcal{D}\eta \exp\left(+\int dx\, dy\, \eta^*(x) g(x,y) \eta(y)\right) = \det g, \qquad (9.85)$$

provided that $\mathcal{D}\eta^* \mathcal{D}\eta = \prod_x (d\eta(x) d\eta^*(x))$.

9.4 EXERCISES

1. Show that eq.(9.11) is equivalent to eq.(7.62).
2. Compute

$$\int dx \frac{\delta^2}{\delta\phi^2(x)}\left(\exp\left(-\int dy\, dz\, \phi^2(y)g(y,z)\phi(z)\right.\right.$$
$$\left.\left.+\int dy\, dz\, ds\, \phi(y)f(y,z)\phi(z)h(z,s)\phi(s)\right)\right).$$

3. Compute the first 5 terms of a functional Taylor expansion of

$$F[A] = \int dx\, dy\, \partial_x A(x)\partial_y A(y) + A^2(x)h(x,y)A^2(y),$$

about the function $A_c(z)$.

4. Show that

$$\left[\exp\left(i\int dx G[-i\delta/\delta J(x)]\right), J(y)\right]$$
$$= \frac{\delta G}{\delta J}[-i\delta/\delta J(y)]\exp\left(i\int dx G[-i\delta/\delta J(x)]\right).$$

5. Solve the functional differential equation,

$$\int dx \frac{\delta}{\delta\phi(x)}\,\phi(x)\,\Psi_n[\phi] = E_n\Psi_n[\phi].$$

6. Find a solution to the functional differential equation,

$$\int dx\, -\left(\frac{\delta^2}{\delta\phi^2(x)} + h(x)\phi^2(x) + J(x)\phi(x)\right)\Psi_n[\phi] = E_n\Psi_n[\phi].$$

7. Compute the fermionic Gaussian functional integral,

$$\int \mathcal{D}\eta^*\,\mathcal{D}\eta\,\exp\left(-\int dx\, dy\, \eta^*(x)g(x,y)\eta(y)\right.$$
$$\left.+\int dx\, \chi^*(x)\eta(x) + \eta^*(x)\chi(x)\right).$$

CHAPTER **10**

Free Fields in the Schrödinger Representation

The Schrödinger representation for field theory is a natural extension of nonrelativistic quantum mechanics familiar from atomic physics. For ordinary quantum mechanics, we start with a Hamiltonian operator and canonically quantize by postulating commutation relations between coordinate or position operators and their conjugate momenta. We reach the coordinate Schrödinger representation by representing the diagonal position operator with its eigenvalues, and use a differential representation of the commutators by replacing the conjugate momenta with derivatives. Coordinate representations of state vectors are called wave functions. The Schrödinger equation becomes a differential equation whose solutions, the eigenfunctions of the Hamiltonian differential operator, represent possible states of the system. In the momentum Schrödinger representation, the momentum operators are assumed to be diagonal and the position operators are replaced by derivatives.

For field theory in the Schrödinger representation, we must only substitute the word "functional" for "function." Differential representations of the canonical commutators are obtained by replacing conjugate momenta with functional derivatives. Coordinate representations of state vectors or elements of Fock space are wave functionals. The Schrödinger equation is a functional differential equation whose solutions, the eigenfunctionals of the Hamiltonian functional differential operator, represent possible states of the system.

We will begin exploring the Schrödinger representation for field theory by recasting the three free field theories into this representation. From this new viewpoint we will obtain the same results as in the operator representation used in the previous chapters.

10.1 FREE SCALAR FIELD THEORY

Recall from chapter 3 that the action for free scalar field theory,

$$\int d^4x\, \mathcal{L}(x) = \frac{1}{2}\int d^4x\, (\partial^\mu\varphi\partial_\mu\varphi - m^2\varphi^2), \tag{10.1}$$

leads to the conjugate field momentum

$$\pi(x) = \frac{\partial\mathcal{L}}{\partial(\partial_t\varphi)} = \dot{\varphi}(x), \tag{10.2}$$

and Hamiltonian,

$$H = \frac{1}{2}\int d^3x\, \pi^2 + |\nabla\varphi|^2 + m^2\varphi^2. \tag{10.3}$$

The operators $\varphi(x)$ and $\pi(x)$ satisfy canonical equal-time commutators,

$$\begin{aligned} &[\varphi(\vec{x},t), \pi(\vec{y},t)] = i\,\delta(\vec{x}-\vec{y}) \\ &[\varphi(\vec{x},t), \varphi(\vec{y},t)] = [\pi(\vec{x},t), \pi(\vec{y},t)] = 0. \end{aligned} \tag{10.4}$$

We will switch to a coordinate Schrödinger representation and work with a basis for Fock space where the operator $\varphi(\vec{x})$, now time independent, is diagonal. Let $|\,\phi\rangle$ be an eigenstate of φ with eigenvalue ϕ.

$$\varphi(\vec{x})|\,\phi\rangle = \phi(\vec{x})|\,\phi\rangle \tag{10.5}$$

Please note that $\varphi(\vec{x})$ is an operator, while $\phi(\vec{x})$ is just an ordinary scalar function. The coordinate representation of the state $|\,\Psi\rangle$, now time dependent, is the wave functional, $\Psi[\phi]$, a functional of the function $\phi(\vec{x})$.

$$\Psi[\phi] = \langle\phi\,|\,\Psi\rangle \tag{10.6}$$

Recall from the previous chapter that the functional derivative satisfies

$$\frac{\delta}{\delta\phi(\vec{x})}\phi(\vec{y}) = \delta(\vec{x}-\vec{y}), \tag{10.7}$$

or, equivalently,

$$\left[\frac{\delta}{\delta\phi(\vec{x})}, \phi(\vec{y})\right] = \delta(\vec{x}-\vec{y}). \tag{10.8}$$

Therefore,

$$\pi(\vec{x}) = -i\frac{\delta}{\delta\phi(\vec{x})} \tag{10.9}$$

constitutes a functional differential representation of the equal-time commutators, eq.(10.4). In terms of the coordinate basis, $|\phi\rangle$,

$$\langle\phi'|\pi(\vec{x})|\phi\rangle = -i\frac{\delta}{\delta\phi(\vec{x})}\delta[\phi-\phi']. \tag{10.10}$$

The differential representation of the field momentum, eq.(10.9), turns the Hamiltonian operator, eq.(10.3), into a functional differential operator,

$$H = \frac{1}{2}\int d^3x \left(-\frac{\delta^2}{\delta\phi^2(\vec{x})} + |\nabla\phi|^2 + m^2\phi^2\right), \tag{10.11}$$

and the Schrödinger equation,

$$i\frac{\partial}{\partial t}|\Psi\rangle = H|\Psi\rangle,$$

into a functional differential equation,

$$i\frac{\partial}{\partial t}\Psi[\phi,t] = \frac{1}{2}\int d^3x \left(-\frac{\delta^2}{\delta\phi^2(\vec{x})} + |\nabla\phi|^2 + m^2\phi^2\right)\Psi[\phi,t]. \tag{10.12}$$

Since the Hamiltonian does not explicitly depend on time, we may separate out the time dependence of the wave functional $\Psi[\phi,t]$ and write

$$\Psi[\phi,t] = e^{-iEt}\,\Psi[\phi]. \tag{10.13}$$

$\Psi[\phi]$ satisfies the time-independent Schrödinger functional equation,

$$\frac{1}{2}\int d^3x - \frac{\delta^2\Psi[\phi]}{\delta\phi^2(\vec{x})} + (|\nabla\phi|^2 + m^2\phi^2)\Psi[\phi] = E\Psi[\phi]. \tag{10.14}$$

Let's solve eq.(10.14) for the lowest energy eigenfunctional, the ground state or vacuum wave functional, $\Psi_0[\phi]$. The form of this equation is similar to eq.(9.23). Since we are looking for the ground state, we may assume that the functional has no nodes and is positive everywhere. Therefore we can write

$$\Psi_0[\phi] = \eta\,\exp(-G[\phi]). \tag{10.15}$$

The unknown functional $G[\phi]$ must satisfy eq.(10.14),

$$\frac{1}{2}\int d^3x\,\frac{\delta G[\phi]}{\delta\phi(\vec{x})}-\left(\frac{\delta G[\phi]}{\delta\phi(\vec{x})}\right)^2=E_0-\frac{1}{2}\int d^3x\,\phi(\vec{x})(-\nabla^2+m^2)\phi(\vec{x}),\tag{10.16}$$

which is nearly identical in form to eq.(9.25). Power counting tells us that $G[\phi]$ should minimally be quadratic in ϕ so that the second functional derivative of G yields a number while the square of the first derivative is still quadratic in ϕ and can be matched with the right-hand side of eq.(10.16). Therefore, we write $G[\phi]$ as a general quadratic in ϕ,

$$G[\phi]=\int d^3x\,d^3y\,\phi(\vec{x})g(\vec{x},\vec{y})\phi(\vec{y}).\tag{10.17}$$

To simplify things, we will take g to be symmetric in $\vec{x}$ and $\vec{y}$. So far, we have found no reason why this isn't true. Plugging eq.(10.17) back into eq.(10.16) and equating like terms on the left- and right-hand sides, we find that

$$\frac{1}{2}\int d^3x\frac{\delta^2 G[\phi]}{\delta\phi^2(\vec{x})}=\int d^3x\,g(\vec{x},\vec{x})=E_0,\tag{10.18}$$

and

$$\begin{aligned}-\frac{1}{2}\int d^3x\left(\frac{\delta G[\phi]}{\delta\phi(\vec{x})}\right)^2&=-2\int d^3x\,d^3y\,d^3z\,\phi(\vec{z})g(\vec{z},\vec{x})g(\vec{x},\vec{y})\phi(\vec{y})\\&=-\frac{1}{2}\int d^3x\,\phi(\vec{x})(-\nabla^2+m^2)\phi(\vec{x}).\end{aligned}\tag{10.19}$$

Therefore, the unknown function $g(\vec{x},\vec{y})$ must satisfy

$$\int d^3x\,g(\vec{z},\vec{x})\,g(\vec{x},\vec{y})=\frac{1}{4}(-\nabla^2+m^2)\delta(\vec{z}-\vec{y}).\tag{10.20}$$

We observe from the right-hand side of eq.(10.20) that the left-hand side must only depend on $\vec{z}-\vec{y}$. Therefore, we may assume that g is translationally invariant, $g(\vec{x},\vec{y})=g(\vec{x}-\vec{y})$. Then, the left-hand side above is just a convolution which suggests that we should Fourier transform eq.(10.20) to solve it. Let the Fourier transform of $g(\vec{x}-\vec{y})$ be $\tilde{g}(\vec{k})$.

$$g(\vec{x},\vec{y})=\int\frac{d^3k}{(2\pi)^3}\,\tilde{g}(\vec{k})\,e^{i\vec{k}\cdot(\vec{x}-\vec{y})}\tag{10.21}$$

The Fourier transform of eq.(10.20) yields the algebraic equation,

$$\tilde{g}^2(\vec{k})=\frac{1}{4}(\vec{k}^2+m^2)\tag{10.22}$$

Thus, $\tilde{g}(\vec{k}) = \sqrt{\vec{k}^2 + m^2}/2 = \omega_k/2$, and

$$g(\vec{x},\vec{y}) = \frac{1}{2}\int \frac{d^3k}{(2\pi)^3}\,\omega_k\, e^{i\vec{k}\cdot(\vec{x}-\vec{y})}. \tag{10.23}$$

At this point we can determine the ground state energy eigenvalue, E_0, from eq.(10.18).

$$\begin{aligned} E_0 = \int d^3x\, g(\vec{x},\vec{x}) &= \frac{1}{2}\int \frac{d^3k}{(2\pi)^3}\omega_k \int d^3x \\ &= \frac{1}{2}\int d^3k\,\omega_k\,\delta^3(0), \end{aligned} \tag{10.24}$$

which agrees with the result obtained in the operator representation in chapter 3, eq.(3.22).

The normalization, η, in eq.(10.15) is determined by the usual requirement that $\langle 0\,|\,0\rangle = 1$. The scalar product or expectation in this representation is a functional integral.

$$\begin{aligned} 1 = \langle 0\,|\,0\rangle &= \int \mathcal{D}\phi\,\langle 0\,|\,\phi\rangle\langle\phi\,|\,0\rangle \\ &= \int \mathcal{D}\phi \Psi_0^*[\phi]\,\Psi_0[\phi] \\ &= \eta^2 \int \mathcal{D}\phi\,\exp\left(-2\int d^3x\,d^3y\,\phi(\vec{x})g(\vec{x}-\vec{y})\phi(\vec{y})\right) \\ &= \eta^2 \frac{(\sqrt{\pi})^\infty}{\sqrt{\det\sqrt{-\nabla^2+m^2}}} \end{aligned} \tag{10.25}$$

This mysterious-looking result can be made to look very familiar by redoing the normalization computation using the wave functional of the Fourier transform of $\phi(\vec{x})$, $\Psi_0[\tilde{\phi}]$. This functional is obtained from $\Psi_0[\phi]$ by the ordinary composition of function(al)s, $\Psi[\phi[\tilde{\phi}]]$. That is, we replace $\phi(\vec{x})$ in $\Psi_0[\phi]$ with $\int d^3k/(2\pi)^3\,\tilde{\phi}(\vec{k})\,\exp(i\vec{k}\cdot\vec{x})$. The result is

$$\Psi_0[\tilde{\phi}] = \eta\,\exp\left(-\frac{1}{2}\int \frac{d^3k}{(2\pi)^3}\,\omega_k\,\tilde{\phi}(\vec{k})\,\tilde{\phi}(-\vec{k})\right). \tag{10.26}$$

Repeating the normalization calculation, this time with $\Psi_0[\tilde{\phi}]$, we have

$$
\begin{aligned}
1 = \langle 0 \,|\, 0 \rangle &= \int \mathcal{D}\phi\, \Psi_0^*[\phi]\, \Psi_0[\phi] \\
&= \int \mathcal{D}\tilde{\phi}\, \Psi_0^*[\tilde{\phi}]\, \Psi_0[\tilde{\phi}] \\
&= \eta^2 \int \mathcal{D}\tilde{\phi}\, \exp\left(-\int \frac{d^3k}{(2\pi)^3}\, \omega_k\, \tilde{\phi}(\vec{k})\, \tilde{\phi}(-\vec{k}) \right) \\
&= \eta^2 \prod_{\vec{k}} \left(\sqrt{\frac{\pi}{\omega_k}} \right),
\end{aligned}
\tag{10.27}
$$

where

$$
\mathcal{D}\tilde{\phi} = \prod_{\vec{k}} \frac{d\tilde{\phi}(\vec{k})}{\sqrt{(2\pi^3)}}.
$$

Thus,

$$
\begin{aligned}
\Psi_0[\tilde{\phi}] &= \left(\prod_{\vec{k}} \left(\frac{\omega_k}{\pi} \right)^{\frac{1}{4}} \right) \exp\left(-\frac{1}{2} \int \frac{d^3k}{(2\pi)^3}\, \omega_k \tilde{\phi}(\vec{k})\, \tilde{\phi}(-\vec{k}) \right) \\
&= \prod_{\vec{k}} \left(\frac{\omega_k}{\pi} \right)^{\frac{1}{4}} \exp\left(-\frac{1}{2}\, \frac{1}{(2\pi)^3} \omega_k \tilde{\phi}^2(|\vec{k}|) \right)
\end{aligned}
\tag{10.28}
$$

$\Psi_0[\tilde{\phi}]$ is just the infinite product of harmonic oscillator ground state wave functions, one wave function for each $\vec{k}$. This result is not surprising since we know that free field Hamiltonians are just an infinite sum of harmonic oscillator Hamiltonians.

From eq.(10.27) and (10.25) we see that

$$
\det \sqrt{-\nabla^2 + m^2} = \prod_{\vec{k}} \omega_k = \prod_{\vec{k}} \sqrt{\vec{k}^2 + m^2}. \tag{10.29}
$$

We can obtain this result without reference to $\Psi_0[\tilde{\phi}]$ by naively applying properties of determinants of finite-dimensional operators to the infinite-dimensional case. Namely, assume that the determinant is the product of the eigenvalues of an operator, and that $\det g^2 = \det g \cdot \det g$ is true in the infinite-dimensional limit. Then $(\det\sqrt{-\nabla^2 + m^2})^2 = \det(-\nabla^2 + m^2)$. For the operator $-\nabla^2 + m^2$ with periodic boundary conditions, we note that

$$
(-\nabla^2 + m^2) e^{-i\vec{k}\cdot\vec{x}} = (\vec{k}^2 + m^2) e^{-i\vec{k}\cdot\vec{x}}.
$$

Thus, $\exp(i\vec{k}\cdot\vec{x})$ is an eigenfunction with eigenvalue $\vec{k}^2+m^2$. Therefore,

$$\det(-\nabla^2+m^2)=\prod_{\vec{k}}(\vec{k}^2+m^2),$$

and

$$\det\sqrt{-\nabla^2+m^2}=\prod_{\vec{k}}\sqrt{\vec{k}^2+m^2}.$$

What about the excited state wave functionals? We could try to solve eq.(10.14) in general terms, but it is easier to construct these functionals by returning to the form of $\Psi_0[\tilde{\phi}]$ as an infinite product of ordinary oscillator ground state wave functions, eq.(10.28). To create an excited state eigenfunctional, we replace the ground state wave function for one (or many) of the mode oscillators in the product in eq.(10.28) with excited state harmonic oscillator wave functions.

For example, let's put one excitation in mode $\vec{k}_1$. Replace the ground state wave function for mode $\vec{k}_1$ in (10.28) with the wave function for the first excited state.

$$\begin{aligned}\Psi_1[\tilde{\phi}] &= \left(\frac{\omega_{k_1}}{4\pi}\right)^{\frac{1}{4}} 2\left(\frac{\omega_{k_1}}{(2\pi)^3}\right)^{\frac{1}{2}}\tilde{\phi}(\vec{k}_1)\exp\left(-\frac{1}{2}\frac{1}{(2\pi)^3}\omega_{k_1}\tilde{\phi}(\vec{k}_1)\tilde{\phi}(-\vec{k}_1)\right)\\ &\quad\prod_{\vec{k}\neq\vec{k}_1}\left(\frac{\omega_k}{\pi}\right)^{\frac{1}{4}}\exp\left(-\frac{1}{2}\frac{1}{(2\pi)^3}\omega_k\tilde{\phi}(\vec{k})\tilde{\phi}(-\vec{k})\right)\\ &= \left(\frac{2\omega_{k_1}}{(2\pi)^3}\right)^{\frac{1}{2}}\tilde{\phi}(\vec{k}_1)\prod_{\vec{k}}\left(\frac{\omega_k}{\pi}\right)^{\frac{1}{4}}\exp\left(-\frac{1}{2}\frac{1}{(2\pi)^3}\omega_k\tilde{\phi}(\vec{k})\tilde{\phi}(-\vec{k})\right)\\ &= \left(\frac{2\omega_{k_1}}{(2\pi)^3}\right)^{\frac{1}{2}}\tilde{\phi}(\vec{k}_1)\Psi_0[\tilde{\phi}],\end{aligned}$$

or

$$\Psi_1[\phi]=\left(\frac{2\omega_{k_1}}{(2\pi)^3}\right)^{\frac{1}{2}}\int d^3y\, e^{-i\vec{k}_1\cdot\vec{y}}\,\phi(\vec{y})\,\Psi_0[\phi]. \qquad (10.30)$$

To verify that eq.(10.30) is an eigenstate and to explicitly compute its eigenvalue, let's plug it back into the time-independent functional Schrödinger equation, (10.14). To begin we have

$$\frac{\delta\Psi_1[\phi]}{\delta\phi(\vec{x})}=\left(\frac{2\omega_{k_1}}{(2\pi)^3}\right)^{\frac{1}{2}}e^{-\vec{k}_1\cdot\vec{x}}\Psi_0[\phi]+\left(\frac{2\omega_{k_1}}{(2\pi)^3}\right)^{\frac{1}{2}}\int d^3y\, e^{-i\vec{k}_1\cdot\vec{y}}\phi(\vec{y})\frac{\delta\Psi_0[\phi]}{\delta\phi(\vec{x})},$$

and

$$\frac{\delta^2\Psi_1[\phi]}{\delta\phi^2(\vec{x})} = 2\sqrt{\frac{2\omega_{k_1}}{(2\pi)^3}}e^{-i\vec{k}_1\cdot\vec{x}}\frac{\delta\Psi_0[\phi]}{\delta\phi(\vec{x})} + \sqrt{\frac{2\omega_{k_1}}{(2\pi)^3}}\int d^3y\, e^{-i\vec{k}_1\cdot\vec{y}}\phi(\vec{y})\frac{\delta^2\Psi_0[\phi]}{\delta\phi^2(\vec{x})}. \tag{10.31}$$

Using the fact that $\Psi_0[\phi]$ is a solution of eq.(10.14) with eigenvalue E_0, we can rewrite the second term on the right-hand side above, after integrating over $\vec{x}$, as

$$\int d^3x \left(\frac{2\omega_{k_1}}{(2\pi)^3}\right)^{\frac{1}{2}} \int d^3y\, e^{-i\vec{k}_1\cdot\vec{y}}\phi(\vec{y})\frac{\delta^2\Psi_0[\phi]}{\delta\phi^2(\vec{x})}$$
$$= -2E_0\Psi_1[\phi] + \int d^3x(|\nabla\phi|^2 + m^2\phi^2)\Psi_1[\phi]. \tag{10.32}$$

Therefore, substituting from eq.(10.31) and eq.(10.32) into eq.(10.14) we have

$$-\left(\frac{2\omega_{k_1}}{(2\pi)^3}\right)^{\frac{1}{2}} \int d^3x\, e^{-\vec{k}_1\cdot\vec{x}}\frac{\delta\Psi_0[\phi]}{\delta\phi(\vec{x})} = (E_1 - E_0)\Psi_1[\phi]. \tag{10.33}$$

Since

$$\frac{\delta\Psi_0[\phi]}{\delta\phi(\vec{x})} = -2\int d^3y\, g(\vec{x},\vec{y})\,\phi(\vec{y})\,\Psi_0[\phi],$$

the left-hand side of eq.(10.33) is

$$\left(\frac{2\omega_{k_1}}{(2\pi)^3}\right)^{\frac{1}{2}} \int \frac{d^3k}{(2\pi)^3}\, d^3x\, d^3y\, e^{-i\vec{k}_1\cdot\vec{x}}\omega_k e^{\vec{k}\cdot(\vec{x}-\vec{y})}\phi(\vec{y})\Psi_0[\phi], \tag{10.34}$$

where we have used the explicit expression for $g(\vec{x},\vec{y})$, eq.(10.23). The integral over $\vec{x}$ above results in $(2\pi)^3\delta^3(k-k_1)$, which makes the $\vec{k}$ integral easy to do. Eq.(10.34) becomes $\omega_{k_1}\ \Psi_1[\phi]$ so $\Psi_1[\phi]$ is an eigenfunctional with $E_1 - E_0 = \omega_{k_1}$.

The momentum operator, P^i, generates infinitesimal spatial displacements and in ordinary quantum mechanics it is represented by $-i\partial/\partial x_i$. In the operator formalism used in chapter 3, the momentum operator acting on the field operator $\varphi(\vec{x},t)$ should have the same effect, namely

$$[P_j, \varphi(\vec{x},t)] = -i\frac{\partial}{\partial x^j}\varphi(\vec{x},t). \tag{10.35}$$

The operator

$$P_j = -\int d^3x\, \varphi(x)\partial_j\pi(x), \tag{10.36}$$

used in conjunction with the equal-time commutators, eq.(10.4), satisfies eq.(10.35). In the Schrödinger representation, P_j becomes the functional differential operator,

$$P_j = i\int d^3x\,\phi(\vec{x})\partial_j\frac{\delta}{\delta\phi(\vec{x})}. \tag{10.37}$$

Using eq.(10.37), we can also verify that $\Psi_1[\phi]$ is a momentum eigenstate. First of all, $\partial/\partial x_j\, g(\vec{x},\vec{y}) = 0$, so the vacuum carries no momentum as expected, $P_j\Psi_0[\phi] = 0$. Then

$$\begin{aligned} P_j\Psi_1[\phi] &= i\int d^3x\phi(\vec{x})\frac{\partial}{\partial x^j}\left(\frac{2\omega_{k_1}}{(2\pi)^3}\right)^{\frac{1}{2}}\int d^3y\,e^{-i\vec{k}_1\cdot\vec{y}}\delta^3(x-y)\Psi_0[\phi] \\ &= (k_1)_j\Psi_1[\phi]. \end{aligned} \tag{10.38}$$

Since $\Psi_1[\phi]$ is an energy eigenstate with energy ω_{k_1} relative to the vacuum, and is also a momentum eigenstate with momentum $\vec{k}_1$, we can clearly use $\Psi_1[\phi]$ as a state describing 1 particle with 4-momentum k_1 and mass m. Once again, we can see that the harmonic oscillator structure of free field theory provides us with a particle interpretation.

We can formalize the particle interpretation by transforming the creation and destruction operators, $a^\dagger(\vec{k})$ and $a(\vec{k})$, into this new representation. From eq.(3.17) we have

$$\begin{aligned} a(\vec{k}) &= \int d^3x\,e^{-ik\cdot x}(\omega_k\varphi(x) + i\pi(x)) \\ a^\dagger(\vec{k}) &= \int d^3x\,e^{-ik\cdot x}(\omega_k\varphi(x) - i\pi(x)). \end{aligned} \tag{10.39}$$

These are easily converted to the Schrödinger representation using the functional derivative representation of $\pi(\vec{x})$, eq.(10.9).

$$\begin{aligned} a(\vec{k}) &= \int d^3x\,e^{i\vec{k}\cdot\vec{x}}\left(\omega_k\phi(\vec{x}) + \frac{\delta}{\delta\phi(\vec{x})}\right) \\ a^\dagger(\vec{k}) &= \int d^3x\,e^{-i\vec{k}\cdot\vec{x}}\left(\omega_k\phi(\vec{x}) - \frac{\delta}{\delta\phi(\vec{x})}\right) \end{aligned} \tag{10.40}$$

We can now verify explicitly that $a(\vec{k})\Psi_0[\phi] = 0$, and that $\Psi_1[\phi] = a^\dagger(\vec{k}_1)/\sqrt{2\omega_{k_1}(2\pi)^3}\,\Psi_0[\phi]$. The other excited state eigenfunctionals (unnormalized) can be constructed by repeatedly applying $a^\dagger(\vec{k}_i)$, eq.(10.40), to $\Psi_0[\phi]$.

$\Psi_1[\phi]$, eq.(10.30), represents a state with 1 scalar particle of mass m, energy ω_{k_1}, and momentum $\vec{k}_1$. Since the particle is in a definite state of energy-momentum, then by the uncertainty principle we have no idea where the particle is located. This is reflected in $\Psi_1[\phi]$ by the fact that $\Psi_1[\phi]$ has no dependence on any spatial coordinate x_i. Similarly, the state

$$\eta\phi(\vec{x}_1)\Psi_0[\phi] \tag{10.41}$$

represents a state where the particle is located at x_1. The wave functional, eq.(10.41), does not depend on any momentum vector so we don't have any idea what the momentum is. Likewise, the wave functional will not be an eigenfunctional of the Hamiltonian or momentum operator.

Although we don't really use the propagator in this representation, we will compute it for the sake of illustration. Suppose that initially the particle is located at $\vec{x}$ at time t and propagates to $\vec{x}'$ at t'. The initial state wave functional is $\eta\phi(\vec{x})\Psi_0[\phi,t]$, while the final state wave functional is $\eta'\phi(\vec{x}')\Psi_0[\phi,t']$. Since the initial state must occur before the final state, then $t' > t$. The contribution to the propagator for this process is the overlap between the two states, now a functional integral.

$$\langle 0\,|\varphi(x')\varphi(x)|\,0\rangle\theta(t'-t) = \int \mathcal{D}\phi\,\phi(\vec{x}')\phi(\vec{x})\Psi_0^*[\phi,t']\Psi_0[\phi,t]$$

$$= \int \mathcal{D}\tilde{\phi}\,\frac{d^3k}{(2\pi)^3}\frac{d^3k'}{(2\pi)^3}\,e^{-i\vec{k}'\cdot\vec{x}'}\tilde{\phi}(\vec{k}')e^{i\vec{k}\cdot\vec{x}}\tilde{\phi}(\vec{k})\Psi_0^*[\tilde{\phi}]\Psi_0[\tilde{\phi}]e^{i\omega_{k'}t'}\,e^{-i\omega_k t}\theta(t'-t)$$

$$= \eta'\eta\int\frac{d^3k}{(2\pi)^3}\frac{d^3k'}{(2\pi)^3}e^{ik'\cdot x'}e^{-ik\cdot x}$$
$$\int\mathcal{D}\tilde{\phi}\,\tilde{\phi}(\vec{k}')\,\tilde{\phi}(\vec{k})\,\exp\left(-\int\frac{d^3q}{(2\pi)^3}\,\omega_q\tilde{\phi}(\vec{q})\tilde{\phi}(-\vec{q})\right)\theta(t'-t). \tag{10.42}$$

Since odd moments of a Gaussian integral vanish, the functional integral in eq.(10.42) above will only be nonzero if $\vec{k}' = \vec{k}$.

$$\int\mathcal{D}\tilde{\phi}\,\tilde{\phi}(\vec{k}')\,\tilde{\phi}(\vec{k})\,\exp\left(-\int\frac{d^3q}{(2\pi)^3}\,\omega_q\,\tilde{\phi}(\vec{q})\tilde{\phi}(-\vec{q})\right) = \frac{(2\pi)^3}{2\omega_k}\,\delta^3(k'-k)\eta^{-2} \tag{10.43}$$

Eq.(10.42) reduces to

$$\langle 0\,|\varphi(x')\varphi(x)|\,0\rangle\theta(t'-t) = \int\frac{d^3k}{(2\pi)^3}\frac{1}{2\omega_k}\,e^{ik\cdot(x-x')}\theta(t'-t). \tag{10.44}$$

Similarly,

$$\langle 0\,|\varphi(x)\varphi(x')|\,0\rangle\theta(t-t') = \eta\eta' \int \frac{d^3k}{(2\pi)^3}\frac{d^3k'}{(2\pi)^3}\, e^{-ik'\cdot x'} e^{ik\cdot x}$$

$$\int \mathcal{D}\tilde{\phi}\; \tilde{\phi}(\vec{k}')\tilde{\phi}(\vec{k}) \quad \exp\left(-\int \frac{d^3q}{(2\pi)^3}\,\omega_q \tilde{\phi}(\vec{q})\tilde{\phi}(-\vec{q})\right)\theta(t-t')$$

$$= \int \frac{d^3k}{(2\pi)^3}\frac{1}{2\omega_k}\, e^{-ik\cdot(x-x')}\theta(t-t'). \tag{10.45}$$

The sum of eqs.(10.44) and (10.45) is the expectation of the time-ordered pair, $T(\varphi(x')\varphi(x))$, the propagator $i\Delta_F(x'-x)$.

$$i\Delta_F(x'-x) = \int \frac{d^3k}{(2\pi)^3}\frac{1}{2\omega_k}\left(e^{ik\cdot(x'-x)}\theta(t'-t) + e^{-ik\cdot(x'-x)}\theta(t-t')\right) \tag{10.46}$$

This agrees with the propagator obtained in the operator representation in chapter 3.

Let us note that we could equally work in the momentum Schrödinger representation by taking a basis where the field momentum operator, $\pi(x)$, is diagonal. Let $|\,\varpi\rangle$ be an eigenstate of π with eigenvalue ϖ.

$$\pi(\vec{x})|\,\varpi\rangle = \varpi(\vec{x})|\,\varpi\rangle \tag{10.47}$$

While $\pi(\vec{x})$ is an operator, $\varpi(\vec{x})$ is just a function. The ground state wave functional in this Schrödinger representation is

$$\langle \varpi\,|\,0\rangle = \Psi_0[\varpi]. \tag{10.48}$$

A differential representation of the equal-time commutators, eq.(10.4), is now provided by taking

$$\varphi(\vec{x}) = i\frac{\delta}{\delta\varpi(\vec{x})}. \tag{10.49}$$

The time-independent functional Schrödinger equation in the momentum representation follows from eq.(10.49) and eq.(10.3).

$$\frac{1}{2}\int d^3x\left(-\frac{\delta}{\delta\varpi(\vec{x})}(-\nabla^2+m^2)\frac{\delta}{\delta\varpi(\vec{x})} + \varpi^2(\vec{x})\right)\Psi[\varpi] = E\,\Psi[\varpi] \tag{10.50}$$

We can use power counting again to obtain the vacuum functional, $\Psi_0[\varpi]$. The potential term, $\varpi^2(\vec{x})$, suggests a Gaussian functional,

$$\Psi_0[\varpi] = \eta\exp\left(-\int d^3x\,d^3y\,\varpi(\vec{x})\,h(\vec{x},\vec{y})\,\varpi(\vec{y})\right). \tag{10.51}$$

This functional will be a solution of eq.(10.50) provided that

$$\int d^3x\,(-\nabla_x^2 + m^2)h(\vec{x},\vec{y})\Big|_{\vec{y}=\vec{x}} = E_0, \tag{10.52}$$

and

$$4\int d^3x\, h(\vec{z},\vec{x})(-\nabla_x^2 + m^2)h(\vec{x},\vec{y}) = \delta(\vec{z}-\vec{y}). \tag{10.53}$$

We can solve eq.(10.53) by Fourier transforming it. The solution is

$$h(\vec{x},\vec{y}) = \frac{1}{2}\int \frac{d^3k}{(2\pi)^3}\, e^{i\vec{k}\cdot(\vec{x}-\vec{y})}\,\frac{1}{\sqrt{\vec{k}^2+m^2}}. \tag{10.54}$$

Using eq.(10.54) we can compute E_0 from eq.(10.52). Since

$$\begin{aligned}(-\nabla_x^2 + m^2)h(\vec{x},\vec{y}) &= \frac{1}{2}\int \frac{d^3k}{(2\pi)^3} e^{-\vec{k}\cdot(\vec{x}-\vec{y})}\sqrt{\vec{k}^2+m^2}\\ &= g(\vec{x},\vec{y}),\end{aligned}$$

then E_0 will agree with our previous result, eq.(10.24).

As you might expect, $\Psi_0[\tilde{\varpi}]$ can be obtained from $\Psi_0[\tilde{\phi}]$ by Fourier transforming each of the individual mode oscillator ground state wave functions. This, together with the fact that $2h(\vec{x},\vec{y})$ is the inverse of $2g(\vec{x},\vec{y})$, that is, $4\int d^3x\, g(\vec{z},\vec{x})\, h(\vec{x},\vec{y}) = \delta(\vec{z}-\vec{y})$, suggests that we may obtain $\Psi_0[\varpi]$ from $\Psi_0[\phi]$ by a functional Fourier transform,

$$\Psi_0[\varpi] = \int \mathcal{D}\phi\, \exp\left(i\int d^3x\, \varpi(\vec{x})\,\phi(\vec{x})\right)\Psi_0[\phi]. \tag{10.55}$$

Since $\Psi_0[\phi]$ is Gaussian, this integral may be done by "completing the square" as discussed in the previous chapter. We do indeed obtain the same $\Psi_0[\varpi]$ as defined by eqs.(10.51) and (10.54) in this way.

10.2 PHOTON FIELD THEORY

In order to canonically quantize the free electromagnetic field we must choose a gauge. In the operator representation, we quantized in both the Coulomb and Lorentz gauges in chapter 5. The quantization in either gauge requires us to modify the canonical procedure. In the Coulomb gauge we modified the canonical equal-time commutators by replacing the δ-function with the transverse δ-function. When we switch into the coordinate functional Schrödinger representation in this gauge, we will have to take the electric field to be a functional directional derivative in the direction of the transverse δ-function in order to differentially represent the modified commutators. This is an added complication we wish to avoid at the moment. In the Lorentz gauge, we are forced to modify the action and to add a constraint to the states. The Hamiltonian is more complicated, and in addition, we must carry along scalar and longitudinal photons and their wave functionals. At the moment, we also wish to avoid this extra work.

A compromise gauge that we will work in is the temporal gauge, $A_0 = 0$. In this gauge we may use the canonical equal-time commutators, and the Hamiltonian is simple, but we still will need to carry longitudinal photons and a constraint on the states.

In the temporal gauge, the Hamiltonian is

$$H = \frac{1}{2}\int d^3x\, \vec{E}^2 + \vec{B}^2, \tag{10.56}$$

where

$$E_i(x) = -\dot{A}_i(x) \qquad \text{and} \qquad \vec{B}(x) = \nabla \times \vec{A}(x). \tag{10.57}$$

The canonical equal-time commutators are

$$[E_i(\vec{x},t), A_j(\vec{y},t)] = i\delta_{ij}\delta(\vec{x}-\vec{y})$$

$$[A_i(\vec{x},t), A_j(\vec{y},t)] = [E_i(\vec{x},t), E_j(\vec{y},t)] = 0. \tag{10.58}$$

In the coordinate functional Schrödinger representation, the operator $A_i(\vec{x})$ is diagonal. Let $|\,a\rangle$ be an eigenstate of $A_i(\vec{x})$ with eigenvalue (eigenfunction) $a_i(\vec{x})$.

$$A_i(\vec{x})|\,a\rangle = a_i(\vec{x})|\,a\rangle \tag{10.59}$$

A_i is an operator, a_i is a function. In this representation,

$$E_i(\vec{x}) = i\frac{\delta}{\delta a_i(\vec{x})} \tag{10.60}$$

is a differential representation of the commutators (10.58). The state $|\,\Psi\rangle$ is represented by the wave functional $\Psi[a] = \langle a\,|\,\Psi\rangle$. $|\,\Psi\rangle$ is an eigenstate of the Hamiltonian, eq.(10.56), if the wave functional, $\Psi[a]$, is a solution of the time-independent functional Schrödinger equation,

$$\frac{1}{2}\int d^3x \left(-\frac{\delta}{\delta\vec{a}(\vec{x})}\cdot\frac{\delta}{\delta\vec{a}(\vec{x})} + \vec{b}(\vec{x})\cdot\vec{b}(\vec{x})\right)\Psi[a] = E\,\Psi[a], \tag{10.61}$$

where $\vec{b}(\vec{x}) = \nabla\times\vec{a}(\vec{x})$.

In the temporal gauge we have set $A_0 = 0$. This means that there is no Hamiltonian equation of motion corresponding to Gauss's law, $\nabla \cdot \vec{E} = 0$. Therefore, we must add this equation as a constraint on the states, $\nabla \cdot \vec{E}|\Psi\rangle = 0$. (Recall that if we try to implement Gauss's law as an operator equation, we end up in the Coulomb gauge.) The wave functionals that satisfy eq.(10.61) must also satisfy the functional Gauss's law constraint,

$$\nabla \cdot \frac{\delta}{\delta \vec{a}(\vec{x})} \Psi[a] = 0. \tag{10.62}$$

To interpret this constraint, let's consider what happens to $\Psi[a]$ under a gauge transformation. When we gauge transform $\vec{a}$ by adding the gradient of an arbitrary function, $a^i(\vec{x}) \to a^i(\vec{x}) - \partial^i \Lambda(\vec{x})$, we induce a change in the functional $\Psi[a]$. Any change in the wave functional due to a change in $\vec{a}$ is given by

$$\delta \Psi[a] = \int d^3x \, \frac{\delta \Psi}{\delta a^i(\vec{x})} \delta a^i(\vec{x}). \tag{10.63}$$

When the change in $\vec{a}(\vec{x})$ is due to a gauge transformation, then $\delta a^i(\vec{x}) = -\partial^i \Lambda(\vec{x})$. Thus, a change in $\Psi[a]$, due to a gauge transformation of $\vec{a}(\vec{x})$, is

$$\begin{aligned} \delta \Psi[a] &= -\int d^3x \, \frac{\delta \Psi}{\delta a^i(\vec{x})} \partial^i \Lambda(\vec{x}) \\ &= \int d^3x \, \nabla \cdot \frac{\delta \Psi}{\delta \vec{a}(\vec{x})} \Lambda(\vec{x}), \end{aligned} \tag{10.64}$$

where the last step is accomplished by an integration by parts. The Gauss's law operator acting on the wave functional has appeared in eq.(10.64). If the wave functional satisfies the Gauss's law constraint, eq.(10.62), then the right-hand side of eq.(10.64) vanishes and the wave functional is invariant under a gauge transformation. On the other hand, if we want the wave functional to be gauge invariant, then the functional must satisfy the Gauss's law constraint because $\Lambda(\vec{x})$ is an arbitrary function. Thus, the Gauss's law constraint requires the physical wave functionals to be gauge invariant. If we know that a wave functional is gauge invariant, then we know that it will satisfy the Gauss's law constraint. Note that when we say "gauge invariant," we mean invariant under time-independent gauge transformations, i.e., $\Lambda(\vec{x})$ is independent of time. We noted earlier that the temporal gauge condition doesn't fully fix the gauge. Any transformation that is time independent leaves $A_0 = 0$. Thus, the constraint insures that the residual gauge symmetry is respected.

We may arrive at the same conclusion by interpreting the action of the constraint on $\Psi[a]$. $\delta/\delta\vec{a}(\vec{x})$ picks out the factors of $\vec{a}$ in $\Psi[a]$, one at a time. $(\nabla \cdot\ )$ then tests whether this factor of $\vec{a}$ is divergence-free. If it is divergence-free, then $\nabla \cdot \delta/\delta\vec{a}(\vec{x})$ will give zero, otherwise not. Thus, the Gauss's law constraint requires that each factor of $\vec{a}$ in $\Psi[a]$ be divergence-free, $\nabla\cdot\vec{a}(\vec{x}) = 0$. But this is precisely the condition that all of the factors of $\vec{a}$ are transverse. The transverse components of $\vec{a}$ are gauge invariant, while the longitudinal components are not. Thus, if $\Psi[a]$ depends only on the transverse components of $\vec{a}$, then it is gauge invariant (and vice versa).

Expectations of gauge invariant operators are functional integrals.

$$\frac{\langle\Psi\,|\mathcal{O}|\,\Psi\rangle}{\langle\Psi\,|\,\Psi\rangle} = \frac{\int \mathcal{D}\vec{a}\,\Psi^*[a]\mathcal{O}(\vec{a})\Psi[a]}{\int \mathcal{D}\vec{a}\,\Psi^*[a]\Psi[a]} \tag{10.65}$$

$\mathcal{D}\vec{a} = \prod_{\vec{x}}\prod_{i=1}^{3} da_i(\vec{x})$. The constraint introduces a small complication in defining the functional integral. The operator $\mathcal{O}$ is gauge invariant under time independent gauge transformations. If it wasn't, its physical meaning would be unclear. Since it is gauge invariant, it only depends on the gauge invariant components of $\vec{a}$, the transverse components, $\vec{a}_T$. Likewise, the physical state wave functionals satisfy the Gauss's law constraint so they are gauge invariant and depend only on $\vec{a}_T$. We are, however, integrating over all of the components of $\vec{a}$, including the gauge dependent, longitudinal component, $\vec{a}_L$. Thus, $\mathcal{D}\vec{a} = \mathcal{D}\vec{a}_L\,\mathcal{D}\vec{a}_T$. Since the integrand does not depend on $\vec{a}_L$, this introduces a rather large infinity,

$$\langle\Psi\,|\mathcal{O}|\,\Psi\rangle = \int \mathcal{D}\vec{a}_L \int \mathcal{D}\vec{a}_T\,\Psi^*[a_T]\,\mathcal{O}(\vec{a}_T)\Psi[a_T]. \tag{10.66}$$

For the Abelian case, this infinity, $\int\mathcal{D}\vec{a}_L$, is easy to identify, separate out, and absorb into the normalization. This factor, $\int\mathcal{D}\vec{a}_L$, is present in the numerator and denominator of eq.(10.65) so it cancels out. We could equally include $\delta[\vec{a}_L]$ in the integrand of eq.(10.66) to accomplish the same thing. In the nonabelian case, such as for Yang-Mills, the separation of the gauge dependent part so that it can be absorbed in the normalization is not so straightforward. The Faddeev-Popov procedure used to accomplish this will be discussed fully in chapter 15.

Let's determine the ground state wave functional, $\Psi_0[a]$, by solving the Schrödinger equation (10.61). We will use a slight variation of the procedure used in the scalar case by solving for $\Psi_0[\tilde{a}]$ directly by Fourier transforming the Schrödinger equation (10.61). Using

$$a_i(\vec{x}) = \int \frac{d^3k}{(2\pi)^3}\,\tilde{a}_i(\vec{k})\,e^{i\vec{k}\cdot\vec{x}} \tag{10.67}$$

and

$$\frac{\delta}{\delta a_i(\vec{x})} = \int \frac{d^3k}{(2\pi)^3}\,\frac{\delta}{\delta\tilde{a}_i(\vec{k})}\,e^{-i\vec{k}\cdot\vec{x}}, \tag{10.68}$$

the Schrödinger equation (10.61) becomes

$$\frac{1}{2}\int\frac{d^3k}{(2\pi)^3}\left(-\frac{\delta}{\delta\tilde{a}_i(\vec{k})}\frac{\delta}{\delta\tilde{a}_i(-\vec{k})}+(\vec{k}\times\tilde{a}(\vec{k}))\cdot(\vec{k}\times\tilde{a}(-\vec{k}))\right)\Psi_0[\tilde{a}] = E_0\Psi_0[\tilde{a}]. \quad (10.69)$$

Note that the conventions we use in the Fourier transforms, eq.(10.67) and (10.68) (i.e. where we put the factors of 2π), require us to take

$$\frac{\delta}{\delta\tilde{a}_i(\vec{p})}\tilde{a}_j(\vec{p}') = (2\pi)^3\,\delta_{ij}\,\delta^3(p-p').$$

To solve eq.(10.69), power counting again suggests that the functional should be Gaussian. Namely, choose

$$\Psi_0[\tilde{a}] = \eta\exp(-G[\tilde{a}]), \quad (10.70)$$

where

$$G[\tilde{a}] = \int\frac{d^3k}{(2\pi)^3}\tilde{g}(\vec{k})\tilde{a}(\vec{k})\cdot\tilde{a}(-\vec{k}). \quad (10.71)$$

As it stands, this is not sufficient. We must not forget the Gauss's law constraint, eq.(10.62), which states that

$$k_i\frac{\delta}{\delta\tilde{a}_i(\vec{k})}G[\tilde{a}] = 0. \quad (10.72)$$

Eq.(10.71) does not satisfy the constraint, but a remedy is not hard to find. If we substitute eq.(10.71) into the constraint, we find a nonvanishing result proportional to $k_i\tilde{a}_i(\vec{k})$. If we make sure that $G[\tilde{a}]$ does not depend on $k_i\tilde{a}(\vec{k})/|\vec{k}|$, then the constraint will be satisfied. This is, of course, just the requirement that $G[\tilde{a}]$ be independent of the longitudinal component of $\tilde{a}$, that is, gauge invariant. Therefore, we fix up eq.(10.71) by removing the longitudinal component from $\tilde{a}$. That is, we replace $\tilde{a}$ in eq.(10.71) with

$$\tilde{a}_i \to \tilde{a}_i - \frac{(\vec{k}\cdot\tilde{a})k_i}{\vec{k}^2}. \quad (10.73)$$

The new functional,

$$\begin{aligned} G[\tilde{a}] &= \int\frac{d^3k}{(2\pi)^3}\,\tilde{g}(\vec{k})\left(\tilde{a}_i(\vec{k}) - \frac{(\vec{k}_j\tilde{a}_j)k_i}{\vec{k}^2}\right)\left(\tilde{a}_i(-\vec{k}) - \frac{(k_j\tilde{a}_j(-\vec{k})k_i}{\vec{k}^2}\right) \\ &= \int\frac{d^3k}{(2\pi)^3}\,\tilde{g}(\vec{k})\left(\tilde{a}_i(\vec{k})\tilde{a}_i(-\vec{k}) - \frac{(k_i\tilde{a}_i(\vec{k}))(k_j\tilde{a}_j(-\vec{k}))}{\vec{k}^2}\right), \end{aligned} \quad (10.74)$$

satisfies the constraint, eq.(10.72).

To actually be a solution of eq.(10.69), the functional (10.74) must satisfy

$$\int \frac{d^3k}{(2\pi)^3} \frac{\delta G[\tilde{a}]}{\delta \tilde{a}_j(\vec{k})} \frac{\delta G[\tilde{a}]}{\delta \tilde{a}_j(-\vec{k})} = \int \frac{d^3k}{(2\pi)^3} (\vec{k} \times \tilde{a}(\vec{k})) \cdot (\vec{k} \times \tilde{a}(-\vec{k})), \quad (10.75)$$

and

$$\frac{1}{2} \int \frac{d^3k}{(2\pi)^3} \left(-\frac{\delta^2 G[\tilde{a}]}{\delta \tilde{a}^2} \right) = E_0. \quad (10.76)$$

From eq.(10.74) we have

$$\frac{\delta G[\tilde{a}]}{\delta \tilde{a}_i(\vec{k})} = 2\tilde{g}(\vec{k}) \left(\tilde{a}_i(-\vec{k}) - \frac{(k_j\, \tilde{a}_j(-\vec{k}))k_i}{\vec{k}^2} \right), \quad (10.77)$$

where we are assuming (unless we run into trouble) that $\tilde{g}(-\vec{k}) = \tilde{g}(\vec{k})$. Using the vector identity

$$(\vec{a} \times \vec{b}) \cdot (\vec{c} \times \vec{d}) = (\vec{a} \cdot \vec{c})(\vec{b} \cdot \vec{d}) - (\vec{a} \cdot \vec{d})(\vec{b} \cdot \vec{c}), \quad (10.78)$$

the right-hand side of eq.(10.75) may be written as

$$\int \frac{d^3k}{(2\pi)^3} \left(\vec{k}^2 \tilde{a}_i(\vec{k}) \tilde{a}_i(-\vec{k}) - (k_i \tilde{a}_i(\vec{k}))(k_j \tilde{a}_j(-\vec{k})) \right). \quad (10.79)$$

Therefore, eq.(10.75) boils down to the requirement that

$$4\frac{\tilde{g}^2(\vec{k})}{\vec{k}^2} = 1 \quad \text{or} \quad \tilde{g}(\vec{k}) = \frac{1}{2}|\vec{k}| = \frac{1}{2}\omega_k. \quad (10.80)$$

Using eqs.(10.80) and (10.74), and the vector identity (10.78), the ground state wave functional can be written as

$$\Psi_0[\tilde{a}] = \eta \exp\left(-\frac{1}{2} \int \frac{d^3k}{(2\pi)^3} \frac{(\vec{k} \times \tilde{a}(\vec{k})) \cdot (\vec{k} \times \tilde{a}(-\vec{k}))}{|\vec{k}|} \right). \quad (10.81)$$

In terms of the transverse components, $\tilde{a}_T$, eq.(10.81) is just the product of harmonic oscillator ground state wave functions,

$$\Psi_0[\tilde{a}] = \eta \prod_{\vec{k}} \exp\left(-\frac{1}{2} \frac{1}{(2\pi)^3} \omega_k\, \tilde{a}_T(\vec{k}) \cdot \tilde{a}_T(-\vec{k}) \right), \quad (10.82)$$

as expected. The result, eq.(10.81), may be transformed back into a functional of $\vec{a}(\vec{x})$.

$$\begin{aligned}\Psi_0[a] &= \eta \exp\left(-\frac{1}{(2\pi)^2}\int d^3x\, d^3y\, \frac{(\nabla\times\vec{a}(\vec{x}))\cdot(\nabla\times\vec{a}(\vec{y}))}{|\vec{x}-\vec{y}|^2}\right)\\ &= \eta \exp\left(-\frac{1}{(2\pi)^2}\int d^3x\, d^3y\, \frac{\vec{b}(\vec{x})\cdot\vec{b}(\vec{y})}{|\vec{x}-\vec{y}|^2}\right)\end{aligned} \tag{10.83}$$

As in the scalar case, the harmonic oscillator nature of the wave functionals lends itself to a particle interpretation. A wave functional representing a state with one photon of momentum $\vec{k}_1$ is obtained from eq.(10.82) by replacing the ground state wave function for mode $\vec{k}_1$ with the harmonic oscillator wave function for the first excited state. If we want the wave functional for a state with two photons of momentum $\vec{k}_1$, then we replace mode $\vec{k}_1$ with the second excited wave function, and so on.

The excited state wave functionals are conveniently produced from $\Psi_0[\tilde{a}]$ or $\Psi_0[a]$ using the functional form of the $\mathbf{a}(\vec{k},\lambda)$ and $\mathbf{a}^\dagger(\vec{k},\lambda)$ operators of eq.(5.28). Using the conventions defined in chapter 5 for the polarization vectors, we can invert eq.(5.28) to give

$$\begin{aligned}\mathbf{a}(\vec{k},\lambda) &= \int d^3x\, e^{-i\vec{k}\cdot\vec{x}}\,\vec{\varepsilon}(\vec{k},\lambda)\cdot(|\vec{k}|\vec{A}(\vec{x}) - i\vec{E}(\vec{x}))\\ \mathbf{a}^\dagger(\vec{k},\lambda) &= \int d^3x\, e^{i\vec{k}\cdot\vec{x}}\vec{\varepsilon}(\vec{k},\lambda)\cdot(|\vec{k}|\vec{A}(\vec{x}) + i\vec{E}(\vec{x})).\end{aligned} \tag{10.84}$$

The functional differential representation quickly follows from eq.(10.60). The wave functional that represents the state containing one photon of momentum $\vec{k}_1$ and polarization $\vec{\varepsilon}(\vec{k}_1,\lambda)$ is

$$\Psi_1[a] = \int d^3x e^{-\vec{k}_1\cdot\vec{x}}\vec{\varepsilon}(\vec{k}_1,\lambda)\cdot\left(|\vec{k}_1|\vec{a}(\vec{x}) - \frac{\delta}{\delta\vec{a}(\vec{x})}\right)\Psi_0[a], \tag{10.85}$$

or

$$\begin{aligned}\Psi_1[\tilde{a}] &= \vec{\varepsilon}(\vec{k}_1,\lambda)\cdot\left(|\vec{k}_1|\left(\tilde{a}(\vec{k}_1) - \frac{(\vec{k}_1\cdot\tilde{a}(\vec{k}_1))\vec{k}_1}{|\vec{k}_1|^2}\right) - \frac{\delta}{\delta\tilde{a}(-\vec{k}_1)}\right)\Psi_0[\tilde{a}]\\ &= \vec{\varepsilon}(\vec{k}_1,\lambda)\cdot\left(|\vec{k}_1|\tilde{a}_T(\vec{k}_1) - \frac{\delta}{\delta\tilde{a}_T(-\vec{k}_1)}\right)\Psi_0[\tilde{a}_T].\end{aligned} \tag{10.86}$$

Using the vacuum wave functional, $\Psi_0[a]$, eq.(10.83), the computation of the propagator $\langle 0\,|T(A^i_T(\vec{x})A^j_T(\vec{y}))|\,0\rangle$ reduces to a Gaussian functional integral that is straightforward to compute. We will not explicitly need the propagator in this representation.

10.3 SPINOR FIELD THEORY

Now let us consider the Schrödinger representation for free fermion fields and fermion wave functionals. Recall from chapter 4 that the Dirac Hamiltonian is

$$H = \int d^3x\, \Psi^\dagger(x)(-i\vec{\alpha}\cdot\nabla + \beta m)\Psi(x), \tag{10.87}$$

where the matrices α_i and β satisfy anticommutators given by eq.(4.2), and where Ψ is a 4-component spinor. The field operator conjugate to $\Psi(x)$ is $i\Psi^\dagger(x)$. Because the quanta of the spinor fields Ψ and $\Psi^\dagger$ are supposed to be fermions, Ψ and $\Psi^\dagger$ must satisfy equal-time anticommutation relations,

$$\begin{aligned}\{\Psi_\alpha(\vec{x},t), \Psi^\dagger_\beta(\vec{y},t)\} &= \delta_{\alpha\beta}\,\delta^3(\vec{x}-\vec{y})\\ \{\Psi_\alpha(\vec{x},t), \Psi_\beta(\vec{y},t)\} &= \{\Psi^\dagger_\alpha(\vec{x},t), \Psi^\dagger_\beta(\vec{y},t)\} = 0.\end{aligned} \tag{10.88}$$

In the coordinate Schrödinger representation, the state $|\,\Phi\rangle$ is represented by the wave functional $\langle\psi\,|\,\Phi\rangle = \Phi[\psi]$. The state $|\psi\rangle$ is an eigenstate of the field operator $\Psi(\vec{x})$ with eigenvalue $\psi(\vec{x})$. Since $\Psi(\vec{x})$ satisfies the anticommutators eq.(10.88) and therefore squares to zero, the eigenvalues, $\psi(\vec{x})$ must be spinors of anticommuting variables or Grassmann functions, $\psi^2_\alpha(\vec{x}) = 0$.

The defining relationship of a functional derivative with respect to a Grassmann function given in chapter 9 allows us to represent $\Psi^\dagger_\beta(\vec{x})$ as the functional derivative,

$$\Psi^\dagger_\beta(\vec{x}) = \frac{\delta}{\delta\psi_\beta(\vec{x})}, \tag{10.89}$$

and also satisfy eq.(10.88). Therefore, the time independent Schrödinger equation is

$$\int d^3x \left(\frac{\delta}{\delta\psi(\vec{x})}(-i\vec{\alpha}\cdot\nabla + \beta m)\psi(\vec{x})\right)\Phi[\psi] = E\Phi[\psi]. \tag{10.90}$$

This Grassmann functional differential equation will be easier to solve in momentum space so we turn to that now and Fourier transform (10.90).

To avoid some clumsy notation like $\delta/\delta d^\dagger_i(\vec{p})$ and inconvenient normalizations, we will slightly modify the plane wave expansion of the field operators $\Psi(\vec{x})$ and $\Psi^\dagger(\vec{x})$ used in chapter 4. Namely, let

$$\begin{aligned}\Psi_\alpha(\vec{x}) &= \sum_{r=1}^{4}\int \frac{d^3p}{\sqrt{(2\pi)^3}}\sqrt{\frac{m}{E}}\,\mathrm{b}(\vec{p},r)w_\alpha(\vec{p},r)e^{i\vec{p}\cdot\vec{x}},\\ \Psi^\dagger_\alpha(\vec{x}) &= \sum_{r=1}^{4}\int \frac{d^3p}{\sqrt{(2\pi)^3}}\sqrt{\frac{m}{E}}\,\mathrm{b}^\dagger(\vec{p},r)w^\dagger_a(\vec{p},r)e^{-i\vec{p}\cdot\vec{x}}.\end{aligned} \tag{10.91}$$

In terms of the spinors used in chapter 4,

$$\begin{aligned} w(\vec{p}, r=1,2) &= u_i(\vec{p}) \quad i=1,2 \\ w(\vec{p}, r=3,4) &= v_i(-\vec{p}) \quad i=1,2, \end{aligned} \tag{10.92}$$

that is,

$$w(\vec{p},1) = \sqrt{\frac{E+m}{2m}} \begin{pmatrix} 1 \\ 0 \\ \frac{p_z}{E+m} \\ \frac{p_+}{E+m} \end{pmatrix} \qquad w(\vec{p},2) = \sqrt{\frac{E+m}{2m}} \begin{pmatrix} 0 \\ 1 \\ \frac{p_-}{E+m} \\ \frac{-p_z}{E+m} \end{pmatrix}$$

$$w(\vec{p},3) = \sqrt{\frac{E+m}{2m}} \begin{pmatrix} \frac{-p_z}{E+m} \\ \frac{-p_+}{E+m} \\ 1 \\ 0 \end{pmatrix} \qquad w(\vec{p},4) = \sqrt{\frac{E+m}{2m}} \begin{pmatrix} \frac{-p_-}{E+m} \\ \frac{p_z}{E+m} \\ 0 \\ 1 \end{pmatrix}. \tag{10.93}$$

The spinors satisfy orthogonality and completeness relations that we will make use of.

$$\begin{aligned} w^\dagger(\vec{p}, r)\, w(\vec{p}, r') &= \frac{E}{m}\,\delta_{rr'} \\ \bar{w}(\varepsilon_r \vec{p}, r)\, w(\varepsilon_{r'} \vec{p}, r') &= \varepsilon_r\, \delta_{rr'} \\ \sum_{r=1}^{4} w_\alpha(\vec{p}, r) w_\beta^\dagger(\vec{p}, r) &= \frac{E}{m}\,\delta_{\alpha\beta}, \end{aligned} \tag{10.94}$$

where $\varepsilon_r = 1$ for $r = 1,2$ and $\varepsilon_r = -1$ for $r = 3,4$.

In order that the field operators $\Psi_\alpha(\vec{x})$ and $\Psi_\beta^\dagger(\vec{x})$ satisfy the anti-commutation relations, eq.(10.88), the operators $\mathbf{b}$ and $\mathbf{b}^\dagger$ must satisfy

$$\{\mathbf{b}(\vec{p}, r), \mathbf{b}^\dagger(\vec{p}', r')\} = \delta_{rr'}\,\delta^3(p - p')$$

$$\{\mathbf{b}(\vec{p}, r), \mathbf{b}(\vec{p}', r')\} = \{\mathbf{b}^\dagger(\vec{p}, r), \mathbf{b}^\dagger(\vec{p}', r')\} = 0. \tag{10.95}$$

A comparison of eq.(10.95) with eqs.(4.36) and (4.37) will show that we changed the normalization of the $\mathbf{b}$ operators used in chapter 4 along with the definition of the $\mathbf{d}, \mathbf{d}^\dagger$ operators. For example,

$$\begin{aligned} \mathbf{b}_1(p) &= \sqrt{(2\pi)^3}\sqrt{\frac{E}{m}}\mathbf{b}(\vec{p}, r=1) \\ \mathbf{d}_2^\dagger(-p) &= \sqrt{(2\pi)^3}\sqrt{\frac{E}{m}}\mathbf{b}(\vec{p}, r=4). \end{aligned} \tag{10.96}$$

The new notation and normalization will allow us to take

$$\mathbf{b}^\dagger(\vec{p},r) = \frac{\delta}{\delta b(\vec{p},r)}.$$

In other words, the expansions of the Grassmann functions ψ (the "eigenvalues" of the operator Ψ) are

$$\begin{aligned}
\psi_\alpha(\vec{x}) &= \sum_{r=1}^{4} \int \frac{d^3p}{\sqrt{(2\pi)^3}} \sqrt{\frac{m}{E}}\, b(\vec{p},r)\, w_\alpha(\vec{p},r)\, e^{i\vec{p}\cdot\vec{x}} \\
\frac{\delta}{\delta\psi_\alpha(\vec{x})} &= \sum_{r=1}^{4} \int \frac{d^3p}{\sqrt{(2\pi)^3}} \sqrt{\frac{m}{E}} \frac{\delta}{\delta b(\vec{p},r)}\, w^\dagger_\alpha(\vec{p},r)\, e^{-i\vec{p}\cdot\vec{x}}.
\end{aligned} \tag{10.97}$$

Using the expansions in eq.(10.97) above, the Schrödinger equation, eq.(10.90), becomes

$$\sum_{r=1}^{4} \int d^3p\, \varepsilon_r\, E_p \frac{\delta}{\delta b(\vec{p},r)}\, b(\vec{p},r)\, \Phi[b] = E\Phi[b],$$

or

$$\sum_{r=1}^{4} \int d^3p \left(-\varepsilon_r\, E_p\, b(\vec{p},r) \frac{\delta}{\delta b(\vec{p},r)} \right) \Phi[b] = E\Phi[b], \tag{10.98}$$

where $E_p = \sqrt{\vec{p}^2 + m^2}$.

Recall from chapter 4 that the physical, stable vacuum is the state with all of the negative energy states occupied. To determine its wave functional, $\Phi_0[b]$, let's first examine the wave function for a single fermion.

For a single fermion, b will destroy it while $\delta/\delta b$ creates it. Let $\Omega_0(b)$ be the wave function of the state $|\,0\rangle$ where the fermion is absent, and $\Omega_1(b)$ be the wave function of the state $|\,1\rangle$, the state when the fermion is present. Then,

$$\begin{aligned}
b\,\langle b\,|\,0\rangle &= b\,\Omega_0(b) = 0 \\
b\langle b\,|\,1\rangle &= b\,\Omega_1(b) = \Omega_0(b) \\
\frac{\delta}{\delta b}\,\langle b\,|\,0\rangle &= \frac{\delta}{\delta b}\Omega_0(b) = \Omega_1(b) \\
\frac{\delta}{\delta b}\,\langle b\,|\,1\rangle &= \frac{\delta}{\delta b}\Omega_1(b) = 0.
\end{aligned} \tag{10.99}$$

Since b is a Grassmann number, $b^2 = 0$, and the first condition in eq.(10.99) suggests that we take $\Omega_0(b) = b$. Having done this, the second condition then requires that $\Omega_1(b) = 1$. These two choices are consistent with the last two conditions.

This choice for $\Omega_0(b)$ and $\Omega_1(b)$ means that the functional operator appearing in the Hamiltonian, eq.(10.98), has a simple interpretation. The operator $\frac{\delta}{\delta b(\vec{p},r)} b(\vec{p},r)$ tests the functional $\Phi[b]$ for the presence of the factor $b(\vec{p},r)$. If it is present, the result is zero. If not, the wave functional is restored to its original state by taking $\frac{\delta}{\delta b(\vec{p},r)}$. Thus, $\frac{\delta}{\delta b(\vec{p},r)} b(\vec{p},r)$ tests to see if the mode $\vec{p}$, spin variable r, is occupied. If so the result of the operator is 1 and $\varepsilon_r E_p$ is added to the energy of the state. If not, the result is zero and nothing happens. The Hamiltonian is an infinite sum of these operators and tests each mode that can be occupied for the presence of a fermion.

Since the Hamiltonian, eq.(10.98), is an infinite sum of functional operators that are decoupled in $\vec{p}$ and r, the eigen wave functional Φ will be an infinite product of single fermion wave functions just given above. We can now easily write down the physical vacuum wave functional, $\Phi_0[b]$. All of the negative energy states ($r = 3, 4$) are filled, thus each negative energy mode is represented by the wave function 1. All of the positive energy states are empty, thus each empty mode ($\vec{p}, r = 1, 2$) is represented by the wave function $b(\vec{p},r)$.

$$\Phi_0[b] = \eta \prod_{r=1}^{2} \prod_{\vec{p}} b(\vec{p},r) \tag{10.100}$$

From the definition of the delta-functional for Grassmann variables given in chapter 9, eq.(9.64), the wave functional, $\Phi_0[b]$, above can be recognized as $\delta[b(r=1)]\,\delta[b(r=2)]$. Following the same reasoning, we can see that the wave functional in terms of the x-space spinors, $\psi(\vec{x})$, is

$$\Phi_0[\psi] = \eta\delta[\psi_+], \tag{10.101}$$

where ψ_+ is the positive energy part of ψ.

From the definition of integration for Grassmann functions, namely,

$$\begin{aligned} \int db(\vec{p},r)\, b(\vec{p}',r') &= \delta_{rr'}\,\delta(\vec{p}-\vec{p}') \\ \int d\psi_\alpha(\vec{x})\, \psi_\beta(\vec{x}') &= \delta_{\alpha\beta}\,\delta(\vec{x}-\vec{x}'), \end{aligned} \tag{10.102}$$

the wave functional $\Phi_0[b]$ will only be normalizable if its adjoint wave functional is taken as

$$\begin{aligned} \Phi_0^\dagger[b] &= \eta^* \prod_{r=3}^{4} \prod_{\vec{p}} b(\vec{p},r) \\ \Phi_0^\dagger[\psi] &= \eta^* \delta[\psi_-], \end{aligned} \tag{10.103}$$

in which case $\eta = 1$.

Excited state wave functionals are easily constructed by filling or emptying the desired modes. If we want to add an electron of momentum $\vec{p}$, we replace the factor of $b(\vec{p}, r = 1 \text{ or } 2)$ with 1. If we want to add a positron of momentum $-\vec{p}$, that is, remove a negative energy electron, we multiply the physical vacuum wave functional by $b(\vec{p}, r = 3, \text{ or } 4)$. For example, the wave functional representing the state with one electron of momentum $\vec{p}$, spin up, is

$$\Phi_{1e}[b] = \prod_{\substack{r=1 \\ \text{except}(\vec{p},1)}}^{2} \prod_{\vec{q}} b(\vec{q}, r), \tag{10.104}$$

while the state with one positron of momentum $\vec{p}$, spin up, is

$$\Phi_{1p}[b] = b(-\vec{p}, 4) \prod_{r=1}^{2} \prod_{\vec{q}} b(\vec{q}, r). \tag{10.105}$$

The computation of the propagator is straightforward. Remembering that the time dependence now resides in the states, the contribution to the propagator for the process of an electron to go from x to x' is

$$\langle 0 \,|\Psi_\alpha(\vec{x}')\Psi_\beta^\dagger(\vec{x})|\, 0\rangle \theta(t' - t).$$

Using eqs.(10.100) and (10.97), this contribution is the Grassmann functional integral

$$\sum_{r'=1}^{4} \sum_{r=1}^{4} \int \mathcal{D}b\, \Phi_0^\dagger[b] \int \frac{d^3p'}{\sqrt{(2\pi)^3}} \sqrt{\frac{m}{E_{p'}}}\, b(\vec{p}', r') w_\alpha(\vec{p}', r') e^{i\vec{p}'\cdot\vec{x}'}$$
$$\int \frac{d^3p}{\sqrt{(2\pi)^3}} \sqrt{\frac{m}{E_p}}\, \frac{\delta}{\delta b(\vec{p}, r)}\, w_\beta^\dagger(\vec{p}, r) e^{-i\vec{p}\cdot\vec{x}} \Phi_0[b]\, e^{-iE_1(t-t')}\theta(t' - t), \tag{10.106}$$

where E_1 is the energy of the state containing one electron in the physical vacuum. Since only the negative energy states are filled in $\Phi_0[b]$, then $\delta/\delta b(\vec{p}, r)\ \Phi_0[b] = 0$ for $r = 3, 4$. Therefore,

$$\frac{\delta}{\delta b(\vec{p}, r)} \Phi_0[b] = \prod_{\substack{s=1 \\ \text{except }(\vec{p},r)}}^{2} \prod_{\vec{q}} b(\vec{q}, s). \tag{10.107}$$

Since the b's anticommute, when we multiply (10.107) by $b(\vec{p}', r')$, we will only get a nonzero result if $\vec{p}' = \vec{p}$, $r' = r$, or if $r' = 3, 4$.

$$b(\vec{p}', r') \frac{\delta}{\delta b(\vec{p}, r)} \Phi_0[b] = \begin{cases} \Phi_0[b] & \text{if } \vec{p}' = \vec{p},\ r' = r \\ \prod_{\substack{s=1 \\ \text{except }(\vec{p},r)}}^{2} \prod_{\vec{q}} b(\vec{p}', r') & \text{if } r' = 3, 4 \end{cases} \tag{10.108}$$

When we multiply this result by the adjoint of the vacuum wave functional, $\Phi_0^\dagger[b]$, the second possibility, where $r' = 3, 4$, vanishes because there will be 2 factors of $b(\vec{p}', r')$ and $b^2(\vec{p}', r') = 0$. This implies that $\vec{p}' = \vec{p}$, $r' = r$. Since the second possibility in (10.108) doesn't conserve momentum, we expect it not to contribute. The functional integral over b is trivial and removes all of the factors of b. Thus,

$$\int \mathcal{D}b\, \Phi_0^\dagger[b]\, b(\vec{p}', r')\, \frac{\delta}{\delta b(\vec{p}, r)}\, \Phi_0[b] = \delta_{rr'}\, \delta^3(p - p'), \tag{10.109}$$

and

$$\begin{aligned} &\langle 0\,|\Psi_\alpha(\vec{x}')\Psi_\beta^\dagger(\vec{x})|\,0\rangle\theta(t'-t) = \\ &= \sum_{r=1}^{2} \int \frac{d^3p}{(2\pi)^3}\left(\frac{m}{E_p}\right) w_\alpha(\vec{p}, r) w_\beta^\dagger(\vec{p}, r) e^{i\vec{p}\cdot(\vec{x}'-\vec{x})} e^{-iE_1(t-t')}\theta(t'-t). \end{aligned} \tag{10.110}$$

For $r = 1, 2$, $w(\vec{p}, r) = u_r(p)$ so we may apply the projection operator, eq.(4.24).

$$\sum_{r=1}^{2} w_\alpha(\vec{p}, r)\, \bar{w}_\beta(\vec{p}, r) = \left(\frac{\not{p} + m}{2m}\right)_{\alpha\beta} \tag{10.111}$$

After we rescale the energy of the vacuum to zero (normal ordering), the electron's ($r = 1, 2$) contribution to the propagator is

$$\langle 0\,|\Psi(\vec{x}')\bar{\Psi}(\vec{x})|\,0\rangle\theta(t'-t) = \int \frac{d^3p}{(2\pi)^3}\,\frac{1}{2E}(\not{p} + m)\theta(t'-t)e^{-ip\cdot(x'-x)}, \tag{10.112}$$

which agrees with the result in chapter 4. The positron's ($r = 3, 4$) contribution is $\langle 0\,|\Psi_\alpha^\dagger(\vec{x})\Psi_\beta(\vec{x}')|\,0\rangle\theta(t - t')$. The functional integral that we must do for this result is

$$\int \mathcal{D}b\; \Phi_0^\dagger[b]\, \frac{\delta}{\delta b(\vec{p}, r)}\, b(\vec{p}', r')\, \Phi_0[b].$$

As in the electron's contribution, this integral will only be nonzero if

$$\frac{\delta}{\delta b(\vec{p}, r)}\, b(\vec{p}', r')\, \Phi_0[b] = \Phi_0[b].$$

Now, however, since $b(\vec{p}', r')$ operates before $\delta/\delta b(\vec{p}, r)$, r' must be 3 or 4. If not, 2 identical factors of b will appear and the expression will vanish. The corresponding projection operator that we will need,

$$\sum_{r=3}^{4} \bar{w}_\alpha(\vec{p}, r) w_\beta(\vec{p}, r) = \left(\frac{\not{p} - m}{2m}\right)_{\beta\alpha}, \qquad (10.113)$$

follows from eq.(4.24). The resulting contribution due to the positron,

$$\langle 0 \,|\bar{\Psi}(\vec{x})\Psi(\vec{x}')|\, 0\rangle\theta(t - t')$$
$$= \int \frac{d^3p}{(2\pi)^3} \frac{1}{2E_p} (\not{p} - m)\theta(t - t')e^{ip\cdot(x' - x)}, \qquad (10.114)$$

agrees with the result from chapter 4 and is subtracted from the electron's contribution, eq.(10.112), to obtain the entire propagator.

10.4 EXERCISES

1. Show by direct functional integration that the state $\Psi_1[\phi]$ given by eq.(10.30) satisfies $\langle\Psi_1\,|\,\Psi_1\rangle = \delta^3(0)$.
2. Using the vacuum wave functional, $\Psi_0[a]$, eq.(10.83), compute the propagator, $\langle 0\,|T(A^i_T(\vec{x})A^j_T(\vec{x}'))|\,0\rangle$, as a functional integral with respect to $\vec{A}_T$. Compare the result with the propagator obtained in the operator representation in chapter 5.
3. What is the most likely configuration of the $\varphi(\vec{x})$ in the vacuum? What is it when there is a source, $J(\vec{x})$, present? Compute the functional integral, $\langle 0\,|\varphi(\vec{x})|\,1\rangle$, where $|\,1\rangle$ is a single-particle eigenstate with momentum $\vec{k}$.
4. Compute $\Psi_0[\varpi]$, eq.(10.51), from $\Psi_0[\phi]$, eq.(10.15), using the functional Fourier transform, eq.(10.55).
5. Quantize the electromagnetic field in the Schrödinger representation in the Coulomb gauge by taking the electric field to be the functional directional derivative in the direction of the transverse δ-function,

$$E_i(\vec{x}) = i\frac{\delta_\lambda}{\delta_\lambda a_i(\vec{x})}.$$

This functional directional derivative is defined by

$$\frac{\delta_\lambda}{\delta_\lambda a_i(\vec{x})}\, a_j(\vec{y}) = \delta^{\rm tr}_{ij}(\vec{x} - \vec{y}),$$

where $\delta_{ij}^{\rm tr}$ is given by eq.(5.25). Find the vacuum wave functional in this gauge and compare with eq.(10.83).

6. What is the vacuum wave functional for the electromagnetic field in the temporal gauge when there is a classical background charge, $\rho(\vec{x})$, present? What are the expected electric and magnetic field configurations in this vacuum?
7. Write the Nonlinear Schrödinger model given in chapter 2 in the Schrödinger representation. What is the ground state wave functional, $\Psi_0[\phi] = \langle \phi \,|\, 0 \rangle$?

CHAPTER **11**

Interacting Fields in the Schrödinger Representation

One of our goals in solving interacting quantum field theories is to calculate cross sections for scattering processes that can be compared with experiment. To compute a cross section, we need to know the S-matrix element corresponding to the scattering process. So, no matter which representation of field theory we work with, in the end we want to know the S-matrix elements. How the S-matrix is calculated will vary from representation to representation.

In the operator representation developed in the earlier chapters, we defined the S-matrix in terms of IN-OUT states and field operators. Elements of the S-matrix were related to the Green's functions of the interacting quantum field theory through a reduction formula. The Green's functions were then computed in a perturbation series.

In the functional Schrödinger representation we do not use a formal reduction formula, nor do we speak about the Green's functions (although we could). In the Schrödinger representation, the dynamics resides in the states, not in the field operators. Since the S-matrix is defined as the overlap between initial and final states, we need only compute the interacting initial and final states and project one onto the other. The interacting states are computed perturbatively using Rayleigh-Schrödinger perturbation theory, which is familiar from nonrelativistic quantum mechanics. To obtain a particle interpretation, we perturb off of the known, exact free field theory states. We can find a "Feynman" diagram interpretation for each of the terms in the series for the states. When the final state is projected onto the initial state, the diagrams for the two states combine to form the same diagrams that arose in the operator representation. This program of computing the S-matrix is entirely equivalent to the operator formalism.

We begin by casting the self-interacting scalar theory, φ^4, into the functional Schrödinger representation. We then perturbatively solve the Schrödinger equation for the interacting vacuum wave functional and interpret the terms. We then consider the 2-body elastic scattering process and compare the results with those obtained in the operator representation.

Rayleigh-Schrödinger perturbation theory presents some difficulties when applied to gauge theories. We resolve these difficulties constructively and solve for the perturbative vacuum wave functional for Yang-Mills to the lowest nontrivial order in the coupling. The same technique used in the Yang-Mills case to find the perturbative wave functional must also be applied to spinor QED, which we do in the last section.

This chapter is long because many of the details of the computations have been left *in*. We have done so under the assumption that solving functional differential equations is an unfamiliar topic. A guide around the more lengthy computations is provided for the first pass through.

11.1 SELF-INTERACTING SCALAR FIELD THEORY

Most of the work in bringing φ^4 theory into the Schrödinger representation was done in the previous chapter when we did the free scalar field theory. The field momentum and equal-time commutators are identical in the free and interacting cases, hence we still take

$$\pi(\vec{x}) = -i\frac{\delta}{\delta\phi(\vec{x})}. \tag{11.1}$$

The interacting Hamiltonian, eq.(7.3), becomes the functional differential operator

$$H = \int d^3x\,\frac{1}{2}\left(-\frac{\delta^2}{\delta\phi^2(\vec{x})} + |\nabla\phi|^2 + m^2\phi^2 + \delta m^2\phi^2\right) + \frac{\lambda_0}{4!}\phi^4(\vec{x}), \tag{11.2}$$

which is just the free field Hamiltonian operator, eq.(10.11), plus an interaction Hamiltonian,

$$H_{\text{int}} = H_1 + H_2 = \int d^3x\,\frac{1}{2}\delta m^2\,\phi^2(\vec{x}) + \frac{\lambda_0}{4!}\phi^4(\vec{x}). \tag{11.3}$$

To determine the interacting vacuum wave functional, let us first apply Rayleigh-Schrödinger perturbation theory. Then, for illustration, we will also obtain the perturbative wave functional by directly solving the Schrödinger equation.

Let $\Psi_N[\phi]$ be the eigenfunctional of the Hamiltonian (11.2) with eigenvalue E_N, $H\Psi_N[\phi] = E_N\Psi_N[\phi]$. Write the Hamiltonian as $H = H_0 + \alpha H_{\rm int}$, where H_0 is the free scalar Hamiltonian, eq.(10.11), and where $H_{\rm int}$ is given above in eq.(11.3). Expand both the wave functional and the energy eigenvalue as a power series in α,

$$\begin{aligned}\Psi_N[\phi] &= \Psi_N^{(0)}[\phi] + \alpha\Psi_N^{(1)}[\phi] + \alpha^2\Psi_N^{(2)}[\phi] + \ldots \\ E_N &= E_N^{(0)} + \alpha E_N^{(1)} + \alpha^2 E_N^{(2)} + \ldots\end{aligned} \tag{11.4}$$

and place the expansion in $H\,\Psi_N[\phi] = E_N\,\Psi_N[\phi]$. Equating both sides of this eigenequation order by order in α, and taking the inner product with $\Psi_M^{(0)}[\phi]$, we obtain as usual,

$$\Psi_N^{(1)}[\phi] = \sum_{M\neq N} \frac{\langle\Psi_M^{(0)}\,|H_{\rm int}|\,\Psi_N^{(0)}\rangle}{E_N^{(0)} - E_M^{(0)}}\Psi_M^{(0)}[\phi], \tag{11.5}$$

and

$$E_N^{(1)} = \langle\Psi_N^{(0)}\,|H_{\rm int}|\Psi_N^{(0)}\rangle, \quad E_N^{(2)} = \sum_{M\neq N} \frac{|\langle\Psi_M^{(0)}\,|H_{\rm int}|\Psi_N^{(0)}\rangle|^2}{E_N^{(0)} - E_M^{(0)}}, \tag{11.6}$$

where the zeroth order wave functionals, $\Psi_M^{(0)}[\phi]$, satisfy $H_0\Psi_M^{(0)}[\phi] = E_M^{(0)}\Psi_M^{(0)}[\phi]$. The expectation values appearing in eqs.(11.5) and (11.6) are now functional integrals.

The interaction Hamiltonian, $H_{\rm int}$, is a polynomial in the field coordinates, $\phi(\vec{x})$. The free field wave functionals are all polynomials in $\phi(\vec{x})$ multiplied by the Gaussian $\Psi_0^{(0)}[\phi]$. Therefore, all of the expectation values that appear in eqs.(11.5) and (11.6) are functional integrals that are moments of the Gaussian $\Psi_0^{*(0)}\Psi_0^{(0)}$. As in chapter 9, to make the computation of the moments easier, we define a generating functional of the moments of $\Psi_0^{*(0)}\Psi_0^{(0)}$ by adding a source term. The generating functional we want is

$$\mathcal{G}[J] = \eta^2 \int \mathcal{D}\phi \exp\left(-2\int d^3x\, d^3y\, \phi(\vec{x})g(\vec{x},\vec{y})\phi(\vec{y}) + \int d^3x\, J(\vec{x})\phi(\vec{x})\right), \tag{11.7}$$

where $g(\vec{x},\vec{y})$ was determined in chapter 10, eq.(10.23). η is the normalization of $\Psi_0^{(0)}[\phi]$ so that $\int \mathcal{D}\phi\, \Psi_0^{*(0)}[\phi]\,\Psi^{(0)}[\phi] = 1$. Equivalently, η^2 is the normalization of the generating functional, $\mathcal{G}[J]$, so that $\mathcal{G}[0] = 1$.

The moments are obtained by functionally differentiating with respect to the source function $J(\vec{x})$.

$$\langle \Psi_0^{(0)}\, |\phi(\vec{x}_1)\cdots\phi(\vec{x}_n)|\, \Psi_0^{(0)}\rangle = \frac{\delta}{\delta J(\vec{x}_1)}\cdots\frac{\delta}{\delta J(\vec{x}_n)}\mathcal{G}[J]\bigg|_{J=0} \tag{11.8}$$

The generating functional $\mathcal{G}[J]$ may be obtained in closed form by completing the squares (see eq.(9.50)). The result is

$$\mathcal{G}[J] = \exp\left(\frac{1}{8}\int d^3x\, d^3y\, J(\vec{x})g^{-1}(\vec{x},\vec{y})J(\vec{y})\right), \tag{11.9}$$

where $g^{-1}(\vec{x},\vec{y})$ is the inverse of $g(\vec{x},\vec{y})$,

$$\int d^3z\, g(\vec{x},\vec{z})g^{-1}(\vec{z},\vec{y}) = \delta^3(x-y). \tag{11.10}$$

Use of the generating functional, $\mathcal{G}[J]$, reduces the computation of the shift in the vacuum energy to first order, $E_0^{(1)}$, to functional differentiation. From eqs.(10.15), (10.17), (11.3), (11.6), (11.8), and (11.9) we have

$$\begin{aligned}
&\langle \Psi_0^{(0)}\, |H_1|\, \Psi_0^{(0)}\rangle \\
&= \int \mathcal{D}\phi\, \Psi_0^{*(0)}[\phi] \int d^3x\, \frac{1}{2}\,\delta m^2\, \phi^2(\vec{x})\Psi_0^{(0)}[\phi] \\
&= \int d^3x\, \frac{1}{2}\,\delta m^2\, \eta^2 \int \mathcal{D}\phi\, \phi^2(\vec{x}) \exp\left(-2\int d^3x\, d^3y\, \phi(\vec{x})g(\vec{x},\vec{y})\phi(\vec{y})\right) \\
&= \int d^3x\, \frac{1}{2}\,\delta m^2\, \frac{\delta^2}{\delta J(\vec{x})^2}\mathcal{G}[J]\bigg|_{J=0} \\
&= \int d^3x \frac{1}{2}\,\delta m^2 \left(\frac{1}{4}g^{-1}(\vec{x},\vec{x}) + \left(\frac{1}{4}\int d^3y\, J(\vec{y})\, g^{-1}(\vec{y},\vec{x})\right)^2\right)\mathcal{G}[J]\bigg|_{J=0} \\
&= \int d^3x\, \frac{1}{8}\,\delta m^2\, g^{-1}(\vec{x},\vec{x})\, \mathcal{G}[0] \\
&= \frac{1}{4}\,\delta m^2 \int \frac{d^3k}{(2\pi)^3}\, \frac{1}{\omega_k} \int d^3x \\
&= \frac{1}{4}\delta m^2 \int d^3k\, \frac{1}{\omega_k}\, \delta^3(0).
\end{aligned} \tag{11.11}$$

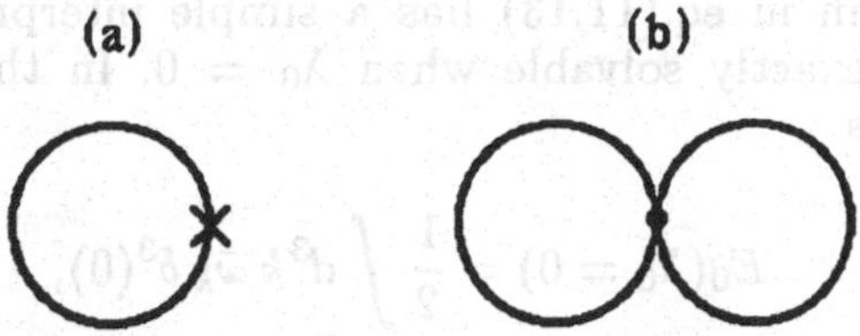

Figure 11.1 Two diagrams representing vacuum bubbles which contribute to the phase shift of the S-matrix in the interacting vacuum, $\langle 0\,|S|\,0\rangle$, or equivalently, the shift in the energy of the vacuum, $E_0^{(1)}$. Both diagrams vanish if the Hamiltonian is normal ordered.

Similarly,

$$\begin{aligned}\langle\Psi_0^{(0)}\,|H_2|\,\Psi_0^{(0)}\rangle &= \int d^3x\,\frac{\lambda_0}{4!}\frac{\delta^4}{\delta J(\vec{x})^4}\mathcal{G}[J]\bigg|_{J=0}\\ &= 3\frac{\lambda_0}{4!}\left(\frac{1}{4}g^{-1}(\vec{x},\vec{x})\right)^2\int d^3x \qquad (11.12)\\ &= \frac{\lambda_0}{32}\left(\int\frac{d^3k}{(2\pi)^3}\frac{1}{\omega_k}\right)^2(2\pi)^3\delta^3(0).\end{aligned}$$

All together,

$$E_0^{(1)} = \left(\frac{\delta m^2}{4}\int\frac{d^3k}{(2\pi)^3}\frac{1}{\omega_k} + \frac{\lambda_0}{32}\left(\int\frac{d^3k}{(2\pi)^3}\frac{1}{\omega_k}\right)^2\right)(2\pi)^3\delta^3(0). \qquad (11.13)$$

In the operator formalism of chapter 7, the comparable calculation of the vacuum energy is the denominator in eq.(7.97), the phase shift of the S-matrix on the interacting vacuum, $\langle 0\,|S|\,0\rangle$. At first order, this phase shift is $-i\langle 0\,|H_{\text{int}}|\,0\rangle$. In chapter 7, however, we normal ordered the Hamiltonian so $\langle 0|:H_{\text{int}}:|0\rangle = 0$. We have not normal ordered here so eq.(11.13) is not zero. We can remove the normal ordering in chapter 7 by allowing for contractions within the H_{int}. The two extra terms corresponding to $E_0^{(1)}$ above are represented by the diagrams in figure 11.1. When we apply the rules in chapter 7 to the diagrams, multiply by i, and do the k_0 integration, we get the same result as eq.(11.13).

The first term in eq.(11.13) has a simple interpretation, since the theory remains exactly solvable when $\lambda_0 = 0$. In this case, the exact vacuum energy is

$$E_0(\lambda_0 = 0) = \frac{1}{2}\int d^3k\, \bar{\omega}_k\, \delta^3(0), \tag{11.14}$$

where $\bar{\omega}_k = \sqrt{k^2 + m_0^2}$. In terms of $\omega_k = \sqrt{k^2 + m^2}$, we have $\bar{\omega}_k = \sqrt{k^2 + m^2 + \delta m^2} = \omega_k\sqrt{1 + \delta m^2/\omega_k^2}$. Expanding $\bar{\omega}_k$ to first order in δm^2, we find $\bar{\omega}_k = \omega_k(1 + \delta m^2/(2\omega_k^2) + \cdots)$. When we insert this into eq.(11.14), we recover eq.(11.11).

To determine the perturbed wave functional, $\Psi_0^{(1)}[\phi]$, via eq.(11.5), we must compute $\int \mathcal{D}\phi\, \Psi_M^{*(0)}[\phi]\, \phi^n(\vec{x})\, \Psi_0^{(0)}[\phi]$ for $n = 2, 4$. The result of the computation is eq.(11.24). The wave functional $\phi^4(\vec{x})\Psi_0^{(0)}[\phi]$ represents a combination of states containing 0, 2, and 4 particles. This can be seen by recalling that $\phi(\vec{x})$ is the eigenvalue of the field operator $\varphi(\vec{x})$ so the state above represents $\varphi(\vec{x})$ operating on vacuum 4 times. Each $\varphi(\vec{x})$ can be decomposed into a creation and destruction operator.

The free field wave functionals, $\Psi_M^{(0)}[\phi]$, are orthonormal, so the only states that will have a nonvanishing overlap with $\phi^4(\vec{x})\Psi_0^{(0)}[\phi]$ are the free states with 0, 2, and 4 particles. Because of the energy denominator in eq.(11.15), we do not need to worry about the state with zero particles. It is easiest to write down the excited state wave functionals in terms of $\tilde{\phi}(\vec{k})$. The states, $\Psi_M^{(0)}[\phi]$, that we need to consider are

$$\begin{aligned}
\Psi_2^{(0)}[\tilde{\phi}] &= \frac{1}{\sqrt{2!}}\left(\sqrt{\frac{2\omega_{k_1}}{(2\pi)^3}}\sqrt{\frac{2\omega_{k_2}}{(2\pi)^3}}\,\tilde{\phi}(\vec{k}_1)\tilde{\phi}(\vec{k}_2) - \delta^3(k_1 + k_2)\right)\Psi_0^{(0)}[\tilde{\phi}] \\
\Psi_4^{(0)}[\tilde{\phi}] &= \frac{1}{\sqrt{4!}}\Bigg[\sqrt{2\omega_{k_1}/(2\pi)^3}\cdots\sqrt{2\omega_{k_4}/(2\pi)^3}\,\tilde{\phi}(\vec{k}_1)\cdots\tilde{\phi}(\vec{k}_4) \\
&\quad - \Big(\sqrt{2\omega_{k_1}/(2\pi)^3}\sqrt{2\omega_{k_2}/(2\pi)^3}\,\tilde{\phi}(\vec{k}_1)\tilde{\phi}(\vec{k}_2)\,\delta^3(k_3 + k_4) \\
&\qquad + \cdots + \sqrt{2\omega_{k_3}/(2\pi)^3}\sqrt{2\omega_{k_4}/(2\pi)^3}\,\tilde{\phi}(\vec{k}_3)\tilde{\phi}(\vec{k}_4)\,\delta^3(k_1 + k_2)\Big) \\
&\quad - \Big(\delta^3(k_1 + k_2)\delta^3(k_3 + k_4) + \delta^3(k_1 + k_3)\delta^3(k_2 + k_4) \\
&\qquad + \delta^3(k_1 + k_4)\delta^3(k_2 + k_3)\Big)\Bigg]\Psi_0^{(0)}[\tilde{\phi}].
\end{aligned} \tag{11.15}$$

The first functional represents a state with one particle in mode $\vec{k}_1$ and one in $\vec{k}_2$, while the second functional represents a state with 4 particles in $\vec{k}_1 \cdots \vec{k}_4$.

Since the excited state wave functionals are written in terms of $\tilde{\phi}$, we might as well compute the overlap as a functional integral over $\tilde{\phi}$.

$$\int \mathcal{D}\phi\, \Psi_M^{*(0)}[\phi] \int d^3x\, \phi^4(\vec{x})\Psi_0^{(0)}[\phi]$$

$$= \int \frac{d^3k_1'}{(2\pi)^3} \cdots \frac{d^3k_4'}{(2\pi)^3} (2\pi)^3\, \delta^3(k_1' + k_2' + k_3' + k_4')$$

$$\int \mathcal{D}\tilde{\phi}\, \Psi_M^{*(0)}[\tilde{\phi}]\, \tilde{\phi}(\vec{k}_1')\, \tilde{\phi}(\vec{k}_2')\, \tilde{\phi}(\vec{k}_3')\, \tilde{\phi}(\vec{k}_4')\, \Psi_0^{(0)}[\tilde{\phi}] \quad (11.16)$$

For $\Psi_M^{(0)}[\tilde{\phi}] = \Psi_2^{(0)}[\tilde{\phi}]$, the product of Gaussian integrals will involve an odd moment that will cause the functional integral to vanish unless $\vec{k}_1$ in $\Psi_2^{(0)}$ is equal to one of the four $\vec{k}_j'$, $\vec{k}_2$ in $\Psi_2^{(0)}$ is equal to one of the three remaining $\vec{k}_j'$, and the last two $\vec{k}_j'$'s add to zero. The contribution from $\vec{k}_1 = \vec{k}_2$ cancels, leaving

$$\int \mathcal{D}\phi\, \Psi_2^{*(0)}[\phi] \int d^3x\, \phi^4(\vec{x})\, \Psi_0^{(0)}[\phi]$$

$$= \frac{6\sqrt{2}}{(2\pi)^3} \int \frac{d^3k_1'}{(2\pi)^3} \cdots \frac{d^3k_4'}{(2\pi)^3} (2\pi)^3\delta^3(k_1' + k_2' + k_3' + k_4')\sqrt{2\omega_{k_1}}\sqrt{2\omega_{k_2}}$$
$$\frac{(2\pi)^3\delta^3(k_1 + k_1')}{2\omega_{k_1}} \frac{(2\pi)^3\delta^3(k_2 + k_2')}{2\omega_{k_2}} \frac{(2\pi)^3\delta^3(k_3' + k_4')}{2\omega_{k_3'}}$$

$$= \frac{6\sqrt{2}}{\sqrt{2\omega_{k_1}}\sqrt{2\omega_{k_2}}}\, \delta^3(k_1 + k_2) \left(\int \frac{d^3k}{(2\pi)^3} \frac{1}{2\omega_k}\right). \quad (11.17)$$

For this case, the sum over the states $\Psi_M^{(0)}[\phi]$ in eq.(11.15), $\sum_M$, amounts to integrating over $\vec{k}_1$ and $\vec{k}_2$. The difference in energy between $E_0^{(0)}$ and $E_2^{(0)}$ is just the energy of the two particles, $\omega_{k_1} + \omega_{k_2}$. Thus, the λ_0 contribution to the perturbed wave functional from $\Psi_2^{(0)}$ is

$$\frac{\lambda_0}{4!}\sum_M \frac{\langle \Psi_2^{(0)} \,|\, \int d^3x\, \phi^4(\vec{x}) | \, \Psi_0^{(0)} \rangle}{E_0^{(0)} - E_M^{(0)}} \Psi_2^{(0)}[\tilde{\phi}]$$

$$= -\frac{\lambda_0}{4}\left(\int \frac{d^3k}{(2\pi)^3}\frac{1}{2\omega_k}\right)\left[\int \frac{d^3k_1}{(2\pi)^3}\frac{1}{2\omega_{k_1}}\tilde{\phi}(\vec{k}_1)\,\tilde{\phi}(-\vec{k}_1)\right.$$

$$\left. - \int \frac{d^3k_1}{(2\pi)^3}\frac{1}{(2\omega_{k_1})^2}(2\pi)^3\delta^3(0)\right]\Psi_0^{(0)}[\tilde{\phi}]. \quad (11.18)$$

The second term in eq.(11.18) that is independent of $\tilde{\phi}(\vec{k}_1)$ and $\tilde{\phi}(\vec{k}_2)$ is a contribution to the λ_0 correction to the normalization of Ψ_0 so as to maintain $\langle \Psi_0 \,|\, \Psi_0 \rangle = 1 + O(\lambda_0^2)$.

The contribution from the 4-particle functional in eq.(11.15) may be determined in a similar manner. We can compute the functional integral that represents the overlap by taking the appropriate eighth derivative of the generating functional, eq.(11.19), or its Fourier transform, or by writing the functional integral over $\tilde{\phi}$ as an infinite product of Gaussians as we did for the 2-particle contribution. Since odd moments of a Gaussian vanish, the integral is zero unless the $\vec{k}_i'$'s match the $\vec{k}_j$'s. The Gaussian integrals for these modes contribute $(2\omega_{k_i})^{-1}$ besides the usual factor that cancels the normalization in $\Psi_0^{(0)}[\tilde{\phi}]$.

$$\int \frac{d\tilde{\phi}(\vec{k}_1)}{\sqrt{(2\pi)^3}}\,\tilde{\phi}(-\vec{k}_1)\,\tilde{\phi}(\vec{k}_1')\,\exp\!\left(-\frac{\omega_{k_1}}{(2\pi)^3}\,\tilde{\phi}^2(|\vec{k}_1|)\right)$$

$$= \left(\frac{\pi}{\omega_{k_1}}\right)^{\frac{1}{2}}\frac{(2\pi)^3}{2\omega_{k_1}}\,\delta^3(k_1 - k_1') \quad (11.19)$$

Thus, the overlap is

$$\langle \Psi_4^{(0)} \,|\, \int d^3x\phi^4(\vec{x}) | \, \Psi_0^{(0)} \rangle = \frac{\sqrt{4!}\;\delta^3(k_1 + k_2 + k_3 + k_4)}{(2\pi)^3\sqrt{2\omega_{k_1}\cdots 2\omega_{k_4}}}. \quad (11.20)$$

The difference in energy between the two states is $E_4^{(0)} - E_0^{(0)} = \sum_i \omega_{k_i}$ and the sum over M is an integral over $\vec{k}_1 \cdots \vec{k}_4$. The contribution from $\Psi_4^{(0)}$ is

$$-\frac{\lambda_0}{4!}\int\frac{d^3k_1}{(2\pi)^3}\cdots\frac{d^3k_4}{(2\pi)^3}\frac{(2\pi)^3\delta^3(k_1+\cdots+k_4)}{\omega_{k_1}+\omega_{k_2}+\omega_{k_3}+\omega_{k_4}}\tilde{\phi}(\vec{k}_1)\cdots\tilde{\phi}(\vec{k}_4)\,\Psi_0^{(0)}[\tilde{\phi}]$$
$$+\frac{\lambda_0}{4}\int\frac{d^3k_1}{(2\pi)^3}\frac{d^3k_3}{(2\pi)^3}\frac{1}{2\omega_{k_3}}\frac{1}{2(\omega_{k_1}+\omega_{k_3})}\tilde{\phi}(\vec{k}_1)\tilde{\phi}(-\vec{k}_1)\,\Psi_0^{(0)}[\tilde{\phi}]$$
$$-3\frac{\lambda_0}{4!}\int\frac{d^3k_1}{(2\pi)^3}\frac{d^3k_3}{(2\pi)^3}\frac{1}{2\omega_{k_1}}\frac{1}{2\omega_{k_3}}\frac{1}{2(\omega_{k_1}+\omega_{k_3})}(2\pi)^3\delta^3(0)\,\Psi_0^{(0)}[\tilde{\phi}]. \tag{11.21}$$

The last term is another λ_0 contribution to the normalization.

To determine the δm^2 contribution to $\Psi_0^{(1)}[\phi]$, we need to compute $\int\mathcal{D}\phi\,\Psi_M^{*(0)}[\phi]\,\phi^2(\vec{x})\,\Psi_0^{(0)}[\phi]$. The wave functional, $\phi^2(\vec{x})\Psi_0^{(0)}[\phi]$, represents a state containing zero or two particles, thus we need only use $\Psi_M^{(0)}=\Psi_2^{(0)}$ of eq.(11.15). The resulting overlap is

$$\langle\Psi_2^{(0)}(k_1,k_2)|\int d^3x\,\phi^2(\vec{x})|\,\Psi_0^{(0)}\rangle=\sqrt{2}\sqrt{2\omega_{k_1}}\sqrt{2\omega_{k_2}}\,\frac{\delta^3(k_1-k_2)}{2\omega_{k_1}\,2\omega_{k_2}}. \tag{11.22}$$

Therefore, the δm^2 contribution is

$$-\frac{\delta m^2}{2}\int\frac{d^3k_1}{(2\pi)^3}\frac{1}{2\omega_{k_1}}\tilde{\phi}(\vec{k}_1)\tilde{\phi}(-\vec{k}_1)\,\Psi_0^{(0)}[\tilde{\phi}]$$
$$+\frac{\delta m^2}{2}\int\frac{d^3k_1}{(2\pi)^3}\frac{1}{(2\omega_{k_1})^2}(2\pi)^3\delta^3(0)\Psi_0^{(0)}[\tilde{\phi}]. \tag{11.23}$$

As with $E_0^{(1)}(\delta m^2)$, eq.(11.11), we may obtain eq.(11.23) by taking $\Psi_0^{(0)}[\tilde{\phi}]$ with $\tilde{g}(k)=1/2\,\tilde{\omega}_k$, which is the exact vacuum functional for $\lambda_0=0$, expand $\tilde{\omega}_k$ around ω_k, and expand the exponential term to obtain the part proportional to δm^2. This piece is identical to the first term in eq.(11.23). The second term maintains the normalization through order δm^2.

All together, the entire perturbed functional to first order is the sum of eqs.(11.18), (11.21), and (11.23).

$$\Psi_0^{(1)}=$$
$$-\frac{\delta m^2}{2}\int\frac{d^3k_1}{(2\pi)^3}\Big(\frac{1}{2\omega_{k_1}}\tilde{\phi}(\vec{k}_1)\tilde{\phi}(-\vec{k}_1)-\frac{1}{(2\omega_{k_1})^2}(2\pi)^3\delta^3(0)\Big)\Psi_0^{(0)}[\tilde{\phi}]$$
$$-\frac{\lambda_0}{4}\int\frac{d^3k}{(2\pi)^3}\int\frac{d^3k_1}{(2\pi)^3}\frac{1}{2\omega_{k_1}}\frac{1}{2(\omega_k+\omega_{k_1})}\tilde{\phi}(\vec{k}_1)\,\tilde{\phi}(-\vec{k}_1)\,\Psi_0^{(0)}[\tilde{\phi}]$$

$$+\frac{\lambda_0}{4}\left(\int\frac{d^3k}{(2\pi)^3}\frac{1}{2\omega_k}\right)\int\frac{d^3k_1}{(2\pi)^3}\frac{1}{(2\omega_{k_1})^2}(2\pi)^3\delta^3(0)\,\Psi_0^{(0)}[\tilde{\phi}]$$
$$-3\frac{\lambda_0}{4!}\int\frac{d^3k_1}{(2\pi)^3}\frac{d^3k_3}{(2\pi)^3}\frac{1}{2\omega_{k_1}}\frac{1}{2\omega_{k_3}}\frac{1}{2(\omega_{k_1}+\omega_{k_3})}(2\pi)^3\delta^3(0)\,\Psi_0^{(0)}[\tilde{\phi}]$$
$$-\frac{\lambda_0}{4!}\int\frac{d^3k_1}{(2\pi)^3}\cdots\frac{d^3k_4}{(2\pi)^3}\frac{(2\pi)^3\delta^3(k_1+\cdots+k_4)}{\omega_{k_1}+\cdots+\omega_{k_4}}\tilde{\phi}(\vec{k}_1)\cdots\tilde{\phi}(\vec{k}_4)\,\Psi_0^{(0)}[\tilde{\phi}] \tag{11.24}$$

For illustration, let us obtain the same results for $E_0^{(1)}$ and $\Psi_0^{(1)}[\phi]$ by directly solving the functional Schrödinger equation. Following eq.(11.44), we will apply the results to 2-body elastic scattering.

To lowest order in λ_0 and δm^2, let us write

$$\Psi_0[\phi]=\eta\exp(-G[\phi]-\delta m^2\,R_1[\phi]-\lambda_0\,R_2[\phi]). \tag{11.25}$$

$G[\phi]$ is the free field quadratic functional found in chapter 10.

$$G[\phi]=\int d^3x\,d^3y\,\phi(\vec{x})\,g(\vec{x},\vec{y})\,\phi(\vec{y}), \tag{11.26}$$

where $g(\vec{x},\vec{y})$ is given by eq.(10.23). $R_1[\phi]$ and $R_2[\phi]$ are unknown functionals that we want to solve for. The results are expressed in the form of eqs.(11.28), (11.30), (11.32), (11.40), and (11.43). We will find that $-\delta m^2\,R_1[\phi]-\lambda_0\,R_2[\phi]$ is just eq.(11.24). The value of directly solving the Schrödinger equation is that it suggests that we write the solution as an exponential. We then see that we can exponentiate eq.(11.24) to obtain contributions to the perturbative wave functional for higher powers of λ_0 and δm^2.

If we place the functional, eq.(11.25), into the Schrödinger equation based on eq.(11.2), and separately collect the terms proportional to λ_0 and δm^2, we find that $R_{1,2}[\phi]$ must satisfy

$$\begin{aligned}\frac{1}{2}\int d^3x\left(\frac{\delta^2R_1}{\delta\phi^2(\vec{x})}-2\frac{\delta G}{\delta\phi(\vec{x})}\frac{\delta R_1}{\delta\phi(\vec{x})}\right)+\phi^2(\vec{x})&=\frac{E_0^{(1)}(\delta m^2)}{\delta m^2}\\ \int d^3x\frac{1}{2}\left(\frac{\delta^2R_2}{\delta\phi^2(\vec{x})}-2\frac{\delta G}{\delta\phi(\vec{x})}\frac{\delta R_2}{\delta\phi(\vec{x})}\right)+\frac{1}{4!}\phi^4(\vec{x})&=\frac{E_0^{(1)}(\lambda_0)}{\lambda_0}.\end{aligned} \tag{11.27}$$

Once again, we apply power counting to guess at trial functionals. For R_1, we can see from eq.(11.27) that we must produce a number with no dependence on ϕ, $E_0^{(1)}(\delta m^2)$, as well as a quadratic functional $\phi^2(\vec{x})$, by differentiation. This is the same situation as in the free field case. If R_1 is quadratic in ϕ, the second derivative will produce a pure number. Since the first derivative of G contains a single factor of ϕ, then the first derivative of a quadratic R_1 combined with the first derivative of G is quadratic. Therefore, we may take

$$R_1[\phi] = \int d^3x\, d^3y\, \phi(\vec{x})\, h_1(\vec{x}, \vec{y})\, \phi(\vec{y}). \tag{11.28}$$

When we plug this functional back into eq.(11.27), we find that the symmetric $h_1(\vec{x}, \vec{y})$ must satisfy

$$4 \int d^3z\, g(\vec{x}, \vec{z})\, h_1(\vec{z}, \vec{y}) = \frac{1}{2}\delta^3(x - y)$$

and

$$E_0^{(1)}(\delta m^2) = \delta m^2 \int d^3x\, h_1(\vec{x}, \vec{x}). \tag{11.29}$$

To solve this, we Fourier transform it and use the known Fourier transform of $g(\vec{x}, \vec{z})$, eq.(10.23). The result is $4(\omega_k/2)\tilde{h}_1(\vec{k}) = 1/2$, hence

$$h_1(\vec{x}, \vec{y}) = \frac{1}{4} \int \frac{d^3k}{(2\pi)^3} \frac{1}{\omega_k} e^{i\vec{k}\cdot(\vec{x}-\vec{y})}, \tag{11.30}$$

and

$$E_0^{(1)}(\delta m^2) = \frac{\delta m^2}{4} \int \frac{d^3k}{(2\pi)^3} \frac{1}{\omega_k} \int d^3x, \tag{11.31}$$

which agrees with eqs.(11.11) and (11.23).

Moving on to $R_2[\phi]$, we see from eq.(11.27) that it also must produce a pure number, $E_0^{(1)}(\lambda_0)/\lambda_0$. Since the first derivative of G contains one factor of ϕ, the only way to produce a number is to let R_2 contain a quadratic functional. The cross derivative of a quadratic R_2 with G will only produce a quadratic functional which can never match the potential, $\phi^4(\vec{x})$. Therefore, R_2 must also include a functional that contains 4 factors of ϕ. So, we take

$$\begin{aligned} R_2[\phi] = &\int d^3x\, d^3y\, \phi(\vec{x})\, h_2(\vec{x}, \vec{y})\phi(\vec{y}) \\ &+ \int d^3x_1 \cdots d^3x_4\, h(\vec{x}_1, \ldots, \vec{x}_4)\phi(\vec{x}_1)\cdots\phi(\vec{x}_4). \end{aligned} \tag{11.32}$$

Because of the symmetric nature of the factors of $\phi(\vec{x}_i)$ in R_2, we may take h_2 and h to be symmetric in their arguments, $h_2(\vec{x}, \vec{y}) = h_2(\vec{y}, \vec{x})$, $h(\vec{x}_3, \vec{x}_4, \vec{x}_2, \vec{x}_1) = h(\vec{x}_1, \vec{x}_2, \vec{x}_3, \vec{x}_4)$, etc.

Plugging eq.(11.32) back into eq.(11.27) and equating terms of like powers of ϕ, we find that h_2 and h must satisfy

$$E_0^{(1)}(\lambda_0) = \lambda_0 \int d^3x\, h_2(\vec{x}, \vec{x}), \tag{11.33}$$

$$4\int d^3x\, d^3y\, d^3z\, g(\vec{x},\vec{z})\, h_2(\vec{z},\vec{y})\, \phi(\vec{x})\, \phi(\vec{y})$$
$$= 6\int d^3x\, d^3x_3\, d^3x_4\, h(\vec{x},\vec{x},\vec{x}_3,\vec{x}_4)\phi(\vec{x}_3)\phi(\vec{x}_4), \tag{11.34}$$

$$8\int d^3x\, d^3x_1 \cdots d^3x_4\, g(\vec{x}_1,\vec{x})\, h(\vec{x},\vec{x}_2,\vec{x}_3,\vec{x}_4)\phi(\vec{x}_1)\cdots\phi(\vec{x}_4)$$
$$= \frac{1}{4!}\int d^3x\, \phi^4(\vec{x}). \tag{11.35}$$

The third equation above, eq.(11.35), determines h. With h, we can find h_2 using the second equation, eq.(11.34). Then, with h_2, we can determine the energy eigenvalue, $E_0^{(1)}(\lambda_0)$, from eq.(11.33).

To untangle the proper relationship between g and h from eq.(11.35) above, we functionally differentiate both sides of eq.(11.35) four times, once each with respect to $\phi(\vec{z}_i)$ for $i = 1, \ldots, 4$. The result is that h must satisfy

$$48\int d^3x\Big(g(\vec{z}_1,\vec{x})h(\vec{x},\vec{z}_2,\vec{z}_3\vec{z}_4) + g(\vec{z}_2,\vec{x})h(\vec{z}_1,\vec{x},\vec{z}_3,\vec{z}_4)$$
$$+ g(\vec{z}_3,\vec{x})h(\vec{z}_1,\vec{z}_2,\vec{x},\vec{z}_4) + g(\vec{z}_4,\vec{x})h(\vec{z}_1,\vec{z}_2,\vec{z}_3,\vec{x})\Big)$$
$$= \delta^3(z_1 - z_2)\, \delta^3(z_1 - z_3)\, \delta^3(z_1 - z_4). \tag{11.36}$$

To solve this equation for h, we write g and h in terms of their Fourier transforms $\tilde{g}$, $\tilde{h}$.

$$g(\vec{x},\vec{y}) = \frac{1}{2}\int \frac{d^3k}{(2\pi)^3}\omega_k\, e^{i\vec{k}\cdot(\vec{x}-\vec{y})}$$
$$h(\vec{x}_1,\vec{x}_2\vec{x}_3,\vec{x}_4) = \int \frac{d^3k_1}{(2\pi)^3}\cdots\frac{d^3k_4}{(2\pi)^3}\,\tilde{h}(\vec{k}_1,\cdots,\vec{k}_4)e^{i\vec{k}_1\cdot\vec{x}_1}\cdots e^{i\vec{k}_4\cdot\vec{x}_4} \tag{11.37}$$

The left-hand side of eq.(11.36) becomes

$$24\int \frac{d^3k_1}{(2\pi)^3}\cdots\frac{d^3k_4}{(2\pi)^3}\, e^{i\vec{k}_1\cdot\vec{z}_1}\cdots e^{i\vec{k}_4\cdot\vec{z}_4}(\omega_{k_1}+\omega_{k_2}+\omega_{k_3}+\omega_{k_4})\,\tilde{h}(\vec{k}_1,\vec{k}_2,\vec{k}_3,\vec{k}_4). \tag{11.38}$$

We can see that the energy denominator of eq.(11.5), $E_M^{(0)} - E_0^{(0)}$, has appeared in eq.(11.38). The right-hand side of eq.(11.36) is conveniently written as

$$\delta^3(z_2 - z_1)\, \delta^3(z_3 - z_1)\, \delta^3(z_4 - z_1)$$
$$= \int \frac{d^3k_2}{(2\pi)^3}\frac{d^3k_3}{(2\pi)^3}\frac{d^3k_4}{(2\pi)^3}\, e^{i\vec{k}_2\cdot(\vec{z}_2-\vec{z}_1)}e^{i\vec{k}_3\cdot(\vec{z}_3-\vec{z}_1)}e^{i\vec{k}_4\cdot(\vec{z}_4-\vec{z}_1)}$$

$$= \int \frac{d^3k_1}{(2\pi)^3} \cdots \frac{d^3k_4}{(2\pi)^3} (2\pi)^3\, \delta^3(k_1+k_2+k_3+k_4)\, e^{i\vec{k}_1\cdot\vec{z}_1} e^{i\vec{k}_2\cdot\vec{z}_2} e^{i\vec{k}_3\cdot\vec{z}_3} e^{i\vec{k}_4\cdot\vec{z}_4}. \tag{11.39}$$

Equating eqs.(11.38) and (11.39) we find

$$\tilde{h}(\vec{k}_1, \vec{k}_2, \vec{k}_3, \vec{k}_4) = \frac{1}{24} \frac{(2\pi)^3 \delta^3(k_1+k_2+k_3+k_4)}{\omega_{k_1}+\omega_{k_2}+\omega_{k_3}+\omega_{k_4}}, \tag{11.40}$$

which is what we found earlier, eq.(11.21).

We now apply the same procedure to eq.(11.34) and differentiate both sides by $\delta/\delta\phi(\vec{z}_1)$ and $\delta/\delta\phi(\vec{z}_2)$.

$$12 \int d^3x\, h(\vec{x}, \vec{x}, \vec{z}_1, \vec{z}_2) = 4 \int d^3x \Big(g(\vec{z}_1, \vec{x})\, h_2(\vec{x}, \vec{z}_2) + g(\vec{z}_2, \vec{x})\, h_2(\vec{x}, \vec{z}_1) \Big) \tag{11.41}$$

After Fourier transforming and using $\tilde{h}$ from eq.(11.40), eq.(11.41) implies that $\tilde{h}_2$ satisfies

$$\begin{aligned} \frac{1}{4} \int \frac{d^3k_1}{(2\pi)^3} \frac{d^3k_2}{(2\pi)^3} \frac{e^{i\vec{k}_1\cdot(\vec{z}_1-\vec{z}_2)}}{\omega_{k_1}+\omega_{k_2}} & \\ = 2 \int \frac{d^3k_1}{(2\pi)^3} \frac{d^3k_2}{(2\pi)^3} & (\omega_{k_1}+\omega_{k_2}) \tilde{h}_2(\vec{k}_1, \vec{k}_2)\, e^{i\vec{k}_1\cdot\vec{z}_1} e^{i\vec{k}_2\cdot\vec{z}_2}. \end{aligned} \tag{11.42}$$

Thus,

$$\tilde{h}_2(\vec{k}_1, \vec{k}_2) = \frac{1}{16} \frac{\delta^3(k_1+k_2)}{\omega_{k_1}} \int \frac{d^3k}{(2\pi)^3} \frac{1}{\omega_k+\omega_{k_1}}. \tag{11.43}$$

Again, this agrees with eq.(11.24). From eq.(11.43) and eq.(11.33) we find that the eigenvalue is

$$\begin{aligned} E_0^{(1)}(\lambda_0) &= \lambda_0 \int \frac{d^3k_1}{(2\pi)^3}\, \tilde{h}_2(\vec{k}_1, -\vec{k}_1) \int d^3x \\ &= \frac{\lambda_0}{16} \int \frac{d^3k}{(2\pi)^3} \frac{d^3k_1}{(2\pi)^3} \frac{1}{\omega_{k_1}} \frac{1}{\omega_k+\omega_{k_1}} \int d^3x \\ &= \frac{\lambda_0}{16} \int \frac{d^3k}{(2\pi)^3} \frac{d^3k_1}{(2\pi)^3} \frac{1}{2} \left(\frac{1}{\omega_{k_1}} \frac{1}{\omega_k+\omega_{k_1}} + \frac{1}{\omega_k} \frac{1}{\omega_k+\omega_{k_1}} \right) \int d^3x \\ &= \frac{\lambda_0}{32} \int \frac{d^3k}{(2\pi)^3} \frac{d^3k_1}{(2\pi)^3} \frac{1}{\omega_k\,\omega_{k_1}} \int d^3x, \end{aligned} \tag{11.44}$$

which agrees with our previous result, eq.(11.12).

Now let us consider the same 2-body elastic scattering process as we did at the end of chapter 7. Suppose that two particles of momentum p_1 and p_2 collide and depart with momenta p_1' and p_2'. The initial state, $|p_1p_2\rangle$, is a state with two particles of momenta p_1 and p_2. The final state is $|p_1'p_2'\rangle$. The amplitude, or S-matrix element, corresponding to this process is $\langle p_1'p_2'|p_1p_2\rangle$. There are two entirely equivalent ways to approach computing this scalar product.

The first way is to determine the perturbative wave functional for the state containing the two free particles. This means that we would compute corrections to $\Psi^{(0)}_{p_1p_2}[\phi]$ by either directly solving the functional Schrödinger equation or using Rayleigh-Schrödinger perturbation theory, eqs.(11.5)-(11.6). The S-matrix element is the functional integral

$$\begin{aligned}S_{\beta\alpha} &= \langle p_1'p_2'\,|p_1p_2\rangle\\ &= \int \mathcal{D}\phi \Psi^*_{p_1'p_2'}[\phi]\,\Psi_{p_1p_2}[\phi]\\ &= \int \mathcal{D}\phi\,(\Psi^{*(0)}_{p_1'p_2'}[\phi] + \Psi^{*(1)}_{p_1'p_2'}[\phi] + \cdots)(\Psi^{(0)}_{p_1p_2}[\phi] + \Psi^{(1)}_{p_1p_2}[\phi] + \cdots).\end{aligned} \tag{11.45}$$

The alternative viewpoint to determine the S-matrix element is to use the functional form of the creation operator, eq.(10.40), and let it operate on the interacting vacuum wave functional, $\Psi_0[\phi]$.

$$\langle\phi\,|p_1p_2\rangle = \Psi_{p_1p_2}[\phi] = \mathrm{a}^\dagger(\vec{p}_2)\mathrm{a}^\dagger(\vec{p}_1)\Psi_0[\phi]. \tag{11.46}$$

$$\begin{aligned}S_{\beta\alpha} &= \int \mathcal{D}\phi\,\Psi_0^*[\phi]\,\mathrm{a}(\vec{p}_1{}')\mathrm{a}(\vec{p}_2{}')\mathrm{a}^\dagger(\vec{p}_2)\mathrm{a}^\dagger(\vec{p}_1)\Psi_0[\phi]\\ &= \int \mathcal{D}\phi(\Psi_0^{*(0)}[\phi] + \Psi_0^{*(1)}[\phi] + \cdots)\\ &\qquad \mathrm{a}(\vec{p}_1{}')\mathrm{a}(\vec{p}_2{}')\mathrm{a}^\dagger(\vec{p}_2)\mathrm{a}^\dagger(\vec{p}_1)(\Psi_0^{(0)}[\phi] + \Psi_0^{(1)}[\phi] + \cdots).\end{aligned} \tag{11.47}$$

This approach is in line with the spirit of the IN-OUT fields.

As long as we have already explicitly computed the perturbative vacuum wave functional to first order, we will take the latter approach. Expanding eq.(11.47) and explicitly including the normalization we have

$$\begin{aligned}S_{\beta\alpha} =&\eta^2 \int \mathcal{D}\phi\,\Psi_0^{*(0)}[\phi]\mathrm{a}(\vec{p}_1{}')\,\mathrm{a}(\vec{p}_2{}')\,\mathrm{a}^\dagger(\vec{p}_2)\,\mathrm{a}^\dagger(\vec{p}_1)\Psi_0^{(0)}[\phi]\\ &+ \eta^2 \int \mathcal{D}\phi\,\Psi_0^{*(0)}[\phi]\mathrm{a}(\vec{p}_1{}')\,\mathrm{a}(\vec{p}_2{}')\,\mathrm{a}^\dagger(\vec{p}_2)\,\mathrm{a}^\dagger(\vec{p}_1)\Psi_0^{(1)}[\phi]\\ &+ \eta^2 \int \mathcal{D}\phi\,\Psi_0^{*(1)}[\phi]\mathrm{a}(\vec{p}_1{}')\,\mathrm{a}(\vec{p}_2{}')\,\mathrm{a}^\dagger(\vec{p}_2)\,\mathrm{a}^\dagger(\vec{p}_1)\Psi_0^{(0)}[\phi] + \dots\,,\end{aligned} \tag{11.48}$$

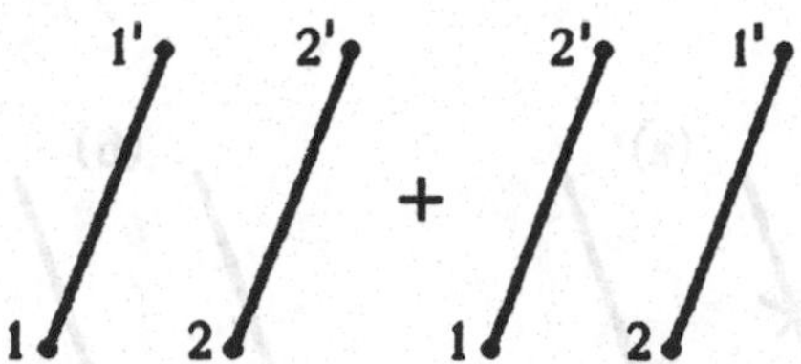

Figure 11.2 Free field contribution to the 2-body elastic scattering S-matrix element. This amplitude is equal to the overlap of the free particle wave functionals.

where

$$\eta^{-2} = \int \mathcal{D}\phi\, (\Psi_0^{*(0)}[\phi] + \Psi_0^{*(1)}[\phi] + \cdots)(\Psi_0^{(0)}[\phi] + \Psi_0^{(1)}[\phi] + \cdots). \quad (11.49)$$

The effect of carrying out the normalization is the equivalent of dividing out the disconnected vacuum bubbles.

Since the creation and destruction operators are naturally written in terms of $\tilde{\phi}$, it will be more convenient to functionally integrate (11.48) over $\tilde{\phi}$. With

$$\begin{aligned} \mathrm{a}(\vec{p}) &= \omega_p\, \tilde{\phi}(\vec{p}) + \frac{\delta}{\delta\tilde{\phi}(-\vec{p})} \\ \mathrm{a}^\dagger(\vec{p}) &= \omega_p\, \tilde{\phi}(\vec{p}) - \frac{\delta}{\delta\tilde{\phi}(-\vec{p})}, \end{aligned} \quad (11.50)$$

the first integral in eq.(11.48), the free field contribution, is

$$16\eta^2 \int \mathcal{D}\tilde{\phi}\, \omega_{p_1'}\omega_{p_2'}\omega_{p_1}\omega_{p_4}\, \tilde{\phi}(-\vec{p}_1)\tilde{\phi}(-\vec{p}_2)\tilde{\phi}(\vec{p}_1')\tilde{\phi}(\vec{p}_2') \exp\left(-\int \frac{d^3p}{(2\pi)^3}\, \omega_p\, \tilde{\phi}(-\vec{p})\tilde{\phi}(\vec{p})\right). \quad (11.51)$$

Since odd moments vanish, we must have $\vec{p}_1 = \vec{p}_1'$, $\vec{p}_2 = \vec{p}_2'$, or $\vec{p}_1 = \vec{p}_2'$, $\vec{p}_2 = \vec{p}_1'$. Therefore, the first term in eq.(11.48) is

$$4\omega_{p_1}\omega_{p_2}\Big((2\pi)^3\,\delta^3(p_1 - p_1')\,(2\pi)^3\,\delta^3(p_2 - p_2') + (2\pi)^3\,\delta^3(p_1 - p_2')\,(2\pi)^3\,\delta^3(p_2 - p_1')\Big), \quad (11.52)$$

which is identical to the S-matrix elements arising from the diagrams in figure 11.2, computed using the methods and rules of chapter 7. See also eq.(7.52).

Figure 11.3 Contributions to the S-matrix arising from the overlap of the free particle wave functionals and the first-order perturbatively corrected functional proportional to δm^2. The result is equivalent to, but not equal to, the operator rules (non normal ordered) applied to diagram (b). The difference is the number represented by the vacuum bubble.

The second and third integrals in eq.(11.48) involve the first correction $\Psi_0^{(1)}[\tilde{\phi}]$, eq.(11.24), of the vacuum wave functional. Let us consider the contribution to eq.(11.48) from the term in $\Psi_0^{(1)}[\tilde{\phi}]$ that is proportional to δm^2. As usual, the nonvanishing contribution from eq.(11.48) will occur when the momenta $\vec{p}_i$, $\vec{p}_i'$, and $\vec{k}_1$ from eqs.(11.48) and (11.24) match to form even moments. The case where $\vec{k}_1$ from the δm^2 term of $\Psi_0^{(1)}[\tilde{\phi}]$ matches one $\vec{p}_i$ and one $\vec{p}_i'$ yields a result that is identical to that of the diagram in figure 11.3(a). When the $\vec{p}_i$ match the $\vec{p}_i'$, we obtain the free field contribution, eq.(11.52), multiplied by a number. This result is equivalent to, but not equal to, the diagram in figure 11.3(b). The difference is in the number represented by the vacuum bubble.

Next, consider the contribution to eq.(11.48) from the quadratic functional in eq.(11.24) that is proportional to λ_0. As above, if k_1 matches one

Figure 11.4 Contribution to the S-matrix arising from the overlap of the free particle wave functionals and the first-order quadratic functional proportional to λ_0. Operator rules applied to (b) yield equivalent but unequal results because of the vacuum bubble.

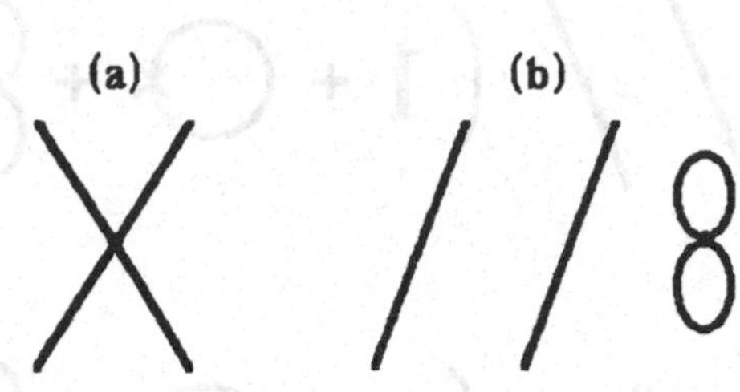

Figure 11.5 Contribution to the S-matrix arising from the projection of the free particle wave functionals onto the quartic portion of the first-order perturbative functional proportional to λ_0. Only (a) contributes to the connected part of S.

p_i and one p_i', the result is identical to that of diagram 11.4(a). If the p_i match the p_i', we again find the free field result multiplied by a number as represented in figure 11.4(b).

Finally, if we consider the remaining part of $\Psi_0^{(1)}$, the quartic functional proportional to λ_0, we have two possibilities represented by the diagrams in figure 11.5. The result given by eq.(11.48) is identical to that of the diagram in figure 11.5(a) and arises when the 4 k_1's match all of the p_i, p_i'. The situation represented in figure 11.5(b) occurs when the p_i's and p_i''s match and is just a number times the free field contribution.

If we sum together the results from eq.(11.48) corresponding to the diagrams in figures 11.2, 11.3(b), 11.4(b), and 11.5(b), we have the free field result, eq.(11.52), multiplied by a number. This sum corresponds to the diagram in figure 11.6. The number multiplying this result differs between the two representations. In the Schrödinger representation, this number is precisely the normalization of the vacuum wave functional $\Psi_0[\phi]$ expanded in a perturbation series. In other words, this number is absorbed in the normalization so the effect of normalizing the wave functionals is to remove the disconnected vacuum bubbles.

All of the processes in figures 11.2-11.5 are disconnected except for figure 11.5(a). This is the only process to this order that contributes to the scattering. Let us explicitly calculate the S-matrix contribution from this part of eq.(11.48) and show that it is identical to the result obtained in chapter 7, eq.(7.131).

Combining the second and third integrals in eq.(11.48), the functional integral that we must calculate is

(a)

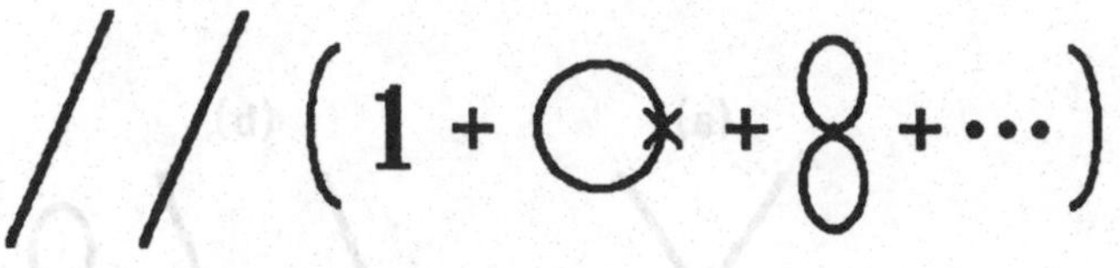

(b) $\eta^{-2} = (1 + \ldots + \ldots + \cdots)$

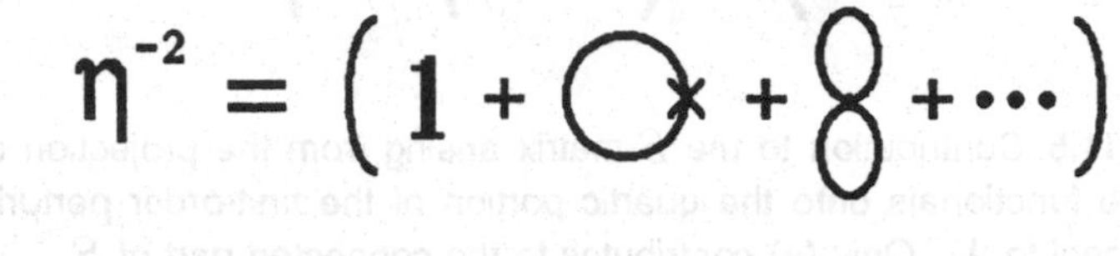

Figure 11.6 (a) Sum of terms to first order that contribute to the disconnected part of the S-matrix element for 2-body elastic scattering. The sum of the vacuum bubbles is just an ordinary number that can be factored out. (b) This sum of vacuum bubbles represents the normalization of the perturbed wave functional. Since the wave functionals must be normalized, all factors arising from vacuum bubbles will cancel and we need only compute overlaps that do not involve such disconnected vacuum processes.

$$-\frac{\lambda_0}{4!}\eta^2 \int \frac{d^3k_1}{(2\pi)^3} \cdots \frac{d^3k_4}{(2\pi)^3} \frac{(2\pi)^3\delta^3(k_1+k_2+k_3+k_4)}{\omega_{k_1}+\omega_{k_2}+\omega_{k_3}+\omega_{k_4}}$$

$$16\omega_{p_1}\cdots\omega_{p_2'} \int \mathcal{D}\tilde{\phi}\, \tilde{\phi}(-\vec{p}_1)\tilde{\phi}(-\vec{p}_2{}')\tilde{\phi}(\vec{k}_1)\cdots\tilde{\phi}(\vec{k}_4)$$

$$\tilde{\phi}(\vec{p}_1)\tilde{\phi}(\vec{p}_2)\Psi_0^{*(0)}[\tilde{\phi}]\Psi_0^{(0)}[\tilde{\phi}]e^{i(\omega_{p_1'}+\omega_{p_2'})t'}e^{-i(\omega_{p_1}+\omega_{p_2})t}, \quad (11.53)$$

where we have explicitly included the time dependence of the states (initial state at t, final state at t'). In order to get a nonzero result, we must have $\vec{p}_1{}' = \vec{k}_1$, $\vec{p}_2{}' = \vec{k}_2$, $\vec{p}_1 = -\vec{k}_3$, $\vec{p}_2 = -\vec{k}_4$, or $\vec{p}_1{}' = \vec{k}_2$, $\vec{p}_2{}' = \vec{k}_1$, $\vec{p}_1 = -\vec{k}_3$, $\vec{p}_2 = -\vec{k}_4$, etc. Since we are integrating over the $\vec{k}_i$'s, all 24 different combinations of matching $\vec{p}_i$ and $\vec{p}_i{}'$ with $\vec{k}_j$ give the same result. After carrying out the functional integration, eq.(11.53) becomes

$$-\lambda_0 \int \frac{d^3k_1}{(2\pi)^3} \cdots \frac{d^3k_4}{(2\pi)^3} \frac{(2\pi)^3\delta^3(k_1+k_2+k_3+k_4)}{\omega_{k_1}+\omega_{k_2}+\omega_{k_3}+\omega_{k_4}}$$

$$16\omega_{p_1}\cdots\omega_{p_2'} \frac{(2\pi)^3\delta^3(p_1'-k_1)}{2\omega_{p_1'}} \frac{(2\pi)^3\delta^3(p_2'-k_2)}{2\omega_{p_2'}}$$

$$\frac{(2\pi)^3\delta^3(p_1+k_3)}{2\omega_{p_1}} \frac{(2\pi)^3\delta^3(p_2+k_4)}{2\omega_{p_2}} e^{-(\omega_{p_1'}+\omega_{p_2'})t'} e^{-i(\omega_{p_1}+\omega_{p_2})t}$$

$$= -\lambda_0 \frac{(2\pi)^3\delta^3(p_1'+p_2'-p_1-p_2)}{\omega_{p_1'}+\omega_{p_2'}+\omega_{p_1}+\omega_{p_2}} e^{i(\omega_{p_1'}+\omega_{p_2'})t'} e^{-i(\omega_{p_1}+\omega_{p_2})t}. \qquad (11.54)$$

Experimental measurements are only made in the asymptotic region, so we want to consider $t' \to \infty$, $t \to -\infty$. Take $T = t' = -t$. The S-matrix contribution corresponding to figure 11.5(a) is

$$\begin{aligned} S_{fi} &= -\lambda_0 \lim_{T\to\infty} \frac{(2\pi)^3\delta^3(p_1'+p_2'-p_1-p_2)}{\omega_{p_1'}+\omega_{p_2'}+\omega_{p_1}+\omega_{p_2}} e^{-(\omega_{p_1'}+\omega_{p_2'}+\omega_{p_1}+\omega_{p_2})T} \\ &= -i\lambda_0\,(2\pi)^3\,\delta^3(p_1'+p_2'-p_1-p_2)\,(2\pi)\,\delta(\omega_{p_1'}+\omega_{p_2'}+\omega_{p_1}+\omega_{p_2}) \\ &= -i\lambda_0(2\pi)^4\delta^4(p_1'+p_2'-p_1-p_2), \end{aligned} \qquad (11.55)$$

where we have used the relation

$$\lim_{T\to\infty} \frac{e^{i\omega T}}{\omega} = 2\pi i\,\delta(\omega). \qquad (11.56)$$

Eq.(11.55) agrees with eq.(7.131). At this point, the procedure for carrying the S-matrix through to a cross section is the same as in chapter 7.

11.2 YANG-MILLS

We are going to consider "pure" Yang-Mills with no dynamical quarks. The action is given in chapter 6, eq.(6.18).

$$\mathcal{L}(x) = -\frac{1}{2}\mathrm{tr}\, F^{\mu\nu}F_{\mu\nu}, \qquad (11.57)$$

where

$$F^{\mu\nu} = \partial^\mu A^\nu - \partial^\nu A^\mu + ig_0[A^\mu, A^\nu], \tag{11.58}$$

is the field-strength tensor, the nonabelian generalization of the electric and magnetic fields. $F^{\mu\nu}$ is the matrix $F_a^{\mu\nu}T^a$, where T^a is a matrix representation of the generators of the group of 2×2 unitary matrices with unit determinant, $SU(2)$. The T^a's satisfy the algebra

$$[T^a, T^b] = i\epsilon^{abc}T_c. \tag{11.59}$$

In the adjoint representation, the index a, referred to as the "color" index, runs from 1 to 3 and we may take $T^a = \sigma^a/2$, where the σ^a are the usual Pauli matrices.

We are going to quantize Yang-Mills in the Schrödinger representation in the temporal gauge, $A_0^a = 0$. We choose the temporal gauge for the same reasons as discussed in the free photon case in the previous chapter. In this gauge, the field momentum is

$$\Pi_a^i(x) = F_a^{0i} = \dot{A}_a^i = -E_a^i, \tag{11.60}$$

the color electric field. The Hamiltonian is

$$H = \frac{1}{2}\int d^3x\,(\vec{E}^a \cdot \vec{E}^a + \vec{B}^a \cdot \vec{B}^a), \tag{11.61}$$

where $B^{ia} = -(1/2)\epsilon^{ijk}F_{jk}^a$ is the color magnetic field. Canonical quantization implies that we take the equal-time commutators,

$$[E_a^i(\vec{x},t), A_b^j(\vec{y},t)] = i\delta^{ij}\delta_{ab}\delta^3(x-y)$$

$$[E_a^i(\vec{x},t), E_b^j(\vec{y},t)] = [A_a^i(\vec{x},t), A_j^b(\vec{x},t)] = 0, \tag{11.62}$$

as our quantum conditions. In the Schrödinger representation where the operator $A_a^i(\vec{x})$ is diagonal with eigenvalue $a_a^i(\vec{x})$, a functional differential representation of the quantum conditions is

$$E_a^j(\vec{x}) = i\frac{\delta}{\delta a_a^j(\vec{x})}. \tag{11.63}$$

The ground state wave functional, a representation of the vacuum state, satisfies the functional Schrödinger equation,

$$\frac{1}{2}\int d^3x \left(-\frac{\delta}{\delta\vec{a}_a(\vec{x})}\cdot\frac{\delta}{\delta\vec{a}_a(\vec{x})} + \vec{b}_a(\vec{x})\cdot\vec{b}_a(\vec{x})\right)\Psi_0[a] = E_0\Psi_0[a], \tag{11.64}$$

where

$$\vec{b}_a(\vec{x}) = \nabla \times \vec{a}_a(\vec{x}) - \frac{g_0}{2}\epsilon_{abc}\vec{a}^b(\vec{x}) \times \vec{a}^c(\vec{x}) \tag{11.65}$$

is the eigenvalue of the diagonal time-independent color magnetic field operator.

As in the Abelian case, there is no Hamiltonian equation corresponding to Gauss's law so we must add this equation as a constraint,

$$D \cdot \vec{E}_a | \Psi\rangle = 0. \tag{11.66}$$

$D^i = \partial^i + ig_0[A^i$, is the gauge covariant derivative. Even in the absence of color charged quarks, the color electric field is not divergence free. From eq.(11.66) we have

$$\nabla \cdot \vec{E} | \Psi\rangle \;=\; -ig_0[A^i, E^i]| \Psi\rangle, \tag{11.67}$$

which shows that the gluons, the quanta of the A^i operators, carry a color charge. As in the Abelian case, the color Gauss's law constraint implies that the wave functionals representing physical states are gauge invariant under time-independent gauge transformations,

$$A^i(\vec{x}) \to U(\vec{x})A^i(\vec{x})U^\dagger(\vec{x}) + \frac{i}{g_0}(\partial^i U(\vec{x}))U^\dagger(\vec{x}), \tag{11.68}$$

where U is an element of the group $SU(2)$. To show this we follow the same development as in the Abelian case and consider the change in a wave functional due to a gauge transformation of $a^i(\vec{x}) = a^i_b(\vec{x})T^b$.

$$\delta\Psi[a] = \int d^3x\, \frac{\delta\Psi}{\delta a^i_a(\vec{x})}\,\delta a^i_a(\vec{x}) \tag{11.69}$$

$\delta a^i_a(\vec{x})$ in eq.(11.69) represents the change in $a^i_a(\vec{x})$ due to an infinitesimal gauge transformation. The infinitesimal version of eq.(11.68) may be obtained by expanding $U = \exp(ig_0\theta_b T^b) = 1 + ig_0\theta_b T^b + \ldots$ and keeping terms to order θ in eq.(11.68). Then

$$\begin{aligned} \delta a^i &= -(\partial^i\theta + ig_0[a^i, \theta]) \\ \delta a^i_a &= -\partial^i\theta_a + g_0\epsilon_{abc}a^i_b\theta_c. \end{aligned} \tag{11.70}$$

Placing this back into eq.(11.69) and integrating the first term by parts we have

$$\begin{aligned} \delta\Psi[a] &= \int d^3x\, \partial^i \frac{\delta\Psi}{\delta a^i_a(\vec{x})}\theta_a + g_0\epsilon^{abc}\frac{\delta\psi}{\delta a^i_a(\vec{x})}a^i_b(\vec{x})\theta_c(\vec{x}) \\ &= \int d^3x\theta_a\left(\partial^i\frac{\delta}{\delta a^i_a(\vec{x})} - g_0\epsilon^{abc}a^i_b\frac{\delta}{\delta a^i_c(\vec{x})}\right)\Psi[a] \\ &= \int d^3x\theta_a(\vec{x})(D\cdot\vec{E})^a\Psi[a]. \end{aligned} \tag{11.71}$$

Since $\theta_a(\vec{x})$ is arbitrary, if the wave functional satisfies the Gauss's law constraint, then the right-hand side vanishes, showing the $\Psi[a]$ is gauge invariant under time-independent gauge transformations. The converse also applies.

In the limit $g_0 \to 0$, the system reduces to 3 uncoupled copies of free photon field theory (pure Abelian gauge theory). Therefore, we already know from chapter 10, eq.(10.83), that the vacuum wave functional in this limit is

$$\begin{aligned}\Psi_0[a, g_0 = 0] &= \Psi_0^{(0)}[a] \\ &= \eta \exp\left(-\frac{1}{(2\pi)^2}\int d^3x\, d^3y\, \frac{\nabla \times \vec{a}_b(\vec{x}) \cdot \nabla \times \vec{a}_b(\vec{y})}{|\vec{x}-\vec{y}|^2}\right),\end{aligned} \tag{11.72}$$

where the sum on the color index, b, runs from 1 to 3. Using this wave functional, let us solve perturbatively to first order in g_0 for the vacuum wave functional that satisfies eq.(11.64). This computation is rather lengthy, but in our usual style, we leave all of the details in. We will find that the standard Rayleigh-Schrödinger perturbation theory fails to give the correct result, because it does not take the constraint into account. We can modify the usual solution to the Rayleigh-Schrödinger theory, eqs.(11.5) and (11.6), to incorporate the constraint. See the Bibliography. Instead, we will resort to directly solving the functional Schrödinger differential equation. We will discover that the constraint provides just the right amount of information to solve the functional equation. The resulting vacuum wave functional to order g_0, $\Psi_0^{(1)}$, is given by eq.(11.93). This technique also works for QED as we will demonstrate in the next section.

If we apply Rayleigh-Schrödinger perturbation theory, then the first order vacuum wave functional is

$$\Psi_0^{(1)}[a] = \sum_{M\neq 0} \frac{\langle \Psi_M^{(0)} | H_{\text{int}} | \Psi_0^{(0)} \rangle}{E_0^{(0)} - E_M^{(0)}} \Psi_M^{(0)}[a], \tag{11.73}$$

with eigenvalue

$$E_0^{(1)} = \langle \Psi_0^{(0)} | H_{\text{int}} | \Psi_0^{(0)} \rangle. \tag{11.74}$$

Eq.(11.73) fails to give the correct form for the perturbed wave functional. Eq.(11.74) and the usual formula for $E_0^{(2)}$, eq.(11.6), are not complete.

The reason that the usual Rayleigh-Schrödinger perturbation theory fails is that it does not incorporate or obey the Gauss's law constraint that we must apply to the wave functionals. It is easy to see that eq.(11.73) generates wave functionals that do not obey the Gauss's law constraint when we recall from chapter 10 that the unperturbed wave functionals, $\Psi_M^{(0)}[a]$, depend only on the transverse part of $\vec{a}_b(\vec{x})$, and satisfy $\nabla \cdot \frac{\delta}{\delta \vec{a}_b(\vec{x})}\Psi_M^{(0)}[a] = 0$. If we apply $\nabla \cdot \frac{\delta}{\delta \vec{a}_b(\vec{x})}$ to eq.(11.73), we see that the right-hand side vanishes, so that $\Psi_0^{(1)}[a]$ generated by eq.(11.73) also depends only on the transverse part of $\vec{a}_b(\vec{x})$. For the nonabelian case, the transverse components are no longer gauge invariant. From the color Gauss's law constraint, eq.(11.67), we see that what we really want is

$$\nabla \cdot \frac{\delta}{\delta \vec{a}_a(\vec{x})}\Psi_0^{(1)}[a] = g_0 \epsilon^{abc} \vec{a}_b(\vec{x}) \cdot \frac{\delta}{\delta \vec{a}_c(\vec{x})}\Psi_0^{(0)}[a]. \tag{11.75}$$

At this point one might be tempted to abandon the temporal gauge and switch to the Coulomb gauge, where there is no constraint placed on the wave functionals. Then we might expect eqs.(11.73) and (11.74) to be valid. We choose the temporal gauge for convenience. The nonabelian Coulomb gauge Hamiltonian contains functional determinants, which are harder to handle. In addition, it is known that these determinants are actually zero, which makes the Coulomb gauge ambiguous (the Gribov ambiguity). While this ambiguity is not present perturbatively, if we contemplate some nonperturbative calculation (for example, a variational calculation) then we must face the problem. Instead, we will stick with the temporal gauge and use the simplicity of the Hamiltonian in this gauge to directly solve the Schrödinger equation for the first perturbative correction to the vacuum functional. We will find that the constraint, eq.(11.75), comes to our rescue to make the problem easier.

The ground state wave functional satisfies eq.(11.64). Since $\Psi_0[a]$ is the lowest energy eigenfunctional of the Schrödinger equation (11.64), we expect that $\Psi_0[a]$ will not possess any nodes, that is, that $\Psi_0[a]$ is positive everywhere. Therefore, we can write $\Psi_0[a]$ as the exponential of a functional. To first order in g_0, let's take

$$\Psi_0[a] = \eta \exp(-G[a] - g_0 F[a] + \cdots), \tag{11.76}$$

and plug it into the Schrödinger equation, keeping terms to first order in g_0. $\Psi_0^{(0)}[a] = \eta \exp(-G[a])$ satisfies

$$\frac{1}{2}\int d^3x \, \frac{\delta}{\delta \vec{a}_a(\vec{x})} \cdot \frac{\delta G[a]}{\delta \vec{a}_a(\vec{x})} - \frac{\delta G[a]}{\delta \vec{a}_a(\vec{x})} \cdot \frac{\delta G[a]}{\delta \vec{a}_a(\vec{x})} + (\nabla \times \vec{a}_a(\vec{x})) \cdot (\nabla \times \vec{a}_a(\vec{x})) = E_0^{(0)}, \tag{11.77}$$

and is given by eq.(11.72). It also satisfies (as it must) the Gauss's law constraint to zeroth order in g_0,

$$\nabla \cdot \frac{\delta}{\delta \vec{a}_a(\vec{x})}\Psi_0^{(0)}[a] = 0. \tag{11.78}$$

To first order in g_0 we find that F must satisfy

$$\int d^3x \frac{1}{2}\frac{\delta}{\delta\vec{a}_a(\vec{x})}\cdot\frac{\delta F[a]}{\delta\vec{a}_a(\vec{x})} - \frac{\delta G[a]}{\delta\vec{a}_a(\vec{x})}\cdot\frac{\delta F[a]}{\delta\vec{a}_a(\vec{x})} - \frac{1}{2}\epsilon^{abc}(\nabla\times\vec{a}_a(\vec{x}))\cdot(\vec{a}_b(\vec{x})\times\vec{a}_c(\vec{x})) = E_0^{(1)}, \quad (11.79)$$

and the constraint,

$$\nabla\cdot\frac{\delta F[a]}{\vec{a}_a(\vec{x})} = \epsilon^{abc}\vec{a}_b(\vec{x})\cdot\frac{\delta G}{\delta\vec{a}_c(\vec{x})}. \quad (11.80)$$

When we apply power counting to F, we see that it must produce a number, $E_0^{(1)}$, when we remove 2 factors of $\vec{a}_a(\vec{x})$ by taking the second derivative. This implies that F should contain a quadratic functional in $\vec{a}_a(\vec{x})$. However, the potential contains 3 factors of $\vec{a}$, each with a different value for the color index. If F is a quadratic functional in $\vec{a}(\vec{x})$, the $\frac{\delta G}{\delta\vec{a}_a}\cdot\frac{\delta F}{\delta\vec{a}_a}$ will also be quadratic, which will never cancel this potential.

The way out of this dilemma is to realize that this means that $E_0^{(1)} = 0$, and that the functional F cannot contain 2 factors of $\vec{a}_a$ with the same color index. To match the potential term, F must contain 3 factors of $\vec{a}$. To avoid a nonzero second functional derivative with respect to $\vec{a}_a$, each factor of $\vec{a}_a$ must carry a different value for the color index. Therefore, we can take

$$F = \epsilon^{abc} f[a^a, a^b, a^c]. \quad (11.81)$$

This choice is consistent with the form of the potential in eq.(11.79) and the right-hand side of the constraint, eq.(11.80), since both are proportional to ϵ^{abc}. With this choice of F, eq.(11.81), we have

$$\int d^3x \frac{\delta}{\delta a^a(\vec{x})}\cdot\frac{\delta F[a]}{\delta a^a(\vec{x})} = 0, \qquad E_0^{(1)} = 0, \quad (11.82)$$

so eq.(11.79) reduces to

$$\int d^3x \frac{\delta G[a]}{\delta\vec{a}^a(\vec{x})}\cdot\frac{\delta F[a]}{\delta\vec{a}^a(\vec{x})} = -\frac{1}{2}\int d^3x\, \epsilon^{abc}(\nabla\times\vec{a}_a(\vec{x}))\cdot(\vec{a}_b(\vec{x})\times\vec{a}_c(\vec{x})). \quad (11.83)$$

At this point, it is easier to solve this functional differential equation in the momentum representation.

$$a_a^i(\vec{x}) = \int\frac{d^3k}{(2\pi)^3}\tilde{a}_a^i(\vec{k})e^{i\vec{k}\cdot\vec{x}}$$
$$\frac{\delta}{\delta a_a^i(\vec{x})} = \int\frac{d^3k}{(2\pi)^3}\frac{\delta}{\delta\tilde{a}_a^i(\vec{k})}e^{-i\vec{k}\cdot\vec{x}} \quad (11.84)$$

This choice implies that

$$\frac{\delta}{\delta \tilde{a}_a^i(\vec{k})} \tilde{a}_b^j(\vec{k}') = (2\pi)^3 \delta_{ab} \delta_{ij} \delta^3(k - k').$$

Eq.(11.83) transforms into

$$\int \frac{d^3k}{(2\pi)^3} \frac{\delta G}{\delta \tilde{a}_a^i(-\vec{k})} \frac{\delta F}{\delta \tilde{a}_a^i(\vec{k})} =$$

$$-i\frac{\epsilon^{abc}}{2} \int \frac{d^3k_1}{(2\pi)^3} \cdots \frac{d^3k_3}{(2\pi)^3} (2\pi)^3 \delta^3(k_1+k_2+k_3)(\vec{k}_1 \times \tilde{a}_a(\vec{k}_1)) \cdot (\tilde{a}_b(\vec{k}_2) \times \tilde{a}_c(\vec{k}_3)). \tag{11.85}$$

From eq.(10.81) we know that

$$\frac{\delta G}{\delta \tilde{a}_a^i(-\vec{k})} = -\frac{(\vec{k} \times (\vec{k} \times \tilde{a}_a(\vec{k}))_i}{|\vec{k}|}, \tag{11.86}$$

so the left-hand side of eq.(11.85) becomes

$$\begin{aligned} \int \frac{d^3k}{(2\pi)^3} \frac{\delta G}{\delta \tilde{a}_a^i(-\vec{k})} \frac{\delta F}{\delta \tilde{a}_a^i(\vec{k})} &= -\int \frac{d^3k}{(2\pi)^3} \frac{(\vec{k} \times (\vec{k} \times \tilde{a}_a(\vec{k})))}{|\vec{k}|} \cdot \frac{\delta F}{\delta \tilde{a}_a(\vec{k})} \\ &= \int \frac{d^3k}{(2\pi)^3} \left(|\vec{k}| \frac{\delta F}{\delta \tilde{a}_a^i(\vec{k})} \tilde{a}_a^i(\vec{k}) \right) - \frac{(\vec{k} \cdot \tilde{a}_a(\vec{k}))(\vec{k} \cdot \delta F/\delta \tilde{a}_a(\vec{k}))}{|\vec{k}|}, \end{aligned} \tag{11.87}$$

where we have made use of the vector identity eq.(10.78). Next, we observe that

$$\int \frac{d^3k}{(2\pi)^3} \frac{\delta F}{\delta \tilde{a}_a(\vec{k})} \cdot \tilde{a}_a(\vec{k}) = F[\tilde{a}] \qquad \text{(no sum on } a) \tag{11.88}$$

so the first term on the right-hand side of eq.(11.85) is virtually just F. This means that if we somehow knew the longitudinal part of F, $\vec{k} \cdot \delta F/\delta \tilde{a}_a(\vec{k})$, then we could add the second term on the right-hand side of eq.(11.87) onto the right-hand side of eq.(11.85) and quickly solve for F in terms of the known functionals. But we do know what the longitudinal part of F is! It is precisely this part that is dictated by the Gauss's law constraint. Using eq.(11.86), the transformed constraint is

$$\vec{k} \cdot \frac{\delta F}{\delta \tilde{a}_a(\vec{k})} = i\epsilon^{abc} \int \frac{d^3k'}{(2\pi)^3} \tilde{a}^c(-(\vec{k} + \vec{k}')) \cdot \frac{(\vec{k}' \times (\vec{k}' \times \tilde{a}^b(\vec{k}')))}{|\vec{k}'|}. \tag{11.89}$$

Using eq.(11.87), we insert eq.(11.89) back into the Schrödinger equation, eq.(11.85). We then have,

$$\int \frac{d^3k}{(2\pi)^3}\,|\vec{k}|\,\frac{\delta F}{\delta \tilde{a}_a(\vec{k})}\cdot \tilde{a}_a(\vec{k})$$
$$= -i\epsilon^{abc}\int \frac{d^3k_1}{(2\pi)^3}\cdots\frac{d^3k_3}{(2\pi)^3}(2\pi)^3\delta^3(k_1+k_2+k_3)$$
$$\left(\frac{1}{2}(\vec{k}_1\times\tilde{a}_a(\vec{k}_1))\cdot(\tilde{a}_b(\vec{k}_2)\times\tilde{a}_c(\vec{k}_3))\right.$$
$$\left.-\frac{(\vec{k}_1\cdot\tilde{a}_a(\vec{k}_1))(\tilde{a}_c(\vec{k}_3)\cdot\vec{k}_2\times(\vec{k}_2\times\tilde{a}_b(\vec{k}_2)))}{|\vec{k}_1||\vec{k}_2|}\right). \quad (11.90)$$

Now we only need to isolate F. On the left-hand side of eq.(11.90) we are summing over the 3 possible values of the color index, $a = 1, 2, 3$.

$$\int \frac{d^3k}{(2\pi)^3}\,|\vec{k}|\,\frac{\delta F}{\delta \tilde{a}_a(\vec{k})}\cdot \tilde{a}_a(\vec{k})$$
$$= \int \frac{d^3k}{(2\pi)^3}\,|\vec{k}|\left(\frac{\delta F}{\delta\tilde{a}_1(\vec{k})}\cdot\tilde{a}_1(\vec{k})+\frac{\delta F}{\delta\tilde{a}_2(\vec{k})}\cdot\tilde{a}_2(\vec{k})+\frac{\delta F}{\delta\tilde{a}_3(\vec{k})}\cdot\tilde{a}_3(\vec{k})\right). \quad (11.91)$$

Recall that F is proportional to ϵ^{abc}, thus $\delta F/\delta\tilde{a}_1(\vec{k})$ does not depend on $\tilde{a}_1(\vec{k})$, $\delta F/\delta\tilde{a}_2(\vec{k})$ does not depend on $\tilde{a}_2(\vec{k})$, and so on. Therefore, we can completely remove the $\tilde{a}$ dependence from eq.(11.90) by functionally differentiating both sides of eq.(11.90) three times, once each with respect to $\tilde{a}_1(\vec{q}_1)$, $\tilde{a}_2(\vec{q}_2)$, and $\tilde{a}_3(\vec{q}_3)$. Let's examine the first term in (11.91) as we functionally differentiate.

$$\frac{\delta}{\delta\tilde{a}_3(\vec{q}_3)}\frac{\delta}{\delta\tilde{a}_2(\vec{q}_2)}\frac{\delta}{\delta\tilde{a}_1(\vec{q}_1)}\left(\int\frac{d^3k}{(2\pi)^3}|\vec{k}|\frac{\delta F}{\delta\tilde{a}_1(\vec{k})}\cdot\tilde{a}_1(\vec{k})\right)$$
$$= \frac{\delta}{\delta\tilde{a}_3(\vec{q}_3)}\frac{\delta}{\delta\tilde{a}_2(\vec{q}_2)}|\vec{q}_1|\frac{\delta F}{\delta\tilde{a}_1(\vec{q}_1)}$$
$$= |\vec{q}_1|\frac{\delta^3 F}{\delta\tilde{a}_3(\vec{q}_3)\delta\tilde{a}_2(\vec{q}_2)\delta\tilde{a}_1(\vec{q}_1)}$$

Similarly,

$$\frac{\delta}{\delta\tilde{a}_3(\vec{q}_3)}\frac{\delta}{\delta\tilde{a}_2(\vec{q}_2)}\frac{\delta}{\delta\tilde{a}_1(\vec{q}_1)}\left(\int\frac{d^3k}{(2\pi)^3}|\vec{k}|\frac{\delta F}{\delta\tilde{a}_2(\vec{k})}\cdot\tilde{a}_2(\vec{k})\right)$$
$$= |\vec{q}_2|\frac{\delta^3 F}{\delta\tilde{a}_3(\vec{q}_3)\delta\tilde{a}_2(\vec{q}_2)\delta\tilde{a}_1(\vec{q}_1)}.$$

Thus, from eq.(11.91), we have

$$\frac{\delta}{\delta\tilde{a}_3(\vec{q}_3)}\cdots\frac{\delta}{\delta\tilde{a}_1(\vec{q}_1)}\int\frac{d^3k}{(2\pi)^3}\frac{\delta F}{\delta\tilde{a}_a(\vec{k})}\cdot\tilde{a}_a(\vec{k})$$
$$=(|\vec{q}_1|+|\vec{q}_2|+|\vec{q}_3|)\frac{\delta^3 F}{\delta\tilde{a}_3(\vec{q}_3)\cdots\delta\tilde{a}_1(\vec{q}_1)}. \quad (11.92)$$

After we differentiate the right-hand side of eq.(11.90) we will also just have a function of $(\vec{q}_1, \vec{q}_2, \vec{q}_3)$. Therefore, to recover F, we divide both sides of the differentiated equation by $(|\vec{q}_1|+|\vec{q}_2|+|\vec{q}_3|)$, then dot both sides by $\tilde{a}_1(\vec{q}_1)\tilde{a}_2(\vec{q}_2)\tilde{a}_3(\vec{q}_3)$, and integrate over $\vec{q}_1$, $\vec{q}_2$, and $\vec{q}_3$. The left-hand side is F,

$$\int\frac{d^3q_1}{(2\pi)^3}\cdots\frac{d^3q_3}{(2\pi)^3}\frac{\delta^3 F}{\delta\tilde{a}_3^i(\vec{q}_3)\delta\tilde{a}_2^j(\vec{q}_2)\delta\tilde{a}_1^k(\vec{q}_1)}\tilde{a}_1^k(\vec{q}_1)\tilde{a}_2^j(\vec{q}_2)\tilde{a}_3^i(\vec{q}_3)=F[\tilde{a}],$$

so

$$F[\tilde{a}]=-i\epsilon^{abc}\int\frac{d^3q_1}{(2\pi)^3}\cdots\frac{d^3q_3}{(2\pi)^3}(2\pi)^3\delta^3(q_1+q_2+q_3)$$
$$\left(\frac{1}{2}\frac{(\vec{q}_1\times\tilde{a}_a(\vec{q}_1))\cdot(\tilde{a}_b(\vec{q}_2)\times\tilde{a}_c(\vec{q}_3))}{|\vec{q}_1|+|\vec{q}_2|+|\vec{q}_3|}\right.$$
$$\left.-\frac{(\vec{q}_1\cdot\tilde{a}_a(\vec{q}_1))(\tilde{a}_c(\vec{q}_3)\cdot(\vec{q}_2\times(\vec{q}_2\times\tilde{a}_b(\vec{q}_2))))}{|\vec{q}_1||\vec{q}_2|(|\vec{q}_1|+|\vec{q}_2|+|\vec{q}_3|)}\right). \quad (11.93)$$

One may explicitly check that F satisfies the Schrödinger equation (11.85) and the constraint, eq.(11.89). The first term in eq.(11.93) is similar in form to the result that would be obtained from Rayleigh-Schrödinger theory, eq.(11.73), except that we are also including the longitudinal component of $\tilde{a}(\vec{k})$. The second term must be added to the first to satisfy the Schrödinger equation and the constraint.

The fact that the "energy denominator," $E_M^{(0)}-E_0^{(0)}=|\vec{q}_1|+|\vec{q}_2|+|\vec{q}_3|$, appears in both terms in F, eq.(11.93), suggests that we may repair the Rayleigh-Schrödinger perturbative formula, eq. (11.73), by adding a nonlocal term to the Hamiltonian of the form

$$-i\epsilon^{abc}\int\frac{d^3k_1}{(2\pi)^3}\frac{d^3k_2}{(2\pi)^3}\frac{d^3k_3}{(2\pi)^3}(2\pi)^3\delta^3(k_1+k_2+k_3)$$
$$\left(\frac{\tilde{a}_c(\vec{k}_3)\cdot(\vec{k}_2\times(\vec{k}_2\times\tilde{a}_b(\vec{k}_2)))(\vec{k}_1\cdot\tilde{a}_a(\vec{k}_1))}{|\vec{k}_1||\vec{k}_2|}\right), \quad (11.94)$$

and by relaxing the Gauss's law constraint on $\Psi_M^{(0)}[\tilde{a}]$, that is, by assuming that $\tilde{a}$ in $\Psi_M^{(0)}[\tilde{a}]$ also contains the longitudinal component. While this will produce the correct $\Psi_0^{(1)}[\tilde{a}]$, it won't fix higher-order corrections. We will have to add more terms to the Hamiltonian to accomplish this.

Our method of solving for $\Psi_0^{(1)}[\tilde{a}]$ by inserting the Gauss's law constraint does tell us how to rewrite the Schrödinger equation so that the solutions automatically satisfy the constraint. To incorporate the constraint into the Schrödinger equation we start with the original functional equation, eq.(11.64), and write

$$\Psi_0[\tilde{a}] = \eta^2 \exp(-(G+Q)), \tag{11.95}$$

where $G[\tilde{a}]$ satisfies eqs.(11.77), (11.78), and (11.86). When we plug this back into eq.(11.64), we will find a term $\nabla \cdot \delta Q/\delta\tilde{a}$ just as we did in solving for F. This term arises from the cross derivative of Q with G, $\delta G/\delta\tilde{a} \cdot \delta Q/\delta\tilde{a}$, when eq.(11.86) is used along with the vector identity, eq.(10.78). We then use the Gauss's law constraint to replace $\nabla \cdot \delta Q/\delta\tilde{a}$ with $ig_0[\tilde{a}^i, \delta/\delta\tilde{a}^i](G+Q)$. Q satisfies the *new* Schrödinger equation

$$\begin{aligned}
&\frac{1}{2}\int \frac{d^3k}{(2\pi)^3}\frac{\delta}{\delta\tilde{a}^a(\vec{k})}\cdot\frac{\delta}{\delta\tilde{a}_a(-\vec{k})}Q - \int\frac{d^3k}{(2\pi)^3}|\vec{k}|\tilde{a}^a(\vec{k})\cdot\frac{\delta Q}{\delta\tilde{a}^a(\vec{k})}\\
&+ i\epsilon^{abc}g_0\int\frac{d^3k_1}{(2\pi)^3}\cdots\frac{d^3k_3}{(2\pi)^3}(2\pi)^3\delta^3(k_1+k_2+k_3)\frac{\vec{k}_1\cdot\tilde{a}^a(\vec{k}_1)}{|\vec{k}_1|}\\
&\qquad\qquad \tilde{a}^c(\vec{k}_3)\cdot\left[\frac{\vec{k}_2\times(\vec{k}_2\times\tilde{a}^b(\vec{k}_2))}{|\vec{k}_2|} - \frac{\delta Q}{\delta\tilde{a}^b(-\vec{k}_2)}\right]\\
&+\frac{1}{2}\int\frac{d^3k}{(2\pi)^3}\tilde{b}^a(-\vec{k})\cdot\tilde{b}^a(\vec{k}) - (\nabla\times\tilde{a}^a(-\vec{k}))\cdot(\nabla\times\tilde{a}^a(\vec{k})) = E_0 - E_0^{(0)}.
\end{aligned} \tag{11.96}$$

The Rayleigh-Schrödinger formula based on this new Hamiltonian will yield the correct results that do satisfy the Gauss's law constraint.

We should mention that there is a known *exact* solution to the eigenfunctional equation for the Yang-Mills vacuum, eq.(11.64). This solution is $\exp(W[a])$, where

$$W[a] = -\frac{1}{2}\int d^3x\,\vec{a}^a(\vec{x})\cdot(\nabla\times\vec{a}^a(\vec{x})) + \frac{g_0}{3}\epsilon^{abc}\vec{a}^a(\vec{x})\cdot(\vec{a}^b(\vec{x})\times\vec{a}^c(\vec{x})). \tag{11.97}$$

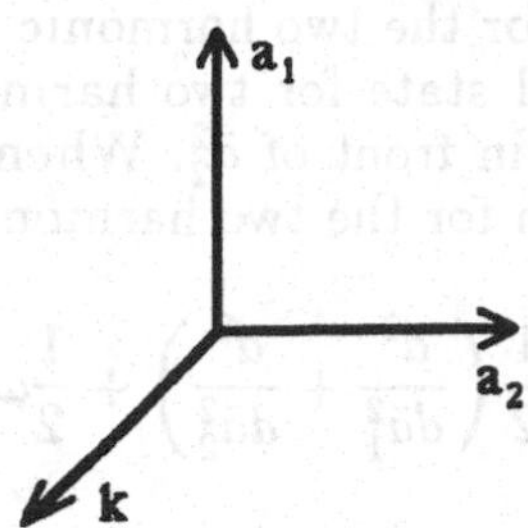

Figure 11.7 Decomposition of the transverse $\tilde{a}$ into 2 orthogonal linear polarizations, $\tilde{a}_1(\vec{k})$ and $\tilde{a}_2(\vec{k})$.

This functional is gauge invariant so it satisfies the Gauss's law constraint. Since

$$\frac{\delta W[a]}{\delta a_a^i(\vec{x})} = -b_a^i(\vec{x}), \qquad \frac{\delta}{\delta \vec{a}_a(\vec{x})} \cdot \frac{\delta W[a]}{\delta \vec{a}_a(\vec{x})} = 0, \tag{11.98}$$

it also satisfies the Schrödinger equation with eigenvalue zero. Why is this not the vacuum wave functional we are seeking? The difficulty is that this functional is not normalizable.

To get an idea what is wrong, let's consider the $g_0 = 0$ limit. In this limit the Yang-Mills system reduces to 3 uncoupled copies of free electrodynamics. Let's look at only one of the three copies and drop the color index on $\vec{a}$. Observe that this $g_0 = 0$ limit of $W[a]$,

$$W_0[a] = W[a, g_0 = 0] = -\frac{1}{2}\int d^3x\, \vec{a}(\vec{x}) \cdot (\nabla \times \vec{a}(\vec{x})), \tag{11.99}$$

also satisfies the Schrödinger equation for free photons, eq.(10.61), with zero eigenvalue and is gauge invariant.

$$\frac{\delta W_0}{\delta \vec{a}(\vec{x})} = -\nabla \times \vec{a}(\vec{x}) = -\vec{b}(\vec{x}), \qquad \frac{\delta^2 W_0}{\delta \vec{a}^2(\vec{x})} = 0, \qquad \nabla \cdot \frac{\delta W_0}{\delta \vec{a}(\vec{x})} = 0. \tag{11.100}$$

For $\vec{a}$ real, the Fourier transform of $W_0[a]$ is

$$W_0[\tilde{a}] = -\frac{i}{2}\int \frac{d^3k}{(2\pi)^3}\, \tilde{a}(\vec{k}) \cdot (\vec{k} \times \tilde{a}(\vec{k})). \tag{11.101}$$

Now let's look at the wave function for one particular mode, $\vec{k}$. Decompose the transverse $\tilde{a}$ into two orthogonal linear polarizations, $\tilde{a}_1(\vec{k})$ and $\tilde{a}_2(\vec{k})$ as in figure 11.7. Then

$$W_0 \sim |\vec{k}|(\tilde{a}_1^2 - \tilde{a}_2^2). \tag{11.102}$$

This wave function for the two harmonic oscillators, $\tilde{a}_1$ and $\tilde{a}_2$, looks like the ordinary ground state for two harmonic oscillators except that the wrong sign appears in front of $\tilde{a}_2^2$. When we run this wave function through the Hamiltonian for the two harmonic oscillators,

$$H = -\frac{1}{2}\left(\frac{d^2}{d\tilde{a}_1^2} + \frac{d^2}{d\tilde{a}_2^2}\right) + \frac{1}{2}\omega^2(\tilde{a}_1^2 + \tilde{a}_2^2), \tag{11.103}$$

where $\omega = ik$, we find that, due to the incorrect sign, the energy eigenvalue is zero. The incorrect sign also makes the wave function non-normalizable.

From another viewpoint, we know that in a classical, linearly polarized, plane electromagnetic wave in the vacuum, the electric and magnetic fields are transverse and orthogonal. If the plane wave is traveling in the z-direction and is electrically polarized in the x-direction, then the magnetic field points in the y-direction only. If we operate on $W_0[a]$ with the electric field we find (eq.(11.100))

$$E_x \exp(W_0[a]) = -ib_x \exp(W_0[a]). \tag{11.104}$$

The functional $\exp(W_0[a])$ appears to associate the wrong component of the magnetic field with the electric field, that is, the wrong polarization of $\vec{b}$ with $\vec{E}$. Instead, we expect that

$$\begin{aligned} E_x\Psi[a] &= -ib_y\Psi[a] \\ &= -i\frac{(\nabla\times\vec{b})_x}{|\nabla|}\Psi[a]. \end{aligned} \tag{11.105}$$

If we require $W_0[a]$ to satisfy eq.(11.105) we recover the correct ground state.

For the case of free photons we know the correct vacuum wave functional, so we know to change the sign in eq.(11.102) so as to associate the correct polarization of the magnetic field with the electric field. The Yang-Mills generalization of $W_0[a]$, eq.(11.97), may also be very close to the correct vacuum wave functional in the sense of a misplaced minus sign or associating the wrong component of the color magnetic field with the (polarized) color electric field. However, as the coupling becomes strong, the notion of polarization that is present in the $g_0 = 0$ limit loses meaning. No one knows just how close the state $\exp(W[a])$ is to the all-important true confining vacuum state. It remains a tantalizing mystery.

11.3 SPINOR QUANTUM ELECTRODYNAMICS

To write down QED in the Schrödinger representation, we must choose a gauge. For the reasons already given in the previous section and in chapter 10 we will work in the temporal gauge, $A_0 = 0$. The Hamiltonian can be obtained from eq.(8.7) by setting $A_0 = 0$.

$$H = \int d^3x \Psi^\dagger(\vec{x})(-i\vec{\alpha}\cdot\nabla + \beta m_0)\Psi(\vec{x}) + \frac{1}{2}(\vec{E}^2(\vec{x}) + \vec{B}^2(\vec{x})) - e_0\Psi^\dagger(\vec{x})(\vec{\alpha}\cdot\vec{A})\Psi(\vec{x}) \tag{11.106}$$

H becomes the Schrödinger functional differential operator upon the introduction of functional derivatives for $\Psi^\dagger(\vec{x})$ and $E^i(\vec{x})$ as given by eqs.(10.89) and (10.60). The result is

$$H = \int d^3x \frac{\delta}{\delta\psi(\vec{x})}(-i\vec{\alpha}\cdot\nabla + \beta m_0)\psi(\vec{x}) + \frac{1}{2}\left(-\frac{\delta}{\delta\vec{a}(\vec{x})}\cdot\frac{\delta}{\delta\vec{a}(\vec{x})} + \vec{b}(\vec{x})\cdot\vec{b}(\vec{x})\right) - e_0\frac{\delta}{\delta\psi(\vec{x})}(\vec{\alpha}\cdot\vec{a}(\vec{x}))\psi(\vec{x}). \tag{11.107}$$

Since there is no Gauss's law Hamiltonian equation of motion in the temporal gauge we must impose it as a constraint. The physical eigenfunctionals of the Hamiltonian, $\Phi[a,\psi]$, must satisfy

$$(\nabla\cdot\vec{E} - e_0\rho)\Phi[a,\psi] = 0, \tag{11.108}$$

that is,

$$\left(i\nabla\cdot\frac{\delta}{\delta\vec{a}(\vec{x})} - e_0\frac{\delta}{\delta\psi_\beta(\vec{x})}\psi_\beta(\vec{x})\right)\Phi[a,\psi] = 0, \tag{11.109}$$

where $\rho = j^0 = \psi^\dagger\psi$ is the charge density, eq.(8.5).

Instead of computing the perturbative vacuum wave functional we are going to calculate to first order in e_0 the wave functional representing a state with just a single electron present. The determination of the vacuum wave functional to the same order is left as an exercise. We will use the perturbative wave functional for the state containing one electron to compute the self-energy of the electron and its shift in mass due to the electromagnetic interaction. In addition, we will use this functional to compute the S-matrix to lowest order for the process of electron-electron elastic scattering, to reproduce the result obtained at the end of chapter 8. Again, directly solving the functional Schrödinger equation for the perturbative wave functional is involved. The technique used to treat the constraint in the previous section for Yang-Mills will also be applied here.

This will be our first nontrivial computation involving Grassmann variables. The resulting wave functional is defined by eq.(11.113) and given by eq.(11.120).

It is easier to solve the functional Schrödinger equation in the momentum representation. Using eqs.(10.67), (10.68), and (10.97), the Hamiltonian, eq.(11.107), becomes $H = H_\gamma + H_e + e_0 H_I$ where

$$H_\gamma = \frac{1}{2}\int \frac{d^3k}{(2\pi)^3}\left(-\frac{\delta}{\delta\tilde{a}_i(\vec{k})}\frac{\delta}{\tilde{a}_i(-\vec{k})} + (\vec{k}\times\tilde{a}(\vec{k}))\cdot(\vec{k}\times\tilde{a}(-\vec{k}))\right),$$

$$H_e = \sum_{r=1}^{4}\int d^3p(\epsilon_r\, E_p \frac{\delta}{\delta b(\vec{p},r)} b(\vec{p},r)), \tag{11.110}$$

and

$$H_I = -\sum_{r=1}^{4}\sum_{r'=1}^{4}\int \frac{d^3k}{(2\pi)^3} d^3p \left(\frac{m_0}{E_{p+k}}\right)^{\frac{1}{2}}\left(\frac{m_0}{E_p}\right)^{\frac{1}{2}}$$
$$w^\dagger(\vec{p}+\vec{k},r')(\alpha\cdot\tilde{a}(\vec{k}))w(\vec{p},r)\frac{\delta}{\delta b(\vec{p}+\vec{k},r')} b(\vec{p},r).$$

Similarly, the Gauss's law operator, $\mathbf{G} = \nabla\cdot\vec{E} - e_0\rho$, becomes

$$\mathbf{G} = \vec{k}\cdot\frac{\delta}{\delta\tilde{a}(\vec{k})} - e_0\sum_{r=1}^{4}\sum_{r'=1}^{4}\int \frac{d^3p}{(2\pi)^3}\left(\frac{m_0}{E_p}\right)^{\frac{1}{2}}\left(\frac{m_0}{E_{p+k}}\right)^{\frac{1}{2}}$$
$$w^\dagger(\vec{p}+\vec{k},r')w(\vec{p},r)\frac{\delta}{\delta b(\vec{p}+\vec{k},r')} b(\vec{p},r). \tag{11.111}$$

Let $\Phi_q[\tilde{a}, b]$ represent the state containing a single electron of momentum $\vec{q}$ and spin up. To lowest order, the wave functional is $\Phi_q^{(0)}[\tilde{a}, b]$ and satisfies $(H_\gamma + H_e)\,\Phi_q^{(0)}[\tilde{a}, b] = E_q\Phi_q^{(0)}[\tilde{a}, b]$ where $E_q = \sqrt{\vec{q}^2 + m_0^2}$. Using the results from the previous chapter, we have

$$\Phi_q^{(0)}[\tilde{a}, b] = \eta\Psi_0[\tilde{a}]\Omega_{(q,1)}[b]$$
$$= \eta\exp\left(-\frac{1}{2}\int\frac{d^3k}{(2\pi)^3}\frac{(\vec{k}\times\tilde{a}(\vec{k}))\cdot(\vec{k}\times\tilde{a}(-\vec{k}))}{|\vec{k}|}\right)\prod_{s=1}^{2}\prod_{\substack{\vec{p}\\ \text{except}(\vec{q},1)}} b(\vec{p},s). \tag{11.112}$$

Rayleigh-Schrödinger perturbation theory, eqs.(11.5) and (11.6), fails to yield the correct energy and perturbative wave functional for the identical reason given in the previous section. The culprit is the Gauss's law constraint. We will be able to solve the Schrödinger equation directly and satisfy the constraint using the same procedure as for the Yang-Mills wave functionals. Namely, we will assume that $\Phi_q^{(0)}[\tilde{a}, b]$ is Gaussian in the longitudinal component of $\tilde{a}$ and directly insert the Gauss's law constraint into the functional differential equation in the obvious place.

Expanding $\Phi_q[\tilde{a}, b]$ in a power series in e_0, $\Phi_q[\tilde{a}, b] = \Phi_q^{(0)}[\tilde{a}, b] + e_0\Phi_q^{(1)}[\tilde{a}, b] + \ldots$, we will assume that

$$\begin{aligned}\Phi_q^{(1)}[\tilde{a}, b] &= \eta F[\tilde{a}, b]\Psi_0[\tilde{a}] \\ &= \eta F[\tilde{a}, b]\exp(-G[\tilde{a}]),\end{aligned} \tag{11.113}$$

where $\Psi_0[\tilde{a}]$ is given in eq.(11.112) above or eq.(10.81). To satisfy the Schrödinger equation to order e_0, $F[\tilde{a}, b]$ must satisfy

$$\begin{aligned}-\frac{1}{2}\int \frac{d^3k}{(2\pi)^3}\frac{\delta}{\delta\tilde{a}(-\vec{k})}\cdot\frac{\delta F}{\delta\tilde{a}(\vec{k})} - 2\frac{\delta G}{\delta\tilde{a}(-\vec{k})}\cdot\frac{\delta F}{\delta\tilde{a}(\vec{k})} \\ +\sum_{r=1}^{4}\int d^3p\,(\epsilon_r E_p \frac{\delta}{\delta b(\vec{p}, r)} b(\vec{p}, r))F[\tilde{a}, b] \\ -\sum_{r=1}^{4}\sum_{r'=1}^{4}\int \frac{d^3k}{(2\pi)^3}d^3p\left(\frac{m_0}{E_p}\right)^{\frac{1}{2}}\left(\frac{m_0}{E_{p+k}}\right)^{\frac{1}{2}} \\ w^\dagger(\vec{p}+\vec{k}, r')(\vec{\alpha}\cdot\tilde{a}(\vec{k}))w(\vec{p}, r)\frac{\delta}{\delta b(\vec{p}+\vec{k}, r')}b(\vec{p}, r)\Omega_{(q,1)}[b] \\ = E_q^{(1)}\Omega_{(q,1)} + E_q F[\tilde{a}, b].\end{aligned} \tag{11.114}$$

$E_q^{(1)}$ is a pure number (no dependence on $\tilde{a}$ or $b(p, r)$), which we expect to vanish. If $E_q^{(1)}$ did not vanish, then $F[\tilde{a}, b]$ must contain a term proportional to $\Omega_{(q,1)}[b]$ to satisfy eq.(11.114). In addition, this term (call it F_0) must not have any dependence on $\tilde{a}(\vec{k})$ (i.e. $\delta F/\delta\tilde{a}_i(\vec{k}) = 0$) or must depend quadratically on $\tilde{a}$ (i.e. $\delta^2 F/\delta\tilde{a}^2(\vec{k})$ = a pure number). If this term in $F[\tilde{a}, b]$ is independent of $\tilde{a}$, then F will not be normalizable so this possibility is excluded. If F is quadratic in $\tilde{a}(\vec{k})$, we will not be able to satisfy eq.(11.114) without introducing a term that is quartic in $\tilde{a}$, which in turn will require a term with 6 factors of $\tilde{a}$ and so on. Such a monstrosity, if even normalizable, can be avoided by taking $E_q^{(1)}$ to be zero. Then we may take F to be linear in $\tilde{a}$.

Another way to see that $E_q^{(1)}$ must vanish is that the interacting portion of the Hamiltonian, H_I, which is linear in $\tilde{a}$, creates or destroys a photon. When this Hamiltonian acts once (order e_0) on the state containing 1 electron, the result is a state containing at least one electron and a photon. We are seeking eigenvalues of the Hamiltonian, thus we want the final state to be one with just one electron and no photons. The Hamiltonian must act again to absorb the extra photon. Thus $E_q^{(1)} = 0$, while $E_q^{(2)}$ will not vanish.

With $E_q^{(1)} = 0$, we may take F to be linear in $\tilde{a}(\vec{k})$ so the first term in eq.(11.114) will vanish. Using the explicit form for $G[\tilde{a}]$, eqs.(10.77) and (10.80), the second term becomes

$$\int \frac{d^3k}{(2\pi)^3}|\vec{k}|\left(\tilde{a}(\vec{k})\cdot\frac{\delta F}{\delta\tilde{a}(\vec{k})}\right)-\left(\frac{(\vec{k}\cdot\tilde{a}(-\vec{k}))(\vec{k}\cdot\delta F/\delta\tilde{a}(\vec{k}))}{|\vec{k}|}\right). \quad (11.115)$$

We are assuming that the second term in eq.(11.115) does not vanish even though Gauss's law to lowest order would indicate so. By keeping this term, however, we must also find the longitudinal part of F, $\vec{k}\cdot\delta F/\delta\tilde{a}(\vec{k})$. As in the Yang-Mills case, the Gauss's law constraint comes to the rescue. At first order in e_0, the Gauss's law constraint, eq.(11.111), says that

$$\vec{k}\cdot\frac{\delta F}{\delta\tilde{a}(\vec{k})}=\sum_{r=1}^{4}\sum_{r'=1}^{4}\int\frac{d^3p}{(2\pi)^3}\left(\frac{m_0}{E_p}\right)^{\frac{1}{2}}\left(\frac{m_0}{E_{p+k}}\right)^{\frac{1}{2}}$$
$$w^\dagger(\vec{p}+\vec{k},r')w(\vec{p},r)\frac{\delta}{\delta b(\vec{p}+\vec{k},r')}b(\vec{p},r)\Omega_{(q,1)}[b]. \quad (11.116)$$

Insert this into eq.(11.115) and eq.(11.115) back into the Schrödinger equation, eq.(11.114). The result is

$$\int\frac{d^3k}{(2\pi)^3}|\vec{k}|\left(\tilde{a}(\vec{k})\cdot\frac{\delta F}{\delta\tilde{a}(\vec{k})}\right)+\sum_{r=1}^{4}\epsilon_r E_k\frac{\delta}{\delta b(\vec{k},r)}b(\vec{k},r)F[\tilde{a},b]-E-q\,F[\tilde{a},b]$$
$$=\sum_{r=1}^{4}\sum_{r'=1}^{4}\int\frac{d^3k}{(2\pi)^3}d^3p\left(\frac{m_0}{E_p}\right)^{\frac{1}{2}}\left(\frac{m_0}{E_{p+k}}\right)^{\frac{1}{2}}$$
$$w^\dagger(\vec{p}+\vec{k},r')\left(\vec{\alpha}\cdot\tilde{a}(\vec{k})+\frac{\vec{k}\cdot\tilde{a}(\vec{k})}{|\vec{k}|}\right)w(\vec{p},r)\frac{\delta}{\delta b(\vec{p}+\vec{k},r')}b(\vec{p},r)\Omega_{(q,1)}[b], \quad (11.117)$$

where we have collected the terms dependent on F on the left-hand side and the remainder on the right. Both the Gauss's law operator and the interaction Hamiltonian have the same functional dependence on b. This allows us to quickly determine the b dependence of $F[\tilde{a}, b]$, because the operator on the left-hand side of eq.(11.117) acting on $F[\tilde{a}, b]$ does not change the functional dependence of F on b. Therefore, $F[\tilde{a}, b]$ must have the same functional dependence on b as appears on the right-hand side of eq.(11.117).

$$F[\tilde{a}, b] \propto \int d^3p\, f[\tilde{a}] \sum_{r=1}^{4} \sum_{r'=1}^{4} \frac{\delta}{\delta b(\vec{p}+\vec{k}, r')} b(\vec{p}, r) \Omega_{(q,1)}[b] \qquad (11.118)$$

When we plug this back into eq.(11.117) and eliminate common factors from both sides we find that $f[\tilde{a}]$ must satisfy

$$\begin{aligned}
&\sum_{r'=1}^{2} \sum_{r=3}^{4} \int \frac{d^3k}{(2\pi)^3} d^3p (|\vec{k}| + E_q + E_{p+k} + E_p - E_q) f[\tilde{a}] \\
&\qquad\qquad \frac{\delta}{\delta b(\vec{p}+\vec{k}, r')} b(\vec{p}, r) \Omega_{(q,1)}[b] \\
&+ \sum_{r'=1}^{2} \int \frac{d^3k}{(2\pi)^3} (|\vec{k}| + E_{q+k} - E_q) f[\tilde{a}] \frac{\delta}{\delta b(\vec{q}+\vec{k}, r')} b(\vec{q}, 1) \Omega_{(q,1)}[b] \\
&= \sum_{r'=1}^{2} \sum_{r=3}^{4} \int \frac{d^3k}{(2\pi)^3} d^3p \left(\frac{m_0}{E_p}\right)^{\frac{1}{2}} \left(\frac{m_0}{E_{p+k}}\right)^{\frac{1}{2}} \\
&\quad w^\dagger(\vec{p}+\vec{k}, r') \left(\vec{\alpha} \cdot \tilde{a}(\vec{k}) + \frac{\vec{k} \cdot \tilde{a}(\vec{k})}{|\vec{k}|} \right) w(\vec{p}, r) \frac{\delta}{\delta b(\vec{p}+\vec{k}, r')} b(\vec{p}, r) \Omega_{(q,1)}[b] \\
&+ \sum_{r'=1}^{2} \int \frac{d^3k}{(2\pi)^3} \left(\frac{m_0}{E_q}\right)^{\frac{1}{2}} \left(\frac{m_0}{E_{q+k}}\right)^{\frac{1}{2}} \\
&\quad w^\dagger(\vec{q}+\vec{k}, r') \left(\vec{\alpha} \cdot \tilde{a}(\vec{k}) + \frac{\vec{k} \cdot \tilde{a}(\vec{k})}{|\vec{k}|} \right) w(\vec{q}, 1) \frac{\delta}{\delta b(\vec{q}+\vec{k}, r')} b(\vec{q}, 1) \Omega_{(q,1)}[b].
\end{aligned} \qquad (11.119)$$

The "energy denominator" factors have appeared on the left-hand side of eq.(11.119). We may move these factors to the right-hand side to solve for $f[\tilde{a}]$ by functionally differentiating eq.(11.119) by $\tilde{a}_i(\vec{k})$, dividing by the energy factors, multiplying by $\tilde{a}_i(\vec{k})$ and integrating over $\vec{k}$. The final result for $F[\tilde{a}, b]$ is

$$F[\tilde{a}, b] = \sum_{r'=1}^{2}\sum_{r=3}^{4}\int \frac{d^3k}{(2\pi)^3}d^3p\left(\frac{m_0}{E_p}\right)^{\frac{1}{2}}\left(\frac{m_0}{E_{p+k}}\right)^{\frac{1}{2}}\frac{1}{E_p + E_{p+k} + |\vec{k}|}$$
$$w^\dagger(\vec{p}+\vec{k}, r')\left(\vec{\alpha}\cdot\tilde{a}(\vec{k}) + \frac{\vec{k}\cdot\tilde{a}(\vec{k})}{|\vec{k}|}\right)w(\vec{p}, r)\frac{\delta}{\delta b(\vec{p}+\vec{k}, r')}b(\vec{p}, r)\Omega_{(q,1)}[b]$$

$$-\sum_{r'=1}^{2}\int \frac{d^3k}{(2\pi)^3}\left(\frac{m_0}{E_q}\right)^{\frac{1}{2}}\left(\frac{m_0}{E_{q+k}}\right)^{\frac{1}{2}}\frac{1}{E_q - E_{q+k} - |\vec{k}|}$$
$$w^\dagger(\vec{q}+\vec{k}, r')\left(\vec{\alpha}\cdot\tilde{a}(\vec{k}) + \frac{\vec{k}\cdot\tilde{a}(\vec{k})}{|\vec{k}|}\right)w(\vec{q}, 1)\Omega_{(q+k,r')}[b]. \quad (11.120)$$

The first term above is essentially independent of $\vec{q}$ and represents a disconnected vacuum process. If we replace $\Omega_{(q,1)}[b]$ by the filled vacuum, the first term becomes the wave functional to order e_0. The second term represents the interaction of the electron with momentum $\vec{q}$ with a photon of momentum $\vec{k}$.

With the use of the identity

$$w^\dagger(\vec{p}+\vec{k}, r')(\vec{k}\cdot\vec{\alpha})w(\vec{p}, r) = (\epsilon_{r'}E_{p+k} - \epsilon_r E_p)w^\dagger(\vec{p}+\vec{k}, r')w(\vec{p}, r), \quad (11.121)$$

one may easily verify that F given in eq.(11.120) does indeed satisfy the Gauss's law constraint, eq.(11.116), and the Schrödinger equation, eq.(11.114). From the form of F, we can see that the usual Rayleigh-Schrödinger expression for F,

$$F = \sum_N \frac{\langle\Phi_N^{(0)}|H_I|\Phi_q^{(0)}\rangle}{E_q - E_N}|\Phi_N^{(0)}\rangle, \quad (11.122)$$

will yield the correct functional if we replace $\vec{\alpha}\cdot\tilde{a}(\vec{k})$ in H_I, eq.(11.110), with $(\vec{\alpha}\cdot\tilde{a}(\vec{k}) + \vec{k}\cdot\tilde{a}(\vec{k})/|\vec{k}|)$, and assume that $\Psi_0[\tilde{a}]$ depends on the longitudinal component of $\tilde{a}(\vec{k})$ in the same manner as the transverse components. The change in H_I given above amounts to adding a non-local, non gauge invariant term to the Hamiltonian of the form

$$-e_0\int\frac{(\nabla\cdot\vec{a}(\vec{x}))\frac{\delta}{\delta\psi(\vec{y})}\psi(\vec{y})}{|\vec{x}-\vec{y}|^2}d^3x d^3y. \quad (11.123)$$

Let us next use the first order functional to compute the energy eigenvalue of the single electron state to second order in e_0. From this we may compute the shift in the electron mass from its bare mass, m_0, due to the electromagnetic interaction. The Rayleigh-Schrödinger expression for the second order correction to the eigenvalue is

$$\begin{aligned} E_q^{(2)} &= \langle \Phi_q^{(0)} \,| H_I |\, \Phi_q^{(1)} \rangle \\ &= \eta^2 \int \mathcal{D}\tilde{a}\,\mathcal{D}b\, \Psi_0^*[\tilde{a}]\Omega^\dagger_{(q,1)}[b] H_I[\tilde{a},b] F[\tilde{a},b]\Psi_0[\tilde{a}]. \end{aligned} \tag{11.124}$$

To compute the correct shift in E_q, we must use the modified Hamiltonian (add eq.(11.123)) and assume $\Psi_0[\tilde{a}]$ depends equally on the longitudinal component of $\tilde{a}(\vec{k})$. An alternative way to find $E_q^{(2)}$ is to use F and solve the functional Schrödinger equation directly. The same method of inserting the Gauss's law constraint must be employed to complete the calculation. Both methods yield the same result, but eq.(11.124) is easier to compute with. The result is eq.(11.128) and eq.(11.130). The resulting shift in the mass of the electron, δm, is given by eq.(11.133), and its cutoff dependence by eq.(11.134).

Let us consider the details of computing eq.(11.124). In order that the functional integral over b not vanish, we must have $H_I[\tilde{a},b]\,F[\tilde{a},b] \propto \Omega_{(q,1)}[b]$. Since

$$\begin{aligned} H_I = &-\sum_{s=1}^{4}\sum_{s'=1}^{4}\int \frac{d^3k'}{(2\pi)^3} d^3p' \left(\frac{m_0}{E_{p'}}\right)^{\frac{1}{2}} \left(\frac{m_0}{E_{p'+k'}}\right)^{\frac{1}{2}} \\ & w^\dagger(\vec{p}'+\vec{k}',s') \left(\vec{\alpha}\cdot\tilde{a}(\vec{k}') + \frac{\vec{k}'\cdot\tilde{a}(\vec{k}')}{|\vec{k}'|}\right) w(\vec{p}',s) \frac{\delta}{\delta b(\vec{p}'+\vec{k}',s')} b(\vec{p}',s'), \end{aligned} \tag{11.125}$$

then we must have

$$\sum_{s=1}^{4}\sum_{s'=1}^{4}\sum_{r'=1}^{2}\sum_{r=3}^{4} \frac{\delta}{\delta b(\vec{p}'+\vec{k}',s')} b(\vec{p}',s) \frac{\delta}{\delta b(\vec{p}+\vec{k},r')} b(\vec{p},r)\Omega_{(q,1)}[b] = \Omega_{(q,1)}[b], \tag{11.126}$$

when H_I acts on the first term in the functional $F[\tilde{a},b]$, eq.(11.120), and

$$\sum_{s=1}^{4}\sum_{s'=1}^{4}\sum_{r'=1}^{2} \frac{\delta}{\delta b(\vec{p}'+\vec{k}',s')} b(\vec{p}',s)\Omega_{(q+k,r')}[b] = \Omega_{(q,1)}[b] \tag{11.127}$$

when H_I acts on the second term in $F[\tilde{a},b]$. Eq.(11.127) is easily satisfied if $p' = q+k$, $s = r'$, and $p'+k' = q$, $s' = 1$. After the functional integral over b is done, the contribution to eq.(11.124) from the second term in $F[\tilde{a}, b]$ is

$$\sum_{r'=1}^{2} \eta^2 \int \mathcal{D}\tilde{a}\Psi_0^*[\tilde{a}]\Psi_0[\tilde{a}] \int \frac{d^3k}{(2\pi)^3}\frac{d^3k'}{(2\pi)^3} d^3p' \left(\frac{m_0}{E_q}\right)^{\frac{1}{2}} \left(\frac{m_0}{E_{q+k}}\right)^{\frac{1}{2}}$$

$$\left(\frac{m_0}{E_{p'}}\right)^{\frac{1}{2}} \left(\frac{m_0}{E_{p'+k'}}\right)^{\frac{1}{2}} \frac{1}{E_q - E_{q+k} - |\vec{k}|}$$

$$w^\dagger(\vec{p}' + \vec{k}', 1)\left(\vec{\alpha}\cdot\tilde{a}(\vec{k}') + \frac{\vec{k}'\cdot\tilde{a}(\vec{k}')}{|\vec{k}'|}\right) w(p', r')$$

$$w^\dagger(\vec{q} + \vec{k}, r')\left(\vec{\alpha}\cdot\tilde{a}(\vec{k}) + \frac{\vec{k}\cdot\tilde{a}(\vec{k})}{|\vec{k}|}\right) w(\vec{q}, 1)\delta^3(p' - q - k)\delta^3(p' + k' - q)$$

which is

$$\sum_{r'=1}^{2} \eta^2 \int \mathcal{D}\tilde{a}\Psi_0^*[\tilde{a}]\Psi_0[\tilde{a}] \int \frac{d^3k}{(2\pi)^3}\frac{m_0}{E_{q+k}}\frac{m_0}{E_q}\frac{1}{E_q - E_{q+k} - |\vec{k}|}$$

$$w^\dagger(\vec{q}, 1)\left(\vec{\alpha}\cdot\tilde{a}(-\vec{k}) - \frac{\vec{k}\cdot\tilde{a}(-\vec{k})}{|\vec{k}|}\right) w(\vec{q} + \vec{k}, r')$$

$$w^\dagger(\vec{q} + \vec{k}, r')\left(\vec{\alpha}\cdot\tilde{a}(\vec{k}) + \frac{\vec{k}\cdot\tilde{a}(\vec{k})}{|\vec{k}|}\right) w(\vec{q}, 1).$$

The Gaussian functional integral is now readily evaluated to yield

$$\sum_{r'=1}^{2} \int \frac{d^3k}{(2\pi)^3}\frac{m_0}{E_{q+k}}\frac{m_0}{E_q}\frac{1}{E_q - E_{q+k} - |\vec{k}|}\frac{1}{2|\vec{k}|}$$

$$w^\dagger(\vec{q}, 1)\left(\alpha_j - \frac{k_j}{|\vec{k}|}\right) w(\vec{q} + \vec{k}, r') w^\dagger(\vec{q} + \vec{k}, r')\left(\alpha_j + \frac{k_j}{|\vec{k}|}\right) w(\vec{q}, 1). \quad (11.128)$$

At first glance it would appear that we can only trivially satisfy eq.(11.126) by taking $s = r' \neq 1$, $p' = p + k$, $p' + k' = p$, and $s' = r$. Since the electron at $\vec{q}$ in $\Omega_{(q,1)}[b]$ is not touched by any b or $\delta/\delta b$, this solution of eq.(11.126) must represent a disconnected vacuum process and therefore represents the shift in vacuum energy at second order in e_0. This shift is unmeasurable and does not contribute to the self-interaction of the electron. Therefore, it would appear that the first term of $F[\tilde{a}, b]$ would not contribute to the shift in energy of the electron, eq.(11.124). However, this is not the shift in the vacuum energy, because $s = r'$ cannot be 1 due to the presence of the electron at $\vec{q}$, so there is a contribution to the self-mass of the electron from this term. To determine this contribution, we anticommute $\delta/\delta b(\vec{p}' + \vec{k}', r')$ with $b(\vec{p}', s)$ in eq.(11.126) and get

$$\sum_{s'=1}^{4}\sum_{r'=1}^{2}\sum_{r=3}^{4}\frac{\delta}{\delta b(\vec{p}' + \vec{k}', s')}\left(\delta^3(p+k-p')\delta_{r's} - \frac{\delta}{\delta b(\vec{p}+\vec{k}, r')}b(\vec{p}', s)\right)$$
$$b(\vec{p}, r)\Omega_{(q,1)}[b] = \Omega_{(q,1)}[b]. \quad (11.129)$$

Now we can easily see that the $\delta^3(p+k-p')\delta_{r's}$ term above yields the vacuum energy while the second term may be satisfied if $s' = r$, $p'+k' = p$, $r' = s = 1$, $p + k = q$, and $p' = q$. Thus, the contribution to eq.(11.124) from the first term in $F[\tilde{a}, b]$ is

$$\sum_{r=3}^{4}\eta^2\int \mathcal{D}\tilde{a}\Psi_0^*[\tilde{a}]\Psi_0[\tilde{a}]\int\frac{d^3k}{(2\pi)^3}\frac{d^3k'}{(2\pi)^3}d^3p\,d^3p'\left(\frac{m_0}{E_{p'}}\right)^{\frac{1}{2}}\left(\frac{m_0}{E_{p'+k'}}\right)^{\frac{1}{2}}$$
$$\left(\frac{m_0}{E_p}\right)^{\frac{1}{2}}\left(\frac{m_0}{E_{p+k}}\right)^{\frac{1}{2}}\frac{1}{E_p + E_{p+k} + |\vec{k}|}$$
$$w^\dagger(\vec{p}' + \vec{k}', r)\left(\vec{\alpha}\cdot\tilde{a}(\vec{k}') + \frac{\vec{k}'\cdot\tilde{a}(\vec{k}')}{|\vec{k}'|}\right)w(\vec{p}', 1)$$
$$w^\dagger(\vec{p}+\vec{k}, 1)\left(\vec{\alpha}\cdot\tilde{a}(\vec{k}) + \frac{\vec{k}\cdot\tilde{a}(\vec{k})}{|\vec{k}|}\right)w(\vec{p}, r)\delta^3(p'+k'-p)\delta^3(p+k-q)\delta^3(p'-q)$$
$$= \sum_{r=3}^{4}\int\frac{d^3k}{(2\pi)^3}\frac{m_0}{E_q}\frac{m_0}{E_{q+k}}\frac{1}{E_q + E_{q+k} + |\vec{k}|}\frac{1}{2|\vec{k}|}$$
$$w^\dagger(\vec{q} + \vec{k}, r)\left(\alpha_j + \frac{k_j}{|\vec{k}|}\right)w(\vec{q}, 1)w^\dagger(\vec{q}, 1)\left(\alpha_j - \frac{k_j}{|\vec{k}|}\right)w(\vec{q} + \vec{k}, r). \quad (11.130)$$

Eqs.(11.128) and (11.130) together represent the shift in energy of the electron due to the electromagnetic interaction to second order in e_0. The contribution from eq.(11.130) is equivalent to that from the diagram in figure 11.8(a) while eq.(11.128) is equivalent to diagram 11.8(b).

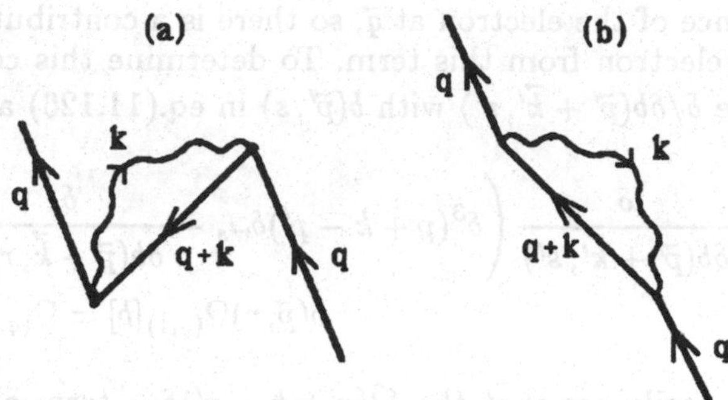

Figure 11.8 Diagrams leading to the shift, δm, of the mass of the electron to order α when $\vec{q} \longrightarrow 0$. Diagram (a) contains one positron in the intermediate state, and arises from H_I acting on the first term of $F[\tilde{a}, b]$ in eq.(11.120). The intermediate state in diagram (b) contains one electron, and arises from H_I acting on the second term of $F[\tilde{a}, b]$.

If we take $\vec{q} \rightarrow 0$ in eq.(11.124), then the shift in the energy eigenvalue for the single electron state is just the shift in the mass of the electron. Thus,

$$\delta m = \lim_{\vec{q}\to 0} e_0^2 \sum_{r=3}^{4} \int \frac{d^3k}{(2\pi)^3} \frac{m_0}{E_q} \frac{m_0}{E_{q+k}} \frac{1}{E_q + E_{q+k} + |\vec{k}|} \frac{1}{2|\vec{k}|}$$
$$w^\dagger(\vec{q}, 1)\Big(\alpha_j - \frac{k_j}{|\vec{k}|}\Big) w(\vec{q} + \vec{k}, r) w^\dagger(\vec{q} + \vec{k}, r)\Big(\alpha_j + \frac{k_j}{|\vec{k}|}\Big) w(\vec{q}, 1)$$

$$+ \lim_{\vec{q}\to 0} e_0^2 \int \frac{d^3k}{(2\pi)^3} \frac{m_0}{E_q} \frac{m_0}{E_{q+k}} \frac{1}{E_q - E_{q+k} - |\vec{k}|} \frac{1}{2|\vec{k}|}$$
$$w^\dagger(\vec{q}, 1)\Big(\alpha_j - \frac{k_j}{|\vec{k}|}\Big) w(\vec{q} + \vec{k}, r') w^\dagger(\vec{q} + \vec{k}, r')\Big(\alpha_j + \frac{k_j}{|\vec{k}|}\Big) w(\vec{q}, 1)$$

$$= e_0^2 \sum_{r=3}^{4} \int \frac{d^3k}{(2\pi)^3} \frac{m_0}{E_k} \frac{1}{m_0 + E_k + |\vec{k}|} \frac{1}{2|\vec{k}|}$$
$$w^\dagger(0, 1)\Big(\alpha_j - \frac{k_j}{|\vec{k}|}\Big) w(\vec{k}, r) w^\dagger(\vec{k}, r)\Big(\alpha_j + \frac{k_j}{|\vec{k}|}\Big) w(0, 1)$$

$$+ e_0^2 \sum_{r'=1}^{2} \int \frac{d^3k}{(2\pi)^3} \frac{m_0}{E_k} \frac{1}{m_0 - E_k - |\vec{k}|} \frac{1}{2|\vec{k}|}$$

$$w^\dagger(0,1)\Big(\alpha_j - \frac{k_j}{|\vec{k}|}\Big) w(\vec{k}, r') w^\dagger(\vec{k}, r') \Big(\alpha_j + \frac{k_j}{|\vec{k}|}\Big) w(0,1). \quad (11.131)$$

To evaluate the first term above, add and subtract the identical term, except sum over r from 1 to 2. Using the completeness relation for the spinors $w(\vec{k}, r)$, eq.(10.94), the spinor string in the first term of eq.(11.131) reduces to

$$\begin{aligned}
\sum_{r=1}^{4} w^\dagger(0,1)\Big(\alpha_j - \frac{k_j}{|\vec{k}|}\Big) w(\vec{k}, r) w^\dagger(\vec{k}, r) \Big(\alpha_j + \frac{k_j}{|\vec{k}|}\Big) w(0,1) \\
= \frac{E_k}{m_0} w^\dagger(0,1)\Big(\alpha_j - \frac{k_j}{|\vec{k}|}\Big)\Big(\alpha_j + \frac{k_j}{|\vec{k}|}\Big) w(0,1) \\
= 2\frac{E_k}{m_0} w^\dagger(0,1) w(0,1) \\
= 2\frac{E_k}{m_0}.
\end{aligned} \quad (11.132)$$

Therefore, eq.(11.131) becomes

$$\delta m = e_0^2 \int \frac{d^3k}{(2\pi)^3} \frac{1}{m_0 + E_k + |\vec{k}|} \frac{1}{|\vec{k}|}$$

$$+ e_0^2 \sum_{r'=1}^{2} \int \frac{d^3k}{(2\pi)^3} \frac{1}{2|\vec{k}|} \frac{m_0}{E_k} \Big(\frac{1}{m_0 - E_k - |\vec{k}|} - \frac{1}{m_0 + E_k + |\vec{k}|} \Big)$$

$$w^\dagger(0,1)\Big(\alpha_j - \frac{k_j}{|\vec{k}|}\Big) w(\vec{k}, r') w^\dagger(\vec{k}, r') \Big(\alpha_j + \frac{k_j}{|\vec{k}|}\Big) w(0,1).$$

Using the projection operator, eq.(10.111), the spinor string reduces to $E_k/m_0 - 2$, thus the shift in the mass is

$$\begin{aligned}
\delta m &= e_0^2 \int \frac{d^3k}{(2\pi)^3} \frac{1}{|\vec{k}|} \Bigg[\frac{1}{m_0 + E_k + |\vec{k}|} \\
&\quad + \frac{1}{2}\Big(1 - \frac{2E_k}{m_0}\Big) \Big(\frac{1}{m_0 - E_k - |\vec{k}|} - \frac{1}{m_0 + E_k + |\vec{k}|} \Big) \Bigg] \\
&= \frac{e_0^2}{4\pi^2} \frac{1}{m_0} \int_0^\infty dk\, k - E_k + \frac{2m_0^2}{E_k} \\
&= \frac{\alpha}{m_0 \pi} \int_0^\infty dk\, k - \sqrt{k^2 + m_0^2} + \frac{2m_0}{\sqrt{k^2 + m_0^2}},
\end{aligned} \quad (11.133)$$

where α is the fine structure constant. While this integral appears to be quadratically divergent, it is easily evaluated to show that the divergence is actually logarithmic. If we cut the integral off at $\Lambda/2$, where $\Lambda/2 \gg m_0$, then

$$\begin{aligned}\delta m &= \frac{\alpha}{m_0\pi}\left(\frac{1}{2}k^2 - \frac{1}{2}(k\sqrt{k^2+m_0^2} + m_0^2\ln(k+\sqrt{k^2+m_0^2}))\right.\\ &\qquad \left.+\, 2m_0^2\ln(k+\sqrt{k^2+m_0^2})\right)\Bigg|_0^{\Lambda/2} \\ &\approx \frac{3}{2}\frac{\alpha m_0}{\pi}\ln\left(\frac{\Lambda}{m_0}\right).\end{aligned} \tag{11.134}$$

We may now give a short preview of renormalization. The mass of the electron to order α is $m = m_0 + \delta m$. m_0 is the "bare" mass of the electron, the mass of the electron in the absence of electromagnetic interactions, that is, the mass of an uncharged electron. The shift in the mass due to the presence of the electromagnetic to lowest nontrivial order in perturbation theory, δm, is infinite. But we know that the physical mass of the electron which includes electromagnetic interactions is finite. To get a finite answer out of perturbation theory we had to cut off the integral, thereby introducing an arbitrary scale, Λ. So what can we do?

Clearly, the physically measurable mass of the electron, m, cannot depend on some arbitrary unphysical cutoff Λ. To make sense of the result above, $m = m_0 + \delta m$ must be independent of Λ. How can we do this? The bare mass, m_0, is *not* physically measurable (requires infinite energy), thus m_0 *can* depend on Λ, $m_0 = m_0(\Lambda)$. Let us choose m_0 to depend on Λ in such a way that m is independent of Λ.

To find such a function $m_0(\Lambda)$, we take $m = m_0 + \delta m(\Lambda)$ to be a constant and invert eq.(11.134) to solve for m_0 in terms of Λ. To order α we have

$$\begin{aligned} m &= m_0 + \delta m \\ &= m_0\left(1 + \frac{3}{2}\frac{\alpha}{\pi}\ln\left(\frac{\Lambda}{m_0}\right)\right). \end{aligned}$$

Thus,

$$\begin{aligned} m_0 &= \frac{m}{1+\frac{3}{2}\frac{\alpha}{\pi}\ln\left(\frac{\Lambda}{m_0}\right)} \\ &= m\left(1 - \frac{3}{2}\frac{\alpha}{\pi}\ln\left(\frac{\Lambda}{m_0}\right) + \cdots\right). \end{aligned} \tag{11.135}$$

At zeroth order in α, $m_0 = m$. To get m_0 to first order in α, we replace

m_0 with m on the right-hand side of eq.(11.135).

$$m_0(\Lambda) = m\left(1 - \frac{3}{2}\frac{\alpha}{\pi}\ln\left(\frac{\Lambda}{m}\right) + O(\alpha^2)\right) \tag{11.136}$$

Using eqs.(11.134) and (11.136) we can indeed verify that to order α,

$$\begin{aligned}\frac{dm}{d\Lambda} &= \frac{d}{d\Lambda}(m_0 + \delta m)\\ &= \frac{dm_0}{d\Lambda}\left(1 + \frac{3\alpha}{2\pi}\ln\left(\frac{\Lambda}{m_0}\right)\right) + m_0\left(\frac{3}{2}\frac{\alpha}{\pi}\left(\frac{1}{\Lambda} - \frac{1}{m_0}\frac{dm_0}{d\Lambda}\right)\right)\\ &= 0 + O(\alpha^2).\end{aligned} \tag{11.137}$$

Alternatively, we could have started with eq.(11.134) and required that $dm/d\Lambda = 0$, so that m would be independent of Λ. This would yield a first order differential equation for m_0 whose solution to order α is eq.(11.136). The physical mass m provides the initial or boundary condition to anchor the trajectory of m_0.

The mass that appears in the Hamiltonian, eq.(11.107), is the bare mass, m_0. We can rewrite the Hamiltonian entirely in terms of the renormalized or physical mass m by substituting in eq.(11.136) for m_0. When we do so we get the Hamiltonian with m instead of m_0 and one extra term,

$$H_{\text{ct}} = -\int d^3x \frac{\delta}{\delta\psi(\vec{x})}\,\beta\left(\frac{3\alpha}{2\pi}\ln\left(\frac{\Lambda}{m}\right)\right)\psi(\vec{x}). \tag{11.138}$$

The extra term that appears is called a counterterm and can be thought of as an additional part of H_I. If we now proceed to carry out the identical calculation of the energy eigenvalue of a single electron to order α except that we include the counterterm in H_I, we find that to zeroth order in α the mass is the physical mass m (instead of m_0). At order α the contribution from the counterterm just cancels the shift, δm, that arises from the original interaction in H_I. Thus, by construction, there is *no* shift in the mass of the electron from its physically measured value m obtained at zeroth order. Since this mass is finite, we have removed one infinity from the theory at order α.

Once we introduce a cutoff, we have changed the theory we are dealing with. To recover the original theory, we must take $\Lambda \to \infty$. Since the renormalized mass m is independent of Λ, it remains fixed and finite in the limit. However, we can see from eq.(11.136) that m_0 diverges as $\Lambda \to \infty$. Does this mean that the bare mass of the electron, the mass in the absence of the electromagnetic interaction, the mass we would measure at momentum scales comparable to Λ, is actually infinite? We cannot say. When we examine the infinities present in QED to second order in perturbation theory (order α) in upcoming chapters, we will find that $\alpha = e_0^2/4\pi$ must also be renormalized, that is, that the bare charge, e_0, must also depend on Λ and does so in such a way that $e_0 \to \infty$ as the cutoff is removed. Since α is scale-dependent, then at some energy scale α must become larger than 1, and there perturbation theory will no longer be valid. When this occurs, we no longer possess a valid prediction of the bare mass, m_0.

As a final application of the wave functional $F[\tilde{a}, b]$, eq.(11.120), let us recompute the S-matrix element for electron-electron elastic scattering at order α. Recall that the S-matrix element is just the overlap between initial and final states. The initial state consists of two electrons with momenta $\vec{p}_1$ and $\vec{p}_2$, and with spin s_1 and s_2. The final state also consists of two electrons, and let's call their momenta $\vec{p}_1'$ and $\vec{p}_2'$. The S-matrix element is

$$S_{2\to 2} = \langle p_1', s_1'; p_2', s_2 \,|\, p_1, s_1; p_2, s_2 \rangle, \tag{11.139}$$

which in the Schrödinger representation is a functional integral over $\tilde{a}$ and b.

At zeroth order in e_0, the state $|\, p_1, s_1; p_2, s_2\rangle$ represents two free electrons, so the wave functional is

$$\langle \tilde{a}, b \,|\, p_1, s_1; p_2, s_2\rangle = \Phi^{(0)}_{p_1 p_2}[\tilde{a}, b] = \eta \Psi_0[\tilde{a}]\Omega_{(p_1,s_1;p_2,s_2)}[b]$$
$$= \eta \exp\left(-\frac{1}{2}\int \frac{d^3k}{(2\pi)^3}\frac{(\vec{k}\times\tilde{a}(\vec{k}))\cdot(\vec{k}\times\tilde{a}(-\vec{k}))}{|\vec{k}|}\right) \underset{\text{except}(\vec{p}_1,s_1),(\vec{p}_2,s_2)}{\textstyle\prod_{s=1}^{2}\prod_{\vec{p}}} b(\vec{p}, s). \tag{11.140}$$

To complete the calculation of $S_{2\to 2}$ to order α we need the first perturbative correction to $\Phi^{(0)}_{p_1 p_2}$. We could find this wave functional by solving the Schrödinger equation and constraint as we did to obtain $F[\tilde{a}, b]$; however, to this order, the only difference between the functional for the 1-electron state, eq.(11.120), and the 2-electron state is just the removal of an extra electron from the fermionic part of the functional. Therefore,

$$\Phi^{(1)}_{p_1 p_2}[\tilde{a}, b] = \eta F_2[\tilde{a}, b]\Psi_0[\tilde{a}], \tag{11.141}$$

where

$$F_2[\tilde{a}, b] = \sum_{r'=1}^{2}\sum_{r=3}^{4}\int \frac{d^3k}{(2\pi)^3}\, d^3p \left(\frac{m_0}{E_p}\right)^{\frac{1}{2}}\left(\frac{m_0}{E_{p+k}}\right)^{\frac{1}{2}} \frac{1}{E_p + E_{p+k} + |\vec{k}|}$$
$$w^\dagger(\vec{p}+\vec{k}, r')\left(\vec{\alpha}\cdot\tilde{a}(\vec{k}) + \frac{\vec{k}\cdot\tilde{a}(\vec{k})}{|\vec{k}|}\right) w(\vec{p}, r)\frac{\delta}{\delta b(\vec{p}+\vec{k}, r')} b(\vec{p}, r)\Omega_{(p_1,s_1;p_2,s_2)}[b]$$

$$-\sum_{r'=1}^{2}\int\frac{d^3k}{(2\pi)^3}\left(\frac{m_0}{E_{p_1}}\right)^{\frac{1}{2}}\left(\frac{m_0}{E_{p_1+k}}\right)^{\frac{1}{2}}\frac{1}{E_{p_1}-E_{p_1+k}-|\vec{k}|}$$
$$w^\dagger(\vec{p}_1+\vec{k},r')\left(\vec{\alpha}\cdot\tilde{a}(\vec{k})+\frac{\vec{k}\cdot\tilde{a}(\vec{k})}{|\vec{k}|}\right)w(\vec{p}_1,s_1)\Omega_{(p_1+k,r';p_2,s_2)}[b]$$
$$+\sum_{r'=1}^{2}\int\frac{d^3k}{(2\pi)^3}\left(\frac{m_0}{E_{p_2}}\right)^{\frac{1}{2}}\left(\frac{m_0}{E_{p_2+k}}\right)^{\frac{1}{2}}\frac{1}{E_{p_2}-E_{p_2+k}-|\vec{k}|}$$
$$w^\dagger(\vec{p}_2+\vec{k},r')\left(\vec{\alpha}\cdot\tilde{a}(\vec{k})+\frac{\vec{k}\cdot\tilde{a}(\vec{k})}{|\vec{k}|}\right)w(\vec{p}_2,s_2)\Omega_{(p_1,s_1;p_2+k,r')}[b].$$
$$(11.142)$$

The final state to the same order is the adjoint of (11.142) with $p_1 \to p_1'$, $p_2 \to p_2'$, $s_1 \to s_1'$, and $s_2 \to s_2'$.

Let us compute only the connected part of $S_{2\to2} = S^c_{2\to2}$ in eq.(11.139). The first term in $F_2[\tilde{a}, b]$ contributes only to disconnected vacuum processes, so we may ignore it. At order α,

$$S^c_{2\to2} = e_0^2\eta^2\int\mathcal{D}\tilde{a}\,\mathcal{D}b\Psi_0^*[\tilde{a}]\Psi_0[\tilde{a}]\,F_2^\dagger[p_1',p_2';\tilde{a},b]F_2[p_1,p_2;\tilde{a},b]. \quad (11.143)$$

We leave the details of the functional integration and the comparison with eq.(8.53) as an exercise.

11.4 EXERCISES

1. Generate the expressions for $\Psi_2^{(0)}[\tilde{\phi}]$ and $\Psi_4^{(0)}[\tilde{\phi}]$ given in eq.(11.15) by applying the raising operator, eq.(11.50), to the vacuum functional. Note that the convention we adopted to define the Fourier transform implies that

$$\frac{\delta}{\delta\tilde{\phi}(\vec{p})}\tilde{\phi}(\vec{p}') = (2\pi)^3\delta^3(p-p').$$

2. Compute the overlap given by eq.(11.20) by direct functional integration.
3. Consider a self-interacting scalar field theory given by

$$S[\phi] = \int d^4x\,\frac{1}{2}(\partial^\mu\phi\partial_\mu\phi - m_0^2\phi^2) - \frac{\nu_0}{3!}\phi^3.$$

   Solve the functional Schrödinger equation for the vacuum wave functional as a power series in ν_0 through order ν_0^2. Compare your answer with the result from Rayleigh-Schrödinger perturbation theory.
4. Show by direct substitution that the functional $F[\tilde{a}]$, eq.(11.93), satisfies the Gauss's law constraint, eq.(11.89).
5. Solve the functional Schrödinger equation for QED for the vacuum wave functional to first order in e_0.
6. Prove the identity, eq.(11.121).
7. Verify by direct substitution that the functional, eq.(11.141), is a solution of the Schrödinger equation and the Gauss's law constraint at order e_0.
8. Complete the functional integral in eq.(11.143) and compare the result with that obtained in the operator representation, eq.(8.53).

CHAPTER **12**

Path Integral Representation of Quantum Mechanics

The final representation of quantum field theory that we will consider, the path integral representation, is usually the least familiar representation of quantum mechanics. Therefore, we will begin by developing the concepts and definitions in the context of ordinary, nonrelativistic point particle quantum mechanics. The generalization to field theory follows in the next chapter.

12.1 CONSTRUCTIVE DEFINITION OF A PATH INTEGRAL IN QUANTUM MECHANICS

Let us consider a single particle in a potential $V(x)$ described by a Hamiltonian, H. Let $|\,\psi(t)\rangle$ be a time-dependent state of this 1-particle system. The probability density for the particle to be located at the space-time point (x,t) is $|\psi(x,t)|^2$ where $\psi(x,t) = \langle x\,|\,\psi(t)\rangle$ is the wave function representing the state $|\,\psi(t)\rangle$ in the basis of position operator eigenvectors, $|\,x\rangle$.

The time-dependent state $|\,\psi(t)\rangle$ satisfies the Schrödinger equation

$$i\frac{\partial}{\partial t}|\,\psi(t)\rangle = \mathbf{H}|\,\psi(t)\rangle. \tag{12.1}$$

The formal solution to (12.1) is

$$|\psi(t)\rangle = e^{-i\mathbf{H}(t-t_0)}|\psi(t_0)\rangle. \tag{12.2}$$

In the position basis this becomes

$$\begin{aligned}\langle x|\psi(t)\rangle = \psi(x,t) &= \langle x|e^{-i\mathbf{H}(t-t_0)}|\psi(t_0)\rangle \\ &= \int dx_1\,\langle x|e^{-i\mathbf{H}(t-t_0)}|x_1\rangle\langle x_1|\psi(t_0)\rangle \\ &= \int dx_1\,G(x,t;x_1,t_0)\psi(x_1,t_0),\end{aligned} \tag{12.3}$$

where we have assumed that the position operator eigenvectors are complete,

$$\mathbf{1} = \int dx_1\,|x_1\rangle\langle x_1|. \tag{12.4}$$

The "time-evolution" operator, $\exp(-i\mathbf{H}(t-t_0))$, evolves the system from time t_0 to t, that is, it propagates the states forward in time from t_0 to t. The Green's function for this Schrödinger equation, $G(x,t;x_1,t_0)$, is the matrix element of the time-evolution operator in the $|x\rangle$ basis. Feynman discovered that $G(x,t;x_1,t_0)$ may be represented by a path integral. Let us constructively define a path integral in this context.

Suppose the particle is located at x_0 at time t_0. Then $|\psi(t_0)\rangle = |x_0\rangle$, that is, $\psi(x_1,t_0) = \langle x_1|\psi(t_0)\rangle = \langle x_1|x_0\rangle = \delta(x_1-x_0)$. The wave function at time t is $\psi(x,t)$, thus the quantum mechanical amplitude for the particle to go from (x_0,t_0) to (x,t) is $\psi(x,t)$. From eq.(12.3) we can see that when $|\psi(t_0)\rangle = |x_0\rangle$, then

$$\begin{aligned}\psi(x,t) &= \int dx_1\,G(x,t;x_1,t_0)\delta(x_1-x_0) \\ &= G(x,t;x_0,t_0).\end{aligned} \tag{12.5}$$

Therefore, $G(x,t;x_0,t_0)$ is the quantum mechanical amplitude for a particle to go from (x_0,t_0) to (x,t). We *do not know* how the particle got from (x_0,t_0) to (x,t). We do not know what path it took. In fact, if we don't try to measure which path it takes, then quantum mechanics tells us that it took all paths. While we can't say which path the particle took, if any, we *can* calculate the amplitude that it took a certain path. The total amplitude, $G(x,t;x_0,t_0)$, will be the sum of all amplitudes that represent equivalent physical processes. Since the physical result of propagating a particle from (x_0,t_0) to (x,t) on any path is the same (we aren't looking in between), then $G(x,t;x_0,t_0)$ is the sum of amplitudes, one amplitude for each path. The total amplitude G will be a sum over paths, weighting

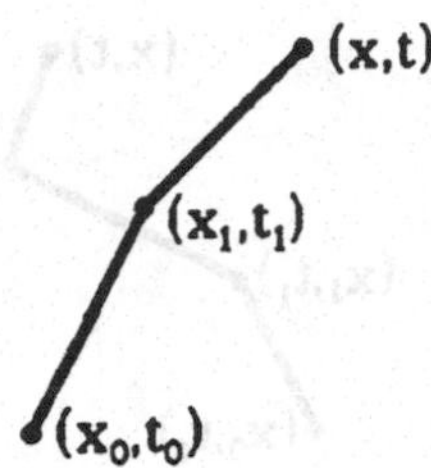

Figure 12.1 The single-particle propagation process broken up into 2 steps. If the straight line between (x_0, t_0) and (x_1, t_1) represents a sum over all paths connecting the two points, and similarly for (x_1, t_1) and (x, t), then when we integrate (x_1, t_1) over all of space-time we are summing over all paths connecting (x_0, t_0) to (x, t).

each path by its quantum mechanical amplitude. This sum over paths is what we call a path integral.

We can construct this path integral and determine the weight or amplitude for each path by breaking up the propagation process into a series of smaller steps. The effect of breaking up the propagation also breaks any path leading from (x_0, t_0) to (x, t) into a collection of smaller pieces.

To begin, consider breaking up the propagation process into 2 steps. The amplitude $G(x, t; x_0, t_0)$ to go from (x_0, t_0) to (x, t) can be written in terms of the amplitude to go from (x_0, t_0) to (x_1, t_1) and the amplitude to go from (x_1, t_1) to (x, t) as

$$G(x, t; x_0, t_0) = \int dx_1\, G(x, t; x_1, t_1)\, G(x_1, t_1; x_0, t_0), \tag{12.6}$$

where $t_0 < t_1 < t$ (figure 12.1). The result, eq.(12.6), follows from inserting eq.(12.4) into eq.(12.3).

$$\begin{aligned} G(x, t; x_0, t_0) &= \langle x\, |e^{-i\mathbf{H}(t-t_0)}|\, x_0\rangle \\ &= \langle x\, |e^{-i\mathbf{H}(t-t_1)}\, e^{-i\mathbf{H}(t_1-t_0)}|\, x_0\rangle \\ &= \int dx_1\, \langle x\, |e^{-i\mathbf{H}(t-t_1)}|\, x_1\rangle\langle x_1\, |e^{-i\mathbf{H}(t_1-t_0)}|\, x_0\rangle \\ &= \int dx_1\, G(x, t; x_1, t_1)\, G(x_1, t_1; x_0, t_0). \end{aligned} \tag{12.7}$$

We may think of the product $G(x, t; x_1, t_1)\, G(x_1, t_1; x_0, t_0)$ as the amplitude for the path of the particle to definitely pass through the space-time point (x_1, t_1) on its way to (x, t) from (x_0, t_0).

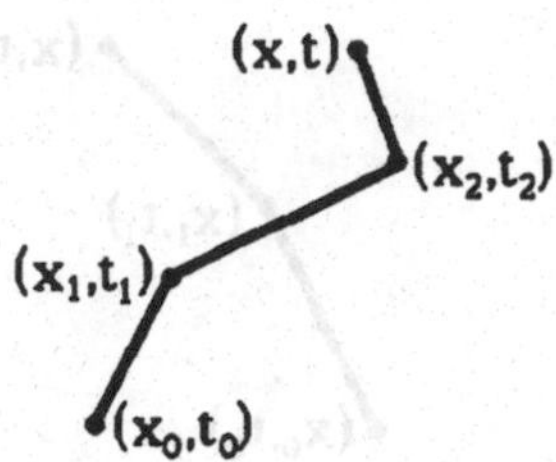

Figure 12.2 The single-particle propagation process broken up into 3 steps. For fixed (x_1, t_1) and (x_2, t_2), this figure represents summing over all paths connecting (x_0, t_0) to (x, t) that pass through (x_1, t_1) and (x_2, t_2).

We can repeat this procedure choosing a t_2 such that $t > t_2 > t_1 > t_0$ and inserting $\int dx_2 \,|\, x_2\rangle\langle x_2 \,|$ into $\langle x \,|\, \exp(-i\mathbf{H}(t - t_1))|\, x_1\rangle$ to break the propagation into 3 steps (figure 12.2). Following the same steps as in eq.(12.7) we have

$$G(x, t; x_0, t_0) = \int dx_2\, dx_1 G(x, t; x_2, t_2) G(x_2, t_2; x_1, t_1) G(x_1, t_1; x_0, t_0), \tag{12.8}$$

thus, $G(x, t; x_2, t_2)\, G(x_2, t_2; x_1, t_1)\, G(x_1, t_1, x_0, t_0)$ is the amplitude for the particle to pass through (x_1, t_1) and (x_2, t_2) on its way from (x, t) from (x_0, t_0).

Now let's repeat this procedure many times and break up the interval between t_0 and t into $N + 1$ equal pieces (figure 12.3). The amplitude to pass through (x_j, t_j) for $j = 1, \ldots, N$, will be the product, $G(x, t; x_N, t_N) \cdots G(x_1, t_1, x_0, t_0)$, while the total amplitude will be

$$G(x, t; x_0, t_0) = \int dx_N \cdots dx_1$$
$$G(x, t; x_N, t_N)\, G(x_N, t_N; x_{N-1}, t_{N-1}) \cdots G(x_1, t_1; x_0, t_0). \tag{12.9}$$

In the limit $N \to \infty$, the product

$$\lim_{N\to\infty} G(x, t; x_N, t_N) \cdots G(x_1, t_1; x_0, t_0) \tag{12.10}$$

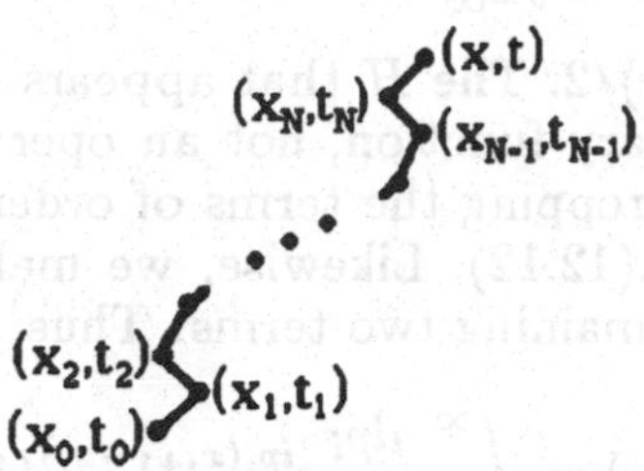

Figure 12.3 The single-particle propagation process broken up into $N+1$ steps. As we integrate over all of the intermediate points (x_i, t_i), $i = 1, \ldots, N$, $t_i < t_{i+1}$, we are summing over all paths connecting (x_0, t_0) to (x, t). In the limit $N \to \infty$, it does not matter if the small propagation step taken between (x_i, t_i) and (x_{i+1}, t_{i+1}) is itself a sum over paths or just a contribution from a single straight path.

will be the amplitude for the particle to take a specific path $x(t)$ between (x_0, t_0) and (x, t). The total amplitude, G, is

$$\begin{aligned} G(x,t;x_0,t_0) &= \lim_{N\to\infty} \int \prod_{i=1}^{N} dx_i\, G(x,t;x_N,t_N)\cdots G(x_1,t_1;x_0,t_0) \\ &\sim \int \mathcal{D}x(t)\cdots, \end{aligned} \tag{12.11}$$

which we can begin to recognize as a functional integral with respect to paths $x(t)$, $\int \mathcal{D}x(t)\cdots$, that is, an integral over paths between (x_0, t_0) and (x, t). We will define what we mean by $\mathcal{D}x(t)$ in a little bit.

To write the integrand in eq.(12.11) in a more useful, compact form, let's examine $G(x_{i+1}, t_{i+1}; x_i, t_i)$ for large N so that $t_{i+1} - t_i = \epsilon$ is small. Since the time interval is small we may expand the exponential in $G(x_{i+1}, t_{i+1}; x_i, t_i)$ and ignore the higher order terms in ϵ to obtain

$$\begin{aligned} G(x_{i+1},t_{i+1};x_i,t_i) &= \langle x_{i+1}\,|\,e^{-i\mathbf{H}(t_{i+1}-t_i)}\,|\,x_i\rangle \\ &= \langle x_{i+1}\,|\,e^{-i\mathbf{H}\epsilon}\,|\,x_i\rangle \\ &= \langle x_{i+1}\,|1 - i\mathbf{H}\epsilon + O(\epsilon^2) + \ldots|\,x_i\rangle \\ &= \langle x_{i+1}\,|\,x_i\rangle - i\epsilon\langle x_{i+1}\,|\,\mathbf{H}\,|\,x_i\rangle + \ldots \end{aligned} \tag{12.12}$$

Next, write the matrix elements on the right-hand side of (12.12) in the momentum representation,

$$\langle x_{i+1}\,|\,x_i\rangle = \delta(x_{i+1} - x_i) = \int_{-\infty}^{\infty} \frac{dp_i}{2\pi} e^{ip_i(x_{i+1}-x_i)}$$

$$\langle x_{i+1} |\mathbf{H}| x_i\rangle = \int_{-\infty}^{\infty} \frac{dp_i}{2\pi} H(p_i, \widehat{x}_i) e^{ip_i(x_{i+1}-x_i)}, \tag{12.13}$$

where $\widehat{x}_i = (x_{i+1} + x_i)/2$. The H that appears on the right-hand side of eq.(12.13) is an ordinary function, not an operator. Since ϵ is small, we make little error by dropping the terms of order ϵ^2 and greater from the right-hand side of eq.(12.12). Likewise, we make little error $(O(\epsilon^2))$ by exponentiating the remaining two terms. Thus,

$$\begin{aligned} G(x_{i+1}, t_{i+1}; x_i, t_i) &= \int_{-\infty}^{\infty} \frac{dp_i}{2\pi} e^{ip_i(x_{i+1}-x_i)}(1 - i\epsilon H(p_i, \widehat{x}_i) + \cdots) \\ &\approx \int_{-\infty}^{\infty} \frac{dp_i}{2\pi} e^{ip_i(x_{i+1}-x_i)}\, e^{-i\epsilon H(p_i,\widehat{x}_i)}. \end{aligned} \tag{12.14}$$

Now the infinite product in the integrand in eq.(12.11), the amplitude to take a particular path, can be easily written as

$$\begin{aligned} &\lim_{N\to\infty} \prod_{i=0}^{N} G(x_{i+1}, t_{i+1}; x_i, t_i) \\ &\quad = \lim_{N\to\infty} \int \prod_{i=0}^{N} \frac{dp_i}{2\pi} \prod_{j=1}^{N} e^{ip_j(x_{j+1}-x_j)}\, e^{-i(t-t_0)H(p_j,\widehat{x}_j)/(N+1)} \\ &= \lim_{N\to\infty} \int \prod_{j=0}^{N} \frac{dp_j}{2\pi} \exp\left(i \sum_{j=0}^{N} p_j(x_{j+1} - x - j) - \left(\frac{t-t_0}{N+1}\right) H(p_j, \widehat{x}_j)\right) \\ &\quad \equiv \int \mathcal{D}p \exp\left(i \int_{t_0}^{t} p\dot{x} - H(p, x)\, d\bar{t}\right), \end{aligned} \tag{12.15}$$

where $x_{N+1} = x$, $t_{N+1} = t$. The entire amplitude is

$$\begin{aligned} &G(x, t; x_0, t_0) = \\ &\lim_{N\to\infty} \int \prod_{j=0}^{N} \frac{dp_j}{2\pi} \prod_{i=1}^{N} dx_i \exp\left(i \sum_{j=0}^{N} p_j(x_{j+1} - x_j) - \left(\frac{t-t_0}{N+1}\right) H(p_j, \widehat{x}_j)\right) \\ &\qquad \equiv \int \mathcal{D}\overline{x}\mathcal{D}p \exp\left(i \int_{t_0}^{t} p\dot{\overline{x}} - H(p, \overline{x}) d\bar{t}\right). \end{aligned} \tag{12.16}$$

The first line of eq.(12.16) provides a constructive definition of a path integral, symbolically written as the second line of (12.16). If H is quadratic in p with no funny extra p dependence, then we can carry out the N Gaussian p integrals and take the limit. To do so we must complete the squares in the exponential. The net result of the functional integration is to replace p with $\dot{x}$ in the exponential (more explicitly, replace p_j with $(x_{j+1} - x_j)/\epsilon$), and to add a normalization factor that is absorbed into the measure $\mathcal{D}\bar{x}(\bar{t})$.

Specifically, if $H = p^2/2 + V(x)$, then completing the squares in eq.(12.14) yields

$$\begin{aligned} G(x_{i+1}, t_{i+1}; x_i, t_i) &= \int_{-\infty}^{\infty} \frac{dp_i}{2\pi} e^{ip_i(x_{j+1}-x_j)} e^{-i\epsilon(p_i^2/2+V(\widehat{x}_i))} \\ &= e^{i(x_{i+1}-x_i)^2/2\epsilon} e^{-i\epsilon V(\widehat{x}_i)} \int_{-\infty}^{\infty} \frac{dp_i}{2\pi} e^{-i\epsilon(p_i-\Delta x_i/\epsilon)^2/2} \\ &= \left(\frac{1}{2\pi i\epsilon}\right)^{1/2} \exp\left(i\epsilon(\frac{\dot{x}_i^2}{2} - V(\widehat{x}_i)\right), \end{aligned} \tag{12.17}$$

where $\dot{x}_i = (x_{i+1} - x_i)/\epsilon$. Eq.(12.16) becomes

$$\begin{aligned} G(x, t; x_0, t_0) &= \int \mathcal{D}\bar{x}(\bar{t}) \exp\left(i \int_{t_0}^{t} L[\dot{\bar{x}}, \bar{x}, \bar{t}]d\bar{t}\right) \\ &= \int \mathcal{D}\bar{x}(\bar{t}) \exp\left(iS[\bar{x}(\bar{t})]\right), \end{aligned} \tag{12.18}$$

where L is the classical Lagrangian and $S[\bar{x}(\bar{t})]$ is the classical action, and where the path integral is functionally integrated only over functions $\bar{x}(\bar{t})$ such that $\bar{x}(t_0) = x_0$ and $\bar{x}(t) = x$, and where

$$\mathcal{D}\bar{x}(\bar{t}) = \lim_{N\to\infty} \prod_{i=1}^{N} d\bar{x}_i \left(\frac{1}{2\pi i\epsilon}\right)^{1/2}, \quad N\epsilon = \text{fixed} = t - t_0. \tag{12.19}$$

Thus, the amplitude to take a particular path is the exponential of i times the classical action associated with the path.

You will have noticed that $\widehat{x}_i = (x_i + x_{i+1})/2$ appeared in eq. (12.13) instead of just x_i. The choice of $\widehat{x}_i$ over x_i amounts to an operator ordering prescription upon quantization. The particular choice of $\widehat{x}_i$ leads to a symmetric operator ordering since x_i and x_{i+1} are treated on equal footing. For example, if the term xp appeared in H, then the ordering upon quantization is ambiguous since $\mathbf{x}$ and $\mathbf{p}$ do not commute. The choice $\mathbf{px}$ leads to placing x_i in eq.(12.13) since

$$\begin{aligned} \langle x_{i+1} | \mathbf{px} | x_i \rangle &= \int \frac{dp_i}{2\pi} \frac{dp_j}{2\pi} \langle x_{i+1} | p_j \rangle \langle p_j | \mathbf{p} | p_i \rangle \langle p_i | \mathbf{x} | x_i \rangle \\ &= \int \frac{dp_i}{2\pi} \frac{dp_j}{2\pi} x_i p_i \langle x_{i+1} | p_j \rangle \langle p_j | p_i \rangle \langle p_i | x_i \rangle \\ &= \int \frac{dp_i}{2\pi} e^{ip_i(x_{i+1}-x_i)} x_i p_i. \end{aligned}$$

The choice of $\widehat{x}_i$ on the right-hand side leads to a symmetric ordering of x and p.

$$\langle x_{i+1} \,|\frac{1}{2}(\mathbf{xp} + \mathbf{px})|\, x_i\rangle = \int \frac{dp_i}{2\pi} e^{ip_i(x_{i+1}-x_i)} p_i \widehat{x}_i$$

As you can see, the object of interest to calculate in the path integral representation is $G(x, t; x_0, t_0)$, the matrix element of the time-evolution operator, which is a path integral. By construction, G is the Green's function for the Schrödinger equation, and therefore propagates quantum mechanical states forward in time. Each representation of quantum mechanics has its particular usefulness, and given the nature of G, the path integral representation is well suited to compute transition amplitudes, vacuum expectation values, and coordinate or momentum representation matrix elements of operators.

The path integral representation of transition amplitudes is a straightforward application of eqs.(12.2), (12.3), and (12.16) or (12.18). If the system is in a state $|\,\psi\rangle$ at time t_0, then it will evolve to a state $|\,\phi\rangle$ at time t, where by eq.(12.2),

$$|\,\phi\rangle = e^{-i\mathbf{H}(t-t_0)}|\,\psi\rangle.$$

The amplitude for the system to be in a state $|\,\chi\rangle$ at time t is $\langle\chi\,|\,\phi\rangle$. Thus, the transition amplitude for the system to go from state $|\,\psi\rangle$ at t_0 to state $|\,\chi\rangle$ at t is

$$\begin{aligned}\langle\chi\,|\,\phi\rangle &= \langle\chi\,|e^{-i\mathbf{H}(t-t_0)}|\,\psi\rangle \\ &= \int dx\, dx_0\, \langle\chi\,|\,x\rangle\langle x\,|e^{-i\mathbf{H}(t-t_0)}|\,x_0\rangle\langle x_0\,|\,\psi\rangle \\ &= \int dx\, dx_0\, \chi^*(x) G(x, t; x_0, t_0)\psi(x_0) \\ &= \int dx\, dx_0 \int \mathcal{D}\bar{x}(\bar{t})\chi^*(x) \exp(iS[\bar{x}(\bar{t})])\, \psi(x_0),\end{aligned} \tag{12.20}$$

where all paths satisfy $\bar{x}(t_0) = x_0$ and $\bar{x}(t) = x$.

To arrive at a path integral representation of the vacuum expectation value of an operator, let us rewrite G in terms of the energy eigenstates, $|\,\phi_n\rangle$, of the Hamiltonian, $\mathbf{H}$. $|\,\phi_n\rangle$ satisfies $\mathbf{H}|\,\phi_n\rangle = E_n|\,\phi_n\rangle$; assume the collection of eigenstates forms a complete set,

$$\sum_n |\,\phi_n\rangle\langle\phi_n\,| = 1. \tag{12.21}$$

Insert this form of 1 into the matrix element expression for G.

$$\begin{aligned}
G(x,t;x_0,t_0) &= \langle x\,|e^{-i\mathbf{H}(t-t_0)}|\,x_0\rangle \\
&= \sum_{n,m}\langle x\,|\,\phi_n\rangle\langle\phi_n\,|e^{-i\mathbf{H}(t-t_0)}|\,\phi_m\rangle\langle\phi_m\,|\,x_0\rangle \\
&= \sum_{n,m}\phi_n(x)e^{-iE_nt}\langle\phi_n\,|\,\phi_m\rangle e^{iE_mt_0}\phi_m^*(x_0) \\
&= \sum_{n} e^{-iE_n(t-t_0)}\phi_n^*(x_0)\phi_n(x)
\end{aligned} \tag{12.22}$$

For convenience, set $t_0 = 0$ and consider the limit $t \to \infty$ (or if you prefer, $it \to \infty$). In this limit, all contributions to G in the sum in eq.(12.22) vanish except for the smallest value, the ground state ($n = 0$), provided there is a gap between E_0 and E_1. Any term in the sum in eq.(12.22) for $n > 0$ will oscillate rapidly as t gets large and will not contribute. In terms of Euclidean (imaginary) time, each term above the ground state will exponentially decay away relative to the ground state as $t \to \infty$. Thus,

$$\lim_{t\to\infty} G(x,t;x,0) = \phi_0^*(x)\phi_0(x), \tag{12.23}$$

and

$$\lim_{t\to\infty} G(x,t;0,0) \approx \phi_0(x). \tag{12.24}$$

The vacuum expectation value of an operator $O(x)$ is

$$\begin{aligned}
\langle O\rangle &= \lim_{t\to\infty}\int dx G(x,t;x,0)O(x) \\
&= \lim_{t\to\infty}\int dx G(x,t;x,-t)O(x) \\
&= \int dx \int \mathcal{D}\bar{x}(t)O(\bar{x})\exp\left(i\int_{-\infty}^{\infty} L(\dot{\bar{x}},\bar{x},t)\,dt\right),
\end{aligned} \tag{12.25}$$

where all the paths that are integrated over satisfy $\bar{x}(-\infty) = x = \bar{x}(\infty)$.

To derive the path integral representation of a time-dependent operator matrix element in the coordinate representation, it is useful to start from the Heisenberg representation. The time-independent Heisenberg state, $|\,xt\rangle$, is related to the time dependent Schrödinger state, $|\,x\rangle$, via $|\,xt\rangle = \exp(i\mathbf{H}t)|\,x\rangle$. (Yes, the notation is lousy, but standard.) Therefore, $G(x,t;x_0,t_0)$ can be written as $\langle xt\,|\,x_0t_0\rangle$. The general matrix element of the time-dependent operator $O(t)$ is

$$\begin{aligned}
&\langle xt\,|\mathbf{O}(\bar{t})|\,x_0t_0\rangle \\
&\quad = \int dx_1\,dx_2\,\langle xt\,|x_2\bar{t}\rangle\langle x_2\bar{t}\,|\mathbf{O}(\bar{t})|\,x_1\bar{t}\rangle\langle x_1\bar{t}\,|x_0t_0\rangle \\
&\quad = \int dx_1\,dx_2 G(x,t;x_2,\bar{t})\langle x_2\,|\mathbf{O}_s|\,x_1\rangle G(x_1,\bar{t};x_0,t_0),
\end{aligned} \tag{12.26}$$

where the $\mathbf{O}_s$ is the operator in the Schrödinger picture. Let us suppose that the operator is diagonal in the position basis,

$$\langle x_2\,|\mathbf{O}_s|\,x_1\rangle = \langle x_2(\bar{t})\,|\mathbf{O}_s|\,x_1(\bar{t})\rangle = O(x_1)\,\delta(x_1 - x_2). \tag{12.27}$$

Then, by using eqs.(12.6) and (12.18), the matrix element (12.26) becomes

$$\langle xt\,|\mathbf{O}(\bar{t})|\,x_0t_0\rangle = \int \mathcal{D}x_1(t_1)\,\exp\left(i\int_{t_0}^{t} L[x_1(t_1)]\,dt_1\right) O[x_1(\bar{t})], \tag{12.28}$$

where all of the paths $x_1(t)$ start at x_0 at t_0 and end at x at t, $x_1(t_0) = x_0$, $x_1(t) = x$. In the limit $t, -t_0 \to \infty$, eq.(12.28) becomes the vacuum expectation value.

$$\langle\mathbf{O}(\bar{t})\rangle = \int \mathcal{D}x_1(t_1)\,\exp\left(i\int_{-\infty}^{\infty} L[x_1(t_1)]\,dt_1\right) O[x_1(\bar{t})] \tag{12.29}$$

Compare this result with eq.(12.25).

Next consider two time-dependent operators, $\mathbf{O}_1(t)$ and $\mathbf{O}_2(t)$. If $t_0 < t_1 < t_2 < t$, then

$$\begin{aligned}
&\langle xt\,|\mathbf{O}_2(t_2)\mathbf{O}_1(t_1)|\,x_0t_0\rangle \\
&\quad = \int \mathcal{D}x_1(\bar{t}) O_2[x_1(t_2)] O[x_1(t_1)]\,\exp\left(i\int_{t_0}^{t} L[x_1(\bar{t})]\,d\bar{t}\right).
\end{aligned} \tag{12.30}$$

On the other hand, if $t_0, t_2 < t_1 < t$, then

$$\begin{aligned}
&\langle xt\,|\mathbf{O}_1(t_1)\mathbf{O}_2(t_2)|\,x_0t_0\rangle \\
&\quad = \int \mathcal{D}x_1(\bar{t}) O_1[x_1(t_1)] O_2[x_1(t_2)]\,\exp\left(i\int_{t_0}^{t} L[x_1(\bar{t})]\,d\bar{t}\right).
\end{aligned} \tag{12.31}$$

Notice that O_1 and O_2, appearing on the right-hand sides of eqs.(12.30) and (12.31) above, are just ordinary functions, not operators, so that their order is immaterial. Thus, the right-hand sides of eqs.(12.30) and (12.31) are identical. The only difference is that $t_1 < t_2$ in eq.(12.30) and $t_2 < t_1$

in eq.(12.31). Reversing the argument, we have

$$\int \mathcal{D}x_1(\bar{t})\, O_1[x_1(t_1)]\, O_2[x_2(t_2)]\, \exp\left(i\int_{t_0}^{t} L[x_1(\bar{t})]\, d\bar{t}\right)$$
$$= \langle xt\,|T(\mathbf{O}_1(t_1)\mathbf{O}_2(t_2))|\, x_0t_0\rangle, \quad (12.32)$$

where T is the usual time-ordering operator. Time-ordered products have natural path integral representations.

Finally, the generalization to multiparticle systems is straightforward. For example, let the Hamiltonian, $H = (p_1^2 + p_2^2)/2 + V(x_1, x_2)$, describe a 2-particle system. The amplitude for particle 1 to arrive at (x,t) from (x_0, t_0) and for particle 2 to arrive at (X,t) from (X_0, t_0) is

$$G(x, X, t; x_0, X_0, t_0)$$
$$= \langle x, X, t\,|e^{-i\mathbf{H}(t-t_0)}|\, x_0, X_0, t_0\rangle$$
$$= \int \mathcal{D}p_1\, \mathcal{D}p_2\, \mathcal{D}x_1\, \mathcal{D}x_2\, \exp\left(i\int_{t_0}^{t} d\bar{t}\;\, p_1\dot{x}_1 + p_2\dot{x}_2 - H(p_1, p_2, \widehat{x}_1, \widehat{x}_2)\right), \quad (12.33)$$

where $x_1(t_0) = x_0$, $x_2(t_0) = X_0$, $x_1(t) = x$, and $x_2(t) = X$. The generalization to an infinite number of particles and degrees of freedom follows in the next chapter.

12.2 FREE PARTICLE PATH INTEGRAL

With the formalism in place, it is time to use the operational definition of a path integral, eq.(12.16), to actually calculate $G(x, t; x_0, t_0)$ for a 1-particle system. The simplest system, a free particle, provides an excellent example since it can be done exactly.

For a free particle of mass m, the Hamiltonian is simply

$$H = \frac{1}{2m}p^2, \quad (12.34)$$

so

$$G(x, t; x_0, t_0) = \int \mathcal{D}p\mathcal{D}\bar{x}\, \exp\left(i\int_{t_0}^{t} (p\dot{\bar{x}} - \frac{1}{2m}p^2)\, d\bar{t}\right), \quad (12.35)$$

where all paths start at (x_0, t_0) and end at (x, t). Our operational definition, eq.(12.16), tells us that we should subdivide the time interval into $N+1$ equal parts, compute the $2N$ integrals and let $N \to \infty$. Let t_i be the time of the ith slice, $\bar{x}_i$ be the spatial coordinate of the path

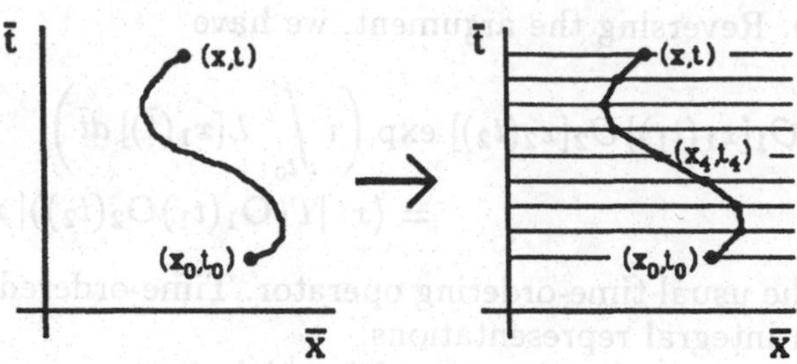

Figure 12.4 To compute a path integral using the constructive definition we discretize space-time in the time direction. Each path is broken up into $N+1$ parts and the contribution from each part is added up. Eventually we let $N \to \infty$.

on the ith slice, $\bar{x}(t_i) = \bar{x}_i$, and let the time interval between slices be $\epsilon = t_{i+1} - t_i = (t - t_0)/N + 1$. Each path is then broken up into $N+1$ parts as in figure 12.4. With this discretization in the time direction, the time derivative is taken as the forward difference,

$$\dot{\bar{x}} = \frac{\bar{x}_{i+1} - \bar{x}_i}{\epsilon}. \tag{12.36}$$

According to eq.(12.16),

$$\begin{aligned} &G(x, t; x_0, t_0) \\ &= \lim_{N\to\infty} \int \prod_{j=0}^{N} \frac{dp_j}{2\pi} \prod_{i=1}^{N} d\bar{x}_i \exp\left(i \sum_{j=0}^{N} \epsilon p_j \left(\frac{\bar{x}_{j+1} - \bar{x}_j}{\epsilon} \right) - \frac{1}{2m} p_j^2 \epsilon \right) \\ &= \lim_{N\to\infty} \int \prod_{i=1}^{N} d\bar{x}_i \left(\frac{m}{2\pi i \epsilon} \right)^{\frac{N+1}{2}} \exp\left(i \sum_{j=0}^{N} \frac{m\epsilon}{2} \left(\frac{\bar{x}_{j+1} - \bar{x}_j}{\epsilon} \right)^2 \right). \end{aligned} \tag{12.37}$$

It is easy to recognize the discretized form of the free particle action,

$$S = \int_{t_0}^{t} d\bar{t}\, \frac{m}{2} \dot{\bar{x}}^2,$$

that arises after the $N+1$ Gaussian integrals over p_j are done.

To complete the calculation we must do the remaining N Gaussian integrals over $\bar{x}_i$ and take $N \to \infty$. Each Gaussian requires completing the squares so let's start with $i = 1$.

$$
\begin{aligned}
&G(x,t;x_0,t_0)\\
&= \lim_{N\to\infty}\left(\frac{m}{2\pi i\epsilon}\right)^{\frac{N+1}{2}} \int \prod_{i=2}^{N} dx_i \exp\left(i\sum_{j=2}^{N}\frac{m\epsilon}{2}\left(\frac{x_{j+1}-x_j}{\epsilon}\right)^2\right) \\
&\qquad \int_{-\infty}^{\infty} dx_1 \exp\left(\frac{im}{2\epsilon}\left((x_2-x_1)^2-(x_1-x_0)^2\right)\right).
\end{aligned}
\tag{12.38}
$$

Completing the squares for x_1 we have

$$
\begin{aligned}
(x_2-x_1)^2+(x_1-x_0)^2 &= 2x_1^2 - 2x_1(x_2+x_0) + (x_2^2+x_0^2)\\
&= 2\left(x_1 - \frac{1}{2}(x_2+x_0)\right)^2 + \frac{1}{2}(x_2-x_0)^2.
\end{aligned}
$$

Thus,

$$
\begin{aligned}
&\int_{-\infty}^{\infty} dx_1 \exp\left(\frac{im}{2\epsilon}\left((x_2-x_1)^2+(x_1-x_0)^2\right)\right)\\
&= \exp\left(\frac{im}{4\epsilon}(x_2-x_0)^2\right)\int_{-\infty}^{\infty} dx_1 \exp\left(\frac{im}{\epsilon}\left(x_1-\frac{1}{2}(x_2+x_0)\right)^2\right)\\
&= \left(\frac{\pi\epsilon}{im}\right)^{\frac{1}{2}} \exp\left(\frac{im}{4\epsilon}(x_2-x_0)^2\right)
\end{aligned}
$$

and

$$
\begin{aligned}
G(x,t;x_0,t_0) &= \lim_{N\to\infty}\left(\frac{m}{2\pi i\epsilon}\right)^{\frac{N+1}{2}}\left(\frac{\pi\epsilon}{im}\right)^{\frac{1}{2}}\\
&\int \prod_{i=2}^{N} dx_i \exp\left(i\sum_{j=2}^{N}\frac{m\epsilon}{2}\left(\frac{x_{j+1}-x_j}{\epsilon}\right)^2\right)\exp\left(\frac{im}{4\epsilon}(x_2-x_0)^2\right).
\end{aligned}
\tag{12.39}
$$

Continuing this process, we next look at the integral over x_2.

$$
\begin{aligned}
&G(x,t;x_0,t_0)\\
&= \lim_{N\to\infty}\left(\frac{m}{2\pi i\epsilon}\right)^{\frac{N+1}{2}}\left(\frac{\pi\epsilon}{im}\right)^{\frac{1}{2}} \int \prod_{i=3}^{N} dx_i \exp\left(i\sum_{j=3}^{N}\frac{m\epsilon}{2}\left(\frac{x_{j+1}-x_j}{\epsilon}\right)^2\right)
\end{aligned}
$$

$$\int_{-\infty}^{\infty} dx_2 \exp\left(\frac{im\epsilon}{2}\left(\frac{x_3 - x_2}{\epsilon}\right)^2\right) \exp\left(\frac{im}{4\epsilon}(x_2 - x_0)^2\right) \tag{12.40}$$

Completing the squares,

$$\begin{aligned}(x_3 - x_2)^2 + \frac{1}{2}(x_2 - x_0)^2 &= \frac{3}{2}x_2^2 - x_2(2x_3 + x_0) + x_3^2 + \frac{x_0^2}{2} \\ &= \frac{3}{2}\left(x_2 - \left(\frac{2x_3 + x_0}{3}\right)\right)^2 + \frac{1}{3}(x_3 - x_0)^2,\end{aligned}$$

so

$$\begin{aligned}\int_{-\infty}^{\infty} dx_2 \exp&\left(\frac{im\epsilon}{2}\left(\frac{x_3 - x_2}{\epsilon}\right)^2\right) \exp\left(\frac{im}{4\epsilon}(x_2 - x_0)^2\right) \\ &= \left(\frac{2 \cdot 2\epsilon\pi}{3im}\right)^{\frac{1}{2}} \exp\left(\frac{im}{2 \cdot (3\epsilon)}(x_3 - x_0)^2\right),\end{aligned}$$

and

$$\begin{aligned}G(x, t; x_0, t_0) = \lim_{N\to\infty} &\left(\frac{m}{2\pi i\epsilon}\right)^{\frac{N+1}{2}} \left(\frac{1}{2}\right)\left(\frac{2\pi\epsilon}{2im}\right)^{\frac{1}{2}} \left(\left(\frac{2}{3}\right)\frac{2\pi\epsilon}{im}\right)^{\frac{1}{2}} \\ &\int \prod_{i=3}^{N} dx_i \exp\left(i\sum_{j=3}^{N} \frac{m\epsilon}{2}\left(\frac{x_{j+1} - x_j}{\epsilon}\right)^2\right) \exp\left(\frac{im}{2 \cdot (3\epsilon)}(x_3 - x_0)^2\right).\end{aligned} \tag{12.41}$$

Now we can see a pattern emerge upon which to do induction. For the result from the next step, the x_4 integral, we replace 3ϵ by 4ϵ and x_3 by x_4 in the exponential above and we will pick up a factor

$$\left(\left(\frac{3}{4}\right)\frac{2\pi\epsilon}{im}\right)^{\frac{1}{2}}$$

in front. The result after the kth integral over x_k will be

$$G(x,t;x_0,t_0) = \lim_{N\to\infty} \left(\frac{m}{2\pi i\epsilon}\right)^{\frac{N+1}{2}}$$

$$\prod_{j=1}^{k}\left(\left(\frac{j}{j+1}\right)\frac{2\pi\epsilon}{im}\right)^{\frac{1}{2}} \int \prod_{i=k+1}^{N} dx_i \exp\left(i \sum_{j=k+1}^{N} \frac{m\epsilon}{2}\left(\frac{x_{j+1}-x_j}{\epsilon}\right)^2\right)$$

$$\exp\left(\frac{im}{2\cdot(k+1)\epsilon}(x_{k+1}-x_0)^2\right). \quad (12.42)$$

After all N integrals are done we finally have

$$G(x,t;x_0,t_0) =$$

$$\lim_{N\to\infty} \left(\frac{m}{2\pi i\epsilon}\right)^{\frac{N+1}{2}} \prod_{j=1}^{N}\left(\left(\frac{j}{j+1}\right)\frac{2\pi\epsilon}{im}\right)^{\frac{1}{2}} \exp\left(\frac{im}{2(N+1)\epsilon}(x_{N+1}-x_0)^2\right). \quad (12.43)$$

But $x_{N+1} = x$, $(N+1)\epsilon = t - t_0$, and

$$\prod_{j=1}^{N}\left(\left(\frac{j}{j+1}\right)\frac{2\pi\epsilon}{im}\right)^{\frac{1}{2}} = \left(\frac{2\pi\epsilon}{im}\right)^{\frac{N}{2}} \prod_{j=1}^{N}\left(\frac{j}{j+1}\right)^{\frac{1}{2}} = \left(\frac{2\pi\epsilon}{im}\right)^{\frac{N}{2}}\left(\frac{1}{N+1}\right)^{\frac{1}{2}}, \quad (12.44)$$

so

$$\begin{aligned} G(x,t;x_0,t_0) &= \lim_{N\to\infty} \left(\frac{m}{2\pi i\epsilon}\right)^{\frac{N+1}{2}} \left(\frac{2\pi\epsilon}{im}\right)^{\frac{N}{2}} \left(\frac{1}{N+1}\right)^{\frac{1}{2}} \exp\left(\frac{im}{2(t-t_0)}(x-x_0)^2\right) \\ &= \lim_{N\to\infty} \left(\frac{m}{2\pi i\epsilon(N+1)}\right)^{\frac{1}{2}} \exp\left(\frac{im}{2(t-t_0)}(x-x_0)^2\right) \\ &= \left(\frac{m}{2\pi i(t-t_0)}\right)^{\frac{1}{2}} \exp\left(\frac{im}{2(t-t_0)}(x-x_0)^2\right). \end{aligned} \quad (12.45)$$

12.3 HARMONIC OSCILLATOR PATH INTEGRAL

The harmonic oscillator was done in the Heisenberg and Schrödinger representations in chapter 2. For comparison, we now solve it again with path integrals. Since the potential is quadratic in x, the path integral is still Gaussian so that it can be computed exactly. We could compute it using brute force in a manner similar to the way we did the free particle case in the previous section. Instead, let's take a different approach and do a little functional calculus.

The Hamiltonian for the harmonic oscillator is

$$H = \frac{1}{2m}p^2 + m\omega^2 x^2,$$

so we may start directly for eq.(12.18).

$$\begin{aligned} G(x,t;x_0,t_0) &= \int \mathcal{D}\bar{x}(\bar{t}) \exp\left(iS[\bar{x}(\bar{t})]\right) \\ &= \int \mathcal{D}\bar{x}(\bar{t}) \exp\left(i\int_{t_0}^{t} d\bar{t}\,\frac{1}{2}m(\dot{\bar{x}}^2 - \omega^2\bar{x}^2)\right), \end{aligned} \tag{12.46}$$

where $\bar{x}(t_0) = x_0$ and $\bar{x}(t) = x$. Consider the functional Taylor expansion of the action functional about a particular path, x_{cl}.

$$S[\bar{x}(\bar{t})] = S[x_{\text{cl}}(\bar{t})] + \left.\frac{\delta S}{\delta \bar{x}}\right|_{\bar{x}=x_{\text{cl}}} (\bar{x}(\bar{t}) - x_{\text{cl}}(\bar{t})) + \frac{1}{2}\left.\frac{\delta^2 S}{\delta \bar{x}^2}\right|_{\bar{x}=x_{\text{cl}}} (\bar{x}(\bar{t}) - x_{\text{cl}}(\bar{t}))^2 \tag{12.47}$$

Since the action is quadratic in the path $\bar{x}(\bar{t})$, the Taylor series vanishes after the third term. Now we want to do what would amount to a stationary phase approximation for a general action and choose $x_{\text{cl}}(\bar{t})$ so that

$$\left.\frac{\delta S}{\delta \bar{x}}\right|_{\bar{x}=x_{\text{cl}}} = 0. \tag{12.48}$$

Since $\delta S/\delta\bar{x} = 0$ is just the equation of motion, then $x_{\text{cl}}(\bar{t})$ is simply the classical path. For a harmonic oscillator that starts at (x_0, t_0) and goes through (x, t), the classical path is

$$x_{\text{cl}}(\bar{t}) = x_0\left(\frac{\sin(\omega(t-\bar{t}))}{\sin(\omega(t-t_0))}\right) + x\left(\frac{\sin(\omega(\bar{t}-t_0))}{\sin(\omega(t-t_0))}\right), \tag{12.49}$$

so that the first term in the Taylor expansion is

$$S[x_{\text{cl}}] = \frac{m}{2}\left(\frac{\omega}{\sin(\omega(t-t_0))}\right)\left((x^2 + x_0^2)\cos(\omega(t-t_0)) - 2xx_0\right). \tag{12.50}$$

Placing eqs.(12.50), (12.48), and (12.47) into eq.(12.46), the expression for G reduces to the path integral

$$G = \exp\left(\frac{im}{2}\left(\frac{\omega}{\sin(\omega(t-t_0))}\right)\left((x^2+x_0^2)\cos(\omega(t-t_0)) - 2xx_0\right)\right)$$
$$\int \mathcal{D}\bar{x}(\bar{t})\, \exp\left(-\frac{im}{2}\int_{t_0}^{t} d\bar{t}\,(\bar{x}-x_{\text{cl}})(\partial_t^2+\omega^2)(\bar{x}-x_{\text{cl}})\right). \tag{12.51}$$

To do the remaining path integral we switch path variables from $\bar{x}(\bar{t})$ to $y(\bar{t}) = \bar{x}(\bar{t}) - x_{\text{cl}}(\bar{t})$. All paths $y(\bar{t})$ are fluctuations about the classical path with the endpoints fixed, that is, $y(t_0) = y(t) = 0$. If we identify the point $y(t)$ with $y(t_0)$, then the paths being integrated over may be considered as periodic with period $t - t_0$. According to eq.(9.47), the remaining path integral in eq.(12.51) is the square root of 1 over a determinant.

$$G(x,t;x_0,t_0) = \left(\det\left(\frac{im}{2\pi}(\partial_t^2+\omega^2)\right)\right)^{-\frac{1}{2}}$$
$$\exp\left(\frac{im}{2}\left(\frac{\omega}{\sin(\omega(t-t_0))}\right)\left((x^2+x_0^2)\cos(\omega(t-t_0)) - 2xx_0\right)\right). \tag{12.52}$$

The determinant of an operator is the product of its eigenvalues. Since the paths over which we integrated were periodic with period $T = t - t_0$, and had fixed endpoints at zero, the eigenfunctions of $\partial_t^2 + \omega^2$ are $\sin(n\pi\bar{t}/T)$, $n = 1, 2, 3, \ldots$ with eigenvalues $\lambda_n = -(n\pi/T)^2 + \omega^2$. Thus,

$$\begin{aligned}\det\left(\frac{im}{2\pi}(\partial_t^2+\omega^2)\right) &= \prod_n \frac{m}{2\pi i}\left(\left(\frac{n\pi}{T}\right)^2 - \omega^2\right) \\ &= \prod_n \left(\frac{m}{2\pi i}\left(\frac{n\pi}{T}\right)^2\right)\prod_n\left(1 - \frac{\omega^2 T^2}{(n\pi)^2}\right).\end{aligned} \tag{12.53}$$

Now observe that if we set $\omega = 0$, we must recover the result of the previous section for the free particle. Thus, we arrive at the curious result,

$$\prod_n \left(\frac{m}{2\pi i}\left(\frac{n\pi}{T}\right)^2\right) = \frac{2\pi i T}{m}. \tag{12.54}$$

The remaining part of eq.(12.53) looks like a polynomial in ωT written in factorized form. If $\omega T = n\pi$ for any $n = 1, 2, 3, \ldots$, then the product is zero. Therefore, the zeroes of this function are $n\pi$ which suggests that the infinite product is related to $\sin(\omega T)$. This infinite product is, in fact,

Euler's famous factorization of $\sin(\omega T)/\omega T$,

$$\frac{\sin(\omega T)}{\omega T} = \prod_n \left(1 - \left(\frac{\omega T}{n\pi}\right)^2\right). \tag{12.55}$$

One may arrive at such an expression using the Mittag-Leffler theorem or, for example, in the following way:

Consider the Fourier series expansion of $\cos(\alpha x)$ on $x \in [-\pi, \pi]$ where α is not an integer.

$$\cos(\alpha x) = \sum_n a_n(\alpha)\cos(nx), \tag{12.56}$$

where

$$\begin{aligned} a_n = \frac{2}{\pi}\int_0^\pi dx\, \cos(\alpha x)\cos(nx) &= \frac{1}{\pi}\left(\frac{\sin((\alpha+n)\pi)}{\alpha+n} + \frac{\sin((\alpha-n)\pi)}{\alpha-n}\right) \\ &= \frac{1}{\pi}(-1)^n \frac{\alpha}{\alpha^2-n^2}\sin(\alpha\pi). \end{aligned} \tag{12.57}$$

Evaluate the series eq.(12.56) at $x = \pi$ and now think of it as a series expansion in terms of α.

$$\cos(\alpha\pi) = \frac{2\alpha\sin(\alpha\pi)}{\pi}\left(\frac{1}{2\alpha^2} + \frac{1}{\alpha^2-1} + \frac{1}{\alpha^2-2^2} + \cdots\right)$$

or

$$\cot(\alpha\pi) - \frac{1}{\pi\alpha} = -\frac{2\alpha}{\pi}\left(\frac{1}{1-\alpha^2} + \frac{1}{2^2-\alpha^2} + \frac{1}{3^2-\alpha^2} + \cdots\right). \tag{12.58}$$

Integrate both sides of eq.(12.58) with respect to α from 0 to x,

$$\begin{aligned} \int_0^x d\alpha\, \cot(\alpha\pi) - \frac{1}{\alpha\pi} &= \frac{1}{\pi}\left(\ln(\sin(\alpha\pi)) - \ln(\pi\alpha)\right)\Big|_0^x \\ &= \frac{1}{\pi}\ln\left(\frac{\sin(\pi x)}{\pi x}\right) \end{aligned}$$

$$\begin{aligned} \int_0^x d\alpha\, -\frac{2\alpha}{n^2-\alpha^2} &= \ln(n^2-\alpha^2)\Big|_0^x \\ &= \ln\left(1-\frac{x^2}{n^2}\right). \end{aligned}$$

Thus,

$$\ln\left(\frac{\sin(\pi x)}{\pi x}\right) = \sum_n \ln\left(1 - \frac{x^2}{n^2}\right). \tag{12.59}$$

The expansion, eq.(12.55), follows from setting $\pi x = \omega T$ and exponentiating both sides of eq.(12.59). All together we have

$$\det\left(\frac{im}{2\pi}(\partial_t^2 + \omega^2)\right) = \frac{2\pi i T}{m}\frac{\sin(\omega T)}{\omega T}, \tag{12.60}$$

so

$$\begin{aligned} G(x,t;x_0,t_0) = &\left(\frac{m\omega}{2\pi i \sin(\omega(t-t_0))}\right)^{\frac{1}{2}} \\ &\exp\left(\frac{im}{2}\left(\frac{\omega}{\sin(\omega(t-t_0))}\right)\left((x^2+x_0^2)\cos(\omega(t-t_0)) - 2xx_0\right)\right). \end{aligned} \tag{12.61}$$

Now let's recover the energy eigenfunctions from the path integral, $G(x,t;x_0,t_0)$, using eq.(12.22). Begin by rewriting G in a form such that an exponential of i times something times $T = t - t_0$ appears in front. This is accomplished by writing the sines and cosines in terms of exponentials and expanding.

$$\begin{aligned} &G(x,T;x_0,0) \\ &= \left(\frac{m\omega}{\pi}\right)^{\frac{1}{2}} e^{-i\omega T/2}\left(1 - e^{-2i\omega T}\right)^{-1/2} \\ &\qquad \exp\left[-\frac{m\omega}{2}\left((x^2+x_0^2)\left(\frac{1+e^{-2i\omega T}}{1-e^{-2i\omega T}}\right) - \frac{4xx_0e^{-i\omega T}}{1-e^{-2i\omega T}}\right)\right] \\ &= \left(\frac{m\omega}{\pi}\right)^{\frac{1}{2}} e^{-i\omega T/2}\left(1 + \frac{1}{2}e^{-2i\omega T} + \cdots\right) \\ &\qquad \exp\left[-\frac{m\omega}{2}\left((x^2+x_0^2)\left(1 + 2e^{-2i\omega T} + 2e^{(-2i\omega T)^2} + \cdots\right)\right.\right. \\ &\qquad\qquad \left.\left. - 4xx_0e^{-i\omega T}\left(1 + e^{-2i\omega T} + e^{(-2i\omega T)^2} + \cdots\right)\right)\right] \\ &= \left(\frac{m\omega}{\pi}\right)^{\frac{1}{2}} e^{-i\omega T/2}\exp\left(-\frac{m\omega}{2}(x^2+x_0^2)\right)\left(1 + \frac{1}{2}e^{-2i\omega T} + \cdots\right) \\ &\qquad \left(1 - m\omega(x^2+x_0^2)e^{-2i\omega T} + \cdots\right)\left(1 + 2m\omega x x_0 e^{-i\omega T} + \cdots\right) \end{aligned} \tag{12.62}$$

In this form it is easy to see the leading term in the $iT \to \infty$ limit.

$$G(x,T;x_0,0) \stackrel{iT\to\infty}{\longrightarrow} e^{-i\omega T/2}\left(\frac{m\omega}{\pi}\right)^{\frac{1}{2}} \exp\left(-\frac{m\omega}{2}(x^2+x_0^2)\right) \\ = e^{-iE_0T}\phi_0^*(x)\phi_0(x_0), \tag{12.63}$$

thus,

$$E_0 = \frac{\omega}{2} \quad \text{and} \quad \phi_0(x) = \left(\frac{m\omega}{\pi}\right)^{1/4} \exp\left(-\frac{m\omega}{2}x^2\right). \tag{12.64}$$

If we expand the remaining series in eq.(12.62) we find that the term with the smallest frequency above $E_0 = \omega/2$ implies that $E_1 = 3\omega/2$, and that

$$\phi_1^*(x)\phi_1(x_0) = 2m\omega x x_0 \phi_0^*(x)\phi_0(x_0),$$

so

$$\phi_1(x) = \sqrt{2m\omega}\, x \left(\frac{m\omega}{\pi}\right)^{1/4} \exp\left(-\frac{m\omega}{2}x^2\right). \tag{12.65}$$

The next highest term implies that $E_2 = 5\omega/2$ and so on.

Forced Harmonic Oscillator

Now add an arbitrary external force, $F(\bar{t})$, on the oscillator. The action functional becomes

$$S[\bar{x}(\bar{t})] = \int_{t_0}^{t} d\bar{t}\, \frac{1}{2}m(\dot{\bar{x}}^2 - \omega^2 x^2) + \bar{x}F(\bar{t}), \tag{12.66}$$

leading to the equation of motion,

$$\frac{\delta S}{\delta \bar{x}} = m(\partial_{\bar{t}}^2 + \omega^2)\bar{x} - F(\bar{t}) = 0. \tag{12.67}$$

Let $x_{\text{cl}}^F(\bar{t})$ be the classical path in the presence of the force that satisfies eq.(12.67) with the usual boundary conditions: $x_{\text{cl}}^F(t_0) = x_0$, and $x_{\text{cl}}^F(t) = x$. The general solution to eq.(12.67) is

$$x_{\text{cl}}^F(\bar{t}) = x_{\text{cl}}(\bar{t}) - \int_{t_0}^{t} dt'\, g(\bar{t},t')F(t'), \tag{12.68}$$

where $x_{\text{cl}}(\bar{t})$ is the classical path in the absence of the force and is given already by eq.(12.49). The Green's function, $g(\bar{t}, t')$ satisfies

$$m(\partial_{\bar{t}}^2 + \omega^2) g(\bar{t}, t') = -\delta(\bar{t} - t'), \tag{12.69}$$

and since $x_{\text{cl}}(\bar{t})$ already satisfies the boundary conditions, then $g(\bar{t}, t')$ must satisfy $g(t, t') = 0 = g(t_0, t')$. An explicit expression for g is

$$g(\bar{t}, t') = \begin{cases} \frac{\sin(\omega(t-\bar{t}))\sin(\omega(t'-t_0))}{m\omega\sin(\omega(t-t_0))} & t' \leq \bar{t} \\ \frac{\sin(\omega(t-t'))\sin(\omega(\bar{t}-t_0))}{m\omega\sin(\omega(t-t_0))} & \bar{t} \leq t'. \end{cases} \tag{12.70}$$

To compute the path integral in the presence of the force,

$$G^F(x, t; x_0, t_0) = \int \mathcal{D}\bar{x}(\bar{t}) \exp\left(i \int_{t_0}^{t} d\bar{t} \, \frac{1}{2} m(\dot{\bar{x}}^2 - \omega^2 x^2) + \bar{x} F(\bar{t})\right), \tag{12.71}$$

we follow the same steps as in the $F = 0$ case and expand S in a functional Taylor series,

$$\begin{aligned} S[\bar{x}(\bar{t})] &= S[x_{\text{cl}}^F] + \left.\frac{\delta S}{\delta \bar{x}}\right|_{\bar{x}=x_{\text{cl}}^F} (\bar{x} - x_{\text{cl}}^F) + \frac{1}{2} \left.\frac{\delta^2 S}{\delta \bar{x}^2}\right|_{\bar{x}=x_{\text{cl}}^F} (\bar{x} - x_{\text{cl}}^F)^2 \\ &= S[x_{\text{cl}}^F] + \frac{1}{2} \left.\frac{\delta^2 S}{\delta \bar{x}^2}\right|_{\bar{x}=x_{\text{cl}}^F} (\bar{x} - x_{\text{cl}}^F)^2. \end{aligned} \tag{12.72}$$

Therefore,

$$\begin{aligned} &G^F(x, t; x_0, t_0) \\ &= \exp(iS[x_{\text{cl}}^F]) \int \mathcal{D}\bar{x}(\bar{t}) \exp\left(\frac{im}{2} \int_{t_0}^{t} d\bar{t} \left.\frac{\delta^2 S}{\delta \bar{x}^2}\right|_{\bar{x}=x_{\text{cl}}^F} (\bar{x} - x_{\text{cl}}^F)^2\right). \end{aligned} \tag{12.73}$$

Since

$$\left.\frac{\delta^2 S(F)}{\delta \bar{x}^2}\right|_{\bar{x}=x_{\text{cl}}^F} = \left.\frac{\delta^2 S(F=0)}{\delta \bar{x}^2}\right|_{\bar{x}=x_{\text{cl}}^F} = \left.\frac{\delta^2 S(F=0)}{\delta \bar{x}^2}\right|_{\bar{x}=x_{\text{cl}}}, \tag{12.74}$$

the path integral in eq.(12.73) leads to the same determinant as in the $F = 0$ case. The final result is

$$
\begin{aligned}
G^F&(x,t;x_0,t_0)\\
&=\left(\det\left(-\frac{im}{2\pi}(\partial_{\bar t}^2+\omega^2)\right)\right)^{-\frac{1}{2}}\exp\left(iS[x_{\rm cl}^F]\right)\\
&=\left(\frac{m\omega}{2\pi i\sin(\omega(t-t_0))}\right)^{\frac{1}{2}}\exp\left(iS[x_{\rm cl}^F]\right)\\
&=\exp\left(\frac{ix}{\sin(\omega(t-t_0))}\int_{t_0}^{t}d\bar t\,\sin(\omega(\bar t-t_0))F(\bar t)\right.\\
&\qquad+\frac{ix_0}{\sin(\omega(t-t_0))}\int_{t_0}^{t}d\bar t\,\sin(\omega(t-\bar t))F(\bar t)\\
&\qquad\left.+\frac{i}{m\omega\sin(\omega(t-t_0))}\int_{t_0}^{t}dt'\int_{t_0}^{t'}d\bar t\,F(t')\sin(\omega(t-t'))\sin(\omega(\bar t-t_0))F(\bar t)\right)\\
&\qquad\quad\cdot G(x,t;x_0,t_0),
\end{aligned}
\tag{12.75}
$$

where $G(x,t;x_0,t_0)$ is the path integral when $F=0$, eq.(12.61).

Alternatively, we can expand the action about $x_{\rm cl}(\bar t)$, the classical path in the absence of the force. In this case we have

$$
\begin{aligned}
S&[\bar x(\bar t)]\\
&=S[x_{\rm cl}]+\left.\frac{\delta S}{\delta\bar x}\right|_{\bar x=x_{\rm cl}}(\bar x-x_{\rm cl})+\frac{1}{2}\left.\frac{\delta^2 S}{\delta\bar x^2}\right|_{\bar x=x_{\rm cl}}(\bar x-x_{\rm cl})^2\\
&=S[x_{\rm cl}]+\int d\bar t\,F(\bar t)(\bar x-x_{\rm cl})\;-\;\frac{m}{2}\int d\bar t\,(\bar x-x_{\rm cl})(\partial_{\bar t}^2+\omega^2)(\bar x-x_{\rm cl}).
\end{aligned}
\tag{12.76}
$$

$S[x_{\rm cl}]$ is equal to eq.(12.50) plus an additional term,

$$
\begin{aligned}
\int_{t_0}^{t}d\bar t\,x_{\rm cl}(\bar t)F(\bar t)=&\frac{x_0}{\sin(\omega(t-t_0))}\int_{t_0}^{t}d\bar t\,\sin(\omega(t-\bar t))F(\bar t)\\
&+\frac{x}{\sin(\omega(t-t_0))}\int_{t_0}^{t}d\bar t\,\sin(\omega(\bar t-t_0))F(\bar t),
\end{aligned}
\tag{12.77}
$$

where we have substituted eq.(12.49) for the classical path. Compare this piece with eq.(12.75). The path integral we must do from eq.(12.76) over paths $y(\bar t)=\bar x(\bar t)-x_{\rm cl}(\bar t)$ is

$$\int \mathcal{D}y(\bar{t}) \exp\left(-\frac{im}{2}\int_{t_0}^{t} d\bar{t}\, y(\bar{t})(\partial_{\bar{t}}^2+\omega^2)y(\bar{t}) + i\int_{t_0}^{t} d\bar{t}\, F(\bar{t})y(\bar{t})\right)$$

$$= \exp\left(\frac{i}{2m}\int_{t_0}^{t} d\bar{t}\int_{t_0}^{t} dt'\, F(\bar{t})\left(\frac{1}{\partial_{\bar{t}}^2+\omega^2}\right)F(t')\right)\int \mathcal{D}y(\bar{t})$$

$$\exp\left(-\frac{im}{2}\int_{t_0}^{t} d\bar{t}\left(y-\frac{F}{m(\partial_{\bar{t}}^2+\omega^2)}\right)(\partial_{\bar{t}}^2+\omega^2)\left(y-\frac{F}{m(\partial_{\bar{t}}^2+\omega^2)}\right)\right), \tag{12.78}$$

where we have completed the square in the exponential. The path integral over $y(\bar{t})$ is the same as eq.(12.51) and is equal to one over the square root of the determinant of the operator $-im/2\pi(\partial_{\bar{t}}^2+\omega^2)$. The operator $1/(\partial_{\bar{t}}^2+\omega^2)$ in eq.(12.78) above is the inverse of $\partial_{\bar{t}}^2+\omega^2$, that is, the Green's function for $\partial_{\bar{t}}^2+\omega^2$, eq.(12.70). Combining eqs.(12.77), (12.78), and (12.60) we recover eq.(12.75),

$$G^F(x,t;x_0,t_0) = \exp\left(i\int_{t_0}^{t} d\bar{t}\, x_{\rm cl}(\bar{t})F(\bar{t})\right)$$

$$\exp\left(\frac{i}{2}\int_{t_0}^{t} d\bar{t}\int_{t_0}^{t} dt'\, F(\bar{t})g(\bar{t},t')F(t')\right)\cdot G(x,t;x_0,t_0). \tag{12.79}$$

$G^F(x,t;x_0,t_0)$ also serves as a generating functional for matrix elements of time-ordered products of the position operator of the harmonic oscillator. Starting from eq.(12.71), functionally differentiate $G^F(x,t;x_0,t_0)$ with respect to $F(t_1)$, where $t_0 < t_1 < t$, then set F to zero.

$$-i\frac{\delta}{\delta F(t_1)}G^F(x,t;x_0,t_0)\Big|_{F=0}$$

$$= -i\frac{\delta}{\delta F(t_1)}\int \mathcal{D}\bar{x}(\bar{t})\exp\left(i\int_{t_)}^{t} d\bar{t}\,\frac{1}{2}m(\dot{\bar{x}}^2-\omega^2\bar{x}^2)+\bar{x}(\bar{t})F(\bar{t})\right)\Big|_{F=0}$$

$$= \int \mathcal{D}\bar{x}(\bar{t})\,\bar{x}(t_1)\exp\left(i\int_{t_0}^{t} d\bar{t}\,\frac{1}{2}m(\dot{\bar{x}}^2-\omega^2\bar{x}^2)\right)$$

$$= \langle x,t\,|\mathbf{X}(t_1)|\,x_0,t_0\rangle, \tag{12.80}$$

where we have used eq.(12.28) to arrive at the last line above. If we perform the same operation on the explicit expression for $G^F(x,t;x_0,t_0)$, eq.(12.79), we find

$$\begin{aligned}\langle x,t\,|\mathbf{X}(t_1)|\,x_0,t_0\rangle &= -i\frac{\delta}{\delta F(t_1)}G^F(x,t;x_0,t_0)\bigg|_{F=0} \\ &= x_{\rm cl}(t_1)G(x,t;x_0,t_0).\end{aligned} \tag{12.81}$$

The second functional derivative of G^F generates the matrix element of the time-ordered product of $\mathbf{X}(t_1)\mathbf{X}(t_2)$.

$$\begin{aligned}&(-i)^2\frac{\delta}{\delta F(t_1)}\frac{\delta}{\delta F(t_2)}G^F(x,t;x_0,t_0)\bigg|_{F=0} \\ &\quad= \langle x,t\,|T(\mathbf{X}(t_1)\mathbf{X}(t_2))|\,x_0,t_0\rangle \\ &\quad= \Big(x_{\rm cl}(t_1)x_{\rm cl}(t_2)+\frac{1}{m}g(t_1,t_2)\Big)G(x,t;x_0,t_0),\end{aligned} \tag{12.82}$$

and so on. In the limit that $t, -t_0 \to \infty$, G^F generates ground state or vacuum expectation values of time-ordered products of $\mathbf{X}(t)$.

12.4 PERTURBATION THEORY

Let's now consider the amplitude, $G(x,t;x_0,t_0)$, for a particle to go from (x_0,t_0) to (x,t) in the presence of a general potential $V(x)$. Only in very special cases will we be able to compute this amplitude exactly. However, often the Hamiltonian, H, for the system can be separated into a part, H_0, plus a potential, $\lambda V(x)$, where we do know the amplitude, $G_0(x,t;x_0,t_0)$, exactly when $H = H_0$. When λ is "small," we can use perturbation theory to find an approximate expression for $G(x,t;x_0,t_0)$ in terms of $G_0(x,t;x_0,t_0)$. To find the path integral version of this, start with the path integral representation of $G(x,t;x_0,t_0)$, eq.(12.18).

$$G(x,t;x_0,t_0) = \int \mathcal{D}\bar{x}(\bar{t})\,\exp\left(i\int_{t_0}^{t} d\bar{t}\,S_0[\bar{x}(\bar{t})] - \lambda V(\bar{x})\right) \tag{12.83}$$

S_0 is the action corresponding to the system with $\lambda = 0$. By starting from eq.(12.18) instead of eq.(12.16) we are assuming that H does not contain any funny dependence on p that would prevent us from carrying out the integral over $p(\bar{t})$ in eq.(12.16). If λ is small, then we can safely expand the exponential, $\exp(-i\lambda \int V[\bar{x}(\bar{t})]\, d\bar{t})$, that appears in eq.(12.83).

$$
\begin{aligned}
&G(x,t;x_0,t_0) \\
&= \int \mathcal{D}\bar{x}(\bar{t}) \exp\left(i\int_{t_0}^{t} d\bar{t} S_0[\bar{x}(\bar{t})]\right) \sum_{n=0}^{\infty} \frac{(-i\lambda)^n}{n!} \left(\int_{t_0}^{t} d\bar{t}\, V[\bar{x}(\bar{t})]\right)^n \\
&= \sum_{n=0}^{\infty} \frac{(-i\lambda)^n}{n!} \int \mathcal{D}\bar{x}(\bar{t}) \left(\int_{t_0}^{t} dt'\, V[\bar{x}(t')]\right)^n \exp\left(i\int_{t_0}^{t} d\bar{t}\, S_0[\bar{x}(\bar{t})]\right) \\
&\equiv \sum_{n=0}^{\infty} G^{(n)}(x,t;x_0,t_0).
\end{aligned}
\tag{12.84}
$$

At $n = 0$ we have $G^{(0)}(x,t;x_0,t_0) = G_0(x,t;x_0,t_0)$. Examine the next term.

$$
\begin{aligned}
G^{(1)}(x,t;x_0,t_0) &= -i\lambda \int \mathcal{D}\bar{x}(\bar{t}) \int_{t_0}^{t} dt_1\, V[\bar{x}(t_1)] \exp\left(i\int_{t_0}^{t} d\bar{t}\, S_0[\bar{x}(\bar{t})]\right) \\
&= -i\lambda \int_{t_0}^{t} dt_1 \int \mathcal{D}\bar{x}(\bar{t})\, V[\bar{x}(t_1)] \exp\left(i\int_{t_0}^{t} d\bar{t}\, S_0[\bar{x}(\bar{t})]\right)
\end{aligned}
\tag{12.85}
$$

As we did in constructing the definition of a path integral earlier in the chapter, now break the propagation process into two pieces. If we insist that the particle go through the space-time point (x_1, t_1) on its way from (x_0, t_0) to (x,t), then, to first order in λ, the amplitude for this process is

$$
-i\lambda G_0(x,t;x_1,t_1)V(x_1,t_1)G_0(x_1,t_1;x_0,t_0), \tag{12.86}
$$

that is, the particle is free to take any path from (x_0, t_0) to (x_1, t_1) where it meets the potential at (x_1, t_1). It is then free to take any path again from (x_1, t_1) to (x, t). To get the total amplitude to first order, we must sum over all x_1 and $t_1 \in [t_0, t]$.

$$
\begin{aligned}
&G^{(1)}(x,t;x_0,t_0) \\
&= \int_{t_0}^{t} dt_1 \int_{-\infty}^{\infty} dx_1\, (-i\lambda) G_0(x,t;x_1,t_1)V(x_1,t_1)G_0(x_1,t_1;x_0,t_0) \\
&= -i\lambda \int_{t_0}^{t} dt_1 \int_{-\infty}^{\infty} dx_1 \int \mathcal{D}\bar{x}(\bar{t}) \exp\left(i\int_{t_1}^{t} d\bar{t}\, S_0[\bar{x}(\bar{t})]\right) V(x_1,t_1) \\
&\quad \int \mathcal{D}y(\bar{t}) \exp\left(i\int_{t_0}^{t_1} d\bar{t}\, S_0[y(\bar{t})]\right),
\end{aligned}
\tag{12.87}
$$

which is, of course, what we mean by eq.(12.85).

Expanding $G(x,t;x_0,t_0)$ in terms of G_0 lends itself easily to a diagrammatic interpretation of the terms. Once again we represent G_0 by a straight line as in figure 12.5(a). If we represent the point where the potential acts with an $\times$, then eq.(12.85) or (12.87) is represented by figure 12.5(b). Clearly, the next diagram of the sequence is figure 12.5(c) which represents $G^{(2)}$. To translate from a diagram back to an expression we start at the top of the diagram and work down, writing the expression down from left to right. Each straight line between (x_i,t_i) and (x_{i+1},t_{i+1}) corresponds to a factor $G^{(0)}(x_{i+1},t_{i+1};x_i,t_i)$ and each $\times$ at (x_i,t_i) corresponds to a factor $-i\lambda V(x_i,t_i)$. The total amplitude is the integral over space-time of all the points where the potential acts. For figure 12.5(c) we have $(t_1 < t_2)$,

$$
\begin{aligned}
&G^{(2)}(x,t;x_0,t_0)\\
&= (-i\lambda)^2 \int_{t_0}^{t} dt_2 \int_{t_0}^{t_2} dt_1\, G_0(x,t;x_2,t_2)V(x_2,t_2)\\
&\qquad\qquad G_0(x_2,t_2;x_1,t_1)V(x_1,t_1)G_0(x_1,t_1;x_0,t_0)\\
&= (-i\lambda)^2 \int_{t_0}^{t} dt_2 \int_{t_0}^{t_2} dt_1 \int \mathcal{D}\bar{x}(\bar{t})\, V[\bar{x}(t_1)]\, V[\bar{x}(t_2)]\, \exp\left(i\int_{t_0}^{t} d\bar{t}\, S_0[\bar{x}]\right)\\
&= \frac{(-i\lambda)^2}{2!} \int_{t_0}^{t} dt_2 \int_{t_0}^{t} dt_1 \int \mathcal{D}\bar{x}(\bar{t})\, V[\bar{x}(t_1)]\, V[\bar{x}(t_2)]\, \exp\left(i\int_{t_0}^{t} d\bar{t}\, S_0[\bar{x}]\right).
\end{aligned}
\tag{12.88}
$$

$G^{(n)}$ will be represented by a diagram with n vertices, $\times$.

$$
\begin{aligned}
G^{(n)}(x,t;x_0,t_0) &= (-i\lambda)^n \int_{t_0}^{t} dt_n \cdots \int_{t_0}^{t_2} dt_1 G_0(x,t;x_n,t_n)\\
&\qquad V(x_n,t_n)\cdots V(x_1,t_1)G_0(x_1,t_1;x_0,t_0)\\
&= \frac{(-i\lambda)^n}{n!} \int_{t_0}^{t} dt_n \cdots \int_{t_0}^{t} dt_1 \int \mathcal{D}\bar{x}(\bar{t})\, V[\bar{x}(t_n)]\cdots V[\bar{x}(t_1)]\\
&\qquad \exp\left(i\int_{t_0}^{t} d\bar{t}\, S_0[\bar{x}(\bar{t})]\right)
\end{aligned}
\tag{12.89}
$$

The perturbative expansion of G above is very similar in structure to the time-ordered exponential expression for the U-matrix, eq.(7.87), encountered in chapter 7. In that case, that expression for U arose from a Neumann series solution to an integral equation, eq.(7.82). We can also

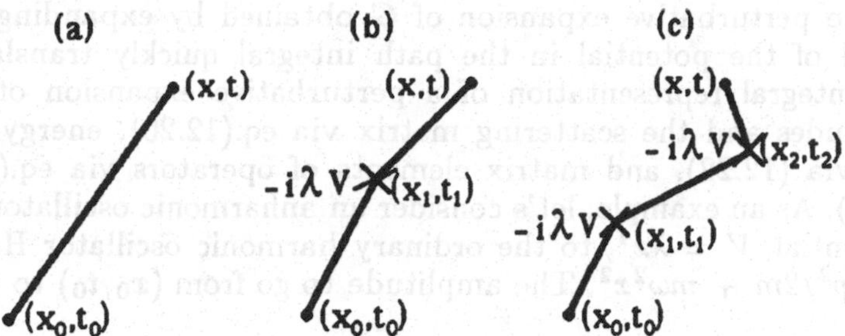

Figure 12.5 Diagrammatic representation of the first three terms of the Neumann series solution to the integral equation (12.90). (a) $G_0(x,t;x_0,t_0)$; (b) $G^{(1)}(x,t;x_0,t_0)$; (c) $G^{(2)}(x,t;x_0,t_0)$.

interpret the sum, eq.(12.84), as a Neumann series and find an integral equation that G satisfies by factoring a

$$-i\lambda \int_{t_0}^{t} dt_n \int dx_n\, G_0(x,t;x_n,t_n)V(x_n,t_n)$$

from the left-hand side of $G^{(n)}$ for $n \geq 1$. The remaining sum can be seen to be just G. Therefore, G satisfies the integral equation,

$$\begin{aligned} G(x,t;x_0,t_0) &\\ = G_0(x,t;x_0,t_0) &\\ - i\lambda \int_{t_0}^{t} dt_n \int dx_n\, G_0(x,t;x_n,t_n)V(x_n,t_n)\Big(G_0(x_n,t_n;x_0,t_0) + \cdots\Big) &\\ = G_0(x,t;x_0,t_0) &\\ - i\lambda \int_{t_0}^{t} dt_n \int dx_n\, G_0(x,t;x_n,t_n)V(x_n,t_n)G(x_n,t_n;x_0,t_0). & \end{aligned} \tag{12.90}$$

By realizing that G_0 is the Green's function for the $\lambda = 0$ Schrödinger equation,

$$i\frac{\partial}{\partial t}G_0(x,t;x_0,t_0) = \mathbf{H}_0 G_0(x,t;x_0,t_0) - i\delta(t-t_0)\delta(x-x_0), \tag{12.91}$$

one can see that the solution to the integral equation (12.90) satisfies

$$i\frac{\partial}{\partial t}G(x,t;x_0,t_0) = \mathbf{H}G(x,t;x_0,t_0) - i\delta(t-t_0)\delta(x-x_0). \tag{12.92}$$

The perturbative expansion of G obtained by expanding the exponential of the potential in the path integral quickly translates into a path integral representation of a perturbative expansion of transition amplitudes and the scattering matrix via eq.(12.20), energy eigenfunctions via (12.22), and matrix elements of operators via eq.(12.25) and (12.28). As an example, let's consider an anharmonic oscillator by adding a potential, $V = \lambda x^4$, to the ordinary harmonic oscillator Hamiltonian, $H_0 = p^2/2m + m\omega^2 x^2$. The amplitude to go from (x_0, t_0) to (x,t) is

$$\begin{aligned} &G(x,t;x_0,t_0) \\ &= \int \mathcal{D}\bar{x}(\bar{t}) \exp\left(i\int_{t_0}^{t} d\bar{t}\ S_0[\bar{x}(\bar{t})] - \lambda\bar{x}^4\right) \\ &= \int \mathcal{D}\bar{x}(\bar{t}) \exp\left(i\int_{t_0}^{t} d\bar{t}\ S_0[\bar{x}(\bar{t})]\right)\left(1 - i\lambda\int_{t_0}^{t} dt'\ \bar{x}^4(t') + \ldots\right) \\ &= G_0(x,t;x_0,t_0) + G^{(1)}(x,t;x_0,t_0) + \ldots, \end{aligned} \tag{12.93}$$

where $G_0(x,t;x_0,t_0)$ is given by eq.(12.61). The first order correction to G,

$$G^{(1)}(x,t;x_0,t_0) = -i\lambda\int_{t_0}^{t} dt' \int \mathcal{D}\bar{x}(\bar{t})\ \bar{x}^4(t') \exp\left(i\int_{t_0}^{t} d\bar{t}\ S_0[\bar{x}(\bar{t})]\right), \tag{12.94}$$

can be computed using the path integral expression for the forced harmonic oscillator as a generating functional.

$$\begin{aligned} &G^{(1)}(x,t;x_0,t_0) \\ &= -i\lambda\int_{t_0}^{t} dt'\ i^4 \frac{\delta^4}{\delta F^4(t')} G^F(x,t;x_0,t_0)\Big|_{F=0} \\ &= -i\lambda\int_{t_0}^{t} dt'\left(x_{\rm cl}^4(t') - i6x_{\rm cl}^2(t')g(t',t') - 3g(t',t')^2\right)G_0(x,t;x_0,t_0) \end{aligned} \tag{12.95}$$

From eq.(12.70),

$$g(t',t') = \frac{\sin(\omega(t-t'))\sin(\omega(t'-t_0))}{m\omega\sin(\omega(t-t_0))},$$

and from eq.(12.49),

$$x_{\rm cl}(t') = x_0\frac{\sin(\omega(t-t'))}{\sin(\omega(t-t_0))} + x\frac{\sin(\omega(t'-t_0))}{\sin(\omega(t-t_0))}.$$

Thus,

$$G^{(1)}(x,t;x_0,t_0) =$$
$$-i\lambda\Bigg[\left(\frac{x^4+x_0^4}{\omega\sin^4(\omega T)}\right)\left(\frac{3}{8}\omega T-\frac{1}{4}\sin^3(\omega T)\cos(\omega T)-\frac{3}{8}\sin(\omega T)\cos(\omega T)\right)$$

$$+\left(\frac{4(x_0^3x+x^3x_0)}{\omega\sin^4(\omega T)}-\frac{i6(x_0^2+x^2)}{m\omega^2\sin^3(\omega T)}\right)$$
$$\left(\frac{\sin^5(\omega T)}{4}-\cos(\omega T)\left(\frac{3}{8}\omega T-\frac{1}{4}\sin^3(\omega T)\cos(\omega T)-\frac{3}{8}\sin(\omega T)\cos(\omega T)\right)\right)$$

$$+\left(\frac{6x^2x_0^2}{\omega\sin^4(\omega T)}-\frac{3}{m^2\omega^3\sin^2(\omega T)}-\frac{i12xx_0}{m\omega^2\sin^3(\omega T)}\right)$$
$$\left(\frac{\omega T}{8}-\frac{3}{16}\sin(2\omega T)+\frac{\omega T}{4}\cos^2(\omega T)\right)\Bigg]G_0(x,t;x_0,t_0), \quad (12.96)$$

where $T=t-t_0$.

Let's extract the order-λ correction to the ground state energy eigenvalue, and the order λ correction to the ground state wave function. To reach the ground state we take the limit $iT\to\infty$ and look at the leading term. Two useful limits to keep in mind are

$$\lim_{iT\to\infty}\frac{1}{\sin(\omega T)}=0 \quad \text{and} \quad \lim_{iT\to\infty}\frac{\cos(\omega T)}{\sin(\omega T)}=i. \quad (12.97)$$

In the limit $iT\to\infty$, $G^{(1)}(x,t;x_0,t_0)$ reduces to

$$-i\lambda\left(\frac{-i}{4}\left(\frac{x^4+x_0^4}{\omega}\right)-i\frac{3}{4}\left(\frac{x_0^2+x^2}{m\omega^2}\right)\right.$$
$$\left.-i\frac{9}{8}\frac{1}{m^2\omega^3}+\frac{3}{4}\frac{\omega T}{m^2\omega^3}\right)e^{-i\omega T/2}\phi_0^{(0)}(x)\phi_0^{(0)*}(x_0). \quad (12.98)$$

Comparing this to the expansion of $\sum_n \exp(-iE_nT)\phi_n(x)\phi_n^*(x_0)$ to first order,

$$\lim_{iT\to\infty}\sum_n e^{-i(E_n^{(0)}+\lambda E_n^{(1)}+\cdots)T}(\phi_n^{(0)*}(x_0)+\lambda\phi_n^{(1)*}(x_0)+\cdots)$$
$$(\phi_n^{(0)}(x)+\lambda\phi_n^{(1)}(x)+\cdots)$$

$$= e^{-iE_0^{(0)}T}\phi_0^{(0)*}(x_0)\phi_0^{(0)}(x)$$
$$+ e^{-iE_0^{(0)}T}(-i\lambda E_0^{(1)}T + \lambda\phi_0^{(1)*}(x_0)\phi_0^{(0)}(x) + \lambda\phi_0^{(1)}(x)\phi_0^{(0)*}(x_0)) + \ldots, \tag{12.99}$$

we find that

$$E_0^{(1)} = \frac{3}{4}\frac{1}{(m\omega)^2},$$

and

$$\lambda\phi_0^{(1)}(x) = -\lambda\left(\frac{1}{4\omega}x^4 + \frac{3}{4}\frac{x^2}{m\omega^2} - \frac{9}{16}\frac{1}{m^2\omega^3}\right)\phi_0^{(0)}(x). \tag{12.100}$$

12.5 EXERCISES

1. Do the functional integral over p in eq.(12.16) for the Hamiltonian,
$$H = \frac{1}{2}m(x)p^2,$$
where $m(x)$ is some arbitrary function of x. Does your result agree with eqs.(12.18)-(12.19)?
2. If a free particle is located at x_0 at time t_0, what is the probability that the particle will be found between x and $x + dx$ at time t? As $t, -t_0 \to \infty$, this probability vanishes. What does this mean? What is the probability at time t if the particle experiences a harmonic oscillator potential instead? What happens as $t - t_0 \to \infty$?
3. Show that the classical action for a harmonic oscillator that starts at (x_0, t_0) and ends at (x, t) is equal to eq.(12.50).
4. Compute the path integral for the action
$$S[\bar{x}(\bar{t})] = \int_{t_0}^{t} d\bar{t}\,\frac{1}{2}m\dot{\bar{x}}^2 + \bar{x}F(\bar{t}).$$
Verify that the function sandwiched between the two factors of F in the exponent of the result is the Green's function for the operator d^2/dx^2. What are the boundary conditions on this Green's function?
5. Show that if the action, $S[x]$, is at most quadratic in $x(t)$, then the path integral is proportional to $\exp(iS[x_{\text{cl}}])$, where x_{cl} is the classical path. Express the constant of proportionality in terms of a determinant.
6. Consider the anharmonic oscillator with potential $V = \nu x^3$. Compute the path integral that represents $G(x, t; x_0, t_0)$ through order ν^2. Determine the lowest order corrections to the ground state energy and wave function.

CHAPTER 13

Path Integrals in Free Field Theory

Using the results in the previous chapter for path integrals in quantum mechanics, we are now ready to generalize the definitions to the simplest field theories, the free fields. Recall that in first quantized systems that the number of particles is fixed, and that the position and momentum, $\vec{x}_i$ and $\vec{p}_i$, of each particle become operators. In the path integral representation we integrate over paths in phase space, $(\vec{x}_i(t), \vec{p}_i(t))$. In a second quantized system, the wave function, $\phi(\vec{x}, t)$, of the first quantized system is elevated to an operator. Therefore, in the path integral representation of a field theory we can expect that we will need to integrate over "paths" in a "phase space" of functions whose coordinates are $(\phi(\vec{x}, t), \pi(\vec{x}, t))$, $\pi(\vec{x}, t)$ being the appropriate field momentum.

Next recall that the quantum theory of free fields amounts to an infinite collection of quantum harmonic oscillators, and that the wave functionals of the free field theory are infinite products of ordinary harmonic oscillator wave functions, one function for each mode. We have explicitly worked out the path integral for one harmonic oscillator so we should expect that the relevant path integral for the free field theory will be an infinite product of harmonic oscillator path integrals. In addition, we found that the forced harmonic oscillator path integral serves as a generating functional for matrix elements of time-ordered products of the position operator, and in particular, for vacuum or ground state expectations of time-ordered products of the position operator if we take $-t_0, t \to \infty$. Therefore, we can expect that the path integral in field theory in the presence of an arbitrary source, $J(\vec{x}, t)$, for "paths" in the function space starting in the distant past and going to the distant future will be a generating functional for time-ordered products of the field $\phi(\vec{x}, t)$ in the vacuum. These time-ordered products or Green's functions will yield S-matrix elements by way of a reduction formula. Let's now recover the results of chapters 3, 4, 5, and 10 by way of path integrals for the three familiar free fields: the scalar field, free photons, and the free spinor field. As can be expected, the path integral for free photons will require some extra attention because of the gauge symmetry. A complete treatment of this problem in the context of perturbation theory will wait until chapter 15. With the free field path integrals in hand we will add field interactions in the next chapter, apply perturbation theory in the same manner as with ordinary path integrals, and recover the diagrammatic representation of the series.

13.1 FREE SCALAR FIELD THEORY

We may constructively define the path integral for free scalar field theory by following the same procedure as in the previous chapter. Let $|\Psi(t)\rangle$ be a state of the system at time t. $|\Psi(t)\rangle$ evolves by the Schrödinger equation,

$$i\frac{\partial}{\partial t}|\Psi(t)\rangle = \mathbf{H}|\Psi(t)\rangle, \tag{13.1}$$

where the Hamiltonian, $\mathbf{H}$, for the free scalar field is given by eq.(10.3). In the $|\phi\rangle$ basis, the wave functional representing $|\Psi(t)\rangle$ is $\langle\phi\,|\,\Psi(t)\rangle = \Psi[(\phi(\vec{x}), t]$, and the Hamiltonian is eq.(10.11). The formal solution to eq.(13.1) is

$$|\Psi(t)\rangle = e^{-i\mathbf{H}(t-t_0)}|\Psi(t_0)\rangle, \tag{13.2}$$

which, in the $|\psi\rangle$ basis, is

$$\Psi[\phi(\vec{x}), t] = \int \mathcal{D}\phi_0 \langle\phi\,|\,e^{-i\mathbf{H}(t-t_0)}\,|\,\phi_0\rangle\Psi[\phi_0(\vec{x}), t_0]. \tag{13.3}$$

Thus, $\langle\phi\,|e^{-i\mathbf{H}(t-t_0)}|\,\phi_0\rangle \equiv G[\phi, t; \phi_0, t_0]$, the time-evolution operator, represents the amplitude for the field in configuration $\phi_0(\vec{x})$ at t_0 to evolve to $\phi(\vec{x})$ at t.

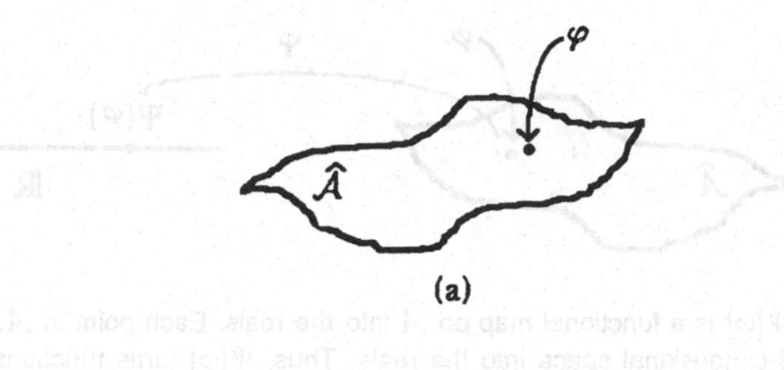

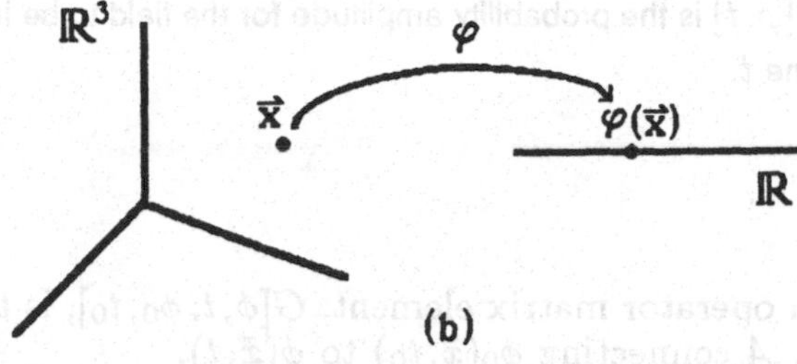

Figure 13.1 $\widehat{\mathcal{A}}$ is the space of all maps from 3-dimensional space into the reals. (a) A point, φ, in $\widehat{\mathcal{A}}$ is (b) a configuration of the field at a given instant, $\varphi : \mathbb{R}^3 \to \mathbb{R}$.

If $\widehat{\mathcal{A}}$ is the space of all maps from 3-dimensional space into the reals, then a point in $\widehat{\mathcal{A}}$ is a configuration of the field at a given instant, i.e. $\phi \in \widehat{\mathcal{A}}$ implies $\phi : \mathbb{R}^3 \to \mathbb{R}$. See figure 13.1.

$\Psi[\phi, t]$ is a functional map on $\widehat{\mathcal{A}}$ into the reals, that is, Ψ turns points in $\widehat{\mathcal{A}}$ into real numbers. $\Psi[\phi, t]$ describes the state of the system at time t. That is, $\Psi[\phi, t]$ is the probability amplitude for the field to be in configuration $\phi(\vec{x})$ at time t. See figure 13.2. $\int \mathcal{D}\phi_0$ in eq.(13.3) above is an integral over $\widehat{\mathcal{A}}$. $G[\phi, t; \phi_0, t_0]$ is the amplitude for the system to start in configuration $\phi_0(\vec{x})$ at t_0 ($\Psi[\phi_1, t_0] = \delta[\phi_1 - \phi_0]$) and end up in configuration $\phi(\vec{x})$ at t. See figure 13.3. To draw the time evolution we can stack up copies of $\widehat{\mathcal{A}}$, labeling each layer by a time t. Call this collection $\mathcal{A}(= \widehat{\mathcal{A}} \times \mathbb{R})$. A point in $\mathcal{A}$ is a mapping of space into the reals at a time t, $(\phi(\vec{x}), t) = \phi(\vec{x}, t)$, that is, a mapping of space-time into the reals. One possible path of evolution of $\phi_0(\vec{x}, t_0)$ to $\phi(\vec{x}, t)$ through $\mathcal{A}$ is given in figure 13.4. In a manner similar to a first quantized system, the

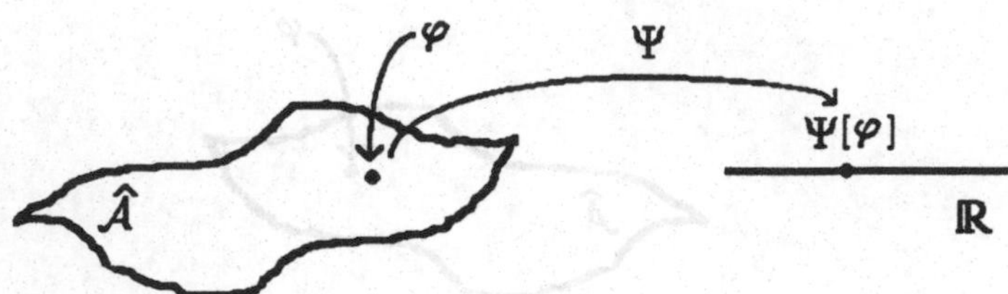

Figure 13.2 $\Psi[\varphi]$ is a functional map on $\widehat{\mathcal{A}}$ into the reals. Each point in $\widehat{\mathcal{A}}$, $\varphi \in \widehat{\mathcal{A}}$, is a map of 3-dimensional space into the reals. Thus, $\Psi[\varphi]$ turns functions, φ, into real numbers. $\Psi[\varphi, t]$ is the probability amplitude for the field to be in the configuration $\varphi(\vec{x}) \in \widehat{\mathcal{A}}$ at time t.

time-evolution operator matrix element, $G[\phi, t; \phi_0, t_0]$, is the sum over all paths through $\mathcal{A}$ connecting $\phi_0(\vec{x}, t_0)$ to $\phi(\vec{x}, t)$.

To obtain the path integral representation of $G[\phi, t; \phi_0, t_0]$ we divide the interval $t - t_0$ into $N + 1$ equal parts as in figure 13.5. Next, we insert $1 = \int \mathcal{D}\phi_i \mid \phi_i\rangle\langle\phi_i \mid$, $\phi_i = \phi(\vec{x}, t_i)$, at each division. The functional integral $\int \mathcal{D}\phi_i$ is over the copy of $\widehat{\mathcal{A}}$ at $t = t_i$. Finally, we let $N \to \infty$.

$$G[\phi, t; \phi_0, t_0] = \lim_{N\to\infty} \int \prod_{i=1}^{N} \mathcal{D}\phi_i \, \langle \phi \, | e^{-i\mathbf{H}(t-t_N)} | \, \phi_N \rangle \cdots \langle \phi_1 \, | e^{-i\mathbf{H}(t_1-t_0)} | \, \phi_0 \rangle \tag{13.4}$$

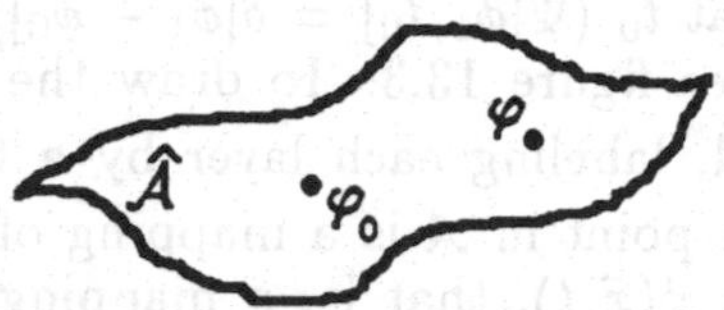

Figure 13.3 The matrix element of the time evolution operator, $G[\phi, t; \phi_0, t_0]$, is the probability amplitude for the system to start at the point φ_0 in $\widehat{\mathcal{A}}$ at t_0, and end at the point φ in $\widehat{\mathcal{A}}$ at time t.

As N gets large, the interval $t_{i+1} - t_i$ gets small, and we can safely expand the exponential that appears in $G[\phi_{i+1}, t_{i+1}; \phi_i, t_i]$.

$$\begin{aligned} G[\phi_{i+1}, t_{i+1}; \phi_i, t_i] &= \langle \phi_{i+1} | e^{-i\mathbf{H}(t_{i+1}-t_i)} | \phi_i \rangle \\ &\approx \langle \phi_{i+1} | 1 - i\mathbf{H}(t_{i+1} - t_i) | \phi_i \rangle \\ &= \delta[\phi_{i+1} - \phi_i] - i\epsilon \langle \phi_{i+1} | \mathbf{H} | \phi_i \rangle, \end{aligned} \tag{13.5}$$

where $\epsilon = t_{i+1} - t_i$. Recall that the δ-functional is an infinite product of δ-functions, one for each point $\vec{x}$ in space.

$$\delta[\phi_{i+1} - \phi_i] = \prod_{\vec{x}} \delta(\phi_{i+1}(\vec{x}) - \phi_i(\vec{x})) \tag{13.6}$$

Next we write each individual δ-function as a Fourier representation,

$$\delta(\phi_{i+1}(\vec{x}) - \phi_i(\vec{x})) = \int_{-\infty}^{\infty} \frac{d\pi_i}{2\pi} \exp\Big(i\pi_i(\vec{x})(\phi_{i+1}(\vec{x}) - \phi_i(\vec{x}))\Big), \tag{13.7}$$

so

$$\begin{aligned} \delta[\phi_{i+1} - \phi_i] &= \prod_{\vec{x}} \int \frac{d\pi_i(\vec{x})}{2\pi} \exp\Big(i\pi_i(\vec{x})(\phi_{i+1}(\vec{x}) - \phi_i(\vec{x}))\Big) \\ &= \int \mathcal{D}\pi_i \exp\left(i \int d^3x\, \pi_i(\vec{x})(\phi_{i+1}(\vec{x}) - \phi_i(\vec{x}))\right), \end{aligned} \tag{13.8}$$

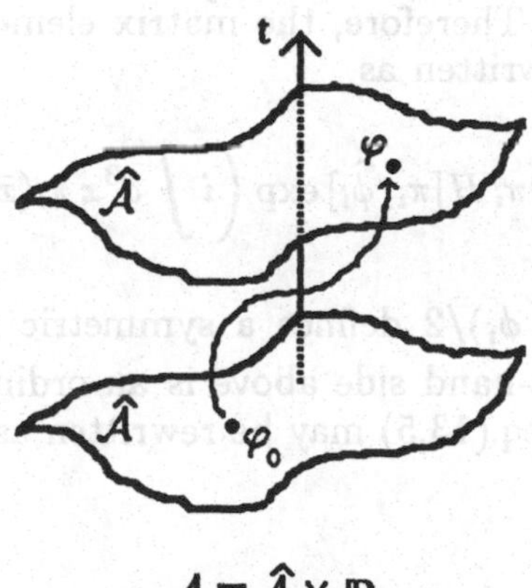

Figure 13.4 $G[\phi, t; \phi_0, t_0]$ may be represented by a path integral through the space $\mathcal{A} = \widehat{\mathcal{A}} \times \mathbb{R}$, i.e. through a space made of copies of $\widehat{\mathcal{A}}$ stacked upon each other, each layer being labeled by the time t. One possible path of evolution of $(\varphi_0(\vec{x}), t) = \varphi(\vec{x}, t)$ is shown. To obtain $G[\phi, t; \phi_0, t_0]$ we must sum over all paths.

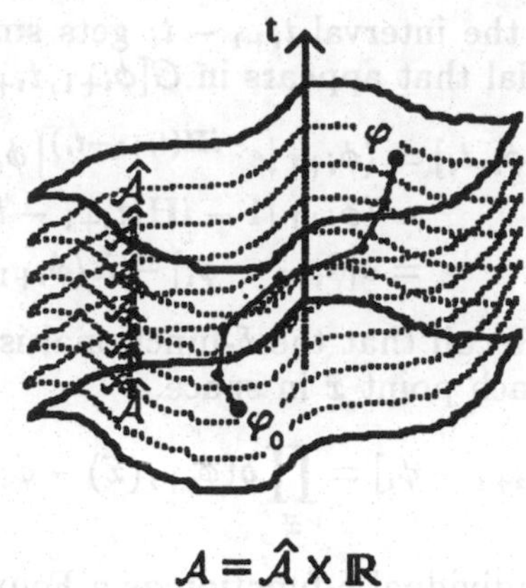

Figure 13.5 Illustration of the constructive definition of the path integral on the space $\mathcal{A}$. The interval $t - t_0$ is broken up into $N + 1$ equal parts, and the functional integral with respect to $\varphi(\vec{x}_i, t_i)$ is performed over $\widehat{\mathcal{A}}$ on each intermediate time slice, t_i. The path integral is the limit $N \to \infty$.

which we recognize as the functional Fourier transform introduced in eq.(10.55). From the Schrödinger representation of the field momentum, eq.(10.9), we can also recognize $\exp(i \int d^3x\, \pi(\vec{x})\phi(\vec{x}))$ as a field momentum eigenfunctional. Therefore, the matrix element of the Hamiltonian can be conveniently written as

$$\langle \phi_{i+1} |\mathbf{H}| \phi_i \rangle = \int \mathcal{D}\pi_i\, H[\pi_i, \widehat{\phi}_i] \exp\left(i \int d^3x\, \pi_i(\vec{x})(\phi_{i+1}(\vec{x}) - \phi_i(\vec{x})) \right), \tag{13.9}$$

where $\widehat{\phi}_i = (\phi_{i+1} + \phi_i)/2$ defines a symmetric operator ordering and $H[\pi_i, \widehat{\phi}_i]$ on the right-hand side above is an ordinary functional, not an operator. Therefore, eq.(13.5) may be rewritten as

$$\begin{aligned} &G[\phi_{i+1}, t_{i+1}; \phi_i, t_i] \\ &= \int \mathcal{D}\pi_i\, (1 - i\epsilon H[\pi_i, \widehat{\phi}_i] + O(\epsilon^2)) \exp\left(i \int d^3x\, \pi_i(\vec{x})(\phi_{i+1}(\vec{x}) - \phi_i(\vec{x})) \right) \\ &\approx \int \mathcal{D}\pi_i \exp\left(i \int d^3x\, \pi_i(\vec{x})(\phi_{i+1}(\vec{x}) - \phi_i(\vec{x})) - \epsilon H[\pi_i, \widehat{\phi}_i] \right), \end{aligned} \tag{13.10}$$

where we assume that ϵ is small enough so that little error is made by replacing $1 - i\epsilon H$ with $\exp(i\epsilon H)$. Placing eq.(13.10) into eq.(13.4), we have

$$\begin{aligned} G[\phi, t; \phi_0, t_0] &= \lim_{N\to\infty} \int \prod_{i=1}^{N} \mathcal{D}\phi_i \prod_{j=0}^{N} \mathcal{D}\pi_j \\ &\quad \exp\left(i\epsilon \sum_{j=0}^{N} \int d^3x \pi_j(\vec{x}) \frac{(\phi_{j+1}(\vec{x}) - \phi_j(\vec{x}))}{\epsilon} - H[\pi_j, \widehat{\phi}_j] \right) \\ &\equiv \int \mathcal{D}\bar{\phi}\mathcal{D}\pi \exp\left(i \int_{t_0}^{t} d\bar{t} \int d^3x \pi(\vec{x})\dot{\bar{\phi}}(\vec{x}) - H[\pi, \bar{\phi}] \right). \end{aligned} \tag{13.11}$$

Since $H[\pi, \widehat{\phi}]$ is quadratic in π, we may easily carry out the functional integration over π by completing the squares. The net result of this is to replace $\pi(\vec{x})$ by $\dot{\bar{\phi}}(\vec{x})$ in the exponential.

$$\begin{aligned} G[\phi, t; \phi_0, t_0] &= \int \mathcal{D}\bar{\phi} \exp\left(i \int_{t_0}^{t} d\bar{t} \int d^3x \mathcal{L}[\dot{\bar{\phi}}, \bar{\phi}, \bar{t}] \right) \\ &= \int \mathcal{D}\bar{\phi} \exp\left(iS[\bar{\phi}(\bar{t})] \right) \end{aligned} \tag{13.12}$$

$\mathcal{L}$ is the classical Lagrangian density and S is the classical action functional, eq.(3.4). In eq.(13.12), $\mathcal{D}\bar{\phi}$ means

$$\mathcal{D}\bar{\phi} = \lim_{N\to\infty} \prod_{i=1}^{N} \prod_{\vec{x}} d\phi(\vec{x}, t_i) \left(\frac{1}{2\pi i\epsilon} \right)^{\frac{1}{2}}, \qquad N\epsilon = t - t_0. \tag{13.13}$$

For most practical calculations, the normalization in eq.(13.13) will be tossed into the overall normalization of the path integral and forgotten.

The path integral, eq.(13.12), is a sum over paths in the space $\mathcal{A} = \widehat{\mathcal{A}} \times \mathbb{R}$, where all paths start at the point $\phi_0(\vec{x}, t_0)$ and end at $\phi(\vec{x}, t)$. When we compare this expression and eq.(13.11) with eq.(12.16) and eq.(12.18) we see that our initial expectation is fulfilled. In a first quantized system, the position and momentum, x and p, are elevated to operators. In the path integral representation we integrate over paths $(x(t), p(t))$ in phase space. In a second quantized system, the field position and momentum are elevated to operators. In the path integral representation we integrate over paths $(\phi(\vec{x}, t), \pi(\vec{x}, t))$ in a functional phase space.

Now let's compute the path integral, eq.(13.12). To minimize the amount of work we must do and to make use of the results in the previous chapter, let us switch coordinates and work with the Fourier transform of $\phi(\vec{x})$,

$$\tilde{\phi}(\vec{k}) = \int d^3x \phi(\vec{x})\, e^{-i\vec{k}\cdot\vec{x}}. \tag{13.14}$$

The action for $\tilde{\phi}(\vec{k},t)$ is

$$S[\tilde{\phi}] = \int_{t_0}^{t} d\bar{t} \int \frac{d^3k}{(2\pi)^3}\frac{1}{2}(\partial_{\bar{t}}\tilde{\phi}(-\vec{k})\partial_{\bar{t}}\tilde{\phi}(\vec{k}) - \omega_k^2\tilde{\phi}(-\vec{k})\tilde{\phi}(\vec{k})), \tag{13.15}$$

where $\omega_k^2 = |\vec{k}|^2 + m^2$. Therefore,

$$G[\tilde{\phi},t;\tilde{\phi}_0,t_0]$$

$$= \int \mathcal{D}\tilde{\phi}\, \exp\left(i\int_{t_0}^{t} d\bar{t} \int \frac{d^3k}{(2\pi)^3}\frac{1}{2}(\partial_{\bar{t}}\tilde{\phi}(-\vec{k})\partial_{\bar{t}}\tilde{\phi}(\vec{k}) - \omega_k^2\tilde{\phi}(-\vec{k})\tilde{\phi}(\vec{k})\right). \tag{13.16}$$

Now we observe that the action, eq.(13.15), is really just the sum of actions of harmonic oscillators, 1 oscillator for each mode $\vec{k}$.

$$S[\tilde{\phi}] \sim -\frac{1}{2}\frac{1}{(2\pi)^3}\sum_{\vec{k}}\int_{t_0}^{t} d\bar{t}\,(\tilde{\phi}(-\vec{k})(\partial_{\bar{t}}^2 + \omega_k^2)\tilde{\phi}(\vec{k}))$$

Since the action is diagonal in $\tilde{\phi}(\vec{k}) = \tilde{\phi}_{\vec{k}}$, we can write the path integral as an infinite product of path integrals, 1 integral for each mode $\vec{k}$.

$$G[\tilde{\phi},t;\tilde{\phi}_0,t_0]$$

$$= \int \mathcal{D}\tilde{\phi}\, \exp\left(\frac{i}{2}\frac{1}{(2\pi)^3}\sum_{\vec{k}}\int_{t_0}^{t} d\bar{t}\, \partial_{\bar{t}}\tilde{\phi}(-\vec{k})\partial_{\bar{t}}\tilde{\phi}(\vec{k}) - \omega_k^2\tilde{\phi}(-\vec{k})\tilde{\phi}(\vec{k})\right)$$

$$= \prod_{\vec{k}}\int \mathcal{D}\tilde{\phi}_{\vec{k}}(\bar{t})\, \exp\left(\frac{i}{2}\frac{1}{(2\pi)^3}\int_{t_0}^{t} d\bar{t}\,(\partial_{\bar{t}}\tilde{\phi}_{\vec{k}}(\bar{t}))^2 - \omega_k^2\tilde{\phi}_{\vec{k}}(\bar{t}))^2\right) \tag{13.17}$$

Comparing this result with eq.(12.46) ($\bar{x} \to \tilde{\phi}_{\vec{k}}$) we can easily see that the path integral above factors into an infinite product of harmonic oscillator path integrals, just as we expected for free fields. Using the result from the previous chapter for a single harmonic oscillator, eq.(12.61), we can quickly write down the result of the functional integration above,

$$G[\tilde{\phi},t;\tilde{\phi}_0,t_0]$$

$$= \prod_{\vec{k}} \left(\frac{\omega_k}{2\pi i \sin(\omega_k(t-t_0))} \right)^{\frac{1}{2}}$$

$$\exp\left(\frac{i}{2}\frac{1}{(2\pi)^3}\left(\frac{\omega_k}{\sin(\omega_k(t-t_0))}\right)\left((\tilde{\phi}^2_{\vec{k}}+\tilde{\phi}^2_{0\vec{k}})\cos(\omega_k(t-t_0)) - 2\tilde{\phi}_{\vec{k}}\tilde{\phi}_{0\vec{k}}\right)\right)$$

$$= \prod_{\vec{k}} \left(\frac{\omega_k}{2\pi i \sin(\omega_k(t-t_0))} \right)^{\frac{1}{2}}$$

$$\exp\left(\frac{i}{2}\int \frac{d^3k}{(2\pi)^3}\frac{\omega_k}{\sin(\omega_k(t-t_0))}\left((\tilde{\phi}^2(\vec{k})+\tilde{\phi}^2_0(\vec{k}))\cos(\omega_k(t-t_0))\right.\right.$$

$$\left.\left. - 2\tilde{\phi}(\vec{k})\tilde{\phi}_0(\vec{k})\right)\right). \qquad (13.18)$$

To recover the ground state wave functional in the Schrödinger representation, eq.(10.28), we take $t - t_0 = T$ to ∞ and look at the leading term. Since

$$\sin(\omega_k T) \to -\frac{i}{2}e^{i\omega_k T}, \qquad \frac{\cos(\omega_k T)}{\sin(\omega_k T)} \to i, \qquad (13.19)$$

as T gets large, then

$$G[\tilde{\phi},\tilde{\phi}_0;T] \longrightarrow \left(\prod_{\vec{k}} \left(\frac{\omega_k}{\pi}\right)^{\frac{1}{2}} e^{-i\omega_k T/2} \right)$$

$$\exp\left(-\int \frac{d^3k}{(2\pi)^3}\frac{\omega_k}{2}(\tilde{\phi}^2(\vec{k})+\tilde{\phi}^2_0(\vec{k}))\right). \qquad (13.20)$$

Therefore, the vacuum energy density is

$$\mathcal{E}_0 = \int d^3k \, \frac{1}{2}\omega_k,$$

and the wave functional is

$$\Psi_0[\tilde{\phi}] = \prod_{\vec{k}} \left(\frac{\omega_k}{\pi}\right)^{\frac{1}{4}} \exp\left(-\frac{1}{2}\frac{\omega_k}{(2\pi)^3}\tilde{\phi}(-\vec{k})\tilde{\phi}(\vec{k})\right), \qquad (13.21)$$

which is the result obtained by solving the functional Schrödinger equation in chapter 10.

To probe the dynamics of the system, we can add an arbitrary external source, $J(\vec{x})$, for $\phi(\vec{x})$. This amounts to adding an arbitrary external force to each of the harmonic oscillators that make up the modes of the field. While such an exercise is not particularly interesting for free fields, it will lead us in the interacting case to a path integral representation of the generating functional of vacuum expectations of time-ordered products of the field. This expression is the goal of this chapter and the following chapter. With this generating functional we can compute the Green's functions of the interacting theory through a perturbation series. The Green's functions can then be sent through a reduction formula to become S-matrix elements and ultimately cross sections as we did in chapter 7.

When we add a source, the action becomes

$$\begin{aligned} S &= \int_{t_0}^{t} d\bar{t} \int d^3x \frac{1}{2}(\partial_\mu \phi(\vec{x},\bar{t})\partial^\mu \phi(\vec{x},\bar{t}) - m^2\phi^2(\vec{x},\bar{t})) + J(\vec{x},\bar{t})\phi(\vec{x},\bar{t}) \\ &= \int_{t_0}^{t} d\bar{t} \int \frac{d^3k}{(2\pi)^3} \left(\frac{1}{2}(\partial_{\bar{t}}\tilde{\phi}(-\vec{k},\bar{t})\partial_{\bar{t}}\tilde{\phi}(\vec{k},\bar{t}) - \omega_k^2\tilde{\phi}(-\vec{k},\bar{t})\tilde{\phi}(\vec{k},\bar{t})) \right. \\ &\qquad \left. + \tilde{J}(-\vec{k},\bar{t})\tilde{\phi}(\vec{k},\bar{t}) \right), \end{aligned} \tag{13.22}$$

so

$$\begin{aligned} &G_J[\tilde{\phi},t;\tilde{\phi}_0,t_0] \\ &= \int \mathcal{D}\tilde{\phi} \exp\left(i\int_{t_0}^{t} d\bar{t} \int \frac{d^3k}{(2\pi)^3}\frac{1}{2}(\partial_{\bar{t}}\tilde{\phi}(-\vec{k},\bar{t})\partial_{\bar{t}}\tilde{\phi}(\vec{k},\bar{t}) \right. \\ &\qquad \left. - \omega_k^2\tilde{\phi}(-\vec{k},\bar{t})\tilde{\phi}(\vec{k},\bar{t})) + \tilde{J}(-\vec{k},\bar{t})\tilde{\phi}(\vec{k},\bar{t}) \right) \\ &= \prod_{\vec{k}} \int \mathcal{D}\tilde{\phi}_{\vec{k}}(\bar{t}) \exp\left(\frac{i}{(2\pi)^3}\int_{t_0}^{t} d\bar{t}\, \frac{1}{2}(\partial_{\bar{t}}\tilde{\phi}_{\vec{k}}^2(\bar{t}) - \omega_k^2\tilde{\phi}_{\vec{k}}^2(\bar{t})) + \tilde{J}_{\vec{k}}(\bar{t})\tilde{\phi}_{\vec{k}}(\bar{t}) \right), \end{aligned} \tag{13.23}$$

which is a product of forced harmonic oscillator path integrals, eq.(12.71). We can use the result of one oscillator, eq.(12.75) or (12.79), to write down G_J.

$$G_J[\tilde{\phi}, t; \tilde{\phi}_0, t_0] = \prod_{\vec{k}} \exp\left(i\frac{1}{(2\pi)^3}\int_{t_0}^{t} d\bar{t}\, \tilde{\phi}_{\rm cl}(\vec{k}, \bar{t})\tilde{J}(-\vec{k}, \bar{t})\right)$$
$$\exp\left(\frac{i}{2}\frac{1}{(2\pi)^3}\int_{t_0}^{t} d\bar{t}\int_{t_0}^{t} dt'\, \tilde{J}(-\vec{k}, \bar{t})g(\bar{t}, t')\tilde{J}(\vec{k}, t')\right) G[\tilde{\phi}, t; \tilde{\phi}_0, t_0], \tag{13.24}$$

where $\tilde{\phi}_{\rm cl}$ is the classical path connecting $\tilde{\phi}_0(\vec{k})$ at t_0 to $\tilde{\phi}(\vec{k})$ at t in the absence of the source,

$$\tilde{\phi}_{\rm cl}(\vec{k}, \bar{t}) = \tilde{\phi}_0(\vec{k})\left(\frac{\sin(\omega_k(t-\bar{t}))}{\sin(\omega_k(t-t_0))}\right) + \tilde{\phi}(\vec{k})\left(\frac{\sin(\omega_k(\bar{t}-t_0))}{\sin(\omega_k(t-t_0))}\right), \tag{13.25}$$

and $g(\bar{t}, t')$ is the Green's function for the harmonic oscillator, eq.(12.70),

$$g(\bar{t}, t') = \begin{cases} \frac{\sin(\omega_k(t-\bar{t}))\,\sin(\omega_k(t'-t_0))}{\omega_k \sin(\omega_k(t-t_0))} & t' \le \bar{t} \\ \frac{\sin(\omega_k(t-t'))\,\sin(\omega_k(\bar{t}-t_0))}{\omega_k \sin(\omega_k(t-t_0))} & \bar{t} \le t'. \end{cases} \tag{13.26}$$

In a scattering experiment, two particles that are essentially free are brought together to react, and the kinematics of the outgoing particles are measured. The region of the reaction is usually very tiny and all measurements are made in the "asymptotic" region, far (in terms of the relevant scales for the particle) in space and time from the interaction. This means that J has compact support and that in practical calculations we are always interested in the limit $t, -t_0 \to \infty$. Only the vacuum state contributions to G_J will survive in this limit, hence the limit, $t, -t_0 \to \infty$ of G_J is called the "vacuum-to-vacuum amplitude in the presence of the source J" and is conventionally denoted by $Z[J]$. Let us explicitly take the limit of G_J in eq.(13.24) to obtain the free field vacuum-to-vacuum amplitude, $Z_0[J]$. Let $t = -t_0 = T$. Consider $\tilde{\phi}_{\rm cl}$ first, eq.(13.25).

$$\begin{aligned} \frac{\sin(\omega_k(t-\bar{t}))}{\sin(\omega_k(t-t_0))} &= \frac{\sin(\omega_k(T-\bar{t}))}{\sin(2\omega_k T)} \\ &= \frac{\cos(\omega_k\bar{t})}{2\cos(\omega_k T)} - \frac{\sin(\omega_k\bar{t})}{2\sin(\omega_k T)} \\ &\xrightarrow{iT\to\infty} 0, \end{aligned} \tag{13.27}$$

since both $\cos(\omega T)$ and $\sin(\omega T)$ go like $\exp(i\omega T)$ as iT gets large. Thus, $\tilde{\phi}_{\rm cl}$ vanishes; the classical vacuum is zero. Next consider the Green's function, $g(\bar{t}, t')$, for $t' \le \bar{t}$.

$$\frac{\sin(\omega_k(t-\bar{t}))\sin(\omega_k(t'-t_0))}{\omega_k\sin(\omega_k(t-t_0))}$$

$$= (\sin(\omega_k T)\cos(\omega_k\bar{t}) - \sin(\omega_k\bar{t})\cos(\omega_k T))\left(\frac{(\sin(\omega_k t')\cos(\omega_k T) + \sin(\omega_k T)\cos(\omega_k t'))}{\omega_k\sin(2\omega_k T)}\right)$$

$$= \frac{1}{2\omega_k}\Bigg(\sin(\omega_k t')\cos(\omega_k\bar{t}) - \sin(\omega_k\bar{t})\cos(\omega_k t') + \cos(\omega_k\bar{t})\cos(\omega_k t')\left(\frac{\sin(\omega_k T)}{\cos(\omega_k T)}\right) - \sin(\omega_k\bar{t})\sin(\omega_k t')\left(\frac{\cos(\omega_k T)}{\sin(\omega_k T)}\right)\Bigg)$$

$$\xrightarrow{iT\to\infty} \frac{1}{2\omega_k}\Big(\sin(\omega_k(\bar{t}-t')) - i\cos(\omega_k(\bar{t}-t'))\Big)$$

$$= \frac{i}{2\omega_k}\,e^{-i\omega_k(\bar{t}-t')}. \tag{13.28}$$

Likewise, if $\bar{t} \leq t'$, then

$$g(\bar{t},t') \longrightarrow \frac{i}{2\omega_k}e^{-i\omega_k(t'-\bar{t})} = \frac{i}{2\omega_k}e^{i\omega_k(\bar{t}-t')}. \tag{13.29}$$

Altogether we have

$$\begin{aligned} g(\bar{t},t') &= \frac{i}{2\omega_k}\left(e^{-i\omega_k(\bar{t}-t')}\theta(\bar{t}-t') + e^{i\omega_k(\bar{t}-t')}\theta(t'-\bar{t})\right) \\ &= -\tilde{\Delta}_F(\vec{k},\bar{t}-t'). \end{aligned} \tag{13.30}$$

When we compare this with the results of chapter 3, we find that $g(\bar{t},t')$ has turned into the spatial Fourier transform of the Feynman propagator, eqs.(3.48) and (3.52). Thus,

$$Z_0[\tilde{J}]$$

$$= Z_0[0]\exp\left(-\frac{i}{2}\int\frac{d^3k}{(2\pi)^3}\int_{-\infty}^{\infty}d\bar{t}\int_{-\infty}^{\infty}dt'\;\tilde{J}(-\vec{k},\bar{t})\tilde{\Delta}_F(\vec{k},\bar{t}-t')\tilde{J}(\vec{k},t')\right),$$

or

$$Z_0[J] = Z_0[0]\exp\left(-\frac{i}{2}\int d^4\bar{x}\int d^4x'\;J(\bar{x})\Delta_F(\bar{x}-x')J(x')\right). \tag{13.31}$$

$Z_0[0]$, the normalization of the vacuum-to-vacuum amplitude, contains the limiting value of $G[\tilde{\phi},T;\tilde{\phi}_0,-T]$ as T gets large. In general, we do not need to know the exact form of $Z_0[0]$ so we do not need to look at $G[\phi,T;\phi_0,-T]$.

Recall from chapter 3 that we built boundary conditions into the Feynman propagator such that positive energy solutions to the Klein-Gordon equation propagate only forward in time while negative energy solutions propagate only backwards in time. We essentially had to add this feature by hand by subtracting a factor $i\epsilon$ in the denominator of Δ_F, eq.(3.52), to displace the poles of the propagator, i.e. we time-ordered the fields. However, the method used here to determine and compute the path integral provided the proper propagator, satisfying the desired boundary conditions automatically. A heuristic argument of how this arises is as follows: The path integral is a sum of phases, that is, there is a factor of i that appears in front of the classical action in the exponent of the integrand. This means that the integrand oscillates horribly as we continuously change the path and there is no apparent reason why the path integral should converge to anything meaningful. To remedy this, we can add a "convergence factor" to the integrand, that is, add a term to the action like $i\int \epsilon\phi^2(x)\, d^4x$, $\epsilon > 0$, so that the path integral becomes

$$G = \int \mathcal{D}\phi \, \exp\left(iS[\phi]\right) \exp\left(-\epsilon \int d^4x \phi^2(x)\right). \tag{13.32}$$

After we compute the integral, then we can get rid of the extra factor by taking $\epsilon \longrightarrow 0^+$. This procedure is equivalent to adding a small imaginary piece to the mass term, $m^2 \longrightarrow m^2 - i\epsilon$. Now if we explicitly calculate the path integral after we take the limit $T \rightarrow \infty$ (instead of before, as we just did) then we will arrive with the correct boundary conditions. So, in some sense, the boundary conditions we want to use to avoid the problem of negative energy solutions are also the ones that make the path integral well defined. Time-ordering arises naturally in defining the path integral.

In the last chapter we showed that the forced harmonic oscillator path integral can be used as a generating functional for time-ordered products of the position operator. By the same reasoning, the vacuum-to-vacuum amplitude in the presence of a source is a generating functional for time-ordered products of the fields. It is the desire to calculate these Green's functions to input into the reduction formulae that makes $Z[J]$ the object of interest to calculate in the path integral representation.

As a matter of illustration, let us quickly compute a couple of time-ordered products using the explicit form of $Z_0[J]$, eq.(13.31).

$$\begin{aligned} \langle 0 \left| T(\phi(x_1)\phi(x_2)) \right| 0\rangle &= \frac{1}{i^2}\frac{\delta}{\delta J(x_1)}\frac{\delta}{\delta J(x_2)}\frac{Z_0[J]}{Z_0[0]}\bigg|_{J=0} \\ &= i\Delta_F(x_1 - x_2) \end{aligned} \tag{13.33}$$

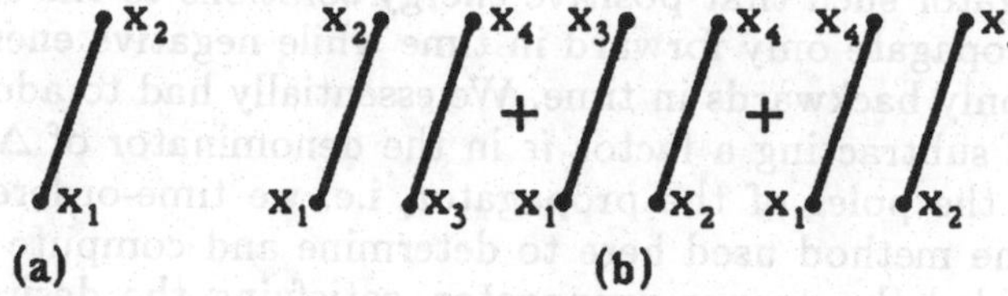

Figure 13.6 Diagrammatic representation of the free field (a) 2-point function, and (b) 4-point function generated by functional differentiation of the vacuum-to-vacuum amplitude, $Z_0[J]$.

$$
\begin{aligned}
&\langle 0\,|T(\phi(x_1)\phi(x_2)\phi(x_3)\phi(x_4))|\,0\rangle \\
&\quad = \frac{1}{i^4}\frac{\delta}{\delta J(x_1)}\frac{\delta}{\delta J(x_2)}\frac{\delta}{\delta J(x_3)}\frac{\delta}{\delta J(x_4)}\frac{Z_0[J]}{Z_0[0]}\bigg|_{J=0} \\
&\quad = i\Big(\Delta_F(x_1-x_2)\Delta_F(x_3-x_4)+\Delta_F(x_1-x_3)\Delta_F(x_2-x_4) \\
&\qquad\qquad + \Delta_F(x_1-x_4)\Delta_F(x_2-x_3)\Big)
\end{aligned}
\tag{13.34}
$$

If we represent the 2-point function, $\langle 0\,|T(\phi(x_1)\phi(x_2))|\,0\rangle$ as a straight line connecting x_1 to x_2, then eqs.(13.33) and (13.34) may be drawn diagrammatically as in figure 13.6. Higher n-point functions are just sums of products of $n/2$ 2-point functions if n is even, and zero if n is odd. Diagrammatically, this is just a set of $n/2$ disconnected lines. Since the field is free, no 2 lines cross to interact.

Define a new functional, $W_0[J]$, from $Z_0[J]$, via

$$Z_0[J] = Z_0[0]\,\exp(iW_0[J]). \tag{13.35}$$

Now note that

$$
\begin{aligned}
i\frac{\delta}{\delta J(x_1)}W_0[J]\bigg|_{J=0} &= 0 \\
i^2\frac{\delta}{\delta J(x_1)}\frac{\delta}{\delta J(x_2)}W_0[J]\bigg|_{J=0} &= \Delta_F(x_1-x_2) \\
i^n\prod_{j=1}^{n}\frac{\delta}{\delta J(x_j)}W_0[J]\bigg|_{J=0} &= 0, \qquad n\geq 3.
\end{aligned}
$$

$W_0[J]$ only generates one diagram, figure 13.6(a). The difference between diagram 13.6(a) and diagram 13.6(b) is that (a) is connected while (b) is not. We can split (b) in half without cutting a solid line and end up with 2 legitimate diagrams. Thus, $W_0[J]$ appears to generate only connected diagrams, at least in free field theory. This remains true for interacting theories, too.

13.2 FREE PHOTON FIELD THEORY

Using the experience gained in the previous section, we can try to compute the vacuum-to-vacuum amplitude starting from an expression similar to eq.(13.12), but where we have already taken $t, -t_0 \to \infty$. We know that we can obtain the correct boundary conditions on the propagator by adding a convergence factor to the action. However, if we recall some of the difficulties we met in quantizing the electromagnetic field in the operator formalism in chapter 5, we will realize that the gauge symmetry must add some new wrinkles to the path integral procedure. The difficulty that the gauge symmetry presents is that we would like to write down $Z_0[J]$ by integrating over all A^μ, $\int \mathcal{D}A^\mu$, yet the action, being gauge invariant, does not depend on all 4 components of the potential. This will lead to a huge infinity in computing the path integral. This divergence arises for the same reason that $\int f(x)\, dx\, dy$ is infinite. If we can somehow split up $\int \mathcal{D}A^\mu$ into an integral over gauge invariant components and gauge dependent components, then we can take the infinite integral over the gauge dependent components and throw it into the normalization and forget it (for example, separate out $\int dy$ from $\int f(x)\, dx\, dy$ and normalize it away). However, it may not be easy to split the components into gauge dependent and gauge invariant parts. This is the case with Yang-Mills. Chapter 15 will deal with a systematic method for accomplishing this, the Faddeev-Popov procedure. In the meantime, we know how to separate gauge dependent and gauge invariant components for the electromagnetic field so we can just separate out the infinite volume. The price for such simplicity will be manifest Lorentz covariance.

The simplest way to get something physically meaningful from the vacuum-to-vacuum amplitude is to only couple the source to the gauge invariant components, that is, the transverse part of A^μ.

$$Z_0[J] = \int \mathcal{D}A^\mu \,\exp\left(i \int d^4x\, \frac{1}{2}(\vec{E}^2 - \vec{B}^2) + \vec{J}\cdot\vec{A}_T\right) \qquad (13.36)$$

$\vec{E}$ and $\vec{B}$ are the usual electric and magnetic fields, $\vec{E} = -\nabla A^0 - \partial_t \vec{A}$, $\vec{B} = \nabla \times \vec{A}$. Any part of the integral $\int \mathcal{D}A^\mu \cdots$ that doesn't depend on $\vec{J}$ will be thrown into $Z_0[0]$. By the Gaussian nature of the integrand, it is easy to see that the integral over A^0 and the longitudinal component, $\vec{A}_L$, will not depend on $\vec{J}$ so we may rewrite eq.(13.36) as

$$Z_0[J] = \mathcal{N} \int \mathcal{D}A^\mu \, \delta[A^0]\delta[\vec{A}_L] \exp\left(i \int d^4x \frac{1}{2}(\vec{E}^2 - \vec{B}^2) + \vec{J}\cdot\vec{A} \right), \tag{13.37}$$

where $\mathcal{N}$ is a normalization that will eventually become $Z_0[0]$. The appearance of the δ-functionals above is easy to interpret. They express a gauge-fixing, here the Coulomb gauge. In arriving at eq.(13.37) from eq.(13.36), the infinite "gauge volume" has quietly disappeared into $Z_0[0]$. The integrals over A^0 and $\vec{A}_L$ are now trivial to carry out and we have

$$Z_0[J] = \mathcal{N} \int \mathcal{D}\vec{A}_T \exp\left(i \int d^4x \frac{1}{2}(\vec{E}_T^2 - \vec{B}^2) + \vec{J}\cdot\vec{A}_T \right), \tag{13.38}$$

where $\vec{E}_T = -\partial_t \vec{A}_T$ is the transverse electric field depending upon $\vec{A}_T$ only. To complete the integration it is easier to switch variables to untangle the "transverseness" from the space components. Since "transverseness" can be easily defined with respect to the momentum vector, $\vec{k}$, we should once again Fourier transform $\vec{A}_T$.

$$\vec{A}_T(x) = \int \frac{d^3k}{(2\pi)^3} \sum_{\lambda=1}^{2} \vec{\varepsilon}(\vec{k},\lambda)\tilde{A}(\vec{k},\lambda,t)e^{i\vec{k}\cdot\vec{x}} \tag{13.39}$$

The polarization vectors, $\vec{\varepsilon}(\vec{k},\lambda)$, were defined in chapter 5 and satisfy eqs.(5.29)-(5.30). Written in terms of the transformed variables, the vacuum-to-vacuum amplitude becomes

$$Z_0[\tilde{J}] = \mathcal{N} \int \mathcal{D}\tilde{A}(\vec{k},\lambda,\bar{t}) \exp\Bigg(i \sum_{\lambda=1}^{2} \int d\bar{t} \int \frac{d^3k}{(2\pi)^3} \left(\partial_{\bar{t}} \tilde{A}_{\vec{k}}(\lambda,\bar{t}) \right)^2 - \vec{k}^2 \tilde{A}_{\vec{k}}^2(\lambda,\bar{t}) + \tilde{J}(-\vec{k},\lambda,\bar{t})\tilde{A}_{\vec{k}}(\lambda,\bar{t}) \Bigg). \tag{13.40}$$

Again we recognize that $Z_0[\tilde{J}]$ is an infinite product of harmonic oscillator path integrals, one for each mode $\vec{k}$ and polarization λ. Compare eq.(13.29) with eq.(13.23). Using eq.(13.24) and the limit of $g(\bar{t},t')$, eq.(13.26), as $\bar{t}, -t' \to \infty$, eq.(13.30), we can compute $Z_0[\tilde{J}]$ as

$$\frac{Z_0[\tilde{J}]}{Z_0[0]} =$$
$$\exp\left(-\frac{i}{2}\sum_{\lambda=1}^{2}\int\frac{d^3k}{(2\pi)^3}\int_{-\infty}^{\infty}d\bar{t}\int_{-\infty}^{\infty}dt'\tilde{J}(-\vec{k},\bar{t},\lambda)\tilde{\Delta}_F(\vec{k},\bar{t}-t')\tilde{J}(\vec{k},t',\lambda)\right) \tag{13.41}$$

where

$$\tilde{\Delta}_F(\vec{k},\bar{t}-t') = -\frac{i}{2\omega_k}\left(e^{-i\omega_k(\bar{t}-t')}\theta(\bar{t}-t') + e^{i\omega_k(\bar{t}-t')}\theta(t'-\bar{t})\right), \tag{13.42}$$

and $\omega_k = |\vec{k}|$. To compare with the results of chapter 5, let us transform the functional $Z_0[\tilde{J}]$ back to $\vec{x}$-space. A careful insertion of some factors of unity in the exponent of eq.(13.41) yields

$$\sum_{\lambda=1}^{2}\int\frac{d^3k}{(2\pi)^3}\tilde{J}(-\vec{k},\bar{t},\lambda)\tilde{\Delta}_F(\vec{k},\bar{t}-t')\tilde{J}(\vec{k},t',\lambda)$$
$$=\sum_{\lambda=1}^{2}\sum_{\lambda'=1}^{2}\sum_{\lambda''=1}^{2}\int\frac{d^3k}{(2\pi)^3}\frac{d^3k'}{(2\pi)^3}\frac{d^3k''}{(2\pi)^3}\delta(\vec{k}'-\vec{k})\delta(\vec{k}+\vec{k}'')$$
$$\tilde{J}(-\vec{k}',\bar{t},\lambda')\delta_{\lambda'\lambda}\tilde{\Delta}_F(\vec{k},\bar{t}-t')\delta_{\lambda\lambda''}\tilde{J}(\vec{k}'',t',\lambda'')$$
$$=\sum_{\lambda=1}^{2}\sum_{\lambda'=1}^{2}\sum_{\lambda''=1}^{2}\int\frac{d^3k}{(2\pi)^3}\frac{d^3k'}{(2\pi)^3}\frac{d^3k''}{(2\pi)^3}\int d^3x\,d^3x'\,e^{i(\vec{k}'-\vec{k})\cdot\vec{x}}\,e^{i(\vec{k}+\vec{k}'')\cdot x'}$$
$$\tilde{J}(-\vec{k}',\bar{t},\lambda')\vec{\varepsilon}(\vec{k}',\lambda')\cdot\vec{\varepsilon}(\vec{k},\lambda)\tilde{\Delta}_F(\vec{k},\bar{t}-t')\vec{\varepsilon}(\vec{k}'',\lambda'')\cdot\vec{\varepsilon}(\vec{k},\lambda)\tilde{J}(\vec{k},t',\lambda)$$
$$=\int d^3x\,d^3x'\,J_i(\vec{x},\bar{t})D^{\rm tr}_{ij}(x-x')J_j(\vec{x}',t'). \tag{13.43}$$

Thus,

$$Z_0[J] = Z_0[0]\exp\left(-\frac{i}{2}\int d^4x\,d^4x'\,J_i(x)D^{\rm tr}_{ij}(x-x')J_j(x')\right), \tag{13.44}$$

where

$$D^{\rm tr}_{ij}(x-x') = -\int\frac{d^3k}{(2\pi)^3}\frac{i}{2\omega_k}\left(e^{-i\omega_k(\bar{t}-t')}\theta(\bar{t}-t') + e^{i\omega_k(\bar{t}-t')}\theta(t'-\bar{t})\right)$$
$$\sum_{\lambda=1}^{2}\varepsilon_i(\vec{k},\lambda)\varepsilon_j(\vec{k},\lambda). \tag{13.45}$$

If we apply the same procedure to eq.(13.45) that took eq.(3.48) to eq.(3.52) (representing the Heaviside function, θ, by the integral eq.(3.49)), then we find that $D^{\rm tr}_{ij}(x-x')$ is identical to the Coulomb gauge propagator of chapter 5, eq.(5.37).

13.3 FREE SPINOR FIELD THEORY

Coordinate representations of spinor fields are anticommuting or Grassmann variables. We cannot rely entirely on the experience we gained in doing the bosonic fields since integration of anticommuting is only a formal operation (e.g. what does convergence mean?). In addition, to guide us in the bosonic case we relied heavily on the harmonic oscillator path integral. We do not presently possess the equivalent of this for the fermionic case. Nonetheless, to proceed let us copy the structures we found in the bosonic case and see how far we get.

The path integral that we are interested in is the time-evolution operator for the Schrödinger equation of the Dirac field, eq.(10.90). In view of eqs.(13.11) and (13.12), we can expect that the relevant path integral is

$$G[\psi, t; \psi_0, t_0] = \int \mathcal{D}\psi^\dagger(x)\mathcal{D}\psi(x)\, \exp\left(i\int_{t_0}^{t} d\bar{t} \int d^3x \bar{\psi}(\vec{x},\bar{t})(i\partial\!\!\!/ - m)\psi(\vec{x},\bar{t})\right), \tag{13.46}$$

where $\psi(x)$ is a 4-component spinor and each component is an anticommuting variable. The biggest difference between eq.(13.46) and its bosonic partners is that the integration is over spaces of anticommuting functions.

As with the photon path integral, let's decouple spin information in a computationally convenient manner by switching to momentum representation variables. If we substitute

$$\psi(\vec{x},t) = \sum_{r=1}^{4}\int \frac{d^3p}{\sqrt{(2\pi)^3}}\left(\frac{m}{E_p}\right)^{\frac{1}{2}} b(\vec{p},r,t)w(\vec{p},r)\, e^{i\vec{p}\cdot\vec{x}}, \tag{13.47}$$

where the momentum spinors are given by eq.(10.93) and $E_p^2 = \vec{p}^2 + m^2$, then the Lagrangian in the exponent of eq.(13.46) becomes

$$\begin{aligned}
&\int d^3x \bar{\psi}(\vec{x},\bar{t})(i\partial\!\!\!/ - m)\psi(\vec{x},\bar{t}) \\
&= \sum_{r=1}^{4}\sum_{r'=1}^{4}\int \frac{d^3p}{\sqrt{(2\pi)^3}}\frac{d^3p'}{\sqrt{(2\pi)^3}}\int d^3x \left(\frac{m}{E_p}\right)^{\frac{1}{2}}\left(\frac{m}{E_{p'}}\right)^{\frac{1}{2}} e^{i(\vec{p}-\vec{p}')\cdot\vec{x}} \\
&\qquad b^*(\vec{p}',r',\bar{t})w^\dagger(\vec{p}',r')\gamma^0(i\gamma^0\partial_{\bar{t}} - \gamma^j p^j - m)w(\vec{p},r)b(\vec{p},r,\bar{t}) \\
&= \sum_{r=1}^{4}\sum_{r'=1}^{4}\int d^3p \frac{m}{E_p}\, b^*(\vec{p},r',\bar{t})w^\dagger(\vec{p},r')(i\partial_{\bar{t}} - \vec{\alpha}\cdot\vec{p} - \gamma^0 m)w(\vec{p},r)b(\vec{p},r,\bar{t}) \\
&= \sum_{r=1}^{4}\int d^3p b^*(\vec{p},r,\bar{t})(i\partial_{\bar{t}} - \epsilon_r E_p)b(\vec{p},r,\bar{t}),
\end{aligned} \tag{13.48}$$

where $\epsilon_r = 1$ for $r = 1, 2$ and -1 for $r = 3, 4$, and where we have made use of the relations (10.94) and (11.121) that the $w(\vec{p}, r)$ satisfy.

Under this change of variables, the path integral factors into an infinite product of single-particle fermionic path integrals, one path integral for each mode $\vec{p}$ and component of spin r.

$$G[b_f, t; b_i, t_0] = \prod_{r=1}^{4} \prod_{\vec{p}} \int \mathcal{D}b^* \mathcal{D}b \exp\left(i \int_{t_0}^{t} d\bar{t} \int d^3p\, b^*(\vec{p}, r, \bar{t})(i\partial_{\bar{t}} - \epsilon_r E_p) b(\vec{p}, r, \bar{t}) \right). \tag{13.49}$$

Let us concentrate on the path integral for a single positive energy mode,

$$G(\eta, t; \eta_0, t_0) = \int \mathcal{D}\eta^*(\bar{t}) \mathcal{D}\eta(\bar{t}) \exp\left(i \int_{t_0}^{t} d\bar{t} \eta^*(\bar{t})(i\partial_{\bar{t}} - E_p)\eta(\bar{t}) \right). \tag{13.50}$$

To compute this path integral we follow the constructive definition and divide the interval t_0 to t into $N+1$ equal parts, compute the integrals on each slice, and then let $N \to \infty$. Let t_i be the time on the ith slice with $t_{N+1} = t$, and $\epsilon = t_{i+1} - t_i$. Also, let $\eta_i = \eta(t_i)$, $\eta_{N+1} = \eta$. Then eq.(13.50) becomes

$$\begin{aligned} G(\eta, t; \eta_0, t_0) &= \lim_{N\to\infty} \int d\eta_0^* \prod_{i=1}^{N} (d\eta_i d\eta_i^*) \prod_{j=0}^{N} \exp\left(\eta_j^* \left(-\frac{\eta_{j+1} - \eta_j}{\epsilon} \epsilon - i\epsilon E_p \eta_j \right) \right) \\ &= \lim_{N\to\infty} \int d\eta_0^* \prod_{i=1}^{N} (d\eta_i d\eta_i^*) \prod_{j=0}^{N} \Big(1 - \eta_j^*((\eta_{j+1} - \eta_j) + i\epsilon E_p \eta_j) \Big) \end{aligned} \tag{13.51}$$

where the ordering of the product of exponentials is such that the contribution of the earlier time slice appears to the right of the later slice. The choice of ordering of the differentials, $d\eta_0^* d\eta_1 d\eta_1^* \cdots d\eta_N d\eta_N^*$, is made in accordance with eq.(9.85) instead of eq.(9.84) because the combination $\eta_j^* \eta_j$ appears with a $+$ sign in the exponential in eq.(13.51).

Using the Grassmann rules of integration, eq.(9.61), the η_j^* integrals are straightforward and serve to eliminate the "1−" from each factor of the product. This leaves $((-1)^{N(N+1)}=1$ for all N),

$G(\eta,t;\eta_0,t_0)=$

$$\lim_{N\to\infty}(-1)^{N+1}\int d\eta_1\cdots d\eta_N\,((\eta-\eta_N)+i\epsilon E_p\eta_N)\cdots((\eta_1-\eta_0)+i\epsilon E_p\eta_0). \tag{13.52}$$

The integral over η_N yields

$$\int d\eta_N(\eta-(1-i\epsilon E_p)\eta_N)(\eta_N-(1-i\epsilon E_p)\eta_{N-1}) = -\eta+(1-i\epsilon E_p)^2\eta_{N-1}. \tag{13.53}$$

Continuing with the η_{N-1} integral we have,

$$\int d\eta_{N-1}(-\eta+(1-i\epsilon E_p)^2\eta_{N-1})(\eta_{N-1}-(1-i\epsilon E_p)\eta_{N-2}) = \eta-(1-i\epsilon E_p)^3\eta_{N-2}. \tag{13.54}$$

It is now easy to see the pattern. By induction we find

$$\int d\eta_1\cdots d\eta_N\,(\eta-(1-i\epsilon E_p)\eta_N)\cdots(\eta_1-(1-i\epsilon E_p)\eta_0) = (-1)^N\eta+(-1)^{N+1}(1-i\epsilon E_p)^{N+1}\eta_0. \tag{13.55}$$

Thus,

$$\begin{aligned} G(\eta,t;\eta_0,t_0) &= \lim_{N\to\infty}(-\eta+(1-i\epsilon E_p)^{N+1}\eta_0) \\ &= -\eta+\lim_{N\to\infty}\left(1-\frac{iE_p(t-t_0)}{N+1}\right)^{N+1}\eta_0 \\ &= -\eta+e^{-iE_p(t-t_0)}\eta_0. \end{aligned} \tag{13.56}$$

In performing the computation of this path integral, we have assumed $t>t_0$. Therefore, implicit in eq.(13.56) is a step function, $\theta(t-t_0)$. In passing to the last line of eq.(13.56) we made use of the definition of the exponential, $\lim_{n\to\infty}(1-x/n)^n=\exp(-x)$. To demonstrate the validity of this representation of the exponential on more familiar ground, let us consider $f_n(x)=(1-x/n)^n$ and denote $f(x)=\lim_{n\to\infty}f_n(x)$. Differentiate f_n to obtain

$$\frac{df_n}{dx}=n\Big(1-\frac{x}{n}\Big)^{n-1}\Big(-\frac{1}{n}\Big)=-f_{n-1}(x). \tag{13.57}$$

In the limit $n \to \infty$, there is no difference between f_n and f_{n-1},

$$\lim_{n\to\infty} \frac{f_n(x)}{f_{n-1}(x)} = \lim_{n\to\infty}\left(1-\frac{x}{n}\right) = 1, \tag{13.58}$$

thus the limit of eq.(13.57) is $f'(x) = -f(x)$. The solution of this equation is $f(x) = A\exp(-x)$, and it follows from $f(0) = \lim_{n\to\infty} f_n(0) = \lim_{n\to\infty} 1 = 1$ that $A = 1$.

Let us check the result, eq.(13.56). In the Schrödinger representation the Schrödinger equation is

$$E_p \frac{d}{d\eta}\eta\psi_n(\eta) = E_n\psi_n(\eta). \tag{13.59}$$

The Grassmann nature of η implies that $\psi_n(\eta) = A + B\eta$. Plugging this into eq.(13.59) yields $B = 0$, $E_pA = E_nA$, or $B \neq 0$, $E_n = 0$. Therefore, there are two states, ψ_0 and ψ_1, with $E_0 = 0$ and $E_1 = E_p$, respectively. The Green's function for the Schrödinger equation $i\partial_t\psi = E_p\partial_\eta\eta\psi$ is

$$G(\eta, t; \eta_0, t_0) = \sum_n (-1)^n e^{-iE_n(t-t_0)}\psi_n(\eta)\psi_n^*(\eta_0), \tag{13.60}$$

where n is the occupation number. Since $\psi_0^*(\eta) = 1$ and $\psi_n^*(\eta_0) = \eta$, then eq.(13.60) is identical to eq.(13.56).

For a single negative energy mode ($r = 3$ or 4),

$$\begin{aligned} G(\xi, t; \xi_0, t_0) &= \int \mathcal{D}\xi^*(\bar{t})\mathcal{D}\xi(\bar{t}) \exp\left(i\int_{t_0}^{t} d\bar{t}\, \xi^*(\bar{t})(i\partial_{\bar{t}} + E_p)\xi(\bar{t})\right) \\ &= -\xi + e^{iE_p(t-t_0)}\xi_0. \end{aligned} \tag{13.61}$$

Once again, the factor $\theta(t - t_0)$, implicit in all path integral calculations in this chapter, has been suppressed. Combining one positive and one negative energy mode with the same $\vec{p}$ we have

$$G = \eta\xi - e^{-iE_p(t-t_0)}\eta_0\xi - e^{iE_p(t-t_0)}\eta\xi_0 + \eta_0\xi_0. \tag{13.62}$$

In the limit $i(t - t_0) \to \infty$, the term $\exp(iE_p(t - t_0))\eta\xi_0$ dominates. This implies that the vacuum state is $\psi_0(\xi, \eta) = \eta$, that is, the positive energy mode is empty and the negative energy mode is occupied. Therefore, when we string all of the path integrals for single modes, eq.(13.56) or eq.(13.61), together to form the free spinor field path integral, eq.(13.49), and take the limit $i(t - t_0) \to \infty$, the resulting vacuum state will be the physical, filled vacuum. The wave functional resulting from the path integral in this limit will agree with the one that we obtained in chapter 10, eq.(10.101), by solving the functional Schrödinger differential equation, eq.(10.98). Therefore, the path integral defined by eq.(13.46), in combination with the rules of integration for anticommuting variables, is the desired generalization of the bosonic path integral to fermionic fields.

In order to build a generating functional, we must add sources to the vacuum-to-vacuum amplitude. We can expect that the appropriate fermionic generating functional is

$$Z_0[\bar{\chi},\chi] = \int \mathcal{D}\psi^*(x)\mathcal{D}\psi(x) \exp\left(i\int d^4x\, \bar{\psi}(i\partial\!\!\!/ - m)\psi + \bar{\chi}\psi + \bar{\psi}\chi\right). \tag{13.63}$$

Using the momentum expansion for $\psi(\vec{x},t)$, eq.(13.47), and that for χ,

$$\chi(\vec{x},t) = \sum_{r=1}^{4}\int \frac{d^3p}{\sqrt{(2\pi)^3}}\left(\frac{m}{E_p}\right)^{\frac{1}{2}} \tilde{\chi}(\vec{p},r,t)w(\vec{p},r)\, e^{i\vec{p}\cdot\vec{x}}, \tag{13.64}$$

the generating functional becomes

$$Z_0[\tilde{\chi}^*,\tilde{\chi}] = \int \mathcal{D}b^*\mathcal{D}b \exp\Bigg(i\int_{-\infty}^{\infty} d\bar{t}\int d^3p\, b^*(\vec{p},r,\bar{t})(i\partial_{\bar{t}} - \epsilon_r E_p)b(\vec{p},r,\bar{t}) + i\epsilon_r\left(\frac{m}{E_p}\right)\left(\tilde{\chi}^*(\vec{p},r,\bar{t})b(\vec{p},r,\bar{t}) + b^*(\vec{p},r,\bar{t})\tilde{\chi}(\vec{p},r,\bar{t})\right)\Bigg), \tag{13.65}$$

which is a product of single-particle path integrals in the presence of an external force. Once again, let us concentrate on a single mode.

$$G[\zeta^*,\zeta] = \int \mathcal{D}\eta^*\mathcal{D}\eta \exp\left(i\int_{-\infty}^{\infty} d\bar{t}\eta^*(\bar{t})(i\partial_{\bar{t}} - E_p)\eta(\bar{t}) + \left(\frac{m}{E_p}\right)\left(\zeta^*\eta + \eta^*\zeta\right)\right) \tag{13.66}$$

We could follow the same procedure as in the sourceless case to compute this path integral (break up the interval into $N+1$ equal parts and take the limit), but that computation is rather involved and tedious. Instead, let us use the fact that the fermionic integration is designed to be translationally invariant. If ζ is an anticommuting variable,

$$\int d\eta = \int d(\eta+\zeta) = 0 \quad \text{and} \quad \int d\eta\, \eta + \zeta = \int d(\eta+\zeta)\, \eta + \zeta = 1.$$

Fourier transforming the anticommuting variables with respect to time,

$$\eta(t) = \int \frac{d\omega}{\sqrt{2\pi}}\, e^{-i\omega t}\tilde{\eta}(\omega),$$

the path integral, eq.(13.66), may be written as

$$G[\tilde{\zeta}^*, \tilde{\zeta}] =$$

$$\int \mathcal{D}\eta^* \mathcal{D}\eta \,\exp\Big(i \int d\omega \tilde{\eta}^*(\omega)(\omega - E_p)\tilde{\eta}(\omega) + \frac{m}{E_p}(\tilde{\zeta}^*(\omega)\tilde{\eta}(\omega) + \tilde{\eta}^*(\omega)\tilde{\zeta}(\omega))\Big)$$

$$= \int \mathcal{D}\eta^* \mathcal{D}\eta \,\exp\bigg(-i \int d\omega \Big(\tilde{\eta}^* + \frac{m}{E_p}\frac{1}{\omega - E_p}\tilde{\zeta}^*\Big)$$

$$(\omega - E_p)\Big(\tilde{\eta} + \frac{m}{E_p}\frac{1}{\omega - E_p}\tilde{\zeta}\Big) - \Big(\frac{m}{E_p}\Big)^2 \tilde{\zeta}^*(\omega)\frac{1}{w - E_p}\tilde{\zeta}(\omega)\bigg)$$

$$= \exp\bigg(-i\Big(\frac{m}{E_p}\Big)^2 \int d\omega \tilde{\zeta}^*(\omega)\frac{1}{w - E_p}\tilde{\zeta}(\omega)\bigg)\, G[\tilde{\zeta}^* = 0, \tilde{\zeta} = 0]. \qquad (13.67)$$

To invert the result, eq.(13.67), back into a functional of $\zeta^*(t)$ and $\zeta(t)$, we need a prescription for handling the pole at $\omega = E_p$, that is, we need to specify boundary conditions on the Green's function, $g_-(\bar{t}, t') = 1/(i\partial_{\bar{t}} - E_p)$. By starting with a path integral where the limit $t, -t_0 \to \infty$ has already been taken, we have lost the proper boundary conditions. We know in the bosonic case that we can get the proper boundary conditions by adding an "$i\epsilon$" convergence factor, but we have no such convergence thing to resort to for fermionic integration. We know from chapters 4 and 10 that we want the positive energy modes to only propagate forward in time. This implies that we should take

$$g_-(t, t') = -i\theta(t - t')e^{-iE_p(t-t')}.$$

However, let us proceed without relying on previous results. One possible approach is to define the path integral in Euclidean space (take time τ to be imaginary $= it$). The pole in the Euclidean Green's function will now be off of the axis and in the proper half-plane. Once the path integral is computed, the result is analytically continued back to real time.

The approach we will take is to stick with real time but reconsider the path integral over a finite time interval from t_0 to t and then take the limit $t, -t_0 \to \infty$ after the integral is computed. Instead of computing eq.(13.66) directly, notice that the Green's function we want to know in eq.(13.67) is the limiting value of the path integral

$$G_2(\eta, t; \eta_0, t_0) = \int \mathcal{D}\eta^* \mathcal{D}\eta\, \eta^*(t_1)\eta(t_2) \exp\left(i \int_{t_0}^{t} d\bar{t}\, \eta^*(i\partial_{\bar{t}} - E_p)\eta\right). \tag{13.68}$$

We will compute this path integral directly. Following the same steps that led to eq.(13.56), we divide the interval $[t_0, t]$ into $N+1$ equal parts. Let t_1 fall on the Jth time slice and t_2 on the Kth. Then,

$$\begin{aligned} G_2 &= \lim_{N\to\infty} \int d\eta_0^* \prod_{i=1}^{N} (d\eta_i d\eta_i^*)\, \eta_J^* \eta_K \prod_{j=0}^{N} \exp\Big(-\eta_j^*(\eta_{j+1} - (1 - i\epsilon E_p)\eta_j)\Big)\Big) \\ &= \lim_{N\to\infty} \int d\eta_0^* \prod_{i=1}^{N} (d\eta_i d\eta_i^*)\, \eta_J^* \eta_K \prod_{j=0}^{N} \Big(1 - \eta_j^*(\eta_{j+1} - (1 - i\epsilon E_p)\eta_j)\Big)\Big), \end{aligned} \tag{13.69}$$

where the ordering is the same as in eq.(13.51). The integrals over η_j^* are straightforward. For $j \neq J$, each integral serves to replace $1 - \eta_j^*(\eta_{j+1} - (1 - i\epsilon E_p)\eta_j)$ by $-\eta_{j+1} - (1 - i\epsilon E_p)\eta_j$. For $j = J$, the factor is replaced by 1 since $\eta_J^* \eta_J^* = 0$. Thus $((-1)^{2N+N(N+1)} = 1)$,

$$\begin{aligned} G_2 = &\lim_{N\to\infty} (-1)^J \int d\eta_1 \cdots d\eta_N\, \eta_K(\eta - (1 - i\epsilon E_p)\eta_N) \cdots \\ &\cdots (\eta_{J+2} - (1 - i\epsilon E_p)\eta_{J+1}) 1 (\eta_J - (1 - i\epsilon E_p)\eta_{J_1}) \cdots (\eta_1 - (1 - i\epsilon E_p)\eta_0). \end{aligned} \tag{13.70}$$

Now we must specify the time-ordering. Let us first do $K > J$, corresponding to $t_2 > t_1$. The first $N - K + 1$ integrals are identical to the computation leading to eq.(13.55), thus

$$\begin{aligned} G_2 = &\lim_{N\to\infty} (-1)^J (-1)^{2(N-K)+1} \int d\eta_1 \cdots d\eta_N (\eta - (1 - i\epsilon E_p)^{N-K+1} \eta_K) \\ &\eta_K(\eta_K - (1 - i\epsilon E_p)\eta_{K-1}) \cdots (\eta_{J+2} - (1 i\epsilon E_p)\eta_{J+1}) 1 \\ &(\eta_J - (1 - i\epsilon E_p)\eta_{J-1}) \cdots (\eta_1 - (1 - i\epsilon E_p)\eta_0). \end{aligned} \tag{13.71}$$

The integral over η_K picks out the factor proportional to η, and leaving a factor of η_{K-1} out in front of the remaining product,

$$\begin{aligned} G_2 = &\lim_{N\to\infty} (-1)^J (-1)^{2(N-K)+2} \int d\eta_1 \cdots d\eta_{K-1}\, \eta (1 - i\epsilon E_p)\eta_{K-1} \\ &(\eta_{K-1} - (1 - i\epsilon E_p)\eta_{K-2}) \cdots 1 \cdots (\eta_1 - (1 - i\epsilon E_p)\eta_0). \end{aligned} \tag{13.72}$$

Each integral over $d\eta_i$, down until $i = J+1$, results in another factor of $(-1)^2(1 - i\epsilon E_p)$, while the integral over η_{J+1} is just -1.

$$G_2 = \lim_{N\to\infty} (-1)^J(-1)^{2(N-K)+2}(-1)^{2(K-J-1)}$$

$$\int d\eta_1 \cdots d\eta_J \, \eta(1-i\epsilon E_p)^{K-J-1}(\eta_J-(1-i\epsilon E_p)\eta_{J-1})\cdots(\eta_1-(1-i\epsilon E_p)\eta_0). \tag{13.73}$$

The remaining integrals just contribute factors of (-1) (total $= (-1)^{J-1}$). Thus, the result for $K > J$ is

$$G_2 = \lim_{N\to\infty} -\eta(1 - i\epsilon E_p)^{K-J-1} = -\theta(t_2 - t_1)e^{-iE_p(t_2-t_1)}\eta. \tag{13.74}$$

For $J > K$ we follow the same procedure to get

$$G_2 = \lim_{N\to\infty} (1 - i\epsilon E_p)^{N-(J-K-1)}\eta_0 = e^{-iE_p(t-t_0)}\, e^{-iE_p(t_2-t_1)}\theta(t_1 - t_2). \tag{13.75}$$

All together we have

$$G_2(\eta,t;\eta_0,t_0) = e^{-iE_p(t_2-t_1)}\left(-\eta\theta(t_2 - t_1) + \eta_0\theta(t_1 - t_2)e^{-iE_p(t-t_0)}\right). \tag{13.76}$$

In the limit that $t, t_0 \to \infty$, the factor proportional to η_0 vanishes leaving eq.(13.74). Since

$$\langle\eta^*(t_1)\eta(t_2)\rangle \;\propto\; (-i)^2\left(\frac{E_p}{m}\right)^2 \frac{\delta}{\delta\zeta(t_1)}\frac{\delta}{\delta\zeta^*(t_2)}G[\zeta^*,\zeta],$$

then the single positive energy mode path integral, eq.(13.66), with the proper boundary conditions is

$$G[\zeta^*,\zeta] = \exp\left(-i\left(\frac{m}{E_p}\right)^2\int_{-\infty}^{\infty} dt\int_{-\infty}^{\infty} dt'\,\zeta^*(t)g_-(t,t')\zeta(t')\right)G[0,0], \tag{13.77}$$

where

$$g_-(t,t') = \theta(t-t')e^{-iE_p(t-t')}. \tag{13.78}$$

For a single negative energy mode, the 2-point function yields

$$\begin{aligned} &G_2[\eta,t;\eta_0,t_0] \\ &= \int \mathcal{D}\eta^*\mathcal{D}\eta\,\eta^*(t_1)\eta(t_2)\exp\left(i\int_{t_0}^{t} d\bar{t}\,\eta^*(\bar{t})(i\partial_{\bar{t}} + E_p)\eta(\bar{t})\right) \\ &= -\eta\theta(t_2 - t_1)e^{iE_p(t_2-t_1)} + \eta_0 e^{iE_p(t-t_0)}\, e^{-iE_p(t_2-t_1)}\theta(t_1 - t_2). \end{aligned} \tag{13.79}$$

In the limit $t, -t_0 \to \infty$, the term proportional to η_0 dominates. Thus,

$$G[\zeta^*, \zeta] = \int \mathcal{D}\eta^* \mathcal{D}\eta \exp\Big(i \int_{-\infty}^{\infty} d\bar{t}\, \eta^*(\bar{t})(i\partial_{\bar{t}} + E_p)\eta(\bar{t}) + \Big(\frac{m}{E_p}\Big)\Big(\zeta^*(\bar{t})\eta(\bar{t}) + \eta^*(\bar{t})\zeta(\bar{t})\Big)\Big)$$
$$= \exp\left(-i \Big(\frac{m}{E_p}\Big)^2 \int_{-\infty}^{\infty} dt \int_{-\infty}^{\infty} dt'\, \zeta^*(t) g_+(t,t')\zeta(t') \right) G[0,0], \tag{13.80}$$

where

$$g_+(t,t') = \theta(t'-t)e^{iE_p(t-t')}. \tag{13.81}$$

Combining the results, eqs.(13.77) and (13.80), the vacuum-to-vacuum amplitude in the presence of sources, eq.(13.65) is

$$Z_0[\tilde{\chi}^*, \tilde{\chi}] =$$
$$Z_0[0,0] \prod_{\vec{p}} \prod_{r=1}^{4} \exp\left(-i \Big(\frac{m}{E_p}\Big)^2 \int_{-\infty}^{\infty} dt \int_{-\infty}^{\infty} dt'\, \tilde{\chi}^*(\vec{p}, r, t) g(t,t',r) \tilde{\chi}(\vec{p}, r) \right) \tag{13.82}$$

where $g(t,t',r)$ is given by eq.(13.78) for $r = 1,2$ and by eq.(13.81) for $r = 3,4$.

Finally, we want to transform the functional dependence on $\tilde{\chi}^*(\vec{p}, r, t)$ and $\tilde{\chi}(\vec{p}, r, t')$ back to $\bar{\chi}(x)$ and $\chi(x')$. We begin by making use of the orthonormality of the w spinors,

$$\epsilon_r \bar{w}(\vec{p}, r) w(\vec{p}, r') = \delta_{rr'}, \tag{13.83}$$

so that

$$Z_0[\tilde{\chi}^*, \tilde{\chi}] = \exp\Big(i \int_{-\infty}^{\infty} dt \int_{-\infty}^{\infty} dt' \int d^3p \Big(\frac{m}{E_p}\Big)^2$$
$$\sum_{r,r'=1}^{2} \tilde{\chi}^*(\vec{p}, r, t) \bar{w}(\vec{p}, r) \theta(t-t') e^{-iE_p(t-t')} w(\vec{p}, r') \tilde{\chi}(\vec{p}, r', t')$$
$$- \sum_{r,r'=3}^{4} \tilde{\chi}^*(\vec{p}, r, t) \bar{w}(\vec{p}, r) \theta(t'-t) e^{iE_p(t'-t)} w(\vec{p}, r') \tilde{\chi}(\vec{p}, r', t') \Big) Z_0[0,0]. \tag{13.84}$$

Next we use

$$(\not{p} - \epsilon_r m) w(\vec{p}, r) = 0$$

to write

$$\sum_{r'=1}^{2}\sum_{r=1}^{2}\bar{w}(\vec{p},r)w(\vec{p},r') = \sum_{r,r'=1}^{4}\bar{w}(\vec{p},r)\left(\frac{\not{p}+m}{2m}\right)w(\vec{p},r'), \tag{13.85}$$

and

$$\sum_{r'=3}^{4}\sum_{r=3}^{4}\bar{w}(\vec{p},r)w(\vec{p},r') = \sum_{r,r'=1}^{4}\bar{w}(\vec{p},r)\left(\frac{\not{p}-m}{2m}\right)w(\vec{p},r'), \tag{13.86}$$

so that we may write

$$\begin{aligned} Z_0[\bar{\chi},\chi] = \exp\Bigg(i\int d^4x\int d^4x'\int d^3p\,\frac{m}{E_p}\sum_{r,r'=1}^{4}\tilde{\chi}^*(\vec{p},r,t)\bar{w}(\vec{p},r)e^{-i\vec{p}\cdot\vec{x}} \\ \left(\frac{\theta(t-t')}{2E_p}(\not{p}+m)e^{-ip\cdot(x-x')} - \frac{\theta(t'-t)}{2E_p}(\not{p}-m)e^{ip\cdot(x-x')}\right) \\ w(\vec{p},r')\tilde{\chi}(\vec{p},r',t')e^{i\vec{p}\cdot\vec{x}'}\Bigg)Z_0[0,0] \quad (13.87) \\ = \exp\left(i\int d^4x\int d^4x'\,\bar{\chi}(x)S_F(x-x')\chi(x')\right)Z_0[0,0], \end{aligned}$$

where $S_F(x-x')$ is the free spinor field propagator we found in chapters 4 and 10, eq.(4.51). As in the bosonic case, $Z_0[\bar{\chi},\chi]$ serves as a generating functional of time-ordered products of the field in the vacuum.

13.4 EXERCISES

1. Show by direct calculation that $G[\tilde{\phi},t;\tilde{\phi}_0,t_0]$, eq.(13.18), satisfies

$$\left(i\frac{\partial}{\partial t}-H\right)G[\tilde{\phi},t;\tilde{\phi}_0,t_0] = \delta[\tilde{\phi}-\tilde{\phi}_0]\delta(t-t_0),$$

   where H is the functional differential operator, eq.(10.11), in the momentum representation. This makes G the Green's functional of the Schrödinger functional differential equation.

2. Show by direct calculation that $G(\eta,t;\eta_0,t_0)$, eq.(13.56), satisfies

$$\left(i\frac{d}{d\eta}-H\right)G(\eta,t;\eta_0,t_0) = -\delta(\eta-\eta_0)\delta(t-t_0),$$

   where H is given by eq.(13.59). Note that the fermionic interpretation of the δ-functional will be required.

CHAPTER 14

Interacting Fields and Path Integrals

In the path integral representation of free field theory we concentrated on computing the vacuum-to-vacuum amplitude in the presence of an arbitrary source. Motivated by the results of the path integral for a forced harmonic oscillator, we demonstrated that this vacuum-to-vacuum amplitude acted as a generating functional for vacuum expectations of time-ordered products of the free fields. We also demonstrated that the free field theory path integrals factored into an infinite product of harmonic oscillator path integrals thereby giving us a particle interpretation.

In the operator representation we found that the key ingredients from an interacting field theory used to compute cross sections via a reduction formula were the vacuum expectations of time-ordered products of the interacting fields. These vacuum expectations were computed as a perturbation series, the first term being the appropriate free field result. Since we perturbed off of a free field, the interacting field inherited the particle interpretation provided by the free fields.

Guided by the development of the path integral for the anharmonic oscillator given at the end of chapter 12, we will translate the goals and results of interacting field theory in the operator representation into the path integral technology. We will accomplish this by defining and perturbatively solving the path integral representing the vacuum-to-vacuum amplitude for the interacting fields with arbitrary sources. A perturbation series for the time-ordered products can then be obtained by repeated functional differentiation of the vacuum-to-vacuum amplitude.

14.1 SELF-INTERACTING SCALAR FIELD THEORY

The objects of interest to calculate from an interacting field theory are the Green's functions, or vacuum expectations of the time-ordered products of the fields. From chapter 7, we know that these expectations are processed through a reduction formula to be transformed into S-matrix elements. The S-matrix elements in turn are used directly to compute cross sections.

Following the lead given by the free field path integral representation, we define the vacuum-to-vacuum amplitude with sources for the self-interacting scalar ϕ^4 theory as

$$Z[J] = \int \mathcal{D}\phi \exp\left(i \int d^4x \, \frac{1}{2}(\partial_\mu\phi\partial^\mu\phi - m_0^2\phi^2) - \frac{\lambda_0}{4!}\phi^4 + J(x)\phi(x)\right). \tag{14.1}$$

We expect, as in the free field case, that $Z[J]$ will generate the desired time-ordered products of the fields in the vacuum. Since we have no idea how to compute $Z[J]$ as it stands, we assume that λ_0 is "small," and follow the treatment of the weakly anharmonic oscillator in chapter 12. That is, we will compute $Z[J]$ perturbatively as a power series in λ_0. This merely means writing $\exp(i\int d^4x\lambda_0\phi^4)$ as a power series.

$$\begin{aligned} Z[J] &= \int \mathcal{D}\phi \exp\left(-i\int d^4x \, \frac{\lambda_0}{4!}\phi^4\right) \\ &\qquad \exp\left(i\int d^4x \, \partial_\mu\phi\partial^\mu\phi - m_0^2\phi^2 + J(x)\phi(x)\right) \\ &= \sum_{n=0}^{\infty} \frac{(-i\lambda_0/4!)^n}{n!} \int \mathcal{D}\phi \left(\int d^4x\phi^4(x)\right)^n \\ &\qquad \exp\left(i\int d^4x\partial_\mu\phi\partial^\mu\phi - m_0^2\phi^2 + J\phi\right) \end{aligned} \tag{14.2}$$

The key step in simplifying the expression above is to notice that the factor $(\int d^4x\phi^4(x))^n$ in the integrand may be replaced by functional differentiation with respect to the source.

$$\left(\int d^4x \, \phi^4(x)\right)^n \longrightarrow \left(\int d^4x \, (-i)^4 \frac{\delta^4}{\delta^4 J(x)}\right)^n \tag{14.3}$$

The functional derivatives may then be brought outside of the integral,

$$Z[J] = \sum_{n=0}^{\infty} \frac{(-i\lambda_0/4!)^n}{n!} \left(\int d^4x \, (-i)^4 \frac{\delta^4}{\delta^4 J(x)} \right)^n$$
$$\int \mathcal{D}\phi \, \exp\left(i \int d^4x \partial_\mu \phi \partial^\mu \phi - m_0^2 \phi^2 + J\phi \right). \quad (14.4)$$

Now, however, we quickly recognize the remaining path integral as the one we already calculated in the previous chapter, the free field vacuum-to-vacuum amplitude, $Z_0[J]$. Thus,

$$\begin{aligned} Z[J] &= \sum_{n=0}^{\infty} \frac{(-i\lambda_0/4!)^n}{n!} \left(\int d^4x \, (-i)^4 \frac{\delta}{\delta^4 J(x)} \right)^n Z_0[J] \\ &= \sum_{n=0}^{\infty} \frac{(-i\lambda_0/4!)^n}{n!} \left(\int d^4x \, (-i)^4 \frac{\delta}{\delta^4 J(x)} \right)^n Z_0[0] \\ &\quad \exp\left(-\frac{i}{2} \int d^4\bar{x} \int d^4x' J(\bar{x}) \Delta_F(\bar{x} - x') J(x') \right). \end{aligned} \quad (14.5)$$

The computation of the vacuum expectations of time-ordered products as a perturbation series is now a piece of cake: just functionally differentiate eq.(14.5).

$$\langle 0 | T(\varphi(x_1) \cdots \varphi(x_n)) | 0 \rangle = (-i)^n \frac{\delta}{\delta J(x_1)} \cdots \frac{\delta}{\delta J(x_n)} \frac{Z[J]}{Z[0]} \bigg|_{J=0} \quad (14.6)$$

For example, if we write the 2-point function as

$$\begin{aligned} &\langle 0 | T(\phi(x_1)\phi(x_2)) | 0 \rangle \\ &\quad = G^{(0)}(x_1, x_2) + \lambda_0 G^{(1)}(x_1, x_2) + \lambda_0^2 G^{(2)}(x_1, x_2) + \cdots, \end{aligned}$$

then

$$\begin{aligned} G^{(0)}(x_1, x_2) &= (-i)^2 \frac{\delta}{\delta J(x_1)} \frac{\delta}{\delta J(x_2)} \\ &\quad \exp\left(-\frac{i}{2} \int d^4x \, d^4x' \; J(x) \Delta_F(x - x') J(x') \right) \bigg|_{J=0} \\ &= i\Delta_F(x_1 - x_2), \end{aligned} \quad (14.7)$$

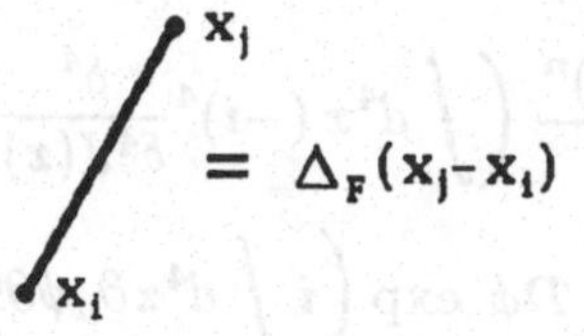

Figure 14.1 Graphical representation of the free field 2-point function or propagator.

the free field result. At first order in λ_0 we have

$$G^{(1)}(x_1, x_2) = (-i)^2 \frac{\delta}{\delta J(x_1)} \frac{\delta}{\delta J(x_2)} \frac{1}{Z[0]}$$
$$(-i) \int d^4y \, \frac{\delta^4}{\delta J^4(y)} \exp\left(-\frac{i}{2} \int d^4x \, d^4x' \; J(x)\Delta_F(x - x')J(x') \right)\Bigg|_{J=0}$$
$$= -\frac{1}{2} \int d^4y \, \Delta_F(y - y)\Delta_F(x_1 - y)\Delta_F(x_2 - y), \tag{14.8}$$

and so on.

We can generate a diagrammatic representation of the series that is identical to the one in chapter 7 (except for the mass shift and normal ordering) by identifying a solid line connecting the points x_i and x_j with each factor of $\Delta_F(x_i - x_j)$ as in figure 14.1. The series representing the 2-point function $G_2(x_1, x_2)$ through $G_2^{(1)}(x_1, x_2)$, eq.(14.8), is drawn in figure 14.2.

$Z[J]$ generates disconnected as well as connected diagrams. As in the free field case, we can define a functional, $W[J]$, that generates only connected diagrams or terms. In fact, the definition is identical to that of the free field case, eq.(13.35).

$$Z[J] = Z[0] \exp(iW[J]) \tag{14.9}$$

In terms of the free field result,

$$W_0[J] = -\frac{1}{2} \int d^4x \, d^4x' \; J(x)\Delta_F(x - x')J(x'), \tag{14.10}$$

and eq.(14.2), we have

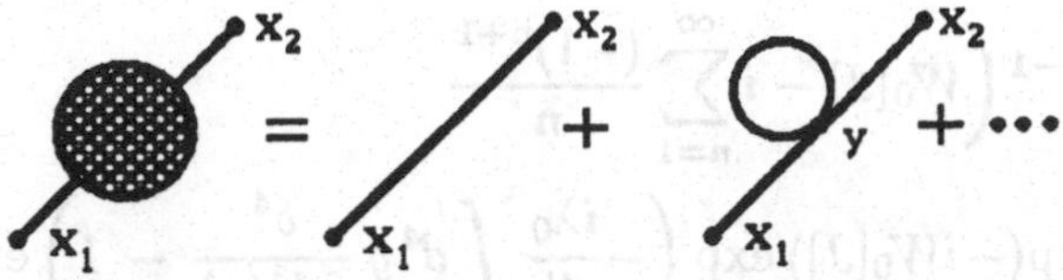

Figure 14.2 Graphical representation of the series expansion of the interacting field 2-point function through first order in λ_0 as generated by $Z[J]$, eq.(14.5). $Z[J]$ generates both connected and disconnected terms.

$$
\begin{aligned}
&\exp(iW[J]) \\
&\quad = Z[0]^{-1}\exp\left(-\frac{i\lambda_0}{4!}\int d^4y\,\frac{\delta^4}{\delta J^4(y)}\right)\exp(iW_0[J]) \qquad (14.11)\\
&\quad = Z[0]^{-1}\exp(iW_0[J])\exp(-iW_0[J]) \\
&\qquad\qquad \exp\left(-\frac{i\lambda_0}{4!}\int d^4y\,\frac{\delta^4}{\delta J^4(y)}\right)\exp(iW_0[J]).
\end{aligned}
$$

Thus,

$$
\begin{aligned}
iW[J] &= Z[0]^{-1}\Bigg(iW_0[J] \\
&\quad + \ln\left(\exp(-iW_0[J])\exp\left(-\frac{i\lambda_0}{4!}\int d^4y\,\frac{\delta^4}{\delta J^4(y)}\right)\exp(iW_0[J])\right)\Bigg) \\
&= Z[0]^{-1}\Bigg(iW_0[J]+ \\
&\ln\left(1+\exp(-iW_0[J])\exp\left(-\frac{i\lambda_0}{4!}\int d^4y\,\frac{\delta^4}{\delta J^4(y)}-1\right)\exp(iW_0[J])\right)\Bigg).
\end{aligned}
\tag{14.12}
$$

Since we are considering λ_0 to be "small," we can use the series expansion of $\ln(1+x)$ to eventually write $W[J]$ as a perturbative expansion in powers of λ_0.

$$
\begin{aligned}
W[J] =& Z[0]^{-1}\Bigg(W_0[J] - i\sum_{n=1}^{\infty}\frac{(-1)^{n+1}}{n} \\
&\left(\exp(-iW_0[J])\exp\left(-\frac{i\lambda_0}{4!}\int d^4y\,\frac{\delta^4}{\delta J^4(y)} - 1\right)\exp(iW_0[J])\right)^n\Bigg) \\
=& Z[0]^{-1}\Bigg(W_0[J] + i\sum_{n=1}^{\infty}\frac{(-1)^n}{n} \\
&\left(\exp(-iW_0[J])\sum_{m=1}^{\infty}\left(\frac{-i\lambda_0}{4!}\right)^m\frac{1}{m!}\left(\int d^4y\,\frac{\delta^4}{\delta J^4(y)}\right)^m \exp(iW_0[J])\right)^n\Bigg)
\end{aligned}
\tag{14.13}
$$

To simplify the expression, insert $1 = \exp(iW_0[J])\exp(-iW_0[J])$ between each factor of $\int d^4y\,\delta^4/\delta J^4(y)$. Then,

$$
\begin{aligned}
W[J] =& \\
Z[0]^{-1}\Bigg(&W_0[J] + i\sum_{m=1}^{\infty}\sum_{n=1}^{m}\sum_{l=0}^{n-1}\frac{(-1)^{n+l}}{n}\,\frac{n!}{l!(n-l)!}\,\frac{1}{m!}\,(n-l)^m\left(\frac{-i\lambda_0}{4!}\right)^m \\
&\left(\exp(-iW_0[J])\left(\int d^4y\,\frac{\delta^4}{\delta J^4(y)}\right)\exp(iW_0[J])\right)^m\Bigg) \\
=& Z[0]^{-1}\Bigg(W_0[J] - \frac{\lambda_0}{4!}\Bigg(\int d^4y\; - 3\Delta_F^2(y-y) \\
&+ i\Delta_F(y-y)\left(\int d^4x\,J(x)\Delta_F(x-y)\right)^2 \\
&+ \left(\int d^4x\,J(x)\Delta_F(x-y)\right)^4\Bigg) + \cdots\Bigg)
\end{aligned}
\tag{14.14}
$$

One can verify directly that $W[J]$ only generates connected terms. For the 2-point function to first order in λ_0 we have

$$
\begin{aligned}
&G^c(x_1, x_2)\\
&= (-i)^2 \frac{\delta}{\delta J(x_1)} \frac{\delta}{\delta J(x_2)} (iW_0[J]) \Big|_{J=0}\\
&= i\Delta_F(x_1 - x_2) - \frac{\lambda_0}{2} \int d^4y\, \Delta_F(y-y)\, \Delta_F(x_1 - y)\, \Delta_F(x_2 - y).
\end{aligned}
\tag{14.15}
$$

We can just as easily generate the expansion of the n-point function in momentum space by Fourier transforming the sources in the generating functionals. For example, the generating functional for the connected Green's functions, $W[J]$, becomes

$$
\begin{aligned}
&W[\tilde{J}] =\\
&Z[0]^{-1} \left(W_0[\tilde{J}] + i \sum_{m=1}^{\infty} \sum_{n=1}^{m} \sum_{l=0}^{n-1} \frac{(-1)^{n+l}}{n} \frac{n!}{l!(n-l)!} \frac{1}{m!} (n-l)^m \left(\frac{-i\lambda_0}{4!} \right)^m \right.\\
&\left(\exp(iW_0[\tilde{J}]) \int \frac{d^4k_1}{(2\pi)^4} \cdots \frac{d^4k_4}{(2\pi)^4} (2\pi)^4\, \delta^4(k_1 + k_2 + k_3 + k_4) \right.\\
&\left.\left. \frac{\delta}{\delta \tilde{J}(k_1)} \cdots \frac{\delta}{\delta \tilde{J}(k_4)} \exp(iW_0[\tilde{J}]) \right)^m \right),
\end{aligned}
\tag{14.16}
$$

where

$$
W_0[\tilde{J}] = -\frac{1}{2} \int \frac{d^4p}{(2\pi)^4} \tilde{J}(-p) \tilde{\Delta}_F(p) \tilde{J}(p). \tag{14.17}
$$

The 2-point function in the momentum representation, $\tilde{G}^c(p_1, p_2)$, to lowest order is

$$
\begin{aligned}
\tilde{G}_c^{(0)}(p_1, p_2) &= (-i)^2 \frac{\delta}{\delta \tilde{J}(p_1)} \frac{\delta}{\delta \tilde{J}(p_2)} \left(-\frac{i}{2} \int \frac{d^4p}{(2\pi)^4} \tilde{J}(-p) \tilde{\Delta}_F(p) \tilde{J}(p) \right)\\
&= i \frac{\delta}{\delta \tilde{J}(p_2)} \left(\tilde{J}(-p_1) \tilde{\Delta}_F(p_1) \right)\\
&= i \tilde{\Delta}_F(p_1) (2\pi)^4 \delta^4(p_1 + p_2),
\end{aligned}
\tag{14.18}
$$

which is eq.(7.122). The higher-order terms and 4-point functions are left as exercises.

14.2 SPINOR QUANTUM ELECTRODYNAMICS

The same methods and ideas used for the scalar field theory of the previous section may equally be applied to QED. We will work in the Coulomb gauge. We learned from the previous chapter that working in the Coulomb gauge means that we couple the source only to the gauge invariant part of the vector potential. By an unimportant redefinition of the normalization in front of $Z[J]$, this is equivalent to inserting the δ-functionals $\delta[\vec{A}_L]\delta[A^0]$ into the functional integral. See eqs. (13.36)-(13.37). This is intuitively appealing since in the absence of any charges the classical fields in the Coulomb gauge satisfy $A^0 = 0$, and $\vec{A}_L = 0$. However, in the presence of charges, we will still have $\vec{A}_L = 0$ in the Coulomb gauge, but the field no longer satisfies $A^0 = 0$ because of Gauss's law. Therefore, in defining the path integral we will couple the source to A^0 in addition to $\vec{A}_T$. This is equivalent to inserting only $\delta[\vec{A}_L]$ into the functional integral. Since the action is

$$S = \int d^4x \, \frac{1}{2}(\vec{E}^2 - \vec{B}^2) + \bar{\psi}(i\not{D} - m_0)\psi, \tag{14.19}$$

where $\not{D} = \not{\partial} + ie_0\not{A}$ is the gauge covariant derivative, then the vacuum-to-vacuum amplitude, complete with sources, is

$$Z[J, \bar{\chi}, \chi] = N \int \mathcal{D}A^\mu \mathcal{D}\bar{\psi}\mathcal{D}\psi\delta[\vec{A}_L] \exp\Big(i \int d^4x \, \frac{1}{2}(\vec{E}^2 - \vec{B}^2) + \bar{\psi}(i\not{\partial} - m_0)\psi - e_0\bar{\psi}\not{A}\psi + J_\mu A^\mu + \bar{\chi}\psi + \bar{\psi}\chi\Big), \tag{14.20}$$

where $\vec{E} = -\nabla A^0 - \partial_t \vec{A}_T$. Now we pull the same maneuver as in the previous section that led to eq.(14.5) (which is embodied in eq.(14.3)), and replace the fields in the interaction portion of the action by functional derivatives.

$$Z[J, \bar{\chi}, \chi] = N \exp\left(-ie_0 \int d^4y \, \gamma^\mu \frac{\delta}{\delta J^\mu(y)} \frac{\delta}{\delta \chi(y)} \frac{\delta}{\delta \bar{\chi}(y)}\right) \int \mathcal{D}A^\mu \, \mathcal{D}\bar{\psi}\mathcal{D}\psi \, \delta[\vec{A}_L] \exp\Big(i \int d^4x \, \frac{1}{2}(\vec{E}^2 - \vec{B}^2) + \bar{\psi}(i\not{\partial} - m_0)\psi + J_\mu A^\mu + \bar{\chi}\psi + \bar{\psi}\chi\Big) \tag{14.21}$$

The remaining integral over ψ and $\bar{\psi}$ in eq.(14.21) is identical to the path integral we did for the free field case in the previous chapter, eq.(13.63).

Using the result, eq.(13.87), we have

$$
\begin{aligned}
Z[J,\bar{\chi},\chi] = N' \exp\left(-ie_0 \int d^4y\, \gamma^\mu \frac{\delta}{\delta J^\mu(y)} \frac{\delta}{\delta \chi(y)} \frac{\delta}{\delta \bar{\chi}(y)}\right) \\
\exp\left(-\int d^4x \int d^4x'\, \bar{\chi}(x) S_F(x-x')\chi(x')\right) \\
\int \mathcal{D}A^\mu\, \delta[\vec{A}_L] \exp\left(i\int d^4x\, \frac{1}{2}(\vec{E}^2+\vec{B}^2) + J_\mu A^\mu\right).
\end{aligned}
\tag{14.22}
$$

The δ-functional in the integral over A^μ guarantees that $\nabla \cdot \vec{A} = 0$, thus the integral over A^0 decouples from the integral over $\vec{A}_T$, the transverse potential. The only apparent coupling between A^0 and $\vec{A}_T$ is in the term $\vec{E}^2$, and this coupling vanishes in the Coulomb gauge.

$$
\begin{aligned}
\int d^4x\, \vec{E}^2 &= \int d^4x (\nabla A^0 + \partial_t \vec{A}_T)^2 \\
&= \int d^4x (\nabla A^0)^2 + 2(\nabla A^0 \cdot \partial_t \vec{A}_T) + (\partial_t \vec{A}_T)^2 \\
&= \int d^4x (\nabla A^0)^2 + (\partial_t \vec{A}_T)^2 - 2(A^0 \partial_t (\nabla \cdot \vec{A}_T)) \\
&= \int d^4x (\nabla A^0)^2 + (\partial_t \vec{A}_T)^2
\end{aligned}
\tag{14.23}
$$

Therefore, the remaining integral in eq.(14.22) can be rewritten as

$$
\begin{aligned}
\int \mathcal{D}A^\mu \delta[\vec{A}_L] \exp\left(i\int d^4x\, \frac{1}{2}(\vec{E}^2+\vec{B}^2) + J_\mu A^\mu\right) \\
= \int \mathcal{D}A^0 \exp\left(i\int d^4x\, -\frac{1}{2}A^0 \nabla^2 A^0 + J_0 A^0\right) \\
\int \mathcal{D}\vec{A}_T \exp\left(i\int d^4x\, \frac{1}{2}(\vec{E}_T^2+\vec{B}^2) + \vec{J}\cdot\vec{A}_T\right).
\end{aligned}
\tag{14.24}
$$

We now recognize the integral over $\vec{A}_T$ as being identical to the free photon Coulomb gauge integral that we computed in the previous chapter, eq.(13.44). The computation of the integral over A^0 is also straightforward: just Fourier transform and complete the squares. Thus, the vacuum-to-vacuum amplitude is

$$
Z[J,\bar{\chi},\chi] = Z[0,0,0] \exp\left(-ie_0 \int d^4y\, \gamma^\mu \frac{\delta}{\delta J^\mu(y)} \frac{\delta}{\delta \chi(y)} \frac{\delta}{\delta \bar{\chi}(y)}\right)
$$

$$\exp\left(-\int d^4x \int d^4x' \bar{\chi}(x) S_F(x-x')\chi(x') + \frac{i}{2} J^j(x) D^{\rm tr}_{jk}(x-x') J^k(x') \right.$$
$$\left. + \frac{i}{2} J^0(x)\Delta_0(x-x')J^0(x') \right), \tag{14.25}$$

where the propagator for A^0 is

$$\begin{aligned} \Delta_0(x-x') &= \int \frac{d^4k}{(2\pi)^4}\frac{1}{|\vec{k}^2|} e^{-ik\cdot(x-x')} \\ &= \frac{\delta(t-t')}{4\pi|\vec{x}-\vec{x}'|}. \end{aligned} \tag{14.26}$$

Recall from chapter 5 that we put $D^{\rm tr}_{jk}(x-x')$ into a more useful form, which separated the covariant from the noncovariant parts by introducing the coordinates of the orthonormal frame in which quantization took place. The result is (eq.(5.40)),

$$D^{\rm tr}_{\mu\nu}(x-x') =$$
$$\int \frac{d^4k}{(2\pi)^4}\frac{e^{-ik\cdot(x-x')}}{k^2+i\epsilon}\left(-g_{\mu\nu} - \frac{k^2\eta_\mu\eta_\nu - (k\cdot\eta)(k_\mu\eta_\nu + k_\nu\eta_\mu) + k_\mu k_\nu}{(k\cdot\eta)^2 - k^2}\right). \tag{14.27}$$

η_μ is a time-like unit vector which in the frame of quantization is (1,0,0,0). The result of the path integral over A^0 is written in this frame. Written in the frame expressed in eq.(14.27), the result of the A^0 integration is

$$J^0(x)\Delta_0(x-x')J^0(x') = J^\mu(x)\eta_\mu\eta_\nu\Delta_0(x-x')J^\nu(x'). \tag{14.28}$$

However, in the frame of quantization we also have $(k\cdot\eta)^2 = k_0^2$ so that $(k\cdot\eta)^2 - k^2 = |\vec{k}|^2$. For $\epsilon \to 0^+$, the term proportional to $\eta_\mu\eta_\nu$ in $D^{\rm tr}_{\mu\nu}(x-x')$ is

$$-\int \frac{d^4k}{(2\pi)^4}\frac{e^{-ik\cdot(x-x')}}{|\vec{k}|^2}\eta_\mu\eta_\nu = -\Delta_0(x-x')\eta_\mu\eta_\nu. \tag{14.29}$$

The A^0 integration contributes a factor which just cancels the $\eta_\mu\eta_\nu$ term in $D^{\rm tr}_{\mu\nu}(x-x')$. Of course, since J^0 couples to $\rho = \gamma^0\bar{\psi}\psi$ in the interaction, we recognize the contribution from the A^0 integration as just the instantaneous Coulomb energy. This piece fits neatly into $D^{\rm tr}_{\mu\nu}(x-x')$ to fill in the $\mu=\nu=0$ component to build the covariant part, $D_{\mu\nu}(x-x')$, the term proportional to $g_{\mu\nu}$ in eq.(14.27). We can combine eq.(14.28) with eq.(14.27) to rewrite the vacuum-to-vacuum amplitude as

$$Z[J,\bar{\chi},\chi] = Z[0,0,0]\exp\left(-ie_0\int d^4y\,\gamma^\mu \frac{\delta}{\delta J^\mu(y)}\frac{\delta}{\delta\chi(y)}\frac{\delta}{\delta\bar{\chi}(y)}\right)$$

$$\exp\left(-\int d^4x\int d^4x'\,\bar{\chi}(x)S_F(x-x')\chi(x') + \frac{i}{2}J^\mu(x)\bar{D}^{\rm tr}_{\mu\nu}(x-x')J^\nu(x')\right), \tag{14.30}$$

where

$$\bar{D}^{\rm tr}_{\mu\nu}(x-x') = \int \frac{d^4k}{(2\pi)^4}\frac{e^{-ik\cdot(x-x')}}{k^2+i\epsilon}\left(-g_{\mu\nu}+\frac{(k\cdot\eta)(k_\mu\eta_\nu+k_\nu\eta_\mu)-k_\mu k_\nu}{(k\cdot\eta)^2-k^2}\right). \tag{14.31}$$

In all practical calculations, we can drop the second term in eq.(14.31) above and replace $\bar{D}^{\rm tr}_{\mu\nu}(x-x')$ with $-g_{\mu\nu}D(x-x')$, the covariant Lorentz gauge propagator, in $Z[J,\bar{\chi},\chi]$, eq.(14.30). This substitution is valid because free photons are transverse, and the electric current is conserved. See the discussion in chapter 8 surrounding figure 8.3.

For small e_0, we arrive at the perturbative expansion of $Z[J,\bar{\chi},\chi]$ by expanding the first exponential of functional derivatives. By direct calculation, one can see that $Z[J,\bar{\chi},\chi]$ generates disconnected as well as connected terms. The generating functional for connected terms only, $W[J,\bar{\chi},\chi]$, is defined in the same way as in the scalar case,

$$Z[J,\bar{\chi},\chi] = Z[0,0,0]\ \exp(iW[J,\bar{\chi},\chi]). \tag{14.32}$$

Let us denote the free field result as $W_0[J,\bar{\chi},\chi]$,

$$W_0[J,\bar{\chi},\chi] =$$

$$-\int d^4x\int d^4x'\Big(\frac{1}{2}J^\mu(x)D_{\mu\nu}(x-x')J^\nu(x') + i\bar{\chi}(x)S_F(x-x')\chi(x')\Big). \tag{14.33}$$

Following the same development as in the scalar case, we arrive at the perturbative expansion for $W[J,\bar{\chi},\chi]$

$$W[J,\bar{\chi},\chi] = Z[0,0,0]^{-1}$$

$$\left(W_0[J,\bar{\chi},\chi] + i\sum_{m=1}^{\infty}\sum_{n=1}^{m}\sum_{l=0}^{n-1}\frac{(-1)^{n+l}}{n}\frac{n!}{l!(n-l)!}\frac{1}{m!}(n-l)^m(-ie_0)^m\right.$$

$$\left.\left(\exp(-iW_0[J,\bar{\chi},\chi])\int d^4y\,\gamma^\mu\frac{\delta}{\delta J^\mu(y)}\frac{\delta}{\delta\chi(y)}\frac{\delta}{\delta\bar{\chi}(y)}\exp(iW_0[J,\bar{\chi},\chi])\right)\right). \tag{14.34}$$

The computation of the first two nonvanishing terms in the 2-point functions for the electron and photon for comparison with the results obtained in the operator formalism in chapter 8 is left as an exercise.

14.3 EXERCISES

1. Compute $Z[0]$ to order λ_0^3 from eq.(14.5) and draw the diagrams corresponding to each term. What is the effect, in terms of diagrams, of dividing by $Z[0]$ in eq.(14.6) or in the normalization of $W[J]$ to $W[0] = 1$ in eq.(14.14)?
2. Derive the perturbative expansion of $W[J]$, eq.(14.14), from the expression, eq.(14.13).
3. Compute the connected 4-point function through order λ_0^3 from $W[J]$, eq.(14.14). Do the same using eq.(14.16) in the momentum representation. Compare with the 4-point function generated by $Z[J]$.
4. Compute the $O(\lambda_0^2)$ contributions to the connected 2-point function using $W[J]$.
5. Write down the generating functional, $W[J]$, for a scalar field theory with $V = \nu_0\varphi^3/3!$. Compute the 2-point function through order ν_0^2, and the 3-point function through order ν_0^3. Draw the diagrammatic representation of each term.
6. Compute $Z[0, 0, 0]$ from eq.(14.30) through order e_0^2. Draw the corresponding diagrams.
7. Compute the e_0^3 contributions to the vertex function and compare with the result in the operator representation.

CHAPTER 15

Yang-Mills and Faddeev-Popov

A difficulty arises in defining the path integral representing the vacuum-to-vacuum amplitude when a local symmetry, such as a gauge symmetry, is present in the classical action. The path integral is even more divergent than usual. The problem is as follows: Suppose we are quantizing a classical field theory with generic gauge dependent field $A_\mu(x)$. The action, $S[A_\mu]$, is gauge invariant. The naive path integral, the path integral suggested by scalar field theory, is $Z[J] = \int \mathcal{D}A_\mu \exp(iS[A_\mu, J_\mu])$. In this path integral we are integrating over all configurations of the field A_μ, and each configuration in the sum is related to an infinite number of others by gauge transformations. However, the integrand is identical for two gauge equivalent configurations, hence the two configurations contribute the same information. This means that by integrating over all configurations, we are computing an infinite number of copies of some integral. Put in a slightly different fashion, $A_\mu(x)$ contains gauge dependent and gauge invariant components. When we integrate over the space of configurations, $\int \mathcal{D}A_\mu$, we are integrating with respect to both gauge invariant and gauge dependent parts. Yet, the action, which is gauge invariant, depends only on the gauge invariant part of A_μ. The integrand for the gauge dependent parts is unity, and so the integral over the gauge dependent parts diverges. This is the extra divergence we want to remove.

One solution to this problem was offered in chapter 13 as a cheap but workable solution for free photons. In that case we coupled the source to the gauge invariant part of the vector potential, A_μ. The integral over the gauge dependent components was independent of the source and casually thrown into the normalization and forgotten. There are three reasons why this isn't satisfactory in general. The first is that while QED is simple enough so that we could easily identify the gauge dependent components, the same is not true for Yang-Mills. Secondly, this method put us in the Coulomb gauge which is not manifestly covariant. What if we wanted or needed to work in a covariant gauge? Finally, how do we switch gauges and know that we have quantized properly to get the same physical results?

A procedure for properly defining the path integral in any acceptable linear gauge was developed by Faddeev and Popov. The idea is to pick a gauge and change coordinates. When we pick a gauge, then some of the configurations A^μ will satisfy the gauge condition, but most will not. Let $\bar{A}^\mu$ denote a configuration that does satisfy the gauge choice. If A^μ is a configuration that does not satisfy the gauge condition ($A^\mu \neq \bar{A}^\mu$), then there is a gauge transformation, U, that takes one from $\bar{A}^\mu$ to A^μ, $A^\mu = U[\bar{A}^\mu]$. The desired change of coordinates is from a general A^μ to the set $(\bar{A}^\mu, U)$. The change of coordinates will involve a functional Jacobian, J, which must be determined, $\int \mathcal{D}A_\mu \to \int \mathcal{D}\bar{A}^\mu \mathcal{D}U\, J$. Since the integrand is gauge invariant, then it cannot depend on U, and the $\int \mathcal{D}U$ integral may be pulled out and placed in the normalization. The integral $\int \mathcal{D}U$ is called the "gauge volume," and this is the infinity which we wish to extract to properly define the path integral. Since we are writing the integral as $\int \mathcal{D}U \int \mathcal{D}\bar{A}^\mu$ of something, we are integrating the "something" over all gauge transformations. Therefore, no matter what gauge we choose by this procedure (provided, of course, that it results in a nonvanishing Jacobian), the end result must be gauge invariant.

The Faddeev-Popov procedure is just a change of coordinates, but since it is for a functional integral, the way the change is actually carried out may at first appear unfamiliar. Instead of jumping directly to Yang-Mills, let us begin with a very simple example. This will allow us to define, in a familiar setting, the notation often used and to introduce the slight variations in which the procedure is practiced. Following the simple example we will consider QED and recover the result of chapter 13 and derive the path integral in the covariant Lorentz gauge. Finally, we will apply the procedure to Yang-Mills where its utility will become most apparent. We will need to apply the procedure again to define the string "path" integral in future chapters.

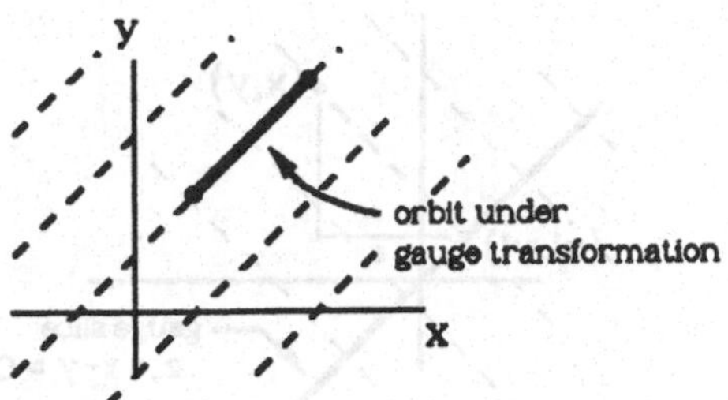

Figure 15.1 The motion of a "configuration," (x, y), in the configuration space, $\mathbb{R}^2$, under the "gauge" transformation defined by eq.(15.3). The path is called a gauge orbit. In the simple example here, the gauge orbits are lines of constant $x - y$.

15.1 A SIMPLE EXAMPLE

Consider the following integral:

$$Z = \int_{-\infty}^{\infty} dx \int_{-\infty}^{\infty} dy \; e^{-(x-y)^2}. \tag{15.1}$$

This integral is divergent since the integrand depends only on the difference between x and y. The size of the divergence is easily determined by switching variables from (x, y) to $z_- = x - y$ and $z_+ = (x + y)/2$. Then,

$$Z = \int_{-\infty}^{\infty} dz_+ \int_{-\infty}^{\infty} dz_- \; e^{-z_-^2} = \sqrt{\pi} \int dz_+. \tag{15.2}$$

The size of the divergence is the volume of the reals, i.e. the 1-dimensional real line.

Now let's cast this problem into a different setting. The integrand is translationally invariant so that the transformation

$$\begin{aligned} x &\to x + a \\ y &\to y + a, \end{aligned} \tag{15.3}$$

where $a \in \mathbb{R}$ leaves the integrand invariant. Let's call the symmetry operation, eq.(15.3), a gauge transformation. The "action," $(x - y)^2$, is gauge invariant. a is an element of the gauge group, the reals. The gauge volume is $\int da$, which is the volume of the real line. A gauge "orbit," the path of a "configuration" (x, y) through $\mathbb{R}^2$ under a gauge transformation, is a line of constant $x - y$ as shown in figure 15.1. Since the action is invariant along a gauge orbit, we can easily see that by integrating over all configurations, the entire (x, y) plane, we are redundantly adding together an infinite number of copies of 1 finite Gaussian integral. We want to develop a procedure for extracting this finite integral from Z.

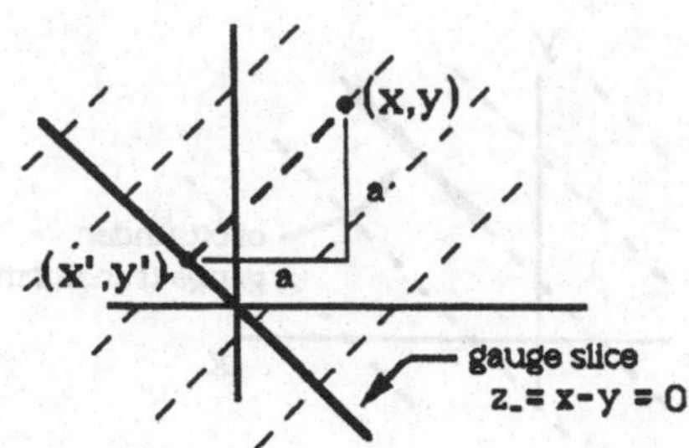

Figure 15.2 The gauge choice $x + y = 0$ defines a "gauge slice" through the configuration space. (x', y') is a configuration on the slice, that is, it satisfies the gauge condition. (x, y) is a gauge equivalent configuration, since both (x, y) and (x', y') reside on the same gauge orbit. a is the gauge transformation that takes us from the slice to (x, y).

In terms of the "gauge" words, it is clear that the transformation in eq.(15.2) just expresses the separation of the configuration (x, y) into its gauge invariant component, z_-, and its gauge dependent component, z_+. To put eq.(15.2) in a form in which we can generalize a little, we can change variables from (x, y) to (z_-, a) instead of (z_-, z_+). Here a is the gauge transformation that takes a point on the line $z_+ = x + y = 0$ to (x, y) as shown in figure 15.2. In other words, we trade the gauge dependent coordinate z_+ for the gauge transformation that takes us to it.

Under this change of variables, Z is

$$Z = \int da \int dz_-\, e^{-z_-^2} = \int da\, \sqrt{\pi}, \tag{15.4}$$

where the gauge volume, $\int da$, now appears explicitly. Now let's rewrite the Gaussian integral above once more.

$$Z = \int da \int dx\, dy\, 2\delta(x + y)\, e^{-(x-y)^2} = \int da\, \sqrt{\pi} \tag{15.5}$$

In this form, we can interpret the δ-function as a gauge-fixing term. That is, we have fixed the gauge such that $x + y = 0$, something analogous to the Coulomb gauge. Let us call the choice of gauge, $x + y = 0$, the gauge slice. The factor of 2 that has appeared is the Jacobian of the change of coordinates from (x, y) to a variable that runs along the gauge slice along with the gauge transformation that takes you off the gauge slice to the

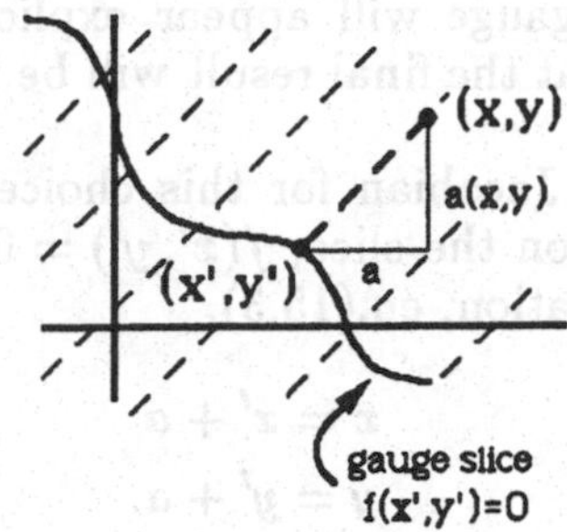

Figure 15.3 Illustration of a general choice of gauge, $f(x, y) = 0$. The desired change of coordinates is from (x, y) to (s, a), s being a variable that runs along the slice, and a being the gauge transformation that runs from the slice to (x, y).

point (x, y).

Suppose we want to make a different gauge choice, $f(x, y) = 0$, as in figure 15.3. What will be the generalization of eq.(15.5)? Let s be a variable that runs along the gauge slice. $\partial/\partial s$ is a vector tangent to f, $df/ds = 0$. Again, let $a(x, y)$ be the gauge transformation that takes the particular point on the gauge slice where the gauge orbit crosses the slice to the point (x, y) off the slice. Change variables in Z from (x, y) to (s, a).

$$Z = \int da\, ds\, J\, e^{-h(s)}, \tag{15.6}$$

where the Jacobian, J is

$$J = \det \begin{vmatrix} \frac{\partial x}{\partial s} & \frac{\partial x}{\partial a} \\ \frac{\partial y}{\partial s} & \frac{\partial y}{\partial a} \end{vmatrix}. \tag{15.7}$$

So far our choice of s is arbitrary in that we can choose s to increment at any rate we wish as we march along the slice. The rate at which s changes along the slice need not be uniform either. For convenience, we will choose the scale of s to be defined by

$$\int dx\, dy\, \delta(f(x, y)) = \int ds. \tag{15.8}$$

The rate of change of this s as we move along the gauge slice will not be uniform unless the gauge choice, f is linear in x and y. The reason this choice is desirable is that we can substitute eq.(15.8) back into eq.(15.6) so that the choice of gauge will appear explicitly in the integral. The Jacobian will insure that the final result will be independent of the choice of gauge.

Let's compute the Jacobian for this choice of s. Let (x', y') be the coordinates of a point on the slice, $f(x', y') = 0$. Then, by the definition of the gauge transformation, eq.(15.3),

$$\begin{aligned} x &= x' + a \\ y &= y' + a. \end{aligned} \tag{15.9}$$

Since x' and y' reside on the slice, they must be functions of s only, $x' = x'(s)$, $y' = y'(s)$. Therefore,

$$\frac{\partial x}{\partial a} = 1 = \frac{\partial y}{\partial a}$$

so

$$J = \left| \frac{\partial x}{\partial s} - \frac{\partial y}{\partial s} \right| = \left| \frac{\partial x'}{\partial s} - \frac{\partial y'}{\partial s} \right| .$$

f is constant along the gauge slice, so $df/ds = 0$.

$$0 = \frac{df}{ds} = \frac{\partial f}{x'} \frac{\partial x'}{\partial s} + \frac{\partial f}{\partial y'} \frac{\partial y'}{\partial s} \tag{15.10}$$

A particular solution to eq.(15.10) that satisfies eq.(15.8) can be found by carrying out the integral on the left-hand side of eq.(15.8) with respect to either x or y. Using the property of the δ-function expressed in eq.(3.14) and integrating eq.(15.8) with respect to y,

$$\int dx\, dy \left| \frac{df}{dy} \right|^{-1}_{y=y'} \delta(y - y') = \int dx \left| \frac{df}{dy} \right|^{-1}_{y=y'} = \int ds,$$

so that

$$\frac{\partial x'}{\partial s} = \pm \left| \frac{\partial f}{\partial y'} \right| .$$

If we choose the + sign, then eq.(15.10) implies

$$\frac{\partial x'}{\partial s} = \frac{\partial f}{\partial y'}, \qquad \frac{\partial y'}{\partial s} = -\frac{\partial f}{\partial x'} . \tag{15.11}$$

Therefore, the Jacobian we want is

$$J = \left| \frac{\partial f}{\partial y'} + \frac{\partial f}{\partial x'} \right| = \left| \frac{\partial f}{\partial y} + \frac{\partial f}{\partial x} \right|_{f=0} , \tag{15.12}$$

and the proper definition of Z in the presence of the symmetry is

$$Z = \int da \int dx\, dy\, \delta(f(x,y)) \left| \frac{\partial f}{\partial y} + \frac{\partial f}{\partial x} \right| e^{-(x-y)^2}. \tag{15.13}$$

We have explicitly separated out the gauge volume for any legitimate choice of gauge ($J \neq 0$), and this divergence may now be thrown into the normalization and forgotten. One convenient way to forget about the divergence is to say that the Z we really want is, by definition,

$$Z = \frac{\int dx\, dy\, e^{-(x-y)^2}}{V_g}, \tag{15.14}$$

where V_g is the gauge volume.

Now let us rederive eq.(15.13) by an easier construction. Let's insert 1 in the form of $\int df\, \delta(f)$ into Z, eq.(15.1).

$$Z = \int dx\, dy \int df\, \delta(f)\, e^{-(x-y)^2} \tag{15.15}$$

If we start on the gauge slice (x', y') where $f = 0$ and begin to move off the gauge slice via a gauge transformation, then the value of f on the gauge orbit is going to change, that is, f is not constant along a gauge orbit (otherwise it would not be a good gauge choice). Therefore, we may consider f to be a function of a, the gauge transformation, $f = f(a)$. So, change variables from f to a in eq.(15.15) above.

$$Z = \int dx\, dy \int da\, \left.\frac{df}{da}\right|_{f=0} \delta(f(a))\, e^{-(x-y)^2} \tag{15.16}$$

However, from eq.(15.9) we have

$$\begin{aligned} \left.\frac{df}{da}\right|_{f=0} &= \frac{\partial f}{\partial x'}\frac{\partial x'}{\partial a} + \frac{\partial f}{\partial y'}\frac{\partial y'}{\partial a} \\ &= \left.\left(\frac{\partial f}{\partial x} + \frac{\partial f}{\partial y}\right)\right|_{f=0}, \end{aligned} \tag{15.17}$$

and we arrive at eq.(15.13) again.

For the sake of convention, let us rewrite eqs.(15.15)-(15.17) using the notation that has become standard for gauge theories. Define

$$\begin{aligned} \Delta_f^{-1} &= \int da\, \delta(f(x(a), y(a))) \\ &= \int df\, \det\left|\frac{da}{df}\right| \delta(f) \\ &= \det\left.\left|\frac{da}{df}\right|\right|_{f=0}. \end{aligned} \tag{15.18}$$

Thus,

$$\Delta_f = \det\left|\frac{df}{da}\right|_{f=0}.$$

Insert $1 = \Delta_f \Delta_f^{-1}$ into eq.(15.1).

$$\begin{aligned} Z &= \int dx\, dy\, \Delta_f\, \Delta_f^{-1}\, e^{-(x-y)^2} \\ &= \int dx\, dy\, \det\left|\frac{df}{da}\right|_{f=0} \int da\, \delta(f)\, e^{-(x-y)^2}, \end{aligned} \tag{15.19}$$

and we return to eq.(15.17).

A quick computational example is in order. Let's take the gauge slice to be $f(x,y) = (x+y)^2 - (x-y) = 0$. According to eqs.(15.12)-(15.13), the Jacobian is $J = 4(x+y)$, and

$$Z = \int da \int dx\, dy\, \delta((x+y)^2 - (x-y))\, 4(x+y)\, e^{-(x-y)^2}. \tag{15.20}$$

Alternatively,

$$\begin{aligned} f(a) = f(x(a), y(a)) &= (x' + a + y' + a)^2 - (x' - a - y' + a) \\ &= 4a^2 + 4a(x' + y') \end{aligned}$$

so

$$\left.\frac{df}{da}\right|_{a=0} = 4(x' + y'),$$

and eq.(15.16) yields the identical result, eq.(15.20).

So far we have always been choosing the gauge slice to satisfy $f = 0$. However, our result is independent of the choice of slice, so we could just as well choose $f = c$, where c is a real constant. We have defined the part of Z that we are interested in to be nonzero only on the gauge slice, but this is not the only option. Our goal, after all, is find a definition of Z that is finite and independent of choice of gauge slice. One way to guarantee independence of slice is add up all of the slices, weighting each slice in such a way that the integral along 1 gauge orbit is finite.

For example, we can add up the contributions from all slices by integrating over c, weighting each slice, $f = c$, by $\exp(-c^2)$. Starting from our single slice definition of Z,

$$Z = \int da \int dx\, dy\, \delta(f(x,y) - c) \left|\frac{df}{da}\right|_{f=c} e^{-(x-y)^2}, \tag{15.21}$$

we can insert $1 = \int dc \exp(-c^2)/\sqrt{\pi}$. The result is

$$\begin{aligned} Z &= \int da \int dx\, dy \int dc\, \delta(f-c) \frac{e^{-c^2}}{\sqrt{\pi}} \left|\frac{df}{da}\right|_{a=0} e^{-(x-y)^2} \\ &= \int da \int dx\, dy \frac{1}{\sqrt{\pi}} e^{-f^2} \left|\frac{df}{da}\right|_{a=0} e^{-(x-y)^2}. \end{aligned} \tag{15.22}$$

We can interpret the result, eq.(15.22), as the addition of a term to the "action," the new action being $(x-y)^2 + f^2$. This new term is called a gauge-fixing term and is analogous to what we did to quantize the electromagnetic field in the Lorentz gauge back in chapter 5.

Finally, before moving on to the electromagnetic field, we want to derive the same result from a slightly different viewpoint, one based on the following fact: If g_{ij} is the metric on the (tangent) space we want to integrate over, then the measure that is invariant under coordinate transformations is $\int d^n x \sqrt{\det g_{ij}}$. This is just another way of interpreting the Jacobian (geometrical).

Recall that ordinary multi-dimensional Gaussian integrals produce 1 over the square root of determinants of operators. If we set up a Gaussian integral such that the metric is the operator in the exponent, then the integral will produce $(\sqrt{\det g_{ij}})^{-1}$, i.e. J^{-1}. The metric is used to measure distances on the space we are integrating over. It defines a scalar product at a point of two vectors tangent to the space at that point. The infinitesimal square distance between a point $\vec{x}$ on the space and a neighboring point $\vec{x} + \delta\vec{x}$ is $||\delta\vec{x}||^2 = \delta x^i g_{ij} \delta x^j$. Therefore, the Gaussian integral that produces the Jacobian we need is

$$\int d\delta\vec{x}\, e^{-||\delta\vec{x}||^2} = \int d\delta\vec{x} \exp\left(-\sum_{i,j+1}^{n} \delta x^i\, g_{ij}\, \delta x^j\right) \propto \frac{1}{\sqrt{\det g_{ij}}}, \tag{15.23}$$

and the Jacobian, J can be defined by

$$J(\vec{x}) \frac{1}{\pi^{n/2}} \int d\delta\vec{x}\, e^{-||\delta\vec{x}||^2} = 1. \tag{15.24}$$

The Gaussian nature of this construction allows us to apply it in the limit $n \to \infty$, that is, to functional integrals.

Let's apply this method to our simple example, eq.(15.1). Let $f(x,y) = 0$ be our choice of gauge and denote a point on the gauge slice as (x', y') (i.e. (x', y') is a solution to $f(x,y) = 0$). Any point off the slice can be reached by eq.(15.9),

$$x = x' + a, \qquad y = y' + a.$$

In the (x, y) coordinate system, the metric is the unit matrix, $g_{ij} = \delta_{ij}$. The square distance between neighboring points is $||\delta\vec{x}||^2 = (\delta x)^2 + (\delta y)^2$, and $\sqrt{\det g_{ij}} = 1$, as expressed in eq.(15.1).

For the sake of variety, let us change coordinates to (x', a) instead of (s, a). Denote the metric in the (x', a) system as h_{ij}. From eq.(15.9), we have

$$\begin{aligned} \delta x &= \delta x' + \delta a \\ \delta y &= \delta y' + \delta a. \end{aligned} \tag{15.25}$$

The gauge choice enters via

$$\delta f = \frac{\partial f}{\partial x'}\delta x' + \frac{\partial f}{\partial y'}\delta y' = 0, \tag{15.26}$$

that is, f is constant along the slice. From eqs.(15.25) and (15.26) we find

$$\delta y' = -\beta\delta x' + \delta a, \qquad \beta \equiv \left(\frac{\frac{\partial f}{\partial x'}}{\frac{\partial f}{\partial y'}}\right) = \left(\frac{\frac{\partial f}{\partial x}}{\frac{\partial f}{\partial y}}\right)_{f=0}. \tag{15.27}$$

Since

$$\begin{aligned} ||\delta\vec{x}||^2 &= (\delta x)^2 + (\delta y)^2 \\ &= (\delta x' + \delta a)^2 + (-\beta\delta x' + \delta a)^2 \\ &= (\delta x' \;\; \delta a)\Big(h_{ij}\Big)\left(\delta x' \;\; \delta a\right)^{\mathrm{T}}, \end{aligned} \tag{15.28}$$

then the metric is

$$h_{ij} = \begin{pmatrix} 1+\beta^2 & 1-\beta \\ 1-\beta & 2 \end{pmatrix}. \tag{15.29}$$

The Jacobian is given by eq.(15.24),

$$\frac{J(x', a)}{\pi}\int d\delta x'\, d\delta a\, \exp\left(-(\delta x'\; \delta a)h_{ij}\begin{pmatrix}\delta x' \\ \delta a\end{pmatrix}\right) = 1, \tag{15.30}$$

from which we obtain

$$J(x', a) = \sqrt{\det h_{ij}} = \sqrt{2(1+\beta^2) - (1-\beta)^2} = 1 + \beta, \tag{15.31}$$

and finally,

$$Z = \int da\, dx'\, (1+\beta)e^{-(x'-y'(x'))^2} = \int da\, \sqrt{\pi}, \tag{15.32}$$

where $f(x', y'(x')) = 0$.

We have spent several pages developing methods to find the Jacobian of a coordinate transformation, even though we know how to compute it by the use of elementary calculus. The purpose in doing so is that these methods may easily be generalized to infinite-dimensional integrals where the Jacobian is now a functional determinant. Our first application in this limit will be to pure QED.

15.2 QED

The vacuum-to-vacuum amplitude, $Z[J]$, expressed as a path integral for QED is naively expected to be

$$Z[J] = \int \mathcal{D}A^\mu \exp(iS[J,A]). \tag{15.33}$$

S is the usual action,

$$S[J,A] = \int d^4x \, \frac{1}{2}(\vec{E}^2 - \vec{B}^2) + J_\mu A^\mu,$$

where $\vec{E} = -\nabla A^0 - \partial_t \vec{A}$, $\vec{B} = \nabla \times \vec{A}$. This action is invariant under the gauge symmetry

$$A^\mu(x) \to A^\mu(x) - \partial^\mu \Lambda(x), \tag{15.34}$$

provided that the source, $J^\mu(x)$, is conserved, $\partial_\mu J^\mu = 0$, which we now assume.

In eq.(15.33), we are integrating over all configurations of the vector potential, A^μ. However, since the action is invariant under (15.34), any two configurations related by a gauge transformation will contribute the same amount to the integral. Since, for any particular configuration, we can find an infinite number of gauge equivalent configurations, the integral eq.(15.33), must be wildly divergent. To get a finite answer, or at least a renormalizable result, we should expect to need to form equivalence classes of gauge equivalent configurations (A^μ/Λ) and integrate only over the classes. In other words, to make sense of eq.(15.33), we should divide by the gauge volume.

$$Z[J] = \frac{\int \mathcal{D}A^\mu \exp(iS[J,A])}{V_{\text{gauge}}}. \tag{15.35}$$

To apply the Faddeev-Popov procedure developed in the previous section to separate out the gauge volume from the numerator in eq.(15.35), we start by picking a gauge. Let us start with the Coulomb gauge, and then redo the procedure in the Lorentz gauge.

In the Coulomb gauge the gauge condition is $F[A] = \nabla \cdot \vec{A} = 0$. On the gauge slice, the vector potential, $\vec{A}$, is transverse. From eq.(15.34) we can see that Λ plays an analogous role to a in the simple example. In accordance with eq.(15.19), we insert 1 into $Z[J]$ to obtain

$$Z[J] = V_{\text{gauge}}^{-1} \int \mathcal{D}A^\mu \int \mathcal{D}\Lambda \; \det \left| \frac{\delta F[A]}{\delta \Lambda} \right|_{F=0} \delta[F[\vec{A}]] \exp(iS[A,J]), \tag{15.36}$$

where $F[\vec{A}] = \nabla \cdot \vec{A}$. Let $\vec{A}_T$ be the transverse potential, the coordinates on the gauge slice. Any $\vec{A}$ off the slice can be written as $= \vec{A}_T - \nabla\Lambda$. Therefore,

$$\frac{\delta F[\vec{A}]}{\delta\Lambda} = \frac{\delta(\nabla\cdot\vec{A})}{\delta\Lambda} = \frac{\delta(\nabla\cdot(\vec{A}_T - \nabla\Lambda))}{\delta\Lambda} = -\nabla^2, \tag{15.37}$$

and

$$Z[J] = V_{\text{gauge}}^{-1}\int \mathcal{D}A^\mu \int \mathcal{D}\Lambda \, \det(\nabla^2)\, \delta[\nabla\cdot\vec{A}] \exp(iS[A,J]).$$

Nothing in the integrand depends upon Λ so the volume $\int\mathcal{D}\Lambda$ may be separated out to cancel the denominator. The Jacobian is also independent of A^μ and may be taken out of the integral and absorbed into the normalization, $Z[0]$. This leaves

$$Z[J] = N\int \mathcal{D}A^\mu \, \delta[\nabla\cdot\vec{A}] \exp(iS[A,J]). \tag{15.38}$$

The δ-functional, $\delta[\nabla\cdot\vec{A}]$, decouples the A^0 integral from the $\vec{A}_T$ integral. The resulting integral is identical to eq.(14.24), and, therefore, is equal to (eq.(14.25))

$$Z[J] = Z[0]\exp\left(-\frac{i}{2}\int d^4x \int d^4x' \, J^j(x) D^{\text{tr}}_{jk}(x-x') J^k(x') \right.$$
$$\left. + J^0(x)\Delta_0(x-x')J^0(x')\right) \tag{15.39}$$

The A^0 integral guarantees that Gauss's law will be satisfied.

$$\langle\nabla\cdot\vec{E}(y)\rangle = \int \mathcal{D}A^\mu \, \delta[\nabla\cdot\vec{A}]\,(\nabla\cdot\vec{E}(y)) \exp(iS[A,J]) = J^0(y) \tag{15.40}$$

Now let's repeat the procedure in the Lorentz gauge. $Z[J]$ is still given by eq.(15.36) except now $F[A] = \partial_\mu A^\mu$. Since the gauge choice is still linear in A^μ, the Jacobian remains independent of A^μ and can be taken outside the integral and absorbed into the normalization.

$$\det\left|\frac{\delta F[A]}{\delta\Lambda}\right|_{F=0} = \det\left|\frac{\delta(\partial_\mu A^\mu)}{\delta\Lambda}\right|_{\Lambda=0} = \det|-\partial_\mu\partial^\mu| \tag{15.41}$$

The volume $\int \mathcal{D}\Lambda$ may equally be separated out of the integral over A^μ and canceled by the denominator. Then

$$Z[J] = \int \mathcal{D}A^\mu \, \delta[\partial \cdot A] \exp(iS[A,J]). \tag{15.42}$$

While perfectly acceptable, this form of $Z[J]$ is not particularly easy to calculate with, since $\partial \cdot A$ does not appear explicitly in the action. A form that is far more manageable, and one that connects with the quantization procedure used in the operator representation in chapter 5, may be obtained by employing the procedure that led to eq.(15.22). This means that we choose the gauge condition to be $F[A] = \partial_\mu A^\mu - c(x)$, $c(x)$ an arbitrary function, and then sum the contribution for each gauge slice (labeled by c), weighting each slice by a Gaussian, $\exp(-\alpha/2 \int d^4x\, c^2(x))$. The Jacobian is the same on each slice and so remains equal to eq.(15.41). After we absorb the Jacobian into the normalization, and cancel the gauge volume, we have

$$\begin{aligned} Z[J] &= N \int \mathcal{D}A^\mu \int \mathcal{D}c\, \delta[\partial_\mu A^\mu - c] \exp\Bigl(i\frac{\alpha}{2}\int d^4x c^2(x)\Bigr) \exp(iS[A,J]) \\ &= N \int \mathcal{D}A^\mu \exp\Bigl(iS[A,J] + i\frac{\alpha}{2}\int d^4x (\partial_\mu A^\mu)^2\Bigr) \\ &= N \int \mathcal{D}A^\mu \exp(iS'[A,J]). \end{aligned} \tag{15.43}$$

The new exponent, $S'[A,J]$, can be thought of as a modified action, and is identical to the action we used in chapter 5, eq.(5.43). The additional piece is called a gauge-fixing term. The integral eq.(15.43) is now a simple Gaussian and is straightforward to compute. For $\alpha = 1$, we get the usual covariant propagator, $D_{\mu\nu}(x - x') = -g_{\mu\nu}\Delta_F(x - x', m = 0)$, eq.(5.64).

$$Z[J] = Z[0] \exp\Bigl(-\frac{i}{2}\int d^4x\, d^4x'\, J^\mu(x) D_{\mu\nu}(x - x') J^\nu(x')\Bigr) \tag{15.44}$$

15.3 YANG-MILLS

The path integral we want to make sense of is the usual vacuum-to-vacuum amplitude,

$$Z[J] = \int \mathcal{D}A_a^\mu \exp(iS[A_a^\mu]), \tag{15.45}$$

where A_a^μ is the color vector potential. The Yang-Mills action is

$$S[A_a^\mu] = -\frac{1}{2}\int d^4x\, \mathrm{tr} F_{\mu\nu}F^{\mu\nu}, \tag{15.46}$$

where the color field strength is

$$F^{\mu\nu} = \partial^\mu A^\nu - \partial^\nu A^\mu + ig_0[A^\mu, A^\nu]. \tag{15.47}$$

A^μ is in the adjoint representation of the gauge group $SU(2)$, thus $A^\mu = A_a^\mu T^a$, where $T^a = \sigma^a/2$, $a = 1,2,3$, are the generators of the gauge group, and satisfy the algebra $[T^a, T^b] = i\epsilon^{abc}T_c$. σ^a may be taken as the usual Pauli matrices, eq.(4.4).

The action, eq.(15.46), is invariant under the gauge transformation

$$A_\mu \to UA_\mu U^\dagger + \frac{i}{g_0}(\partial_\mu U)U^\dagger, \tag{15.48}$$

where

$$U(x) = \exp(ig_0\theta^a(x)T^a) \tag{15.49}$$

is an element of the gauge group $SU(2)$. Since we are integrating over all configurations in eq.(15.45), we must again form equivalence classes of configurations and integrate only over the classes $= A^\mu/U$. Thus, we divide Z by the gauge volume and pick one representative A_a^μ from each equivalence class by choosing a gauge slice. The coordinate change necessary to cancel the gauge volume involves changing from a general configuration A_a^μ to $(A_a^{\mu\prime}, U)$, where $A_a^{\mu\prime}$ satisfies the gauge choice and lies on the gauge slice, and where U is the gauge transformation that takes one off the slice at $A_a^{\mu\prime}$ to reach A_a^μ. This is illustrated in figure 15.4. The coordinates that we will use to characterize $U(x)$ are the $\theta^a(x)$ in eq.(15.49) above. The gauge slice is characterized by $\theta^a(x) = 0$. This construction will be adequate only for a perturbative evaluation of Z since only the $U(x)$ that are continuously connected to the identity, δ_{ij}, can be generated this way. (Configurations A_a^μ that have a nontrivial winding number are left out of the integral by this construction.)

With these coordinates and gauge choice, $F[A]$, Z becomes (in accordance with eq.(15.19))

$$Z = V_{\text{gauge}}^{-1} N \int \mathcal{D}A_a^\mu \int \mathcal{D}\theta^b(x)\, \delta[F[A]]\, \det\left|\frac{\delta F}{\delta\theta}\right|_{F=0} \exp(iS[A]). \tag{15.50}$$

Let us work in the covariant Lorentz gauge,

$$F[A] = F^a T^a = \partial_\mu A_a^\mu(x) T^a = \partial_\mu A^\mu(x), \tag{15.51}$$

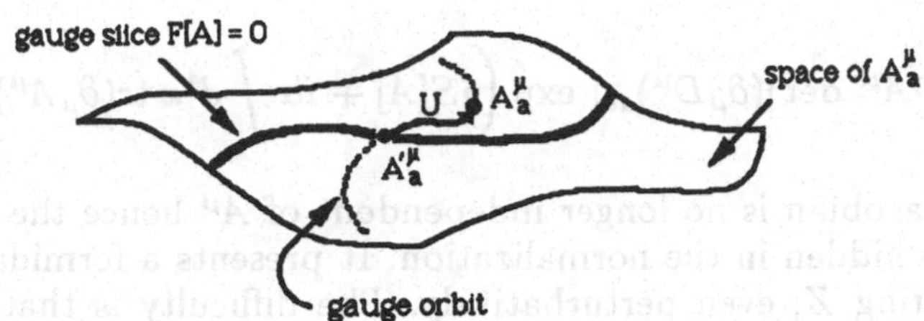

Figure 15.4 Illustration of the change of coordinates in Z leading to eq.(15.50). A gauge choice is made. Configurations $A^{\mu\prime}_a$ that satisfy the gauge conditions, $F[A] = 0$, fall on the gauge slice. Two configurations that are gauge equivalent lie on the same gauge orbit. A general configuration A^μ_a is decomposed into $(A^{\mu\prime}_a, U)$, where $A^{\mu\prime}_a$ is the point where the gauge orbit, on which A^μ_a resides, intersects the gauge slice. U is the gauge transformation relating $A^{\mu\prime}_a$ to A^μ_a.

where the gauge condition is now a matrix since A^μ is one. With this choice, an explicit computation of the Jacobian yields

$$\Delta^F = \det \left| \frac{\delta F}{\delta \theta} \right|_{F=0} = \det |M_{ab}|_{F=0}, \tag{15.52}$$

where

$$\begin{aligned} M_{ab} = \frac{\delta F_a(x)}{\delta \theta^b(y)} &= \int d^4 z \, \frac{\delta F_a(x)}{\delta A^\mu_c(z)} \frac{\delta A^\mu_c(z)}{\delta \theta^b(y)} \bigg|_{\theta^b=0} \\ &= \int d^4 z \, \partial_\mu \delta_{ac} \, \delta^4(x-z) (D^\mu)_{cb} \, \delta^4(z-y) \\ &= (\partial_\mu D^\mu)_{ab} \, \delta^4(x-y), \end{aligned} \tag{15.53}$$

where

$$(D^\mu)_{ab} = \partial_\mu \delta_{ab} + i g_0 \epsilon_{abc} A^{\mu c}$$

is the gauge covariant derivative. Δ^F is gauge invariant so the gauge volume, $\int \mathcal{D}\theta^b(x)$, may be separated out of the integral over A^μ_a to cancel the denominator. Hence, in the Lorentz gauge,

$$Z = \int \mathcal{D}A^\mu \, \delta[\partial_\mu A^\mu] \det |(\partial_\mu D^\mu)_{ab}| \exp(iS[A]). \tag{15.54}$$

As in the $g_0 \to 0$ limit, QED, $\partial_\mu A^\mu$ does not appear explicitly in $S[A]$, so it is more convenient computationally to add together the contribution from all slices labeled by $c^a(x)$, each slice weighted by the Gaussian

$\exp(i\alpha/2 \int d^4x \,\mathrm{tr}\, c^2(x))$. Following the same steps that led to eq.(15.43) we find that

$$Z = \int \mathcal{D}A^\mu \det |(\partial_\mu D^\mu)_{ab}| \exp\left(iS[A] + i\alpha \int d^4x \,\mathrm{tr}(\partial_\mu A^\mu)^2\right). \quad (15.55)$$

The Jacobian is no longer independent of A^μ hence the determinant cannot be hidden in the normalization. It presents a formidable obstacle to computing Z, even perturbatively. The difficulty is that even a perturbative expansion of the determinant leads to nonlocal (noncontact) interactions, and the machinery that we have developed so far is not adequate for the task.

The solution to this problem is to find a representation of the determinant that leads to local interactions. Recall from chapter 9 that the result of a fermionic Gaussian integral over complex anticommuting functions is the determinant of the operator sandwiched between the fields.

$$\det M = \int \mathcal{D}\eta^* \mathcal{D}\eta \exp\left(-\int d^4x \, d^4x' \, \eta^*(x) M(x - x')\eta(x')\right) \quad (15.56)$$

Therefore, the Jacobian, Δ^F, eqs.(15.52)-(15.53) may be represented as

$$\begin{aligned}
&\det |M_{ab}|_{F=0} \\
&= \int \mathcal{D}\eta^{*a} \mathcal{D}\eta^a \exp\left(-\int d^4x \, d^4x' \, \eta^{*a}(x) M_{ab}(x - x')\eta^b(x')\right) \\
&= \int \mathcal{D}\eta^{*a} \mathcal{D}\eta^a \exp\left(-\int d^4x \, d^4x' \, \eta^{*a}(x) \, \partial_\mu D^\mu_{ab} \delta^4(x - x') \, \eta^b(x')\right) \\
&= \int \mathcal{D}\eta^{*a} \mathcal{D}\eta^a \exp\left(-\int d^4x \,\mathrm{tr} \eta^* \partial_\mu \partial^\mu \eta + i g_0 \eta^{*a} \epsilon_{abc} \partial_\mu (A^{\mu c} \eta^b)\right).
\end{aligned} \quad (15.57)$$

Eq.(15.57) itself resembles a vacuum-to-vacuum amplitude for a field theory of fermions coupled to an external field, $A^{\mu c}$. However, the kinetic term in the action above corresponds to a bosonic field.

The fictitious anticommuting fields, $\eta^{*a}(x)$ and $\eta^a(x)$, originally introduced independently by Feynman and by De Witt, are called Faddeev-Popov ghosts. Please note that the term "ghosts" is used in two very different ways in field theory. One way is as above, as a representation of a Jacobian. The other way the word is used is to describe states whose norm have the wrong sign, and therefore the potential to destroy unitarity instead of restore it. One such example is provided by the states of scalar photons encountered in quantizing the electromagnetic field in the operator formalism in chapter 5. Without the constraint, $\langle \partial \cdot A \rangle = 0$,

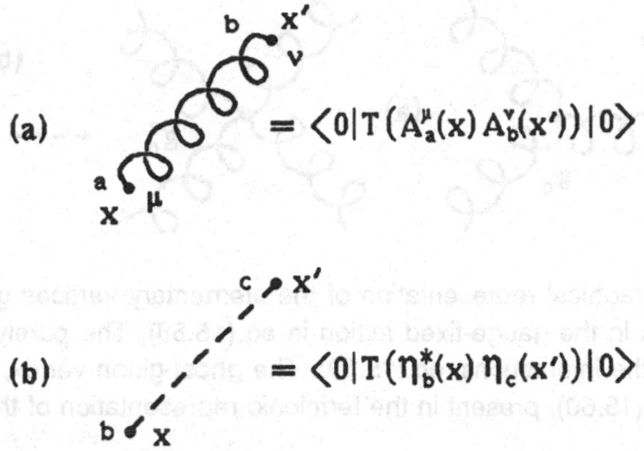

(a) $= \langle 0|T(A^\mu_a(x)\,A^\nu_b(x'))|0\rangle$

(b) $= \langle 0|T(\eta^*_b(x)\,\eta_c(x'))|0\rangle$

Figure 15.5 Graphical representation of the free field 2-point functions for (a) the gluons, and (b) the ghosts.

these ghosts would render the theory useless for lack of unitarity and (consequently) a probability interpretation.

When we add sources to Z, eq.(15.55), after eq.(15.57) is substituted in for the determinant, we must add sources for the Faddeev-Popov ghosts, too. Thus, the Yang-Mills vacuum-to-vacuum amplitude in the presence of sources in the covariant Lorentz gauge is

$$
\begin{aligned}
Z[J,\chi^*,\chi] = N\int \mathcal{D}A^\mu_a\,\mathcal{D}\eta^{*b}\,\mathcal{D}\eta^b\ \exp\Bigg(i\int d^4x\ &\frac{1}{2}\,\mathrm{tr}F^{\mu\nu}F_{\mu\nu} + \alpha\,\mathrm{tr}(\partial_\mu A^\mu)^2 \\
&- i\,\mathrm{tr}(\partial_\mu\eta^*\partial^\mu\eta) + ig_0\epsilon_{abc}\eta^{*a}A^{\mu c}\partial_\mu\eta^b \\
&+ \mathrm{tr}(J^\mu A_\mu) + \mathrm{tr}(\chi^*\eta) + \mathrm{tr}(\eta^*\chi)\Bigg).
\end{aligned}
\tag{15.58}
$$

To perturbatively evaluate eq.(15.58), we replace any term in the action that is not linear or quadratic in the fields with functional derivatives with respect to the appropriate source. After the replacement, these terms may be pulled outside of the integral. The remaining integral is now strictly quadratic in the fields and may be computed exactly.

The interacting terms are

$$
\begin{aligned}
&\int d^4x\ \frac{1}{2}g_0\epsilon_{abc}(\partial^\mu A^\nu_a - \partial^\nu A^\mu_a)A^b_\mu A^c_\nu, \\
&\int d^4x\ -\frac{1}{4}g_0^2\epsilon_{abc}\epsilon_{ade}A^\mu_b A^\nu_c A^d_\mu A^e_\nu,
\end{aligned}
\tag{15.59}
$$

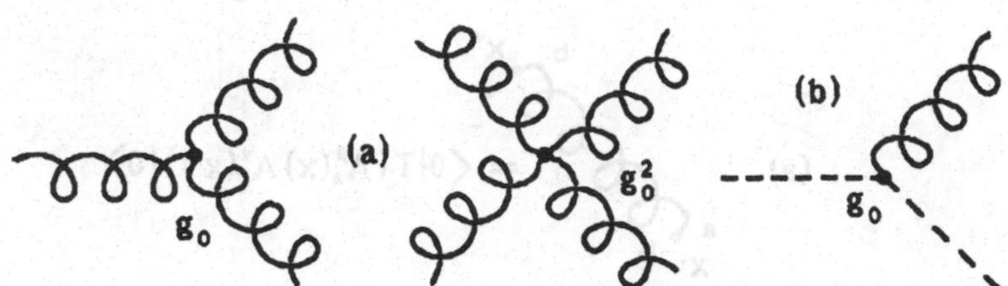

Figure 15.6 Graphical representation of the elementary vertices generated by the interacting terms in the gauge-fixed action in eq.(15.58). The purely gluonic vertices in (a) arise from the interactions, eq.(15.59). The ghost-gluon vertex, (b), arises from the interaction, eq.(15.60), present in the fermionic representation of the Jacobian.

from the $\mathrm{tr}F^{\mu\nu}F_{\mu\nu}$ piece, and

$$i\int d^4x\; g_0\epsilon_{abc}\eta^{*a}A^{\mu}_{c}\partial_{\mu}\eta^{b}, \tag{15.60}$$

from the Jacobian. The integral over quadratic pieces with sources will determine the propagators for each of the fields. This computation is left as an exercise. If we represent the gluon propagator for the A^{μ}_{a} field as a curly line as in figure 15.5(a), and the propagator for the ghost field as a dashed line, figure 15.5(b), then the diagrammatic vertices generated by

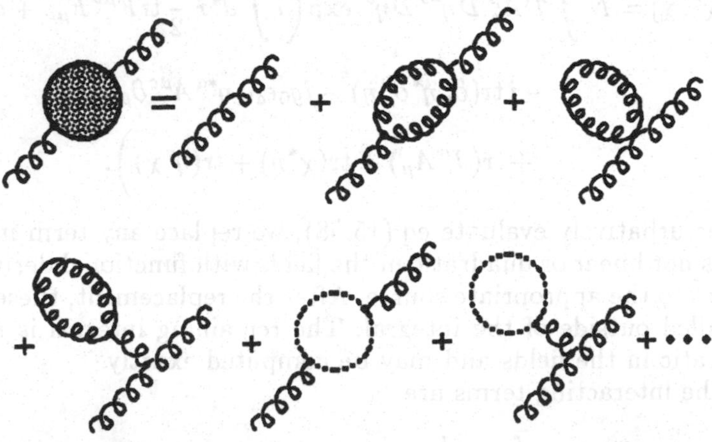

Figure 15.7 Diagrammatic representation of the perturbative expansion of the full, interacting 2-point function for the gluons through order g_0^2.

the interacting terms are represented in figure 15.6(a) for eq.(15.59), and figure 15.6(b) for eq.(15.60). The perturbative expansion to order g_0^2 of the 2-point function for the gluons will look diagrammatically like figure 15.7.

We could equally well introduce the ghost field representation of the Jacobian in the path integral for $Z[J]$ for QED. Since $\delta A^\mu/\delta\Lambda$ is independent of A^μ, the ghost fields will never couple to A^μ so we will not see them in any calculation involving photons. However, since $\delta A_a^\mu/\delta\theta^b$ does depend on A_a^μ, does this mean that we must introduce ghosts in every gauge? No, let's consider the axial gauge defined by $F[A] = A_z^a = 0$. In this gauge, we have

$$\begin{aligned}\int d^4z\,\frac{\delta F_a(x)}{\delta A_c^\mu(z)}\,\frac{\delta A_c^\mu(z)}{\delta\theta^b(y)} &= \int d^4z\,\delta_{ac}\delta_{\mu z}\delta^4(x-z)\,(D^\mu)_{ab}\,\delta^4(z-y)\\ &= (D^z)_{ab}\delta^4(x-y).\end{aligned} \tag{15.61}$$

However, if $A_z^a = 0$ for $a = 1, 2, 3$, then $D_{ab}^z = \partial^z\delta_{ab}$. Since the Jacobian is independent of A_a^μ, there is no need to introduce ghosts. When such a situation arises, the gauge is referred to as "ghost-free."

15.4 EXERCISES

1. Carry out the integral, eq.(15.20).
2. Compute the Jacobian, $J(s, a)$, for the simple example Z, eq.(15.1), via the "metric" method, eq.(15.24), for the gauge choice $f(x, y) = 0$, where a is defined by eq.(15.9), and where s is defined by eq.(15.8). Substitute in eq.(15.8) for $\int ds$ to arrive at eq.(15.13).
3. Consider the vacuum-to-vacuum amplitude for QED, $Z[J]$, given by eq.(15.35). Choose the Coulomb gauge by taking the gauge condition to be $F[A] = \nabla\cdot\vec{A} - c(x) = 0$. Add together all gauge slices (labeled by c) by weighting each slice by the Gaussian, $\exp(-i/2\int d^4x\, c^2(x))$ and inserting

   $$1 = N\int \mathcal{D}c\,\exp\left(-\frac{i}{2}\int d^4x\, c^2(x)\right)$$

   into the integral. The result will be an integral similar to eq.(15.43). Compute this integral to find the Coulomb gauge photon propagator and show that it is equal to $D_{\mu\nu}^{\text{tr}}$, eq.(14.27). Note that we did not need to introduce polarization vectors, nor modify any commutators.
4. Do the integral in eq.(15.40) to verify the result.

5. **Compute the Lorentz gauge photon propagator for eq.(15.43) for $\alpha \neq 1$.**
6. **Show that the gauge orbits intersect the gauge slice at right angles in the Coulomb gauge.**
7. **Use the metric method to define the vacuum-to-vacuum amplitude for QED in the Coulomb gauge. Use the x and y components of the transverse potential, $\vec{A}_T$, as the coordinates of the gauge slice. The result will not be identical to eq.(15.38). How are they related?**
8. **Determine the free field gluon and ghost propagator in the Lorentz gauge by computing the $g_0 = 0$ limit of $Z[J, \chi^*, \chi]$, eq.(15.58).**
9. **Develop a perturbative expansion of $Z[J, \chi^*, \chi]$ as a power series in g_0 from eq.(15.58).**
10. **Compute the terms appearing in figure 15.7 for the gluonic 2-point function in the momentum representation.**

CHAPTER 16

Hiding the Infinities

16.1 OVERVIEW

To make sense of a quantum field theory we must not only be able to solve it (at least approximately), but also we must be able to interpret the results. We are able to solve the free field theories exactly, and their harmonic oscillator structure lends itself to a particle interpretation. This in turn provides an identification of the parameter in the action that looks like a mass with the actual mass of the particle.

When we add interactions, we can no longer solve the resulting interacting quantum theory (except for a few special cases), so we must be satisfied with an approximate solution. The approximate solution is usually a perturbative expansion of some sort, where the lowest order term is the free quantum theory or the interacting classical theory. Our confidence in the validity of the series is based on what we believe to be the identification of a "small" expansion parameter, usually a coupling constant or $\hbar$, and much of the way we interpret the results relies on the interpretation of the lowest order term.

If the perturbative solution of an interacting field theory is to be useful and practical, then the higher order corrections should be progressively smaller in comparison to the lower order terms, at least for a while. It is therefore surprising that often when we expect that the perturbative solution should be valid, the first quantum corrections to the classical field theory are not only not smaller, but infinite. The first impulse is to abandon the series and maybe even the theory. However, we must first examine how we are interpreting the results. There exists a reinterpretation of the parameters and components appearing in many of the physically interesting actions such that the results of the series are finite.

The systematic "removal" of the divergences is called the process of renormalization. The basic idea behind renormalization is simple. We calculate a term in a series for a Green's function which turns out to be divergent. This will imply that some physical quantity, such as the mass of a particle, will be divergent. Or, perhaps, we calculate the corrections to the energy eigenvalue of some state as we did in chapter 11 and we find that it diverges. The first thing we do is to *change* the original theory in such a way that everything is finite. This modification is called regularizing the theory. Let the regularization be described by Λ such that in the limit $\Lambda \to \infty$, we recover the original theory. Now we go back and recalculate the Green's function or physical quantities of interest. For finite Λ everything is finite, but physically measurable quantities now depend on an unphysical arbitrary parameter(s) Λ. Having added an unphysical Λ, we must somehow remove it from physically measurable quantities. What do we have available? There are parameters in the original action which we are used to thinking of as related to masses and couplings (charges). However, when we add interactions, the identification is usually functionally more complicated. If we give up the notion that they *must* be identified somehow with the physical masses and charges, then they are just arbitrary parameters that appear in the action. But, by being arbitrary, these parameters *may* depend on Λ without physical consequence. Therefore, we *choose* them to depend on Λ in such a way as to make physically measurable quantities independent of Λ. Assuming that this may be done, then we may take the limit $\Lambda \to \infty$ safely to get back to the original theory. The physical quantities produced by the renormalized theory will be finite. The parameters appearing in the original action will be divergent. However, we cannot directly measure these parameters so the fact that they are infinite has no direct experimental consequence.

For concreteness, let's assume we have a theory described by an action that has two parameters, m_0 and g_0. In the classical limit of the theory, we have physical interpretations of m_0 and g_0, say, the mass and charge of the particle, respectively. Now we calculate two physical processes from which the true mass and charge, m and g, may be measured experimentally, and we find that the two expressions diverge. Next, we regularize the theory and recompute the two processes. The resulting mass and charge will be finite, but dependent on the two parameters, m_0 and g_0, and on the unphysical regulator, Λ.

$$\begin{aligned} m &= m(m_0, g_0, \Lambda) \\ g &= g(m_0, g_0, \Lambda) \end{aligned} \tag{16.1}$$

We can experimentally measure m and g so we know in principle what values belong on the left-hand side above. To make sense of the regularized theory, we now must somehow make m and g independent of Λ. To do this, we assume that m_0 and g_0 are functions of Λ, and that they depend on Λ in such a way so that m and g do not depend on Λ. Thus, the particular $m_0(\Lambda)$ and $g_0(\Lambda)$ we want will satisfy

$$\begin{aligned} \frac{dm}{d\Lambda} &= 0 = \frac{\partial m}{\partial m_0}\frac{\partial m_0}{\partial \Lambda} + \frac{\partial m}{\partial g_0}\frac{\partial g_0}{\partial \Lambda} + \frac{\partial m}{\partial \Lambda} \\ \frac{dg}{d\Lambda} &= 0 = \frac{\partial g}{\partial m_0}\frac{\partial m_0}{\partial \Lambda} + \frac{\partial g}{\partial g_0}\frac{\partial g_0}{\partial \Lambda} + \frac{\partial g}{\partial \Lambda}. \end{aligned} \tag{16.2}$$

The coupled equations are first order and require one "initial" condition each to specify a unique trajectory. The initial values are supplied by the experimentally measured or renormalized mass and charge, m and g. The particular Λ at which the identification is made depends on the details of the experimental measurement and its relation to the theory, the so-called renormalization prescription. More about this in a moment.

It is, of course, possible that just $m_0(\Lambda)$ and $g_0(\Lambda)$ above will not be enough freedom to render the entire theory finite in the limit $\Lambda \to \infty$. We may have to add more interactions and parameters to the original action and apply the same procedure to make the entire theory finite in this limit. (In general, the fields themselves in the original action will also be multiplicatively scaled with a Λ-dependent parameter, but this does not require the addition of new interactions.) However, as long as the terms that we must ultimately add are local interactions and are finite in number, then we say the theory is renormalizable. As you can intuitively see, renormalizability is a nontrivial constraint on the possible quantum theories we might consider.

In practice, the coupled differential equations are not so easy to solve. Instead of trying to solve them in the form (16.2), we follow a constructive procedure. Again, for concreteness, let's assume that we are dealing with a theory with parameters m_0 and g_0, and field ϕ_0, and that the Hamiltonian, H, can be written as $H_0 + g_0 V$, where H_0 is the free field Hamiltonian, and $V(\phi_0)$ is the interaction. The mass term in H_0 will look like $1/2 \int d^3x\, m_0^2 \phi_0^2$. In the Schrödinger representation, the energy eigenfunctional equation,

$$\left(H_0' + \frac{1}{2}\int d^3x\, m_0^2\phi_0^2 \;+\; g_0 V(\phi_0)\right)\Psi_n[\phi_0] = E_n \Psi_n[\phi_0], \tag{16.3}$$

may be solved perturbatively as a series expansion in g_0 in a similar manner to what we did in chapter 11. In this way, we compute corrections to the energy eigenvalue, E_n, of the state as a series in g_0. To compute the mass of a particle in the presence of the interactions, we calculate the energy eigenvalue of a state with a single particle. In general, E_1, to first order in g_0, will depend on m_0, g_0, and the momentum of the particle, $\vec{p}$.

$$E_1 = E_1^{(0)} + g_0 E_1^{(1)} = \sqrt{\vec{p}^2 + m_0^2} + g_0 E_1^{(1)}(\vec{p}, m_0^2), \tag{16.4}$$

thus,

$$\begin{aligned} m &= m_0 + g_0 E_1^{(1)}(0, m_0) \\ &= m_0 + \delta m(m_0, g_0). \end{aligned} \tag{16.5}$$

Suppose that we found that the correction to E_1 at order g_0, $E_1^{(1)}(\vec{p}, m_0)$, is divergent. We regularize the theory and recompute E_1 to the same order in g_0 to get

$$m = m_0(\Lambda) + \delta m(m_0, g_0, \Lambda). \tag{16.6}$$

Now we observe that to zeroth order in g_0, that $m = m_0$, and $g = g_0$. Therefore, to first order we may replace m_0 with m, and g_0 with g in δm.

$$m = m_0(\Lambda) + \delta m(m, g, \Lambda) + O(g_0^2). \tag{16.7}$$

The errors we make in replacing m_0 with m, and g_0 with g, are of higher order in g_0 than we are working at, i.e. order g_0^2 and higher. Now we can quickly solve for m_0 algebraically.

$$m_0(\Lambda) = m - \delta m(m, g, \Lambda) \tag{16.8}$$

Since we are assuming that m is finite and constant, the resulting $m_0(\Lambda)$ will be just the right function of Λ to absorb the Λ dependence that we are seeking. That is, if we differentiate eq.(16.7) with respect to Λ, we will find that m is independent of Λ to the order in g_0 that we are working at.

$$\frac{dm}{d\Lambda} = 0 + O(g_0^2) \tag{16.9}$$

To begin to construct a Hamiltonian that contains only finite parameters, and gives finite results, we replace the divergent "bare" mass, m_0, with the finite renormalized mass, m, via eq.(16.8).

$$\begin{aligned} H_{\text{ren}} = {H_0}' &+ \frac{1}{2}\int d^3x\, m^2 \phi_0^2 + g_0 V(\phi_0) \\ &- \int d^3x\, m\, \delta m(m, g, \Lambda)\phi_0^2 + \frac{1}{2}\int d^3x\, \delta m^2(m, g, \Lambda)\phi_0^2 \end{aligned} \tag{16.10}$$

Two new terms have appeared, one at $O(g)$ and one at $O(g^2)$. These terms are called "counterterms." Reference to the bare mass m_0 has disappeared. If we recompute the E_1 to order g using the *new* Hamiltonian, H_{ren}, we obtain

$$E_1 = \sqrt{\vec{p}^2 + (m - \delta m)^2} + gE_1^{(1)}(\vec{p}, m, \Lambda), \tag{16.11}$$

and at $\vec{p} = 0$,

$$m = (m - \delta m(m, g, \Lambda)) + gE_1^{(1)}(0, m, \Lambda) = m. \tag{16.12}$$

The eigenvalue to $O(g)$ is, by construction, finite. In the limit, $\Lambda \to \infty$, it remains fixed since it is independent of Λ. However, we are taking the difference between 2 infinite quantities to keep it that way. In general, we must also add additional counterterms to render all physical quantities finite at order g.

Note that the finite parameter m that we are inserting into the Hamiltonian to replace m_0 is assumed to be the experimentally measured rest mass of a single, isolated particle. For computational convenience, we chose the 3-momentum, $\vec{p}$, to be zero when we made the identification between eigenvalue and mass. We could choose a different momentum $\vec{q}$ without consequence. We still get the same rest mass because the assumption that the measurement is made of an isolated particle implies that the particle must be "on mass shell," that is, that the 4-momentum squared, $p^2 = E_p^2 - \vec{p}^2$, is equal to the invariant mass, m^2.

The momentum or energy scale at which the relation between the quantity computed and an experimentally measured parameter is identified (which is related to the initial conditions for $\partial m_0/\partial\Lambda$ at order g) is called the subtraction point. The subtraction point is arbitrary so we are not stuck renormalizing "on shell." It may be more convenient to perform the subtraction "off shell," $p^2 = \mu^2$, $\mu^2 \neq$ the experimentally measured square mass. The difference between the two renormalization "prescriptions" is a finite renormalization. Since the subtraction point is arbitrary and unphysical, then the final answers we compute had better not depend on the subtraction scale, μ^2. Therefore, in the same spirit that led to eq.(16.2), $d/d\mu^2$ of physical quantities had better be zero. The physical quantities can be related to a particular subset of renormalized inverse Green's functions. When $d/d\mu^2$ is applied to this subset, the resulting coupled differential equations are known as the "renormalization group equations." We will return to this in more detail in chapter 18.

The analysis of the single-particle energy eigenvalue presented above follows the computation done with the electron mass in QED at the end of chapter 11. One can also see that we have already added the mass renormalization counterterm to the interacting theories in chapters 7, 8, and 11, in anticipation that m_0 would need to acquire a Λ dependence.

The substitution of the measured mass m in the action for the divergent parameter m_0 is also convenient for comparison with scattering experiments.

Now let's recast the same procedure into the context of the Green's functions to connect with the operator formalism and path integral viewpoint. In particular, let's consider the inverse, G_2^{-1}, of the 2-point function, $G_2 = \langle 0\,|T(\varphi_0(x_1)\varphi_0(x_2))|\,0\rangle$. In momentum space, the free field inverse propagator is

$$(G_2^{(0)})^{-1}(p) = -i(p^2 - m_0^2), \tag{16.13}$$

which vanishes when $p^2 = m_0^2$. By the particle interpretation, this is the invariant square mass of the free particle. That is, particles propagate at masses set by the poles of the 2-point function.

The 2-point function for the interacting theory can be perturbatively computed as a power series in g_0,

$$G_2(p^2) = G_2^{(0)}(p^2) + g_0 G_2^{(1)}(p^2) + \dots . \tag{16.14}$$

The corrections, $G_2^{(n)}(p^2)$ normally arise naturally in the form

$$G_2^{(n)}(p^2) = G_2^{(0)}(p^2)\Sigma_n(p^2)G_2^{(0)}(p^2) \tag{16.15}$$

so that, to order g_0, we may write

$$G_2(p^2) = \frac{i}{p^2 - m_0^2 - g_0\Sigma_1(p^2)} \tag{16.16}$$

or

$$G_2^{-1}(p^2) = -i(p^2 - m_0^2 - g_0\Sigma_1(p^2)) + O(g_0^2). \tag{16.17}$$

Once again we find that $\Sigma_1(p^2, m_0)$ is divergent so we regularize and recompute G_2^{-1}.

$$G_2^{-1}(p^2) = -i(p^2 - m_0^2 - g_0\Sigma_1(p^2, m_0, \Lambda)) + O(g_0^2) \tag{16.18}$$

We know experimentally that the particle should propagate at its measured mass m. We must now choose a renormalization prescription, i.e. a subtraction point. Following the free field result and the fact that the measurements of mass of isolated particles are done "on shell," we take $G_2(m^2) = 0$. This leads to

$$m^2 = m_0^2(\Lambda) - g_0\Sigma_1(m^2, m_0^2, \Lambda). \tag{16.19}$$

As in the previous example, to order g_0 we may replace g_0 with g and m_0 with m, leaving

$$m^2 = m_0^2(\Lambda) - g\Sigma_1(m^2, m^2, \Lambda) + O(g_0^2),$$

or

$$\begin{aligned} m_0^2(\Lambda) &= m^2 + g\Sigma_1(m^2, m^2, \Lambda) \\ &= m^2 + \delta m^2. \end{aligned} \tag{16.20}$$

When we substitute this back into the action, we now have a mass term, $-1/2 \int d^4x\, m^2\phi_0^2$, where the parameter m really is the physical mass, but we also have a new interaction generated by the counterterm, $1/2 \int d^4x \delta m^2\, \phi_0^2$. The new interaction introduces a new vertex which must appear in the diagrams representing the series expansion of G_2. Traditionally, the mass counterterm vertex is represented by an $\times$ as shown in figure 7.12.

The subtraction (16.19) and subsequent counterterm will guarantee that the pole of the 2-point function will occur at m, thus eliminating 1 infinity. However, the propagator G_2 is also used off shell when it occurs as an internal line in a diagram. So, even though we have subtracted the divergence $g\Sigma_1(m^2, m^2, \Lambda)$, we must still eliminate $g\Sigma_1(p^2, m^2, \Lambda)$ for $p^2 \neq m^2$. This means that we want to place some condition on $G_2^{-1}(p^2)$ that involves p^2. Again, following the lead of the free field result, we additionally require

$$\frac{\partial}{\partial p^2} G_2^{-1}(p^2) = -i \tag{16.21}$$

on mass shell. The mass counterterm guarantees that the pole of the propagator will occur at a finite value, but does not fix the residue at the pole. The new requirement above does that.

When we apply eq.(16.21) to eq.(16.18), we find

$$\begin{aligned} \frac{\partial}{\partial p^2} G_2^{-1}(p^2)\bigg|_{p^2=m^2} &= \frac{\partial}{\partial p^2}\left(-i\left(p^2 - m_0^2 - g_0\Sigma_1(p^2, m_0^2, \Lambda)\right)\right)_{p^2=m^2} \\ &= -i\left(1 - g_0\frac{\partial}{\partial p^2}\Sigma_1(p^2, m_0^2, \Lambda)\right)_{p^2=m^2} \end{aligned} \tag{16.22}$$

$\partial/\partial p^2\, \Sigma_1$ does not vanish on shell; in fact, it usually diverges. To absorb this divergence, we must introduce a Λ-dependent scale factor, $Z(\Lambda)$, via

$$G_2^{-1}(p^2) = -iZ(\Lambda)(p^2 - m_0^2 - g_0\Sigma_1(p^2, m_0^2, \Lambda)), \tag{16.23}$$

and require

$$Z(\Lambda)\left(1 - g\frac{\partial}{\partial p^2}\Sigma_1(p^2, m^2, \Lambda)\bigg|_{p^2=m^2}\right) = 1. \tag{16.24}$$

We can reproduce the first term in eq.(16.23) by adding a counterterm of the form $(Z(\Lambda) - 1)\int d^4x\, \partial_\mu\phi_0\partial^\mu\phi_0$, and the second term by scaling the mass counterterm we have already considered, and so on. However, it is much easier just to realize that the field ϕ_0 is also not directly physically measurable. This allows ϕ_0 to possess a Λ dependence via a scale. Therefore, if we define the renormalized field ϕ as $\sqrt{Z(\Lambda)}\phi_0$, eq.(16.23) is generated by the usual action, written in terms of ϕ instead of ϕ_0, along with the mass counterterm already introduced, also written in terms of ϕ, $1/2 \int d^4x\, \delta m^2 \phi^2$. This scaling is called "wave function renormalization" and is the source of the factors of $\sqrt{Z}$ that appear in the definition of the IN–OUT fields, reduction formulae, and Feynman rules in chapters 7 and 8.

The charge, g_0, must also be renormalized. Which inverse n-point function is used depends upon how g is defined experimentally and measured. When we continue to the next order in g_0, $O(g_0^2)$, we compute quantities using the adjusted action, the action that includes the counterterms generated by the lower order results.

The renormalized action, the action written in terms of the renormalized quantities, ϕ, m, and g, and containing the counterterms, will produce finite Green's functions to whatever order in g you have renormalized to. For the example we have just considered, the renormalized 2-point function to order g can be written as

$$G_{2R}^{-1}(p^2) = -i(p^2 - m^2 - \Sigma_R(p^2)), \tag{16.25}$$

where the renormalization prescription requires that Σ_R satisfy (on shell)

$$\Sigma_R(m^2) = 0, \qquad \text{and} \qquad \frac{\partial}{\partial p^2}\Sigma_R\bigg|_{p^2=m^2} = 0. \tag{16.26}$$

As we stated earlier, the subtraction point is arbitrary. We could just as well take the renormalization prescription off shell to $p^2 = \mu^2$. The mass and charge that we wish to renormalize to must be dependent on the subtraction point used, $m = m(\mu^2)$, $g = g(\mu^2)$, in such a way so as to result in the same experimentally measured mass and charge on shell. The scaling $Z(\Lambda)$ must also depend on μ^2 so that the renormalized Green's functions done at μ^2 will produce the same Green's function when written entirely in terms of bare quantities. The Green's functions written in terms of bare quantities cannot depend on μ^2. But since μ^2 is arbitrary, the renormalized Green's functions must not either. The statement of this fact, $d/d\mu^2\, G_{nR}^{-1} = 0$, is called the renormalization group equation. We will discuss this equation in more detail in chapter 18.

When we examine which diagrams in the series for the Green's functions are divergent, we find that they all contain at least one loop. To get a divergence, we have to have at least one independent 4-momentum traveling through an internal line or propagator over which we must integrate. The most common divergence comes from the behavior of the integral at large internal momenta or small distances. It turns out that the number of integrals over independent momenta to be done is equal to the number of loops in the diagram. Therefore, you need a loop to get a divergence, but, of course, not all loops must produce divergences.

Let's see how it arises that the number of loops tracks the number of independent internal momenta. For definiteness, consider a self-interacting scalar field theory with $V(\varphi) = \varphi^n$. Each vertex will then have n propagators or lines emerging from it. Consider a connected diagram with E external legs, and I internal lines. Only one end of an external leg terminates at a vertex, while both ends of internal lines are connected to vertices. Therefore, the number of vertices, v, in the diagram is

$$v = \frac{E + 2I}{n}. \qquad (16.27)$$

If there are v vertices in the diagram, then there must be at least $v - 1$ internal lines to connect all of the vertices together, otherwise the diagram will be disconnected. Any additional internal line added must form a loop since it will connect two vertices already connected by the initial $v - 1$ internal lines. In fact, each additional internal line will form an additional loop. Therefore, the number of loops, L, is

$$L = I - (v - 1). \qquad (16.28)$$

Recall from the Feynman rules in chapter 7 that we associate an integral, $\int d^4k_i/(2\pi)^3$ for each internal line carrying momentum K_i^μ. However, we must conserve momentum at each vertex so there are v δ-functions of the momenta entering each vertex. Momentum conservation at each vertex will determine only $v-1$ of the internal momenta since one δ-function will ultimately express overall momentum conservation of the entire diagram. After the $v - 1$ δ-functions are integrated over, there will be $I - (v - 1)$ leftover integrals to do over the $I - (v - 1)$ remaining independent internal momenta. Therefore, the number of loops is equal to the number of independent internal momenta.

We can estimate how badly a diagram might diverge by observing that each internal line propagator will contribute a factor, $|k_i|^{-2}$, to the integrand of the integral represented by the diagram, while each internal momentum that must be integrated over will contribute $|k|^d$ to the measure, where d = number of space-time dimensions. The superficial degree of divergence, D, is the difference between the exponents.

$$D = dL - 2I \qquad (16.29)$$

If $D = 0$, the integral will probably diverge like $\ln(\Lambda)$, where Λ is the largest momentum allowed, and if $D > 0$, the divergence will probably go like Λ^D, etc. If we substitute eq.(16.28) into eq.(16.29) and use eq.(16.27)

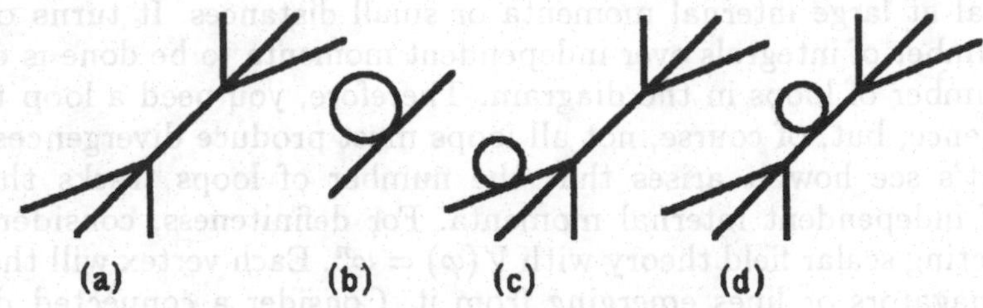

Figure 16.1 (a): Tree level 6-point function which is finite, $D = -2$. (b): One-loop contribution to the 2-point function which is divergent, $D = 2$. (c): Diagram in (b) attached as an external leg to (a), and (d): replacing an internal line in (a). Both (c) and (d) are divergent, yet $D = -2$.

to eliminate I, we find that the superficial degree of divergence of a diagram with E external legs and v vertices in 4 space-time dimensions is

$$D = 4 - E + (n-4)v. \tag{16.30}$$

The self-interacting theory we have considered so far is $V(\varphi_0) = \varphi_0^4$, i.e. $n = 4$, in which case $D = 4 - E$. Interestingly, whether a diagram superficially diverges or not depends only on the number of external legs. There are only 2 possibilities for superficial divergence: $E = 2$, and $E = 4$. The 2-point function may potentially diverge as Λ^2 (it does) while the 4-point function may potentially diverge as $\ln(\Lambda)$ (it does). We have already seen that the renormalization of the 2-point function generates mass counterterms, but of the form of interaction already present in the action. Renormalization of the 4-point function will generate a counterterm proportional to φ_0^4 plus wave function renormalization, again already present in the original action. We can now begin to see why this theory might be renormalizable.

You will notice that we used the phrase "will probably diverge like" above. There is no guarantee that the divergence will be described by D. For example, any diagram representing a term in the series for the 6-point function ($E = 6$) will have $D < 0$, but will not necessarily converge. As a specific example, we can take the ordinary tree level diagram for the 6-point function, figure 16.1(a), which is finite, and attach a divergent 2-point term, figure 16.1(b), to one of the legs. The resulting diagram, figure 16.1(c), will still have $D < 0$, but will not converge. Alternatively,

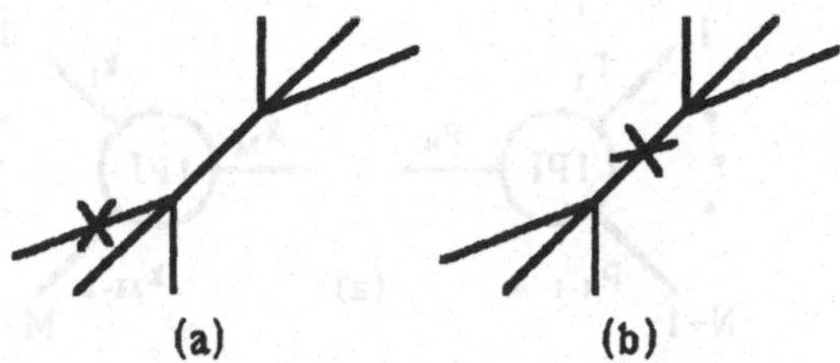

Figure 16.2 Diagrams generated by the addition of the mass counterterm that correspond to the diagrams in figure 16.1(c) and (d). When added to 16.1(c) and (d), a finite result is obtained.

we may add a loop to the internal line as in figure 16.1(d) and reach the same conclusion. However, one gets the feeling that if we add counterterms that renormalize the 2-point function, we will also make the diagrams in figure 16.1(c) and (d) finite, too. This is, in fact, true. For example, after we have renormalized at 1-loop, we will have correspondingly two new diagrams generated by the counterterm vertex proportional to δm^2, figure 16.2. The counterterm vertex can appear wherever a loop can appear, hence the sum of the two always cancels and leaves a finite result. This means that there is a subset of diagrams, which if made finite by renormalization, will render the remaining diagrams, not belonging to the subset, finite, too. The diagrams belonging to this subset are called proper diagrams.

A proper diagram is a truncated, connected, 1-particle irreducible diagram. A 1-particle irreducible (1PI) diagram is one which remains

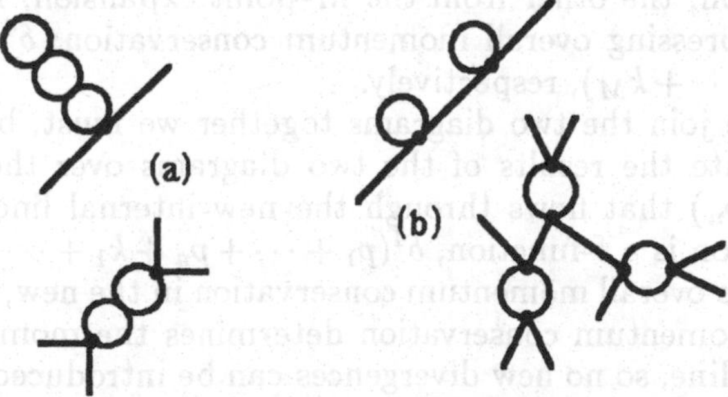

Figure 16.3 (a) Examples of 1-particle irreducible diagrams. The diagrams remain connected after 1 arbitrary internal line is cut. (b) Examples of 1-particle reducible diagrams.

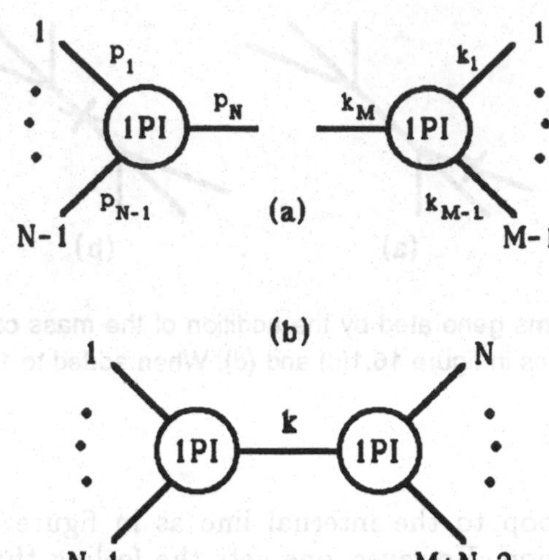

Figure 16.4 (a) Two 1-particle irreducible (1PI) diagrams joined together by identifying 1 external leg each to form a 1-particle reducible diagram in (b). (b) The momentum flowing through the new internal line, k, is determined completely by momentum conservation.

connected when an arbitrary internal line is cut. Diagrams with internal lines but without loops clearly are never 1PI. Diagrams in figure 16.3(a) are 1PI while those in 16.3(b) are not. Diagram 16.1(b) is 1PI while (a), (c), and (d) are not.

Why are 1PI diagrams the important ones? Consider a 1-particle reducible diagram built from two 1PI diagrams as in figure 16.4. In figure 16.4(a) we have two 1PI diagrams, one a term in the N-point function expansion, the other from the M-point expansion. Both contain δ-functions expressing overall momentum conservation, $\delta^4(p_1 + \cdots + p_N)$ and $\delta^4(k_1 + \cdots + k_M)$, respectively.

When we join the two diagrams together we must, by the Feynman rules, integrate the results of the two diagrams over the momentum k $(= k_m = -p_n)$ that flows through the new internal line. The result of this integration is a δ-function, $\delta^4(p_1 + \cdots + p_n + k_1 + \cdots + k_m - 1)$, that represents the overall momentum conservation in the new, larger diagram. Therefore, momentum conservation determines the momentum k in the new internal line, so no new divergences can be introduced by joining two 1PI diagrams in this way. If the 1PI diagrams have been made finite by renormalization, then the new diagram will be finite, too. Any diagram is either 1PI or not. If it is not, then by breaking internal lines that do

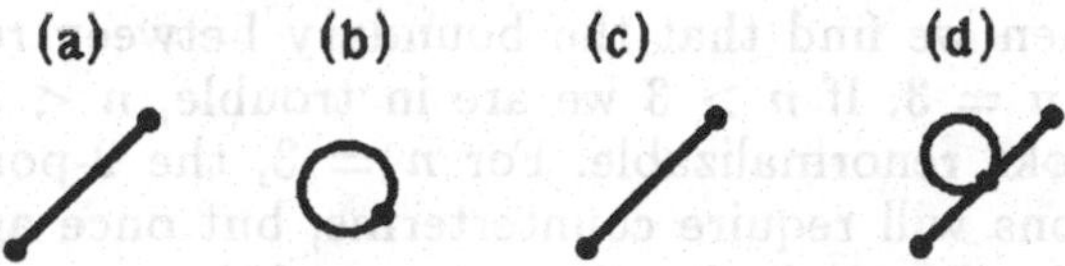

Figure 16.5 "Proper vertices" resulting from the truncation of the diagrams in figure 16.1. Note that diagrams 16.1(a) and (c) contain the same proper vertex.

disconnect pieces, we will eventually break the diagram into 1PI pieces. If each piece is made finite, then the whole works put back together again will be finite.

A truncated diagram is one with its external legs removed. When we truncate the diagrams in figure 16.1, we end up with the diagrams in figure 16.5. Note that truncation of figure 16.1(a) and 16.1(c) results in the same "proper vertex." Why do we only care about the truncated versions of 1PI diagrams? Recall that the effect of the reduction formula is to truncate the legs of any diagram by acting with the operator, $1/\sqrt{Z}$ $(\partial^\mu \partial_\mu + m^2)$, on the n-point function and taking the result on shell. Since the renormalized mass appears in the operator above, any self-energy part hanging on an external leg, as in figure 16.1(c), is also chopped. Also note that the part of the 2-point function to $O(g)$ that had to be subtracted in our example, eq.(16.23), is Σ_1. By eqs.(16.14) and (16.15), we can see that Σ_1 represents the truncated diagram for the 2-point function at order g.

The proper diagrams represent the terms in a perturbative expansion of proper functions. Any Green's function can be built up from the proper functions. To make things finite, we worry only about the proper parts. The generating functional of the proper functions is called the effective action. Given the role of the proper functions, it is easy to see why it is desirable to compute the effective action. We will consider the effective action in more detail in chapter 18 and use it to renormalize the self-interacting scalar theory $\lambda\phi^4$ at 1-loop.

In ϕ^4 theory, if subtractions are made to the 2-point and 4-point functions to render them finite, then any diagram will be finite. To accomplish this, we do not need to generate counterterms that are functionally different from the terms already present in the action. How special is ϕ^4? Consider a potential, $V = \phi^n$ with $n > 4$ in 4 space-time dimensions. From eq.(16.30), we can see that there will be a diagram for each E-point function that will have $D < 0$. This is a sure sign that we will need an infinite number of different counterterms beyond what is already in the action to perform the subtraction, and the theory will be nonrenormalizable. For $n < 4$ there appears to be no problem. If we look at 6 space-time dimensions, then we find that the boundary between renormalizable or not occurs at $n = 3$. If $n > 3$ we are in trouble, $n < 3$ is no problem, and $n = 3$ looks renormalizable. For $n = 3$, the 2-point, 4-point, and 6-point functions will require counterterms, but once again they are already present in the original action, or are of the renormalizable variety. The field ϕ has dimensions of 1 over length $(1/l)$ in 4 space-time dimensions. Therefore, the coupling λ in $\lambda\phi^4$ is dimensionless, ν in $\nu\phi^3$ has units of $1/l$, and ρ in $\rho\phi^6$ has units of l^2. In 6 space-time dimensions, the field ϕ has dimension of $1/l^2$. This makes ν dimensionless, and means λ has units of l^2. The boundary between renormalizable and nonrenormalizable apparently occurs when the classical coupling is dimensionless. Couplings with units of length that are positive are to be avoided. If such a coupling crops up as a counterterm, then trouble has begun.

Since loops play an important role in renormalization, it is convenient to work with an expansion of the vacuum-to-vacuum amplitude of Green's functions in terms of the number of loops. Certainly this expansion just reorders the diagrams or terms contained in a coupling constant expansion. However, the loop expansion has an obvious physical interpretation in that it is an expansion in powers of $\hbar$. The lowest order terms, the tree diagrams (no loops), are independent of $\hbar$. They correspond to a perturbative solution of the classical field theory. There is no quantum mechanics involved other than the motivation to write down a classical field theory using spinors, etc. It is also no surprise that the tree diagrams are finite. The 1-loop terms are proportional to $\hbar$, and are the first quantum corrections to the classical field theory, and so on.

To see that an expansion in powers of $\hbar$ is an expansion in loops, consider $Z[J]$ for the scalar field described by the action,

$$S[\varphi] = \int d^4x\, \frac{1}{2}(\partial_\mu\varphi\partial^\mu\varphi - m^2\varphi^2) - V(\varphi), \tag{16.31}$$

but this time put the factor of $1/\hbar$ in front explicitly.

$$\begin{aligned} Z[J] &= \int \mathcal{D}\phi \, \exp\left(\frac{i}{\hbar}S[\phi] + i\int d^4x \, J(x)\phi(x)\right) \\ &= \exp\left(-\frac{i}{\hbar}V\Big(-i\frac{\delta}{\delta J}\Big)\right) \\ &\quad \int \mathcal{D}\phi \, \exp\left(\frac{i}{\hbar}\int d^4x \, \frac{1}{2}(\partial^\mu\phi\partial_\mu\phi + m^2\phi^2) + \hbar J\phi\right) \\ &= \exp\left(-\frac{i}{\hbar}V\Big(-i\frac{\delta}{\delta J}\Big)\right) \\ &\quad \exp\left(\frac{i\hbar}{2}\int d^4x \, d^4x' \, J(x)\Delta_F(x-x')J(x')\right) \end{aligned} \tag{16.32}$$

$-i/\hbar\, V(-i\delta/\delta J)$ generates a vertex in a diagram; therefore each vertex present contributes a factor of $\hbar^{-1}$. Each internal line represents $\Delta_F(x - x')$, and therefore contributes a factor of $\hbar$. Each external leg also contributes a factor of $\hbar$. Therefore, a diagram with E external legs, I internal legs, and v vertices will be proportional to $\hbar^{E+I-v}$, which, from eq.(16.28), is $\hbar^{E-1+L}$. If we restrict our attention to truncated diagrams, then $E = 0$, and the diagram is proportional to $\hbar^{L-1}$. The factor $1/\hbar$ that appears in front of tree diagrams simply reflects that the action we started from is multiplied by $1/\hbar$ (that is, a factor of $1/\hbar$ will also appear in front of $W[J]$). This factor of $1/\hbar$ in the classical limit may be absorbed into the source of the classical field equation. Then, the E-point function, even in the classical limit, will have a factor of $\hbar^E$ in front. Therefore, if we leave the legs on the diagram, we get the same result.

In the next chapter we will renormalize QED at 1-loop. We will concentrate on extracting the consequences of the results in view of the discussion in this chapter. One consequence of renormalization that we have not discussed yet will appear with respect to processes in QED involving axial vector currents coupled to the usual vector currents. At the classical level the Lagrangian is chirally invariant. There is no distinction between left-handedness and right-handedness. Parity is conserved. The classical action is invariant under global rotations generated by γ_5. Renormalization of diagrams describing processes that connect axial vector-to-vector currents require regularization. The regularized theory is *not* invariant under γ_5 rotations. Therefore, the chiral symmetry is broken by the renormalization process. Such symmetry-breaking quantum corrections are called anomalies. We will compute the chiral anomaly at 1-loop in chapter 17. We will encounter anomalies again in string theory where the conformal anomaly plays a central role.

16.2 EXERCISES

1. What is the superficial degree of divergence for a scalar theory in 2 space-time dimensions for $V(\phi) = \phi^n$? Note that $D \geq 0$ only for diagrams that have no vertex or 1 vertex. These divergences may be eliminated simply by normal ordering. Such a theory is called "super-renormalizable." This is another example of the desirable properties that occur in 2-d. Strings are one way to elevate the nice properties of 2-dimensional theories to higher space-time dimensions. In this case, the 2-d theory lives on the world-sheet of the string.

CHAPTER **17**

Renormalization of QED at 1-Loop

Our goal is to make finite predictions from 1-loop QED by suitable redefinitions of the parameters m_0, e_0, A_0^μ, and ψ_0 that appear in the classical action.

17.1 PRELIMINARIES

We have already written down three 1-loop graphs that are divergent: figures 8.8, 8.9, and 8.10. How many more can we expect? To get an idea, let's compute the superficial degree of divergence, D, when we have a mixture of fermions and bosons.

Recall that the superficial degree of divergence is the difference between the number of factors of internal momenta, k_i, being integrated over in the numerator versus the number in the denominator. The factors of k_i in the numerator arise from the volume elements $d^d k_i$ in d dimensions, thus there are d factors of k_i for each internal momentum. Recall from the previous chapter that the number of independent internal momenta to integrate over is equal to the number of loops, L, thus the total number of factors of k_i in the numerator is $d\,L$. The factors of k_i in the denominator arise from the internal propagators, i.e. the internal lines. We have one factor of k_i for each fermion propagator, and one factor of

k_i^2 for each boson propagator. Thus, the total number of factors of k_i in the denominator is $I_f + 2I_b$, where I_f is the number of internal fermion lines in the graph and I_b is the number of internal boson lines. Altogether, the superficial degree of divergence is

$$D = dL - I_f - 2I_b. \tag{17.1}$$

Now consider a diagram with V vertices. To produce a connected diagram we must connect the vertices like a dot-to-dot puzzle using $V-1$ internal lines. Any extra internal lines left lying around must now form loops when connected to any two vertices. Thus,

$$L = I_b + I_f + 1 - V. \tag{17.2}$$

There is only one vertex in QED, and it connects 2 fermionic propagators to one bosonic line. Both ends of an internal line must be connected to a vertex, while only one end of an external leg terminates at a vertex. Since one and only one boson line must terminate at each vertex, then

$$V = 2I_b + E_b, \tag{17.3}$$

where E_b is the number of external bosonic legs present. By identical reasoning, since two and only two fermionic lines must terminate at each vertex, then

$$2V = 2I_f + E_f. \tag{17.4}$$

In 4 dimensions, we may express D entirely in terms of E_b and E_f by substituting eq.(17.2) into eq.(17.1) and eliminating V using eq.(17.3) plus 3/2 times eq.(17.4).

$$D = 4 - E_b - \frac{3}{2}E_f \tag{17.5}$$

The diagram will be superficially divergent if $D > 0$, which means that $E_f \leq 2$. Since E_f must be even, this leaves $E_f = 0$ and 2. When $E_f = 0$, then we may have $E_b = 2, 3, 4$. The diagram with $E_b = 3$ must vanish by Furry's theorem (i.e. charge conjugation invariance - see chapter 8). This leaves $E_b = 2$, the vacuum polarization diagram, figure 17.1(a), with $D = 2$, and $E_b = 4$ which represents light scattering off of light, figure 17.1(b), with $D = 0$. When $E_f = 2$, we may only have $E_b = 0$ and 1. $E_b = 0$ is the electron self-energy diagram we have already encountered, figure 17.1(c), with $D = 1$, while $E_b = 1$ is the vertex correction, figure 17.1(d), with $D = 0$.

As we have shown in chapter 11, the electron self-energy diagram, figure 17.1(c), is only logarithmically divergent even though $D = 1$. The vacuum polarization and vertex correction, figures 17.1(a) and (d), also are logarithmically divergent. Each will generate counterterms of the form

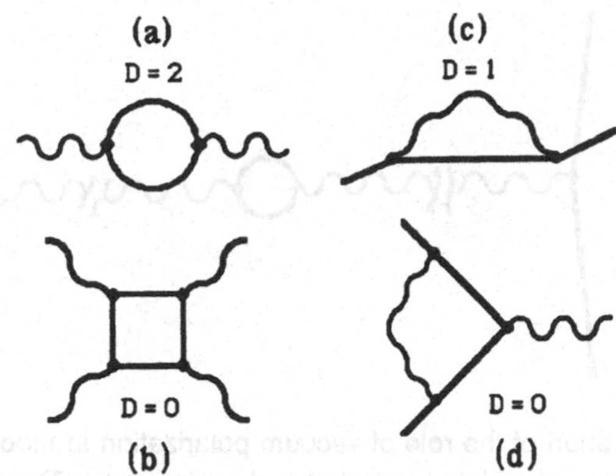

Figure 17.1 Superficially divergent diagrams in QED at 1-loop: (a) vacuum polarization, $D = 2$. (b) photon-photon scattering, $D = 0$. (c) electron self-energy, $D = 1$. (d) vertex correction, $D = 0$.

already present in the original classical action. However, the light-light scattering diagram, figure 17.1(b), is a different matter. It is superficially logarithmically divergent. The difficulty is that, if divergent, it would generate a new counterterm involving the product of 4 photon fields, a completely new term not present classically. This is an unhealthy prospect. Fortunately, this diagram is actually convergent due to gauge invariance.

Given the divergent diagrams, what can we expect in terms of counterterms? From the electron self-energy diagram we can expect a mass shift counterterm so as to fix the pole of the electron propagator at its physically measured mass. This won't fix the value of the residuc at the pole, thus we can also expect a counterterm that renormalizes or rescales the fields ψ_0 and $\bar{\psi}_0$. Together, both renormalizations generate counterterms that look like a rescaling of the gauge covariant Dirac action.

Since there is no mass term for the photon, we can expect that the vacuum polarization diagram will *not* require us to introduce a mass shift counterterm to keep the pole of the photon propagator at zero. It would be very unhealthy for the theory if such a shift was required because the mass term is not gauge invariant. We will need a counterterm to fix the residue of the photon propagator pole. Such a term will rescale A_0^μ and will look like $F^{\mu\nu}F_{\mu\nu}$. Such a rescaling of A_0^μ will also affect the vertex term. We may interpret this as a rescaling of the electric charge because the residue of the pole of the photon propagator as $k^2 \to 0$ is measurably related to the coefficient of the static $1/r$ Coulomb potential felt by two identical, widely separated sources (e.g. electrons). The role of the corrected photon propagator is given by the diagram in figure 17.2. The two electrons are

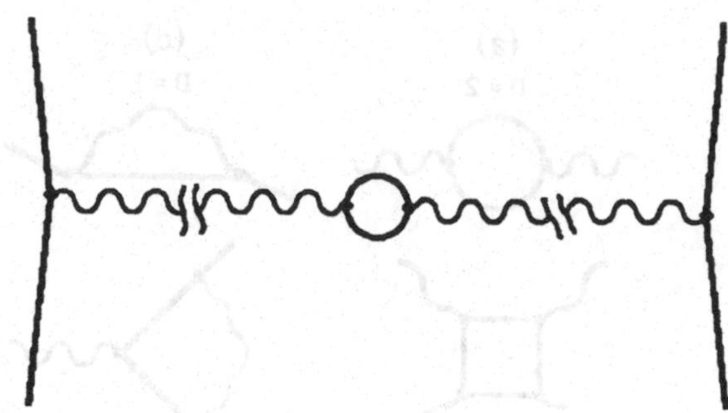

Figure 17.2 Illustration of the role of vacuum polarization in modifying Coulomb's law. Here the 2 electrons are widely separated and nearly static. The momentum exchange in scattering is small. This puts the photon nearly on shell. The modification of Coulomb's law can be interpreted as a rescaling of the renormalized, physically measurable electric charge. This makes the electric charge scale-dependent and leads to the notion of a running coupling constant.

widely separated so the momentum transfer via the photon is very tiny. This puts the photon nearly on mass shell.

The vertex correction will generate a counterterm that looks like the vertex itself except with a different scaling. In summary, we expect to generate a δm and 3 "wave function" renormalizations: Z_2 for ψ_0 and $\bar{\psi}_0$, Z_3 for A_0^μ, and Z_1 for γ^μ. The electric charge e_0 is rescaled by the shifts of the fields and γ^μ in the interaction term in the action.

17.2 REGULATORS

Having identified the divergent diagrams, we can now begin the renormalization process. The first step is to regularize the theory in some way such that all quantities are finite. In all 3 diagrams (figure 17.3), we are integrating over the momentum that flows around the loop. The integrals diverge because the propagators in the loops do not vanish quickly enough as the loop momentum (p in figure 17.3(a), and k in figures 17.3(b) and (c)) gets large. Since it is an ultraviolet divergence (large momentum), it is the short-distance behavior of the theory that is not properly behaved. Figure 17.4 is the vacuum polarization diagram written in x-space. The loop introduces the factor $(S_F(x-y))^2$ into the integrand. $S_F(x-y)$ is a distribution, and the product of 2 distributions is not well defined. The integral over x and y diverges as x approaches y.

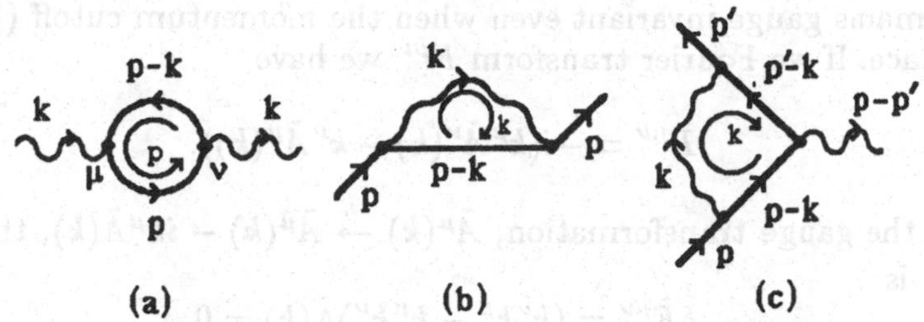

Figure 17.3 Loops represent integrals over the internal momentum flowing in the loop as illustrated above. In each case the integral diverges because the propagators in the loop do not vanish quickly enough as the loop momentum (p in (a), k in (b),(c)) gets large.

One obvious way to regularize the diagrams is to prevent $x - y$ from going to zero, that is, make the interaction vertex fuzzy at x and y to smoothe things out. Equivalently, we can just cut off the loop momentum integral at some large but finite value of momentum. This is what we did in chapter 11 when we computed the perturbative shift in the energy eigenvalue of the single-electron state (which is equivalent to the shift of the pole in the electron propagator due to diagram 17.3(b)). While such a regularization is straightforward, it is not desirable here because it breaks the gauge symmetry. The regularized theory is no longer a gauge theory.

To see why this is so, consider the field equation,

$$\partial_\mu F^{\mu\nu} = j^\nu. \tag{17.6}$$

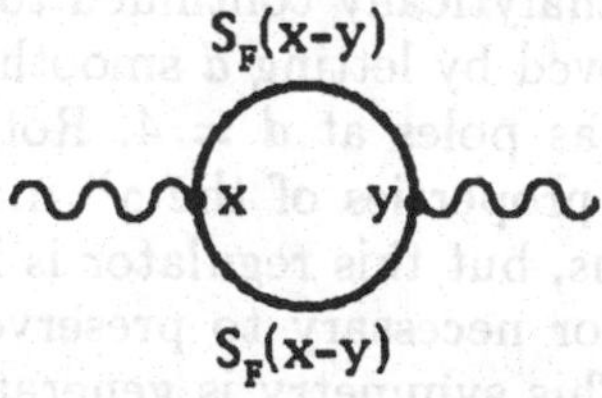

Figure 17.4 Vacuum polarization diagram written in x-space. The loop introduces a factor of $S_F(x-y)^2$, a product of 2 distributions, that is not necessarily well defined. The integral over x and y diverges as x approaches y.

$F^{\mu\nu}$ remains gauge invariant even when the momentum cutoff ($|k^\mu| < \Lambda$) is in place. If we Fourier transform $F^{\mu\nu}$ we have

$$\tilde{F}^{\mu\nu} = -i(k^\mu \tilde{A}^\nu(k) - k^\nu \tilde{A}^\mu(k)).$$

Under the gauge transformation, $\tilde{A}^\mu(k) \to \tilde{A}^\mu(k) - ik^\mu\tilde{\Lambda}(k)$, the change in $\tilde{F}^{\mu\nu}$ is

$$\delta\tilde{F}^{\mu\nu} = (k^\nu k^\mu - k^\mu k^\nu)\tilde{\Lambda}(k) = 0.$$

Since the Fourier transform of $F^{\mu\nu}$ and the gauge transformation are locally defined in k^μ-space, that is, they involve only a single point in k^μ-space, the symmetry is preserved when a cutoff is applied. However, for the charged fermionic currents, $j^\nu = \bar{\psi}(x)\gamma^\nu\psi(x)$, this is not true. Since the current is the product of two fields, the Fourier transform is a convolution, and therefore, involves an integral over all of k^μ space. If we cut off this convolution and transform back, the current is no longer locally defined nor gauge invariant. After regularizing in this way, the left-hand side of eq.(17.6) is still gauge invariant, but the right-hand side is not.

There are several popular ways to regularize in a gauge invariant manner. A momentum cutoff can be implemented by defining the gauge theory on discretized space-time. Usually a uniform discretization is used resulting in a regular lattice. The lattice spacing, a^μ, provides a natural momentum cutoff at $k^\mu = \pi/a^\mu$. The regulator is removed by letting the lattice spacing go to zero. As with any momentum cutoff, the lattice regulator does break rotational invariance. Also, difficulties arise when Dirac fermions are placed on a lattice.

For a given diagram, the superficial degree of divergence may be decreased simply by lowering the dimension of space-time. Dimensional regularization amounts to computing divergent diagrams in an arbitrary (smaller) space-time dimension where they are convergent. The result of the integration is analytically continued to non-integer dimensions, d. The regulator is removed by letting d smoothly go back to 4. Ultraviolet divergences show up as poles at $d = 4$. Rotational invariance is maintained. The essential properties of the γ^μ matrices can be continued to non-integer dimensions, but this regulator is inappropriate for situations where it is desirable or necessary to preserve chiral symmetries involving Dirac fermions. This symmetry is generated by the γ_5 matrix whose properties change in different dimensions.

A gauge invariant regulator special to Abelian gauge theories that does not suffer the difficulties mentioned above is the Pauli-Villars regularization. Pauli-Villars regularization amounts to adding auxiliary spinor fields with large masses. In addition, the photon field is also modified (which is why this regulator is special to QED). Gauge invariance is maintained because the fields are minimally coupled to the photons with the gauge covariant derivative. In order to cancel all of the divergences, the auxiliary spinors may have to acquire some strange properties: e.g. some of the spinor fields may be bosonic. The regulator is removed by letting the auxiliary masses go to ∞. The infinite-mass particles will then decouple leaving the original theory.

In the remainder of this section we will first employ Pauli-Villars regularization to compute the electron self-energy diagram, figure 17.3(b). We will then recompute this diagram along with the vacuum polarization and vertex correction using dimensional regularization. In the last section we will use Pauli-Villars to compute the chiral anomaly.

To regularize QED via Pauli-Villars we add auxiliary massive spinor fields and modify the photon field. At 1-loop, the modification to the photon field is sufficient to regularize the electron self-energy diagram and the vertex correction (figures 17.3(b) and (c)) and only 1 additional spinor field (satisfying Bose statistics) is required to regularize the vacuum polarization diagram (figure 17.3(a)). At higher loops we will need 3 auxiliary spinor fields (1 fermionic and 2 bosonic).

The regularized action density in the Lorentz gauge (also referred to as the "Feynman gauge") is

$$\mathcal{L}_{\text{reg}} = -\frac{1}{4}F^{\mu\nu}\left(\frac{\partial_\sigma\partial^\sigma + M^2}{M^2}\right)F_{\mu\nu} - \frac{1}{2}\partial_\mu A^\mu\left(\frac{\partial_\sigma\partial^\sigma + M^2}{M^2}\right)\partial_\nu A^\nu$$
$$+ \bar{\psi}(i\partial\!\!\!/ - m_0 - e_0 A\!\!\!/)\psi + \sum_{i=1}^{N}\bar{\psi}_i(i\partial\!\!\!/ - M_i - e_0 A\!\!\!/)\psi_i. \tag{17.7}$$

The regularization is gauge invariant because the spinor fields are minimally coupled to the photon field using the gauge covariant derivative, $D\!\!\!\!/ = \partial\!\!\!/ + ie_0 A\!\!\!/$, and because $F_{\mu\nu}$ is gauge invariant. This regularization is special to QED because the field strength is invariant. Pauli-Villars will not maintain gauge invariance in Yang-Mills or QCD because the color $F_{\mu\nu}$ is only gauge covariant, not invariant, thus the higher derivatives that are added will break the gauge invariance of the action.

To determine the modified photon propagator we examine the piece of the action that is quadratic in the photon field. Fourier transforming this piece we have

$$\int d^4x\, \mathcal{L}_{\text{reg}}(A^2) \longrightarrow$$

$$\frac{1}{4}\int \frac{d^4k}{(2\pi)^4}(k_\mu \tilde{A}_\nu - k_\nu \tilde{A}_\nu)\left(\frac{k^2 - M^2}{M^2}\right)(k^\mu \tilde{A}^\nu - k^\nu \tilde{A}^\mu)$$
$$+\frac{1}{2}\int \frac{d^4k}{(2\pi)^4}(k_\mu \tilde{A}^\mu)\left(\frac{k^2 - M^2}{M^2}\right)(k_\nu \tilde{A}^\nu)$$
$$= \frac{1}{2}\int \frac{d^4k}{(2\pi)^4} k^2 \left(\frac{k^2 - M^2}{M^2}\right)\tilde{A}^\mu \tilde{A}_\mu. \tag{17.8}$$

Therefore, the regularized photon propagator is

$$D^{\text{reg}}_{\mu\nu}(k) = g_{\mu\nu}\,\frac{M^2}{k^2(k^2 - M^2)} = -g_{\mu\nu}\left(\frac{1}{k^2} - \frac{1}{k^2 - M^2}\right). \tag{17.9}$$

For $M \to \infty$, we recover the original photon and its propagator, eq.(5.64).

What are the effects of the extra spinors? Whenever a closed fermion loop appears in a diagram we must sum over the contributions of each spinor field to the loop. If the spinor is a true fermion (obeying Fermi-Dirac statistics, i.e. anticommutators), then one Feynman rule states that each closed loop introduces a factor of -1. Therefore, the effects of identical fermionic and bosonic spinors traveling around a loop cancel. To regularize fermionic loops, we add auxiliary spinor fields, some bosonic, some fermionic, and arrange the masses of the fields and their statistics to cancel the contribution of the loop at large loop momentum.

To illustrate this, let's consider, qualitatively, the vacuum polarization diagram in figure 17.3(a). The interesting part of the diagram, the fermion loop, for the original electron is proportional to

$$\int \frac{d^4p}{(2\pi)^4}\,\text{tr}\left(\frac{1}{\not p - m_0 + i\epsilon}\gamma_\mu \frac{1}{\not p - \not k - m_0 + i\epsilon}\gamma_\nu\right). \tag{17.10}$$

To this we must add the contribution of the auxiliary spinors,

$$\sum_{i=1}^{N} c_i \int \frac{d^4p}{(2\pi)^4}\,\text{tr}\left(\frac{1}{\not p - M_i + i\epsilon}\gamma_\mu \frac{1}{\not p - \not k - M_i + i\epsilon}\gamma_\nu\right), \tag{17.11}$$

where $c_i = 1$ for fermions and $c_i = -1$ for bosons.

The spinor propagators may be rewritten as

$$\frac{1}{\not p - m + i\epsilon} = \frac{1}{\not p - m + i\epsilon}\frac{\not p + m}{\not p + m} = \frac{\not p + m}{p^2 - m^2 + i\epsilon}. \tag{17.12}$$

Using the trace identities in the exercises of chapter 4, expression (17.10) becomes

$$\int \frac{d^4p}{(2\pi)^4}\frac{1}{p^2 - m_0^2 + i\epsilon}\frac{1}{(p-k)^2 - m_0^2 + i\epsilon}\,\mathrm{tr}\left((\not p + m_0)\gamma_\mu(\not p - \not k + m_0)\gamma_\nu\right)$$

$$= 4\int \frac{d^4p}{(2\pi)^4}\frac{1}{p^2 - m_0^2 + i\epsilon}\frac{1}{(p-k)^2 - m_0^2 + i\epsilon}$$
$$(p_\mu(p-k)_\nu + p_\nu(p-k)_\mu - (p^2 - m^2 - p\cdot k)g_{\mu\nu}). \tag{17.13}$$

Eq.(17.11) reduces to an expression identical to eq.(17.13) with m_0^2 replaced by M_i^2 and multiplied by c_i. For large $|p|$, eq.(17.13) goes like

$$\int \frac{d^4p}{(2\pi)^4}\frac{1}{|p|^4}(2p^2 - m_0^2). \tag{17.14}$$

The extra spinors at large $|p|$ will contribute

$$\sum_{i=1}^{N} c_i \int \frac{d^4p}{(2\pi)^4}\frac{1}{|p|^4}(2p^2 - M_i^2). \tag{17.15}$$

To regularize, we want eq.(17.14) plus (17.15) to vanish. This means that c_i and M_i must satisfy

$$1 + \sum_{i=1}^{N} c_i = 0 \quad \text{and} \quad m_0^2 + \sum_{i=1}^{N} c_i M_i^2 = 0. \tag{17.16}$$

One particular solution to the constraints which involves the minimal number of additional spinors is

$$c_1 = 1,\ c_2 = c_3 = -1,\ c_i = 0 \quad \text{for } i \geq 4$$

$$M_1^2 = m_0^2 + 2M^2, \quad M_2^2 = M_3^2 = m_0^2 + M^2. \tag{17.17}$$

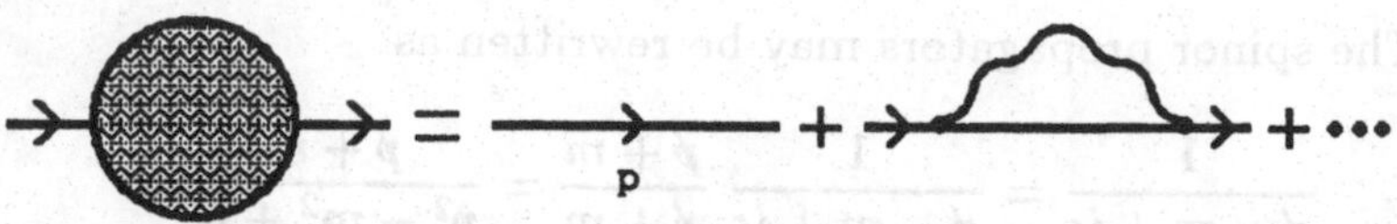

Figure 17.5 The electron self-energy diagram is the second term (the first "radiative" correction) in a perturbative expansion of the full electron propagator.

Due to gauge invariance, the 1-loop vacuum polarization diagram only diverges logarithmically. The piece involving m_0^2 in eq.(17.14) is canceled by part of the rest of the diagram. Therefore, we need only the first constraint in eq.(17.16) which is easily satisfied with 1 extra spinor field acting as a boson with $c_1 = -1$ and $M_i^2 = M^2$.

Now let us move on to compute the regularized electron self-energy diagram. We will leave the actual subtraction and renormalization for the next section. The self-energy diagram is the second term in a perturbative expansion of the full electron propagator as illustrated in figure 17.5. Using the Feynman rules given in chapter 8 and the momentum and vertex labels of figure 17.6, the self-energy diagram is equal to

$$\frac{i}{\not p - m_0 + i\epsilon}\left(-i\Sigma(p)\right)\frac{i}{\not p - m_0 + i\epsilon}, \tag{17.18}$$

where

$$\Sigma(p) = (-ie_0)^2 \int \frac{d^4k}{(2\pi)^4}\frac{1}{k^2 + i\epsilon}\gamma_\mu \frac{i}{\not p - \not k - m_0 + i\epsilon}\gamma^\mu. \tag{17.19}$$

The interesting part of the diagram, $\Sigma(p)$, is ultraviolet divergent so we must regularize it by modifying the photon propagator. However, the derivative of $\Sigma(p)$ is also infrared ($k^\mu \to 0$) divergent when the external particles are "on mass shell" ($p^2 = m^2$ for electrons and $k^2 = 0$ for photons). This divergence arises from the fact that the photon is massless and is unphysical. We will return to discuss it later. For now, we can regularize it by giving the photon a tiny mass, λ. Using the regularized photon propagator, eq.(17.9), $\Sigma(p)$ becomes

$$\Sigma^{\text{reg}}(p) = (-ie_0)^2 \int \frac{d^4k}{(2\pi)^4}\left(\frac{1}{k^2 - \lambda^2 + i\epsilon} - \frac{1}{k^2 - M^2 + i\epsilon}\right)$$
$$\gamma_\mu \frac{i}{\not p - \not k - m_0 + i\epsilon}\gamma^\mu, \tag{17.20}$$

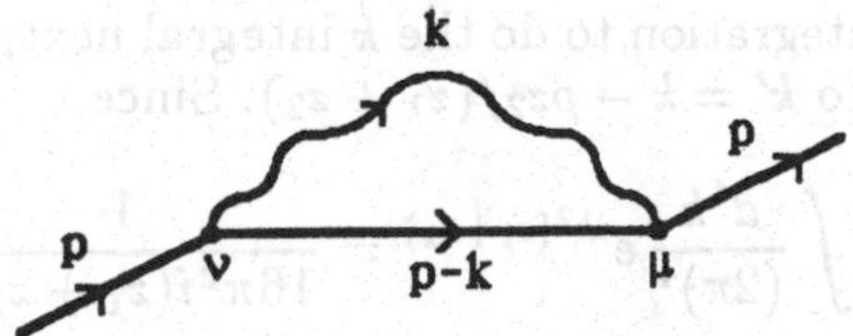

Figure 17.6 Momentum/vertex labels used in the expression for the loop, $\Sigma(p)$, in eq.(17.19) below.

where $M^2 \gg \lambda^2$. Let us just concentrate on the first term of the photon propagator. After we have evaluated it, we can merely substitute M^2 for λ^2 to get the second piece.

To evaluate the integral, we reach into Feynman's bag of tricks. First we elevate the propagator denominators into exponentials.

$$\begin{aligned}\frac{i}{\not p - \not k - m_0 + i\epsilon} &= \frac{i(\not p - \not k + m_0)}{(p-k)^2 - m_0^2 + i\epsilon} \\ &= (\not p - \not k + m_0)\int_0^\infty dz\; e^{iz((p-k)^2 - m_0^2 + i\epsilon)},\end{aligned} \tag{17.21}$$

and,

$$\frac{i}{k^2 - \lambda^2 + i\epsilon} = \int_0^\infty dz\; e^{iz(k^2 - \lambda^2 + i\epsilon)}. \tag{17.22}$$

The first term of eq.(17.20) becomes

$$-i(-ie_0)^2 \int \frac{d^4k}{(2\pi)^4}\int_0^\infty dz_1 \int_0^\infty dz_2 \\ e^{i(k^2-\lambda^2-i\epsilon)z_1}\; e^{i((p-k)^2 - m_0^2 + i\epsilon)z_2}\gamma_\mu(\not p - \not k + m_0)\gamma^\mu. \tag{17.23}$$

Using the commutator for the γ matrices, it is easy to show that

$$\gamma_\mu \not p \gamma^\mu = -2\not p, \tag{17.24}$$

so that $\gamma_\mu(\not p - \not k + m_0)\gamma^\mu = -2\not p + 2\not k + 4m_0$.

Next, we complete the square in the exponentials with respect to k,

$$\begin{aligned}&\exp(ik^2 z_1)\; \exp(i(p-k)^2 z_2) \\ &\quad = \exp(ip^2 z_2)\; \exp(i(z_1+z_2)k^2 - 2i(p\cdot k)z_2) \\ &\quad = \exp(ip^2 z_2)\; \exp\left(i\left(k - \frac{pz_2}{z_1+z_2}\right)^2 (z_1+z_2)\right)\exp\left(-i\frac{p^2 z_2^2}{z_1+z_2}\right),\end{aligned} \tag{17.25}$$

flip the order of integration to do the k integral next, and shift integration variables from k to $k' = k - pz_2/(z_1 + z_2)$. Since,

$$\int \frac{d^4k}{(2\pi)^4} e^{ik^2(z_1+z_2)} = \frac{1}{16\pi^2 i(z_1+z_2)^2},$$

$$\int \frac{d^4k}{(2\pi)^4} \not{k}\, e^{ik^2(z_1+z_2)} = 0, \tag{17.26}$$

then eq.(17.23) becomes

$$\left(\frac{e_0^2}{4\pi}\right)\frac{1}{2\pi}\int_0^\infty dz_1 \int_0^\infty dz_2 \left(2m_0 - \frac{\not{p}z_1}{z_1+z_2}\right)\frac{1}{(z_1+z_2)^2} \exp\left(i\left(\frac{p^2 z_1 z_2}{z_1+z_2} - m_0^2 z_2 - \lambda^2 z_1\right)\right). \tag{17.27}$$

Next, insert the identity

$$1 = \int_0^\infty \frac{d\beta}{\beta}\,\delta\left(1 - \frac{z_1+z_2}{\beta}\right), \tag{17.28}$$

and redefine z_1 to βz_1 and z_2 to βz_2. We have

$$\frac{\alpha_0}{2\pi}\int_0^\infty dz_1 \int_0^\infty dz_2 \left(2m_0 - \frac{\not{p}z_1}{z_1+z_2}\right)\frac{1}{(z_1+z_2)^2}\,\delta(1-(z_1+z_2)) \int_0^\infty \frac{d\beta}{\beta}\exp\left(i\beta\left(\frac{p^2 z_1 z_2}{z_1+z_2} - m_0^2 z_2 - \lambda^2 z_1\right)\right), \tag{17.29}$$

where $\alpha_0 = e_0^2/4\pi$ is the (bare) fine structure constant. The δ-function makes the integral over z_1 easy to do. Switching variables from z_2 to $z = (1 - z_2)$, we obtain

$$\Sigma^{\text{reg}}(p) = \frac{\alpha_0}{2\pi}\int_0^1 dz\,(2m_0 - z\not{p}) \int_0^\infty \frac{d\beta}{\beta}\exp(i\beta I(p, m_0, \lambda)) - \exp(i\beta I(p, m_0, M)), \tag{17.30}$$

where

$$I(p, m_0, \lambda) = p^2 z(1-z) - m_0^2(1-z) - \lambda^2 z + i\epsilon. \tag{17.31}$$

The integral over β can be done,

$$\int_0^\infty \frac{d\beta}{\beta}\exp(ia\beta) - \exp(ib\beta) = \ln\left(\frac{a}{b}\right), \tag{17.32}$$

so, *finally*,

$$\Sigma^{\rm reg}(p) = \frac{\alpha_0}{2\pi}\int_0^1 dz(2m_0 - z\not{p})\ln\left(\frac{M^2 z}{m_0^2(1-z)+\lambda^2 z - p^2 z(1-z) - i\epsilon}\right), \tag{17.33}$$

where we have assumed that M^2 is large enough that the rest of the terms in the numerator of the log can be neglected. $M^2 \rightarrow \infty$ removes the regulator, so we can see that $\Sigma(p)$ is logarithmically divergent.

The $\lambda \rightarrow 0$ limit may safely be taken when $p^2 < m_0^2$ in which case

$$\Sigma^{\rm reg}(p) = \frac{\alpha_0}{2\pi}\left((2m_0 - \frac{\not{p}}{2})\ln\left(\frac{M^2}{m_0^2}\right) + 2m_0\left[1 + \frac{m_0^2 - p^2}{p^2}\ln\left(1 - \frac{p^2}{m_0^2}\right)\right]\right.$$
$$\left. - \frac{1}{2}\not{p}\left[\frac{3}{2} + \frac{m_0^4 - p^4}{p^4}\ln\left(1 - \frac{p^2}{m_0^2}\right) + \frac{m_0^2}{p^2}\right]\right). \tag{17.34}$$

We will return to $\Sigma^{\rm reg}(p)$ in the next section to do the subtraction and complete the renormalization of the electron propagator at 1-loop.

Let us return now to $\Sigma(p)$, eq.(17.19), and now employ dimensional regularization. To dimensionally regularize we compute the integral in $2n$ dimensions. The electric charge is dimensionless in 4 space-time dimensions, but acquires a dimension of $(mass)^{1/2}$ in 3 space-time dimensions, and so on. For our convenience, we will write the dimensionful electric charge as a dimensionless number times a mass μ to the appropriate power. Thus, e_0 in eq.(17.19) is replaced by $e_0\,\mu^{2-n}$. The dimensionally regularized $\Sigma(p)$ is

$$\Sigma^{\rm reg}(p) = (-ie_0\mu^{2-n})^2\int\frac{d^{2n}k}{(2\pi)^{2n}}\frac{1}{k^2+i\epsilon}\gamma_\mu\frac{i}{\not{p} - \not{k} - m_0 + i\epsilon}\gamma^\mu. \tag{17.35}$$

Once again we elevate the denominators into exponentials by applying eqs.(17.21) and (17.22).

$$\Sigma^{\rm reg}(p) = (-ie_0\mu^{2-n})^2\int\frac{d^{2n}k}{(2\pi)^{2n}}\int_0^\infty dz_1\int_0^\infty dz_2$$
$$\exp(i(k^2+i\epsilon)z_1)\exp(i((p-k)^2 - m_0^2 + i\epsilon)z_2)\gamma_\mu(\not{p} - \not{k} + m_0)\gamma^\mu. \tag{17.36}$$

The trace of a product of γ matrices is dimension-dependent,

$$\gamma_\mu\gamma^\mu = 2n, \qquad \text{and,} \qquad \gamma^\mu\gamma^\nu\gamma_\mu = -2 + 2(2-n)\gamma^\nu, \tag{17.37}$$

thus the product of γ matrices in eq.(17.36) reduces to

$$-(2-2(2-n))(\not p-\not k)+2nm_0.$$

Again we complete the squares in the exponential, a computation identical to eq.(17.25). Flipping the order of integration, and shifting the integration variable, we now have

$$\Sigma^{\rm reg}(p)=-i(-ie_0\mu^{2-n})^2\int_0^\infty dz_1\int_0^\infty dz_2\exp\left(ip^2\frac{z_1z_2}{z_1+z_2}-m_0^2z_2\right)$$

$$\int\frac{d^{2n}k}{(2\pi)^{2n}}\left(2nm_0-(2-2(2-n))\left(\not p\frac{z_1}{z_1+z_2}-\not k\right)\right)\exp(ik^2(z_1+z_2)). \tag{17.38}$$

The Gaussian integral in $2n$ dimensions is

$$\begin{aligned}\int\frac{d^{2n}k}{(2\pi)^{2n}}\exp(iak^2)&=\frac{1}{(2\pi)^{2n}}\left(\frac{\pi}{ia}\right)^{\frac{1}{2}}\left(\frac{\pi}{-ia}\right)^{\frac{2n-1}{2}}\\&=\frac{1}{(4\pi a)^n}\frac{-1}{(-1)^{1/2}}\frac{1}{(-i)^n}\\&=-\frac{1}{(4\pi a)^n}\frac{\exp(i\pi n/2)}{i}.\end{aligned} \tag{17.39}$$

Odd moments vanish so eq.(17.38) reduces to

$\Sigma^{\rm reg}(p)=$

$$-\left(\frac{e_0^2}{4\pi}\right)\frac{1}{4\pi}\exp(i\pi n/2)\int_0^\infty dz_1\int_0^\infty dz_2\left(2nm_0-(2-2(2-n))\not p\frac{z_1}{z_1+z_2}\right)$$

$$\frac{(4\pi\mu^2(z_1+z_2))^{2-n}}{(z_1+z_2)^2}\exp\left(ip^2\frac{z_1z_2}{z_1+z_2}-m_0^2z_2\right). \tag{17.40}$$

The identity, eq.(17.28), is inserted and z_1, z_2 are rescaled by β, $z_1\to\beta z_1$, $z_2\to\beta z_2$. The integral over z_2 is easily done leaving

$$\Sigma^{\rm reg}(p)=-\left(\frac{e_0^2}{4\pi}\right)\frac{1}{2\pi}\exp(i\pi n/2)\int_0^1dz\,(nm_0-(1-(2-n))\not pz)$$

$$\int_0^\infty\frac{d\beta}{\beta}(4\pi\mu^2\beta)^{2-n}\exp\Big(i\beta(p^2z(1-z)-m_0^2(1-z))\Big). \tag{17.41}$$

We now think of n as being continuous and substitute $\varepsilon = 2 - n$. The regulator is removed by taking $n \to 2$ from below, that is, $\varepsilon \to 0^+$. Using

$$\int_0^\infty \frac{dx}{x} x^\varepsilon e^{iqx} = \Gamma(\varepsilon) q^{-\varepsilon} \exp(-i\varepsilon\pi/2), \tag{17.42}$$

the integral over β may be done.

$$\Sigma^{\text{reg}}(p) = \left(\frac{e_0^2}{4\pi}\right) \frac{1}{2\pi} \exp(-i\pi\varepsilon)\Gamma(\varepsilon)$$
$$\int_0^1 dz \left((2-\varepsilon)m_0 - (1-\varepsilon)\not{p}z\right) \left(\frac{4\pi\mu^2}{p^2 z(1-z) - m_0^2(1-z)}\right)^\varepsilon \tag{17.43}$$

The expression diverges as ε goes to zero because of the presence of the Γ function. To determine the size of the divergence and the finite piece, expand $\Sigma^{\text{reg}}(p)$ about $\varepsilon = 0$. The expansion of $\Gamma(\varepsilon)$ about $\varepsilon = 0$ is

$$\Gamma(\varepsilon) = \frac{1}{\varepsilon} - \gamma + O(\varepsilon), \tag{17.44}$$

where γ is the Euler-Mascheroni constant $(= 0.577\cdots)$. Since

$$a^\varepsilon = 1 + \varepsilon \ln(a) + O(\varepsilon^2), \tag{17.45}$$

then

$$\Sigma^{\text{reg}}(p) = \frac{\alpha_0}{2\pi}(1 - i\varepsilon\pi + O(\varepsilon^2))\left(\frac{1}{\varepsilon} - \gamma + O(\varepsilon)\right)$$
$$\int_0^1 dz((2-\varepsilon)m_0 - (1-\varepsilon)\not{p}z)\left(1 + \varepsilon \ln\left(\frac{4\pi\mu^2}{p^2 z(1-z) - m_0^2(1-z)}\right) + O(\varepsilon^2)\right)$$
$$= \frac{\alpha_0}{2\pi}\Bigg[(2m_0 - \frac{\not{p}}{2})\frac{1}{\varepsilon} - (i\pi + \gamma)(2m_0 - \frac{\not{p}}{2}) - m_0 + \not{p}/2$$
$$+ \int_0^1 dz(2m_0 - \not{p}z) \ln\left(\frac{4\pi\mu^2}{p^2 z(1-z) - m_0^2(1-z)}\right) + O(\varepsilon)\Bigg]. \tag{17.46}$$

Compare this result with the Pauli-Villars regularization, eq.(17.34). The ultraviolet divergent part is the same if we identify $1/\varepsilon$ with $\ln(M^2/m_0^2)$.

Both eq.(17.34) and eq.(17.46) are finite as $p^2 \to m_0^2$; however, the derivative, $\partial/\partial\not{p}\ \Sigma^{\rm reg}(p)$ at $\not{p} = m_0$ is not. This is the infrared divergence that arises when particles are taken "on shell" and at least one of them is massless. The divergence is eliminated in eq.(17.33) by allowing the photon to have a small mass, λ. We can do the same thing in eq.(17.46), or we may start from eq.(17.43) and do the z integral before taking the limit $\varepsilon \to 0$. In this case, the infrared divergence will manifest itself as a pole in ε. Therefore, unlike Pauli-Villars, dimensional regularization will also regularize the infrared difficulties without recourse to a fictitious photon mass. However, it may be computationally more convenient just to add the artificial photon mass.

Now let us turn our attention to the vacuum polarization diagram, figure 17.3(a). It is the second term in the expansion of the photon propagator. The interesting part, the fermion loop, was given in eq.(17.10).

$$\Pi^{\mu\nu} = -(-ie_0)^2 \int \frac{d^4p}{(2\pi)^4} \operatorname{tr}\left[\left(\frac{i}{\not{p} - m_0 + i\epsilon}\right)\gamma^\mu\left(\frac{i}{\not{p} - \not{k} - m_0 + i\epsilon}\right)\gamma^\nu\right] \tag{17.47}$$

The extra minus sign in front is associated with the presence of the fermion loop (Feynman rule number 6 in chapter 8). The trace in 4 dimensions is worked out in eq.(17.13). To regularize, we reduce the space-time dimension.

$$\begin{aligned}
\Pi^{\mu\nu}_{\rm reg}(k) & \\
&= -(-ie_0m^{2-n})^2 \int \frac{d^{2n}p}{(2\pi)^{2n}} \frac{i}{p^2 - m_0^2 + i\epsilon} \frac{i}{(p-k)^2 - m_0^2 + i\epsilon} \\
&\quad \operatorname{tr}\Big((\not{p} + m_0)\gamma^\mu(\not{p} - \not{k} + m_0)\gamma^\nu\Big) \\
&= (e_0\mu^{2-n})^2 \int \frac{d^{2n}p}{(2\pi)^{2n}} \frac{i}{p^2 - m_0^2 + i\epsilon} \frac{i}{(p-k)^2 - m_0^2 + i\epsilon} \\
&\quad 2^n\Big(p^\mu(p-k)^\nu + p^\nu(p-k)^\mu - ((p^2 - m_0^2) - p\cdot k)g^{\mu\nu}\Big)
\end{aligned} \tag{17.48}$$

Next, use eq.(17.21) to elevate the denominators as was done for $\Sigma^{\rm reg}(p)$ and complete the squares in the exponential.

$$\begin{aligned}
\Pi^{\mu\nu}_{\rm reg}(k) = (e_0\mu^{2-n})^2 \int_0^\infty dz_1 \int_0^\infty dz_2 \int \frac{d^{2n}p}{(2\pi)^{2n}} \\
2^n\Big(p^\mu(p-k)^\nu + p^\nu(p-k)^\mu - ((p^2 - m_0^2) - p\cdot k)g^{\mu\nu}\Big)
\end{aligned}$$

$$\exp\left(i(z_1+z_2)\left(p - \frac{z_2k}{z_1+z_2}\right)^2\right)\exp\left(i\frac{z_1z_2k^2}{z_1+z_2} - i(m_0^2 - i\epsilon)(z_1+z_2)\right). \tag{17.49}$$

After shifting integration variables ($p \to p - z_2\, k/(z_1+z_2)$), the Gaussian moments can be computed using eq.(17.39).

$$\begin{aligned} \int \frac{d^{2n}p}{(2\pi)^{2n}}\, p^2 \exp(iap^2) &= -\frac{n}{(4\pi a)^n}\,\frac{1}{a}\,\exp(in\pi/2), \\ \int \frac{d^{2n}p}{(2\pi)^{2n}}\, p^\mu p^\nu \exp(iap^2) &= -\frac{g^{\mu\nu}}{(4\pi a)^n}\,\frac{1}{2a}\,\exp(in\pi/2) \end{aligned} \tag{17.50}$$

As usual, the odd moments vanish. The first integral above is easily produced from eq.(17.39) by differentiation with respect to ia. The result of the second integral may also be obtained by differentiation of eq.(17.39).

$$\int \frac{d^{2n}p}{(2\pi)^{2n}} \exp(ia(p-k)^2) = -\frac{1}{(4\pi a)^n}\,\frac{\exp(in\pi/2)}{i},$$

thus,

$$\int \frac{d^{2n}p}{(2\pi)^{2n}} \exp(iap^2 - 2iap\cdot k) = -\exp(-iak^\rho g_{\rho\sigma} k^\sigma)\,\frac{1}{(4\pi a)^n}\,\frac{\exp(in\pi/2)}{i}.$$

Eq.(17.50) follows from

$$\begin{aligned} \frac{1}{(-2ia)^2}\frac{\partial}{\partial k_\mu}\frac{\partial}{\partial k_\nu}\int \frac{d^{2n}p}{(2\pi)^{2n}} \exp(iap^2 - 2iap\cdot k)\Big|_{k=0} \\ = \int \frac{d^{2n}p}{(2\pi)^{2n}}\, p^\mu p^\nu \exp(iap^2). \end{aligned}$$

The result of the Gaussian integration is

$$\begin{aligned} \Pi^{\mu\nu}_{\text{reg}}(k) &= (e_0\mu^{2-n})^2\frac{\exp(in\pi/2)}{(4\pi)^n}\int_0^\infty dz_1 \int_0^\infty dz_2\, \frac{2^n}{(z_1+z_2)^n} \\ &\left(\left(\frac{n-1}{(z_1+z_2)} + im_0^2 - ik^2\frac{z_1 z_2}{(z_1+z_2)^2}\right)g^{\mu\nu} - 2i\frac{z_1 z_2}{(z_1+z_2)^2}(k^\mu k^\nu - g^{\mu\nu}k^2)\right) \\ &\exp\left(ik^2\frac{z_1 z_2}{z_1+z_2} - i(m_0^2 - i\epsilon)(z_1+z_2)\right). \end{aligned} \tag{17.51}$$

The fermion loop, $\Pi^{\mu\nu}(k)$, is the current-current Green's function, $\langle 0\,|T(j^\mu(x) j^\nu(x'))|\,0\rangle$, $j^\mu(x) = \bar\psi(x)\gamma^\mu\psi(x)$, in the momentum representation. Global gauge invariance of the action leads to current conservation, $\partial_\mu j^\mu = 0$. This implies that $k_\mu \Pi^{\mu\nu}(k) = 0$. The second factor in eq.(17.51) automatically satisfies this condition, but at first sight, it appears that the first term, proportional only to $g^{\mu\nu}$, does not. The regularization is suppose to preserve the gauge symmetry; therefore this first term must vanish, and indeed it does. To see this, we rescale z_1 and z_2 by β, $z_1 \to \beta z_1$, and $z_2 \to \beta z_2$, and then notice that the first term above may be generated by differentiation with respect to β.

$$-\beta\frac{\partial}{\partial\beta}\frac{1}{\beta^{n-1}}\int_0^\infty dz_1\,dz_2\,\frac{1}{(z_1+z_2)^{n+1}}\exp\left(i\beta\left(k^2\frac{z_1 z_2}{z_1+z_2}-im_0^2(z_1+z_2)\right)\right)$$

$$= \int_0^\infty dz_1\int_0^\infty dz_2\,\frac{\beta^{2-n}}{(z_1+z_2)^n}\left(\frac{n-1}{\beta(z_1+z_2)}+im_0^2-ik^2\frac{z_1 z_2}{(z_1+z_2)^2}\right)$$
$$\exp\left(i\beta\left(k^2\frac{z_1 z_2}{z_1+z_2}-im_0^2(z_1+z_2)\right)\right). \quad (17.52)$$

Now rescale z_1 and z_2 on the left-hand side above by β^{-1}, $z_i \to z_i/\beta$. All factors of β to the right of $\partial/\partial\beta$ disappear and the right-hand side vanishes. Thus, $k_\mu\Pi^{\mu\nu}(k)=0$.

Following the treatment of Σ^{reg}, we next insert the identity, eq.(17.28), into the remaining part of eq.(17.51), and rescale z_1 and z_2 by β. After integrating over z_2 we have

$$\Pi^{\mu\nu}_{\text{reg}}(k) = -i\alpha_0\frac{2}{\pi}\exp(in\pi/2)(k^\mu k^\nu - g^{\mu\nu}k^2)\int_0^1 dz\,z(1-z)$$
$$\int_0^\infty\frac{d\beta}{\beta}(2\pi\mu^2\beta)^{2-n}\exp(i\beta(k^2(1-z)z-(m_0^2-i\epsilon))). \quad (17.53)$$

Substituting $\varepsilon = 2-n$ and using eq.(17.42) we finally end at

$$\Pi^{\mu\nu}_{\text{reg}}(k) = i\frac{2}{\pi}\alpha_0\exp(-i\varepsilon\pi)\Gamma(\varepsilon)(k^\mu k^\nu - g^{\mu\nu}k^2)$$
$$\int_0^1 dz\,z(1-z)\left(\frac{2\pi\mu^2}{k^2z(1-z)-m_0^2}\right)^\varepsilon. \quad (17.54)$$

Expanding about $\varepsilon = 0$ yields

$$\Pi^{\mu\nu}_{\text{reg}}(k) = i\frac{2}{\pi}\alpha_0(k^\mu k^\nu - g^{\mu\nu}k^2)$$
$$\left(\frac{1}{6\varepsilon}-i\frac{\pi}{6}-\frac{\gamma}{6}+\int_0^1 dz\,z(1-z)\ln\left(\frac{2\pi\mu^2}{k^2z(1-z)-m_0^2}\right)\right)+O(\varepsilon).$$
$$(17.55)$$

Note that $\Pi^{\mu\nu}$ involves no photon propagators, hence it is infrared finite.

Finally, we come to the vertex correction, figure 17.3(c), the most computationally intensive of the three divergent 1-loop diagrams. This is because three internal propagators are involved. Employing the notation used in chapter 8, the fermion-boson loop is $(-ie_0)\Gamma^\mu(p,p')$, eq.(8.43), where the momentum labels are those of figure 17.3(c).

$$(-ie_0)\Gamma^\mu(p,p') =$$

$$(-ie_0)^3 \int \frac{d^4k}{(2\pi)^4} \frac{-i}{k^2+i\epsilon} \left(\gamma_\nu \frac{i}{\not p' - \not k - m_0 + i\epsilon} \gamma^\mu \frac{i}{\not p - \not k - m_0 + i\epsilon} \gamma^\nu \right). \tag{17.56}$$

Again, to regularize, we switch to $2n$ dimensions. Elevating all three propagators we have

$$(-ie_0)\Gamma^\mu_{\text{reg}}(p,p') = -(-ie_0\mu^{2-n})^3 \int \frac{d^{2n}k}{(2\pi)^{2n}} \int_0^\infty dz_1\, dz_2\, dz_3$$

$$(\gamma_\nu(\not p' - \not k + m_0)\gamma^\mu(\not p - \not k + m_0)\gamma^\nu)$$

$$\exp\Big(i(z_1(k^2+i\epsilon) + z_2((p'-k)^2 - m_0^2 + i\epsilon) + z_3((p-k)^2 - m_0^2 + i\epsilon)) \Big). \tag{17.57}$$

By power counting, the only part of the integrand that can lead to a divergence when we integrate over k^μ is proportional to $\gamma_\nu \not k \gamma^\mu \not k \gamma^\nu$. Let us separate this piece out and call it $\Gamma^{\mu(a)}$, and the rest $\Gamma^{\mu(b)}$. Upon completing the squares, shifting k to $k - (z_2p' + z_3p)/(z_1+z_2+z_3)$, and doing the Gaussian integration (using eq.(17.50)) we get

$$-ie_0\Gamma^{\mu(a)}_{\text{reg}}(p,p')$$

$$= \frac{1}{2}\frac{1}{(4\pi)^n}(-ie_0\mu^{2-n})^3 \int_0^\infty dz_1\, dz_2\, dz_3 \frac{\gamma_\nu\gamma_\sigma\gamma^\mu\gamma^\sigma\gamma^\nu}{(z_1+z+2+z_3)^{n+1}}$$

$$\exp\left(i\left(\frac{(z_1+z_3)z_2p'^2 - 2z_2z_3p\cdot p' + (z_1+z_2)z_3p^2 - (z_2+z_3)m_0^2}{(z_1+z_2+z_3)} \right) \right), \tag{17.58}$$

and

$$-ie_0\Gamma^{\mu(b)}_{\text{reg}}(p,p') = -\frac{i}{(4\pi)^2}(-ie_0)^3\int_0^\infty dz_1\,dz_2\,dz_3\,\frac{1}{(z_1+z_3+z_3)^3}$$
$$\gamma_\nu((z_1+z_3)\not p' - z_3\not p + m_0(z_1+z_2+z_3))$$
$$\gamma^\mu((z_1+z_2)\not p - z_2\not p' + m_0(z_1+z_2+z_3))\gamma^\nu$$
$$\exp\left(i\left(\frac{(z_1+z_3)z_2p'^2 - 2z_2z_3p\cdot p' + (z_1+z_2)z_3p^2 - (z_2+z_3)m_0^2}{(z_1+z_2+z_3)}\right)\right). \tag{17.59}$$

Multiple application of eq.(17.37) to $\Gamma^{\mu(a)}_{\text{reg}}$ reduces $\gamma_\nu\gamma_\sigma\gamma^\mu\gamma^\sigma\gamma^\nu$ to $(2-2n)^2\gamma^\mu$. Next we insert the 3-dimensional version of the identity, eq.(17.28),

$$1 = \int_0^\infty \frac{d\beta}{\beta}\,\delta\left(1 - \frac{z_1+z_2+z_3}{\beta}\right), \tag{17.60}$$

rescale z_1, z_2, z_3 by β and do the z_1 integral. The integral over β is done using eq.(17.42). This leaves

$$\Gamma^{\mu(a)}_{\text{reg}}(p,p') = \frac{\alpha_0}{2\pi}\exp(-i\varepsilon\pi)(1-\varepsilon)^2\gamma^\mu\Gamma(\varepsilon)\int_0^1 dz_2\int_0^{1-z_2} dz_3$$
$$\left(\frac{4\pi\mu^2}{(1-z_2)z_2p'^2 - 2z_2z_3p\cdot p' + (1-z_3)z_3p^2 - m_0^2(z_2+z_3)}\right)^\varepsilon, \tag{17.61}$$

or, expanding about $\varepsilon = 0$,

$$\Gamma^{\mu(a)}_{\text{reg}}(p,p') = \alpha_0\gamma^\mu\left(\frac{1}{4\pi}\frac{1}{\varepsilon} - \frac{i}{4} - \frac{1}{2\pi} - \frac{\gamma}{4\pi} + \frac{1}{2\pi}\int_0^1 dz_2\int_0^{1-z_2} dz_3\right.$$
$$\left.\ln\left(\frac{4\pi\mu^2}{(1-z_2)z_2p'^2 - 2z_2z_3p\cdot p' + (1-z_3)z_3p^2 - m_0^2(z_2+z_3)}\right) + O(\varepsilon)\right). \tag{17.62}$$

Eq.(17.60) may also be introduced into $\Gamma^{\mu(b)}_{\text{reg}}(p,p')$ to reduce it to

$$\Gamma^{\mu(b)}_{\text{reg}}(p,p') = \frac{\alpha_0}{4\pi}\int_0^1 dz_2\int_0^{1-z_2} dz_3$$
$$\frac{\gamma_\nu((1-z_2)\not p' - z_3\not p + m_0)\gamma^\mu((1-z_3)\not p - z_2\not p' + m_0)\gamma^\nu}{(1-z_2)z_2p'^2 - 2z_2z_3p\cdot p' + (1-z_3)z_3p^2 - m_0^2(z_2+z_3)}. \tag{17.63}$$

As with $\partial/\partial\not p\ \Sigma(p)$, $\Gamma^\mu(p,p')$ is infrared divergent on mass shell. Now let us take the results of the regularized divergent diagrams to complete renormalization at 1-loop by generating counterterms and applying a renormalization prescription. We will then examine the results.

17.3 RENORMALIZATION

Having identified the form and size of the divergences in the 3 primitively divergent diagrams, next we construct counterterms to subtract them off. Let's begin with the electron propagator.

To order e_0^2, the regularized electron propagator is

$$\frac{i}{\not p - m_0} + \frac{i}{\not p - m_0}(-i\Sigma^{\rm reg}(p))\frac{i}{\not p - m_0} = \frac{i}{\not p - m_0 - \Sigma^{\rm reg}(p)} + O(e_0^4), \tag{17.64}$$

where from eq.(17.46) we have

$$\Sigma^{\rm reg}(p) = \frac{e_0^2}{4\pi}\frac{1}{2\pi}\Big((2m_0 - \not p/2)\frac{1}{\varepsilon} + \text{finite terms}\Big). \tag{17.65}$$

Following the discussion in chapter 16, we want the renormalized propagator to look like the free field propagator. Therefore, we take

$$Z_2(\not p - m_0 - \Sigma^{\rm reg}(p)) = \not p - m + \text{finite terms}, \tag{17.66}$$

where m is the renormalized mass which is finite. Equating terms proportional to $\not p$ on both sides of eq.(17.66) determines Z_2.

$$Z_2\Big(1 + \frac{\alpha_0}{4\pi}\frac{1}{\varepsilon}\Big)\not p = \not p, \tag{17.67}$$

which implies that

$$Z_2 = \frac{1}{1 + \frac{\alpha_0}{4\pi}\frac{1}{\varepsilon}} = 1 - \frac{\alpha_0}{4\pi}\frac{1}{\varepsilon} + O(\alpha_0^2), \tag{17.68}$$

where $\alpha_0 = e_0^2/4\pi$ is the bare fine structure constant. The shift in the mass is now determined by equating the terms proportional to the mass,

$$Z_2\Big(m_0 + m_0\frac{\alpha_0}{\pi}\frac{1}{\varepsilon}\Big) = m, \tag{17.69}$$

or

$$\Big(1 - \frac{\alpha_0}{4\pi}\frac{1}{\varepsilon}\Big)m_0\Big(1 + \frac{\alpha_0}{\pi}\frac{1}{\varepsilon}\Big) + O(\alpha_0^2) = m,$$

that is,

$$\begin{aligned} m &= m_0 + \delta m \\ &= m_0 + \frac{3}{4\pi}\alpha_0 m_0 \frac{1}{\varepsilon} + O(\alpha_0^2). \end{aligned} \tag{17.70}$$

From eq.(17.70) we can see that $\delta m = 3/4\pi\, \alpha_0 m_0/\varepsilon$, but also that to the same order in α_0 we may replace m_0 in δm with m. So, to order α_0,

$$\delta m = \frac{3}{4\pi}\alpha_0\, m \frac{1}{\varepsilon} + O(\alpha_0^2). \tag{17.71}$$

Next we want to rewrite the regularized bare Lagrangian density,

$$\mathcal{L} = -\frac{1}{4}(\partial_\mu A_{0\nu} - \partial_\nu A_{0\mu})(\partial^\mu A_0^\nu - \partial^\nu A_0^\mu) - \frac{1}{2}(\partial_\mu A_0^\mu)^2$$
$$+ \bar{\psi}_0(i\partial\!\!\!/ - m_0)\psi_0 - e_0\bar{\psi}_0 A\!\!\!/_0\psi_0, \tag{17.72}$$

such that, to 1-loop in perturbation theory, the electron propagator is

$$\frac{i}{p\!\!\!/ - m + \text{finite terms}}, \tag{17.73}$$

which is finite. With the aid of eq.(17.66) we can accomplish this by substituting $m - \delta m$ for m_0, and rescaling the spinor fields by $Z_2^{1/2}$. If we define the renormalized spinor fields $\psi, \bar{\psi}$ in terms of the bare fields $\psi_0, \bar{\psi}_0$ by

$$\psi_0 = \sqrt{Z_2}\psi, \qquad \bar{\psi}_0 = \sqrt{Z_2}\bar{\psi}, \tag{17.74}$$

then the Lagrangian density,

$$\mathcal{L} = -\frac{1}{4}(\partial_\mu A_{0\nu} - \partial_\nu A_{0\mu})(\partial^\mu A_0^\nu - \partial^\nu A_0^\mu) - \frac{1}{2}(\partial_\mu A_0^\mu)^2$$
$$+ Z_2\bar{\psi}(i\partial\!\!\!/ - m)\psi + Z_2\delta m\bar{\psi}\psi - e_0 Z_2\bar{\psi}A\!\!\!/_0\psi, \tag{17.75}$$

where δm is given by eq.(17.71) and Z_2 by (17.68), will produce eq.(17.73). The finite parts of eq.(17.73) will be uniquely determined by the renormalization prescription (choice of subtraction point) which we will return to after generating the remaining counterterms.

Moving on to the regularized photon propagator, we have, to order e_0^2,

$$\frac{-ig_{\mu\nu}}{k^2} + \frac{-ig_{\mu\rho}}{k^2}\,\Pi^{\rho\sigma}_{\text{reg}}(k)\,\frac{-ig_{\sigma\nu}}{k^2}, \tag{17.76}$$

where, from eq.(17.55), we have

$$\Pi^{\rho\sigma}_{\text{reg}}(k) = \alpha_0\,\frac{2i}{\pi}(k^\rho k^\sigma - g^{\rho\sigma}k^2)\Big(\frac{1}{6\varepsilon}\Big) + \text{finite terms}. \tag{17.77}$$

As with the electron propagator, we would like the renormalized photon propagator to look like the free field result. For the moment ignore the longitudinal part of $\Pi^{\rho\sigma}_{\text{reg}}(k)$ (the part proportional to $k^\rho k^\sigma$). The remaining piece proportional to $g^{\rho\sigma}$ fits nicely into eq.(17.76), so that we can write the photon propagator as

$$\frac{-ig_{\mu\nu}}{k^2(1+\Pi_{\text{reg}}(k))}, \tag{17.78}$$

where $\Pi^{\rho\sigma}_{\text{reg}}(k) \equiv -ig^{\rho\sigma}\, k^2\, \Pi_{\text{reg}}(k)$. This means that we want to define Z_3 by

$$Z_3 k^2(1+\Pi_{\text{reg}}(k)) = k^2 + \text{finite terms}. \tag{17.79}$$

Therefore, eq.(17.77) implies that

$$Z_3 = 1 - \frac{\alpha_0}{3\pi}\frac{1}{\varepsilon} + O(\alpha_0^2), \tag{17.80}$$

and we define the renormalized vector potential, A^μ, in terms of the bare potential, A^μ_0, by

$$A^\mu_0 = \sqrt{Z_3}A^\mu. \tag{17.81}$$

However, because of the presence of the longitudinal $k^\rho k^\sigma$ term in eq.(17.77), the Lagrangian density, (17.75), with A^μ_0 replaced by $Z_3^{1/2}A^\mu$, does not generate the photon propagator,

$$\frac{-ig_{\mu\nu}}{k^2 + \text{finite terms}}, \tag{17.82}$$

at 1-loop. There is no panic involved, because the extra piece generated that doesn't fit into eq.(17.82) is the longitudinal piece. Since the photon propagator will always be attached to at least 1 vertex, and attached to each vertex is a conserved current, j^μ, (i.e. $k^\mu j_\mu = 0$), then the extra piece always projects away.

The presence of the extra piece in $\Pi^{\rho\sigma}_{\text{reg}}(k)$ proportional to $k^\rho k^\sigma$ and the lack of it in the free-field photon propagator tells us that we would have been better off if we had somehow started with a $k^\rho k^\sigma$ term in the free-field no-loop propagator. This longitudinal term is missing even though we have a gauge-fixing term, $\lambda_0/2(\partial_\mu A^\mu_0)^2$, in $\mathcal{L}$, eq.(17.75), because we started from the Feynman (version of the Lorentz) gauge, $\lambda_0 = 1$. We now get the mild surprise that we cannot consistently maintain $\lambda_0 = 1$ after renormalization. In other words, λ_0 is also renormalized. Since all of the physical results are independent of λ_0 due to current conservation, no problems arise from shifts in λ.

The proper thing to do is to start over with an action that includes the gauge-fixing term with $\lambda_0 \neq 1$ and recalculate the 1-loop divergent graphs. The same counterterms will be required; in particular, the same Z_3, eq.(17.80), arises. The remaining divergence in the longitudinal part of eq.(17.77) may then be properly absorbed by defining the renormalized λ in terms of the bare λ_0 by $\lambda_0 = Z_\lambda \lambda$, and taking

$$Z_\lambda = 1 + \frac{\alpha_0}{3\pi}\frac{1}{\varepsilon}. \tag{17.83}$$

Rewriting the Lagrangian density $\mathcal{L}$, eq.(17.75), in terms of the renormalized photon field using eq.(17.81), we now have

$$\begin{aligned}\mathcal{L} = &-\frac{1}{4}Z_3(\partial_\mu A_\nu - \partial_\nu A_\mu)(\partial^\mu A^\nu - \partial^\nu A^\mu) - \frac{1}{2}Z_3(\partial_\mu A^\mu)^2 \\ &+ Z_2\bar{\psi}(i\partial\!\!\!/ - m)\psi + Z_2\delta m\bar{\psi}\psi - e_0 Z_2\sqrt{Z_3}\bar{\psi}A\!\!\!/\psi\end{aligned} \tag{17.84}$$

This Lagrangian density does not produce a completely finite photon propagator at 1-loop since the longitudinal part of the vacuum polarization is not completely canceled. However, this is of no physical consequence, because this piece is always projected out by current conservation. If we start from the Lagrangian density

$$\begin{aligned}\mathcal{L} = &-\frac{1}{4}(\partial_\mu A_{0\nu} - \partial_\nu A_{0\mu})(\partial^\mu A_0^\nu - \partial^\nu A_0^\mu) - \frac{\lambda_0}{2}(\partial_\mu A_0^\mu)^2 \\ &+ \bar{\psi}_0(i\partial\!\!\!/ - m_0)\psi_0 - e_0\bar{\psi}_0 A\!\!\!/_0\psi_0,\end{aligned} \tag{17.85}$$

and apply eqs.(17.74), (17.81), and (17.83), and substitute for m_0 with eq.(17.70), then the new Lagrangian density,

$$\begin{aligned}\mathcal{L} = &-\frac{1}{4}Z_3(\partial_\mu A_\nu - \partial_\nu A_\mu)(\partial^\mu A^\nu - \partial^\nu A^\mu) - \frac{\lambda}{2}Z_3 Z_\lambda(\partial_\mu A^\mu)^2 \\ &+ Z_2\bar{\psi}(i\partial\!\!\!/ - m)\psi + Z_2\delta m\bar{\psi}\psi - e_0 Z_2\sqrt{Z_3}\bar{\psi}A\!\!\!/\psi,\end{aligned} \tag{17.86}$$

does result in a completely finite photon propagator at 1-loop.

Finally, we want to remove the divergence associated with the vertex correction by scaling γ^μ. The full radiative vertex is Λ^μ (see figure 8.10). Thus, to order α_0,

$$\begin{aligned}\Lambda^\mu_{\text{reg}} &= \gamma^\mu + \Gamma^\mu_{\text{reg}} \\ &= \gamma^\mu + \frac{\alpha_0}{4\pi}\frac{1}{\varepsilon}\gamma^\mu + \text{finite terms.}\end{aligned} \tag{17.87}$$

If we replace γ^μ in the vertex in the action density, eq.(17.86), with $Z_1\gamma_\mu$, where

$$Z_1 = 1 - \frac{\alpha_0}{4\pi}\frac{1}{\varepsilon}, \tag{17.88}$$

then, to order α_0, the new full radiative vertex generated is

$$Z_1\Lambda^{\mu}_{\text{reg}} = \left(1 - \frac{\alpha_0}{4\pi}\frac{1}{\varepsilon}\right)\left(1 + \frac{\alpha_0}{4\pi}\frac{1}{\varepsilon}\right)\gamma^{\mu} + \text{finite terms} = \gamma^{\mu} + \text{finite terms,} \tag{17.89}$$

which is finite. To implement this subtraction without modifying $\mathcal{L}$, we multiply γ_μ by $1 = Z_1/Z_1$.

$$\begin{aligned}\mathcal{L} = &-\frac{1}{4}Z_3(\partial_\mu A_\nu - \partial_\nu A_\mu)(\partial^\mu A^\nu - \partial^\nu A^\mu) - \frac{\lambda}{2}Z_3 Z_\lambda(\partial_\mu A^\mu)^2 \\ &+ Z_2\bar{\psi}(i\partial\!\!\!/ - m)\psi + Z_2\delta m\bar{\psi}\psi - \left(e_0\frac{Z_2}{Z_1}\sqrt{Z_3}\right)Z_1\bar{\psi}A\!\!\!/\psi\end{aligned} \tag{17.90}$$

It is now apparent how the finite renormalized electric charge, e, should be defined in terms of the bare charge. Namely,

$$e = e_0\frac{Z_2\sqrt{Z_3}}{Z_1} = e_0\sqrt{Z_3} = e_0\left(1 - \frac{\alpha_0}{6\pi}\frac{1}{\varepsilon} + O(\alpha_0^2)\right). \tag{17.91}$$

The bare charge is rescaled only by Z_3 here because $Z_1 = Z_2$. Since the bare charge and the renormalized charge are equal at order α^0, then to first order in α (the renormalized fine structure constant $e^2/4\pi \approx 1/137$), we may invert eq.(17.91) to get

$$e_0 = e\left(1 + \frac{\alpha}{6\pi}\frac{1}{\varepsilon} + O(\alpha^2)\right). \tag{17.92}$$

From the form of eq.(17.92), we can also see that we may replace α_0 with α in our expressions for δm, Z_1, Z_2, Z_3, and Z_λ, and still have results valid to order α. The change resulting from $\alpha_0 \to \alpha$ is of order α^2 or higher.

When written with the renormalized charge e, the action density, eq.(17.90), produces finite results at 1-loop that are independent of the regularization parameter (except the inconsequential piece of the vacuum polarization). The regulator may be safely removed.

Let us summarize what we have so far. We started with the bare action, the action written in terms of bare fields and couplings. This action produced divergences at 1-loop when written strictly in terms of bare quantities. Next we regularized to determine where the divergences arise, and discovered that if we allowed the bare fields and couplings to depend on the regulating parameter, then the bare fields and couplings could absorb the regulator dependence. This means that the results at 1-loop are finite even when the regulator is removed. The price for this, of course, is that the bare quantities diverge when the regulator is removed. The minimal renormalization performed here to absorb regulator dependence into bare quantities resulted in the renormalized action

$$\mathcal{L}_{\text{ren}} = -\frac{1}{4}Z_3(\partial_\mu A_\nu - \partial_\nu A_\mu)(\partial^\mu A^\nu - \partial^\nu A^\mu) - \frac{1}{2}Z_3(\partial_\mu A^\mu)^2 + Z_2\bar{\psi}(i\not{\partial} - m)\psi + Z_2\delta m\bar{\psi}\psi - e\, Z_1\bar{\psi}\not{A}\psi, \tag{17.93}$$

where

$$\begin{aligned} \delta m &= \frac{3}{4\pi}\alpha\, m\frac{1}{\varepsilon} + O(\alpha^2) \\ Z_1 &= 1 - \frac{\alpha}{4\pi}\frac{1}{\varepsilon}O(\alpha^2) \\ Z_2 &= 1 - \frac{\alpha}{4\pi}\frac{1}{\varepsilon}O(\alpha^2) \\ Z_3 &= 1 - \frac{\alpha}{3\pi}\frac{1}{\varepsilon}. \end{aligned} \tag{17.94}$$

This action density produces finite physical results independent of ε at 1-loop. However, this action density is, by definition, *no* different from the classical action density we started from. The only difference is that the action density above is written in a different functional coordinate system than the bare classical action (but on the same function space). Thus, the classical action is OK, but the unfortunate part is that the quantum action only appears in its classical form in a singular functional coordinate system (the "bare" fields and couplings). Because the transformation out of the singular bare coordinate system is itself singular (the Z's), we cannot unambiguously change coordinates, so we must regularize and introduce renormalized parameters (i.e. m and e) that must be measured experimentally.

Often renormalization is done without the explicit introduction of bare parameters. One starts with the classical action written in terms of renormalized coordinates (renormalized fields and couplings), and then adds counterterms to subtract divergences. The counterterms may be displayed explicitly by adding and subtracting the classical action in $\mathcal{L}_{\text{ren}}$, eq.(17.93).

$$\mathcal{L}_{\text{ren}} = -\frac{1}{4}F^{\mu\nu}F_{\mu\nu} - \frac{1}{2}(\partial_\mu A^\mu)^2 + \bar{\psi}(i\not{\partial} - m)\psi - e\bar{\psi}\not{A}\psi + \mathcal{L}_{\text{ct}}, \tag{17.95}$$

where the counterterms are

$$\mathcal{L}_{\text{ct}} = -\frac{1}{4}(Z_3 - 1)F^{\mu\nu}F_{\mu\nu} - \frac{1}{2}(Z_3 - 1)(\partial_\mu A^\mu)^2 + (Z_2 - 1)\bar{\phi}(i\not{\partial} - m)\psi + Z_2\delta m\,\bar{\psi}\psi - e(Z_1 - 1)\bar{\psi}\not{A}\psi. \tag{17.96}$$

The bare parameters are introduced at the end to demonstrate that the action will take on the appearance of the classical action when expressed in terms of the bare parameters.

The renormalization performed so far, resulting in the action density, eq.(17.93), and the Z's, eq.(17.94), was "minimal." We have not yet specified the renormalization prescription. The finite terms left over are ambiguous and arbitrary, and depend on arbitrary parameters such as μ. To complete the renormalization and extract the physics, we must uniquely define the finite terms. However, before we march on to that, let us turn our attention, for the moment, to the fact that, so far, $Z_1 = Z_2$.

The equality of Z_1 and Z_2 is no accident, and is due to gauge invariance. The formal development that leads to $Z_1 = Z_2$, which is valid to all orders in perturbation theory, results in the Ward-Takahashi identities. A diagrammatic motivation for the identities relevant here, eq.(8.45) and eq.(8.48), was given in chapter 8. Let us heuristically derive the identity from a different viewpoint. A proper derivation of all of the identities can be found in the references in the Bibliography.

We briefly introduced the effective action, $\Gamma[\bar{\psi}_c, \psi_c, A_c^\mu]$, in the last chapter as the generating functional of the proper functions or inverse Green's functions. $\Gamma[\bar{\psi}_c, \psi_c, A_c^\mu]$ is the Legendre transform of the generating functional $W[\chi, \bar{\chi}, J]$, eq.(14.32), and that was introduced in chapter 14. The subscript on the fields in Γ indicates that they are treated as classical fields. The computation and use of the effective action will be developed in the next chapter.

To lowest order (no loops), the effective action is just the classical action,

$$\Gamma[\bar{\psi}_c, \psi_c, A_c^\mu] = \int d^4x \; -\frac{1}{4}F_c^{\mu\nu}F_{c\,\mu\nu} + \bar{\psi}_c(i\partial\!\!\!/ - m)\psi_c - e\bar{\psi}_c A\!\!\!/_c\psi_c.$$

To motivate the form of Γ when loops are included, write $i\partial\!\!\!/ - m$ as $1/(1/(i\partial\!\!\!/ - m))$ which is 1 over the free spinor field 2-point function, $S_F(x-y)$. Thus, $i\partial\!\!\!/ - m \to S_F^{-1}(x-y)$. Let $S^{-1}(x-y)$ be the full inverse fermion 2-point function, $\Lambda_\mu(x,y,z)$ the full radiative vertex or 3-point function, and $D_{\mu\nu}^{-1}(x-y)$ the full inverse photon propagator. Then the effective action starts out as

$$\begin{aligned}
&\Gamma[\bar{\psi}_c, \psi_c, A_c^\mu] \\
&= \int d^4x\, d^4y\, \frac{1}{2}A_c^\mu(x)D_{\mu\nu}^{-1}(x-y)A_c^\nu(y) + \bar{\psi}_c(x)S^{-1}(x-y)\psi_c(y) \\
&\quad - e\int d^4x\, d^4y\, d^4z\, \bar{\psi}_c(x)A_c^\mu(y)\Lambda_\mu(x,y,z)\psi_c(z) + \cdots
\end{aligned} \tag{17.97}$$

The effective action must be gauge invariant, otherwise the theory is mathematically inconsistent and physical results depend on the coordinate system used to define phase. The gauge invariance of Γ means that under a gauge transformation, $\delta\Gamma = 0$. The variation of Γ can be written as

$$\delta\Gamma = \int d^4x\, \delta\bar{\psi}_c \frac{\delta\Gamma}{\delta\bar{\psi}_c} + \frac{\delta\Gamma}{\delta\psi_c}\delta\psi_c + \frac{\delta\Gamma}{\delta A_c^\mu}\delta A_c^\mu.$$

Under an infinitesimal local gauge transformation, $\delta\psi = i\theta(z)\psi$, $\delta\bar{\psi} = -i\theta(z)\bar{\psi}$, and $\delta A^\mu = 1/e\,\partial^\mu\theta(z)$. Thus, the variation in Γ under a gauge transformation is

$$\delta\Gamma = 0 = \int d^4z \left(-i\bar{\psi}_c \frac{\delta\Gamma}{\delta\bar{\psi}_c} + i\frac{\delta\Gamma}{\delta\psi_c} + \frac{1}{e}\,\partial^\mu \frac{\delta\Gamma}{\delta A_c^\mu} \right)\theta(z), \qquad (17.98)$$

where we have integrated the last term by parts. Since $\theta(z)$ is arbitrary, then the remaining factor in the integrand must vanish. When we apply eq.(17.98) to Γ, eq.(17.97), which we can do just by performing a gauge transformation in eq.(17.97), we find that the one term proportional to A^ν will vanish if the full inverse propagator for the photon is transverse,

$$\partial^\mu D^{-1}_{\mu\nu}(x-y) = 0. \qquad (17.99)$$

Recall that we did indeed find this to be the case at 1-loop when we found that $\Pi_{\mu\nu}(k)$ is proportional to $k^\mu k^\nu - g^{\mu\nu}k^2$. Now we know that the same thing must be true at higher orders.

Next, the terms proportional to $\bar{\psi}_c\psi_c$ in eq.(17.98) will vanish if

$$-iS^{-1}(x-y)\delta(z-x) + iS^{-1}(x-y)\delta(z-y) - \frac{\partial}{\partial z^\mu}\Lambda^\mu(x,z,y) = 0. \quad (17.100)$$

If we Fourier transform this identity, then

$$S-1(p') - S^{-1}(p) = (p'-p)^\mu \Lambda_\mu(p', p'-p, p). \qquad (17.101)$$

Recalling that the renormalized propagator is defined by eq.(17.66) using Z_2, and the renormalized vertex by eq.(17.89) using Z_1, then eq.(17.101) implies that we must have $Z_1 = Z_2$. Eq.(8.48) and the limit $p' \to p$, eq.(8.45), then follow.

So far we have seen by direct calculation only that the divergent parts of Z_1 and Z_2 are equal. However, gauge invariance dictates that the finite parts must be equal too, and that the equality must hold at all orders of perturbation theory.

One consequence of $Z_1 = Z_2$ can be seen from eq.(17.91). The electric charge is only renormalized by the factor Z_3, the factor that arose as a result of rescaling the photon field by the vacuum polarization. Thus,

$$eA^\mu = e_0 A_0^\mu. \qquad (17.102)$$

The renormalization of the charge does not depend on what conserved current the photon is coupled to. This means that the electromagnetic field is coupled universally (via the gauge covariant derivative) to all charged particles. As long as the bare charge is identical, the renormalized charge of two different kinds of particles will also be the same. Without this universality, we would have a very difficult time explaining why the renormalized charge of the electron should be the same as the proton, the muon, and so on, as required by experiment.

The renormalization we have done so far, eqs.(17.93) and (17.94), is just the "bare bones," the minimal amount to produce finite answers. We have not yet specified the subtraction point or renormalization prescription. The finite terms in the regularized expressions for the self-energy, vacuum polarization, and vertex correction are arbitrary, and will not be uniquely defined until a renormalization prescription is chosen. Let us turn our attention to that now, beginning with the vacuum polarization.

The transverse part of the regularized vacuum polarization to order α is

$$
\begin{aligned}
&-ig^{\mu\nu}k^2\Pi_{\rm reg}(k)\\
&= -g^{\mu\nu}k^2\,\frac{2\alpha}{\pi}\left(\frac{1}{6\varepsilon}-\frac{i\pi}{6}-\frac{\gamma}{6}+\int_0^1 dz\,(1-z)z\ln\left(\frac{2\pi\mu^2}{k^2z(1-z)-m^2}\right)\right)\\
&\qquad + O(\varepsilon^2)\\
&= -ig^{\mu\nu}k^2\,\frac{2\alpha}{\pi}\left(\frac{1}{6\varepsilon}-\frac{\gamma}{6}+\int_0^1 dz(1-z)z\ln\left(\frac{2\pi\mu^2}{k^2z(1-z)-m^2}\right)\right)+O(\varepsilon^2).
\end{aligned}
\tag{17.103}
$$

The arbitrariness of the finite piece is reflected in the explicit appearance of an arbitrary mass scale, μ. The most popular subtraction point is a $k^2 = 0$, where the photon is on mass shell. This means that we want the finite part of $\Pi_{\rm reg}(k)$ to vanish when $k^2 = 0$. Thus, the finite part is uniquely specified by defining the renormalized $\Pi(k)$ (call it $\Pi_R(k)$) to be

$$
\begin{aligned}
\Pi_R(k) &= \Pi_{\rm reg}(k) - \Pi_{\rm reg}(0)\\
&= \frac{2\alpha}{\pi}\int_0^1 dz\,(1-z)z\ln\left(\frac{m^2}{m^2-k^2(1-z)z}\right).
\end{aligned}
\tag{17.104}
$$

The dependence on the cutoff disappears. This in turn means that the renormalized photon propagator, $D_R^{\mu\nu}(k)$, is

$$
D_R^{\mu\nu}(k) = \lim_{\varepsilon\to 0}\frac{-ig^{\mu\nu}}{k^2(1+\Pi_{\rm reg}(k)-\Pi_{\rm reg}(0))} = \frac{-ig^{\mu\nu}}{k^2(1+\Pi_R(k))}. \tag{17.105}
$$

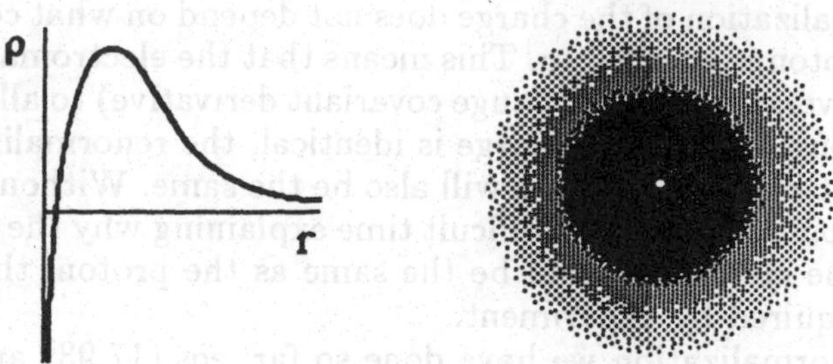

Figure 17.7 Qualitative representation of the charge distribution surrounding a bare electron at $r = 0$ in the vacuum due to vacuum polarization. The measurement of the charge contained in the sphere of radius $r > 0$ is significantly smaller than the bare charge due to screening. The magnitude of the measured charge also depends on r which leads to a scale-dependent electric charge.

The residue at the pole $k^2 = 0$ is identical to the free field result. From eq.(17.105) and eq.(17.79), we can now specify the finite part of Z_3. Z_3 is defined by $Z_3(1 + \Pi_{\text{reg}}(k)) = 1 + \Pi_R(k)$, thus

$$Z_3 = 1 - \frac{\alpha}{3\pi}\left(\frac{1}{\varepsilon} - \gamma + \ln\left(\frac{2\pi\mu^2}{m^2}\right)\right) + O(\alpha^2). \tag{17.106}$$

Now let's take a look at what we have. The finite renormalized electric charge is related to the bare charge by eq.(17.91), $e = \sqrt{Z_3}e_0$. From eq.(17.106), we can see that the bare charge is larger than the renormalized charge. This means that the vacuum *screens* electric charge. A restatement of this fact is that the so-called β-function of the charge is positive. For a fixed bare charge e_0 (independent of μ), $\partial e_0/\partial\mu = 0$. Then, eq.(17.106) implies that, to order α, that

$$\beta(e) = -\mu\frac{\partial e}{\partial\mu} = \frac{e\alpha}{3\pi} = \frac{e^3}{12\pi^2}. \tag{17.107}$$

An intuitive physical interpretation is represented in figures 17.7 and 17.8. The quantum vacuum is not empty and, in fact, is a busy place. The introduction of a bare electron polarizes the vacuum in much the same way as a classical charge polarizes a dielectric medium. In the vacuum, quantum fluctuations result in the formation of virtual electron-positron pairs which eventually disappear by annihilation. Close by the bare charge, the virtual positrons spend more time near the charge before annihilation, while the virtual electrons are repulsed and spend more time away. The

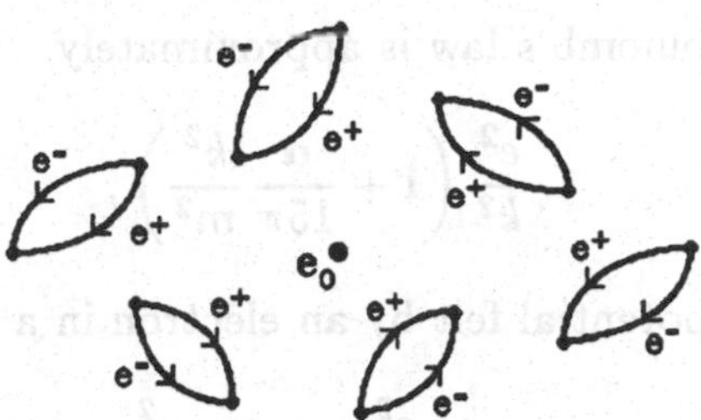

Figure 17.8 Qualitative representation of the virtual positron cloud that surrounds the bare electron. As virtual electron-positron pairs are formed from vacuum fluctuations, the virtual positrons are attracted to the bare electron while the virtual electrons are repulsed. This results in a screening of the bare charge by the vacuum.

net result is that there is a cloud of virtual positrons that surrounds each bare charge, screening the charge and reducing it from the bare charge to the observable renormalized charge.

The "structure" of the virtual particle cloud surrounding the bare charge is contained in $\Pi_R(k)$. As we mentioned earlier and illustrated in figure 17.2, the additional factor, $1+\Pi_R(k)$, in the denominator of the renormalized photon propagator leads to a modification of Coulomb's law. The modification in Coulomb's law results from the use of the renormalized photon propagator in the electron-electron scattering amplitude that we computed in chapter 8. The modification results in a rescaling of the *renormalized* electric charge by

$$e^2 \to \frac{e^2}{1+\Pi_R(k)}. \tag{17.108}$$

This means the observable renormalized charge is momentum-dependent and that QED predicts that the electric charge is not a constant, but that it depends on what energy scale the measurement is made at. The momentum-dependent charge or fine structure constant as given above is referred to as a "running coupling constant."

For static situations, $k_0 = 0$ and $k^2 = -\vec{k}^2$. The tree level electron-electron scattering amplitude (i.e. the photon propagator) leads to the usual Coulomb's law, $e^2/\vec{k}^2$. The quantum corrections modify this to $e^2/\vec{k}^2(1+\Pi_R(-\vec{k}))$. When $|\vec{k}|$ is small (i.e. $k^2 \ll m^2$), then

$$\begin{aligned}\Pi_R(-\vec{k}) &= \frac{2\alpha}{\pi}\int_0^1 dz\,(1-z)z\,\ln\left(1-\frac{k^2}{m^2}z(1-z)+\cdots\right)\\ &\simeq -\frac{\alpha}{15\pi}\frac{k^2}{m^2},\end{aligned} \tag{17.109}$$

so the corrected Coulomb's law is approximately

$$\frac{e^2}{k^2}\left(1+\frac{\alpha}{15\pi}\frac{k^2}{m^2}\right),$$

and the Coulomb potential felt by an electron in a hydrogen atom is

$$V(r) \simeq -\frac{e^2}{4\pi r} - \frac{\alpha}{15\pi}\frac{e^2}{m^2}\delta^3(r). \tag{17.110}$$

In the Dirac theory of the hydrogen atom (i.e. solution of the first quantized Dirac equation in the presence of the ordinary Coulomb potential), the $2S_{1/2}$ and $2P_{1/2}$ levels are degenerate. The modification to the Coulomb energy above splits the two levels, lowering the $2S_{1/2}$ level. This is in agreement with the virtual particle cloud picture. The angular momentum of the electron in the $2S_{1/2}$ state is lower than the $2P_{1/2}$ state hence the electron can spend more time near the nucleus. The square of the wave function is larger near $r = 0$ for the $2S_{1/2}$ state. This allows the electron to spend more time inside the cloud surrounding the proton where the screening is reduced. As the electron sees more of the bare charge of the proton (and vice versa) the binding becomes tighter and the energy of the state drops. The vacuum polarization predicts that the difference in energy between the two states corresponds to a frequency of about 27 MHz. The measured shift, the Lamb shift, is quite a bit larger in the opposite direction, near 1057 MHz. The vacuum polarization effect is swamped by a contribution from the vertex correction; nevertheless it must be included to get QED to agree with experiment.

As k^2 grows, the running coupling constant, $e/\sqrt{1+\Pi_R(k)}$, grows. This implies that as we measure the charge at higher and higher energy scales, we are allowed to probe further into the virtual particle cloud and begin to resolve the bare charge. At large space-like separations, $-k^2 \gg m^2$, the square charge is approximately

$$e^2(k^2) \approx \frac{e^2(m^2)}{\left(1-\frac{e^2(m^2)}{12\pi^2}\ln\left(-\frac{k^2}{m^2}\right)\right)}. \tag{17.111}$$

Now observe that eq.(17.111) is nearly identical to the solution of the β-function, eq.(17.107),

$$\frac{1}{e^2(\mu)} - \frac{1}{e^2(\mu_0)} = -\frac{1}{12\pi^2}\ln\left(\frac{\mu^2}{\mu_0^2}\right), \tag{17.112}$$

which is, of course, no coincidence. As we approach energy scales comparable to the regulator scale, the renormalized charge begins to look like the bare charge. Another way to illustrate this is to write out the photon propagator at large $-k^2$ in terms of the bare charge.

$$\begin{aligned} D_R^{\mu\nu}(k) &= -i\frac{g^{\mu\nu}}{k^2}\left(1+\frac{\alpha_0}{3\pi}\ln\left(-\frac{k^2}{m^2}\right)\right)\left(1-\frac{\alpha_0}{3\pi}\frac{1}{\varepsilon}\right) \\ &= -i\frac{g^{\mu\nu}}{k^2}\left(1+\frac{\alpha_0}{3\pi}\ln\left(-\frac{k^2}{m^2}\right)\right)\left(1-\frac{\alpha_0}{3\pi}\ln\left(\frac{M^2}{m^2}\right)\right) \end{aligned} \tag{17.113}$$

In the second step above we have replaced $1/\varepsilon$ in Z_3 with the Pauli-Villars equivalent, $\ln(M^2/m^2)$. When $-k^2 = M^2$ (the momentum transferred in scattering is equal to the cutoff), the finite part cancels Z_3 leaving just the bare charge. As the regulator is removed, the cancellation occurs at infinite energy.

Since there is experimental evidence that the electric charge is scale-dependent, does this mean that there really may be an infinite bare electric charge in the center of the cloud? Who can say, for $e^2/4\pi$ becomes $\gg 1$ at an energy long before $M^2 \to \infty$, and perturbation theory can no longer be trusted.

Moving on to the electron propagator, once again we would like to choose the subtraction point to be where the external particles are on mass shell. This would mean that the pole of the electron propagator would occur at $p^2 = m^2$ and that the residue at the pole would be equal to i, just as in the free field propagator.

The renormalized electron propagator will be

$$\frac{i}{\not p - m - \Sigma_R(p)}, \tag{17.114}$$

so, in order that the pole reside at $p^2 = m^2$ when the particle is on shell, then

$$\Sigma_R(p)\Big|_{\not p = m} = 0. \tag{17.115}$$

In order for the residue at $\not p = m$ to be i, then we must also have

$$\frac{\partial}{\partial \not p}\Sigma_R(\not p)|_{\not p = m} = 0. \tag{17.116}$$

In terms of the regularized $\Sigma^{\text{reg}}(p)$, eqs.(17.115) and (17.116) will be satisfied by the subtraction,

$$\Sigma_R(\not p) = \Sigma^{\text{reg}}(\not p) - \Sigma^{\text{reg}}(m) - (\not p - m)\frac{\partial}{\partial \not p}\Sigma^{\text{reg}}(\not p)|_{\not p = m}. \tag{17.117}$$

Now we run into a difficulty we mentioned earlier. When we apply the subtraction to $\Sigma^{\text{reg}}(p)$, eq.(17.46),

$$\Sigma^{\text{reg}}(p) = \frac{\alpha}{2\pi}\Bigg(\Big(2m - \frac{\not{p}}{2}\Big)\Big(\frac{1}{\varepsilon} - \gamma - 1\Big) + \int_0^1 dz(2m - \not{p}z)\ln\Big(\frac{4\pi\mu^2}{m^2(1-z) - p^2(1-z)z}\Big)\Bigg), \quad (17.118)$$

we find that the resulting renormalized $\Sigma_R(\not{p})$ is not finite but logarithmically divergent. The source of the divergence is the infrared (small momentum) limit of the integral over the loop momentum, and the divergence occurs because the photon is massless.

The choice of the subtraction point $\not{p} = m$ makes the renormalized electron propagator look like a free field propagator for a particle of mass m when the electron is on shell. This is in congruence with the asymptotic IN–OUT state formalism. Any particle reaching the asymptotic region must be on mass shell and propagating at its physical mass. In the asymptotic region away from the interaction, the particles are essentially free. However, when massless particles are present, the associated force has an infinite range, so technically, there is no asymptotic region. It should not be surprising that we might have difficulty forcing the electron propagator to look like the free particle result on mass shell.

The divergence is unphysical and arises from the particular expansion we chose. The expansion parameter is the fine structure "constant," $\alpha = e^2/4\pi$, or $\alpha = e^2/\hbar c$ in ordinary units. The zeroth order term of the expansion of the electron propagator is the free field result which can be thought of as classical (no loops). The next term arises at 1-loop, and we are trying to shape it into the classical form when the electron is isolated. The choice of expansion, however, is not well suited to take the classical limit $\hbar \to 0$. The divergence arises from a poor choice of expansion parameter. If we expand in some other parameter where the classical limit is not singular, or if we add up all of the higher order terms in the expansion in powers of α, then the infrared divergence goes away.

Now consider the scattering of an electron in an external field. The scattering amplitude will be an expansion in powers of α. The first quantum correction, order α, will involve the vertex correction. The vertex correction is also infrared divergent on shell. However, we know that any change in the electron's momentum results in the radiation of photons. In fact, since the photon is massless, the energy of a single photon can be arbitrarily small, thus the scattering of the electron can, and will, result in the emission of a very large number of low-frequency photons. Since the emission of a photon requires the addition of a vertex, such processes are not included in our expansion unless we include many higher order powers of α. In other words, the involvement of a lot of photons approaches a classical state, and the expansion in powers of α is not suitable to describe classical states unless a very large number of terms are included. If we sum up the series the divergence goes away.

On the practical side, what can we do at this point? Since the subtraction point is arbitrary, we can choose a subtraction point that is off shell. Then the divergence goes away. Alternatively, we can give the photon a small fictitious mass, λ, as we did in the Pauli-Villars regularized computation of $\Sigma(p)$. The mass acts as an infrared cutoff since a photon of arbitrarily small energy no longer can exist. The results will depend on λ, so some care must be exercised in interpreting them. For example, if we consider some scattering process and ask what the probability is that no photons are radiated above some energy k_{min}, or only 1 photon with energy above k_{min}, or only 2, etc., then, when we add together the amplitudes for all equivalent processes, the terms dependent on λ will cancel, leaving something physically meaningful. If dimensional regularization is used, we may leave the photon massless provided all of the integrals are done before the expansion about $\varepsilon = 0$ is performed. The infrared divergences will appear as poles and the poles will cancel when the amplitudes for all physically equivalent processes are added together.

Returning to the electron propagator, let us keep the subtraction point on mass shell at $\not{p} = m$, and add a fictious photon mass. The addition of the photon mass changes eq.(17.118) to

$$\Sigma_{\text{reg}}(\not{p}) = \frac{\alpha}{2\pi}\Big(\Big(2m - \frac{\not{p}}{2}\Big)\Big(\frac{1}{\varepsilon} - \gamma - 1\Big) + m + \int_0^1 dz\,(2m - \not{p}z)\ln\Big(\frac{4\pi\mu^2}{m^2(1-z) + \lambda^2 z - p^2 z(1-z)}\Big)\Big) \tag{17.119}$$

Application of the subtraction, eq.(17.117), yields

$$\Sigma_R(\not{p}) = \frac{\alpha}{2\pi}\Big(-\int_0^1 dz\,(2m - \not{p}z)\ln(m^2(1-z) + \lambda^2 z - p^2 z(1-z)) + \int_0^1 dz\,(2m - \not{p}z)\ln(m^2(1-z)^2 + \lambda^2 z) - (\not{p} - m)2m^2\int_0^1 dz\,\frac{(2-z)z(1-z)}{m^2(1-z)^2 + \lambda^2 z}\Big). \tag{17.120}$$

The first two integrals are finite in the limit $\lambda \to 0$; therefore we set $\lambda = 0$ in them. The last integral diverges logarithmically with $\lambda \to 0$. We separate out and save this term, and then take $\lambda \to 0$ in the remaining, finite piece. The result is

$$\Sigma_R(\not p) = \frac{\alpha}{2\pi}\Bigg(\frac{3}{2}\not p - m - \not p\frac{m^2}{2p^2} + \left(2m(m^2-p^2) - \frac{\not p}{2}\frac{m^4-p^4}{p^4}\right)\ln\left(1-\frac{p^2}{m^2}\right) + (\not p - m)\ln\left(\frac{\lambda^2}{m^2}\right)\Bigg). \tag{17.121}$$

Note that if $p^2 > m^2$, then $\Sigma_R(p)$ becomes complex.

Finally, we consider the subtraction of the vertex function, $\Lambda_\mu(p',p)$. Keeping in mind the motivation for choosing the renormalization prescription for the electron and photon propagators, we fix the renormalized vertex function by the desire to make it look like the classical result, γ_μ, when the particles are on shell. Thus,

$$\Gamma^\mu_R(p',p) = \Gamma^\mu_{\text{reg}}(p,p') - \Gamma^\mu_{\text{reg}}(p,p)\Big|_{\not p = m}. \tag{17.122}$$

Since the subtraction point is on shell, we must be on the lookout for infrared divergences, and add a photon mass, if necessary, to regulate them.

When we computed $\Gamma^\mu_{\text{reg}}(p',p)$, we broke it up into 2 parts. The first part, $\Gamma^{(a)\mu}_{\text{reg}}(p',p)$, eq.(17.62), is ultraviolet divergent when the regulator is removed, while the second part, $\Gamma^{(b)\mu}_{\text{reg}}(p',p)$, eq.(17.63), is ultraviolet finite. $\Gamma^{(a)\mu}_{\text{reg}}(p',p)$ is infrared convergent, hence there is no need to introduce a photon mass in it. Using the prescription above, eq.(17.122), the renormalized version of this is

$$\Gamma^{(a)\mu}_R(p',p) = \gamma^\mu \frac{\alpha}{2\pi}\int_0^1 dz_2 \int_0^{1-z_2} dz_3 \ln\left(\frac{m^2(z_2+z_3)^2}{m^2(z_2+z_3) - (1-(z_2+z_3))(p'^2 z_2 + p^2 z_3)}\right). \tag{17.123}$$

The computation of, and extraction of physical results from $\Gamma^{(b)\mu}_R(p',p)$ is straightforward but rather algebraically involved due to the string of γ matrices that appears in the numerator. We have in mind extracting the QED correction to the magnetic moment of the electron, a famous result first derived by J. Schwinger in 1948. For this, we only need the combination $\bar u(p')\Gamma^{(b)\mu}_R(p',p)u(p)$, where $u(p)$ and $\bar u(p')$ are Dirac momentum spinors, eq.(4.21). Since these spinors satisfy

$$(\not p - m)u(p) = 0, \quad \bar u(p)(\not p - m) = 0,$$

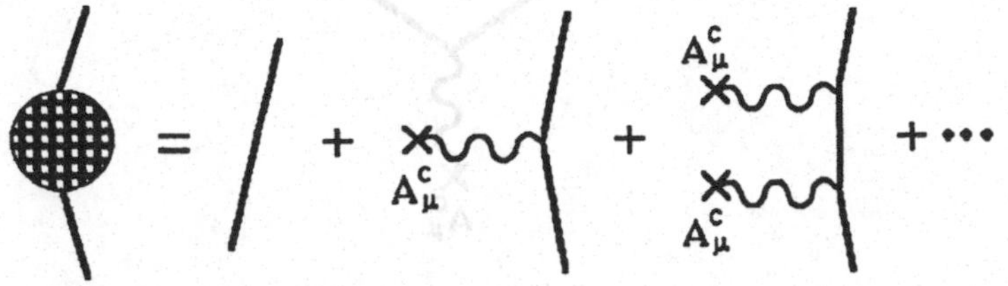

Figure 17.9 Diagrammatic expansion of the Dirac propagator in a classical background electromagnetic field, A^c_μ.

the net effect of sandwiching $\Gamma^\mu_R(p',p)$ between them is to put the fermions on mass shell. Thus, the renormalization prescription states that

$$\bar{u}(p')\Gamma^\mu_R(p',p)u(p) = 0$$

when $p'^\mu - p^\mu = 0$.

Why do we only need $\bar{u}(p')\Gamma^\mu_R(p',p)u(p)$? Consider measuring the magnetic moment, μ, of the electron by "scattering" it in a uniform magnetic field, $\vec{B}$. The macroscopic magnetic field is essentially classical; therefore we think of the operator A^μ as representing quantum fluctuations about a classical background, and write $A_\mu = A^q_\mu + A^c_\mu$. We will drop the q notation off the fluctuating part. In the presence of the classical background, the interaction term in the action becomes

$$e_0 \int d^4x\, \bar{\psi}_0(x)(\not{A} + \not{A}^c)\psi_0(x). \tag{17.124}$$

The classical background introduces a new vertex. For the moment, let $\not{A} = 0$. This leaves us a quantized Dirac field in the presence of a classical external field, A^c_μ. The action is strictly quadratic in ψ_0, hence the vacuum-to-vacuum amplitude is $\det(i\not{\partial} - m - e\not{A}^c(x))$. A perturbative expansion of the electron propagator in this background as a power series in e_0 (or A^c_μ) is represented diagrammatically in figure 17.9. The first order term, figure 17.11(a), is simply a representation of the new vertex, $e_0\not{A}^c$, thus this diagram is equal to

$$\frac{i}{\not{p}' - m_0 + i\epsilon}\, e_0\not{A}^c(p-p')\, \frac{i}{\not{p} - m_0 + i\epsilon}. \tag{17.125}$$

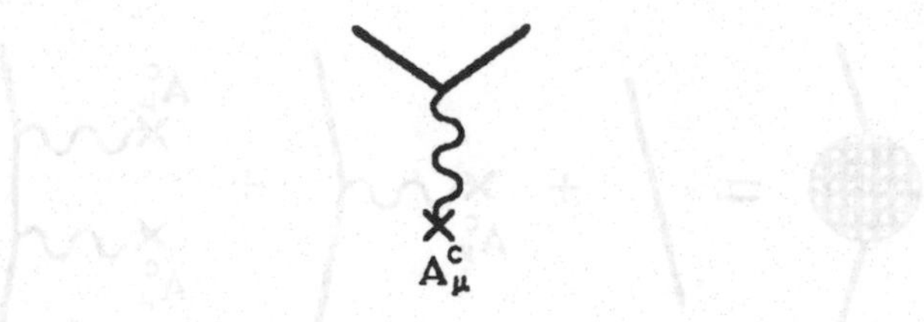

Figure 17.10 Diagrammatic representation of the lowest order term in the amplitude for pair creation in an external electromagnetic field.

When turned on its side, as in figure 17.10, this diagram represents the lowest order term in the amplitude for pair creation in an external electromagnetic field (which actually is a purely electric effect).

The wavy lines in figures 17.9 and 17.11(a) do not represent real or virtual photons, since, strictly speaking, there is no fluctuating electromagnetic field present. The symbolic utility in representing the vertex $\gamma^\mu A^c_\mu(x)$ in this way arises when we reinstate the fluctuating field, A_μ, and consider quantum corrections to figure 17.11(a). The 1-loop diagrams are given in figure 17.11(b)-(e). For example, in figure 17.11(b), the fermion loop is not present as a correction to the photon propagator, because the wavy line emanating from the external field is not a photon. The contribution of the fermion loop is, however, identical to the vacuum polarization contribution to the photon propagator. Similarly, the loop in figure 17.11(e) is identical to the vertex correction.

The corrections to the electron propagator represented by figures 17.11(c) and (d) simply tell us to replace m_0 by m in eq.(17.125) above. Therefore, the sum of the diagrams in figure 17.11 that represents the amplitude for scattering of an electron in an external macroscopic electromagnetic field is

$$\frac{i}{\not p' - m + i\epsilon} e A^c_\mu \left(\gamma^\mu + \Gamma^\mu_R + \Pi^{\mu\nu}_R \frac{g_{\nu\sigma}\gamma^\sigma}{(p'-p)^2 + i\epsilon} \right) \frac{i}{\not p - m + i\epsilon}. \quad (17.126)$$

When we process this through the reduction formula, eq.(8.22), the external legs are removed and replaced by momentum spinors $\bar{u}(p')$ and $u(p)$. Therefore, the vertex correction contribution to the scattering amplitude is $\bar{u}(p')\Gamma^\mu_R(p',p)u(p)$. To determine the QED correction to the magnetic moment of the electron, we need only examine this combination.

In order to determine what part of $\Gamma^\mu(p',p)$ represents a correction to the magnetic moment, let us return to the Dirac equation. Classically, we know that the energy associated with a magnetic moment, $\vec{\mu}$, in a

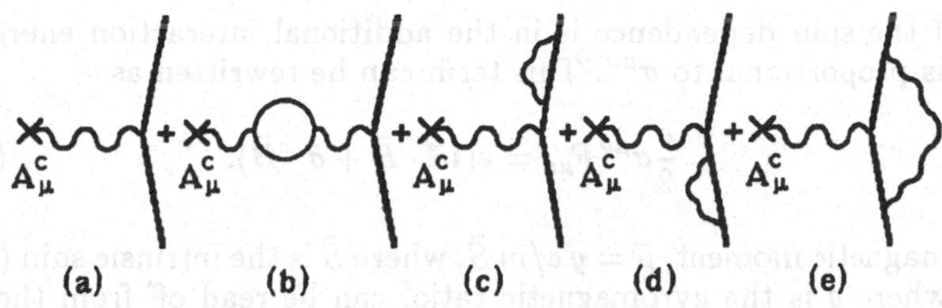

Figure 17.11 (b)-(e): One-loop quantum radiative corrections to the propagator of an electron through a classical external background electromagnetic field, A^c_μ, to first order in the background field (a).

magnetic field is $-\vec{\mu}\cdot\vec{B}$. From quantum mechanics we know that the magnetic moment of a particle is proportional to its intrinsic spin. Recall from chapter 4 that one requirement we imposed in finding a satisfactory relativistic wave equation that is first order in derivatives is that the equation should be the "square root" of the Klein-Gordon equation. If we apply the Dirac operator twice, we should end up with the Klein-Gordon equation. This requirement led to the commutation relations for the Dirac matrices, eq.(4.2). From the correspondence principle, the Klein-Gordon operator can be expressed as a relativistic energy-mass-momentum relation. Let us now follow the same procedure again, except this time we will start with the Dirac equation that is minimally coupled to an external field,

$$(i\partial\!\!\!/ - m - eA\!\!\!/)\psi = 0. \tag{17.127}$$

Application of the operator $i\partial\!\!\!/ + m - eA\!\!\!/$ to both sides yields

$$((i\partial\!\!\!/ - eA\!\!\!/)^2 - m^2)\psi = 0. \tag{17.128}$$

By use of the anticommutators, $\{\gamma^\mu, \gamma^\nu\} = 2g^{\mu\nu}$, we have

$$\gamma^\mu\gamma^\nu = g^{\mu\nu} - i\sigma^{\mu\nu}, \qquad \sigma^{\mu\nu} \equiv \frac{i}{2}[\gamma^\mu, \gamma^\nu]. \tag{17.129}$$

Thus,

$$\begin{aligned}(i\partial\!\!\!/ - eA\!\!\!/)^2 &= (i\partial_\mu - eA_\mu)(i\partial_\nu - eA_\nu)\gamma^\mu\gamma^\nu \\ &= (i\partial - eA)^2 - \frac{e}{2}\sigma^{\mu\nu}F_{\mu\nu},\end{aligned}$$

which means that eq.(17.128) is

$$\left((i\partial - eA)^2 - \frac{e}{2}\sigma^{\mu\nu}F_{\mu\nu} - m^2\right)\psi = 0. \tag{17.130}$$

All of the spin dependence is in the additional interaction energy term that is proportional to $\sigma^{\mu\nu}$. This term can be rewritten as

$$\frac{e}{2}\sigma^{\mu\nu}F_{\mu\nu} = e(i\vec{\alpha}\cdot\vec{E}+\vec{\sigma}\cdot\vec{B}). \tag{17.131}$$

The magnetic moment, $\vec{\mu} = g\,e/m\,\vec{S}$, where $\vec{S}$ is the intrinsic spin ($=\vec{\sigma}/2$), and where g is the gyromagnetic ratio, can be read off from the second piece above as $g = 2$.

Returning to the vertex correction contribution to the scattering of an electron in an external field, figure 17.11(e), the piece of $\bar{u}(p')\Gamma^{\mu}_{R}(p',p)u(p)$ $A^{c}_{\mu}(p'-p)$ that can be interpreted as a radiative correction to the gyromagnetic ratio is proportional to $\sigma^{\mu\nu}(p'-p)_{\nu}A^{c}_{\mu}$ (the Fourier transform of $\sigma^{\mu\nu}F_{\mu\nu}$). Now that we know what we are looking for, let's go back to $\Gamma^{(b)\mu}_{\rm reg}$, eq.(17.63), and extract this piece.

The monster in this computation is the numerator of $\Gamma^{(b)\mu}_{\rm reg}$,

$$\gamma^{\nu}(\not{p}'(1-z_2)-z_3\not{p}+m)\gamma^{\mu}((1-z_3)\not{p}-z_2\not{p}'+m)\gamma_{\nu}. \tag{17.132}$$

To reduce this to a manageable form we use the identities,

$$\gamma^{\nu}\not{a}\not{b}\not{c}\gamma_{\nu} = -2\not{c}\not{b}\not{a},\quad \gamma^{\nu}\not{a}\not{b}\gamma_{\nu} = 4a\cdot b,\quad \text{and}\quad \gamma^{\nu}\not{a}\gamma_{\nu} = -2\not{a}, \tag{17.133}$$

and then commute the remaining $\not{p}'$'s and $\not{p}$'s so that $\not{p}'$ appears on the very left and $\not{p}$ on the very right. Since the numerator is sandwiched between $\bar{u}(p')$ and $u(p)$, we may then use the Dirac equation to replace them with m. With the fermions on shell, eq.(17.132) reduces to

$$-\gamma^{\mu}\left(2m^2\left(-2+2(z_2+z_3)+(z_2+z_3)^2\right)+2(p'-p)^2(1-z_2)(1-z_3)\right)$$
$$+4imz_2\left(1-(z_2+z_3)\right)(p'-p)_{\nu}\sigma^{\mu\nu}. \tag{17.134}$$

Note that the first term does not vanish when $p = p'$. Therefore, it is subtracted away when the renormalization prescription, eq.(17.122), is applied.

When the fermions are on shell, $p^2 = p'^2 = m^2$, and $2(p\cdot p') = 2m^2$ $-(p'-p)^2$. Therefore, eq.(17.134) and eq.(17.63) imply that

$$\bar{u}(p')\Gamma^{\mu}_{R}(p',p)u(p) = \bar{u}(p')\left(\frac{\alpha}{2\pi}\right)\int_0^1 dz_2\int_0^{1-z_2} dz_3$$
$$\frac{(p'-p)^2(1-z_2)(1-z_3)\gamma^{\mu}+2imz_2(1-z_2-z_3)(p'-p)_{\nu}\sigma^{\mu\nu}}{m^2(z_2+z_3)^2+\lambda^2(1-z_2-z_3)-(p'-p)^2z_2z_3}\,u(p), \tag{17.135}$$

where we have included a small photon mass to regulate any infrared difficulties. However, the piece we are interested in, the term proportional to $\sigma^{\mu\nu}$ is infrared finite. When $(p'-p)^2 \ll m^2$ (small momentum transfer), this part is

$$\begin{aligned}\bar{u}(p')\left(\frac{\alpha}{2\pi}\right)\frac{2i}{m}(p'-p)_\nu\sigma^{\mu\nu}\int_0^1 dz_2\int_0^{1-z_2} dz_3\,\frac{z_2(1-z_2-z_3)}{(z_2+z_3)^2}\,u(p)\\ = \bar{u}(p')\frac{i}{2m}\left(\frac{\alpha}{2\pi}\right)(p'-p)_\nu\sigma^{\mu\nu}u(p). \quad (17.136)\end{aligned}$$

Comparing this result with eqs.(17.126) and (17.131), we see that the radiative vertex correction implies that the gyromagnetic ratio, g, to order α, is

$$g = 2\left(1+\frac{\alpha}{2\pi}\right). \quad (17.137)$$

In the same limit, $(p'-p)^2 \ll m^2$, one can show that the first term in eq.(17.135) reduces to

$$\frac{\alpha}{3\pi}\frac{(p'-p)^2}{m^2}\gamma^\mu\left(\ln\left(\frac{m}{\lambda}\right)-\frac{3}{8}\right). \quad (17.138)$$

In this same limit, eq.(17.109) states that the vacuum polarization reduces to

$$-\frac{\alpha}{15\pi}\frac{(p'-p)^2}{m^2}. \quad (17.139)$$

Thus the scattering amplitude, eq.(17.126), for small momentum transfers is

$$eA^c_\mu\left[\gamma^\mu\left(1+\frac{\alpha}{3\pi}\frac{(p'-p)^2}{m^2}\left(\ln\left(\frac{m}{\lambda}\right)-\frac{3}{8}-\frac{1}{5}\right)\right)+\frac{i}{2m}\frac{\alpha}{2\pi}\sigma^{\mu\nu}(p'-p)_\nu\right]. \quad (17.140)$$

We can just as well think of A^c_μ as arising from the Coulomb field around a proton, in which case eq.(17.140) represents radiative corrections to the spectrum of the Dirac hydrogen atom. We have already briefly mentioned the vacuum polarization's contribution to the Lamb shift of the $2S_{1/2}$ state from the $2P_{1/2}$ state. The dominant part of the shift is due to the vertex correction. The details may be found in the references in the Bibliography.

17.4 THE CHIRAL ANOMALY

An "anomaly" occurs when a symmetry present in the classical system is broken upon quantization by the quantum corrections. This definition is a bit limited since it assumes that a meaningful classical limit exists, and that we are able to identify it. In this sense the identification of anomalies relies on perturbation theory. However, anomalies are not believed to be an artifact of perturbation theory. There are several examples where nonperturbative anomalies can be identified. Anomalies are not just annoying little nuisances, either. They can render a quantum field theory mathematically inconsistent or nonrenormalizable, or they can come to the rescue.

As we saw in the previous section, the classical action can be identical to the quantum action written in terms of the bare functional coordinates. This bare coordinate system is rather singular, and we must go through the song-and-dance routine of renormalization to rewrite the action in a finite functional coordinate system. During the process of renormalization we generate counterterms, and it is possible that some counterterms may violate a symmetry that was apparently present in the bare action. Such counterterms would be anomalies. Alternatively, the presence of divergences and the need to renormalize force us to regularize as an intermediate step. The presence of an anomaly depends on the fact that it is not possible to find a regulator that preserves all of the symmetries present in the classical or bare action. Likewise, if there is a good way to regularize a potentially anomalous diagram, then there is no anomaly present.

We have already met one anomaly. The classical action is scale-invariant. Regularization necessarily breaks scale invariance since the regularizing cutoff introduces a scale. There is no way to regularize without introducing such a scale. This anomaly manifests itself in the renormalized theory as a scale-dependent electric charge, a scale dependence not apparent in the bare action.

Continuous symmetries lead to conserved currents. The breakdown of a symmetry by an anomaly is signaled by the failure to maintain current conservation after regularization and renormalization. An anomaly is fatal to a theory when the symmetry and associated current conservation are necessary to maintain mathematical consistency or are required for renormalizability. For example, the gauge symmetry in a gauge theory leads to a conserved "gauge" current. In electrodynamics, the gauge current is the vector current, $j^\mu = \bar{\psi}\gamma^\mu\psi$. If there is no way to regularize the theory and maintain gauge current conservation, then the quantum theory will not be renormalizable and will not be mathematically consistent since physical results will depend on an arbitrary choice of gauge (i.e. coordinate system). As we have seen, we do have a way to regularize QED and maintain current conservation; therefore there is no gauge anomaly in electrodynamics. However, as we are about to see, there is a breakdown of chiral symmetry in the presence of gauge fields coupled to conserved vector currents. Actually, this chiral anomaly represents an ambiguity in the theory. It is possible to maintain axial (chiral) current conservation at the price of gauge invariance since the conserved axial current is not gauge invariant. Alternatively, we may define a gauge invariant axial current, but this current is not conserved. We cannot maintain both gauge invariance and axial current conservation. Since the gauge symmetry is crucial, the choice between the two alternatives is easy to make. The chiral symmetry is not necessary for renormalizability, so who needs it anyway?

In fact, the presence of the chiral anomaly has physical consequences. We will return to these in a moment after stating the result of the computation of the anomaly. The details of the computation of the anomaly will be saved for last.

Recall from chapter 4 that chirality transformations are generated by the γ_5 matrix, $\gamma_5 = i\gamma^0\gamma^1\gamma^2\gamma^3$. Thus,

$$\psi \to \exp(i\theta\gamma_5)\psi, \qquad \psi^\dagger \to \psi^\dagger \exp(-i\theta\gamma_5), \tag{17.141}$$

represents a global chiral transformation. In order to find the associated current we can employ the same device we used in chapter 2, and look at the transformation of the action under a local chiral transformation ($\theta = \theta(x)$). Since γ_5 squares to 1 and anticommutes with the other Dirac matrices,

$$\begin{aligned} S &= \int d^4x\, \bar{\psi}(i\not{D} - m)\psi \\ &\to \int d^4x\, \bar{\psi}\, e^{i\theta(x)\gamma_5}(i\not{\partial} - m - e_0\not{A})e^{i\theta(x)\gamma_5}\psi \\ &= S + \int d^4x\, \theta(x)\left[\partial_\mu(\bar{\psi}\gamma^\mu\gamma_5\psi) - 2im\bar{\psi}\gamma_5\psi\right] + O(\theta^2), \end{aligned} \tag{17.142}$$

where we have employed integration by parts. The terms in the brackets describe a conservation relation,

$$\partial_\mu j_5^\mu = 2mj_5, \tag{17.143}$$

where

$$j_5^\mu = \bar{\psi}\gamma^\mu\gamma_5\psi \tag{17.144}$$

is the axial vector current or axial current or pseudovector current, and

$$j_5 = i\bar{\psi}\gamma_5\psi \tag{17.145}$$

is the pseudoscalar current. Eq.(17.143) states that the axial current is conserved and the chiral symmetry is unbroken when the fermions are massless, $m = 0$. This result is expected, of course, since the projection of spin onto the direction of motion is undefined for a massive particle at rest. However, we will find at 1-loop that the axial current is not conserved even when $m = 0$. That is, when we try to verify eq.(17.143) at 1-loop, we instead find

$$\partial_\mu j_5^\mu = 2mj_5 - \frac{\alpha}{2\pi} F_{\mu\nu}\, {}^*F^{\mu\nu}, \tag{17.146}$$

where ${}^*F^{\mu\nu}$ is the dual of $F^{\mu\nu}$ and is defined by eqs.(5.4) and (5.5). (${}^*F^{\mu\nu}$ is proportional to $\vec{E} \cdot \vec{B}$.) The extra piece in eq.(17.146) not present in eq.(17.143) is "the" anomaly. Note that the axial current defined in eq.(17.144) is gauge invariant and eq.(17.146), which is also gauge invariant, expresses the fact that it is no longer conserved, even in the $m = 0$ limit. On the other hand, we can define an axial current that does satisfy the classical conservation relation, eq.(17.143). It is

$$j_5^{\mu\prime} = \bar{\psi}\gamma^\mu\gamma_5\psi + \frac{\alpha}{2\pi}\epsilon^{\mu\nu\sigma\rho} A_\nu F_{\sigma\rho}. \tag{17.147}$$

However, $j_5^{\mu\prime}$ is clearly no longer gauge invariant due to the explicit appearance of the vector potential. The two alternatives represent the ambiguity introduced by the anomaly.

If we replace γ_5 by 1 in eq.(17.141) and follow the same development, we end up with the usual vector current conservation relation. At 1-loop we explicitly checked that the vector current remained conserved. This amounted to verifying that the Ward-Takahashi identities remained valid in the regularized theory at 1-loop. The Ward-Takahashi identities are an expression of the gauge invariance of the effective action. If the validity of the Ward-Takahashi identities cannot be maintained, then the gauge symmetry is broken. A chiral Ward identity may also be derived for the axial vertex $\gamma^\mu\gamma_5$. At 1-loop, the axial Ward identity acquires an extra term, the anomaly. The anomalous term is proportional to i indicating that the anomaly in the effective action is a nontrivial, A^μ field-dependent phase resulting from performing the functional path integral.

Now back to the physical consequences of eq.(17.146). Even though eq.(17.146) was derived in the context of spinor QED, the existence of the anomaly is dependent on the quantum numbers of the fields (i.e. what group representation they fall into), and not on the particular details of the form of the interaction. An equation very similar to eq.(17.146) will arise in a theory with vector and axial currents.

Eq.(17.146) shows that there is a coupling between the divergence of an axial current and the product of 2 vector potentials. Physically this means that there is an amplitude for the decay of an axial current into 2 photons. Let us consider the decay of the neutral π meson into 2 photons, $\pi \to 2\gamma$, represented diagrammatically in figure 17.12. The pions have

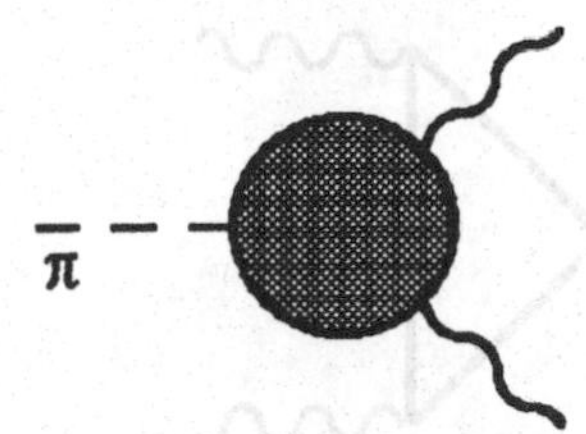

Figure 17.12 Diagrammatic representation of the decay amplitude of a neutral pion into 2 photons.

an intrinsic parity of -1; therefore they are pseudoscalars. Pseudoscalars share the same quantum numbers as divergences of axial currents. In fact, we believe that the pions are up and down quark-antiquark bound states, so in some sense, the π^0 meson carries a weak hadronic flavor current which is part axial. Therefore, the chiral anomaly should influence the $\pi^0 \to 2\gamma$ decay.

With the assumption that the π^0 is a bound quark-antiquark pair, the lowest order contribution to the $\pi^0 \to 2\gamma$ amplitude is the triangle diagram in figure 17.13(a). The fermions traveling in the triangular loop are the constituent quarks that make up the π^0. Let $\phi(x)$ be a scalar interpolating field that creates a state with a single π^0 out of the vacuum. The Partially Conserved Axial Current (PCAC) hypothesis states that $\phi(x)$ is directly proportional to the divergence of an axial current.

$$\partial_\mu A_5^\mu(x) = m_\pi^2 f_\pi \phi(x) \tag{17.148}$$

Here, m_π is the mass of the pion and f_π is the amplitude associated with the decay of a pion to a muon and its neutrino (and therefore may be determined by experiment). Note the similarity between eq.(17.148) and eq.(17.143). Under PCAC, the pion-quark-quark vertex in figure 17.13(a) is an axial vertex, $\gamma^\mu\gamma_5$. The other two radiative vertices are ordinary vector couplings, γ^μ. It is this AVV triangle, figure 17.13(b), that gives rise to the anomaly.

If there was no anomaly, and if PCAC, eq.(17.148), was the whole story, then vector current conservation and the small magnitude of the pion mass would strongly suppress the $\pi^0 \to 2\gamma$ decay mode. However, it is observed that the π^0 decays into 2 photons 99% of the time. The anomaly comes to the rescue. Instead of scrapping PCAC, we modify it using the anomaly (i.e. calculate the triangle diagram for the relevant phenomenological action). In computing the total amplitude associated with figure 17.13(a), we must sum the fermion loop over all of the different types of quarks that can contribute to the pion. In order to get this

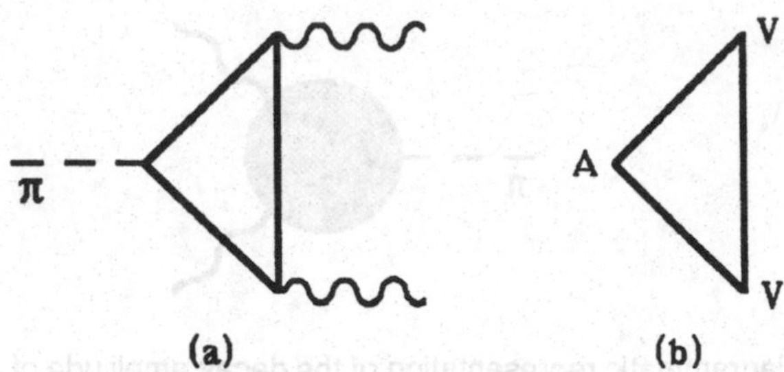

Figure 17.13 (a) Lowest order term in the expansion of the $\pi \to 2\gamma$ decay amplitude. The fermions traveling in the triangular loop are the constituent quarks that make up the neutral pion. (b) Under the PCAC hypothesis, the pion-quark-quark vertex is an axial vertex, $\gamma_\mu\gamma_5$. The 2 other radiative vertices are ordinary vector vertices.

amplitude to agree with experiment (to within 10%) we must assume that each flavor of quark comes in three colors. Thus the $\pi^0 \to 2\gamma$ decay gives some evidence for the existence of anomalies and for the need for 3 quark colors.

The chiral anomaly also comes to aid in the resolution of the so-called $U(1)$ problem in QCD. Eq.(17.147) defines a conserved current that is not gauge invariant. However, the total chiral charge,

$$Q_5{}' = \int d^3x\, j_5^{0\prime}(x), \tag{17.149}$$

is gauge invariant. This means that $Q_5{}'$ is a constant of the motion and that just a little bit of the original chiral symmetry remains intact. In QCD, the axial current, $J_5^\mu = \bar{\psi}^a\gamma^\mu\gamma_5\psi^a$, where a is the color index, is also anomalous.

$$\partial_\mu J_5^\mu = 2im\bar{\psi}^a\gamma_5\psi_a - \frac{g^2}{8\pi^2}\,\mathrm{tr}(F_{\mu\nu}{}^*F^{\mu\nu}), \tag{17.150}$$

where $F^{\mu\nu}$ is now the color field strength. Since $F^{\mu\nu}$ and its dual are gauge covariant, the trace above is gauge invariant. As in the Abelian case, we may also define a conserved axial current,

$$J_5^{\mu\prime} = J_5^\mu - \frac{g^2}{4\pi^2}\epsilon^{\mu\nu\rho\sigma}A_\nu^a\Big(\partial_\rho A_\sigma^a - \partial_\sigma A_\rho^a - \frac{2}{3}g\epsilon_{abc}A_\rho^b A_\sigma^c\Big). \tag{17.151}$$

Note the relationship between the second term above and the functional, $W[a]$, eq.(11.97). This current is, of course, no longer gauge invariant. How about the chiral charge, $Q_5{}'$? It was originally thought that this charge was gauge invariant. The consequence of this little bit of extra symmetry is that the pion should have a light partner which clearly it does not. This is the $U(1)$ problem. However, it turns out that $Q_5{}'$ is not gauge invariant. Without this extra symmetry, the light partner of the pion can become much more massive and the problem goes away.

Why is $Q_5{}'$ no longer invariant as in the Abelian case (in 4 dimensions)? $SU(3)$ gauge transformations can be characterized by a winding number, n. A gauge transformation of vanishing winding number may be continuously deformed to the identity gauge transformation, the unit matrix. Gauge transformations with nonzero winding number cannot. $Q_5{}'$ is only gauge invariant under gauge transformations of zero winding number, and all gauge transformations in QED in 4 dimensions may be deformed to the identity. $Q_5{}'$ is not gauge invariant under topologically nontrivial transformations.

The winding number for Yang-Mills gauge transformations ($SU(2)$) arises in 4 dimensions because the $SU(2)$ group manifold is the 3-sphere, S^3. This happens to be identical to compactified 4-dimensional spacetime (compactified = all points at infinity are identified as a single point at infinity). Thus, an $SU(2)$ gauge transformation is a mapping of S^3 onto itself. As in the case of mappings of the circle onto itself, these mappings can be classified topologically by the number of times the map wraps the 3-sphere onto itself. This classification by wrapping survives for higher-dimensional groups $SU(n)$, $n \geq 3$.

So far the chiral anomaly appears to be friendly. This is so only because in the two cases above the axial current is not coupled to the gauge field (photon or gluon). If it was, then the anomaly turns nasty and destroys the renormalizability and mathematical consistency. In the Weinberg-Salam model or electroweak theory, the gauge fields *are* coupled to axial currents. Here the anomaly presents a real danger, and the theory must be arranged so that the anomaly cancels. This will occur if there are the same number of leptons as there are flavors of quark. So far we know of 6 leptons, but only 5 flavors. This is why we anxiously await the discovery of the "top" quark.

Now let us return to the computation of the anomaly in the AVV triangle diagram, figure 17.14. The μ vertex is an axial vertex, $\gamma^\mu\gamma_5$. The other two vertices are vectorial, γ^ν and γ^ρ. Using the momentum labels of figure 17.14(a), the loop of this diagram is equal to

$$-(-ie)^2 \int \frac{d^4p}{(2\pi)^4} \operatorname{tr}\left(\gamma_\mu\gamma_5 \frac{i}{\not{p} - \not{k}_2 - m}\gamma_\rho \frac{i}{\not{p} - m}\gamma_\nu \frac{i}{\not{p} + \not{k}_1 - m}\right). \tag{17.152}$$

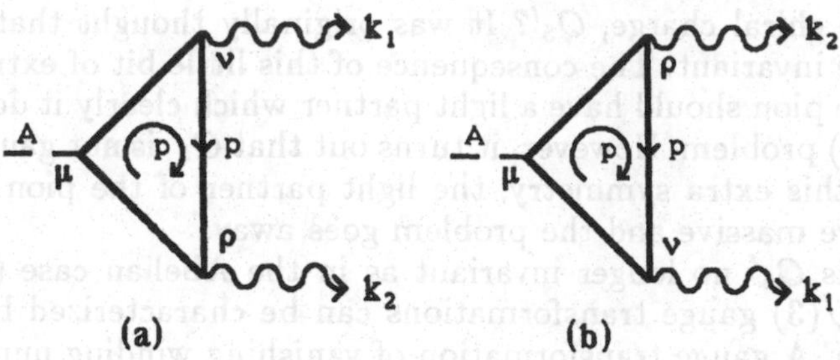

Figure 17.14 Momentum-vertex labels used in computing the chiral triangle anomaly. The photons carry away momenta k_1 and k_2. A momentum p travels around the triangular loop. The Bose symmetry of the photons implies that the total amplitude is the sum of diagrams (a) and (b).

To comply with the Bose symmetry of the photons, we must also include figure 17.14(b), where k_1 and k_2 have been exchanged along with ν and ρ. Thus, the total amplitude corresponding to the AVV triangle is

$$\begin{aligned}T_{\mu\nu\rho} = &-ie^2 \int \frac{d^4p}{(2\pi)^4} \frac{\mathrm{tr}(\gamma_\mu\gamma_5(\not{p} - \not{k}_2 + m)\gamma_\rho(\not{p} + m)\gamma_\nu(\not{p} + \not{k}_1 + m))}{((p+k_1)^2 - m^2)(p^2 - m^2)(p-k_2)^2 - m^2))} \\ &+ ie^2 \int \frac{d^4p}{(2\pi)^4} \frac{\mathrm{tr}(\gamma_\mu\gamma_5(\not{p} - \not{k}_1 + m)\gamma_\nu(\not{p} + m)\gamma_\rho(\not{p} + \not{k}_2 + m))}{((p+k_2)^2 - m^2)(p^2 - m^2)(p-k_1)^2 - m^2))}.\end{aligned} \tag{17.153}$$

Figure 17.14 and eq.(17.153) represent the 3-current correlation function, $T_{\mu\nu\rho} = \langle 0 | T(j^5_\mu j_\nu j_\rho) | 0 \rangle$. Vector current conservation implies that $\partial^\nu T_{\mu\nu\rho} = \partial^\rho T_{\mu\nu\rho} = 0$. If eq.(17.143) holds, then

$$\partial^\mu T_{\mu\nu\rho} = \langle 0 | T(\partial^\mu j^5_\mu j_\nu j_\rho) | 0 \rangle = 2m \langle 0 | T(j^5 j_\nu j_\rho) | 0 \rangle \equiv 2m T_{\nu\rho}.$$

Using the momentum labels in figure 17.14, we therefore want to compute $(k_1 + k_2)^\mu T_{\mu\nu\rho}(k_1, k_2)$.

The first step is to reduce the awful-looking trace that appears in eq.(17.153). In addition, we want to identify $2mT_{\nu\rho}$. $(k_1 + k_2)^\mu$ dotted into the first trace in eq.(17.153) (figure 17.14(a)) is

$$\mathrm{tr}\left((\not{k}_1 + \not{k}_2)\gamma_5(\not{p} - \not{k}_2 + m)\gamma_\rho(\not{p} + m)\gamma_\nu(\not{p} + \not{k}_1 + m)\right). \tag{17.154}$$

We could just compute this straightforwardly using the fact that

$$\mathrm{tr}(\gamma_5) = \mathrm{tr}(\gamma_5\gamma_\mu\gamma_\nu) = 0, \qquad \mathrm{tr}(\gamma_5\gamma_\mu\gamma_\nu\gamma_\rho\gamma_\sigma) = 4i\epsilon_{\mu\nu\rho\sigma}, \tag{17.155}$$

and that the trace of an odd number of γ matrices vanishes. However, using the property $\not{a}\not{a} = a^2$, the form of eq.(17.154) presents us with a way to break up the trace so as to reduce the number of factors that appear in the denominator. As an added bonus, $T_{\nu\rho}$ will pop into view.

What we want to do is to rewrite the term $\not{k}_1 + \not{k}_2$ such that factors of $\not{p} + \not{k}_1 - m$ and $\not{p} - \not{k}_2 - m$ appear. We will then commute these factors such that they will appear in conjunction with $\not{p} + \not{k}_1 + m$ and $\not{p} - \not{k}_2 + m$ respectively. They will then cancel a factor in the denominator.

This is easy to do. All we do is add and subtract $\not{p}$ and m.

$$\begin{aligned}\not{k}_1 + \not{k}_2 &= \not{p} + \not{k}_1 - m - \not{p} + \not{k}_2 + m \\ &= (\not{p} + \not{k}_1 - m) - (\not{p} - \not{k}_2 - m)\end{aligned}$$

Since γ_5 anticommutes with γ_μ, we have

$$\begin{aligned}(\not{k}_1 + \not{k}_2)\gamma_5 &= ((\not{p} + \not{k}_1 - m) - (\not{p} - \not{k}_2 - m))\gamma_5 \\ &= (\not{p} + \not{k}_1 - m)\gamma_5 + \gamma_5(\not{p} - \not{k}_2 + m) \qquad (17.156) \\ &= (\not{p} + \not{k}_1 - m)\gamma_5 + \gamma_5(\not{p} - \not{k}_2 - m) + 2m\gamma_5.\end{aligned}$$

Note that in order to bring $\not{p} - \not{k}_2 - m$ onto $\not{p} - \not{k}_2 + m$ in the trace, we had to commute γ_5. In order to get m to appear with the right sign we had to add and subtract an additional $2m$. This last factor will make $2mT_{\nu\rho}$ explicitly appear. Using eq.(17.156), the trace, eq.(17.154), is

$$\begin{aligned}&((p + k_1)^2 - m^2))\mathrm{tr}(\gamma_5(\not{p} - \not{k}_2 - m)\gamma_\rho(\not{p} + m)\gamma_\nu) \\ &\quad +((p - k_2)^2 - m^2)\mathrm{tr}(\gamma_5\gamma_\rho(\not{p} + m)\gamma_\nu(\not{p} + \not{k}_1 + m)) \\ &\quad +2m\mathrm{tr}(\gamma_5(\not{p} - \not{k}_2 + m)\gamma_\rho(\not{p} + m)\gamma_\nu(\not{p} + \not{k}_1 + m)). \qquad (17.157)\end{aligned}$$

An identical procedure is used on the second trace in eq.(17.153) to yield

$$\begin{aligned}(k_1 + k_2)^\mu T_{\mu\nu\rho}(k_1, k_2) &= 2mT_{\nu\rho}(k_1, k_2) \\ &+ ie^2 \int \frac{d^4p}{(2\pi)^4} \left(\frac{\mathrm{tr}(\gamma_5(\not{p} - \not{k}_2 - m)\gamma_\rho(\not{p} + m)\gamma_\nu)}{(p^2 - m^2)((p - k_2)^2 - m^2)} \right. \\ &\qquad \left. + \frac{\mathrm{tr}(\gamma_5\gamma_\rho(\not{p} + m)\gamma_\nu(\not{p} + \not{k}_1 + m))}{(p^2 - m^2)((p + k_1)^2 - m^2)} \right) \\ &+ ie^2 \int \frac{d^4p}{(2\pi)^4} \left(\frac{\mathrm{tr}(\gamma_5(\not{p} - \not{k}_1 - m)\gamma_\nu(\not{p} + m)\gamma_\rho)}{(p^2 - m^2)((p - k_1)^2 - m^2)} \right. \\ &\qquad \left. + \frac{\mathrm{tr}(\gamma_5\gamma_\nu(\not{p} + m)\gamma_\rho(\not{p} + \not{k}_2 + m))}{(p^2 - m^2)((p + k_2)^2 - m^2)} \right).\end{aligned}$$

(17.158)

If we shift integration variables in the last term of the second integral from p to $p-k_2$, then the denominator, $(p^2-m^2)((p+k_2)^2-m^2)$, will become $((p-k_2)^2-m^2)(p^2-m^2)$, which is identical to the denominator of the first term in the first integral. Under this shift of integration variable, the trace becomes

$$\begin{aligned}\mathrm{tr}(\gamma_5\gamma_\nu(\not{p}+m)\gamma_\rho(\not{p}+\not{k}_2+m)) &\longrightarrow \mathrm{tr}(\gamma_5\gamma_\nu(\not{p}-\not{k}_2+m)\gamma_\rho(\not{p}+m))\\ &= -\mathrm{tr}(\gamma_\nu\gamma_5(\not{p}-\not{k}_2+m)\gamma_\rho(\not{p}+m))\\ &= -\mathrm{tr}(\gamma_5(\not{p}-\not{k}_2+m)\gamma_\rho(\not{p}+m)\gamma_\nu),\end{aligned}$$

which is identical to the trace of the first term in the first integral except for the sign. Therefore, these two terms cancel. If we shift p to $p+k_1$ in the first term of the second integral, it will appear to cancel the second term in the first integral. Thus, at first sight, it appears that we have verified eq.(17.143) and there is no anomaly. However, the integrals are linearly divergent so we are not allowed to shift integration variables and reach such a conclusion. A divergence? That means we must regularize the expressions. To avoid difficulties with the definition of γ_5, we will use Pauli-Villars. This means that we subtract from eq.(17.158) an identical expression except that m is replaced by M. The regulator is removed by taking M to infinity.

Once eq.(17.158) is regulated, then all of the integrals are finite so we are allowed to shift integration variables. The form of the regulated integrals does not change so the same argument above will work, and the result is that all of the extra integrals cancel. Thus,

$$(k_1+k_2)^\mu T^{\mathrm{reg}}_{\mu\nu\rho}(k_1,k_2) = 2mT_{\nu\rho}(k_1,k_2,m) - 2MT_{\nu\rho}(k_1,k_2,M), \quad (17.159)$$

and the fate of the axial current conservation relation, eq.(17.143), rests on the result of $\lim_{M\to\infty} 2MT_{\nu\rho}(k_1,k_2,M)$, to which we now turn.

$$\begin{aligned}&2MT_{\nu\rho}(k_1,k_2,M)\\ &\quad= 2Me^2\int\frac{d^4p}{(2\pi)^4}\frac{\mathrm{tr}(\gamma_5(\not{p}-\not{k}_2+M)\gamma_\rho(\not{p}+M)\gamma_\nu(\not{p}+\not{k}_1+M))}{((p+k_1)^2-M^2)(p^2-M^2)((p-k_2)^2-M^2)}.\\ &\qquad + k_1\leftrightarrow k_2,\ \nu\leftrightarrow\rho \quad (17.160)\end{aligned}$$

The term in the trace proportional to M^3 vanishes because $\mathrm{tr}(\gamma_5\gamma_\rho\gamma_\nu)=0$. The terms proportional to M^2 and M^0 vanish because they involve an odd number of γ matrices. This leaves the term proportional to M, and this piece is independent of p due to the appearance of the antisymmetric tensor, $\epsilon_{\mu\rho\nu\sigma}$. Thus,

$$2MT_{\nu\rho}(k_1,k_2,M) = 8M^2e^2 i\epsilon_{\mu\rho\nu\sigma}k_2^\mu k_1^\sigma(I(k_1,k_2)+I(k_2,k_1)), \quad (17.161)$$

where

$$I(k_1,k_2)=\int\frac{d^4p}{(2\pi)^4}\,\frac{1}{((p+k_1)^2-M^2)(p^2-M^2)((p-k_2)^2-M^2)}. \tag{17.162}$$

To put this into a more manageable form, we elevate the denominator to exponentials using eq.(17.22), complete the squares, and do the Gaussian integral. This yields

$$I(k_1,k_2)=\frac{1}{16\pi^2}\int_0^\infty dz_1\int_0^\infty dz_2\int_0^\infty dz_3\,\frac{1}{(z_1+z_2+z_3)^2}$$
$$\exp\left(-i\frac{(z_1k_1-z_3k_2)^2}{z_1+z_2+z_3}\right)\exp\left(i(z_1k_1^2+z_3k_2^2)\right)\exp\left(-iM^2(z_1+z_2+z_3)\right). \tag{17.163}$$

Next we insert the identity, eq.(17.60), and rescale z_1, z_2, and z_3 by β. The integral over z_2 is easily done, leaving

$$I(k_1,k_2)=\frac{1}{16\pi^2}\int_0^1 dz_1\int_0^{1-z_1}dz_3\int_0^\infty d\beta$$
$$\exp\left(-i\beta(z_1k_1-z_3k_2)^2\right)\exp\left(i\beta(z_1k_1^2+z_3k_2^2)\right)\exp\left(-i\beta M^2\right). \tag{17.164}$$

The integral over β is also elementary. However, before we do it, let's put the photons on shell ($k_1^2=k_2^2=0$, which implies $(k_1-k_2)^2=-(k_1+k_2)^2$). The result of both operations is

$$I(k_1,k_2)=-\frac{i}{16\pi^2}\int_0^1 dz_1\int_0^{1-z_1}dz_3\,\frac{1}{M^2-z_1z_3(k_1+k_2)^2}. \tag{17.165}$$

As M gets large,

$$I(k_1,k_2)\rightarrow-\frac{i}{32\pi^2}\frac{1}{M^2}.$$

Thus, the finite axial current anomaly is

$$\lim_{M\to\infty}2MT_{\nu\rho}(k_1,k_2,M)=\frac{2}{\pi}\alpha\,\epsilon_{\nu\rho\mu\sigma}k_1^\mu k_2^\sigma, \tag{17.166}$$

and

$$\partial^\mu T_{\mu\nu\rho}=2mT_{\nu\rho}+\frac{2\alpha}{\pi}\epsilon_{\nu\rho\mu\sigma}k_1^\mu k_2^\sigma. \tag{17.167}$$

Eq.(17.143) fails to hold at 1-loop.

What happens at higher loops? Are there more chiral anomalies? No, no more, the reason being that there are methods of regularization that respect both gauge invariance and the chiral symmetry. Typically, the regularization is implemented by adding terms to the action that contain higher powers of the gauge covariant derivative acting on the fermions and bosons. This is what we did to the photon in the Pauli-Villars regularization. The higher powers of the covariant derivative increase the power of momentum that appears in the denominator of the propagator, thus decreasing the superficial degree of divergence. However, the added covariant derivatives also introduce momentum-dependent factors at new vertices, which increase the superficial degree of divergence. The net result is that regularization is achieved at 2-loops and higher. Thus, there are no higher-loop chiral anomalies present.

17.5 EXERCISES

1. Generalize the expression (17.5) for the superficial degree of divergence, D, in d space-time dimensions when N_b boson lines and N_f fermion lines terminate at each vertex. Explore what happens in 2 and 4 dimensions.
2. Draw all 2-loop diagrams in QED for which the superficial degree of divergence, D, is ≥ 0. What can we expect in terms of new counterterms?
3. Evaluate the vacuum polarization diagram, figure 17.3(a), and the vertex correction, figure 17.3(c), using Pauli-Villars regularization and compare with the dimensionally regularized results.
4. Carry out the computation of the dimensionally regularized $\Sigma^{\text{reg}}(p)$ where a tiny photon mass, λ, has been included.
5. Show that the light-light scattering diagram, figure 17.1(b), is finite.
6. Starting from the Lagrangian density, eq.(17.85), compute the free-field photon propagator (exercise 3 in chapter 5). Using the result, compute the vacuum polarization diagram, figure 17.3(a). Show that the divergence in the longitudinal part of the vacuum polarization may be absorbed by a rescaling of λ_0 as given by Z_λ in eq.(17.83).
7. For a fixed and finite μ_0, show that eq.(17.112) implies that $e^2(\mu)$ diverges at a finite μ (Landau ghost). If $\alpha(\mu_0) = 1/137$ when $\mu_0 = 1$, at what μ does this occur? Should we worry about this singularity?
8. Determine the finite parts of Z_2 using the renormalization prescription in the text.
9. Verify the identities eq.(17.133).
10. Verify that $Q_5{}'$, eq.(17.149), is gauge invariant.

11. Show that the matrices M,

$$M = \lambda_0 \mathbf{1} + i \sum_{j=1}^{3} \lambda_j \sigma_j,$$

where $\lambda_0, \ldots, \lambda_3$ are real numbers, and where σ_j are the usual Pauli matrices, form a matrix representation of the group $SU(2)$ if $\lambda_0^2 + \lambda_1^2 + \lambda_2^2 + \lambda_3^2 = 1$. Conclude that the group manifold of $SU(2)$ ("coordinatized" by $\lambda_0, \ldots, \lambda_3$) is the 3-sphere.

CHAPTER **18**

The Effective Action

The effective action, the generating functional of proper functions or 1-particle irreducible graphs, is defined, developed, and computed in the context of the self-interacting scalar field theory, φ^4. We have at our disposal the three representations of quantum field theory, and the computation of the 1-loop effective potential is carried through by the appropriate method in each representation. The results are used to renormalize φ^4 at 1-loop. The independence of the resulting physics from the renormalization prescription or subtraction point is then formally explored through the use of the renormalization group equations.

18.1 THE EFFECTIVE ACTION

Consider the vacuum expectation value of the scalar field operator, $\varphi(x)$, when there is an external source, $J(x)$, present.

$$\phi_c(x) = \frac{\langle 0\,|\varphi(x)|\,0\rangle_J}{\langle 0\,|\,0\rangle_J} \tag{18.1}$$

$\phi_c(x)$ is just an ordinary function, and for all intents and purposes, we can think of it as a classical field. One might be tempted to inquire whether there is a classical field theory to go along with this classical field. That is, is there an action, $\Gamma[\phi_c]$, such that the $\phi_c(x)$ that results from the operation of the right-hand side of eq.(18.1) is also the solution to the classical field equation,

$$\frac{\delta\Gamma[\phi_c]}{\delta\phi_c(x)} = -J(x), \tag{18.2}$$

generated by the action $\Gamma[\phi_c]$? The answer is yes, and this action is called the effective action.

To get an idea of how to relate the effective action to the generating functionals that we have already developed, let us consider the classical action for φ^4 theory and begin to think of it as a contribution to the effective action.

$$\Gamma[\phi_c] \sim \int d^4x\, \frac{1}{2}\phi_c(x)(\partial_\mu\partial^\mu - m_0^2)\phi_c(x) - \frac{\lambda_0}{4!}\phi_c^4(x) \tag{18.3}$$

The resulting classical field equation, eq.(18.2), is

$$-\frac{\delta\Gamma[\phi_c]}{\delta\phi_c(x)} = (\partial_\mu\partial^\mu + m_0^2)\phi_c(x) + \frac{\lambda_0}{3!}\phi_c^3(x) = J(x). \tag{18.4}$$

Let us solve eq.(18.4) as a perturbation series in λ_0. Plug

$$\phi_c(x) = \phi_c^{(0)}(x) + \lambda_0\phi_c^{(1)}(x) + \lambda_0^2\phi_c^{(2)}(x) + \ldots \tag{18.5}$$

back into the field equation, eq.(18.4), and equate terms of equal powers of λ_0. The resulting equations are

$$\begin{aligned}
(\partial^\mu\partial_\mu + m_0^2)\phi_c^{(0)}(x) &= J(x)\\
(\partial^\mu\partial_\mu + m_0^2)\phi_c^{(1)}(x) &= -\frac{1}{3!}(\phi_c^{(0}(x))^3\\
(\partial^\mu\partial_\mu + m_0^2)\phi_c^{(2)}(x) &= -\frac{1}{2!}(\phi_c^{(0)}(x))^2\phi_c^{(1)}(x),
\end{aligned} \tag{18.6}$$

and so on. The solution to the first equation can be written formally as

$$\phi_c^{(0)}(x) = \int d^4x\, G_2^{(0)}(x, x_1)J(x_1), \tag{18.7}$$

where $G_2^{(0)}(x, x_1)$ is the (free-field) 2-point Green's function,

$$G_2^{(0)}(x, x_1) = \frac{1}{\partial^\mu\partial_\mu + m_0^2}. \tag{18.8}$$

Using eqs.(18.7) and (18.8), the solutions to the remaining equations in (18.6) can be written as

$$\begin{aligned}\phi_c^{(1)}(x) &= -\frac{1}{3!}\int d^4y_1\, G_2^{(0)}(x,y_1)\left(\phi_c^{(0)}(y_1)\right)^3\\ &= -\frac{1}{3!}\int d^4y_1\, d^4x_1\, d^4x_2\, d^4x_3\, G_2^{(0)}(x,y_1)G_2^{(0)}(y_1,x_1)\\ &\qquad G_2^{(0)}(y_1,x_2)G_2^{(0)}(y_1,x_3)\, J(x_1)\, J(x_2)\, J(x_3),\end{aligned} \tag{18.9}$$

$$\begin{aligned}\phi_c^{(2)}(x) &= -\frac{1}{2!}\int d^4y_2\, d^4y_1\, d^4x_1\, d^4x_2\, G_2^{(0)}(x,y_2)G_2^{(0)}(y_2,x_1)\\ &\qquad G_2^{(0)}(y_2,x_2)G_2^{(0)}(y_2,y_1)\,\phi_c^{(1)}(y_1)\, J(x_1)\, J(x_2)\\ &= \frac{1}{2!}\frac{1}{3!}\int d^4y_2\, d^4y_1\, d^4x_1\cdots d^4x_5 G_2^{(0)}(x,y_2)G_2^{(0)}(y_2,x_1)\\ &\qquad G_2^{(0)}(y_2,x_2)G_2^{(0)}(y_2,y_1)G_2^{(0)}(y_1,x_3)G_2^{(0)}(y_1,x_4)\\ &\qquad G_2^{(0)}(y_1,x_5)\, J(x_1)\cdots J(x_5),\end{aligned} \tag{18.10}$$

and so on. Combining eqs.(18.7)-(18.10), we have

$$\begin{aligned}\phi_c(x) = \int d^4x_1\, G_2^{(0)}(x,x_1)J(x_1)&\\ + \int d^4x_1\cdots d^4x_3\,\frac{1}{3!}G_4^{(0)}(x,x_1,x_2,x_3)J(x_1)J(x_2)J(x_3)&\\ + \int d^4x_1\cdots d^4x_5\,\frac{1}{5!}\,G_6^{(0)}(x,x_1,x_2,x_3,x_4,x_5)\, J(x_1)\cdots J(x_5) + \ldots&\end{aligned} \tag{18.11}$$

where

$$\begin{aligned}&G_4^{(0)}(x_1,x_2,x_3,x_4)\\ &\quad = -\lambda_0\int d^4y_1\, G_2^{(0)}(x_1,y_1)\, G_2^{(0)}(x_2,y_1)\, G_2^{(0)}(x_3,y_1)\, G_2^{(0)}(x_4,y_1),\end{aligned} \tag{18.12}$$

etc. The series eq.(18.11) can be represented diagrammatically by figure 18.1. Alternatively, the n-point Green's function, $G_n^{(0)}$, can be represented

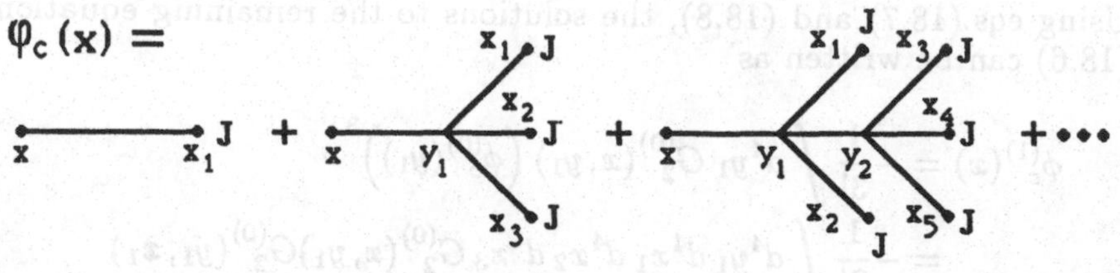

Figure 18.1 Diagrammatic representation of a perturbative solution to a classical field theory driven by a source J. All diagrams are connected and contain no loops.

as in figure 18.2. The series generated involves only connected tree-level (no-loop) diagrams, as can be expected.

The functional form appearing on the right-hand side above is familiar in that it looks like the definition of a generating functional for connected tree-level Green's functions, except that it has been differentiated once. Thus, we can rewrite eq.(18.11) as

$$\phi_c(x) = \frac{\delta W_{\text{tree}}[J]}{\delta J(x)}, \tag{18.13}$$

where

$$W_{\text{tree}}[J] = \int d^4x_1\, d^4x_2\, J(x_1)\, G_2^{(0)}(x_1, x_2)\, J(x_2) + \frac{1}{4!}\int d^4x_1 \cdots d^4x_4\, G_4^{(0)}(x_1, x_2, x_3, x_4)\, J(x_1) \cdots J(x_4) + \dots . \tag{18.14}$$

If we include the (higher-order) quantum corrections, that is, the loop terms, then $W_{\text{tree}}[J]$ will just turn into $W[J]$, eq.(14.14), the natural log of the vacuum-to-vacuum amplitude, $Z[J]$, eq.(9.9), and

$$\phi_c(x) = \frac{\delta W[J]}{\delta J(x)}. \tag{18.15}$$

Eq.(18.15) follows naturally from eq.(18.1) and the path integral definition of $Z[J]$, or the operator representation definition, eq.(9.9), so this is also not surprising. The point of this exercise is to induce the feeling that $W[J]$ and $\Gamma[\phi_c]$ essentially contain the same information. If we know one, we can generate, and therefore know, the other. We started above with the classical action $\Gamma[\phi_c]$ and ended up with $W_{\text{tree}}[J]$. We can reverse the process, so we should be able to define the effective action for the quantum

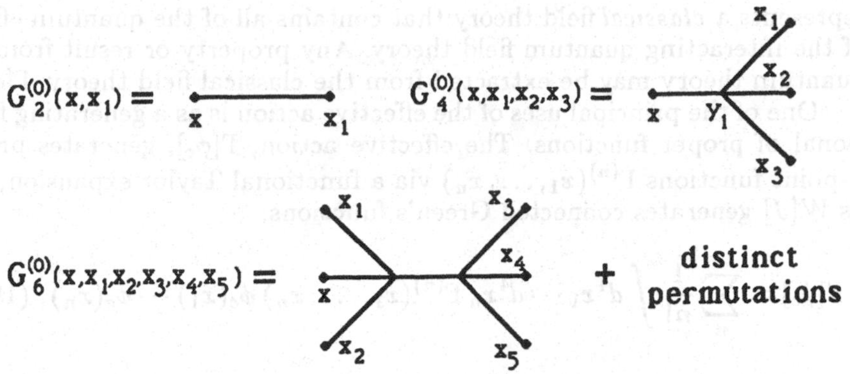

Figure 18.2 Tree-level n-point Green's functions that appear in the perturbative solution of the classical field theory, $\phi_c(x)$. The n-point Green's functions are coefficients in a functional Taylor expansion of the generating functional, $W_{\rm tree}[J]$.

theory in terms of $W[J]$.

Eq.(18.2) and eq.(18.5),

$$\phi_c(x) = \frac{\delta W[J]}{\delta J(x)}, \qquad J(x) = -\frac{\delta\Gamma[\phi_c]}{\delta\phi_c(x)},$$

together form a relationship that is similar to the Hamiltonian and Lagrangian,

$$\dot{x} = \frac{\partial H(p,x)}{\partial p}, \qquad p = \frac{\partial L(\dot{x},x)}{\partial \dot{x}}. \tag{18.16}$$

We know that the Hamiltonian and Lagrangian are related by the Legendre transformation, $H = p\dot{x} - L$; therefore, the effective action and the connected generating functional must also be related by a (functional) Legendre transform.

$$\Gamma[\phi_c] = W[J] - \int d^4x\, J(x)\phi_c(x) \tag{18.17}$$

Since $\Gamma[\phi_c]$ does not depend on $J(x)$, $\delta\Gamma[\phi_c]/\delta J = 0$, eq.(18.15) follows from eq.(18.17) by differentiation of eq.(18.17) with respect to $J(x)$. Likewise, eq.(18.2) follows from eq.(18.17) by differentiation with respect to $\phi_c(x)$, since $W[J]$ is independent of ϕ_c.

We have already seen that the effective action at tree level is just the classical action. The full effective action generated from the $W[J]$ with all loops is the quantum generalization of the classical action. $\Gamma[\phi_c]$

represents a *classical* field theory that contains all of the quantum effects of the interacting quantum field theory. Any property or result from the quantum theory may be extracted from the classical field theory, $\Gamma[\phi_c]$.

One of the principal uses of the effective action is as a generating functional of proper functions. The effective action, $\Gamma[\phi_c]$, generates proper n-point functions $\Gamma^{(n)}(x_1, \ldots, x_n)$ via a functional Taylor expansion, just as $W[J]$ generates connected Green's functions.

$$\Gamma[\phi_c] = \sum_n \frac{1}{n!} \int d^4x_1 \cdots d^4x_n \, \Gamma^{(n)}(x_1, \ldots, x_n) \, \phi_c(x_1) \cdots \phi_c(x_n) \quad (18.18)$$

A perturbative expansion of a proper function can be represented by proper diagrams. A proper diagram is a connected, truncated (external legs removed), 1-particle irreducible (1PI) diagram. The effective action becomes an important tool in the renormalization process because the quantum field theory will be rendered ultraviolet finite by renormalization if and only if the proper functions are made ultraviolet finite. To renormalize, we need only concentrate on $\Gamma[\phi_c]$. An ultraviolet finite $\Gamma[\phi_c]$ will generate an ultraviolet finite $W[J]$.

To see that $\Gamma[\phi_c]$ does generate proper functions, start by functionally differentiating eq.(18.15) with respect to $\phi_c(y)$,

$$\frac{\delta}{\delta\phi_c(y)}\phi_c(x) = \delta^4(x-y) = \frac{\delta}{\delta\phi_c(y)} \frac{\delta W[J]}{\delta J(x)}. \quad (18.19)$$

The particular $\phi_c(x)$ that results from eq.(18.15) is certainly dependent on the source used (i.e. $\phi_c(x)$ is a functional of $J(x)$). The same argument for $J(x)$ can be made via eq.(18.2). With this in mind, we next apply the functional chain rule to eq.(18.19),

$$\frac{\delta}{\delta\phi_c(y)} = \int d^4z \, \frac{\delta J(z)}{\delta\phi_c(y)} \frac{\delta}{\delta J(z)}, \quad (18.20)$$

which results in

$$\delta^4(x-y) = \int d^4z \, \frac{\delta J(z)}{\delta\phi_c(y)} \frac{\delta^2 W}{\delta J(z)\delta J(x)}. \quad (18.21)$$

We can use eq.(18.2) to write $\delta J/\delta\phi_c$ above in terms of a second functional derivative of Γ. This results in

$$\delta^4(x-y) = \int d^4z \, \frac{\delta^2\Gamma}{\delta\phi_c(y)\delta\phi_c(z)} \frac{\delta^2 W}{\delta J(z)\delta J(x)}. \quad (18.22)$$

If we set $J = \phi_c = 0$, we can view this as an infinite-dimensional matrix equation. The δ-function is the unit matrix which means that

$$\left.\frac{\delta^2\Gamma}{\delta\phi_c(y)\delta\phi_c(z)}\right|_{\phi_c=0}$$

is the inverse matrix of

$$\left.\frac{\delta^2 W}{\delta J(z)\delta J(x)}\right|_{J=0}.$$

In terms of the functional Taylor expansions of W and Γ,

$$\frac{\delta^2 W}{\delta J(z)\delta J(x)} = G^{(2)}(z,x), \qquad \frac{\delta^2\Gamma}{\delta\phi_c(y)\delta\phi_c(z)} = \Gamma^{(2)}(y,z). \tag{18.23}$$

Thus,

$$\Gamma^{(2)}(y,z) = (G^{(2)}(y,z))^{-1}. \tag{18.24}$$

If we Fourier transform, this becomes $\tilde{\Gamma}^{(2)}(p) = (\tilde{G}^{(2)}(p))^{-1}$. We can write $\tilde{G}^{(2)}(p)$ as

$$\tilde{G}^{(2)}(p) = \frac{1}{p^2 - m^2 - \Sigma(p)}, \tag{18.25}$$

in which case

$$\tilde{\Gamma}^{(2)}(p) = p^2 - m^2 - \Sigma(p). \tag{18.26}$$

$\tilde{\Gamma}^{(2)}(p)$ involves only a single $\Sigma(p)$ whereas $\tilde{G}^{(2)}(p)$ involves many powers of $\Sigma(p)$,

$$\begin{aligned}\tilde{G}^{(2)}(p) = {} & \frac{1}{p^2 - m^2} + \frac{1}{p^2 - m^2}\Sigma(p)\frac{1}{p^2 - m^2} \\ & + \frac{1}{p^2 - m^2}\Sigma(p)\frac{1}{p^2 - m^2}\Sigma(p)\frac{1}{p^2 - m^2} + \ldots\end{aligned} \tag{18.27}$$

The terms in $\tilde{G}^{(2)}(p)$ that involve more than one $\Sigma(p)$ are represented by 1-particle reducible diagrams as in figure 18.3. We may interpret eq.(18.26) as $-\Sigma(p)$ being the 1PI quantum correction to the proper 2-point function, $p^2 - m^2$. $\Sigma(p)$ in eq.(18.26) does not have any 2-point Green's functions, $1/(p^2 - m^2)$, attached to it; therefore the diagrams

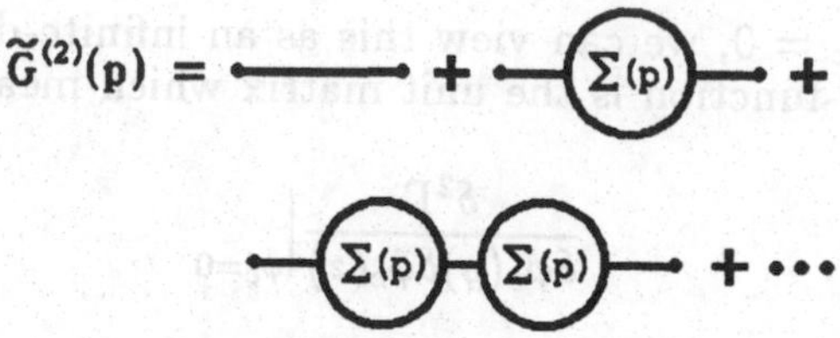

$$\tilde{G}^{(2)}(p) = \text{———} + \text{—}(\Sigma(p))\text{—} + \text{—}(\Sigma(p))(\Sigma(p))\text{—} + \cdots$$

Figure 18.3 Diagrammatic representation of the 2-point function, $\tilde{G}^{(2)}(p)$, in terms of 1-particle irreducible (1PI) pieces, $\Sigma(p)$. Each diagram in the collection $\Sigma(p)$ is 1PI and truncated.

representing it are truncated. The "connectedness" of $\Sigma(p)$ follows from $W[J]$ generating only connected Green's functions.

In order to investigate the 3-point proper function, we differentiate eq.(18.22) with respect to $\phi_c(s)$ again and set $\phi_c = J = 0$. In the process of differentiation, we again apply the chain rule, eq (18.20), and eqs.(18.2) and (18.15). The result is

$$\frac{\delta}{\delta\phi_c(s)}\delta^4(x-y) = 0 = \int d^4z \frac{\delta}{\delta\phi_c(s)}\left(\frac{\delta^2\Gamma}{\delta\phi_c(y)\delta\phi_c(z)}\frac{\delta^2 W}{\delta J(z)\delta J(x)}\right)$$

$$= \int d^4z \frac{\delta^3\Gamma}{\delta\phi_c(s)\delta\phi_c(y)\delta\phi_c(z)}\frac{\delta^2 W}{\delta J(z)\delta J(x)} + \int d^4z \frac{\delta^2\Gamma}{\delta\phi_c(y)\delta\phi_c(z)}\int d^4z_1 \frac{\delta J(z_1)}{\delta\phi_c(s)}\frac{\delta^3 W}{\delta J(z_1)\delta J(z)\delta J(x)}$$

$$= \int d^4z \frac{\delta^3\Gamma}{\delta\phi_c(s)\delta\phi_c(y)\delta\phi_c(z)}\frac{\delta^2 W}{\delta J(z)\delta J(x)} + \int d^4z\, d^4z_1 \frac{\delta^2\Gamma}{\delta\phi_c(y)\delta\phi_c(z)}\frac{\delta^2\Gamma}{\delta\phi_c(s)\delta\phi_c(z_1)}\frac{\delta^3 W}{\delta J(z_1)\delta J(z)\delta J(x)}. \tag{18.28}$$

When J and ϕ_c are set to zero this becomes

$$0 = \int d^4z\, \Gamma^{(3)}(s,y,z)\, G^{(2)}(z,x) + \int d^4z\, d^4z_1\, \Gamma^{(2)}(y,z)\, \Gamma^{(2)}(s,z_1)\, G^{(3)}(x,z,z_1). \tag{18.29}$$

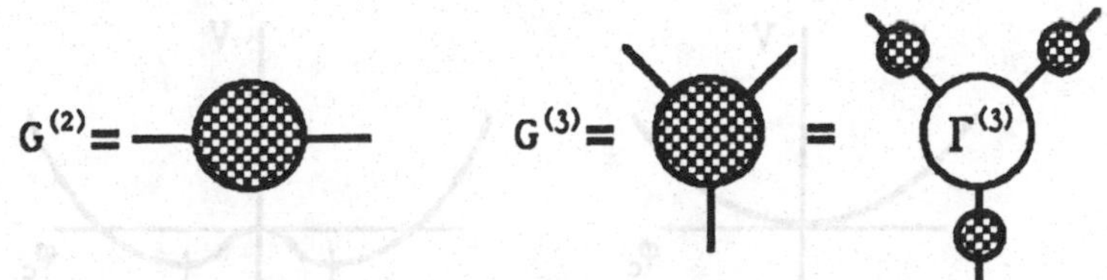

Figure 18.4 Diagrammatic representation of eq.(18.31) where the 3-point function, $G^{(3)}$, is written in terms of the proper 3-point vertex and 2-point Green's function, $G^{(2)}$.

We can solve for $\Gamma^{(3)}$ by applying $\int d^4x\Gamma^{(2)}(x,q)$ to eq.(18.29) above, and then use eq.(18.22). The result is

$$\Gamma^{(3)}(s,y,q) = \int d^4z\, d^4z_1\, d^4x\, \Gamma^{(2)}(y,z)\, \Gamma^{(2)}(s,z_1)\, \Gamma^{(2)}(q,x)\, G^{(3)}(x,z,z_1). \tag{18.30}$$

Clearly, $\Gamma^{(3)}$ represents the connected, truncated 3-point function since $G^{(3)}$ is connected, and each $\Gamma^{(2)}$ above chops off one of the external legs. $\Gamma^{(3)}$ is automatically 1PI, because it is a 3-point function.

The 4-point proper function may be examined by differentiating eq. (18.28) again. We will leave that as an exercise. Let us note that the exercise above also shows us how to recover the Green's functions from the proper functions. Inverting eq.(18.30), we have

$$G^{(3)}(x_1,x_2,x_3) = \int d^4s\, d^4y\, d^4q\, G^{(2)}(x_1,s)\, G^{(2)}(x_2,y)\, G^{(2)}(x_3,q)\, \Gamma^{(3)}(s,y,q), \tag{18.31}$$

which we can represent as the diagram in figure 18.4. The diagrammatic representation of the 3-point Green's function in terms of the 3-point proper function easily motivates why the proper functions are also called proper vertices.

As a simple example, return to the tree-level effective action, i.e. the classical action. From eq.(18.3) and eq.(18.18) we generate

$$\begin{aligned}
\Gamma^{(2)}_{\text{tree}}(x_1,x_2) &= -(\partial_\mu\partial^\mu + m_0^2)\delta^4(x_1-x_2) \\
\Gamma^{(3)}_{\text{tree}}(x_1,x_2,x_3) &= 0 \\
\Gamma^{(4)}(x_1,x_2,x_3,x_4) &= -\lambda_0\delta^4(x_1-x_2)\,\delta^4(x_1-x_3)\,\delta^4(x_1-x_4) \\
\Gamma^{(n)}_{\text{tree}}(x_1,\ldots,x_n) &= 0 \qquad \text{for} \quad n \geq 5.
\end{aligned} \tag{18.32}$$

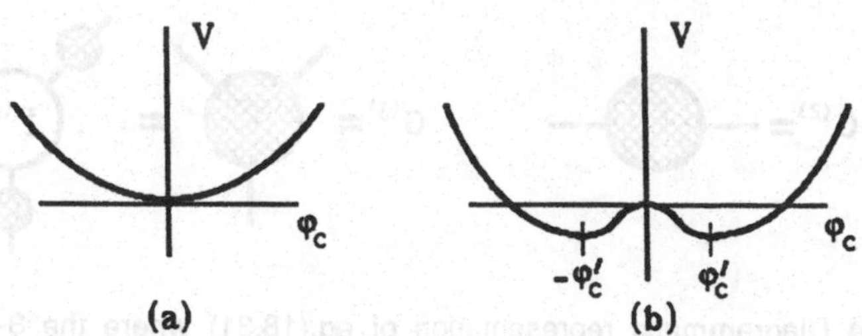

Figure 18.5 Quadratic plus quartic potential, eq.(18.34). (a) When the quadratic contribution is greater than zero ($m_0^2 > 0$), there is just one stable minimum at $\phi_c = 0$. (b) When the quadratic contribution is less than zero ($m_0^2 < 0$), $\phi_c = 0$ becomes an unstable local extremum while 2 stable minima appear at $\pm\phi_c'$.

From figure 18.1 or 18.2 it is easy to see that the n-point tree-level functions for $n > 4$ are not 1PI, which confirms that the proper n-point functions vanish at tree level for $n > 4$.

There is another common way to expand the effective action, this time in powers of momentum, i.e. space-time derivatives of ϕ_c.

$$\Gamma[\phi_c] = \int d^4x \left(-V[\phi_c] - \frac{1}{2} Z[\phi_c](\partial_\mu \phi_c)^2 + \cdots \right) \tag{18.33}$$

The lowest order term, which we can separate out by setting $\phi_c(x) =$ constant, is a potential, the effective potential. The effective potential may be used to determine the stable vacua of the quantum theory, or to determine if the quantum vacuum we are using is stable. We do so by computing ϕ_c from the chosen vacuum candidate and seeing if ϕ_c is at an actual stable minimum of the effective potential $V[\phi_c]$.

To illustrate the idea let's return to the tree-level effective action, eq.(18.3). The potential is

$$V_{\text{tree}}[\phi_c] = \frac{1}{2} m_0^2 \phi_c^2 + \frac{\lambda_0}{4!} \phi_c^4. \tag{18.34}$$

If $m_0^2 > 0$, then the minimum of V occurs at $\phi_c = 0$, and this point is stable (figure 18.5(a)). However, if $m_0^2 < 0$, $\phi_c = 0$ becomes an unstable local extremum of V. There are now two stable minima occurring at

$$\phi_c{}' = \pm \sqrt{\frac{6|m_0^2|}{\lambda_0}}. \tag{18.35}$$

This theory possesses 2 stable vacua that are degenerate (figure 18.5(b)).

The potential (18.34) is symmetric under the transformation $\phi_c \to -\phi_c$. When $m_0^2 > 0$, the vacuum, $\phi_c = 0$, is also symmetric under $\phi_c \to -\phi_c$. However, when $m_0^2 < 0$, the symmetry operation takes us from one vacuum to the other, thus the particular vacuum state chosen (either + or $-\phi_c{}'$) does not respect or reflect the symmetry present in the action. This is an example of a phenomenon called "spontaneous breakdown of symmetry." The symmetry is not really broken. The resulting physics does not care which degenerate vacuum you choose. "Spontaneous breakdown of symmetry" just refers to the fact that the vacuum, a solution of the field equations, does not possess a symmetry that is present in the action.

If we change to a new field variable, $\tilde{\phi}_c$, that is expected to be zero in the true vacuum, then $\tilde{\phi}_c = \phi_c - \phi_c{}'$. The potential written in terms of the new variable is

$$V = \frac{\lambda_0}{6}\phi_c'^2\,\tilde{\phi}_c^2 + \frac{\lambda_0}{6}\phi_c'\,\tilde{\phi}_c^3 + \frac{\lambda_0}{4!}\tilde{\phi}_c^4. \tag{18.36}$$

The mass term (the term proportional to $\tilde{\phi}_c^2$) now appears with a plus sign, as it should, indicating that $\tilde{\phi}_c = 0$ is a stable vacuum. The classical mass squared of the particle is $\lambda_0\phi_c'^2/6$. A cubic interaction has also appeared. Note that the theory still possesses the same symmetry; it is just difficult to recognize it in the $\tilde{\phi}_c$ coordinate system. (Coleman calls this "secret symmetry.")

Once again it is possible that the quantum corrections to the classical potential will spontaneously break a symmetry. Quantum corrections may produce a nonzero vacuum expectation value of the field where the classical vacuum is zero. To find out, we compute the quantum corrections to the effective potential. One example where this does happen is massless scalar electrodynamics.

Next consider the case when $\phi_c = \phi_1 + i\phi_2$ is complex, and suppose the classical potential is

$$V_{\text{tree}}[\phi_c] = -\frac{1}{2}m_0^2|\phi_c|^2 + \frac{\lambda_0}{4!}|\phi_c|^4. \tag{18.37}$$

The entire action is symmetric under the global continuous symmetry,

$$\phi_c \to e^{i\alpha}\phi_c, \qquad \phi_c^* \to e^{-i\alpha}\phi_c, \tag{18.38}$$

which amounts to a rotation of angle α in the (ϕ_1, ϕ_2) plane. This theory possesses an infinite number of degenerate vacua, parametrized by an angle θ,

$$\phi_c' = \sqrt{\frac{6m_0^2}{\lambda_0}}\,e^{i\theta}.$$

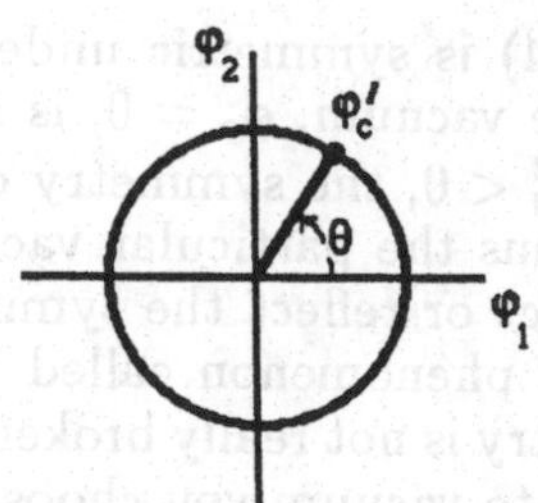

Figure 18.6 The infinite number of degenerate classical vacua of the potential in eq.(18.37) fall on a circle in the (ϕ_1, ϕ_2) plane. We may therefore parametrize the vacua by the angle θ.

See figure 18.6.

The entire effective action, $\Gamma[\phi_1, \phi_2]$, will be invariant under the transformation (18.38). Thus,

$$\delta\Gamma = \int d^4x \, \frac{\delta\Gamma[\phi_1, \phi_2]}{\delta\phi_1(x)} \, \delta\phi_1(x) + \frac{\delta\Gamma[\phi_1, \phi_2]}{\delta\phi_2(x)} \, \delta\phi_2(x) = 0, \qquad (18.39)$$

where $\delta\phi_1$ and $\delta\phi_2$ result from the symmetry operation, eq.(18.38). For small rotations α, this amounts to

$$\delta\phi_1 = -\alpha\phi_2, \qquad \delta\phi_2 = \alpha\phi_1, \qquad (18.40)$$

so eq.(18.39) reads

$$\int d^4x \, -\frac{\delta\Gamma}{\delta\phi_1(x)} \phi_2(x) + \frac{\delta\Gamma}{\delta\phi_2(x)} \phi_1(x) = 0. \qquad (18.41)$$

Differentiate this equation once with respect to $\phi_2(y)$. This results in

$$\int d^4x \, -\frac{\delta^2\Gamma}{\delta\phi_2(y)\delta\phi_1(x)} \phi_2(x) - \frac{\delta\Gamma}{\delta\phi_1(x)} \delta^4(x-y) + \frac{\delta^2\Gamma}{\delta\phi_2(y)\delta\phi_1(x)} \phi_1(x) = 0. \qquad (18.42)$$

Now we want to evaluate this identity in the sourceless vacuum. We must choose one of the degenerate vacua and without loss of generality we can choose $\theta = 0$, that is, $\phi_1' = \sqrt{6m_0^2/\lambda_0}$ and $\phi_2' = 0$. With this choice, the first term in eq.(18.42) above that is proportional to ϕ_2' will vanish. The second term will also vanish since eq.(18.2) tells us that it is proportional to a source $J_1(x)$, that is zero. This leaves

$$\phi_1' \int d^4x \frac{\delta^2 \Gamma}{\delta \phi_2'(y) \delta \phi_2'(x)} = 0. \tag{18.43}$$

Since we are taking ϕ_c to be a constant, ϕ_c', then eq.(18.43) only involves the effective potential. Since ϕ_1' is not zero, and since the effective action generates proper functions, then eq.(18.43) states that the 2-point proper function for the ϕ_2 particle vanishes in the limit of vanishing momentum p.

$$\lim_{p \to 0} \tilde{\Gamma}_2^{(2)}(p) = 0 \tag{18.44}$$

But $\tilde{\Gamma}^{(2)}(p)$ is the inverse of the propagator of the ϕ_2 particle. Thus, eq.(18.44) states that this propagator has a pole at $p = 0$. This means that the ϕ_2 particle is massless. Such massless scalar particles that arise from the spontaneous breakdown of a continuous global symmetry are called Goldstone bosons. In this case the ϕ_1 particle has the same mass as in the real ϕ_c case. We can think of it as arising from quantum fluctuations in the $|\phi_1|$ direction meeting a rising potential in either direction, thus limiting the range of interaction. Fluctuations in the θ direction see a flat potential at the bottom of the "sombrero-shaped" potential. Therefore, the interaction is long-ranged and the resulting particle is massless. Since fluctuations in the θ direction correspond to the ϕ_2 direction at our chosen vacuum (ϕ_1', ϕ_2'), the ϕ_2 particle will be massless.

In the next three sections we will return to the φ^4 theory for a real scalar field and compute the effective potential at 1-loop, once from each representation. Since there is no infinite wave function renormalization at 1-loop in this theory, the effective potential at 1-loop can be used to renormalize the theory at 1-loop.

The method by which the computation of the potential is made varies between representations. In the operator representation (the next section), one defines $W[J]$ as the log of $Z[J]$, where $Z[J]$ is expressed as eq.(9.9). A perturbative expansion of the effective potential amounts to a set of proper diagrams where the incoming momenta all vanish (because the potential is the lowest order term in the momentum expansion of the effective action). The 1-loop proper diagrams with vanishing incoming momenta are computed and summed to yield the result.

In the path integral representation, $Z[J]$ is defined via the vacuum-to-vacuum amplitude as the path integral eq.(14.1). $W[J]$ again is proportional to the log of $Z[J]$. We first compute $W[J]$ to 1-loop from $Z[J]$, and then use the Legendre transform to find the potential. Since we are considering an expansion in loops, which we know from chapter 16 is an expansion in powers of $\hbar$, then the appropriate way to compute $Z[J]$ to

1-loop turns out to be a saddle-point or stationary phase approximation of the path integral. The result of computing $Z[J]$ directly from the path integral by this method is in the form of a functional determinant. From this determinant we must extract $W[J]$ in order to determine the potential.

In the Schrödinger representation, the appropriate method to find the static effective action of the potential is by solving the appropriate functional Schrödinger equation, or by a variational procedure. Both methods arise, because the effective potential is the minimum expected energy per unit volume in the set of states where the expected value of the field is ϕ_c. The vacuum state will minimize this energy density, hence we solve the appropriate Schrödinger equation or use a variational calculation.

To see this in more detail, consider looking variationally for the vacuum or ground state when there are no sources present. This means that we want to find the state, $|\phi_0\rangle$, that minimizes the energy density, $\langle\phi_0|\mathcal{H}|\phi_0\rangle$. The state must be normalizable, hence we have the constraint $\langle\phi_0|\phi_0\rangle = 1$. In order to minimize the energy density in the presence of the constraint, we introduce a Lagrange multiplier (call it E) and minimize $\langle\phi_0|H - E|\phi_0\rangle$ subject to no constraint.

Now we want to think about what happens when there are sources present so that we can actually deal with the effective action. With sources present, the Hamiltonian density becomes $\mathcal{H} - J(x)\varphi(x)$. We now want to minimize $\langle\phi_0|H - E - \int J\varphi|\phi_0\rangle$. In this form, $J(x)$ looks like another Lagrange multiplier. Thus, with a source present, we are minimizing $\langle\phi_0|\mathcal{H}|\phi_0\rangle$ subject to another constraint involving $\langle\phi_0|\varphi(x)|\phi_0\rangle$. However, we know that in the true vacuum in the presence of a source, that $\langle\phi_0|\varphi(x)|\phi_0\rangle = \phi_c(x)$. Therefore, the static effective action, $\Gamma_{\text{static}}[\phi_c]$, is minus the minimum of the energy density, $\langle\phi_0|\mathcal{H}|\phi_0\rangle$ in a state $|\phi_0\rangle$ subject to the constraints, $\langle\phi_0|\phi_0\rangle = 1$ and $\langle\phi_0|\varphi(x)|\phi_0\rangle = \phi_c(x)$. If we take ϕ_c to be a constant, then the constrained minimum energy density will produce the effective potential.

Once we have computed the effective potential at 1-loop, we may use it to renormalize φ^4 at 1-loop by observing that the momentum tree-level 2- and 4-point functions, eq(18.32), say that

$$\tilde{\Gamma}^{(2)}_{\text{tree}}(0) = m_0^2, \quad \frac{\partial\tilde{\Gamma}^{(2)}_{\text{tree}}}{\partial p^2} = 1, \quad \tilde{\Gamma}^{(4)}_{\text{tree}}(0) = -\lambda_0. \tag{18.45}$$

Therefore, we may take

$$\left.\frac{d^2V}{d\phi_c^2}\right|_{\phi_c=\phi'_c} = m^2, \quad \left.\frac{d^4V}{d\phi_c^4}\right|_{\phi_c=\phi'_c} = \lambda, \quad Z_\phi(\phi'_c) = 1, \tag{18.46}$$

as definitions of the renormalized mass, coupling, and field. The particular choice for ϕ'_c amounts to a renormalization prescription or choice of subtraction point.

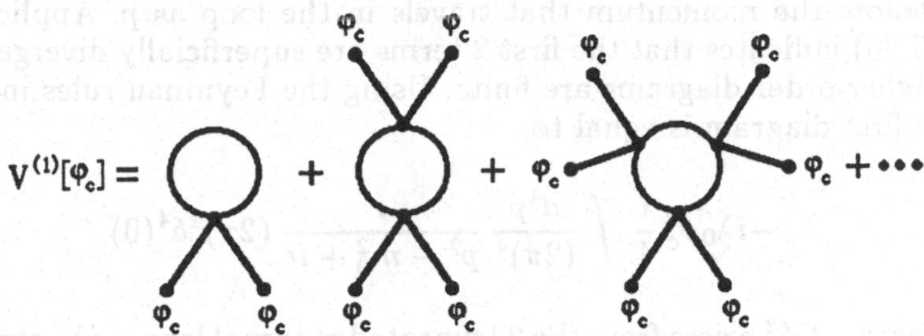

Figure 18.7 Diagrammatic expansion of the 1-loop effective potential, $V^{(1)}[\phi_c]$. A factor of ϕ_c is attached to each external leg to remind us that there are no propagators associated with the truncated legs.

18.2 φ^4 AT 1-LOOP . . . FROM THE OPERATOR REPRESENTATION

In the operator representation, $Z[J]$ is given by the expression (9.9),

$$Z[J] = \langle 0 \,|T \exp\left(i \int d^4x \, J(x)\varphi(x)\right) |\, 0\rangle, \tag{18.47}$$

where the field $\varphi(x)$ above is an operator. The generating functional of connected Green's functions, $W[J]$, is defined by $Z[J] = \exp(iW[J])$. The formal series expansion of the exponential leads to the Taylor expansion form of the generating functional where the coefficients of the expansion are the time-ordered products of the field in the vacuum or Green's functions. $W[J]$ may be put into an identical form where the coefficients are the connected Green's functions. A perturbative expansion of these coefficients leads to a series represented by diagrams, and the Feynman rules are used to evaluate each diagram. The formal development encompassed in eqs.(18.22)-(18.31) relates the connected Green's functions to the coefficients in a Taylor expansion of the effective action, the proper functions. This in turn leads to the identification of the perturbative expansion of the proper functions with an expansion in proper diagrams (connected, truncated, 1PI). If the effective action is expressed as an expansion in powers of momentum, then the lowest order term independent of p is the effective potential. If we expand about $p = 0$, then a perturbative expansion of the effective potential is represented by proper diagrams where the incoming momenta on the truncated legs all vanish. If we expand in powers of $\hbar$, then the term in the expansion of the potential to order $\hbar$ is the collection of all 1-loop proper diagrams with vanishing external momenta. This series is represented diagrammatically in figure 18.7. The notation ϕ_c is attached to each external leg to remind us that the legs are actually truncated so that no propagator appears for any external leg.

Denote the momentum that travels in the loop as p. Application of eq.(16.30) indicates that the first 2 terms are superficially divergent while the higher-order diagrams are finite. Using the Feynman rules in chapter 7, the first diagram is equal to

$$-i\lambda_0 \phi_c^2 \frac{1}{4} \int \frac{d^4p}{(2\pi)^4} \frac{i}{p^2 - m_0^2 + i\epsilon} (2\pi)^4 \delta^4(0). \qquad (18.48)$$

The factor of ϕ_c^2 arises from the 2 truncated external legs, $-i\lambda_0$ comes from the single vertex, $\int d^4p/(2\pi)^4$ from the single-loop momentum, $i/(p^2 - m_0^2 + i\epsilon)$ for propagation around the loop, and $(2\pi)^4\delta^4(0)$ from overall momentum conservation. The δ-function is evaluated at zero because each external momentum vanishes when we are dealing with the effective potential. This factor of $(2\pi)^4\delta^4(0)$ is simply the 4-volume $\int d^4x$, and it appears as such because vanishing external momenta are equivalent to taking ϕ_c = constant. This means that the resulting energy density is independent of x and can be factored out of the energy to leave $\int d^4x$. We will leave this factor out of what follows, keeping in mind that the result is an energy density.

The factor of 1/4 in front of eq.(18.48) is a notorious symmetry factor which we obtain as follows: Let n be the number of vertices. Then there are $2n$ external ϕ_c lines and the diagram is a 1-loop contribution to the $2n$-point function, $\Gamma^{(2n)}$, the coefficient of the $2n^{\text{th}}$ term of the Taylor expansion of $\Gamma[\phi_c]$. In order to obtain its contribution to $\Gamma[\phi_c]$, we must multiply it by $1/(2n)!$ (see eq.(18.18)).

Next, we multiply the result of 1 diagram by its ordinary symmetry factor, s_n, which usually reflects how well the factors of 1/4! have been canceled at each vertex. Finally, we must multiply the result by the number of topologically distinct diagrams. This is determined by labeling each ϕ_c leg with an external momentum, p_i, and counting the number of distinct ways this can be done. In summary, the total symmetry factor is

$$s = \frac{1}{(2n)!} s_n \cdot d_n, \qquad (18.49)$$

where s_n is the symmetry factor of an individual diagram, and d_n is the number of distinct diagrams.

For the 2-legged diagram that represents eq.(18.48), $n = 1$. There is just 1 distinct diagram, $d_1 = 1$, because if we switch momentum labels on the 2 legs, the 2 labels are still connected to the same vertex. The ordinary symmetry factor, s_n, is identical to the symmetry factor in the 1-loop contribution to the 2-point Green's function, $G_2^{(1)}(x_1, x_2)$. For $G_2^{(1)}$

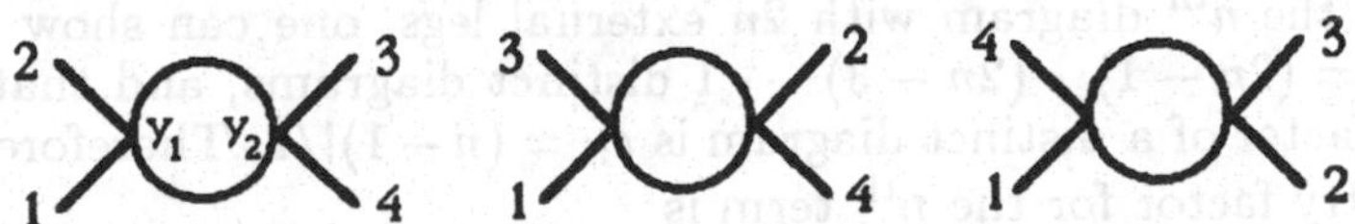

Figure 18.8 The 3 distinct diagrams associated with the ϕ_c^4 term in the expansion of the 1-loop effective potential.

we need to Wick or normal order

$$\langle 0\,|T(\varphi(x_1)\varphi(x_2)\mathcal{H}_I(y))|\,0\rangle \propto \langle 0\,|T(\varphi(x_1)\varphi(x_2)\varphi^4(y))|\,0\rangle,$$

as we did in chapter 7, except that we are not assuming that $\mathcal{H}_I$ is already normal ordered. There are 4 ways to contract $\varphi(x_1)$ with one of the $\varphi(y)$'s. Once this is done, there remain 3 ways to contract $\varphi(x_2)$ with the 3 remaining $\varphi(y)$'s. Thus, there are $4 \times 3 = 12$ identical contributions. However, $\mathcal{H}_I(y)$ contributes a factor of $1/4!$, thus the symmetry factor is $s_1 = 4 \times 3/4! = 1/2$. Plugging this into eq.(18.49) we see that the total symmetry factor for the 2-legged diagram is

$$s = \frac{1}{2}\cdot\frac{1}{2}\cdot 1 = \frac{1}{4}.$$

The next diagram in figure 18.7 is $n = 2$ and is equal to

$$(-i\lambda_0\phi_c)^2\frac{1}{16}\int\frac{d^4p}{(2\pi)^4}\left(\frac{i}{p^2 - m_0^2 + i\epsilon}\right)^2. \tag{18.50}$$

For this case, there are 3 distinct diagrams as illustrated in figure 18.8. To determine the symmetry factor of a single distinct diagram, we must count the number of equivalent ways that $\varphi(x_1)$, $\varphi(x_2)$, $\varphi(x_3)$, and $\varphi(x_4)$ can be contracted with 4 $\varphi(y_1)$'s at one vertex and 4 $\varphi(y_2)$'s at the other. Let's look at the first diagram in figure 18.8. Suppose that we label the vertex that $\varphi(x_1)$ connects to as y_1. Then $\varphi(x_1)$ can only contract with the $\varphi(y_1)$'s and has a choice of 4. Once this is done, $\varphi(x_2)$ may pick from the remaining 3. $\varphi(x_3)$ can choose from 4 $\varphi(y_2)$'s, and $\varphi(x_4)$ from the remaining 3. There are 2 different ways to contract the 2 remaining $\varphi(y_1)$'s with the 2 remaining $\varphi(y_2)$'s to form the loop. Altogether this makes $4\times 3\times 4\times 3\times 2$ equivalent contractions. Each vertex contributes a factor of $1/4!$, so the symmetry of a distinct diagram is $4\times3\times4\times3\times2/4!\times4!$ $= 1/2$ and the total symmetry factor for eq.(18.50) is

$$s = \frac{1}{4!}\cdot\frac{1}{2}\cdot 3 = \frac{1}{16}.$$

For the nth diagram with $2n$ external legs, one can show that there are $d_n = (2n-1)\cdot(2n-3)\cdots 1$ distinct diagrams, and that the symmetry factor of a distinct diagram is $s_n = (n-1)!/2$. Therefore, the total symmetry factor for the nth term is

$$s = \frac{1}{(2n)!}\frac{(n-1)!}{2}\cdot(2n-1)\cdot(2n-3)\cdots 1 = \frac{1}{2n}\cdot\left(\frac{1}{2}\right)^n, \quad (18.51)$$

and the nth diagram contributes

$$\frac{1}{2n}\left(\frac{1}{2}\right)^n(-i\lambda_0\phi_c^2)^n\int\frac{d^4p}{(2\pi)^4}\left(\frac{i}{p^2-m_0^2+i\epsilon}\right)^n \quad (18.52)$$

to the effective potential. The effective potential at 1-loop, which is $\Gamma[\phi_c]$ when ϕ_c is a constant, is the sum over n,

$$\begin{aligned} V^{(1)}[\phi_c] &= -\frac{i}{2}\sum_{n=1}^{\infty}\frac{1}{n}\left(\frac{\lambda_0\phi_c^2}{2}\right)^n\int\frac{d^4p}{(2\pi)^4}\left(\frac{1}{p^2-m_0^2+i\epsilon}\right)^n \\ &= -\frac{i}{2}\int\frac{d^4p}{(2\pi)^4}\ln\left(1-\frac{\lambda_0\phi_c^2/2}{p^2-m_0^2+i\epsilon}\right). \end{aligned} \quad (18.53)$$

The factor of i that has appeared in eq.(18.53) above is due to the convention of defining the log of $Z[J]$ as $iW[J]$.

Now let us use the result (18.53) to renormalize φ^4 at 1-loop. From eq.(18.32) we have $\tilde{\Gamma}^{(2)}_{\text{tree}}(p^2) = p^2 - m_0^2$, which suggests the renormalization prescription

$$\tilde{\Gamma}^{(2)}(m^2) = 0, \qquad \left.\frac{d\tilde{\Gamma}^{(2)}}{dp^2}\right|_{p^2=m^2} = 1, \quad (18.54)$$

where m is the physical, renormalized, finite mass. The second condition above fixes the residue of the 2-point Green's function to be 1 on shell. To accomplish this, we may need to introduce a wave function renormalization, Z_ϕ, to rescale the bare field, φ_0, into the renormalized field φ via $\varphi_0 = \sqrt{Z_\phi}\varphi$.

Also, from eq.(18.32) we have $\tilde{\Gamma}^{(4)}_{\text{tree}}(p_1,p_2,p_3,p_4) = -\lambda_0$. This suggests that we define the renormalized coupling, λ, to be

$$\left.\tilde{\Gamma}^{(4)}(p_1,p_2,p_3,p_4)\right|_{\text{on shell}} = -\lambda \quad (18.55)$$

when the external momenta are on shell.

This choice of renormalization prescription or on shell subtraction is inconvenient, because we do not yet know the 1-loop effective action; we only know the potential. The subtraction point is arbitrary, and we are free to change it. In fact, to make use of the effective potential, we must change the subtraction point to $p = 0$ in eq.(18.54) and $p_i = 0$ in eq.(18.55). Therefore, we take $\tilde{\Gamma}^{(2)}(0) = -m^2$ and $\tilde{\Gamma}^{(4)}(0) = -\lambda$ as definitions of the renormalized mass and coupling. In terms of the effective potential this is

$$\frac{d^2}{d\phi_c^2} V[\phi_c]\bigg|_{\phi_c=0} = m^2, \qquad \frac{d^4}{d\phi_c^4} V[\phi_c]\bigg|_{\phi_c=0} = \lambda. \tag{18.56}$$

Since we are only dealing with the effective potential, we automatically obtain $Z_\phi(0) = 1$. The difference between eq.(18.56) and eqs.(18.54) and (18.55) is a finite renormalization.

The effective potential to 1-loop is eq.(18.53) combined with the tree-level potential,

$$V[\phi_c] = \frac{1}{2} m_0^2 \phi_c^2 + \frac{\lambda_0}{4!}\phi_c^4 - \frac{i}{2}\int \frac{d^4p}{(2\pi)^4} \ln\left(1 - \frac{\lambda_0 \phi_c^2/2}{p^2 - m_0^2 + i\epsilon}\right). \tag{18.57}$$

Application of eq.(18.56) yields

$$m^2 = m_0^2 + \frac{1}{2}\lambda_0 \int \frac{d^4p}{(2\pi)^4} \frac{i}{p^2 - m_0^2 + i\epsilon}, \tag{18.58}$$

and

$$\lambda = \lambda_0 + \frac{3}{2} i \lambda_0^2 \int \frac{d^4p}{(2\pi)^4} \left(\frac{1}{p^2 - m_0^2 + i\epsilon}\right)^2. \tag{18.59}$$

We may replace m_0^2 in the propagators in eqs.(18.58) and (18.59) above with m^2, since this change is of order $\hbar^2$ or 2-loops and higher. The same holds true in the 1-loop contribution to V in eq.(18.57). For the same reason, we may replace λ_0 with λ in eq.(18.58) and λ_0^2 with λ^2 in eq.(18.59). We may do the same in the 1-loop contribution to V in eq.(18.57). Upon this substitution, we can easily solve for m_0^2 in eq.(18.58), and for λ_0 in eq.(18.59).

$$\begin{aligned} m_0^2 &= m^2 - \frac{\lambda}{2}\int \frac{d^4p}{(2\pi)^4} \frac{i}{p^2 - m^2 + i\epsilon}, \\ \lambda_0 &= \lambda - \frac{3}{2} i \lambda^2 \int \frac{d^4p}{(2\pi)^4} \left(\frac{1}{p^2 - m^2 + i\epsilon}\right)^2 \end{aligned} \tag{18.60}$$

We may use eq.(18.60) to completely rewrite $V[\phi_c]$ in terms of finite, renormalized quantities.

$$V[\phi_c] = \frac{1}{2}m^2\phi_c^2 + \frac{\lambda}{4!}\phi_c^4 - \frac{i}{2}\int \frac{d^4p}{(2\pi)^4} \ln\left(1 - \frac{(\lambda\phi_c^2/2)^2}{p^2 - m^2 + i\epsilon}\right)$$
$$-\frac{1}{4}\lambda\phi_c^2 \int \frac{d^4p}{(2\pi)^4} \frac{i}{p^2 - m^2 + i\epsilon} - i\frac{\lambda^2}{16}\phi_c^4 \int \frac{d^4p}{(2\pi)^4} \left(\frac{1}{p^2 - m^2 + i\epsilon}\right)^2 \tag{18.61}$$

$V[\phi_c]$ is now completely finite. We can demonstrate this by regularizing it and noting that the result of integration is independent of the cutoff. Since there is no gauge symmetry present, we do not have to be fancy about our choice of regulator. A simple cutoff of momentum at $|\vec{p}| = \Lambda$, where Λ is large, will suffice.

When the p^0 integration in eq.(18.61) is performed, $V[\phi_c]$ becomes

$$V[\phi_c] = \frac{1}{2}m^2\phi_c^2 + \frac{\lambda}{4!}\phi_c^4 + \frac{1}{2}\int \frac{d^3p}{(2\pi)^3}\left(\sqrt{\vec{p}^2 + m^2 + \lambda\phi_c^2/2} - \sqrt{\vec{p}^2 + m^2}\right)$$
$$+ \frac{\lambda\phi_c^2}{8}\int \frac{d^3p}{(2\pi)^3} \frac{1}{\sqrt{\vec{p}^2 + m^2}} - \frac{\lambda^2\phi_c^4}{64}\int \frac{d^3p}{(2\pi)^3} \frac{1}{(\vec{p}^2 + m^2)^{3/2}}. \tag{18.62}$$

The regularization cuts off the remaining integrals at $|\vec{p}| = \Lambda$. Each integral is elementary and can be done explicitly. For example,

$$\int_0^\Lambda \frac{d^3p}{(2\pi)^3}\sqrt{\vec{p}^2 + a^2}$$
$$= \frac{1}{2\pi^2}\left(\frac{p}{4}(p^2 + a^2)^{\frac{3}{2}} - \frac{a^2}{8}p(p^2 + a^2)^{\frac{1}{2}} - \frac{a^4}{8}\ln(p + \sqrt{p^2 + a^2})\right)\Bigg|_0^\Lambda$$
$$= \frac{1}{2\pi^2}\left(\frac{\Lambda^4}{4} + \frac{1}{4}a^2\Lambda^2 + \frac{1}{32}a^4 - \frac{a^4}{8}\ln\left(\frac{2\Lambda}{a}\right) + \mathcal{O}\left(\frac{a^6}{\Lambda^2}\right)\right). \tag{18.63}$$

The resulting potential is

$$V[\phi_c] = \frac{1}{2}m^2\phi_c^2 + \frac{\lambda}{4!}\phi_c^4$$
$$+ \frac{1}{4\pi^2}\left(\frac{1}{16}\left(m^2 + \frac{\lambda\phi_c^2}{2}\right)^2 \ln\left(1 + \frac{\lambda\phi_c^2}{2m^2}\right) - \frac{1}{32}\lambda m^2\phi_c^2 - \frac{3}{128}\lambda^2\phi_c^4\right). \tag{18.64}$$

Eqs.(18.58) and (18.59) may be used to re-express the bare action in terms of a finite renormalized mass and a coupling. Two counterterms will appear that will insure that the Green's functions are finite at 1-loop.

Let's consider what happens if $m^2 = 0$. One can easily see that it is not possible to take this limit in eq.(18.64). When $m^2 = 0$, the particle is massless and on shell at $p^2 = 0$. This means that the subtraction point we used in fixing eqs.(18.58), (18.59), and (18.64) is now on shell so we can expect to get infrared divergences. The easiest way out of this dilemma is to change the subtraction procedure so that it does not occur on shell. Since the subtraction point is arbitrary, there are no physical consequences in doing so. Backing up to eq.(18.57), the effective potential when $m_0^2 = 0$ is

$$V[\phi_c] = \frac{\lambda_0}{4!}\phi_c^4 - \frac{i}{2}\int \frac{d^4p}{(2\pi)^4} \ln\left(1 - \frac{\lambda_0\phi_c^2/2}{p^2 - i\epsilon}\right). \tag{18.65}$$

The minimum of $V[\phi_c]$ still occurs at $\phi_c = 0$. Instead of using eq.(18.56), we follow Coleman and Weinberg and define the renormalized coupling as

$$\lambda_M = \left.\frac{d^4V}{d\phi_c^4}\right|_{\phi_c = M}, \tag{18.66}$$

where the mass scale M is nonzero but arbitrary. If we follow the same regularization procedure as before, and make use of eq.(18.63), we find

$$V[\phi_c] = \frac{\lambda_0}{4!}\phi_c^4$$
$$+\frac{1}{4\pi^2}\left(\frac{1}{4}\left(\frac{\lambda_0\phi_c^2}{2}\right)\Lambda^2 + \frac{1}{32}\left(\frac{\lambda_0\phi_c^2}{2}\right)^2 - \frac{1}{16}\left(\frac{\lambda_0\phi_c^2}{2}\right)^2 \ln\left(\frac{8\Lambda^2}{\lambda_0\phi_c^2}\right)\right). \tag{18.67}$$

First we make sure that the renormalized mass vanishes and apply

$$\left.\frac{d^2V}{d\phi_c^2}\right|_{\phi_c=0} = 0. \tag{18.68}$$

This subtracts the quadratically divergent term. Now we apply eq.(18.66). The result is

$$\lambda_M = \lambda_0 + \frac{1}{4\pi^2}\left(\frac{3}{16}\lambda_0^2 - \frac{3}{8}\lambda_0^2\ln(4\Lambda^2) + \frac{\lambda_0^2}{64}\left(24\ln\left(\frac{\lambda_0 M^2}{2}\right) + 100\right)\right). \tag{18.69}$$

We invert this to solve for λ_0 in terms of λ_M by replacing λ_0 with λ_M everywhere except the zeroth order term. Such a substitution is valid to 1-loop, since it induces changes only at higher loops. We can do the same thing to $V[\phi_c]$. Now we use the expression of λ_0 in terms of λ_M to completely write $V[\phi_c]$ in terms of λ_M only. The cutoff dependence disappears leaving

$$V[\phi_c] = \frac{\lambda_M}{4!}\phi_c^4 + \frac{1}{64\pi^2}\left(\frac{\lambda_M \phi_c^2}{2}\right)^2 \left(\ln\left(\frac{\phi_c^2}{M^2}\right) - \frac{25}{6}\right). \tag{18.70}$$

The need to avoid infrared divergences apparently forces us to introduce an arbitrary mass scale, M, which potentially can show up in physical results. However, M has no effect on the physics. If we pick a different scale, M', the renormalized value of the coupling, λ'_M, will also change. If we start over at eq.(18.69), we end up with the same functional form for $V[\phi_c]$ as in eq.(18.70), the only difference being that $\lambda_M \to \lambda'_M$ and $M \to M'$. Alternatively, if we apply

$$\lambda'_M = \left.\frac{d^4 V[\phi_c]}{d\phi_c^4}\right|_{\phi_c = M'} \tag{18.71}$$

directly to eq.(18.70), we find

$$\lambda'_M = \lambda_M + \frac{3}{32\pi^2}\lambda_M^2 \ln\left(\frac{M'^2}{M^2}\right). \tag{18.72}$$

If we take this and use it to replace λ_M in eq.(18.70) with λ'_M, we find that V becomes

$$V[\phi_c] = \frac{\lambda'_M}{4!}\phi_c^4 + \frac{1}{64\pi^2}\left(\frac{\lambda'_M \phi_c^2}{2}\right)^2 \left(\ln\left(\frac{\phi_c^2}{M'^2}\right) - \frac{25}{6}\right) + \mathcal{O}({\lambda_M}^3). \tag{18.73}$$

The important point is that, to 1-loop, M has disappeared from eq.(18.73), and that eq.(18.73) is of the same form as eq.(18.70). That is, $V[\phi_c]$ is form-invariant under a change of scale M, $V[\phi_c, \lambda_M, M] = V[\phi_c, \lambda'_M, M']$. No matter what M we pick, we will get the same physics.

18.3 ... FROM THE PATH INTEGRAL REPRESENTATION

In the path integral representation, $Z[J]$ is written as the path integral,

$$\begin{aligned} Z[J] &= \mathcal{N} \int \mathcal{D}\phi \, \exp(iS[\phi, J]) \\ &= \mathcal{N} \int \mathcal{D}\phi \, \exp\left(i \int d^4x \, \frac{1}{2}\left((\partial_\mu \phi)^2 - m_0^2 \phi^2 \right) - \frac{\lambda_0}{4!}\phi^4 + J\phi \right). \end{aligned} \tag{18.74}$$

The normalization, $\mathcal{N}$, is chosen so that $Z[0] = 1$.

$$\mathcal{N}^{-1} = \int \mathcal{D}\phi \, \exp\left(i \int d^4x \, \frac{1}{2}\left((\partial_\mu \phi)^2 - m_0^2 \phi^2 \right) - \frac{\lambda_0}{4!}\phi^4 \right) \tag{18.75}$$

We will determine the effective potential at 1-loop by computing $Z[J]$ to 1-loop, and performing the Legendre transform, eqs.(18.15) and (18.17). To do so, we need the loop expansion of eq.(18.74). In chapter 16 we showed that the loop expansion is identical to an expansion in powers of $\hbar$, or a semi-classical expansion. We also know that the tree-level or no-loop result of eq.(18.74) generates the classical field theory Green's functions, and that the resulting tree-level effective action is the classical action. Now consider expanding the classical action in eq.(18.74) in a functional Taylor series about $\phi = \phi_c$. The first term in the expansion is simply $S[\phi_c] = \Gamma_{\text{tree}}[\phi_c]$. Since the classical field ϕ_c satisfies

$$\left. \frac{\delta S}{\delta \phi} \right|_{\phi = \phi_c} = 0, \tag{18.76}$$

the second term in the functional Taylor expansion will vanish. Together, these suggest that a stationary phase approximation to eq.(18.74) is equivalent to a loop expansion.

To pursue that stationary phase approximation of $Z[J]$, we Taylor expand the classical action in eq.(18.74) about the classical field solution, ϕ_c. To lowest nontrivial order, we truncate the expansion after the third term, the term that is quadratic in the fields. This results in a Gaussian approximation to the path integral, eq.(18.74), which we can compute exactly.

$$\begin{aligned} S[\phi, J] &= S[\phi_c, J] + \left. \frac{\delta S}{\delta \phi} \right|_{\phi = \phi_c} (\phi - \phi_c) + \frac{1}{2} \left. \frac{\delta^2 S}{\delta \phi^2} \right|_{\phi = \phi_c} (\phi - \phi_c)^2 + \cdots \\ &\approx S[\phi_c, J] + \frac{1}{2} \left. \frac{\delta^2 S}{\delta \phi^2} \right|_{\phi = \phi_c} (\phi - \phi_c)^2 \end{aligned} \tag{18.77}$$

Using this approximation, $Z[J]$ becomes

$$Z[J] \approx \mathcal{N} \int \mathcal{D}\phi \, \exp\left(iS[\phi_c, J] + \frac{i}{2} \frac{\delta^2 S}{\delta \phi^2}\bigg|_{\phi=\phi_c} (\phi - \phi_c)^2 \right). \tag{18.78}$$

Since the first term is independent of the integration field, ϕ, it may be pulled from the integral. After shifting integration variables, $\phi \to \phi - \phi_c$, the remaining Gaussian integral yields the usual square root of a determinant as in eq.(9.48).

$$Z[J] \approx \mathcal{N} \exp(iS[\phi_c, J]) \det{}^{-\frac{1}{2}} \left(\frac{\delta^2 S}{\delta \phi^2}\bigg|_{\phi=\phi_c} \right) \tag{18.79}$$

A straightforward calculation gives

$$-\frac{\delta^2 S[\phi, J]}{\delta\phi(x)\phi(y)}\bigg|_{\phi=\phi_c} = \left(\partial_\mu \partial^\mu + m_0^2 + \frac{\lambda_0}{2}\phi_c^2 \right) \delta^4(x-y). \tag{18.80}$$

We apply the same approximation to the normalization factor, $\mathcal{N}$. From eq.(18.11) we can see that $\phi_c = 0$ when $J = 0$ so the first term in the functional expansion of $S[\phi, J = 0]$ also vanishes, $S[\phi_c = 0, 0] = 0$. From eq.(18.80) we have

$$-\frac{\delta^2 S[\phi, 0]}{\delta\phi(x)\phi(y)}\bigg|_{\phi=\phi_c=0} = (\partial_\mu \partial^\mu + m_0^2)\, \delta^4(x-y), \tag{18.81}$$

so

$$\mathcal{N} = \det{}^{\frac{1}{2}} \left((\partial_\mu \partial^\mu + m_0^2)\, \delta^4(x-y) \right), \tag{18.82}$$

and eq.(18.79) becomes

$$Z[J] \approx \exp(iS[\phi_c, J]) \det{}^{\frac{1}{2}} \left(\frac{(\partial_\mu \partial^\mu + m_0^2)\delta^4(x-y)}{(\partial_\mu \partial^\mu + m_0^2 + \lambda_0 \phi_c^2(x)/2)\delta^4(x-y)} \right). \tag{18.83}$$

$1/(\partial_\mu \partial^\mu + m_0^2)$ is just the free-field Green's function, $G_2^{(0)}(x-y)$. Therefore,

$$\frac{1}{(\partial_\mu \partial^\mu + m_0^2)} (\partial_\mu \partial^\mu + m_0^2 + \lambda_0 \phi_c^2(x)/2) = \delta^4(x-y) + G_2^{(0)}(x-y)\, \frac{\lambda_0}{2} \phi_c^2(y), \tag{18.84}$$

and,

$$Z[J] \approx \exp(iS[\phi_c, J]) \det{}^{-\frac{1}{2}} \left(1 + \frac{\lambda_0}{2}\, G_2^{(0)}(x-y)\phi_c^2(y) \right). \tag{18.85}$$

To extract $W[J]$ from $Z[J]$ above, we note that the determinant of any matrix M can be written as

$$\det M = \exp(\mathrm{tr} \ln M). \tag{18.86}$$

Therefore,

$$W[J] \approx S[\phi_c, J] + \frac{i}{2} \mathrm{tr} \ln \left(1 + \frac{\lambda_0}{2} G_2^{(0)}(x-y)\phi_c^2(y)\right). \tag{18.87}$$

In order to see that the term in eq.(18.87) that appears in addition to the classical action is proportional to $\hbar$ (i.e. that it is the 1-loop contribution), we reinstate $\hbar$ explicitly into eq.(18.78).

$$Z[J] \approx \mathcal{N} \exp(\frac{i}{\hbar} S[\phi_c, J]) \int \mathcal{D}\phi \exp\left(\frac{i}{\hbar} \frac{1}{2} \frac{\delta^2 S}{\delta \phi^2}\bigg|_{\phi=\phi_c} (\phi - \phi_c)^2\right) \tag{18.88}$$

After shifting integration variables $\phi \to \phi - \phi_c$, we can completely absorb the $\hbar$ dependence in the Gaussian integral by scaling the shifted field by $\hbar^{1/2}$, $\phi \to \hbar^{1/2}\phi$. Thus, the determinant in eq.(18.79) is independent of $\hbar$. Since

$$Z[J] = \exp\left(\frac{i}{\hbar} W[J]\right),$$

then

$$W[J] \approx S[\phi_c, J] + \frac{i}{2} \hbar \,\mathrm{tr} \ln \left(1 + \frac{\lambda_0}{2} G_2^{(0)}(x-y)\phi_c^2(y)\right). \tag{18.89}$$

In order to see that the terms neglected in the Taylor expansion of S, eq.(18.77), are of order $\hbar^2$ or higher in $W[J]$, we reinstate them in the functional integral, and scale the shifted field as we did above. This is left as an exercise.

Now that we have $W[J]$ to 1-loop, we can Legendre transform it to obtain the effective action to 1-loop. To order $\hbar$, this is simple, because J explicitly appears only in the source term and this is just the piece that is subtracted off in eq.(18.17). Thus,

$$\Gamma[\phi_c] = S[\phi_c, 0] + \frac{i}{2} \hbar \,\mathrm{tr} \ln \left(1 + \frac{\lambda_0}{2} G_2^{(0)}(x-y)\phi_c^2(y)\right) + \mathcal{O}(\hbar^2). \tag{18.90}$$

If J had made an explicit appearance in $W[J]$ besides the source term, then we would have to solve $\phi_c = \delta W/\delta J$ for J as a function of ϕ_c, $J = J[\phi_c]$, in order to eliminate J in favor of ϕ_c to complete the Legendre transform.

The computation of the trace appearing in eq.(18.90) is complicated by the fact that the (infinite-dimensional) matrix is not diagonal. If, however, we just want the effective potential, then we may set $\phi_c =$ constant, and the matrix becomes diagonal in momentum space. Therefore, the 1-loop contribution to the effective potential is

$$V^{(1)}[\phi_c] = -\frac{i}{2}\hbar \int \frac{d^4p}{(2\pi)^4} \ln\left(1 - \frac{\lambda_0 \phi_c^2/2}{p^2 - m_0^2 + i\epsilon}\right). \tag{18.91}$$

This is identical to the result we obtained in the operator representation, eq.(18.53), so the analysis and discussion of renormalization proceeding from this point are the same as in the previous section.

One attractive feature of the path integral computation of the effective potential is that the log arose "naturally" as a result of writing a determinant as an exponential. There was no need to compute symmetry factors for an infinite set of diagrams or to sum the series.

18.4 ... FROM THE SCHRÖDINGER REPRESENTATION

In the Schrödinger representation, we use the alternative, but equivalent, definition of the effective action that was given at the end of the first section. The static effective action is minus the minimum expected energy in a set of normalized states, where the expectation of the field operator in this set of states is equal to $\phi_c(x)$.

$$\Gamma_{\text{static}}[\phi_c] = -\min\langle \psi_c | H | \psi_c \rangle, \tag{18.92}$$

where the states $|\psi_c\rangle$ satisfy the constraints

$$\langle \psi_c | \psi_c \rangle = 1, \qquad \text{and} \qquad \langle \psi_c | \varphi(x) | \psi_c \rangle = \phi_c(x). \tag{18.93}$$

By introducing Lagrange multipliers, E and $J(x)$, this is equivalent to finding the minimum of

$$\min\langle \psi | H - E - \int d^3x J(x)\varphi(x) | \psi \rangle \tag{18.94}$$

without any additional constraints. While any eigenstate of the Hamiltonian $H' = H - \int d^3x J(x)\varphi(x)$ will minimize eq.(18.94), only the vacuum state of H' minimizes eq.(18.92). Thus, minimizing eq.(18.92) is equivalent to solving the functional Schrödinger equation for the vacuum wave functional, $|\psi_{c0}\rangle$, of the Hamiltonian H',

$$\left(H - \int d^3x J(x)\varphi(x)\right) | \psi_{c0} \rangle = E_0 | \psi_{c0} \rangle. \tag{18.95}$$

The energy eigenvalue E_0 is a functional of the source, $J(x)$, $E_0 = E_0[J]$. Application of the Feynman-Hellmann theorem yields

$$\begin{aligned}\frac{\delta E_0[J]}{\delta J(y)} &= \langle \psi_{c0} | \frac{\delta}{\delta J(y)} H' | \psi_{c0} \rangle \\ &= -\langle \psi_{c0} | \varphi(y) | \psi_{c0} \rangle = -\phi_c(y).\end{aligned} \tag{18.96}$$

Eq.(18.96) implies that the static effective action is the Legendre transform of the vacuum energy,

$$\Gamma_{\text{static}}[\phi_c] = -E_0[J] - \int d^3y\, J(y)\phi_c(y). \tag{18.97}$$

That is, the static effective action is $-E_0[J]$ written in terms of ϕ_c only,

$$\Gamma_{\text{static}}[\phi_c] = E_0[0] - E_0[J[\phi_c]], \tag{18.98}$$

where we use eq.(18.96) to solve for $J(x)$ in terms of $\phi_c(x)$. Note that we have normalized $\Gamma_{\text{static}}[\phi_c]$ in eq.(18.98) so that $\Gamma_{\text{static}}[\phi_c = 0] = 0$. This normalization is identical to the one used in the path integral and operator representations ($Z[J = 0] = 1$ implies $\Gamma[\phi_c = 0] = 0$).

Let us illustrate the method contained in eqs.(18.95)-(18.98) by computing the static effective action for free scalar field theory. For a free field theory, the functional Schrödinger equation (18.95) is (see eq.(10.14))

$$\begin{aligned}\frac{1}{2}\int d^3x \left(-\frac{\delta^2}{\delta\phi(\vec{x})^2} + \phi(\vec{x})(-\nabla^2 + m_0^2)\phi(\vec{x}) - 2J(\vec{x})\phi(\vec{x}) \right) \Psi_0[\phi, J] \\ = E_0[J]\Psi_0[\phi, J].\end{aligned} \tag{18.99}$$

To solve this equation, we complete the squares and shift the field $\phi(\vec{x})$.

$$\begin{aligned}&\phi(\vec{x})(-\nabla^2 + m_0^2)\phi(\vec{x}) - 2J(\vec{x})\phi(\vec{x}) \\ &= \left(\phi - \frac{J}{-\nabla^2 + m_0^2}\right)(-\nabla^2 + m_0^2)\left(\phi - \frac{J}{-\nabla^2 + m_0^2}\right) - \frac{J^2}{(-\nabla^2 + m_0^2)}.\end{aligned} \tag{18.100}$$

$h(x, y) = (-\nabla^2 + m_0^2)^{-1}$ is not diagonal, hence $J/(-\nabla^2 + m_0^2)$ is really $\int d^3y\, h(x, y)J(y)$. For convenience, we will use the former notation.

If we shift the field coordinate to $\phi'(\vec{x}) = \phi(\vec{x}) - J/(-\nabla^2 + m_0^2)$, then eq.(18.99) becomes

$$\begin{aligned}\frac{1}{2}\int d^3x \left(-\frac{\delta^2}{\delta\phi'(\vec{x})^2} + \phi'(\vec{x})(-\nabla^2 + m_0^2)\phi'(\vec{x}) - \frac{J^2}{-\nabla^2 + m_0^2} \right) \Psi_0[\phi', J] \\ = E_0[J]\Psi_0[\phi', J].\end{aligned} \tag{18.101}$$

The piece quadratic in the source and independent of $\phi'(\vec{x})$ that explicitly appeared in completing the squares, eq.(18.100), is the source-dependent contribution to the energy. The ϕ'-dependent functional differential operator is identical to eq.(10.14), thus the solution is

$$\begin{aligned}\Psi_0[\phi', J] &= \eta \exp\left(-\frac{1}{2}\int d^3x\, d^3y\, \phi'(\vec{x}) f(\vec{x}-\vec{y})\phi'(\vec{y})\right) \\ &= \eta \exp\left(-\frac{1}{2}\int d^3x\, d^3y \left(\phi(\vec{x}) - \frac{J}{(-\nabla^2+m_0^2)}(\vec{x})\right)\right. \\ &\qquad \left. f(\vec{x}-\vec{y})\left(\phi(\vec{y}) - \frac{J}{(-\nabla^2+m_0^2)}(\vec{y})\right)\right),\end{aligned} \tag{18.102}$$

where

$$f(\vec{x}-\vec{y}) = \sqrt{-\nabla^2+m_0^2} \equiv \int \frac{d^3k}{(2\pi)^3}\, e^{i\vec{k}\cdot(\vec{x}-\vec{y})}\sqrt{\vec{k}^2+m_0^2}, \tag{18.103}$$

and where the energy eigenvalue is

$$E_0[J] = \frac{1}{2}\int d^3x\, f(0) - \frac{1}{2}\int d^3x\, d^3y\, J(\vec{x})\frac{1}{-\nabla^2+m_0^2}J(\vec{y}). \tag{18.104}$$

To determine the effective action, we use eq.(18.96) to rewrite $E_0[J]$ in terms of $\phi_c(\vec{x})$. From eq.(18.96) and eq.(18.98) we have

$$\phi_c(\vec{x}) = -\frac{\delta E_0[J]}{\delta J(\vec{x})} = +\frac{J}{-\nabla^2+m_0^2} = +\int d^3y\, h(\vec{x}-\vec{y})J(\vec{y}), \tag{18.105}$$

and

$$\Gamma_{\text{static}}[\phi_c] = -\frac{1}{2}\int d^3x\, \phi_c(\vec{x})(-\nabla^2+m_0^2)\phi_c(\vec{x}), \tag{18.106}$$

which is identical to the classical result as expected. We can also use eq.(18.105) to express the vacuum wave functional in the presence of the source, $\Psi[\phi, J]$, in terms of ϕ_c.

$$\Psi_0[\phi,\phi_c] = \eta \exp\left(-\frac{1}{2}\int d^3x\, d^3y\, (\phi(\vec{x})-\phi_c(\vec{x}))f(\vec{x}-\vec{y})(\phi(\vec{y})-\phi_c(\vec{y}))\right). \tag{18.107}$$

Now add the quartic interaction, $\lambda_0/4! \int d^3x\, \varphi^4(\vec{x})$. The static effective action may be found by the same method, but we must resort to perturbation theory (or some other approximation) to solve the resulting functional Schrödinger equation, eq.(18.95).

Let's compute the $\mathcal{O}(\lambda_0)$ contribution. From Rayleigh-Schrödinger perturbation theory we know that the shift in the energy is simply the expectation of the interaction in the free-field functional, eq.(18.107).

$$
\begin{aligned}
E_0^{(1)} &= \frac{\lambda_0}{4!}\langle \Psi_{c0}| \int d^3x \phi^4(\vec{x})| \Psi_{c0}\rangle \\
&= \frac{\lambda_0}{4!}\int d^3x \int \mathcal{D}\phi\, \phi^4(\vec{x}) \\
&\qquad \exp\left(-\int d^3y\, d^3z\, (\phi(\vec{y})-\phi_c(\vec{y}))f(\vec{y},\vec{z})(\phi(\vec{z})-\phi_c(\vec{z}))\right)
\end{aligned}
\tag{18.108}
$$

This functional integral is readily computed using the Gaussian moment-generating functional, $\mathcal{G}[j]$, eq.(11.7).

$$
\begin{aligned}
\mathcal{G}[j] &= \eta^2 \int \mathcal{D}\phi \exp\left(-\int d^3x\, d^3y\, \phi(\vec{x})f(\vec{x},\vec{y})\phi(\vec{y}) + \int d^3x\, j(\vec{x})\phi(\vec{x})\right) \\
&= \exp\left(\frac{1}{4}\int d^3x\, d^3y\, j(\vec{x})f^{-1}(\vec{x},\vec{y})j(\vec{y})\right)
\end{aligned}
\tag{18.109}
$$

Shifting variables in eq.(18.108) from ϕ to $\phi-\phi_c$, we have

$$
\begin{aligned}
E_0^{(1)} &= \frac{\lambda_0}{4!}\eta^2\int d^3x \int \mathcal{D}\phi\,(\phi+\phi_c)^4 \exp\left(-\int d^3y\, d^3z\, \phi(\vec{y})f(\vec{y}-\vec{z})\phi(\vec{z})\right) \\
&= \frac{\lambda_0}{4!}\int d^3x \left(\frac{\delta^4}{\delta j^4(\vec{x})} + 6\phi_c^2(\vec{x})\frac{\delta^2}{\delta j^2(\vec{x})} + \phi_c^4(\vec{x})\right)\mathcal{G}[j]\Bigg|_{j=0} \\
&= \frac{\lambda_0}{4!}\int d^3x\, \frac{3}{4}(f^{-1}(0))^2 + 3\phi_c^2(\vec{x})f^{-1}(0) + \phi_c^4(\vec{x}).
\end{aligned}
\tag{18.110}
$$

Therefore, the static effective action to order λ_0 is

$$
\Gamma_{\text{static}}[\phi_c] = \int d^3x -\frac{1}{2}\phi_c(\vec{x})(-\nabla^2+m_0^2)\phi_c(\vec{x}) - \frac{\lambda_0}{4!}\phi_c^4(\vec{x}) - \frac{\lambda_0}{8}\phi_c^2(\vec{x})f^{-1}(0).
\tag{18.111}
$$

The additional piece above beyond the classical action does not involve any derivatives on ϕ_c, hence it is part of the effective potential. It is identical to the 2-legged 1-loop result, eq.(18.48), obtained in the operator representation.

$$
\begin{aligned}
\frac{\lambda_0}{4!}\phi_c^2 f^{-1}(0) &= \frac{\lambda_0}{8}\phi_c^2 \int \frac{d^3k}{(2\pi)^3}\frac{1}{(\vec{k}^2+m_0^2)^{1/2}} \\
&= -i\frac{\lambda_0}{4}\phi_c^2 \int \frac{d^4k}{(2\pi)^4}\frac{1}{k^2-m_0^2}
\end{aligned}
\tag{18.112}
$$

Since the two results are identical, the mass renormalization equation, eq.(18.58), as determined by eq.(18.56), will also be the same.

We can use the mass renormalization equation to rewrite the functional Schrödinger equation in terms of the renormalized mass, m. The new Schrödinger equation, complete with a counterterm, is

$$\left(\int d^3x \; -\frac{1}{2}\left(\frac{\delta^2}{\delta\phi^2(\vec{x})} + \phi(\vec{x})(-\nabla^2 + m^2)\phi(\vec{x})\right) + \frac{\lambda_0}{4!}\phi^4(\vec{x}) - \frac{\lambda_0}{8}\left(\int \frac{d^3k}{(2\pi)^3}\frac{1}{(\vec{k}^2+m^2)^{1/2}}\right)\phi^2(\vec{x})\right)\Psi_0 = E_0\Psi_0.$$

The counterterm guarantees that the eigenvalues will be finite at $\mathcal{O}(\lambda_0)$.

Let's next compute the 1-loop effective potential, this time as a variational calculation. First we need a trial wave functional that satisfies the constraint, $\langle\psi_c|\varphi(\vec{x})|\psi_c\rangle = \phi_c$. Since we will only look for the potential, ϕ_c can be taken as a constant. The Gaussian wave functional we computed for the free-field result, eq.(18.107), satisfies the constraint no matter what $f(\vec{x}-\vec{y})$ is, hence we can use it by treating $f(\vec{x},\vec{y})$ as an unknown variational parameter. With this trial wave functional, we compute $E[f] = \langle\psi_c|H|\psi_c\rangle$ as a functional of $f(\vec{x},\vec{y})$. We then find the function f that minimizes $E[f]$ by solving $\delta E/\delta f = 0$.

Recall that the loop expansion is also an expansion in powers of $\hbar$. In order to extract the 1-loop result, we will need to explicitly keep track of the $\hbar$'s. Putting the $\hbar$'s back into the wave functional, Hamiltonian, and moment-generating functional, we have

$$\Psi_c[\phi,\phi_c,f] = \eta\exp\left(-\frac{1}{2}\frac{1}{\hbar}\int d^3x\, d^3y\,(\phi(\vec{x})-\phi_c)f(\vec{x}-\vec{y})(\phi(\vec{y})-\phi_c)\right)$$

$$H = \int d^3x \; -\frac{\hbar^2}{2}\frac{\delta^2}{\delta\phi^2(\vec{x})} + \frac{1}{2}\phi(\vec{x})(-\nabla^2+m_0^2)\phi(\vec{x}) + \frac{\lambda_0}{4!}\phi^4(\vec{x}) \quad (18.113)$$

$$\mathcal{G}[j] = \exp\left(\frac{\hbar}{4}\int d^3x\, d^3y\, j(\vec{x})f^{-1}(\vec{x}-\vec{y})j(\vec{y})\right).$$

The computation of the expectation of H in the trial state follows along the same lines as $E_0^{(1)}$, eq.(18.110). For convenience, let's work in the momentum representation.

$$\begin{aligned}\mathcal{E}[\tilde{f},\phi_c] &= \langle\Psi_c|\mathcal{H}|\Psi_c\rangle \\ &= \frac{\hbar}{4}\int\frac{d^3k}{(2\pi)^3}\tilde{f}(\vec{k}) + \frac{\hbar}{4}\int\frac{d^3k}{(2\pi)^3}\frac{\vec{k}^2+m_0^2}{\tilde{f}(\vec{k})} + \frac{1}{2}m_0^2\phi_c^2 \\ &\quad + \frac{\lambda_0}{4!}\left(\frac{3\hbar^2}{4}\left(\int\frac{d^3k}{(2\pi)^3}\frac{1}{\tilde{f}(\vec{k})}\right)^2 + 3\hbar\phi_c^2\int\frac{d^3k}{(2\pi)^3}\frac{1}{\tilde{f}(\vec{k})} + \phi_c^4\right)\end{aligned} \quad (18.114)$$

Note that the terms independent of $\hbar$ form the classical potential, and are also independent of $\tilde{f}$. Minimizing the energy density to first order in $\hbar$ we have

$$\frac{\delta \mathcal{E}}{\delta \tilde{f}(p)} = 0 = \frac{\hbar}{4} - \frac{\hbar}{4}\frac{\vec{p}^2 + m_0^2}{\tilde{f}^2(\vec{p})} - \frac{\lambda_0 \hbar}{8}\phi_c^2 \frac{1}{\tilde{f}^2(\vec{p})}, \tag{18.115}$$

which implies

$$\tilde{f}(\vec{p}) = \sqrt{\vec{p}^2 + m_0^2 + \frac{\lambda_0 \phi_c^2}{2}}. \tag{18.116}$$

Plugging this back into $\mathcal{E}[\tilde{f}, \phi_c]$, eq.(18.114), we get

$$\mathcal{E}[\phi_c] = \frac{1}{2}m_0^2\phi_c^2 + \frac{\lambda_0}{4!}\phi_c^4 + \frac{\hbar}{2}\int \frac{d^3k}{(2\pi)^3}\sqrt{\vec{k}^2 + m_0^2 + \frac{\lambda_0 \phi_c^2}{2}} + \mathcal{O}(\hbar^2). \tag{18.117}$$

Application of eq.(18.98) to the energy yields the effective potential to 1-loop.

$$V[\phi_c] = \left(\frac{1}{2}m_0^2\phi_c^2 + \frac{\lambda_0}{4!}\phi_c^4 + \frac{\hbar}{2}\int \frac{d^3k}{(2\pi)^3}\sqrt{\vec{k}^2 + m_0^2 + \frac{\lambda_0 \phi_c^2}{2}} - \frac{\hbar}{2}\int \frac{d^3k}{(2\pi)^3}\sqrt{\vec{k}^2 + m_0^2}\right)\int d^3x \tag{18.118}$$

Comparison of this result with eq.(18.62) verifies that we have obtained the entire 1-loop result. It is not entirely surprising that the 1-loop term leads to a Gaussian wave functional. In the path integral representation, the 1-loop term is identical to the stationary phase approximation. This approximation is also Gaussian in the fluctuating fields.

To arrive at the 1-loop potential we threw away the term in $\mathcal{E}[\tilde{f}, \phi_c]$ that was proportional to $\hbar^2$. What happens if we keep it? Then $\delta\mathcal{E}/\delta\tilde{f} = 0$ implies that

$$\tilde{f}(\vec{p}) = \sqrt{\vec{p}^2 + m_0^2 + \frac{\lambda_0 \phi_c^2}{2} + \frac{\lambda_0 \hbar}{4}\int \frac{d^3k}{(2\pi)^3}\frac{1}{\tilde{f}(\vec{k})}}. \tag{18.119}$$

The extra piece that has appeared is independent of $\vec{p}$ and, therefore, is just a (divergent) constant. If we define the constant as

$$G(\phi_c) = \int \frac{d^3k}{(2\pi)^3}\frac{1}{\tilde{f}(\vec{k})},$$

then we may use eq.(18.119) to derive an integral equation for $G(\phi_c)$.

$$G(\phi_c) = \int \frac{d^3k}{(2\pi)^3} \frac{1}{\left(\vec{k}^2 + m_0^2 + \lambda_0\phi_c^2/2 + \lambda_0\hbar G(\phi_c)/4\right)^{1/2}} \tag{18.120}$$

When eq.(18.119) is inserted back into $\mathcal{E}[\tilde{f}, \phi_c]$, eq.(18.114), the resulting effective potential (density) is

$$V(\phi_c) = \frac{1}{2}m_0^2\phi_c^2 + \frac{\lambda_0}{4!}\phi_c^4 + \frac{\hbar}{2}(F(\phi_c) - F(0)) - \frac{\lambda_0\hbar^2}{32}(G^2(\phi_c) - G^2(0)), \tag{18.121}$$

where

$$F(\phi_c) \equiv \int \frac{d^3k}{(2\pi)^3} \left(\vec{k}^2 + m_0^2 + \lambda_0\phi_c^2/2 + \lambda_0\hbar G(\phi_c)/4\right)^{1/2}, \tag{18.122}$$

a result first obtained by Barnes and Ghandour. This potential defines the following renormalized mass:

$$\begin{aligned} m^2 = \left.\frac{d^2V}{d\phi_c^2}\right|_{\phi_c=0} &= m_0^2 + \frac{\lambda_0\hbar}{4}G(0) \\ &= m_0^2 + \frac{\lambda_0\hbar}{4}\int \frac{d^3k}{(2\pi)^3} \frac{1}{\sqrt{\vec{k}^2 + m^2}}. \end{aligned} \tag{18.123}$$

This is identical to the 1-loop result except that $m_0^2 \to m^2$ in the integral. In other words, the higher-loop terms contained in eq.(18.123) are just those that result from replacing m_0^2 with m^2 in the 1-loop term. The renormalized charge,

$$\lambda = \left.\frac{d^4V}{d\phi_c^4}\right|_{\phi_c=0} = \lambda_0 \frac{1 - \lambda_0\hbar I(m)/4}{1 + \lambda_0\hbar I(m)/8}, \tag{18.124}$$

where

$$I(m) = \int \frac{d^3k}{(2\pi)^3} \frac{1}{(\vec{k}^2 + m^2)^{3/2}}, \tag{18.125}$$

contains other higher-loop effects. It is not at all clear that the resulting renormalized effective potential is actually finite or independent of any cutoff.

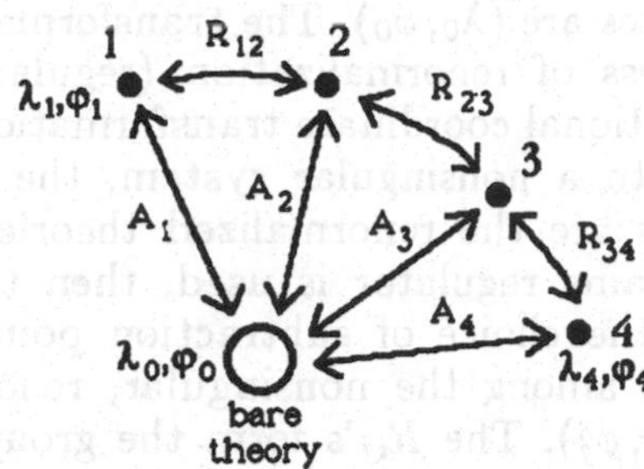

Figure 18.9 Schematic illustration of the relationship between a bare theory (open circle) and several renormalized versions (solid dots). A_1 through A_4 represent the process of renormalization (regularization and subtraction). R_{ij} transform one renormalized theory to another. The R_{ij}'s form a group traditionally called the "renormalization group." The "bare" theory and all four renormalized theories represent the same underlying physical theory.

18.5 THE RENORMALIZATION GROUP

In the process of renormalization, we regularize the bare theory. It doesn't seem to matter which regulator is chosen. Then, it doesn't seem to matter which renormalization prescription or subtraction point is chosen. As long as the regulator is consistent with the desired internal symmetries, we end up with the same physical renormalized theory at the end of the process. This means that there is just one underlying physical theory involved, and this theory is invariant under different renormalization schemes.

We know how to write down the physical theory we are interested in. Unfortunately, we only know how to write it down in a singular functional coordinate system (bare couplings and fields). The underlying physical consequences of the theory are hidden by the poor choice of functional coordinates. In the process of renormalization, we define a functional coordinate transformation from the singular bare system to a nonsingular functional coordinate system, the renormalized couplings and fields. The physical theory we are interested in is invariant under this functional coordinate transformation. The physical results of the theory are independent of which functional coordinate system is chosen. The set of coordinate transformations between the finite functional coordinate systems, or renormalized coordinates, forms a group called the renormalization group. The physical theory is invariant under this group.

The statements above are illustrated schematically in figure 18.9. The open circle labeled "bare theory" represents the theory written in the bare functional coordinate system. For purposes of illustration, let's say those functional coordinates are (λ_0, ϕ_0). The transformations, A_1 through A_4, represent the process of renormalization (regularization plus subtraction). They are functional coordinate transformations (the Z's) that map the bare system into a nonsingular system, the renormalized system, (λ_i, ϕ_i). The results are the renormalized theories, the solid circles, 1 through 4. If the same regulator is used, then the difference between A_1 through A_4 is the choice of subtraction point or scale. The R_{ij}'s are transformations among the nonsingular, renormalized coordinates, $R_{ij} : (\lambda_i, \phi_i) \to (\lambda_j, \phi_j)$. The R_{ij}'s form the group that is traditionally called the renormalization group. Let us emphasize that the "bare theory" and all four "renormalized theories" represent the *same* physical theory.

To illustrate this viewpoint, reconsider the 1-loop result of the effective potential for massless φ^4 theory, eq.(18.70). We found that $V(\phi_c, \lambda_M, M)$ was form-invariant. Under a change of subtraction point $M \to M'$, λ_M changed in such a way that the result was $V(\phi_c, \lambda'_M, M')$, i.e. the same functional form with λ_M replaced by λ'_M and M replaced by M'. The transformation from M to M' is one of the R_{ij}'s of figure 18.9 and is an element of the renormalization group. The form invariance of the potential under this transformation is a reflection of the fact that we are dealing with only one underlying physical theory.

The form invariance of V under a change in M can be expressed as the differential equation,

$$\frac{dV}{dM} = 0. \tag{18.126}$$

It is easy to see why V must satisfy eq.(18.126). It simply states that V is independent of M. M is the scale at which the subtraction was performed. It is arbitrary so we are free to choose any nonzero value. Since it is arbitrary, it cannot be physical. Hence, anything physical, like the potential, cannot depend on it.

As it stands in eq.(18.70), V depends on M explicitly and implicitly through λ_M. Therefore, eq.(18.126) can be written as

$$\left(\frac{\partial}{\partial M} + \frac{\partial \lambda_M}{\partial M}\frac{\partial}{\partial \lambda_M}\right) V(\phi_c, \lambda_M, M) = 0. \tag{18.127}$$

The dependence of V and λ_M on M is often logarithmic. Thus, it is traditional to express eq.(18.126) as $dV/d\ln M = 0$, that is,

$$\left(M\frac{\partial}{\partial M} + M\frac{\partial \lambda_M}{\partial M}\frac{\partial}{\partial \lambda_M}\right) V(\phi_c, \lambda_M, M) = 0. \tag{18.128}$$

This is a special case of the renormalization group equation. The coefficient in front of the derivative with respect to the coupling is called the β-function for the coupling. It describes the flow of the coupling under a change of (subtraction) scale. We may determine the β-function at 1-loop for the massless theory via eq.(18.72). If we treat M' and λ'_M as fixed, then

$$\beta = M\frac{\partial \lambda_M}{\partial M} = \frac{3}{16\pi^2}\lambda_M^2 + \mathcal{O}(\lambda_M^3). \tag{18.129}$$

Compare this result with the β-function for the electric charge in QED, eq.(17.107).

In order to obtain the standard form of the renormalization group equation (which will be eq.(18.133)), we let M be the mass scale at which the renormalized coordinates λ and ϕ_c are defined, i.e. the subtraction point, and note that the entire effective action must be independent of M,

$$M\frac{d}{dM}\Gamma[\phi_c] = 0. \tag{18.130}$$

Although not present in φ^4 at 1-loop, the field ϕ_c is rescaled at higher loops by wave function renormalization via $\phi_c = Z_\phi^{-1/2}\phi_{c0}$. Therefore, when we write out eq.(18.130), we must not forget that Γ depends on M implicitly through $Z_\phi(M)$, too. Thus, eq(18.130) can be written as

$$\left(M\frac{\partial}{\partial M} + \beta\frac{\partial}{\partial \lambda} + \gamma\int d^4x\,\phi_c\frac{\delta}{\delta\phi_c}\right)\Gamma[\phi_c] = 0, \tag{18.131}$$

where the coefficient,

$$\gamma = \frac{1}{2}M\frac{\partial \ln Z_\phi}{\partial M}, \tag{18.132}$$

is called the anomalous dimension. If we expand Γ in a Taylor series in the field, then the n-point proper function must satisfy

$$\left(M\frac{\partial}{\partial M} + \beta\frac{\partial}{\partial \lambda} - n\gamma\right)\tilde{\Gamma}^{(n)}(p_1,\ldots,p_n;\lambda,M) = 0. \tag{18.133}$$

No mention of perturbation theory is made in arriving at eq.(18.131) or eq.(18.133). There is no reason not to believe that they are valid beyond perturbation theory. The power in the renormalization group equation is that it is a linear partial differential equation for the Green's functions whose general solution in terms of the coefficients β and γ can be found. Of course, we do not know ahead of time what the coefficients β and γ are. If we did, we would have solved the theory. We can compute them perturbatively, but the range of their validity is limited by the size of the coupling.

Another use of the renormalization group equation is that it allows us to investigate the asymptotic behavior of $\tilde{\Gamma}^{(n)}$ for large momenta, p_i. In order to see how this might come about, we want to re-express eq.(18.133) as a scaling equation.

From the classical action we see that the physical dimension of the field φ is inverse length or a mass dimension of 1. Therefore, in 4 space-time dimensions, the physical mass dimension of $\tilde{\Gamma}^{(n)}$ must be $4-n$. If we uniformly scale the momenta in $\tilde{\Gamma}^{(n)}$ by s, i.e. $p_i \rightarrow sp_i$, but we keep the subtraction mass scale, M, fixed, then

$$\tilde{\Gamma}^{(n)}(sp_i;\lambda,M) = s^{4-n}\tilde{\Gamma}^{(n)}(p_i;\lambda,M/s). \tag{18.134}$$

This implies that

$$s\frac{\partial}{\partial s}\tilde{\Gamma}^{(n)}(sp_i;\lambda,M)$$
$$= (4-n)s^{4-n}\tilde{\Gamma}^{(n)}(p_i;\lambda,M/s) + s^{4-n}\, s\frac{\partial}{\partial s}\tilde{\Gamma}^{(n)}(p_i;\lambda,M/s). \tag{18.135}$$

Let $x = M/s$. Then

$$\frac{\partial}{\partial x} = \frac{\partial s}{\partial x}\frac{\partial}{\partial s} \quad \text{implies} \quad x\frac{\partial}{\partial x} = -s\frac{\partial}{\partial s}.$$

But

$$x\frac{\partial}{\partial x} = M\frac{\partial}{\partial M},$$

so eq.(18.135), with the aid of eq.(18.134), says

$$\left(M\frac{\partial}{\partial M} + s\frac{\partial}{\partial s}\right)\tilde{\Gamma}^{(n)}(sp_i;\lambda,M) = (4-n)\tilde{\Gamma}^{(n)}(sp_i;\lambda,M). \tag{18.136}$$

The renormalization group equation, eq.(18.133), tells us how $\tilde{\Gamma}^{(n)}$ changes under a change in M without reference to the p_i's. Hence, we can use it to eliminate the $M\,\partial/\partial M$ that appears in eq.(18.136). The result is

$$\left(s\frac{\partial}{\partial s} - \beta\frac{\partial}{\partial\lambda} - (4-n(\gamma+1))\right)\tilde{\Gamma}^{(n)}(sp_i;\lambda,M) = 0. \tag{18.137}$$

This is the desired scaling equation.

The interpretation of eq.(18.137) is straightforward. The classical action for massless ($m_0 = 0$) φ^4 theory is scale-invariant. The coupling is dimensionless in 4 space-time dimensions. Quantization breaks the scale invariance, because we are forced to introduce a mass scale M, the scale at which the renormalized coupling is defined. In order to eliminate M from the physical results, λ becomes a function of M, and no longer is

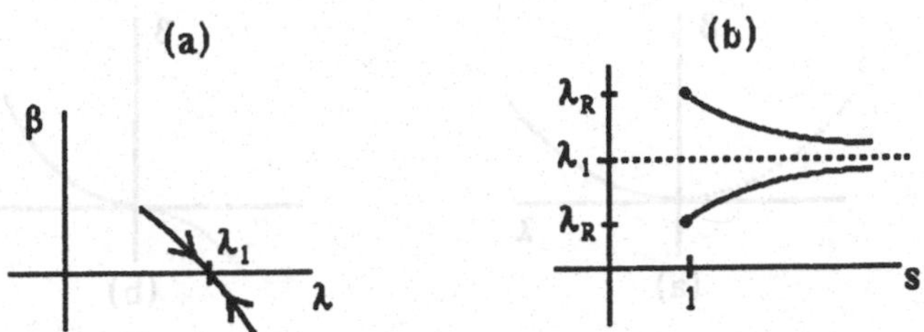

Figure 18.10 (a) Attractive ultraviolet fixed point at $\lambda = \lambda_1$ ($\beta(\lambda_1) = 0 \Rightarrow$ fixed point, $\beta'(\lambda_1) < 0 \Rightarrow$ attractive). (b) Since the fixed point is attractive, $\lambda_R \to \lambda_1$ as s goes from 1 to ∞ no matter whether $\lambda_R > \lambda_1$ or $\lambda_R < \lambda_1$.

scale-invariant. How λ scales is given by the β function. In addition, the field φ no longer scales with its classical value. It scales as if it had an anomalous dimension, $1 + \gamma$.

As we stated earlier, the power in this approach is that the solution to the equation may be found. To do so we treat λ as a function of s. That is, $\lambda(s)$ is the solution to

$$s\frac{\partial \lambda}{\partial s} = \beta(\lambda(s)), \qquad \text{where} \qquad \lambda(s=1) = \lambda_R. \tag{18.138}$$

The solution to eq.(18.137) is

$$\begin{aligned}\tilde{\Gamma}^{(n)}(sp_i;\lambda_R,M) &= s^{4-n}\exp\left(-n\int_1^s \frac{ds'}{s'}\gamma(\lambda(s'))\right)\tilde{\Gamma}^{(n)}(p_i;,\lambda(s),M)\\ &= s^{4-n}\exp\left(-n\int_{\lambda_R}^{\lambda}\frac{d\lambda}{\beta(\lambda)}\gamma(\lambda)\right)\tilde{\Gamma}^{(n)}(p_i;\lambda(s)).\end{aligned} \tag{18.139}$$

The lack of scale invariance is now apparent. However, ...

18.6 THE β-FUNCTION AND ASYMPTOTIC FREEDOM

... following reasoning due to Wilson, suppose that the β-function has a zero in it for some nonzero value of λ. The zeroes of the β-function are called fixed points. If the β-function is zero at $\lambda = \lambda_1$, then the coupling remains fixed at λ_1 as we change s. Suppose also that $d\beta/d\lambda < 0$ at the fixed point, $\lambda = \lambda_1$. Then, as s increases, λ will flow from λ_R to λ_1. We call this fixed point an ultraviolet attractive fixed point. It is "ultraviolet," because it is a fixed point for $s \to \infty$, and it is attractive, because $\lambda_R \to \lambda_1$ as s goes from 1 to ∞ no matter whether $\lambda_R > \lambda_1$ or $\lambda_R < \lambda_1$. See figure

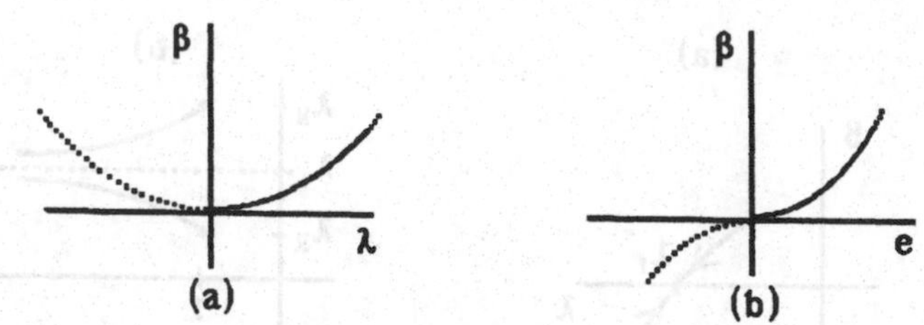

Figure 18.11 Perturbative behavior of the β-function for (a) scalar $\lambda\varphi^4$ theory and (b) QED. In both cases the trivial fixed point at zero coupling is infrared attractive. This implies that perturbation theory is valid at small mass scales or large distances.

18.10. As $s \to \infty$ and we approach the fixed point, the β-function near the fixed point will look like a straight line. That is, near the fixed point,

$$\beta(\lambda) \sim \lambda_1 - \lambda. \tag{18.140}$$

If $\gamma(\lambda)$ doesn't vanish at λ_1, then, near the fixed point, $\gamma \sim \gamma(\lambda_1) \equiv \gamma_1$. Near this fixed point, the exponential factor in eq.(18.139) that ruins the scaling becomes

$$\exp\left(-n \int_{\lambda_R}^{\lambda} \frac{d\lambda}{\beta(\lambda)} \gamma(\lambda)\right) \longrightarrow \exp\left(-n\gamma_1 \int^{\lambda_1} \frac{d\lambda}{\lambda_1 - \lambda}\right). \tag{18.141}$$

From eq.(18.138), we also have near the fixed point that

$$\int \frac{d\lambda}{\lambda_1 - \lambda} \sim \ln(s), \tag{18.142}$$

so together with eq.(18.141), the scaling relation for $\tilde{\Gamma}^{(n)}$ for large s, eq.(18.139), is

$$\tilde{\Gamma}^{(n)}(sp_i; \lambda_R, M) \sim s^{4-n(1+\gamma(\lambda_1))}\tilde{\Gamma}^{(n)}(p_i; \lambda_1, M). \tag{18.143}$$

In this case, $\tilde{\Gamma}^{(n)}(sp_i; \lambda_R, M)$ scales again. The renormalization group analysis has shown us that scale invariance is recovered for large momenta if the β-function develops an attractive fixed point.

Clearly the behavior of the β-function is of interest. So far we only know the β-function perturbatively, and then only the first term. For φ^4 theory, this is given by eq.(18.129), and plotted in figure 18.11(a). For QED, the lowest term in β is given by eq.(17.107), and plotted in figure 18.11(b). In both cases, the trivial fixed point at $\lambda = 0$ or $e = 0$ is infrared

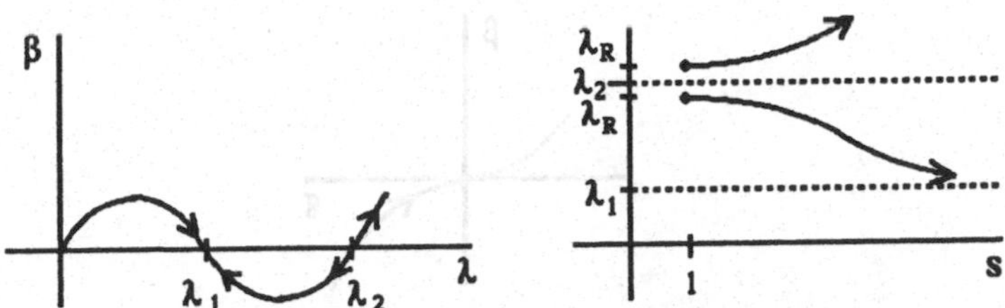

Figure 18.12 Illustration of what happens if the β-function contains multiple zeroes. The fixed point at λ_1 is attractive, while the fixed point at λ_2 is repulsive.

attractive or stable (infrared attractive meaning $\lambda \to 0$ or $e \to 0$ as $s \to 0$). This means that perturbation theory, which remains valid as long as the coupling is small, is valid at small mass scales or large distances.

We can only speculate what happens outside of perturbation theory. If the β-function never develops a zero, then the coupling just keeps on increasing as $s \to \infty$, i.e. at short distances. If the β-function rolls over and develops just one zero, then the situation is exactly as we have already discussed, and is contained in figure 18.10(b). Note, however, that if $\lambda_1 < 1$, then perturbation theory is valid at short distances. If, in addition, λ_R is $< \lambda_1$, then perturbation theory is valid everywhere.

The β-function could contain another zero as depicted in figure 18.12. If $\lambda_R < \lambda_2$, then the coupling will flow to λ_1 at large mass scales. If $\lambda_R > \lambda_2$, then it will flow to the next zero or it will increase forever if there are no zeroes.

So far we have only considered theories where the β-function is positive for small coupling. What happens if we reverse this? Suppose that for a coupling g, $\beta(g) < 0$ for small g as in figure 18.13. Then the trivial zero of β at zero coupling becomes an ultraviolet attractive fixed point. This means that the coupling gets small at small distances or large energies, and perturbation theory is valid at large energies. This situation is called asymptotic freedom, since the asymptotic limit is nearly a free field theory. The only known field theories that possess this property in 4 space-time dimensions are nonabelian gauge theories (which can be coupled to a limited number of fermions). Asymptotic freedom offers the hope to reconcile strong coupling at low energies with the scaling behavior seen (Bjorken scaling) in experiments at high energies. The experiments involve deep inelastic electroproduction where a high energy electron is scattered off of a nucleon. The scattering can be thought of as accomplished by the exchange of a virtual photon that is far off shell. This virtual photon is able to probe very short distance scales within the nucleon. The scaling seen is consistent with the picture that the nucleon

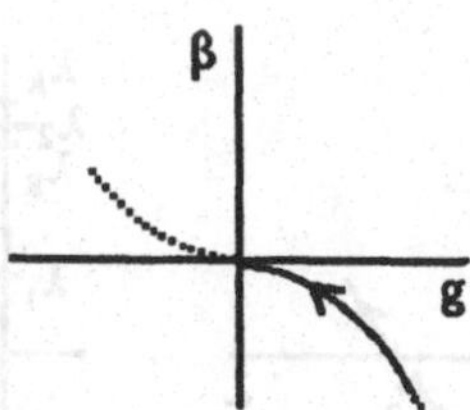

Figure 18.13 β-function for a theory that is "asymptotically free." Here the trivial zero of the β-function at zero coupling is ultraviolet attractive. Perturbation theory is valid at large energies or small distances.

is made up of point-like partons (quarks), and that the partons interact with the incoming virtual photon as if they were free from one another.

So far we have only treated the massless self-interacting scalar field. If the scalar field possesses a bare mass, the renormalization group equations may easily be generalized to include the mass. Once again, we start with eq.(18.130). Now, however, the renormalized mass must also implicitly depend on the subtraction scale, M, thus eq.(18.133) becomes

$$\left(M\frac{\partial}{\partial M} + \beta\frac{\partial}{\partial \lambda} + M\frac{\partial m}{\partial M}\frac{\partial}{\partial m} - n\gamma\right)\tilde{\Gamma}^{(n)}(p_i;\lambda,m,M) = 0. \quad (18.144)$$

From this we may also derive a scaling equation. The mass complicates things in that it is more difficult to solve the equations. By dimensional analysis, each coefficient can depend on λ and m/M. The added dependence on m/M is what slows us down.

The renormalization group equation coefficients $(\beta,\gamma,\ldots)$ are renormalization prescription dependent. There is a renormalization prescription, called the mass independent prescription, invented by 't Hooft and Weinberg, which leads to a β-function and anomalous dimension for the mass and field that are independent of m/M. This prescription essentially says to set the finite parts of the counterterms to zero. Now the renormalization group equation may be solved as before.

At large momenta or scales much larger than the mass, m, we naively expect the theory to act like a massless one. The precise statement of "act like" is contained in the Callan-Symanzik equation. This equation is similar to the renormalization group equation except that it studies the variation of $\tilde{\Gamma}^{(n)}$ with respect to the renormalized mass when the subtraction scale is 0, $M = 0$. For large momenta, the Callan-Symanzik equation becomes identical, in functional form, to eq.(18.133), except that M is replaced by the renormalized mass, m.

$$\left(m\frac{\partial}{\partial m} + \beta(\lambda)\frac{\partial}{\partial \lambda} - n\gamma \right) \Gamma^{(n)}_{\text{asympt}} = 0 \tag{18.145}$$

The analysis now proceeds as in the massless case. The statement that scale invariance should appear in a massive theory at large momentum scales because the masses can be neglected is only true if the β-function develops a fixed point as discussed earlier. Even then, the scaling is not the same as dimensional analysis would predict. The lack of scaling at large momenta indicates that the theory well remembers that there is some renormalized mass involved, because this mass sets the scale of λ and the field. However, the actual value of the mass is nearly forgotten, since any change in it can be compensated by a change in λ and the scale of the field.

18.7 EXERCISES

1. Show that eq.(18.15) follows from eq.(18.1) and the path integral definition of $Z[J]$. The generator of connected Green's functions, $W[J]$, appears in eq.(18.15) instead of $Z[J]$. What is in eq.(18.1) that makes this happen?
2. Draw the first few diagrams in the expansion of $\Sigma(p)$, the 1PI quantum corrections to the proper 2-point function.
3. Obtain an expression for the 4-point proper function, $\Gamma^{(4)}$, by differentiating eq.(18.28) once more. Invert the expression to write the 4-point Green's function in terms of the proper vertices. Draw a diagram to represent the resulting expression for $G^{(4)}$.
4. Higgs mechanism: Consider a charged scalar field with a quadratic and a quartic potential coupled to an Abelian gauge theory (the Abelian Higgs model). The Lagrangian density is

$$\mathcal{L} = -\frac{1}{4}F^{\mu\nu}F_{\mu\nu} + (\partial_\mu - ieA_\mu)\varphi^*(\partial^\mu + ieA^\mu)\varphi - \mu^2|\varphi|^2 - \lambda|\varphi|^4.$$

   Take $\mu^2 < 0$ and $\lambda > 0$. Show that $\mathcal{L}$ is invariant under the local transformation, $\varphi \to \exp(i\alpha(x))\varphi$. Find the expectation of $|\varphi|$ in the ground state. Note that this expectation is independent of x. Write $\varphi = (\varphi_1 + i\varphi_2)/\sqrt{2}$, where φ_1 and φ_2 are real-valued, and spontaneously break the symmetry by making a choice for the ground state expectation of $|\varphi|$ where $\langle\varphi_2\rangle = 0$ ($\theta = 0$ in figure 18.6). Shift the field φ_1 to φ_1' so that $\langle\varphi_1'\rangle = 0$, and rewrite $\mathcal{L}$ in terms of φ_1' and φ_2. Verify that the mass term φ_2^2 has disappeared suggesting that φ_2 is a Goldstone boson as expected. Also identify a mass-like term for the gauge potential, A^μ. Now for the important step. Show that the φ_2 field can be completely gauged away. That is, show that the φ_2 terms can be absorbed in a redefinition of A^μ that amounts to a gauge transformation. This means that the Goldstone boson has disappeared. Note that the resulting Lagrangian implies that the gauge field has acquired a longitudinal polarization (a mass). What is this mass and what is the mass of the Higgs boson, φ_1'? Does the mass term for the gauge field break the gauge symmetry? Count the number of degrees of freedom in the original Lagrangian and compare with the final Lagrangian. This mechanism also works for nonabelian gauge fields.

5. Using the results in sections 2, 3, or 4, determine $\phi_c(x)$ as a functional of $J(x)$ to order λ. Draw a diagrammatic representation of the result and compare with figure 18.1.

6. Extract the renormalized 2-point and 4-point Green's functions at 1-loop for the field $\varphi(x)$ from the effective potential.

II. Strings

CHAPTER **19**

Basic Ideas and Classical String Theory

19.1 OVERVIEW

The quantum theory in which the fundamental constituents are one-dimensional extended objects is called string theory. Strings have a variety of fundamental applications in theoretical physics. Presently, the foremost application of string theory is to the unification of all fundamental forces and elementary particles. Here the basic idea is that all matter is made up of very tiny strings, so that while a particle such as an electron looks like a point at 10^{-13} cm, if we magnify it to scales resolving 10^{-33} cm, then we will see that it is extended like a string (figure 19.1). Unification is achieved in that all particles are made of a single kind of string. The different types of particles we see are just different excitations of the same string. If we excite the string one way we get an electron, another way we get a photon, etc. There is no obvious *a priori* compelling reason to switch from point particles to nearly invisible strings. However, one mode of oscillation of the string is a massless, spin 2 state that can be identified as the graviton. Thus, string theory necessarily contains quantum gravity. This alone would not be an improvement over point particle versions of quantum gravity except that it is possible to define a perturbative expansion of string theory around a fixed, classical background space-time that is finite and anomaly-free at 1-loop (with all indications that this holds at

Figure 19.1 Application of string theory to fundamental particles and forces. All matter is made up of strings. The strings are very tiny, so that (a) at "low" energies all particles look like points and ordinary point particle theories provide an adequate description. (b) Only when we magnify to scales resolving 10^{-33} cm do the particles appear to be extended like a string.

higher loops), thus avoiding the severe difficulties that plague the point particle theories.

There are several arguments of intuitive appeal as to why strings yield improved ultraviolet behavior. The first compares the interaction of 2 point particles to form a single particle with the string equivalent, where 2 strings join to form a single string. The two are represented diagrammatically in figure 19.2. When a point particle moves through space-time, it produces a 1-dimensional world-line. When a string travels through space-time, it sweeps out a world-sheet. Therefore, the string equivalent to the 3-particle diagram is a surface.

When two point particles interact on contact, the interaction necessarily occurs at one space-time point. This point of interaction, the vertex, is a Lorentz invariant notion since it looks the same in different Lorentz frames. However, when two strings join, the point of contact or interaction is no longer Lorentz invariant. The space-time point of interaction will appear to be different to two observers in different frames. In figure 19.2, the person in the frame where the axes are labeled 1 will observe that the two strings made contact at the point labeled 1 on the surface. The person in frame 2, however, will observe the point of contact as point 2 on the surface. In some sense, the point of interaction has been fuzzed out in the string diagram. The divergences arise in point particle theories when 2 vertices get arbitrarily close to one another. In the string case we do not have the notion of two sharp vertices approaching one another, since a single vertex cannot be localized. The purpose of regularizing point particle theories to make them finite (with, say, an ultraviolet momentum cutoff) is just to make the vertex fuzzy. Since the string diagram is a surface, string theory "naturally" accomplishes this.

Another reason point particle theories have difficulty with gravity is that the gravitational coupling constant, Newton's constant, has a positive dimension ($(\text{length})^2$ when $\hbar = c = 1$). This makes the quantized theory unrenormalizable. L-loop diagrams with differing numbers of vertices can

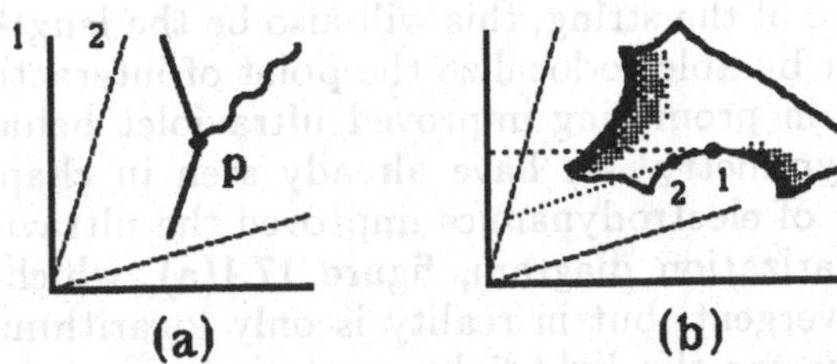

Figure 19.2 (a) Contact interactions of point particles necessarily occur at one space-time point, p. This point of interaction, the vertex, is a Lorentz invariant notion, since it looks the same in different frames. (b) World-sheet representing the joining of two strings to form a single string. Since the world-sheet is a surface, the point of interaction is no longer Lorentz invariant. An observer in frame 1 will see the interaction occur at point 1 on the surface, while the observer in frame 2 will see the interaction occur at point 2.

no longer be added together to cancel the nonrenormalizable behavior, because the different diagrams no longer have the same dimensions. A similar situation arises in the weak interaction where the old, effective 4-fermion theory of Fermi is nonrenormalizable because the coupling has positive dimension. The 4-fermion vertex is replaced by the exchange of the massive gauge bosons, $W^{\pm}$ and Z^0, in the Weinberg-Salam model. The coupling becomes dimensionless and the result is improved ultraviolet behavior and the theory becomes renormalizable.

String theory modifies point particle gravity at short distances by the exchange of massive string states. In addition, the string introduces a new coupling constant, the string tension, T. The string tension T has units of $(length)^{-2}$ when $\hbar = c = 1$, and so introduces a characteristic length squared, L^2. In ordinary units, this length is conveniently defined as $L = \sqrt{c\hbar/\pi T}$. Since string theory is a theory of quantum gravity, the characteristic length in the theory must be on the order of the Planck length, $L_p = \sqrt{G\hbar/c^3}$, the only length that can be constructed from the gravitational constant, $\hbar$, and the speed of light. For the string, however, $\hbar = c = 1$ are not *natural* units. When $\hbar = c = 1$, mass has the units of inverse length. As we shall see, reparametrization invariance on the world-sheet of the string will want to treat length and momentum (i.e. mass when $c = 1$) on the same footing, i.e. the same units. The natural choice of units for the string is $c = 1$ and for the string tension to be dimensionless, $T = 1/\pi$. In these units, $\hbar$ will have the dimensions of length squared and the gravitational constant, $G = (L_p/L)^2$, will be *dimensionless.* When we start to resolve near the characteristic length scale, the stringiness of string theory will become apparent, and the exchange of massive string states will become important. Since $L \sim L_p$ is also the characteristic size of the string, this will also be the length scale at which we will no longer be able to localize the point of interaction.

Another aid in promoting improved ultraviolet behavior in a quantum theory is symmetry. We have already seen in chapter 17 how the gauge symmetry of electrodynamics improved the ultraviolet behavior of the vacuum polarization diagram, figure 17.1(a), which is superficially quadratically divergent, but in reality is only logarithmically divergent. The same is true for the light-light scattering diagram, figure 17.1(b), which is superficially logarithmically divergent, but turns out to be finite. In general, the more symmetry we add, the better the ultraviolet behavior. For example, the incorporation of supersymmetry into point particle gravity made the (supergravity) theory less divergent. The supersymmetry allows for the cancellation of bad behavior by superpartner fermions and bosons. String theory contains yet more symmetry, so we can expect the possibility of ultraviolet convergence.

As if a consistent quantum gravity theory wasn't enough motivation to pursue strings, they also necessarily accommodate Yang-Mills-type interactions. The allowable unifying groups are determined by mathematical consistency, instead of being selected by hand. Of the possibilities, there are several that contain $SU(3) \times SU(2) \times U(1)$ of the standard model (electroweak + QCD). This means that string theory has the potential to describe all fundamental particles and forces presently known.

As the string propagates through space-time, it sweeps out a surface, its world-sheet. The result of first quantizing the string is an ordinary 2-dimensional point particle quantum field theory on the surface. The classical (super)string action is invariant under (super)conformal transformations on the world-sheet. Consistent quantization requires that the resulting 2-dimensional point particle field theory on the world-sheet retain this symmetry. Such field theories are called (super)conformal field theories. First quantized string theory, therefore, consists of the study of conformal field theories on Riemann surfaces. The applications of conformal field theory go beyond strings. Any 2-dimensional statistical system near a second-order phase transition should be describable by a conformally invariant theory. Near such a phase transition, fluctuations of the fields are correlated over long distances and occur equally on all scales. The system loses memory of the underlying lattice so that it may be described by a continuum field theory. Since no scale is preferred, it becomes scale-invariant, i.e. conformally invariant. Since the underlying lattice becomes unimportant, many systems may exhibit the same universal behavior, even though the lattices may be quite different. The infinitesimal 2-dimensional conformal transformations are generated by the centrally extended Virasoro algebra. The coefficient of the central extension, c, characterizes the particular realization of the conformal symmetry that statistical system possesses,

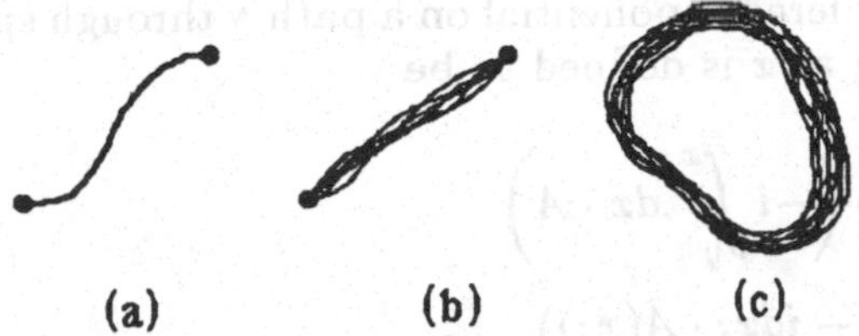

Figure 19.3 (a) Original motivation for the development of string theory. Here 2 quarks are bound together by a string to form a meson. (b) QCD equivalent. Quarks are bound by exchanges of gluons. The gluons themselves are charged, hence the color electric flux lines attract each other to form a narrow flux tube as the quarks are separated. (c) A ring of color electric flux forming a pure glu state (a gluball). In each case, the flux tubes look like strings of finite thickness.

and, therefore, labels the universality classes (but not uniquely). The conformal symmetry is sufficiently large in 2 dimensions that knowledge of the particular realization of the symmetry determines the scaling dimensions of the fields and scaling operators, and the critical exponents.

String theories were originally formulated in the late '60s and early '70s to describe hadron physics. In this case, the string is used to connect quarks together to form the bound quark states that are hadrons (figure 19.3(a)). The typical length of this string is 10^{-13} cm. The abundance of anomalies in 4 space-time dimensions, the persistence of massless states in the string spectrum, and the invention of QCD led to the abandonment of this approach. However, the string picture must still be qualitatively correct. Within QCD, quark bound states are held together by gluons, or flux lines of color electric fields. Since they are thought to confine, and because the color electric flux lines attract each other, a narrow flux tube of glu must connect the bound quarks when their separation grows larger than about 10^{-13} cm (figure 19.3(b)). Similarly, states of pure glu, such as a gluball, form rings of electric flux since individual gluons are also confined (figure 19.3(c)). In each case, the flux tubes look like strings of finite thickness. However, one can argue that a string theory is still the appropriate and desirable description for hadron physics, i.e. gauge fields.

First of all, a proof of confinement is lacking and appears to be very difficult to do in the usual vector potential representation of gauge theory. One criterion signaling confinement has been provided by K. Wilson. He has shown that if the vacuum expectation of the trace of a closed loop path-ordered exponential of the vector potential goes like the exponential of the area of the loop, then the theory confines. Such path-ordered exponentials of the vector potential on closed loops are called Wilson loops.

The path-ordered exponential on a path γ through space-time starting at y and ending at x is defined to be

$$\begin{aligned}\mathcal{W}[\gamma] &= P\exp\left(-i\int_y^x dx\cdot A\right)\\ &\equiv \prod_j (1 - idx_j\cdot A(x_j))\\ &= 1 - i\int_y^x dx_1\cdot A(x_1) - \frac{1}{2}\int_y^x dx_2\int_y^{x_2} dx_1\, A(x_2)A(x_1) + \ldots,\end{aligned} \tag{19.1}$$

where all integrals are taken along the path γ, and where dx_j is tangent to the path γ at the point x_j on the path. Under a gauge transformation, U, the path-ordered exponential transforms as

$$\mathcal{W}[\gamma] \longrightarrow U(x)\mathcal{W}[\gamma]U^\dagger(y). \tag{19.2}$$

When the path γ forms a closed loop so that $y = x$, then the trace is gauge invariant and is known as the Wilson loop, $W[\gamma] = \operatorname{tr}\mathcal{W}[\gamma]$.

When we work in the vector potential representation of the gauge theory, the Wilson loop is an infinite product or, equivalently, an infinite series, which makes computations very difficult. A more natural description of a gauge theory is in terms of Wilson loops only. If the Wilson loop is taken as the fundamental object, then we will not have to face the infinite series representation of it, eq.(19.1). Since they are gauge invariant, in principle, they contain all that is necessary to fully describe the quantum theory. A Wilson loop, however, is nothing more than a closed string (of glu) of infinitesimal width.

There are more reasons for wanting to formulate Yang-Mills or QCD as a string theory of Wilson loops than just a convenient representation for proving confinement. Polyakov has argued that gauge fields are analogous to chiral fields (sigma models) on loops. Chiral fields are completely integrable in 2 dimensions and the hope is that this remains so when they are defined on a loop space. In fact, Polyakov has shown that on the classical level in 3-dimensional Yang-Mills, there exists a set of conserved currents in loop space to make it integrable. If this is true in 4 dimensions at the quantum level, then an exact solution for Yang-Mills in this representation may be found.

Even if this is not so, it has been shown that the large N limit of an $SU(N)$ gauge theory is equivalent to some string theory. The solutions to this string theory could be used to approximately solve Yang-Mills or QCD in a $1/N$ expansion.

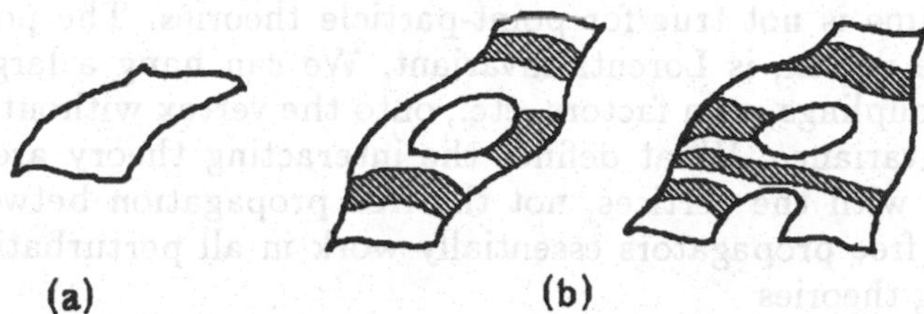

Figure 19.4 (a) Surface or world-sheet representing the free propagation of a string. (b) String diagrams describing the interaction of several strings. On the surface we cannot locally tell the difference between a surface describing a free string (as in (a)) and a surface describing interacting strings. The local, shaded regions on the surfaces in (b) are all topologically identical to (a).

The type of string theory described above is in a "noncritical" space-time dimension where the conformal anomaly is present. So far, the construction of such a theory has been elusive. We will take up this topic in more detail in chapter 25 on noncritical dimensions.

Finally, let us mention that string theory will find an application any time a problem can be formulated as a sum over random surfaces (such as the 3-dimensional Ising model). The propagation of a quantum mechanical particle can be represented as a path integral that sums over all of the particle's possible world-lines connecting the initial and final points. A string sweeps out a world-sheet as it moves through space-time, thus the quantum mechanical propagation of a string can be represented by a sum over all possible world-sheets connecting the initial and final string configurations. The sum over all world-sheets is a sum over random surfaces, hence any sum over surfaces can be interpreted as the propagation of some string.

It is difficult to construct consistent relativistic quantum theories of extended objects. A reflection of this fact is that we know of only a few consistent string theories. String theory is almost unique. Interestingly, the near-uniqueness of string theory arises because the theory of *free* strings is nearly unique, and not because of interactions. String interactions are in a sense almost trivial (perturbatively), since they really only dictate the topology of the string diagram. The string diagram remains a surface no matter how many interactions occur, and *locally* on the surface, we cannot tell the difference between a surface describing the free propagation of a string and a surface describing the splitting and joining of several strings. This is illustrated in figure 19.4. Equivalently, as we have already discussed in the context of figure 19.2, we are unable to pinpoint where a string interaction occurred, because it will look different in two different Lorentz frames.

The same is not true for point-particle theories. The point of interaction, the vertex, is Lorentz invariant. We can hang a large variety of different couplings, spin factors, etc., onto the vertex without spoiling the Lorentz invariance. What defines the interacting theory are the factors associated with the vertices, not the free propagation between vertices. The same free propagators essentially work in all perturbatively defined interacting theories.

In contrast to point particle theories, we cannot construct a consistent string theory in any arbitrary space-time dimension. For a given string theory, there is a maximum allowable space-time dimension, called the critical dimension, above which it appears impossible to make the theory consistent. The difficulty here is the conformal anomaly. Consistent string theories are conformally invariant on the string world-sheet. There is no known method to cancel or eliminate the conformal anomaly above the critical dimension.

The critical dimension is essentially determined by the number of local supersymmetries in the theory that reside on the string world-sheet. (Supersymmetry is a symmetry between particles of differing statistics or spin. It unites fermions and bosons into one group representation. Supersymmetry was discovered theoretically in string theory by Gervais and Sakita in the early '70s. It was applied to point particle theories by Wess and Zumino a couple of years later.) Conformally invariant string actions can be constructed when there are $N = 0$, 1, 2, and 4 local 2-dimensional supersymmetries on the string world-sheet. The critical dimension for no supersymmetry ($N = 0$) is $D = 26$. For 1 supersymmetry, it is $D = 10$, while $N = 2$ requires $D = 4$, and $N = 4$ implies $D = -2$. Clearly $N = 4$ is out. $N = 2$ is apparently unphysical since two of the four dimensions must be time-like. Therefore, we need only consider $N = 0$ and 1.

Strings can be either open (topologically equivalent to the unit interval) with free ends, or closed (topologically equivalent to the circle). In addition, they may be oriented or unoriented. Strings interact by joining and splitting at just one point on the string to preserve causality. Open string theories necessarily include closed strings, because the interaction of 2 strings joining ends to form one string also allows a single string to self-interact and join both of its free ends to form a closed string. The converse is not true. Closed string theories need not include open strings.

Let us consider open string theories first. For an open string, left-moving modes must be identical to right-moving modes since a left-moving mode turns into a right-mover upon reflection at a free end. For $N = 0$, the string is purely bosonic and is called the "Veneziano model." The critical dimension is $D = 26$. This theory is not consistent in the critical dimension in that it contains a tachyon. The theory also is not finite.

Gauge interactions can be added to open strings by attaching group "charges" (factors that transform under the desired group) or Chan-Paton factors to the "ends" of the string. Ultimately this means that group theoretic factors multiply the scattering amplitudes of an open string theory without the charges. $SO(n)$ and $Sp(n)$ are possible gauge groups when the string is unoriented, while $U(n)$ is possible for oriented strings.

The addition of gauge charges does not make the open bosonic string consistent. However, this is not true when supersymmetry is present. The open superstring theory with $N = 1$ world-sheet supersymmetry is called "type I superstring theory." The critical dimension is 10 and the strings are unoriented. There is a consistent truncation of the spectrum that produces space-time supersymmetry as well. The tachyon, which lacks a space-time superpartner, is eliminated by the truncation. Group charges may be added to the ends of the string as with the purely bosonic string. Classically, the same groups are allowed. However, with the addition of fermions, chiral anomalies may appear rendering the theory mathematically inconsistent. There is a unique, anomaly-free, "theory" in the critical dimension when the gauge group is taken to be $SO(32)$.

Moving on to closed string theories, the left- and right-moving modes no longer need to be identical. There are no free ends to add gauge charges to, so gauge interactions must be added in a different way. The gauge charge is essentially distributed along the string by compactification. We will return to this in a moment.

If both left- and right-moving modes lack supersymmetry, then the string is purely bosonic and $D = 26$. The same problems are present (tachyon, etc.) as with the open bosonic string. If both left- and right-moving modes possess $N = 1$ local world-sheet supersymmetry, then the superstring theory is called "type II." The critical dimension is 10 and the strings are oriented. In the critical dimension there is no gauge group (because there are no compactified dimensions) and there are two space-time supersymmetries. In the type IIa theory, the space-time supersymmetries are of opposite chirality, while in a type IIb theory they are of the same chirality. The lack of a gauge group in $D = 10$ means that the "theory" is trivially chiral anomaly-free (and less interesting).

When the right-moving modes are world-sheet supersymmetric, but the left-moving modes are not, the string theory is called heterotic. In order to obtain a "theory" in the critical dimension of the right-moving modes ($D = 10$), 16 of the 26 bosonic degrees of freedom of the left-moving modes must be compactified. The compactified degrees of freedom can be interpreted as internal symmetries of the theory, and result in introducing a gauge interaction. Tachyon-free, consistent "theories" result when the compactification is done in such a way that the gauge group is $E_8 \times E_8$, spin(32) / Z_2, or $SO(16) \times SO(16)$. The compactification that results in the first two gauge groups also produces space-time supersymmetry, while the $SO(16)$ "theory" does not.

You will notice above that quotes sometimes appear on the word "theory." This is because the word "theory" is often used in place of "semi-classical expansion." There are only three known superstring theories: type I, type II, and heterotic. Presently, the only way these quantum theories have been defined is by perturbative or semi-classical (loop) expansions about classical solutions. The expansions only make sense if the classical background solution is chosen correctly. If the background solution resides in $D = 10$ Minkowski space-time, then there is only one known classical solution for type I superstring theory that produces a mathematically consistent semi-classical expansion, and this classical solution has $SO(32)$ as its gauge group. However, we often refer to this solution as "$SO(32)$ type I superstring theory," when it is really an expansion about a particular classical solution. When the word "theory" appears above in quotes, it really means "semi-classical expansion about a classical solution of the theory." "Anomaly-free theories" refer to mathematically consistent expansions, or classical background solutions which produce consistent expansions. Similarly, if the background space-time is 10-dimensional Minkowski space, then we know of only three classical solutions that produce consistent, anomaly-free expansions. The solutions are labeled by the gauge groups they require: $E_8 \times E_8$, spin(32) / Z_2, and $SO(16) \times SO(16)$. From now on we will often substitute "theory" for a particular semi-classical expansion.

While string theory is almost unique, the solutions to the theory are certainly not. In order to describe the observed particles and forces with theories formulated in the critical dimension, we must compactify 6 of the remaining 10 dimensions. If we insist that there be an unbroken $N = 1$ supersymmetry on the 4 noncompact dimensions, then the compactified 6-dimensional manifold must be a Calabi-Yau space or a 6-dimensional orbifold.

In arriving at 4 space-time dimensions in this way using the heterotic string, we first compactify 16 bosonic degrees of freedom, and then compactify 6 of the superstring degrees of freedom (i.e. 6 left-movers and 6 right-movers). However, we could just as well start with the 26 left-movers and 10 right-movers and compactify 22 left-movers and 6 right-movers all at once to arrive at $D = 4$. It is possible to carry out the compactification in such a way that many new gauge symmetries are possible that lead to consistent expansions. It is even possible to cleverly compactify the type II string to get realistic gauge groups (e.g. $SU(3) \times SU(2) \times U(1)$) in 4 dimensions. It may even be possible to obtain consistent expansions from the purely bosonic theory by compactification.

We know many consistent semi-classical expansions of the three superstring theories, some of which give reasonable phenomenology. However, we have no way of knowing which expansion, if any, corresponds to the true vacuum. In defining a perturbative expansion, we must make a choice for what the background metric on space-time will be. String theory is suppose to be a theory of quantum gravity, so the propagation of the string through space-time should determine what the space-time metric is. We don't really want the space-time metric to appear at all in the formulation of the theory. We want it to be something we calculate from the theory.

Having to make a choice for the background metric of space-time means that general coordinate invariance is spontaneously broken. In addition, the desirable background solutions all have unbroken space-time supersymmetry so that all of the terms in the perturbative expansion will be finite. The mass scale for the string theory is the mass associated with the natural length scale. As we discussed earlier, the length scale must be on the order of the Planck length, thus the mass scale is on the order of the Planck mass, $M_p = \sqrt{\hbar c/G}$, which is very large. The only string states that we will be able to observe at ordinary energies are the particles corresponding to the massless string states. Any massive state will have a mass at least on the order of M_p. If supersymmetry is unbroken, then the massless states are truly (identically) massless. We know that all observable particles are not massless, and that the world at ordinary energies is not supersymmetric. Therefore, it is impossible to do true phenomenology from string perturbation theory alone. (When we mentioned that some candidate vacuum states gave reasonable phenomenology, the supersymmetry breaking effects had to be added by hand.) The choice of vacuum state and symmetry breaking effects must be nonperturbative effects. Presently, we have no nonperturbative formulation of string theory. *All* we really know about string theory we have learned from perturbative expansions. However, we really don't know what we are expanding. We do not yet know the fundamental principles upon which string theory is based. In other words, the perturbative expansions define a first quantized string, but what we really need to understand the necessary symmetry breaking is a second quantized theory, a string field theory. If string theories correctly describe reality, then an explanation of why the cosmological constant is zero or nearly zero after supersymmetry breaking presumably resides in the second quantized theory.

There is no guarantee that the string perturbation series is valid since the expansion parameter, which is not a *free* parameter, is not necessarily small. Even though each term is finite, the series can still diverge. String theory may be strongly coupled, but there is still hope that a nonperturbative solution may be found. Two-dimensional field theories enjoy special properties that are absent or rarely seen in higher-dimensional theories. We met one of those properties in chapter 2 in discussing the Nonlinear Schrödinger model. This theory is integrable. It possesses an infinite set of conserved currents, so an exact solution could be found. Strings are a clever way of elevating the special properties of 2-dimensional field theories into higher dimensions. Since string theories contain a huge amount of symmetry, integrability may survive the transition. This suggestion has been made by Friedan and Shenker in their nonperturbative formulation of string theory.

Finally, superstring theories apparently contain no adjustable dimensionless parameters. If nature is stringy, then $\hbar$ is *not* a fundamental constant of nature, an observation first made by Veneziano. Recall that strings introduce a new coupling, the string tension, T. Natural units for the string do not have $\hbar = 1$, but have $T = 1/\pi$. The string tension, $\hbar$, and the speed of light can be combined to form a length. However, $\hbar$ and T always appear together as a ratio and never by themselves. This means that all we have done by introducing T is to change $\hbar$ into a length. Therefore, neither T nor $\hbar$ is a fundamental dimensionful constant of nature. The length associated with them is. Therefore, string theories require only two constants of nature: c, the speed of light, for relativity, and a length, L, for quantization. Why is a length necessary for quantization? As we are about to see, the string action is the area of the world-sheet swept out by the string. In order to quantize this action via a path integral, we need to change exp(action) into a phase. Usually we do this by dividing the action by $\hbar$, since the units of action and $\hbar$ are identical. In order to accomplish this for strings we needed to divide by $\hbar$ *and* multiply by the string tension, T. We could accomplish the same thing just by dividing by L^2. There is no need to introduce $\hbar$ and T.

Why is not the same thing true for point particles? The action for a point particle is the length of its world-line. Why not just divide by L to quantize and forget about $\hbar$? Point particle theories contain several mass parameters, and therefore, adjustable dimensionless parameters (e.g. the ratios of the mass parameters). If we introduce a length L, the one mass scale will be fixed, but the mass ratios still explicitly appear and are not fixed. In this case we need more fundamental constants. Because of its units, it is convenient to introduce $\hbar$ for quantization, and let $\hbar$, c, and the cutoff fix the parameters.

In the following chapters we will concentrate on the first quantization of the bosonic string. Even though it is not a consistent theory, it certainly provides a natural laboratory upon which to develop string perturbation theory and the techniques used. Many of the difficulties encountered in quantizing superstrings also arise in the bosonic theory.

We will begin by quantizing in several gauges in the operator and Schrödinger representations. The main emphasis, however, will be the path integral quantization as pioneered by Polyakov. At the very end we will briefly touch upon superstrings. Before we begin quantization, though, let us examine the classical bosonic string action.

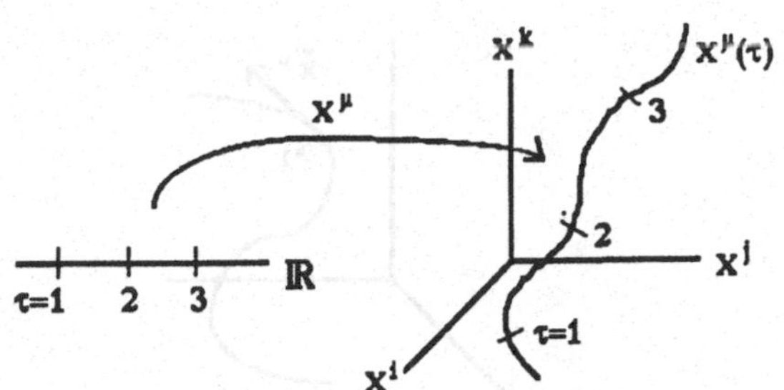

Figure 19.5 Trajectory or world-line of a particle through space-time. x^μ is a mapping from the real line into space-time. The trajectory is parametrized by τ.

19.2 CLASSICAL STRING THEORY

In order to motivate the form of the classical string action, let us first look at the action of a relativistic point particle.

The trajectory of a particle through space-time is its world-line. Let $x^\mu(\tau)$ denote the position of the world-line in space-time. See figure 19.5. $\dot{x}^\mu = \partial_\tau x^\mu$ is a vector tangent to the world-line (at $\tau = \tau_1$ in figure 19.6). It represents the velocity of the particle with respect to the coordinate τ that parametrizes the world-line.

τ is not a measure of proper time for the particle unless the tangent vector is of unit length.

$$\dot{x}^\mu \dot{x}_\mu = 1 \quad \Longleftrightarrow \quad \tau \text{ is proper time.} \tag{19.3}$$

When the world-line is parametrized by the particle's proper time, then the momentum of the particle is $p^\mu = m\dot{x}^\mu$. m is the rest mass of the particle, since $p^\mu p_\mu = p^2 = m^2 \dot{x}^\mu \dot{x}_\mu = m^2$, by virtue of eq.(19.3). If τ is some parametrization other than proper time, we can still express the momentum of the particle in terms of the tangent vector, $\dot{x}^\mu$, simply as

$$p^\mu = \frac{m\dot{x}^\mu}{\sqrt{(\dot{x}^\nu)^2}}. \tag{19.4}$$

Note that eq.(19.4) will satisfy the mass shell condition, $p^2 - m^2 = 0$. Eq.(19.4) represents the equation of motion of a relativistic particle of momentum p^μ and rest mass m.

In an action, $S = \int d\tau\, L(\dot{x}^\mu, x^\mu, \tau)$, where the momentum is conjugate to x^μ, the Lagrangian will satisfy $p^\mu = -\partial L/\partial \dot{x}_\mu$. This is simple to integrate,

$$L = -p_\mu \dot{x}^\mu. \tag{19.5}$$

In order that this action (via Hamilton's principle) generate the same world-lines as the integral curves or solution of eq.(19.4), the momentum

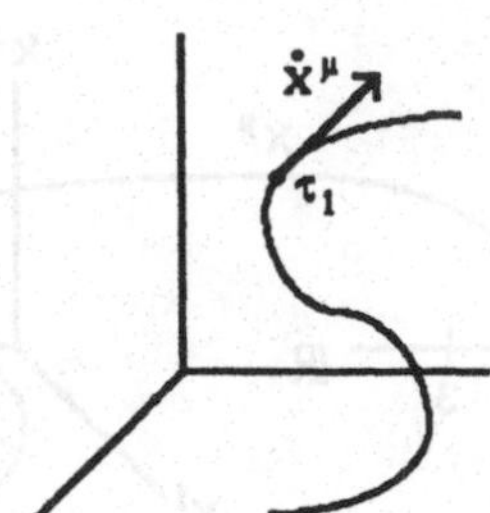

Figure 19.6 $\dot{x}^\mu(\tau_1)$ is a vector tangent to the world-line at space-time point $x^\mu(\tau_1)$.

in eq.(19.5) must satisfy the mass shell condition. Clearly, if we substitute in eq.(19.4) for p_μ above, then this will be true. Therefore, the action for a relativistic particle of rest mass m is

$$S = -m\int d\tau\,\sqrt{\dot{x}_\mu \dot{x}^\mu}. \tag{19.6}$$

Recall that τ is any parametrization of the world-line, and is not necessarily proper time. Any parametrization will do, since the action is reparametrization invariant. To see this, change parametrization in eq.(19.6) from τ to τ'. Let $x^{\mu\prime} = \partial_{\tau'} x^\mu$, and $\dot{x}^\mu = \partial_\tau x^\mu$. By the chain rule we have $\dot{x}^\mu = (d\tau'/d\tau)x^{\mu\prime}$ and $d\tau = (d\tau/d\tau')d\tau'$, hence

$$d\tau\,\sqrt{\dot{x}^\mu \dot{x}_\mu} = d\tau'\left(\frac{d\tau}{d\tau'}\right)\sqrt{\dot{x}^\mu \dot{x}_\mu} = d\tau'\,\sqrt{x^{\mu\prime} x'_\mu}.$$

The action, S, is reparametrization invariant because it represents a geometrical object. In terms of the metric on space-time, the distance squared, $(ds)^2$, separating neighboring points is

$$(ds)^2 = g_{\mu\nu}\,dx^\mu\,dx^\nu. \tag{19.7}$$

If we explicitly introduce the metric into the integrand of eq.(19.6), we have

$$\begin{aligned} d\tau\,\sqrt{\dot{x}^\mu \dot{x}_\mu} &= d\tau\sqrt{g_{\mu\nu}\dot{x}^\mu \dot{x}^\nu} \\ &= d\tau\sqrt{g_{\mu\nu}\frac{dx^\mu}{d\tau}\frac{dx^\nu}{d\tau}} \\ &= \sqrt{g_{\mu\nu}\,dx^\mu\,dx^\nu} \\ &= ds. \end{aligned} \tag{19.8}$$

Thus, the action, eq.(19.6), is the rest mass times the length of the world-line. The classical path connecting 2 space-time points is the world-line of extremal length.

The action written in the form eq.(19.6) may not be the most convenient form to use since it involves a square root. This is especially true when we consider quantization or the massless limit, $m \to 0$. We may get rid of the square root and arrive at a classically equivalent action that is quadratic in $\dot{x}^\mu$ at the cost of adding another function of the parametrization, τ. To find this new form of the action, we return to the original action written explicitly in terms of the momentum, eq.(19.5). The form (19.5) was unsatisfactory because the momentum appearing in it satisfied the mass shell constraint, $p^2 = m^2$. In order to satisfy the constraint, we substituted in eq.(19.4). However, we may keep the form eq.(19.4) and satisfy the constraint by adding a Lagrange multiplier.

$$L = -p_\mu \dot{x}^\mu + \frac{1}{2}\lambda(\tau)(p^2 - m^2) \tag{19.9}$$

To arrive at the desired form, we eliminate p_μ. Since the classical trajectory will be the minimum of the action, we can determine p_μ in terms of $\dot{x}_\mu$ and λ through

$$\frac{\delta L}{\delta p_\mu} = 0 = -\dot{x}^\mu + \lambda p^\mu \quad \text{or} \quad p^\mu = \frac{\dot{x}^\mu}{\lambda}, \tag{19.10}$$

which is just the equation of motion for p_μ. Substituting eq.(19.10) back into eq.(19.9) and into the action, we obtain

$$S = -\frac{1}{2}\int d\tau \, \frac{1}{\lambda(\tau)}(\dot{x}^\mu \dot{x}_\mu) + \lambda(\tau)\, m^2. \tag{19.11}$$

Note that the Lagrange multiplier function, $\lambda(\tau)$, like p_μ, is nondynamical, since $\dot{\lambda} = d\lambda/d\tau$ does not appear in the action.

Eq.(19.11) is classically equivalent to eq.(19.6). To see this, we determine λ in terms of $\dot{x}^\mu$ by solving its equation of motion, $\delta S/\delta\lambda = 0$. The result is then substituted back into eq.(19.11).

As we are about to see, the classical bosonic string action may also be written in two equivalent forms similar to eqs.(19.6) and (19.11) above. The string version of eq.(19.6), which involves a square root, is called

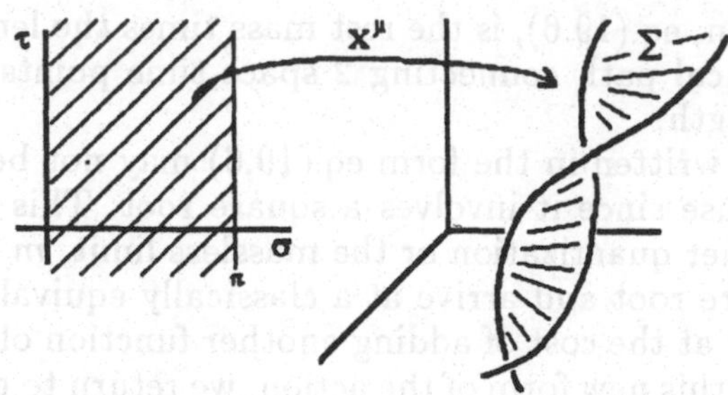

Figure 19.7 The $x^\mu(\sigma, \tau)$ space-time coordinates of the world-sheet of the string, Σ, form a mapping of the 2-dimensional string parameter space (σ, τ) into space-time. σ, which denotes position along the string, traditionally runs from zero to π.

the Nambu-Goto action, while the Lagrange multiplier string version of eq.(19.11) is called the Polyakov action.

Let σ denote position along the string. Traditionally, the range of σ is taken to be $0 \leq \sigma \leq \pi$. Let τ parametrize the time evolution of the string, and let $x^\mu(\sigma, \tau)$ denote where the string world-sheet, Σ, sits in space-time. The x^μ space-time coordinates of the world-sheet form a mapping of the 2-dimensional string parameter space (σ, τ) into space-time as illustrated in figure 19.7.

The action for a relativistic point particle is proportional to the length of its world-line. The string action is the natural higher-dimensional generalization of this result. Namely, the action for a relativistic string is proportional to the area of its world-sheet.

In order to get an explicit expression for the action, let's start by deriving an integral expression for the area of a curved surface or world-sheet Σ that is endowed with a metric. For notational convenience, let ξ^0 $(= \tau)$ and ξ^1 $(= \sigma)$ be coordinates on the surface. We may think of the coordinate system as etching a set of lines on the surface, lines of constant ξ^0 and ξ^1 (figure 19.8(a)). The area of the surface is the sum of the areas of tiny parallelograms formed by vectors tangent to lines of constant ξ^0 and constant ξ^1 (figures 19.8(b) and (c)). $d\xi^0$ and $d\xi^1$ represent displacements in coordinates, not actual distances. The metric on the surface, g_{ab}, tells us how to convert coordinate displacement into actual distances. If ds is the actual distance corresponding to coordinate displacements $(d\xi^0, d\xi^1)$, then $(ds)^2 = g_{ab}\, d\xi^a\, d\xi^b$, for $a, b = 0, 1$. Thus, the actual length of coordinate displacement $d\xi^0$ is

$$||d\xi^0|| = \sqrt{g_{00}}\, d\xi^0, \tag{19.12}$$

and that for $d\xi^1$ is

$$||d\xi^1|| = \sqrt{g_{11}}\, d\xi^1. \tag{19.13}$$

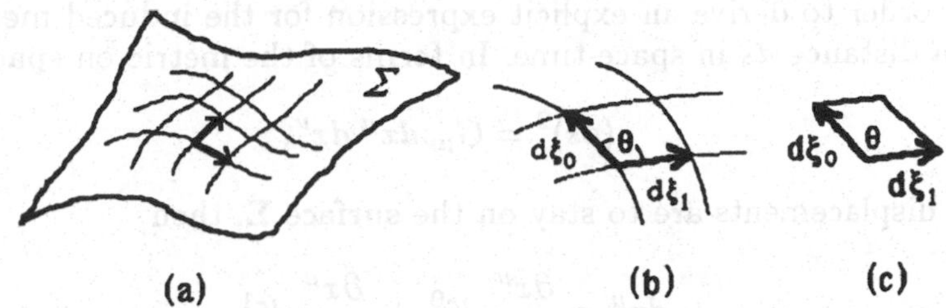

Figure 19.8 (a) A coordinate system (ξ^0, ξ^1) can be thought of as etching a set of lines on the surface Σ, lines of constant ξ^0 and ξ^1. (b) At each point on the surface there are 2 displacement vectors, ∂_{ξ^0} and ∂_{ξ^1}, tangent to the lines of constant ξ^0 and ξ^1. (c) The area of the surface Σ is the sum of the areas of tiny parallelograms formed by the tangent vectors.

The area of the little parallelogram in figure 19.8(c) is

$$\begin{aligned}
||d\xi^0 \times d\xi^1|| &= ||d\xi^0||\,||d\xi^1||\sin(\theta) \\
&= \sqrt{||d\xi^0||^2\,||d\xi^1||^2(1-\cos^2(\theta))} \\
&= \sqrt{||d\xi^0||^2\,||d\xi^1||^2 - (||d\xi^0\cdot d\xi^1||)^2} \qquad (19.14)\\
&= \sqrt{|g_{00}g_{11} - (g_{01})^2|}\, d\xi^0\, d\xi^1 \\
&= \sqrt{|\det g|}\, d\xi^0\, d\xi^1
\end{aligned}$$

where we used the fact that the metric is a symmetric tensor in the last step. Often $|\det g|$ is just written as $|g|$. Thus, the area of the surface, the sum of the areas of the parallelograms, is

$$\text{Area}(\Sigma) = \int d\xi^0\, d\xi^1\, \sqrt{|g|} = \int d\sigma\, d\tau \sqrt{-g}. \qquad (19.15)$$

The minus sign appears at the end of eq.(19.15), because we are assuming σ is space-like, and τ is time-like. This leads to an indefinite metric on the surface whose det is less than zero. Thus, $-g > 0$. If we rotate to Euclidean space (time $\rightarrow$ i time), the minus sign will disappear.

Next, we must get an explicit expression for the metric of the world-sheet Σ that sits in space-time, figure 19.7. Space-time itself has a metric, $G_{\mu\nu}$ (Greek indices will refer to space-time, Latin indices refer to the surface). The fact that this world-sheet sits in space-time means that we can use the space-time metric to measure distances on the surface. That is, mapping the surface into space-time means that the surface picks up a metric, the so-called induced metric. It is the induced metric that belongs in eq.(19.15) above.

In order to derive an explicit expression for the induced metric, consider a distance ds in space-time. In terms of the metric on space-time,

$$(ds)^2 = G_{\mu\nu} dx^\mu dx^\nu . \tag{19.16}$$

If the displacements are to stay on the surface Σ, then

$$dx^\mu = \frac{\partial x^\mu}{\partial \xi^0} d\xi^0 + \frac{\partial x^\mu}{\partial \xi^1} d\xi^1 . \tag{19.17}$$

However, if the displacement stays on the surface, then we may also write eq.(19.16) in terms of the induced metric on the surface,

$$(ds)^2 = h_{ab}\, d\xi^a\, d\xi^b . \tag{19.18}$$

If we plug eq.(19.17) back into eq.(19.16), and compare the result with eq.(19.18), then we find that the induced metric is

$$h_{ab} = G_{\mu\nu} \frac{\partial x^\mu}{\partial \xi^a} \frac{\partial x^\nu}{\partial \xi^b} . \tag{19.19}$$

Substituting eq.(19.19) into eq.(19.15), and returning to our original (σ, τ) notation, we find that the action for the string is

$$S = -T \int d\sigma\, d\tau \sqrt{(\dot{x}^\mu \cdot x'_\mu)^2 - (\dot{x}^\mu)^2 (x'_\nu)^2} , \tag{19.20}$$

where

$$\dot{x}^\mu = \frac{\partial x^\mu}{\partial \tau} \quad \text{and} \quad x'_\mu = \frac{\partial x_\mu}{\partial \sigma} , \tag{19.21}$$

and where the constant of proportionality, $-T$, is minus the string tension. As we mentioned earlier, the string tension is introduced to give the expression (19.20) the units of action. Eq.(19.20) is called the Nambu-Goto action. Compare it to the point particle analog, eq.(19.6). Being a geometrical object (an area), eq.(19.20) is reparametrization invariant, i.e. invariant under coordinate transformations on the world-sheet, $\sigma' = \sigma'(\sigma, \tau)$, $\tau' = \tau'(\sigma, \tau)$.

The action S is sometimes inconvenient to work with because of the square root. This is especially so when we contemplate "path" integral quantization. There exists another form of the action that is classically equivalent but is just quadratic in the derivatives of x^μ. This form is called the Polyakov action, and is

$$S = -\frac{T}{2} \int d\sigma\, d\tau \sqrt{-g}\, g^{ab}\, \partial_a x^\mu \partial_b x_\mu . \tag{19.22}$$

Here g_{ab} is treated as an independent variable, and therefore, is the intrinsic metric on the surface Σ, and not the induced metric. g_{ab} is identified as the induced metric, h_{ab}, eq.(19.19), only as the solution to the classical equation of motion for g_{ab}.

One way to arrive at eq.(19.22) is to start from eqs.(19.20) and (19.19), and examine the equation of motion. For a metric h_{ab}, the inverse is

$$h^{ab} = \frac{-1}{-h}\begin{pmatrix} h_{11} & -h_{01} \\ -h_{10} & h_{00} \end{pmatrix}. \tag{19.23}$$

Since $-h = h_{01}h_{10} - h_{00}h_{11}$, the variation of the determinant is

$$\delta(-\det h_{ab}) = h_{10}\delta h_{01} + h_{01}\delta h_{10} - h_{11}\delta h_{00} - h_{00}\delta h_{11}. \tag{19.24}$$

If we compare eq.(19.24) with the components of $(-h)h^{ab}$ in eq.(19.23), we see that eq.(19.24) can be written as

$$\delta(-\det h_{ab}) = \delta(-h) = (-h)h^{ab}\,\delta h_{ab}. \tag{19.25}$$

Therefore, the variation of the action, eq.(19.20) is

$$\begin{aligned} \delta S &= -T\int d\sigma\, d\tau\, \delta(\sqrt{-h}) \\ &= -\frac{T}{2}\int d\sigma\, d\tau\, \delta(\sqrt{-h})h^{ab}\,\delta h_{ab}. \end{aligned} \tag{19.26}$$

The variation in h_{ab} is due to a variation in δx^μ,

$$\delta h_{ab} = \partial_a \delta x^\mu\, \partial_b x_\mu + \partial_a x_\mu\, \partial_b \delta x^\mu. \tag{19.27}$$

When we plug this back into eq.(19.26) and integrate by parts, we find that the equations of motion are

$$\partial_a(\sqrt{-h}\, h^{ab}\, \partial_b x_\mu) = 0. \tag{19.28}$$

Now, if we treat the metric as independent of x^μ (and call it g_{ab} to reflect this), then eq.(19.28) is easy to integrate.

$$\frac{\delta S}{\delta x^\mu} = -T\partial_a(\sqrt{-g}\, g^{ab}\, \partial_b x_\mu) \tag{19.29}$$

implies

$$S = -\frac{T}{2}\int d\sigma\, d\tau \sqrt{-g}\, g^{ab}\, \partial_a x^\mu \partial_b x_\mu,$$

which is the Polyakov action, eq.(19.22).

We know by construction that if $g_{ab} = h_{ab}$, then the Polyakov action will turn back into the Nambu-Goto action. (The step linking eq.(19.22) to eq.(19.20) for $g_{ab} = h_{ab}$ is eq.(19.27).) Thus, to show that the 2 actions are classically equivalent, we need only show that the classical solution of eq.(19.22) for g_{ab} yields h_{ab}. This is easy to do because g_{ab} is nondynamical. The classical equation for g_{ab} is $\delta S/\delta g_{ab} = 0$. This in turn implies that the stress-energy tensor, T_{ab}, vanishes.

$$\begin{aligned} T_{ab} &= -\frac{2}{T}\sqrt{-g}\,\frac{\delta S}{\delta g^{ab}} \\ &= \partial_a x^\mu \partial_b x_\mu - \frac{1}{2} g_{ab} g^{cd} \partial_c x^\mu \partial_d x_\mu \\ &= 0 \end{aligned} \tag{19.30}$$

The solution to eq.(19.30) is $g_{ab} = h_{ab}$, eq.(19.19).

If we compare eq.(19.22) with its point particle analog, eq.(19.11), we can see that g_{ab} plays the role of a Lagrange multiplier. Since the variation of the action eq.(19.22) with respect to g_{ab} implies that the stress-energy tensor vanishes, then g_{ab} is the Lagrange multiplier associated with the constraint, $T_{ab} = 0$.

Since the action eq.(19.22) is still a geometric object, it retains reparametrization invariance. The action is invariant under a differentiable change of coordinates, $\sigma \to \tilde{\sigma}(\sigma, \tau)$, $\tau \to \tilde{\tau}(\sigma, \tau)$. The collection of all differentiable reparametrizations forms a group called the group of diffeomorphisms of the world-sheet. However, this action possesses another symmetry. It is also invariant under Weyl transformations, or conformal rescalings of the metric, $g_{ab}(\sigma, \tau) \to \exp(\phi(\sigma, \tau))\, g_{ab}(\sigma, \tau)$.

Before we move on to first quantizing the actions, eqs.(19.20) and (19.22), let's solve the classical equations for an open string, starting from the Nambu-Goto action. The particular form of the solution is based on seminar notes by F. Thiess.

Consider the effect on the action by an infinitesimal variation of the world-sheet swept out by the string, where the initial string configuration, $x^\mu(\sigma, \tau_1)$, and final string configuration, $x^\mu(\sigma, \tau_2)$, are held fixed. The variation in the action is

$$\delta S = \int_{\tau_1}^{\tau_2} d\tau\, \frac{\partial L}{\partial x'_\mu}\, \delta x^\mu \bigg|_{\sigma=0}^{\sigma=\pi} - \int_{\tau_1}^{\tau_2} d\tau \int_0^\pi d\sigma \left(\frac{\partial}{\partial \tau}\frac{\partial L}{\partial \dot{x}_\mu} + \frac{\partial}{\partial \sigma}\frac{\partial L}{\partial x'_\mu} \right) \delta x^\mu. \tag{19.31}$$

The principle of least action tells us to set this variation to zero. Thus, the equation of motion for the string is

$$\frac{\partial}{\partial \tau}\frac{\partial L}{\partial \dot{x}_\mu} + \frac{\partial}{\partial \sigma}\frac{\partial L}{\partial x'_\mu} = 0, \tag{19.32}$$

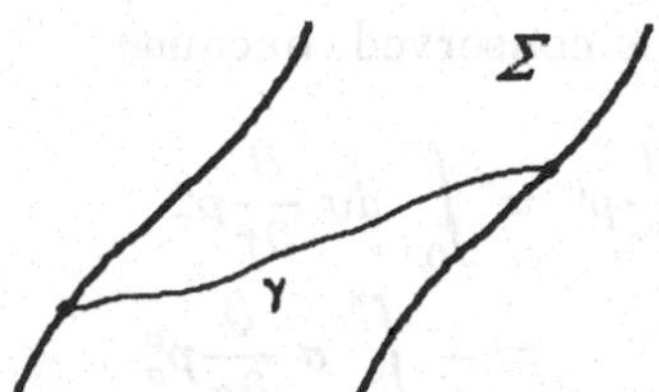

Figure 19.9 γ is a curve that reaches from one edge of the world-sheet, Σ, defined by $\sigma = 0$, to the other edge, $\sigma = \pi$. The total momentum of the string is the integral of the momentum current along any such curve γ.

and the boundary condition at the edges of the string is

$$\left.\frac{\partial L}{\partial x'_\mu}\right|_{\sigma=\pi} - \left.\frac{\partial L}{\partial x'_\mu}\right|_{\sigma=0} = 0. \tag{19.33}$$

The momentum currents conjugate to $\dot{x}_\mu$ and x'_μ are

$$p^\mu_\tau = -\frac{\partial L}{\partial \dot{x}_\mu}, \qquad p^\mu_\sigma = -\frac{\partial L}{\partial x'_\mu}, \tag{19.34}$$

respectively. In terms of the momentum currents, the equations of motion, eq.(19.32), read

$$\frac{\partial}{\partial\tau}p^\mu_\tau + \frac{\partial}{\partial\sigma}p^\mu_\sigma = 0, \tag{19.35}$$

and the boundary condition, eq.(19.33), translates into

$$p^\mu_\sigma(\sigma = \pi) - p^\mu_\sigma(\sigma = 0) = 0. \tag{19.36}$$

This states that the amount of momentum flowing out one end of the string is equal to the amount flowing in the other. For an open string, it is physically reasonable to say that no momentum can flow out of the ends of the string, so that we may take $p^\mu_\sigma(\sigma = \pi) = p^\mu_\sigma(\sigma = 0) = 0$.

The total momentum of the string is the integral of the momentum current along any curve γ that reaches from one edge of the world-sheet, defined by $\sigma = 0$, to the other edge, defined by $\sigma = \pi$ (figure 19.9). Since the net flow out of the ends of the string is zero (eq.(19.36)), then

$$p^\mu = \int_\gamma d\sigma\, p^\mu_\tau + d\tau\, p^\mu_\sigma = \int_0^\pi d\sigma\, p^\mu_\tau. \tag{19.37}$$

The total momentum is conserved, because

$$\begin{aligned}\frac{d}{d\tau}p^{\mu} &= \int_0^{\pi} d\sigma\,\frac{\partial}{\partial\tau}p^{\mu}_{\tau}\\ &= -\int_0^{\pi}\sigma\,\frac{\partial}{\partial\sigma}p^{\mu}_{\sigma}\\ &= p^{\mu}_{\sigma}(\sigma=0) - p^{\mu}_{\sigma}(\sigma=\pi)\\ &= 0.\end{aligned} \tag{19.38}$$

We applied the equations of motion in going from the first line to the second, and the last line follows from eq.(19.36).

In terms of the Nambu-Goto action, eq.(19.20), we have

$$\begin{aligned}p^{\mu}_{\tau} &= T\frac{(\dot{x}\cdot x')x^{\mu\prime} - (x')^2\dot{x}^{\mu}}{\sqrt{(\dot{x}\cdot x')^2 - (x')^2(\dot{x})^2}}\\ p^{\mu}_{\sigma} &= T\frac{(\dot{x}\cdot x')\dot{x}^{\mu} - (\dot{x})^2 x^{\mu\prime}}{\sqrt{(\dot{x}\cdot x')^2 - (x')^2(\dot{x})^2}}.\end{aligned} \tag{19.39}$$

Compare these with the point particle analog, eq.(19.4). The equation of motion, in terms of the derivatives of x^{μ}, follows from eqs.(19.35) and (19.39). From the form of p^{μ}_{τ} and p^{μ}_{σ}, it is simple to see that they satisfy

$$\begin{aligned}p^{\mu}_{\tau}\cdot x'_{\mu} &= 0 \qquad p^{\mu}_{\tau}p_{\tau\mu} + T^2(x')^2 = 0\\ p^{\mu}_{\sigma}\cdot \dot{x}_{\mu} &= 0 \qquad p^{\mu}_{\sigma}p_{\sigma\mu} + T^2(\dot{x})^2 = 0.\end{aligned} \tag{19.40}$$

As in the point particle case, the components of the string momentum are not all independent. Note that if we apply the boundary conditions, $p^{\mu}_{\sigma} = 0$ at $\sigma = 0$ or π, to the last constraint, we have

$$(\dot{x})^2 = 0 \quad \text{at} \quad \sigma = 0 \text{ or } \sigma = \pi. \tag{19.41}$$

This says that the tangent vectors at the edges of the world-sheet Σ formed by the ends of the string are light-like. This means that the ends of the string move at the speed of light.

The equations of motion as determined by eqs.(19.35) and (19.39) are horrible and nearly unsolvable. To get anywhere, we use the reparametrization invariance of the Nambu-Goto action to make a choice of coordinates $(\tilde{\sigma},\tilde{\tau})$ or gauge such that the equations of motion simplify and become tractable.

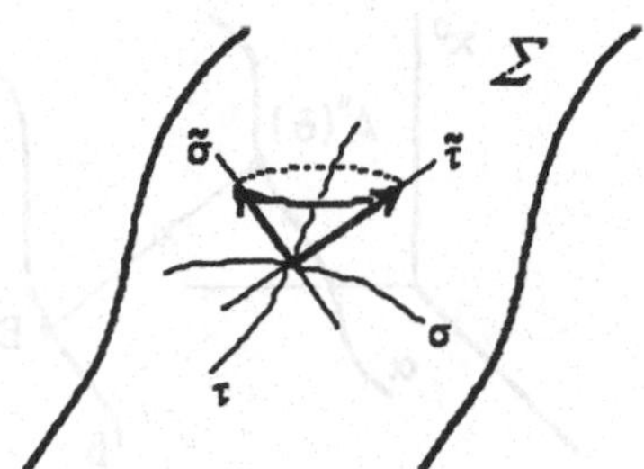

Figure 19.10 World-sheet "light-cone" coordinates $(\tilde{\sigma}, \tilde{\tau})$. At each point (σ, τ) on the surface, there are 2 light-like vectors in the forward direction formed by the intersection of the forward light-cone with the surface. Lines of constant $\tilde{\sigma}$ and $\tilde{\tau}$ are defined to be integral curves of the 2 light-like vector fields.

The particular choice we want to make $(\tilde{\sigma}(\sigma,\tau), \tilde{\tau}(\sigma,\tau))$ is defined as follows: σ is space-like, and τ is time-like, hence there are 2 light-like vectors in the forward direction at each point $x^\mu(\sigma,\tau)$ on the surface Σ. These 2 vectors are defined as tangent to the rays formed by the intersection of the forward light-cone with the surface Σ at $x^\mu(\sigma,\tau)$. This is illustrated in figure 19.10. The coordinate $\tilde{\sigma}$ is defined to be constant along the integral curves of one light-like vector, and the coordinate $\tilde{\tau}$ is defined to be constant along the integral curves of the other light-like vector. Therefore, the lines of constant $\tilde{\sigma}$ and $\tilde{\tau}$ are light-like paths in space-time.

By construction, in this coordinate system,

$$\dot{x}_\mu = \frac{\partial x_\mu}{\partial \tilde{\tau}} \quad \text{and} \quad x'_\mu = \frac{\partial x_\mu}{\partial \tilde{\sigma}}$$

are both light-like, $(\dot{x})^2 = (x')^2 = 0$. In this gauge, the action simplifies considerably. The equation of motion becomes

$$\frac{\partial^2 x^\mu(\tilde{\sigma},\tilde{\tau})}{\partial\tilde{\sigma}\partial\tilde{\tau}} = 0, \tag{19.42}$$

whose general solution is

$$x^\mu(\tilde{\sigma},\tilde{\tau}) = \frac{A^\mu(\tilde{\sigma}) + B^\mu(\tilde{\tau})}{2}. \tag{19.43}$$

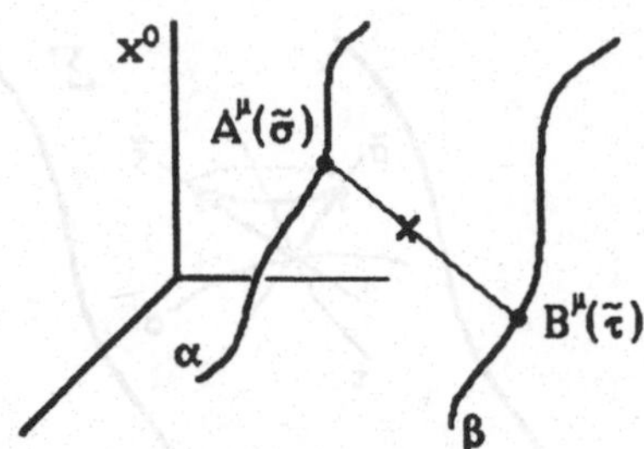

Figure 19.11 Curves $A^\mu(\tilde{\sigma})$ and $B^\mu(\tilde{\tau})$ are light-like trajectories in space-time. Eq. (19.43) implies that the string world-sheet is the locus of mean positions (e.g. marked $\times$) of all pairs of points where one point is from curve α and the other from β.

From this solution we have

$$\dot{x}^\mu = \frac{1}{2}\partial_{\tilde{\tau}} B^\mu \quad \text{and} \quad (x^\mu)' = \frac{1}{2}\partial_{\tilde{\sigma}} A^\mu. \tag{19.44}$$

Since both $\dot{x}^\mu$ and $x^{\mu\prime}$ are light-like vectors, then the curves $A^\mu(\tilde{\sigma})$ and $B^\mu(\tilde{\tau})$ must both be light-like trajectories in space-time. Thus, the string world-sheet is the locus of mean positions of all pairs of points, one point on the curve α, defined as $x^\mu = A^\mu(\tilde{\sigma})$, and the other point on curve β, defined as $x^\mu = B^\mu(\tilde{\tau})$. See figure 19.11.

Let the free ends of the string, at $\sigma = 0$ and $\sigma = \pi$, map out light-like paths $x^\mu = B_0^\mu$, and $x^\mu = B_1^\mu$, respectively. If we map these back into the $(\tilde{\sigma}, \tilde{\tau})$ plane, it will look something like figure 19.12, where $b_0(\tilde{\sigma}) = \tilde{\tau}$ is B_0^μ and $b_1(\tilde{\sigma}) = \tilde{\tau}$ is B_1^μ. The shaded region in between the curves is the part of the $(\tilde{\sigma}, \tilde{\tau})$ plane that is mapped into the world-sheet Σ in space-

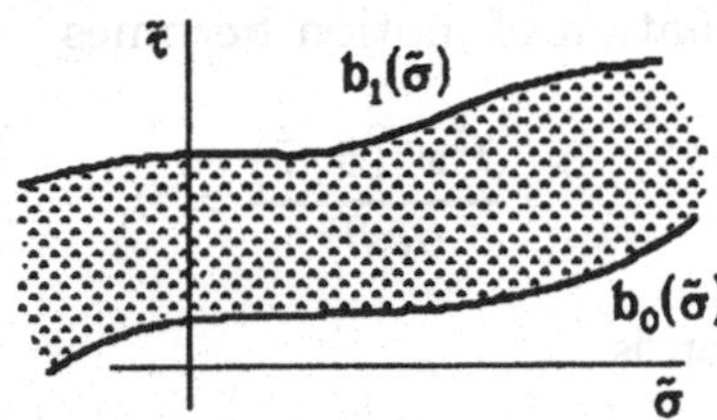

Figure 19.12 Inverse image of the light-like paths, $x^\mu = B_0^\mu$ and $x^\mu = B_1^\mu$, that are the trajectories of the free ends of the string ($\sigma = 0$ and $\sigma = \pi$). The shaded region between $b_0(\tilde{\sigma})$ and $b_1(\tilde{\sigma})$ is the portion of the $(\tilde{\sigma}, \tilde{\tau})$ plane that is mapped into the world-sheet in space-time.

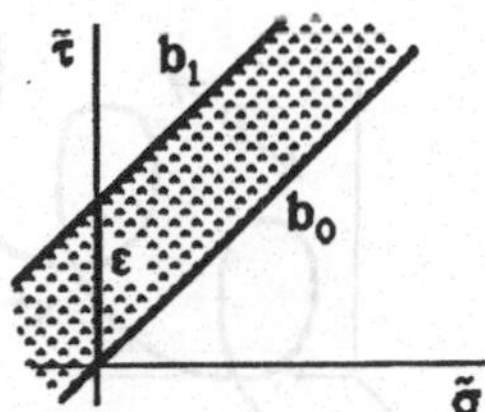

Figure 19.13 Using the residual reparametrization invariance, we can rescale $\tilde{\sigma}$ and $\tilde{\tau}$ such that the curve b_0 in the $(\tilde{\sigma}, \tilde{\tau})$ plane that is mapped onto the light-like trajectory of one free end of the string becomes a straight line of unit slope (i.e. $\sigma = \tilde{\sigma} - \tilde{\tau}$). By eq.(19.46), this means that the curve b_1 that maps into the trajectory of the other free end is also a straight line of unit slope.

time. In this coordinate system, the boundary conditions on the free ends, eq.(19.33), are

$$\dot{x}^\mu \, d\tilde{\tau} = x^{\mu\prime} \, d\tilde{\sigma} \quad \text{or} \quad \partial_{\tilde{\sigma}} A^\mu \, d\tilde{\sigma} = \partial_{\tilde{\tau}} B^\mu \, d\tilde{\tau}. \tag{19.45}$$

This implies that the curves $b_0(\tilde{\sigma})$ and $b_1(\tilde{\sigma})$ that define the edges of the string satisfy

$$\begin{aligned} \partial_{\tilde{\sigma}} A^\mu &= \partial_{\tilde{\tau}} B^\mu \, \frac{d\tilde{\tau}}{d\tilde{\sigma}} \\ &= \partial_{\tilde{\tau}} B^\mu \, b_0'(\tilde{\sigma}) \\ &= \partial_{\tilde{\tau}} B^\mu \, b_1'(\tilde{\sigma}). \end{aligned} \tag{19.46}$$

Given an A^μ and a B^μ, we can use the residual reparametrization invariance to rescale $\tilde{\sigma}$ and $\tilde{\tau}$ so that $\partial_{\tilde{\sigma}} A^0 = 1$ and $\partial_{\tilde{\tau}} B^0 = 1$. If we set $\mu = 0$ in eq.(19.46), this says that $b_0' = b_1' = 1$, so the lines representing the trajectories of the edges of the string are straight lines at 45° in the $(\tilde{\sigma}, \tilde{\tau})$ plane (see figure 19.13). We may also translate $\tilde{\sigma}$ so that one edge of the surface goes through $(\tilde{\sigma} = 0, \tilde{\tau} = 0)$.

However, eq.(19.46) holds for all μ, hence on the line $b_0(\tilde{\sigma})$, we have

$$\partial_{\tilde{\sigma}} A^\mu(\tilde{\sigma}) = \partial_{\tilde{\tau}} B^\mu(\tilde{\tau}) = \partial_{\tilde{\sigma}} B^\mu(\tilde{\sigma}). \tag{19.47}$$

This implies that one curve is a translate of the other,

$$A^\mu(\tilde{\sigma}) - B^\mu(\tilde{\sigma}) = D^\mu. \tag{19.48}$$

Define

$$Q^\mu(\tilde{\sigma}) = B^\mu(\tilde{\sigma}) + \frac{1}{2} D^\mu. \tag{19.49}$$

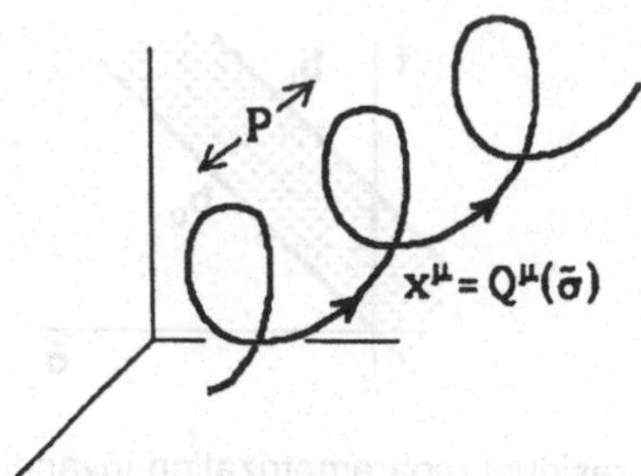

Figure 19.14 Any classical solution of the string equations for an open string can be generated by a single light-like helix, $Q^\mu(\tilde\sigma)$. The periodic vector of the helix, p^μ, is the total momentum of the string. The world-sheet Σ swept out by the string is the locus of mean positions of all pairs of points on the helix.

The general solution, eq.(19.43), can then be written as

$$x^\mu(\tilde\sigma,\tilde\tau)=\frac{Q^\mu(\tilde\sigma)+Q^\mu(\tilde\tau)}{2}. \tag{19.50}$$

Now, $x^{\mu\prime}(\tilde\sigma) = \partial_{\tilde\sigma}Q^\mu(\tilde\sigma)$. If we start at eq.(19.47) again, except apply the analysis to the line $b_1(\tilde\sigma) = \tilde\sigma+\varepsilon$, we get the same result, namely, $x^{\mu\prime}(\tilde\sigma+\varepsilon) = \partial_{\tilde\sigma}Q^\mu(\tilde\sigma) = x^{\mu\prime}(\tilde\sigma)$. This implies that $Q^\mu(\tilde\sigma)$ is a light-like curve in space-time that is periodic,

$$Q^\mu(\tilde\sigma+\varepsilon) = Q^\mu(\tilde\sigma)+p^\mu. \tag{19.51}$$

That is, the curve $Q^\mu(\tilde\sigma)$ is a light-like helix as illustrated in figure 19.14. As you might expect from the choice of notation, the periodic vector of the helix, p^μ, can be shown to be the total momentum, eq.(19.37), of the string.

Therefore, any classical solution to the string equations for an open string is generated by a single light-like helix,

$$x^\mu = Q^\mu(\tilde\sigma), \qquad -\infty<\tilde\sigma<\infty, \tag{19.52}$$

where Q^μ satisfies eq.(19.51), and where p^μ is the total momentum of the string. The world-sheet Σ swept out by the string is the locus of mean positions of all pairs of points on the helix,

$$x^\mu(\tilde\sigma,\tilde\tau)=\frac{Q^\mu(\tilde\sigma)+Q^\mu(\tilde\tau)}{2}. \tag{19.53}$$

19.3 EXERCISES

1. Show that the action eq.(19.11) is classically equivalent to eq.(19.6).
2. Show that the Nambu-Goto action, eq.(19.20), is reparametrization invariant.
3. Show that the Polyakov action, eq.(19.22), is classically equivalent to the Nambu-Goto action, eq.(19.20).
4. Derive eq.(19.39) from the Nambu-Goto action, eq.(19.20), and verify that p^{μ}_{τ} and p^{μ}_{σ} satisfy the constraints, eq.(19.40).
5. Consider a classical string solution generated by a light-like helix, Q^{μ}, with period vector p^{μ}. Show that p^{μ} is also the total momentum of the string.
6. Write down an explicit expression for a light-like helix Q^{μ}. Verify that the x^{μ} generated by it through eq.(19.50) satisfies the original equations of motion, eq.(19.39). Sketch the world-sheet Σ generated by your light-like helix.

CHAPTER **20**

First Quantization

20.1 GAUGE-FIXING AND CONSTRAINTS

When we first quantize a classical point particle system, the positions and momenta of the (fixed number of) particles become operators. The same is true for strings. When we first quantize a string, its string position, $x^\mu(\sigma,\tau)$, and momentum density, $p^\mu_\tau(\sigma,\tau)$, become operators. However, since the string is an extended object, its position depends on two variables. Upon quantization, the string position operator, $x^\mu(\sigma,\tau)$, may also be viewed as a quantized field of some ordinary point particle theory. Thus, the first quantization of a string results in a 2-dimensional second quantized point particle quantum field theory on the world-sheet of the string.

As we have already discussed in the previous chapter, the classical bosonic string action is invariant under diffeomorphisms or reparametrizations of the world-sheet and conformal rescalings of the intrinsic metric on the world-sheet. Since the process of first quantizing this string is equivalent to second quantizing a point particle theory on the world-sheet that possesses gauge symmetries, then we can expect to encounter the same kinds of difficulties in carrying out the quantization as we did with point particle gauge theories in earlier chapters. We can also expect that the same methods and techniques developed for point particle field theories to overcome these difficulties may also be applied to strings.

The analog model for this chapter is chapter 5 on quantization of the free electromagnetic field. The presence of the "gauge" symmetry means that not all of the field components, $x^\mu(\sigma, \tau)$, are dynamical. In addition, we will want to impose equal τ commutators as quantum conditions. An equal τ commutator is not well defined in the presence of reparametrization invariance, since we are free to change τ at will. Therefore, in order to quantize, we will need to choose a "gauge." The choice of gauge will lead to constraints that must be imposed on the quantized system. If the constraints are solved for and imposed as operator equations, then (usually) only the dynamical degrees of freedom will be quantized and the resulting theory will be manifestly unitary. This is what happens when we quantize the string in the light-cone gauge, and is the analog of quantizing the electromagnetic field in the Coulomb gauge. When we quantize in this way, we destroy manifest Lorentz covariance so we must check that the resulting quantized theory is Lorentz invariant. Unlike any point particle analogs, the bosonic string theory will only be Lorentz invariant if, among other things, the dimension of space-time is 26, the critical dimension of the bosonic string.

If we insist on maintaining manifest Lorentz covariance through the quantization process, then we necessarily quantize nondynamical degrees of freedom. The constraints can no longer be implemented as operator equations, because they will conflict with the equal-time commutators. Instead, they must be implemented on the states. In addition, the equal-time commutators will involve the space-time metric. Since the metric is not positive or negative definite, ghost states are introduced which can destroy unitarity. Altogether, this is what happens when we quantize the string in the covariant gauge, and is the analog of quantizing the electromagnetic field in the Lorentz gauge. When the constraints are implemented correctly on the states, then presumably the ghost states decouple and the theory is unitary. However, in string theory this will be true only if the space-time dimension is less than or equal to 26, the critical dimension.

In both gauges, the need to make the quantized theory Lorentz invariant and unitary requires that the string ground state in the critical dimension be a tachyon. This leads to divergences and means that the quantized bosonic string is inconsistent.

As we have already seen, the equations of motion, eqs.(19.35) and (19.39), arising from the Nambu-Goto action, eq.(19.20), are complicated and intractable. In order to proceed, we use the reparametrization invariance to choose a gauge in which the equations are tractable. One possible choice of (σ, τ) leads to

$$\dot{x} \cdot x' = 0 \qquad \text{and} \qquad (x')^2 + (\dot{x})^2 = 0. \tag{20.1}$$

With this choice, the momentum currents simplify considerably,

$$p^{\mu}_{\tau} = T\dot{x}^{\mu}, \quad p^{\mu}_{\sigma} = -Tx^{\mu\prime}. \tag{20.2}$$

The equation of motion, eq.(19.35), turns into the wave equation,

$$\ddot{x}_{\mu} - x''_{\mu} = 0, \tag{20.3}$$

whose general solution,

$$x_{\mu} = A(\sigma - \tau) + B(\sigma + \tau), \tag{20.4}$$

is well known. The choice of gauge is Lorentz covariant, since we are assuming that space-time is Minkowski ($G_{\mu\nu} = \eta_{\mu\nu}$), and since the constraints, eq.(20.1), are invariant.

To interpret the constraints, and to see that it is possible to make such a choice, let's return to the Polyakov action, eq.(19.22), where issues of world-sheet symmetries are more transparent.

The Polyakov action is invariant under 2 world-sheet symmetries: diffeomorphisms of the world-sheet and conformal rescalings of the metric, g_{ab}. Let us use these symmetries to fix the metric as much as possible to make a choice of gauge. The diffeomorphisms or reparametrizations of the world-sheet provide us with a way of fixing 2 degrees of freedom. The Weyl invariance or conformal rescaling adds a third. On the other hand, a 2×2 symmetric tensor has only 3 degrees of freedom, so the symmetries allow us to fully fix g_{ab} to anything we want.

A choice that is manifestly covariant is $g_{ab} = \eta_{ab}$, the 2-dimensional Minkowski metric. With this choice, the Polyakov action becomes

$$S = -\frac{T}{2}\int d\sigma\, d\tau\, (\dot{x}^{\mu})^2 - (x^{\mu\prime})^2, \tag{20.5}$$

which clearly generates the same equation of motion, eq.(20.3).

Recall from the previous chapter that the solution to the equation of motion for g_{ab} is that g_{ab} equals the induced metric, h_{ab}, eq.(19.19).

$$h_{ab} = G_{\mu\nu}\frac{\partial x^{\mu}}{\partial \xi^{a}}\frac{\partial x^{\nu}}{\partial \xi^{b}}$$

If $g_{ab} = \eta_{ab}$, then $g_{\sigma\tau} = g_{\tau\sigma} = 0$. In terms of the induced metric, this is

$$\dot{x}^{\mu}x'_{\mu} = 0. \tag{20.6}$$

In addition, $g_{ab} = \eta_{ab}$ also implies that $g_{\sigma\sigma} = -g_{\tau\tau}$, or

$$(\dot{x}^{\mu})^2 + (x^{\mu\prime})^2 = 0, \tag{20.7}$$

when expressed in terms of the induced metric. This demonstrates that the choice eq.(20.1) is possible.

The equation of motion for g_{ab} generated from the Polyakov action also implies that the stress-energy tensor,

$$T_{ab} = -\frac{2}{T}\frac{1}{\sqrt{-g}}\frac{\delta S}{\delta g^{ab}}, \tag{20.8}$$

must vanish. From eq.(19.27) and eq.(20.8) we have

$$T_{ab} = \partial_a x^\mu \partial_b x_\mu - \frac{1}{2} g_{ab} g^{cd} \partial_c x^\mu \partial_d x_\mu. \tag{20.9}$$

If we use the gauge freedom to "gauge" away g_{ab}, then we have lost eq.(20.8) and $T_{ab} = 0$ as an equation of motion. This means that $T_{ab} = 0$ must be imposed as a constraint. (The same situation arises in electrodynamics when we gauge away A_0. In this case, Gauss's law is lost as an equation of motion and must be imposed as a constraint as we did in chapter 10.) If we plug $g_{ab} = \eta_{ab}$ into eq.(20.9), we find that the constraint $T_{ab} = 0$ is

$$\begin{aligned} T_{01} = T_{10} = \dot{x}^\mu\, x'_\mu = 0 \\ T_{00} = T_{11} = \frac{1}{2}(\dot{x}_\mu^2 + {x'_\mu}^2) = 0. \end{aligned} \tag{20.10}$$

Therefore, if we use the gauge-fixed action, eq.(20.5), then we must impose eq.(20.1) as constraints to insure that the stress-energy tensor vanishes.

In the process of quantization, we must make a choice of how we want to implement the constraints. If we insist on maintaining Lorentz covariance, then the constraints must be implemented as weak constraints on the states to avoid being in conflict with the commutators. In addition, the equal-time commutators, $[x^\mu(\sigma,\tau), p^\nu_\tau(\sigma',\tau)]$, must be proportional to the space-time metric $\eta^{\mu\nu}$, thus negative norm states will be introduced. We will follow the consequences of this path in the latter section of this chapter.

On the other hand, if we give up manifest Lorentz covariance, then we can use the remaining residual gauge freedom to solve the constraints and implement them as operator equations, thereby only quantizing the dynamical degrees of freedom. This choice of gauge is called the light-cone gauge and we will follow this procedure in the next section.

20.2 LIGHT-CONE GAUGE

So far we have used the reparametrization and Weyl invariance to completely specify the intrinsic metric in the Polyakov action, eq.(19.22), into the form $g_{ab} = \eta_{ab}$. The gauge elimination of the metric means that the metric equation of motion, which states that the stress-energy tensor must vanish, now becomes a constraint. In terms of the space-time components of the world-sheet, the constraints are

$$\begin{aligned} \dot{x}^{\mu}\, x'_{\mu} &= 0 \\ \dot{x}^{\mu}\dot{x}_{\mu} + x^{\mu\prime}\, x'_{\mu} &= 0. \end{aligned} \tag{20.11}$$

We will assume that space-time is Minkowski ($G_{\mu\nu} = \eta_{\mu\nu}$). When we apply these gauge constraints to the Nambu-Goto action, we find that the square root disappears. The resulting action in both cases is eq.(20.5).

The equation of motion resulting from this gauge-fixed action is simply the wave equation,

$$\left(\frac{\partial^2}{\partial\sigma^2} - \frac{\partial^2}{\partial\tau^2}\right)x^{\mu} = 0. \tag{20.12}$$

The general solution is

$$x^{\mu}(\sigma,\tau) = x^{\mu}_R(\tau-\sigma) + x^{\mu}_L(\tau+\sigma). \tag{20.13}$$

For open strings, the free boundary conditions are

$$\partial_{\sigma}x^{\mu}(\sigma=0) = \partial_{\sigma}x^{\mu}(\sigma=\pi) = 0, \tag{20.14}$$

while, for closed strings we merely require periodicity,

$$x^{\mu}(\sigma,\tau) = x^{\mu}(\sigma+\pi,\tau). \tag{20.15}$$

Since the equation of motion, eq.(20.12), is linear, we may write the general solution, eq.(20.13), as a superposition of plane wave solutions to eq.(20.12). For the closed string, the periodic solution satisfying eq.(20.15) is

$$\begin{aligned} x^{\mu}_R(\tau-\sigma) &= \frac{1}{2}\bar{x}^{\mu} + \frac{1}{2\pi T}p^{\mu}(\tau-\sigma) + \frac{i}{2}\frac{1}{\sqrt{\pi T}}\sum_{n\neq 0}\frac{1}{n}\alpha^{\mu}_n \exp(-2in(\tau-\sigma)) \\ x^{\mu}_L(\tau+\sigma) &= \frac{1}{2}\bar{x}^{\mu} + \frac{1}{2\pi T}p^{\mu}(\tau+\sigma) + \frac{i}{2}\frac{1}{\sqrt{\pi T}}\sum_{n\neq 0}\frac{1}{n}\tilde{\alpha}^{\mu}_n \exp(-2in(\tau+\sigma)). \end{aligned} \tag{20.16}$$

The factor of $1/n$ that appears above is there for convenience. For the open string, we must satisfy eq.(20.14). This requires us to identify $\widetilde{\alpha}_n^\mu$ with α_n^μ. For this case, we have

$$x^\mu(\sigma,\tau) = \bar{x}^\mu + \frac{1}{\pi T}p^\mu\tau + i\frac{1}{\sqrt{\pi T}}\sum_{n\neq 0}\frac{1}{n}\,\alpha_n^\mu\, e^{-in\tau}\ \cos(n\sigma). \qquad (20.17)$$

In both eq.(20.16) and eq.(20.17), $\bar{x}^\mu$ and p^μ are the centers of position and momentum of the string. Note that in both cases we must have

$$\alpha_{-n}^\mu = \alpha_n^{\mu *}, \qquad \widetilde{\alpha}_{-n}^\mu = \widetilde{\alpha}_n^{\mu *}, \qquad (20.18)$$

in order that x^μ be real.

When we quantize, $x^\mu(\sigma,\tau)$ and its conjugate momentum,

$$p_\tau^\mu = -\frac{\partial L}{\partial \dot{x}_\mu} = T\dot{x}^\mu, \qquad (20.19)$$

will become operators and satisfy canonical quantum equal-τ commutators. This will, in turn, imply that the Fourier coefficients above, α_n^μ and $\widetilde{\alpha}_n^\mu$, will also become operators. The expansion, eqs.(20.16) and (20.17) will provide us with a particle interpretation exactly as it did with free point particle theories in chapters 2 through 5. However, before we go and elevate x^μ and p_τ^μ to operators, we must deal with the constraints. Eqs.(20.16) and (20.17) solve the equations of motion, eq.(20.12), but they do not, as yet, satisfy the constraints, eq.(20.11). In this section, we will implement the constraints as operator equations. The result is quantization in the "light-cone" gauge.

The constraints, eq.(20.11), do not fully fix the gauge. There is residual reparametrization invariance. The constraints tell us that the 2 vectors, $\dot{x}^\mu(\sigma,\tau)$ and $x^{\mu\prime}(\sigma,\tau)$, that are tangent to the string world-sheet at the point (σ,τ) are orthogonal. However, the constraint does not tell us what direction the 2 vectors point. There are lots of orthogonal coordinate systems that can be placed on the world-sheet. A particular choice of orthogonal coordinate system will fully fix the gauge. A convenient choice will allow us to easily solve the constraints, thereby eliminating the nondynamical degrees of freedom.

To motivate the choice, let's examine the residual reparametrization freedom left by the constraints, eq.(20.11). Suppose that (σ,τ) and $(\widetilde{\sigma},\widetilde{\tau})$ are two orthogonal coordinate systems on the world-sheet that satisfy the constraints. How are the two related? By the chain rule we have

$$\begin{aligned}\frac{\partial x^\mu}{\partial \tau} &= \frac{\partial x^\mu}{\partial \widetilde{\tau}}\frac{\partial \widetilde{\tau}}{\partial \tau} + \frac{\partial x^\mu}{\partial \widetilde{\sigma}}\frac{\partial \widetilde{\sigma}}{\partial \tau}\\ \frac{\partial x^\mu}{\partial \sigma} &= \frac{\partial x^\mu}{\partial \widetilde{\tau}}\frac{\partial \widetilde{\tau}}{\partial \sigma} + \frac{\partial x^\mu}{\partial \widetilde{\sigma}}\frac{\partial \widetilde{\sigma}}{\partial \sigma}.\end{aligned} \qquad (20.20)$$

The left-hand side will satisfy the constraints provided that the derivatives of x^μ on the right-hand side do, and that

$$\frac{\partial\widetilde{\sigma}}{\partial\tau}=\frac{\partial\widetilde{\tau}}{\partial\sigma},\qquad\frac{\partial\widetilde{\sigma}}{\partial\sigma}=\frac{\partial\widetilde{\tau}}{\partial\tau}.\tag{20.21}$$

This in turn implies that

$$\frac{\partial}{\partial\tau}\frac{\partial\widetilde{\sigma}}{\partial\tau}=\frac{\partial}{\partial\tau}\frac{\partial\widetilde{\tau}}{\partial\sigma}=\frac{\partial}{\partial\sigma}\frac{\partial\widetilde{\tau}}{\partial\tau}=\frac{\partial}{\partial\sigma}\frac{\partial\widetilde{\sigma}}{\partial\sigma},\tag{20.22}$$

or

$$\left(\frac{\partial^2}{\partial\tau^2}-\frac{\partial^2}{\partial\sigma^2}\right)\widetilde{\sigma}=0.\tag{20.23}$$

Similarly,

$$\left(\frac{\partial^2}{\partial\tau^2}-\frac{\partial^2}{\partial\sigma^2}\right)\widetilde{\tau}=0.\tag{20.24}$$

Notice that $\widetilde{\sigma}(\sigma,\tau)$ and $\widetilde{\tau}(\sigma,\tau)$ satisfy the same equations as $x^\mu(\sigma,\tau)$. This means that we can identify $\widetilde{\sigma}$ or $\widetilde{\tau}$ with one of the x^μ's. That is, we can perform a reparametrization such that one of the coordinates on the world-sheet is identical to one of the space-time coordinates. The choice of which x^μ to identify with τ that leads to the light-cone gauge is as follows:

Let x^0 be the time-like dimension of space-time. For D-dimensional space-time, this leaves $D-1$ spatial dimensions. Define light-cone coordinates x^+ and x^- as

$$\begin{aligned}x^+&=\frac{1}{\sqrt{2}}(x^0+x^{D-1})\\x^-&=\frac{1}{\sqrt{2}}(x^0-x^{D-1}).\end{aligned}\tag{20.25}$$

In these coordinates, the scalar product of two tangent vectors v^μ and w^μ is

$$v\cdot w=\eta_{\mu\nu}v^\mu w^\nu=v^+w^-+v^-w^+-v^iw^i,\tag{20.26}$$

where index i runs from 1 to $D-2$. The light-cone gauge points τ along x^+,

$$x^+=\bar{x}^++\frac{1}{\pi T}p^+\tau.\tag{20.27}$$

With this choice, we can now use the constraints to solve for x^-. Recall that in a general gauge we have the constraint on p^μ_τ, eq.(19.40),

$$p^\mu_\tau\cdot x'_\mu=0.\tag{20.28}$$

Combining this with eq.(20.19) results in the orthonormality constraint. Applying this to x^+ above, $p_\tau^+ = p^+/\pi$ and $x^{+\prime} = 0$, thus

$$\begin{aligned} p_\tau^\mu \cdot x_\mu' &= p_\tau^+ x^{-\prime} + p_\tau^- x^{+\prime} - p_\tau^i x^{i\prime} \\ &= \frac{p^+ x^{-\prime}}{\pi} - p_\tau^i x^{i\prime} \\ &= 0, \end{aligned}$$

or

$$x^{-\prime} = \frac{\pi}{p^+} p_\tau^i x^{i\prime} = \frac{\pi T}{p^+} \dot{x}_i x_i'. \tag{20.29}$$

In addition, we also have

$$p_\tau^\mu p_{\tau\mu} + T^2 (x')^2 = 0, \tag{20.30}$$

which follows from eq.(19.40). From eq.(20.19) we see that this is just the other constraint in eq.(20.11). Eq.(20.30) implies

$$2p_\tau^+ p_\tau^- - p_\tau^i p_\tau^i = -T^2 (2x_+' x_-' - x_i' x_i'),$$

or

$$p_\tau^- = \frac{\pi T^2}{2p^+} \left((\dot{x}^i)^2 (x_i')^2 \right). \tag{20.31}$$

Thus, one of the constraints, eq.(20.28), gives us $x^{-\prime}$, while the other, eq.(20.30), tells us $\dot{x}^-$. We can neatly combine the two constraints, eqs.(20.29) and (20.31), to form

$$\dot{x}^- + x^{-\prime} = \frac{\pi T}{2p^+} (\dot{x}_i + x_i')^2. \tag{20.32}$$

For now, let us concentrate on the open string. We will come back to the closed string in a little bit. Let us substitute the mode expansion for x^μ, eq.(20.17) into eq.(20.32), and solve for α_n^-. Specifically,

$$x^- = \bar{x}^- + \frac{p^-}{\pi T}\tau + \frac{i}{\sqrt{\pi T}} \sum_{n \neq 0} \frac{1}{n} \alpha_n^- \, e^{-in\tau} \cos(n\sigma) \tag{20.33}$$

yields

$$\dot{x}^- + x^{-\prime} = \frac{p^-}{\pi T} + \frac{1}{\sqrt{\pi T}} \sum_{n \neq 0} \alpha_n^- \, e^{-in\tau} \, e^{-in\sigma}. \tag{20.34}$$

We can compactify this expression slightly by defining

$$\alpha_0^- = \frac{p^-}{\sqrt{\pi T}}. \tag{20.35}$$

Then,

$$\dot{x}^- + x^{-\prime} = \frac{1}{\sqrt{\pi T}} \sum_{n=-\infty}^{\infty} \alpha_n^- \, e^{-in\tau} \, e^{-in\sigma}. \tag{20.36}$$

Similarly, if we take

$$\alpha_0^i = \frac{p^i}{\sqrt{\pi T}}, \tag{20.37}$$

then

$$(\dot{x}_i + x_i')^2 = \frac{1}{\pi T} \sum_{m=-\infty}^{\infty} \sum_{l=-\infty}^{\infty} \alpha_m^i \alpha_l^i \, e^{-i(m+l)\tau} \, e^{-i(m+l)\sigma}. \tag{20.38}$$

The combined constraint, eq.(20.32), then requires

$$\alpha_n^- = \frac{\sqrt{\pi T}}{2p^+} \sum_{m=-\infty}^{\infty} \alpha_m^i \alpha_{n-m}^i. \tag{20.39}$$

Thus, the constraints and gauge-fixing have allowed us to specify x^+, and determine α_n^-, i.e. $x^{-\prime}$ and p^-. Therefore, the dynamical degrees of freedom are $\bar{x}^-$, p^+, and α_n^i.

The square mass of the string is $M^2 = p^\mu p_\mu = 2p^+p^- - p^i p^i$. Combining eqs.(20.35), (20.37), and (20.39), we find

$$M^2 = \pi T \left(\sum_{m=-\infty}^{\infty} \alpha_m^i \alpha_{-m}^i - (\alpha_0^i)^2 \right) = \pi T \sum_{m \neq 0} \alpha_m^i \alpha_{-m}^i. \tag{20.40}$$

How about the Hamiltonian? The covariant Lagrangian is given by eq.(20.5).

$$L = -\frac{T}{2} \int d\sigma \, (\dot{x}_\mu)^2 - (x_\mu')^2 \tag{20.41}$$

In the light-cone gauge, we specify x^+ and eliminate x^-. The x^i satisfy the same wave equation as the x^μ in the covariant gauge. Thus the light-cone gauge action is

$$S_{\text{lc}} = \frac{T}{2} \int d\sigma \, d\tau \, (\dot{x}^i)^2 - (x_i')^2. \tag{20.42}$$

The Legendre transform of the light-cone Lagrangian yields the light-cone gauge Hamiltonian.

$$\begin{aligned} H_{\rm lc} &= \frac{T}{2}\int d\sigma(\dot{x}^i)^2 + (x_i')^2 \\ &= \frac{1}{2T}\int d\sigma(p_\tau^i)^2 + T^2(x_i')^2 \end{aligned} \tag{20.43}$$

Using eq.(20.31) and (20.43), we can also express $H_{\rm lc}$ as

$$H_{\rm lc} = \frac{p^+}{\pi T}\int d\sigma\, p_\tau^- = \frac{p^+ p^-}{\pi T}, \tag{20.44}$$

and, in terms of the mode expansion and solution for x^-, eqs.(20.35) and (20.39), the Hamiltonian is

$$H_{\rm lc} = \frac{1}{2}\sum_{m=-\infty}^{\infty} \alpha_m^i \alpha_{-m}^i. \tag{20.45}$$

Now we are finally ready to quantize. The $x^i(\sigma,\tau)$ and $p_\tau^i(\sigma,\tau)$ become operators and satisfy the canonical equal-τ commutators,

$$\begin{aligned} [x^i(\sigma,\tau), p_\tau^j(\sigma',\tau)] &= i\delta^{ij}\delta(\sigma-\sigma') \\ [x^i(\sigma,\tau), x^j(\sigma',\tau)] = [p_\tau^i(\sigma,\tau), p_\tau^j(\sigma',\tau)] &= 0 \\ [\bar{x}^-, p^+] &= -i \\ [\bar{x}^-, x^i] = [\bar{x}^-, p^j] = [p^+, x^i] = [p^+, p^i] &= 0. \end{aligned} \tag{20.46}$$

As we have already mentioned, upon quantization the Fourier coefficients, α_n^i, become operators, just as in the point particle case. In order to determine their commutation relations, we plug the mode expansion for x^i and p_τ^i into eq.(20.46). This results in

$$[\alpha_n^i, \alpha_m^j] = n\delta^{ij}\delta_{n+m}. \tag{20.47}$$

As expected, we can interpret the result of quantization as a collection of quantum harmonic oscillators if we identify

$$\alpha_{-n}^i = (\alpha_n^i)^\dagger, \qquad \text{for} \quad n > 0. \tag{20.48}$$

Note the correspondence with eq.(3.18) for the free scalar field.

Upon quantization, the Hamiltonian suffers ordering ambiguities since α_{-n}^i no longer commutes with α_n^i. To resolve the ambiguity, we normal order the α_n^i's (α_n^i to the right of $(\alpha_n^i)^\dagger$) in the Hamiltonian, and explicitly separate out a divergent constant that represents the sum of the zero-point energies of the oscillators.

$$H = \frac{1}{2}\sum_{m=-\infty}^{\infty} : \alpha_m^i \alpha_{-m}^i : - a \tag{20.49}$$

From eq.(20.47) we have

$$a = -\frac{D-2}{2}\sum_{n=1}^{\infty} n. \tag{20.50}$$

Note that we must do the same thing for the operator α_0^- since $p^+\alpha_0^-$ is proportional to p^+p^-, which is the Hamiltonian.

The string ground state or vacuum, $|0;p\rangle$, is defined by

$$\begin{aligned} \alpha_n^i|0;p\rangle &= 0 \quad \text{for} \quad n > 0 \\ p_{op}^j|0;p\rangle &= p^j|0;p\rangle. \end{aligned} \tag{20.51}$$

We will often write $|0;p\rangle$ as $|0\rangle$ in situations where the overall momentum of the string is unimportant. Excited string states are created by application of the mode creation operators, $(\alpha_n^i)^\dagger = \alpha_{-n}^i$, $n > 0$, to the vacuum state, $|0\rangle$.

The mass squared of each string state is the eigenvalue of the operator M^2 as defined in eq.(20.40).

$$\begin{aligned} M^2 &= 2p^+p^- - p^ip^i \\ &= 2\pi T H - \pi T \alpha_0^i \alpha_o^i \\ &= \pi T\left(\sum_{m\neq 0} : \alpha_m^i \alpha_{-m}^i : -2a\right) \\ &= 2\pi T(N - a), \end{aligned} \tag{20.52}$$

where

$$N = \sum_{m=1}^{\infty} \alpha_{-m}^i \alpha_m^i \tag{20.53}$$

is the usual number operator that counts the number of excitations in a state, N_m, weighted by the oscillator number. The eigenvalues of N are $\sum_{m=1}^{\infty} mN_m$.

Form eq.(20.52), we can immediately see the importance of the constant a. The mass squared of the ground state is $(N = 0)$

$$M^2|0\rangle = -2\pi Ta|0\rangle. \tag{20.54}$$

Values of $a > 0$ lead to disaster since they result in a tachyonic vacuum. Unfortunately, as we shall see, this cannot be avoided.

In the light-cone gauge, we only quantize the dynamical degrees of freedom. There is no possibility of negative normed or ghost states to spoil the probability interpretation. The theory in this gauge is manifestly unitary. However, in order to quantize only the dynamical degrees of freedom, we had to break manifest Lorentz covariance. This means that we must check that the quantized theory is still Lorentz invariant. We must check that quantization has not broken the Lorentz symmetry. What could quantization do? Quantization introduces operator ordering ambiguities in classical expressions which must be resolved. If the ordering is somehow unimportant, then we can be assured that all is well.

One way to check whether a Lorentz anomaly has appeared is to compare the quantized theory with quantization in a different gauge that is manifestly covariant. If the two theories produce identical results, then the Lorentz symmetry is intact in the light-cone gauge, even though it is not manifestly so. Another way to check is to look at the generators of the Lorentz symmetry, which are now operators, and see if they still satisfy the required algebra (commutators).

Before we move on to examine the Lorentz generators in more detail, we can get a hint that Lorentz covariance in the light-cone gauge is a nontrivial requirement that is quite restrictive. First we examine the lowest excited states, the states containing 1 transverse excitation of oscillator number 1,

$$\alpha^i_{-1}|0\rangle.$$

This set of states forms a $D-2$ component vector and is transversely polarized. Under a Lorentz transformation, the new state still only involves the same set of transverse excitations, and no longitudinal excitations are generated. This is just like the photon in the Coulomb gauge in 4 dimensions. The lack of longitudinal modes implies that this must be a massless state if Lorentz invariance is to hold. From eq.(20.52), we can easily see that the mass of this state is

$$M^2 = 2\pi T(1-a).$$

In order to be massless, $a = 1$. If $a = 1$, then the vacuum is a tachyon. In addition, eq.(20.50) indicates that $a = 1$ may restrict the possible dimensions of space-time. In order to know, we need to evaluate the regularized value of the sum,

$$A = \sum_{n=1}^{\infty} n.$$

There are several methods available to determine A_{reg}. One particular choice is called "ζ-function regularization," which analytically continues the ζ-function to define A as $\zeta(-1)$. Another method, using functional directional derivatives, in the Schrödinger representation will be pursued at the end of this section. The result of regularization yields $A_{\text{reg}} = -1/12$ which implies that D must be 26.

The Lorentz generators are

$$M^{\mu\nu} = \int_0^\pi d\sigma\, M_\tau^{\mu\nu} = \int_0^\pi d\sigma\, x^\mu p_\tau^\nu - x^\nu p_\tau^\mu. \tag{20.55}$$

$M_\tau^{\mu\nu}$ is the τ component of the conserved current associated with the Lorentz symmetry of the string action. It may be found by performing an infinitesimal Lorentz transformation, $\delta x^\mu = \delta\Lambda^{\mu\nu} x_\nu$, in the action and examining the piece that is first order in $\delta\Lambda^{\mu\nu}$. Since this is a symmetry operation, $\delta S = 0$, so the current is conserved. $M^{\mu\nu}$ is the "charge" associated with the current. Classically, the generators satisfy the Lorentz algebra,

$$[M^{\mu\nu}, M^{\sigma\rho}] = i\eta^{\nu\sigma} M^{\mu\rho} - i\eta^{\mu\sigma} M^{\nu\rho} - i\eta^{\nu\rho} M^{\mu\sigma} + i\eta^{\mu\rho} M^{\nu\sigma}. \tag{20.56}$$

In the covariant gauge, $M^{\mu\nu}$ is

$$M^{\mu\nu} = T \int_0^\pi d\sigma\, x^\mu \dot{x}^\nu - x^\nu \dot{x}^\mu. \tag{20.57}$$

When we plug in the mode expansion, eq.(20.17), this turns into

$$M^{\mu\nu} = \bar{x}^\mu p^\nu - \bar{x}^\nu p^\mu - i \sum_{n=1}^\infty \frac{1}{n}(\alpha_{-n}^\mu \alpha_n^\nu - \alpha_{-n}^\nu \alpha_n^\mu). \tag{20.58}$$

When we switch to the light-cone gauge, the α_n^+'s go away. α_n^- is given by eq.(20.39), and nothing funny happens to the α_n^i's. Since the transverse α_n^i's are left alone, we do not expect to have any trouble with the M^{ij} generators. Indeed, no operator ordering ambiguities arise after quantization, because M^{ij} is antisymmetric and the α_n^i's that appear in conjunction in M^{ij} commute with one another.

However, M^{-i} is another matter. M^{-i} does suffer ordering ambiguities because α_n^-, eq.(20.39), is quadratic in the α_n^i's, making M^{-i} cubic in the α_n^i's. The transverse α's that appear in conjunction in M^{-i} no longer commute with one another so they must be normal ordered. The normal ordering, which involves introducing a, eq.(20.50), has the potential to spoil the commutation relations. It is not surprising that M^{-i} may cause trouble, because it is a generator of Lorentz transformations that affect x^+. Any rotation of the direction of x^+ will take us out of the light-cone gauge. To restore the gauge, we must reparametrize after the rotation. The net result of this combination is that an infinitesimal Lorentz transformation acts nonlinearly on x^+ in the light-cone gauge, opening the way for potential ordering anomalies.

From eq.(20.56), we can see that, classically, the M^{-i} generators should commute,

$$[M^{-i}, M^{-j}] = 0. \tag{20.59}$$

However, a straightforward, but somewhat lengthy calculation shows that, after quantization,

$$[M^{-i}, M^{-j}] = \frac{2}{(p^+)^2} \sum_{m=1}^{\infty} \left[m\left(1 - \frac{1}{24}(D-2)\right) + \frac{1}{m}\left(\frac{1}{24}(D-2) - a\right)\right] (\alpha^i_{-m}\alpha^j_m - \alpha^j_{-m}\alpha^i_m). \tag{20.60}$$

In order to maintain Lorentz invariance, the right-hand side of eq.(20.60) must vanish. Therefore, the critical space-time dimension is 26 and $a = 1$.

The computation leading to eq.(20.60) is aided with the commutator for α^-_n.

$$[\alpha^-_n, \alpha^-_m] = \frac{\sqrt{\pi T}}{p^+}(n-m)\alpha^-_{n+m} + \frac{1}{(p^+)^2}\left(\frac{D-2}{12}(n^3 - n) + 2an\right)\delta_{n+m} \tag{20.61}$$

This algebra is known as the Virasoro algebra with a central extension (the δ_{n+m} term). We will examine this algebra in a little more detail in the next section.

Before returning to the closed string, let's summarize the lowest lying states of the open string. The vacuum state, $|0\rangle$, is a tachyon with $M^2 = -2\pi T$. The only massless state is the spin 1 vector $\alpha^i_{-1}|0\rangle$ with 24 polarizations. The first states with positive mass, $M^2 = 2\pi T$, are $\alpha^i_{-2}|0\rangle$ and $\alpha^i_{-1}\alpha^j_{-1}|0\rangle$, which number 324 in all. Together they form something that looks like spin 2. At the next level we have $M^2 = 4\pi T$, which can be obtained by $\alpha^i_{-3}|0\rangle$, $\alpha^i_{-2}\alpha^j_{-1}|0\rangle$, and $\alpha^i_{-1}\alpha^j_{-1}\alpha^k_{-1}|0\rangle$. There are a grand total of 3200 possibilities! As you can see, there are many particles. But, remember that the mass scale is near the Planck mass; thus, in a tachyon-free theory, only the massless states can be expect to be seen.

The analysis of the closed string follows along nearly the same lines as the open string. The mode expansions for the left- and right-movers are given by eq.(20.16). In the light-cone gauge, x^+ still satisfies eq.(20.27). This implies that

$$\alpha^+_n = \tilde{\alpha}^+_n = 0, \tag{20.62}$$

for $n \neq 0$. The combined constraint equation (20.32) is still valid, of course. Notice that we could have subtracted eq.(20.29) from eq.(20.31) to obtain

$$\dot{x}^- - x^{-\prime} = \frac{\pi T}{2p^+}(\dot{x}_i - x_i')^2. \tag{20.63}$$

We will need both eqs.(20.32) and (20.63) to determine x^-. In fact, for

$$x^- = \bar{x}^- + \frac{1}{\pi T}p^-\tau + \frac{i}{2}\frac{1}{\sqrt{\pi T}}\sum_{n\neq 0}\frac{1}{n}\left(\alpha_n^- e^{-2in(\tau-\sigma)} + \widetilde{\alpha}_n^- e^{-2in(\tau+\sigma)}\right), \tag{20.64}$$

we get

$$\begin{aligned} \dot{x}^- + x^{-\prime} &= \frac{1}{\pi T}p^- + \frac{2}{\sqrt{\pi T}}\sum_{n\neq 0}\widetilde{\alpha}_n^- e^{-2in(\tau+\sigma)} \\ &= \frac{\alpha_0^-}{\sqrt{\pi T}} - \frac{\widetilde{\alpha}_0^-}{\sqrt{\pi T}} + \frac{2}{\sqrt{\pi T}}\sum_{n=-\infty}^{\infty}\widetilde{\alpha}_n^- e^{-2in(\tau+\sigma)}, \end{aligned} \tag{20.65}$$

and

$$\dot{x}^- - x^{-\prime} = \frac{\widetilde{\alpha}_0^-}{\sqrt{\pi T}} - \frac{\alpha_0^-}{\sqrt{\pi T}} + \frac{2}{\sqrt{\pi T}}\sum_{n=-\infty}^{\infty}\alpha_n^- e^{-2in(\tau-\sigma)}, \tag{20.66}$$

where we have defined

$$\alpha_0^- + \widetilde{\alpha}_0^- = \frac{p^-}{\sqrt{\pi T}}. \tag{20.67}$$

Plugging these into the 2 constraint equations (20.32) and (20.63), we find the following solutions:

$$\alpha_0^- = \widetilde{\alpha}_o^-,$$

$$\widetilde{\alpha}_l^- = \frac{\sqrt{\pi T}}{p^+}\sum_{n=-\infty}^{\infty}\widetilde{\alpha}_n^i\,\widetilde{\alpha}_{l-n}^i, \qquad \alpha_l^- = \frac{\sqrt{\pi T}}{p^+}\sum_{n=-\infty}^{\infty}\alpha_n^i\alpha_{l-n}^i. \tag{20.68}$$

Therefore, the Hamiltonian is

$$H = \frac{p^+p^-}{\pi T} = \sum_{n=-\infty}^{\infty}\alpha_n^i\alpha_{-n}^i + \widetilde{\alpha}_n^i\widetilde{\alpha}_{-n}^i, \tag{20.69}$$

and the square mass is

$$M^2 = p_\mu p^\mu = 2p^+p^- - p^ip^i$$
$$= 4\pi T \sum_{n=1}^{\infty} \alpha^i_n \alpha^i_{-n} + \widetilde{\alpha}^i_n \widetilde{\alpha}^i_{-n}. \qquad (20.70)$$

When we quantize, the commutators, eq.(20.46), imply

$$[\alpha^i_n, \alpha^j_m] = [\widetilde{\alpha}^i_n, \widetilde{\alpha}^j_m] = n\delta^{ij}\delta_{n+m}$$
$$[\alpha^i_n, \widetilde{\alpha}^j_m] = 0, \qquad (20.71)$$

so that α^i_{-n} and $\widetilde{\alpha}^i_{-n}$ for $n > 0$ act as creation operators, and α^i_n and $\widetilde{\alpha}^i_n$ act as destruction operators. Quantization introduces operator ordering ambiguities in α^-_0, $\widetilde{\alpha}^-_0$, H, and M^2, so we normal order just as in the open string analysis.

$$\alpha^-_0 = \frac{\sqrt{\pi T}}{p^+} \sum_{n=-\infty}^{\infty} : \alpha^i_n \alpha^i_{-n} : -a$$
$$\widetilde{\alpha}^-_0 = \frac{\sqrt{\pi T}}{p^+} \sum_{n=-\infty}^{\infty} : \widetilde{\alpha}^i_n \widetilde{\alpha}^i_{-n} : -a \qquad (20.72)$$
$$H = \sum_{n=-\infty}^{\infty} : \alpha^i_n \alpha^i_{-n} : + : \widetilde{\alpha}^i_n \widetilde{\alpha}^i_{-n} : -2a$$
$$M^2 = 4\pi T(N + \widetilde{N} - 2a),$$

where

$$N = \sum_{m=1}^{\infty} \alpha^i_{-m}\alpha^i_m, \qquad \widetilde{N} = \sum_{m=1}^{\infty} \widetilde{\alpha}^i_{-m}\widetilde{\alpha}^i_m. \qquad (20.73)$$

As in the open string theory, the requirement that the light-cone gauge theory be Lorentz invariant is that $a = 1$ and $D = 26$. Thus, the closed string vacuum state, $|\,0\rangle$, that satisfies

$$\alpha^i_n|\,0\rangle = \widetilde{\alpha}^i_n|\,0\rangle = 0, \qquad n > 0, \qquad (20.74)$$

is a scalar tachyon of square mass $M^2 = -8\pi T$. The solution to the constraint equations, eq.(20.68), requires that $N = \widetilde{N}$, so states $\alpha^i_{-n}|\,0\rangle$., $\widetilde{\alpha}^i_{-n}|\,0\rangle$ are forbidden. The next allowable state has 1 right-moving excitation, balanced by 1 left-moving excitation,

$$\alpha^i_{-1}\widetilde{\alpha}^j_{-1}|\,0\rangle. \qquad (20.75)$$

There are $(24)^2$ different states, and all are massless, $M^2 = 0$. From this collection, we can make a state that is the symmetric and traceless combination,

$$\left(\frac{1}{2}(\alpha^i_{-1}\widetilde{\alpha}^j_{-1} + \alpha^j_{-1}\widetilde{\alpha}^i_{-1}) - \alpha^i_{-1}\widetilde{\alpha}^i_{-1}\right)|0\rangle. \tag{20.76}$$

This state transforms as a massless spin 2 particle and is identified with the graviton. The scalar trace state,

$$\alpha^i_{-1}\widetilde{\alpha}^i_{-1}|0\rangle, \tag{20.77}$$

is called the dilaton, and finally there is an antisymmetric combination,

$$(\alpha^i_{-1}\widetilde{\alpha}^j_{-1} - \alpha^j_{-1}\widetilde{\alpha}^i_{-1})|0\rangle. \tag{20.78}$$

The remaining states in the spectrum of the closed string are massive, $M^2 > 0$.

So far we have not mentioned whether the open or closed strings are oriented or not. They are, in fact, oriented because unoriented strings obey an additional symmetry, $\sigma \leftrightarrow -\sigma$. Oriented closed strings are called the "extended Shapiro-Virasoro model," and the spectrum contains all of the states mentioned above. For closed strings, the restriction of non-orientability, $\sigma \leftrightarrow -\sigma$, means that all states must be symmetric under the exchange $\alpha^i_n \leftrightarrow \widetilde{\alpha}^i_n$. Therefore, the antisymmetric combination, eq.(20.78), is not allowed. Unoriented closed bosonic strings are called the "restricted Shapiro-Virasoro model."

The free open string does not contain the graviton. When we add interactions, an open string can self-interact to close by joining ends. Thus, the interacting open string theory necessarily contains a closed string sector, and thereby the graviton.

Functional Schrödinger Representation

From the Hamiltonian, eq.(20.43), and quantum conditions, eq.(20.46), we can immediately go to the functional Schrödinger representation by taking

$$p^j_\tau(\sigma) = -i\frac{\delta}{\delta x^j(\sigma)}. \tag{20.79}$$

The eigen wave functionals, $\Psi[x^i(\sigma)]$, will satisfy the string Schrödinger equation,

$$\frac{1}{2T}\int d\sigma\left(-\frac{\delta^2}{\delta x^j(\sigma)^2} + T^2(\partial_\sigma x^j(\sigma))^2\right)\Psi[x^i(\sigma)] = E\Psi[x^i(\sigma)]. \tag{20.80}$$

Instead of solving eq.(20.80) directly, we will determine the functional differential representation of the α_n^i's and solve the resulting differential equation, eq.(20.51), for the vacuum functional. The remaining eigenfunctionals can then easily be generated by application of the differential operators α_{-n}^i to the vacuum wave functional.

In the Schrödinger representation, the τ dependence resides in the wave functionals, $\Psi[x^i(\sigma), \tau]$, and not the operators. Thus, the position and momentum density of the string, $x^\mu(\sigma)$ and $p_\tau^\mu(\sigma)$, are independent of τ. For the open string, the appropriate mode expansion is

$$\begin{aligned} x^j(\sigma) &= \bar{x}^j + \frac{i}{\sqrt{\pi T}} \sum_{n\neq 0} \frac{1}{n} \alpha_n^j \cos(n\sigma) \\ p_\tau^j(\sigma) &= \frac{p^j}{\pi} + \frac{\sqrt{T}}{\sqrt{\pi}} \sum_{n\neq 0} \alpha_n^j \cos(n\sigma). \end{aligned} \tag{20.81}$$

We may still define

$$\alpha_0^j = \frac{p^j}{\sqrt{\pi T}}.$$

In the light-cone gauge, x^+ is identified with the τ coordinate, and $x^-(\sigma)$ is eliminated by solving the combined constraints (eqs.(20.29) and (20.30)),

$$\frac{p_\tau^-}{T} + (x^-)' = \frac{\pi}{2p^+T}(p_\tau^i + Tx^{i\prime})^2. \tag{20.82}$$

When $x^-(\sigma)$ is expanded in the same way as eq.(20.81), the solution to the constraint (20.82) is identical to the operator representation result, eq.(20.39).

The usual differential representation of the equal-τ commutators, eq.(20.46), leads to eq.(20.79) and

$$p^j = -i\frac{\delta}{\delta \bar{x}^j}. \tag{20.83}$$

This implies that the momentum density mode expansion, eq.(20.81), can be written as

$$\frac{\delta}{\delta x^j(\sigma)} = \frac{1}{\pi}\frac{\delta}{\delta \bar{x}^j} + i\frac{\sqrt{T}}{\sqrt{\pi}} \sum_{n\neq 0} \alpha_n^j \cos(n\sigma). \tag{20.84}$$

The mode expansions, eq.(20.81), are easily inverted to yield

$$\alpha_m^j = \int_0^\pi d\sigma \left(\frac{1}{\sqrt{\pi T}}\, p_\tau^j(\sigma) - im\frac{\sqrt{T}}{\sqrt{\pi}}\, x^j(\sigma) \right) \cos(m\sigma). \tag{20.85}$$

Substitution of eq.(20.79) in for the momentum density yields the desired differential representation of the α^j_m's. For $m > 0$ we have

$$\begin{aligned}\alpha^j_m &= -i\int_0^\pi d\sigma\left(\frac{1}{\sqrt{\pi T}}\frac{\delta}{\delta x^j(\sigma)} + m\frac{\sqrt{T}}{\sqrt{\pi}}\,x^j(\sigma)\right)\cos(m\sigma)\\ (\alpha^j_m)^\dagger &= -i\int_0^\pi d\sigma\left(\frac{1}{\sqrt{\pi T}}\frac{\delta}{\delta x^j(\sigma)} - m\frac{\sqrt{T}}{\sqrt{\pi}}\,x^j(\sigma)\right)\cos(m\sigma)\\ \alpha^j_0 &= -\frac{i}{\sqrt{\pi T}}\frac{\delta}{\delta \bar{x}^j}.\end{aligned} \tag{20.86}$$

Since the vacuum is defined to satisfy $\alpha^j_m|0;p\rangle = 0$, $p^j_{op}|0;p\rangle = p^j|0;p\rangle$, then the vacuum wave functional, $\Psi_0[x^j(\sigma)]$, satisfies the functional differential equations,

$$\begin{aligned}\int_0^\pi d\sigma\,\cos(m\sigma)\left(\frac{1}{\sqrt{\pi T}}\frac{\delta}{\delta x^j(\sigma)} + m\frac{\sqrt{T}}{\sqrt{\pi}}\,x^j(\sigma)\right)\Psi_0[x^j(\sigma)] = 0,\\ -i\frac{\delta}{\delta\bar{x}^j}\Psi_0[x^j(\sigma)] = p_j,\end{aligned} \tag{20.87}$$

for any integer $m > 0$. It is convenient to define Fourier-transformed string position coordinates,

$$\widetilde{x}^j_m = \sqrt{\frac{2}{\pi}}\int_0^\pi d\sigma\,x^j(\sigma)\cos(m\sigma), \tag{20.88}$$

in which case eq.(20.87) becomes

$$\left(\frac{1}{\sqrt{T}}\frac{\delta}{\delta\widetilde{x}^j_m} + m\sqrt{T}\,\widetilde{x}^j_m\right)\Psi_0[\widetilde{x}^j_m] = 0. \tag{20.89}$$

For a given m, the solution is

$$\Psi_0(\widetilde{x}^j_m) = \exp\left(-\frac{T}{2}\,m\,\widetilde{x}^{j\,2}_m\right), \tag{20.90}$$

the harmonic oscillator ground state, thus

$$\begin{aligned}\Psi_0[\widetilde{x}^j] &= \eta\exp(ip_j\bar{x}_j)\prod_{m=1}^{\infty}\exp\left(-\frac{T}{2}\,m\,\widetilde{x}^{j\,2}_m\right)\\ &= \eta\exp(ip_j\bar{x}_j)\,\exp\left(-\frac{T}{2}\sum_{m=1}^{\infty} m\,\widetilde{x}^{j\,2}_m\right),\end{aligned} \tag{20.91}$$

or,

$\Psi_0[x^j(\sigma)]$

$$= \eta \exp(ip_j\bar{x}_j)\exp\left(-\frac{T}{2}\int d\sigma\, d\sigma' x^j(\sigma)s(\sigma,\sigma')x^j(\sigma')\right)$$
$$= \exp\left(i\frac{p^l}{\pi}\int_0^\pi d\sigma\, x^j(\sigma)\right)\exp\left(-\frac{T}{2}\int d\sigma\, d\sigma' x^j(\sigma)s(\sigma,\sigma')x^j(\sigma')\right), \tag{20.92}$$

where

$$s(\sigma,\sigma') = \frac{2}{\pi}\sum_{m=1}^{\infty} m\cos(m\sigma)\,\cos(m\sigma'). \tag{20.93}$$

Note that the center-of-mass coordinates of the string enter Ψ_0 as a phase, reflecting the fact that the string is free. Clearly we could easily get the same result, eqs.(20.92) and (20.93), by introducing $\widetilde{x}^j$, eq.(20.88), into the Schrödinger equation (20.80). The transformation equation (20.88) diagonalizes the functional differential operator, and makes it easy to see that all eigenfunctionals are infinite products, one for each m, of ordinary harmonic oscillator wave functions.

When we run eq.(20.92) through the Schrödinger equation, eq.(20.80), we find that

$$E_0 = \frac{p_j^2}{2\pi T} + \frac{D-2}{2}\sum_{m=1}^{\infty} m, \tag{20.94}$$

in agreement with eq.(20.50). The square mass operator, M^2, may also be directly expressed as a differential operator. From eqs.(20.52), (20.81), and (20.83), we have

$$M^2 = \pi\int d\sigma\left(-\frac{\delta^2}{\delta x^j(\sigma)^2} + T^2(\partial_\sigma x^j(\sigma))^2\right) + \frac{\delta^2}{\delta\bar{x}^{j2}}. \tag{20.95}$$

Direct application of eq.(20.95) to eq.(20.92) results in

$$M^2\Psi_0[x^j(\sigma)] = \left((2\pi T)\frac{D-2}{2}\sum_{m=1}^{\infty} m\right)\Psi_0[x^j(\sigma)], \tag{20.96}$$

which agrees with eq.(20.54). Excited state wave functionals can be found directly from eq.(20.92) by application of $\alpha_m^{j\,\dagger}$, eq.(20.86).

As an application, let's compute the renormalized value of the zero-point energy, $\sum_{n=1}^{\infty} n$. In order to do so, we must regularize the string theory. We will accomplish this using functional directional derivatives.

The regularized theory is defined by replacing the functional derivatives in the Schrödinger equation, eq.(20.80), by the functional directional derivative,

$$\frac{\delta_\lambda}{\delta_\lambda x^j(\sigma)} = \sqrt{\frac{2}{\pi}} \sum_{m=1}^{\infty} \cos(m\sigma) f_m(\lambda) \frac{\delta}{\delta \widetilde{x}_m^j}. \tag{20.97}$$

The direction in function space is taken along a δ-sequence. Thus,

$$\frac{\delta_\lambda}{\delta_\lambda x^j(\sigma)} x^i(\sigma') = \delta^{ij} \delta_\lambda(\sigma - \sigma'), \tag{20.98}$$

where

$$\lim_{\lambda \to 0} \delta_\lambda(\sigma - \sigma') = \delta(\sigma - \sigma'). \tag{20.99}$$

The ground state eigenfunctional of the new, regularized Schrödinger equation is ($p^j = 0$)

$$\Psi_0[\widetilde{x}^j, \lambda] = \eta \exp\left(-\frac{T}{2} \sum_{m=1}^{\infty} \frac{m}{f_m(\lambda)} \widetilde{x}_m^{j\,2} \right), \tag{20.100}$$

and the vacuum energy is

$$E_0(\lambda) = \frac{D-2}{2} \sum_{m=1}^{\infty} f_m(\lambda) m. \tag{20.101}$$

We are free to choose the regulator $f_m(\lambda)$ for our convenience, so let's take

$$f_m(\lambda) = e^{-m\lambda}. \tag{20.102}$$

The vacuum energy becomes

$$E_0 = \frac{D-2}{2} \sum_{m=1}^{\infty} m e^{-m\lambda}, \tag{20.103}$$

which can be explicitly summed to give

$$\begin{aligned} E_0 &= \frac{D-2}{2} \left(-\frac{d}{d\lambda} \right) \sum_{m=1}^{\infty} (e^{-\lambda})^m \\ &= \frac{D-2}{2} \left(-\frac{d}{d\lambda} \right) \left(\frac{e^{-\lambda}}{1 - e^{-\lambda}} \right) \\ &= \frac{D-2}{2} \left(\frac{e^{-\lambda}}{(1 - e^{-\lambda})^2} \right). \end{aligned} \tag{20.104}$$

In order to extract the finite part of eq.(20.104) in the limit $\lambda \to 0$, expand E_0 in a power series in λ,

$$E_0 = \frac{D-2}{2} \sum_{n=-\infty}^{\infty} a_n \lambda^n. \tag{20.105}$$

From eq.(20.104), we have

$$\sum_n \left(\lambda - \frac{\lambda^2}{2} + \frac{\lambda^3}{6} + \cdots\right)^2 a_n \lambda^n = 1 - \lambda + \frac{\lambda^2}{2} + \cdots. \tag{20.106}$$

Thus, $a_{-2} = 1$, $a_{-1} = 0$, $a_0 = -1/12$, so

$$E_0(\lambda) = \frac{D-2}{2}\left(\frac{1}{\lambda^2} - \frac{1}{12} + \mathcal{O}(\lambda^2)\right). \tag{20.107}$$

The first term in eq.(20.107) is quadratically divergent as $\lambda \to 0$, and must be eliminated by a suitable subtraction. To accomplish this, add a counterterm to the Hamiltonian so that the ground state wave function becomes

$$\Psi_0^R[\tilde{x}^j, \lambda] = \eta \exp\left(-\frac{T}{2}\sum_{m=1}^{\infty}\left(\frac{m}{f_m(\lambda)} - 4m^2\lambda\right)\tilde{x}_m^{j\,2}\right). \tag{20.108}$$

Note that the counterterm will apparently vanish in the limit, and that $\Psi_0^R[\tilde{x}^j, \lambda]$ remains normalizable with our choice for $f_m(\lambda)$, eq.(20.102). With the addition of the counterterm, the ground state energy becomes

$$E_0^R = \frac{D-2}{2}\sum_{m=1}^{\infty}\left(me^{-m\lambda} - 4\lambda m^2 e^{-2m\lambda}\right). \tag{20.109}$$

The contribution to the energy from the counterterm is

$$\begin{aligned} -4\lambda \frac{D-2}{2}\sum_{m=1}^{\infty} m^2 e^{-2m\lambda} &= -\lambda \frac{D-2}{2}\left(\frac{d^2}{d\lambda^2}\sum_{m=1}^{\infty} e^{-2m\lambda}\right) \\ &= -4\lambda \frac{D-2}{2}\left(\frac{e^{-2\lambda} + e^{-4\lambda}}{(1-e^{-2\lambda})^3}\right) \\ &= -\frac{D-2}{2}\left(\frac{1}{\lambda^2} + \mathcal{O}(\lambda)\right). \end{aligned} \tag{20.110}$$

Thus,

$$E_0^R = \frac{D-2}{2}\left(-\frac{1}{12}+\mathcal{O}(\lambda)\right). \tag{20.111}$$

The ground state energy is now finite in the limit $\lambda \to 0$, and is identical to the result stated earlier. Since $E_0^R < 0$ when $D > 2$, the ground state is a tachyon.

Observe that the subtraction, eq.(20.110), does not modify the finite piece of eq.(20.107). The series $\sum_m \exp(-2m\lambda)$ would have to contain a term $(\lambda \ln \lambda - \lambda)$ to do so, which it certainly does not. In fact, it is easy to see that the same comment applies when $m^2\lambda$ in the wave functional eq.(20.108) is replaced by

$$\frac{2^{n+1}}{(n!)}\lambda^{n-1}m^n,$$

where $n > 1$. In general, there exists no acceptable subtraction that will modify the finite term in eq.(20.107), hence it is physically meaningful.

20.3 COVARIANT GAUGE

The goal of quantizing in a covariant gauge is to maintain manifest Lorentz covariance. As we saw in the point particle case, the price of doing so is the quantization of nondynamical degrees of freedom and the introduction of ghost states. Constraints arose that had to be applied to the states. The physical point particle states satisfying the constraints were ghost-free. A similar situation occurs for strings. Unlike the point particle case, however, the constrained free string states are not ghost-free unless the dimension of space-time, D, is ≤ 26, and the normal ordering constant, a, introduced in the previous section, is ≤ 1.

In order to maintain manifest Lorentz covariance upon quantization, the equal-τ commutators must be

$$\begin{aligned}
&[x^\mu(\sigma,\tau), p^\nu_\tau(\sigma',\tau)] = -i\eta^{\mu\nu}\delta(\sigma-\sigma')\\
&[x^\mu(\sigma,\tau), x^\nu(\sigma',\tau)] = [p^\mu_\tau(\sigma,\tau), p^\nu_\tau(\sigma',\tau)] = 0.
\end{aligned} \tag{20.112}$$

In particular, the metric on space-time, $\eta_{\mu\nu}$, must appear on the right-hand side of the first commutator above, and all components of x^μ and p^μ_τ must appear on the left. The constraints, eq.(20.10), demonstrate that some of the x^μ and p^μ_τ are nondynamical. Since all components of x^μ and p^ν_τ must appear in eq.(20.112), then some nondynamical degrees of freedom are going to be quantized. In addition, the indefinite Minkowski metric appears on the right-hand side, so the $\mu = \nu = 0$ equal-τ commutator will have the wrong sign on the right-hand side. This introduces ghost states.

We arrive at the covariant gauge by using the reparametrization invariance on the string world-sheet to set the intrinsic metric on the world-sheet to the 2-dimensional Minkowski metric, $g_{ab} = \eta_{ab}$. However, in doing so, we lose the Hamiltonian equation of motion for the metric. Since the metric equation of motion, $\delta S/\delta g_{ab} = 0$, states that the stress-energy tensor,

$$T_{ab} = -\frac{2}{T}\frac{1}{\sqrt{-g}}\frac{\delta S}{\delta g^{ab}}, \tag{20.8}$$

must vanish, then we must impose $T_{ab} = 0$ as a constraint in order to recover the original theory. The constraint, $T_{ab} = 0$, in terms of x^μ is given by eq.(20.10).

In the previous section we used the residual reparametrization invariance to solve the constraints and implement them at the operator level. Presently, manifest covariance dictates the form of the equal-τ commutators as eq.(20.112). We cannot implement the constraints as operator equations, because they are in conflict with the commutators, eq.(20.112). This means that we must apply the constraints to the states. We cannot, however, simply define an acceptable physical state, $|phys\rangle$, via $T_{ab}|phys\rangle = 0$. As in the point particle case, this implementation of the constraints is also in conflict with the equal-τ commutators. We must settle for a realization of the constraints as averages in the physical states,

$$\langle phys\,|T_{ab}|\,phys\rangle = 0. \tag{20.113}$$

Eq.(20.113) will be satisfied if we take as physical states those annihilated by the positive frequency or destruction part of T_{ab},

$$T^{+}_{ab}|\,phys\rangle = 0. \tag{20.114}$$

In the best of all cases, the physical states satisfying eq.(20.114) will be ghost-free. However, as we have already stated, this turns out not to be true (for the free string) unless $D \leq 26$, and $a \leq 1$.

With the choice $g_{ab} = \eta_{ab}$, the equation of motion for x^μ is the wave equation, eq.(20.12). The general solution is given by eq.(20.13), and the boundary conditions by eq.(20.14) or eq.(20.15). The equation of motion is linear so we may write the solution as a superposition of plane waves. The mode expansion for the right- and left-movers of the closed string, $x^\mu_R(\tau - \sigma)$ and $x^\mu_L(\tau + \sigma)$, is given by eq.(20.16). The mode expansion of x^μ for the open string is eq.(20.17).

Upon quantization, the x^μ and p^μ_τ operators satisfy eq.(20.112). From eq.(20.112), it follows that we also have

$$[\bar{x}^\mu, p^\nu] = -i\eta^{\mu\nu}. \tag{20.115}$$

When we plug the mode expansion into the equal-τ commutators, eq.(20.112), we find that mode operators satisfy

$$[\alpha_m^\mu, \alpha_n^\nu] = -m\,\delta_{m+n}\,\eta^{\mu\nu} \tag{20.116}$$

for the open string and for the right-movers on the closed string. For the closed string, we also have

$$\begin{aligned} [\widetilde{\alpha}_m^\mu, \widetilde{\alpha}_n^\nu] &= -m\,\delta_{m+n}\,\eta^{\mu\nu} \\ [\alpha_m^\mu, \widetilde{\alpha}_n^\nu] &= 0. \end{aligned} \tag{20.117}$$

Just as in the previous section for the light-cone gauge, we can interpret the results of quantization as a collection of harmonic oscillators if we take α_m^μ and $\widetilde{\alpha}_m^\mu$ as lowering or destruction operators for $m > 0$, and as raising or creation operators for $m < 0$. The string ground state, $|0;p\rangle$, will satisfy

$$\begin{aligned} \alpha_m^\mu|0;p\rangle = \widetilde{\alpha}_m^\mu|0;p\rangle &= 0 \qquad \text{for } m > 0 \\ p_{op}^\mu|0;p\rangle &= p^\mu|0;p\rangle. \end{aligned} \tag{20.118}$$

However, the appearance of $\eta^{\mu\nu}$ on the right-hand side of eqs.(20.116) and (20.117) means that the $\mu = \nu = 0$ commutator will be reversed from the other commutators. This introduces ghosts since $(m > 0)$,

$$\langle 0;p\,|\alpha_m^0\alpha_{-m}^0|\,0;p\rangle = \langle 0;p\,|[\alpha_m^0, \alpha_{-m}^0]|\,0;p\rangle = -m. \tag{20.119}$$

The ghost states will not be a problem if they are excluded from the physical states by the constraints. Since it is easier to express the states of the string in terms of the mode operators, then in order to apply the constraints to the states, we need to write them in terms of the modes. This amounts to Fourier transforming the components of the stress-energy tensor. However, it turns out to be more convenient to apply the constraints as the mode expansion of linear combinations of the stress-energy components. Let's examine the case for the closed string first.

x_R^μ is a right-moving solution and depends only on the combination $\tau - \sigma$. Thus, $(\partial_\tau + \partial_\sigma)x_R^\mu = 0$, or

$$\dot{x}_R^\mu = -x_R^{\mu\prime}. \tag{20.120}$$

Similarly, $(\partial_\tau - \partial_\sigma)x_L^\mu = 0$, so

$$\dot{x}_L^\mu = x_L^{\mu\prime}. \tag{20.121}$$

When we apply eqs.(20.120) and (20.121) to the constraint equations, eq.(20.10), we find that

$$\begin{aligned} T_{01} = T_{10} = \dot{x}^\mu x'_\mu &= \dot{x}_R^2 + \dot{x}_L^2 = 0 \\ T_{00} = T_{11} = \frac{1}{2}(\dot{x}^2 + (x')^2) &= \dot{x}_L^2 - \dot{x}_R^2 = 0, \end{aligned} \tag{20.122}$$

which together imply that

$$\dot{x}_R^2 = \dot{x}_L^2 = 0. \tag{20.123}$$

If we define the coefficients of the mode expansion of $\dot{x}_R^2$ via

$$\dot{x}_R^2 = -\frac{2}{\pi T} \sum_{m=-\infty}^{\infty} L_m \, e^{-2im(\tau-\sigma)}, \tag{20.124}$$

and define α_0^μ as

$$\alpha_0^\mu = \frac{p^\mu}{2\sqrt{\pi T}}, \tag{20.125}$$

then

$$L_m = -\frac{1}{2} \sum_{n=-\infty}^{\infty} \alpha_n \cdot \alpha_{m-n}. \tag{20.126}$$

Similarly, if

$$\dot{x}_L^2 = -\frac{2}{\pi T} \sum_{m=-\infty}^{\infty} \widetilde{L}_m e^{-2im(\tau+\sigma)}, \tag{20.125a}$$

then

$$\widetilde{L}_m = -\frac{1}{2} \sum_{n=-\infty}^{\infty} \widetilde{\alpha}_n \cdot \widetilde{\alpha}_{m-n}. \tag{20.126a}$$

The constraint equation on the physical states, eq.(20.114), now becomes

$$\begin{aligned} L_m | \, phys \rangle &= 0 \\ \widetilde{L}_m | \, phys \rangle &= 0 \qquad \text{for} \quad m > 0. \end{aligned} \tag{20.127}$$

Note the similarity of L_m and $\widetilde{L}_m$ to α_m^- and $\widetilde{\alpha}_m^-$ of the light-cone gauge in the previous section, eq.(20.68). The $m = 0$ case requires a little more attention, because, once again, there is no operator ordering ambiguity unless $m = 0$. We resolve the ambiguity by defining L_0 with a specific ordering. As in the previous section, we choose normal ordering in which case

$$L_0 = -\frac{1}{2}\alpha_0^2 - \sum_{n=1}^{\infty} \alpha_{-n} \cdot \alpha_n. \tag{20.128}$$

A similar expression holds for $\widetilde{L}_0$. Because of the ordering ambiguity, the classical constraint, $L_0 = \widetilde{L}_0 = 0$ is replaced by

$$(L_0 - a)|phys\rangle = 0 = (\widetilde{L}_0 - a)|phys\rangle, \tag{20.129}$$

where a is the (arbitrary) normal ordering constant.

From eq.(20.125), we see that L_0 and $\widetilde{L}_0$ involve $p^\mu p_\mu = M^2$, the square mass of the string state. Therefore, the constraints, eq.(20.129), are the mass shell conditions for the state,

$$M^2 = -8\pi T\Big(a + \sum_{n=1}^{\infty} \alpha_{-n} \cdot \alpha_n\Big) = -8\pi T\Big(a + \sum_{n=1}^{\infty} \widetilde{\alpha}_{-n} \cdot \widetilde{\alpha}_n\Big). \tag{20.130}$$

Compare this to the closed string light-cone result, eq.(20.72). Note that eq.(20.130) will only be consistent and yield the same square mass for a given state if

$$\sum_{n=1}^{\infty} \alpha_{-n} \cdot \alpha_n |phys\rangle = \sum_{n=1}^{\infty} \widetilde{\alpha}_{-n} \cdot \widetilde{\alpha}_n |phys\rangle,$$

that is,

$$(L_0 - \widetilde{L}_0)|phys\rangle = 0. \tag{20.131}$$

This additional constraint is the only communication between the left- and right-movers of the free closed string. $L_0 - \widetilde{L}_0$ is independent of p^μ, thus such an operation does not involve translation of the string in space-time. Instead, one can show that $L_0 - \widetilde{L}_0$ generates rigid rotations of the closed string, and the constraint, eq.(20.131), reflects the fact that it doesn't matter where we choose the origin of the σ coordinate on the closed string.

The quantum L_m's satisfy the Virasoro algebra with a central extension,

$$[L_m, L_n] = (m-n)L_{m+n} + \frac{D}{12}(m^3 - m)\delta_{m+n}. \tag{20.132}$$

The $\widetilde{L}_m$'s satisfy an identical algebra. The first term on the right-hand side above arises at the classical level. The central extension, which depends on the dimension of space-time, arises from ordering ambiguities.

The direct computation of the anomaly by substitution of eq.(20.126) into the commutator $[L_m, L_n]$ is tedious. Instead, there is a standard method to arrive at eq.(20.132) using the Jacobi identity,

$$[L_k,[L_m,L_n]]+[L_m,[L_n,L_k]]+[L_n,[L_k,L_m]]=0. \tag{20.133}$$

If we write the central term as $A(m)\delta_{m+n}$, then eq.(20.133) implies that

$$\Big((m-n)A(k)+(n-k)A(m)+(k-m)A(n)\Big)\delta_{n+k+m}=0,$$

or

$$(n-m)A(n+m)+(2n+m)A(m)-(2m+n)A(n)=0, \tag{20.134}$$

where we have used the fact that $A(m)=-A(-m)$, which follows from the exchange of n and m in the commutator $[L_m, L_n]$. Set $n=2m$, and assume $m \neq 0$. Then eq.(20.134) becomes

$$A(3m)+5A(m)-4A(2m)=0. \tag{20.135}$$

If we assume $A(m)$ is a polynomial in m, $A(m) = \sum_p a_p m^p$, then eq.(20.135) implies that

$$3^p+5-2^{p+2}=0 \tag{20.136}$$

whenever $a_p \neq 0$. It is not hard to see that the only two solutions to eq.(20.136) are $p=1$ and $p=3$, since 3^p is much greater than 2^{p+2} for $p \geq 4$, and the difference monotonically increases as p increases. Thus,

$$A(m)=a_1 m+a_3 m^3. \tag{20.137}$$

To find a_1 and a_3, we merely evaluate $A(m)$ directly from $[L_m, L_{-m}]$ for 2 values of m. For $m=1$, we have

$$\langle 0\,|[L_1,L_{-1}]|\,0\rangle=0,$$

thus, $A(1)=0$ and $a_1=-a_3$. For $m=2$, we have

$$\langle 0\,|[L_2,L_{-2}]|\,0\rangle=\frac{1}{2}\eta_{\mu\nu}\eta^{\mu\nu}=\frac{1}{2}D, \tag{20.139}$$

so $a_3 = D/12$, and we arrive at eq.(20.132).

Returning to the open string, the L_m's are defined as the mode expansion of the combination of the two constraints in eq.(20.10).

$$(\dot{x}+x')^2=-\frac{2}{\pi T}\sum_{n=-\infty}^{\infty}L_m e^{-im(\tau+\sigma)} \tag{20.140}$$

This yields

$$L_m = -\frac{1}{2}\sum_{n=-\infty}^{\infty}\alpha_n \cdot \alpha_{m-n}, \tag{20.141}$$

where

$$\alpha_0^\mu = \frac{p^\mu}{\sqrt{\pi T}}. \tag{20.142}$$

Once again, operator ordering ambiguities only arise when $m = 0$, where we adopt normal ordering. Thus, the constraint on the states reads

$$\begin{aligned} L_m | \, phys\rangle &= 0 \qquad \text{for} \quad m > 0 \\ (L_0 - a)| \, phys\rangle &= 0 \qquad \text{for} \quad m = 0. \end{aligned} \tag{20.143}$$

The latter constraint yields the mass shell condition,

$$M^2 = -2\pi T\Big(a + \sum_{n=1}^{\infty}\alpha_{-n}\cdot\alpha_n\Big). \tag{20.144}$$

The L_m's satisfy the same Virasoro algebra, eq.(20.132), as in the closed string case. It is left as an exercise to verify that the light-cone gauge mass spectrum and covariant gauge spectrum are identical in the open and closed sectors when $a = 1$ and $D = 26$.

Since we have used all of the α_n^μ's in quantization, then no ordering ambiguities exist in the Lorentz generators, $M^{\mu\nu}$, eq.(20.58), and covariance is maintained through the quantization process. In addition, $[L_m, M^{\mu\nu}] = 0$, so the constraints are Lorentz invariant and the physical states will form Lorentz multiplets. However, we have introduced ghost states, so we must explicitly check whether unitarity remains intact after quantization. It is possible to prove that the physical states are ghost-free for the free string when $a \leq 1$ and $D \leq 26$, as was first done independently by Brower, and by Goddard and Thorn. When interactions are included, the ghost-free conditions are further restricted to $a = 1$ when $D = 26$, or $a < 1$ when $D \leq 25$. The proof is nontrivial and we will omit the details. The interested reader is referred to the Bibliography. Instead, we will show that it is necessary for a to be ≤ 1 and for D to be ≤ 26 for physical states to be ghost-free by explicitly constructing physical states that are ghosts when $a > 1$ or $D > 26$.

Let us begin by considering an open string state consisting of a single excitation with polarization k^μ above a string vacuum of momentum p^μ. The state with the smallest mass is $|\,\phi\rangle = k\cdot\alpha_{-1}|\,0;p\rangle$. The norm of this state is

$$\langle\phi\,|\,\phi\rangle = \langle 0;p\,|(k\cdot\alpha_1)(k\cdot\alpha_{-1})|\,0;p\rangle = -k^2\langle 0;p\,|\,0;p\rangle,$$

and is therefore a ghost when the polarization vector is time-like, $k^2 > 0$. This state must satisfy

$$\begin{aligned} L_m|\,\phi\rangle &= 0 \\ (L_0 - a)|\,\phi\rangle &= 0, \end{aligned} \tag{20.145}$$

in order to be physical. Since,

$$[L_m, \alpha_n^\mu] = -n\,\alpha_{m+n}^\mu, \tag{20.146}$$

$|\,\phi\rangle$ automatically satisfies eq.(20.145) for $m > 1$. $L_1|\,\phi\rangle = 0$ implies that the polarization vector must be orthogonal to the string momentum, $p \cdot k = 0$. Application of the mass shell condition yields

$$p^2 = 2\pi T(1-a).$$

In order to make $|\,\phi\rangle$ a ghost state, we must have $k^2 > 0$. However, since $p \cdot k = 0$, $k^2 > 0$ is allowable only if $p^2 < 0$. Therefore, physical ghost states exist when $a > 1$.

To see that physical ghost states exist for $D > 26$, consider the state

$$|\,\phi\rangle = (\alpha_{-1}\cdot\alpha_{-1} + \beta\alpha_0\cdot\alpha_{-2} + \gamma(\alpha_0\cdot\alpha_{-1})^2)|\,0;p\rangle. \tag{20.147}$$

For $a = 1$, $|\,\phi\rangle$ will satisfy the physical state conditions when

$$p^2 = 2\pi T, \qquad \beta = \frac{D-1}{5}, \qquad \gamma = -\frac{D+4}{10}. \tag{20.148}$$

In this case, the norm of $|\,\phi\rangle$ is

$$\langle\phi\,|\,\phi\rangle = \frac{2}{25}(D-1)(26-D), \tag{20.149}$$

and, therefore, is a ghost when $D > 26$. The details of the computation are left as an exercise.

20.4 EXERCISES

1. Using the mode expansion, verify eqs.(20.47), (20.49), and (20.50).
2. Derive an expression for the Lorentz current generators, $M_\tau^{\mu\nu}$, by performing an infinitesimal Lorentz transformation in the string actions, eqs.(19.20) and (19.22). Show that the Lorentz generators, $M^{\mu\nu}$, are conserved, and verify eq.(20.58).
3. What counterterm (expressed as a functional of $x^j(\sigma)$) must be added to the Hamiltonian so that the ground state wave functional becomes eq.(20.108)? Verify that eq.(20.108) is normalizable.

4. Take the δ-sequence in eq.(20.98) to be

$$\delta_\lambda(\sigma-\sigma') = \begin{cases} \frac{1}{2\lambda} & |\sigma-\sigma'| < \lambda \\ 0 & |\sigma-\sigma'| > \lambda. \end{cases}$$

What is $f_m(\lambda)$? Compute $E_0(\lambda)$, and expand E_0 as a power series in λ, eq.(20.105), to see that the λ-independent term is identical to that in eq.(20.107). Why is this not an acceptable regulator?

5. In the covariant gauge, show that implementation of the constraints as operator equations or as $T_{ab}|\,phys\rangle = 0$ is in conflict with the equal-τ commutators. Also show that eq.(20.115) follows from eq.(20.112).

6. Derive the Hamiltonian in the covariant gauge.

7. Show that the L_m's and the $\widetilde{L}_m$'s are the Fourier coefficients of the components of the stress-energy tensor in world-sheet light-cone coordinates, $\sigma^\pm = \tau \pm \sigma$.

8. Verify that the mass spectrum in the covariant gauge is identical to the spectrum in the light-cone gauge for the ground state and first few excited levels when $a = 1$ and $D = 26$.

9. Verify eq.(20.146) and show that $[L_m, M^{\mu\nu}] = 0$.

10. Show that the state $|\,\phi\rangle$ given by eq.(20.147) will satisfy the physical state conditions for $a = 1$, if p^2, β, and γ are given by eq.(20.148). Compute the norm of $|\,\phi\rangle$.

CHAPTER **21**

The Mathematics of Surfaces

The path integral introduced in chapter 12 is a representation of the 2-point Green's function or propagator of a quantum mechanical point particle. For propagation from $(\vec{x}_0, t_0)$ to $(\vec{x}, t)$, the path integral viewpoint is a sum over all possible paths or world-lines of the particle starting at $(\vec{x}_0, t_0)$ and ending at $(\vec{x}, t)$. The contribution of each path to the sum is weighted by the exponential of the classical action evaluated on the particular world-line. If interactions are included, then the possible paths or world-lines to be summed over involve loops and vertices. For example, the perturbative evaluation of the path integral for a particle with cubic self-interaction is represented diagrammatically in figure 21.1(a).

When the point particle is replaced by a string, the "paths" connecting the initial string configuration with the final configuration are world-sheets swept out by the string. Thus, the "path" integral representation of string propagation is a sum over all possible world-sheets, i.e. a sum over random surfaces. The contribution of each surface to the sum is again weighted by the exponential of the classical action for the particular surface. When interactions are added, the surfaces to be summed over include those with holes or handles. A perturbative evaluation of the path integral for a closed string is illustrated diagrammatically in figure 21.1(b). The expansion is in terms of the genus of the surface, that is, in

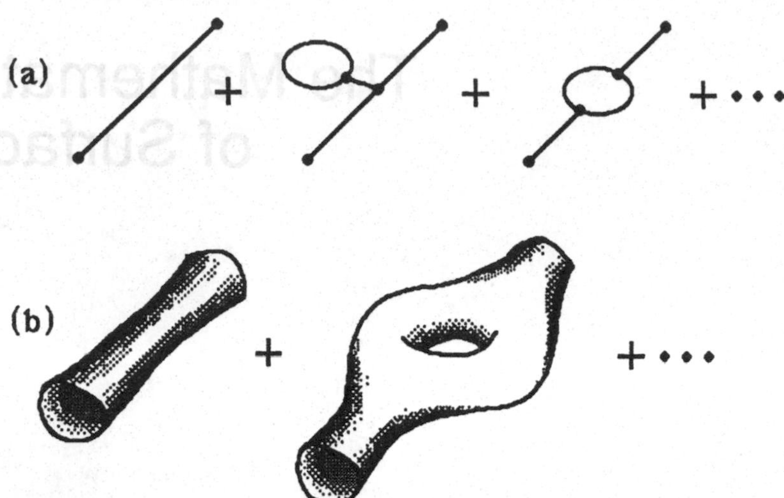

Figure 21.1 Transition from point particles to strings. (a) Perturbative evaluation of a point particle propagator with cubic self-interaction. Paths are world-lines with loops and vertices. Diagrams in the expansion can be labeled by the number of loops. (b) Perturbative evaluation of a closed string propagator. "Paths" are world-sheets or surfaces, so the path integral is a sum over surfaces. Each term in the series can be labeled by the number of holes and handles.

terms of the number of handles. Note that there are 2 distinct 1-loop contributions to the point particle propagator, but only 1 distinct 1-handle surface in the string propagator.

In the Polyakov version of the classical bosonic string action, eq. (19.22), the intrinsic metric on the string world-sheet is treated as an independent variable. Therefore, the surfaces appearing in the "path" integral sum are endowed with a metric. The metric provides us with a notion of orthogonality or rotation by 90°, and therefore with a way to locally define the imaginary "i" and complex numbers. Thus, the metric on the surface canonically defines an "almost complex structure." For the special case of 2 dimensions, this can be automatically extended to a complex structure. Each surface appearing in the sum possesses a complex structure. A connected surface with a complex structure is called a Riemann surface. The path integral viewpoint of string propagation is a sum over random Riemann surfaces.

As discussed in the previous two chapters, the Polyakov action is invariant under reparametrizations of the coordinates on the world-sheet and under Weyl or conformal rescalings of the metric. Therefore, in the absence of anomalies, any 2 conformally equivalent Riemann surfaces contribute identically to the sum. In order to make this sum well-defined, we do not want to sum over all surfaces, but only sum over conformal equivalence classes of Riemann surfaces (Faddeev-Popov procedure). The conformal equivalence classes are identical to the equivalence classes of complex structures on Riemann surfaces. The space of conformal equivalence classes or complex structures for compact surfaces of genus g with n punctures is called moduli space, $\mathcal{M}_{g,n}$ (i.e. each point in a moduli space is a set of conformally equivalent surfaces). Thus, as long as there are no anomalies, the contribution to the string path integral for a given genus g reduces to an integral over the (finite-dimensional) moduli space, $\mathcal{M}_{g,n}$.

The purpose of this chapter is to decode the mathematical statements in the previous 2 paragraphs. In the next chapter we will explicitly compute the genus 0 contribution of compact surfaces with no punctures to this sum (the genus 0 contribution to the partition function). In chapter 23, we will consider the higher-genus contributions.

21.1 MANIFOLDS, STRUCTURES, AND MORPHISMS

Manifolds are generalizations of Euclidean spaces. Locally (i.e. in the neighborhood of a point in the manifold), they look like a Euclidean space. To "look like" something requires a comparison, and this comparison is accomplished using mappings between the neighborhoods and Euclidean space. If the mappings possess the appropriate properties, then we can transfer such things as limits and calculus from Euclidean space to the manifold. In order to give the precise definition of a manifold, we must first define a topology and topological spaces.

A *topology* is a *structure* added to an arbitrary point set which enables one to define a convergent sequence and to define a continuous function in a general setting. Recall the usual ϵ-δ definition of continuity of a function f on the real line: $f : \mathbb{R} \to \mathbb{R}$ is continuous at x_0 if for any positive number ϵ, there exists a positive number δ such that if $|y - x_0| < \delta$, then $|f(y) - f(x_0)| < \epsilon$. The setting of this definition is the real line, a Euclidean space. The definition also involves a positive definite metric. We can generalize this definition a little by considering mappings between spaces with a metric. That is, let X, Y be two spaces with metrics d_1 and d_2, respectively. Then, $f : X \to Y$ is continuous at $x_0 \in X$ if for any $\epsilon > 0$, there exists $\delta > 0$, such that if $d_1(x, x_0) < \delta$, then $d_2(f(x), f(x_0)) < \epsilon$. However, the notion of continuity does not depend on a metric. For example, consider the two curves in figure 21.2. There is no mapping between curve (a) and (b) that is continuous everywhere, and we don't need a metric to tell us this.

A more general setting is provided by noting that the definition on the real line involves open intervals $(x_0 - \delta, x_0 + \delta)$ and $(f(x_0) - \epsilon, f(x_0) + \epsilon)$, which are open (sub)sets. In the case of a metric space, $d_1(x, x_0) < \delta$ defines an open ball, $B_\delta(x_0)$, in X. An open subset V of X is a subset

Figure 21.2 There is no continuous mapping between curves (a) and (b) because of the discontinuous "break" in (b). No need to rely on a metric to arrive at this result. Curves (a) and (b) are topologically inequivalent.

where if $x \in V$, then there is a $\delta > 0$ such that the open ball $B_\delta(x)$ is entirely contained in V. Thus, the 2 definitions of continuity involve open sets.

A *topology*, $\mathcal{T}$, on a set X is defined to be a specified family of open subsets on X satisfying the following 3 properties:

1. The empty set, $\emptyset$, and the space X belong to $\mathcal{T}$.
2. The union of any number (possibly infinite) of open subsets belonging to $\mathcal{T}$ is also in $\mathcal{T}$.
3. The intersection of any finite number (not infinite) of open subsets in $\mathcal{T}$ also belongs to $\mathcal{T}$.

The set X together with a topology $\mathcal{T}$ is called a topological space, and the set X is said to have a topological structure.

In general, a given set X can have many different topologies. For example, consider the set $X = \{x, y\}$ consisting of 2 points. There are 4 possible topologies:

1. $\emptyset, \{x, y\}$
2. $\emptyset, \{x, y\}, \{x\}, \{y\}$
3. $\emptyset, \{x, y\}, \{x\}$
4. $\emptyset, \{x, y\}, \{y\}$

The first topology, which includes $\emptyset$ and X only, is called the indiscrete topology. The second topology, in which all subsets are taken to be open, is called the discrete topology.

When the space X has a metric, then there is a topology induced by the metric. The open subsets that form this topology are the subsets satisfying the definition of an open set given previously (V is open if at each point $x \in V$, there is an open ball, $B_\delta(x)$, entirely contained in V).

An additional important property of a topology is whether or not it contains open sets such that any two distinct points in the space X lie in non-overlapping open sets. If they do, then the topological space $(X, \mathcal{T})$

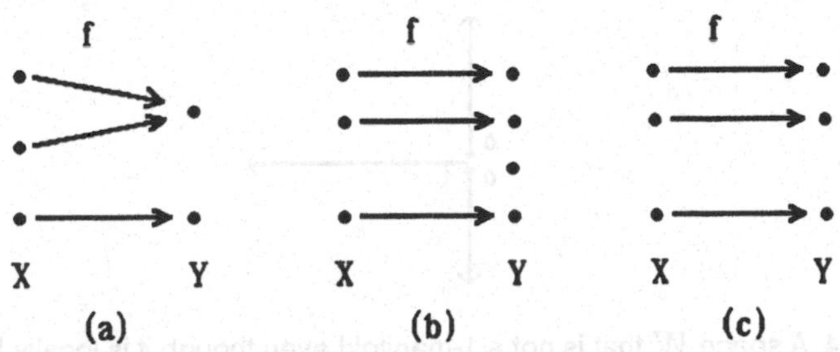

Figure 21.3 Illustration of surjective, injective, and bijective. (a) The map $f : X \to Y$ is surjective or *onto* because at least 1 element of X is mapped onto every element of Y. (b) f is injective or one-to-one (1-1) because no 2 distinct elements of X are mapped to the same element of Y ($f(x_0) = f(x) \Rightarrow x_0 = x$). (c) f is bijective or a 1-1 correspondence because it is 1-1 and onto. The map in (a) is not injective, while the map in (b) is not surjective.

is called a *Hausdorff space*. Restated, X is Hausdorff if for each distinct pair of points x_1, x_2 in X, there exist disjoint open sets $O(x_1)$ and $O(x_2)$, $O(x_1) \cap O(x_2) = \emptyset$, where $x_1 \in O(x_1)$ and $x_2 \in O(x_2)$. In the previous example where X consists of 2 points, only the second topology forms a Hausdorff space. In fact, it is clear that the discrete topology is always Hausdorff.

A metric space with the metric-induced topology is also always a Hausdorff space. If x and y are 2 distinct points, then the open balls $B_\delta(x)$ and $B_\delta(y)$ where $0 < \delta < d(x,y)/2$ are non-overlapping. (Recall that the metric d is assumed to be positive so that $d(x,y) > 0$ when $x \neq y$ and $d(x,y) = 0 \Leftrightarrow x = y$.)

The topology can also be used to define whether the topological space is *connected* or not. A topological space $(X, \mathcal{T})$ is disconnected if there are two non-empty subsets, U and V, of X that are open, $U \in \mathcal{T}$, $V \in \mathcal{T}$, and cover X, $U \cup V = X$, but are disjoint, $U \cap V = \emptyset$. A topological space is connected if it is not disconnected.

Now let us return to the question of continuous mappings. Let $f : X \to Y$ be a function between 2 topological spaces $(X, \mathcal{T}_X)$, and $(Y, \mathcal{T}_Y)$. If $x \in X$, then $f(x) \in Y$ is the *image* of x under f. Let U be an open set in X (i.e. $U \in \mathcal{T}_X$). The image of U under f is the subset $V = f(U) \subset Y$, the range of f with domain U. If V is a subset of Y, then the inverse image of V under f (i.e. $f^{-1}(V)$) is the subset of points in X that are mapped into V by f. The mapping f is defined to be *continuous* when the inverse image of any open set is open. That is, f is continuous if $U = f^{-1}(V) \in \mathcal{T}_X$ when $V \in \mathcal{T}_Y$. This definition agrees with the ϵ-δ definition stated earlier when X and Y are metric spaces with the metric-induced topology.

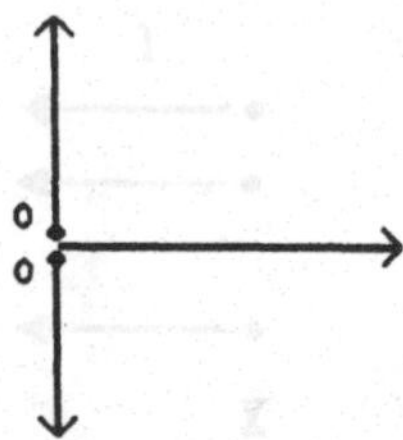

Figure 21.4 A space W that is not a 1-manifold even though it is locally homeomorphic to $\mathbb{R}$, because W is not Hausdorff. The 2 distinct zeroes never lie in non-overlapping open sets.

We should now ask the question of when two topological spaces, $(X, \mathcal{T}_X)$, and $(Y, \mathcal{T}_Y)$, are equivalent. The topological spaces X and Y are equivalent or *homeomorphic* when there is a bijective (1-1 and onto, see figure 21.3) mapping, $f : X \to Y$ that is continuous along with its inverse. When a bijective map and its inverse are continuous, it is called a *homeomorphism*.

Homeomorphism is the first of the morphisms we will meet and it is a bijective mapping that preserves the topological structure. The two spaces involved are homeomorphic or topologically equivalent, because we cannot distinguish between them topologically. A topological result on one space can be transferred to the other via the homeomorphic map. We may think of the homeomorphism as distorting one topological space into another. In a moment, we will consider differentiable structures. The bijective map preserving the differentiable properties is called a diffeomorphism. If a group structure is preserved, then it is called a homomorphism. In the case of other algebraic structures, the mappings are called isomorphisms.

We can now give the formal definition of a topological manifold. An *n-manifold* or n-dimensional manifold M is a Hausdorff topological space such that each point $x \in M$ has an open neighborhood that is homeomorphic with an open subset of n-dimensional Euclidean space, $\mathbb{R}^n$. The "locally looks like" that we mentioned at the beginning of this section is technically defined as open neighborhoods homeomorphic to open subsets of $\mathbb{R}^n$. When we say "surface," we mean a 2-manifold.

The Hausdorff requirement, which is not implied in the "locally looks like," is necessary to eliminate pathological spaces that satisfy the homeomorphic requirement but are not reasonable generalizations of $\mathbb{R}^n$. For example, consider the space W which is a union of subsets in the plane $\mathbb{R}^2$ consisting of the open interval $(0, \infty)$ along the positive x-axis, the half-closed interval $[0, \infty)$ running along the positive y-axis, and $(-\infty, 0]$ running along the negative y-axis. The space is drawn schematically in figure 21.4. $W = (0, \infty) \cup (-\infty, 0] \cup [0, \infty)$. Note that zero appears as 2 distinct points. For any subsets not containing zero, the usual open intervals will serve as open sets. Open sets containing zero must be a union of the open interval $(0, x)$ on the positive x-axis with $[0, y)$ on the positive y-axis or with $(-y, 0]$ on the negative y-axis. Since all open sets are (unions of) open intervals, this space is locally homeomorphic to the real line $\mathbb{R}$. However, W is not an intuitive generalization of the real line. W is not a 1-manifold because it is not Hausdorff. The 2 distinct zeroes never lie in non-overlapping open sets.

If M is an n-manifold, then we know that an open neighborhood surrounds each point in M and there is a homeomorphism between this open neighborhood and an open subset of $\mathbb{R}^n$. Let U_x be an open neighborhood of $x \in M$, let ϕ_x be the homeomorphism, and let V_x be the open subset of $\mathbb{R}^n$, $\phi_x : U_x \rightarrow V_x \subset \mathbb{R}^n$. The homeomorphism ϕ_x and its domain are sometimes called a *chart* at $x \in M$. Since ϕ_x is a homeomorphism, it has a nice continuous inverse, ϕ_x^{-1}, which we can use to place coordinates on the manifold in U_x by borrowing the usual coordinates on $V_x \subset \mathbb{R}^n$, and mapping them back onto U_x. For this reason, the combination (U_x, ϕ_x) is called a *coordinate patch*. A collection of open neighborhoods, $\{U_a, a \in A\}$ (A is some index set), is said to *cover* M if the union of all of the neighborhoods is M, $\cup_{a \in A} U_a = M$. A collection of coordinate patches or charts that covers the manifold M is called an *atlas*.

Now suppose that we contemplate doing calculus on a manifold M. We can accomplish this by doing calculus on $\mathbb{R}^n$ and mapping the results, etc. back over to a coordinate neighborhood on M using a coordinate patch. However, to do this, we will require additional structure on the manifold M. Calculus involves differentials which won't necessarily be mapped correctly by continuous functions alone. To get the results back to the manifold or between manifolds, we will need our charts and coordinate patches to be continuously differentiable.

Thus, we want to build a C^∞ *atlas*. A function $f : U \rightarrow \mathbb{R}^n$ is C^∞ on the open set V if it has continuous partial derivatives of all orders and (possibly mixed) types. $\phi : U \rightarrow \mathbb{R}^n$ is C^∞ if each of its components is. A C^∞ atlas on M is an open cover of M with coordinate patches homeomorphic to $\mathbb{R}^n$, and an additional requirement placed on overlapping charts. When 2 patches, (U_1, ϕ_1) and (U_2, ϕ_2), of a C^∞ atlas overlap, $U_1 \cap U_2 \neq \emptyset$, then the 2 charts must be C^∞-compatible. This means that the composition $\phi_2 \circ \phi_1^{-1} : \mathbb{R}^n \rightarrow \mathbb{R}^n$ and $\phi_1 \circ \phi_2^{-1} : \mathbb{R}^n \rightarrow \mathbb{R}^n$ must be C^∞. The composite mappings, $\phi_2 \circ \phi_1^{-1} : \mathbb{R}^n \rightarrow \mathbb{R}^n$ and $\phi_1 \circ \phi_2^{-1} : \mathbb{R}^n \rightarrow \mathbb{R}^n$, are called *transition functions* for the obvious reason. See figure 21.5. Thus, a C^∞ atlas on M is a collection of C^∞-compatible charts that cover the manifold M. (Composition of homeomorphic functions is always homeomorphic, which is why we did not need an additional compatibility requirement for a topological atlas.)

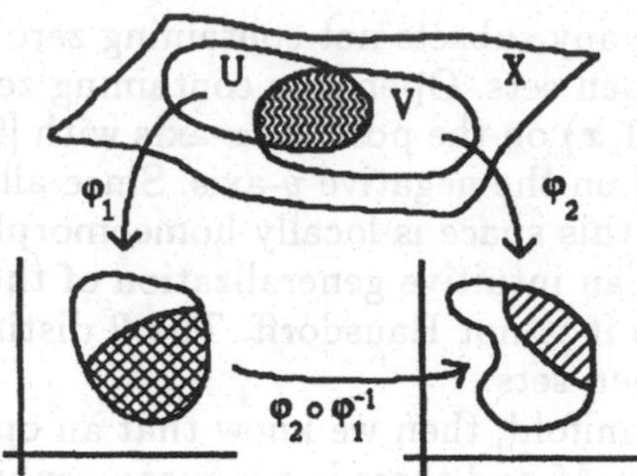

Figure 21.5 (U, ϕ_1) and (V, ϕ_2) are 2 coordinate patches on X. They are C^∞-compatible when the transition functions, $\phi_2 \circ \phi_1^{-1} : \mathbb{R}^n \to \mathbb{R}^n$ and $\phi_1 \circ \phi_2^{-1} : \mathbb{R}^n \to \mathbb{R}^n$, are C^∞ (differentiable). The domain and range of the transition function illustrated are the shaded regions in $\mathbb{R}^n$.

Two C^∞ atlases $\mathcal{A}_1$ and $\mathcal{A}_2$ are equivalent if $\mathcal{A}_1 \cup \mathcal{A}_2$ is also a C^∞ atlas (i.e. all new transition functions are C^∞). A *differentiable structure* on M is an equivalence class of C^∞ atlases on M. There is a maximal atlas for each differentiable structure on M which is the union of all the equivalent atlases in the differentiable structure. A *differentiable manifold*, also called a C^∞ manifold or a smooth manifold, is a topological manifold with a differentiable structure. A single manifold may admit more than one differentiable structure.

The transition functions of an atlas, $\psi_{ij} = \phi_j \circ \phi_i^{-1} : \mathbb{R}^n \to \mathbb{R}^n$, are just ordinary functions on $\mathbb{R}^n$. They transform the coordinates of one overlapping patch into the other in the region of overlap (e.g. ψ_{12} in figure 21.5 transforms the coordinates on U into those on V). Let x_i be the coordinates on the domain of ψ_{ij} and y_i those on the range of ψ_{ij}, that is, $\psi_{ij} : (x_1, \ldots, x_n) \mapsto (y_1, \ldots, y_n)$. The usual Jacobian of ψ_{ij} is $J_{ij} = \partial(y_1, \ldots, y_n)/\partial(x_1, \ldots, x_n)$. ψ_{ij} is a C^∞ function and ψ_{ji} is its C^∞ inverse, thus $J_{ij}J_{ji} = 1$ so $J_{ij} \neq 0$ and $J_{ji} \neq 0$. If $J_{ij} > 0$ for all points in the coordinate patch overlap, then ψ_{ij} is said to be *orientation-preserving.* If ψ_{ij} is orientation-preserving, then so will be ψ_{ji}. A C^∞ atlas is *orientable* if all of the transition functions are orientation-preserving. The differentiable manifold M is oriented if the differentiable structure is oriented. Intuitively, in the case of 2-manifolds, a surface is oriented if it is 2-sided, and non-oriented if it is 1-sided. The cylinder is oriented, but the Möbius strip is not. Any connected, orientable manifold has just 2 orientations. (The definition of orientability is purely topological and does not depend on the existence of a differentiable structure. As we shall see, the definition above is adequate to show that Riemann surfaces are orientable.)

Suppose $f : M \to \mathbb{R}^n$ is a function on manifold M into Euclidean space $\mathbb{R}^n$. Let (U_x, ϕ_x) be a coordinate patch of a C^∞ atlas on M. The function $f \circ \phi_x^{-1} : \mathbb{R}^n \to \mathbb{R}^n$ is the coordinate expression for f. $f \circ \phi_x^{-1}$ is just an ordinary function on Euclidean space, hence we can apply the usual results of calculus to it. f is said to be differentiable (C^∞) on the coordinate patch if $f \circ \phi_x^{-1}$ is differentiable in the ordinary sense. f is differentiable on M if it is differentiable on every coordinate patch.

Suppose $f : M \to N$ is a function from differentiable manifold M to differentiable manifold N. Let (V_y, ψ_y) be a coordinate patch of a C^∞ atlas on N, $\psi_y : V_y \to \mathbb{R}^m$. The composite function, $\psi_y \circ f \circ \phi_x^{-1} : \mathbb{R}^n \to \mathbb{R}^m$, again is just an ordinary function between 2 Euclidean spaces. f is differentiable if $\psi_y \circ f \circ \phi_x^{-1}$ is differentiable in the ordinary sense for any combination of coordinate patches.

When are 2 differentiable manifolds, M and N, equivalent? They are equivalent or *diffeomorphic* when there is a bijective C^∞ map $f : M \to N$ between them whose inverse, $f^{-1} : N \to M$, is also C^∞. Such a map is called a *diffeomorphism*. We cannot distinguish between 2 diffeomorphic manifolds on the basis of properties that concern only their differentiable structures. For low-dimensional topological manifolds ($n \leq 3$), 2 different differentiable structures (i.e. 2 incompatible C^∞ atlases) are usually still diffeomorphic. In fact, in 2 dimensions, surfaces are diffeomorphic if and only if they are homeomorphic.

So far we have discussed topology and not geometry. To bring in geometry, we must add an additional structure on a differentiable manifold, a *geometric structure*. A differentiable manifold with a geometric structure is called a Riemannian manifold, and the resulting geometry is called Riemannian geometry. (A surface with a geometric structure, i.e. a geometric surface, is not called a Riemann surface. We will define a Riemann surface in a moment.)

To describe a geometric structure, we first must define spaces tangent to the manifold at each point. Let $\gamma : \mathbb{R} \to M$ be a map from the real line (parametrized by t) into a differentiable manifold M. Let $x_0 = \gamma(t_0) \in M$ sit in a coordinate patch (U_{x_0}, ϕ_{x_0}). γ is a differentiable curve if $\phi_{x_0} \circ \gamma : \mathbb{R} \to \mathbb{R}$ is differentiable for any patch. The velocity vector of the curve γ, $d\gamma/dt$, is a vector tangent to the curve. At x_0, this vector is also tangent to M. The space of all velocity vectors at x_0 is a linear vector space tangent to M at x_0, and is called the *tangent space of M at x_0*, denoted $T_{x_0}(M)$. It is a linear approximation of the manifold at x_0 (so it has the same dimension as M). The collection of all tangent spaces at each $x \in M$ is called the *tangent bundle*, $T(M)$.

On each tangent space, we can define an inner product or dot product of two vectors. This inner product is a functional that assigns a real number to a pair of vectors in the tangent space. This function is bilinear, symmetric, and positive definite.

A *vector field* on M is a choice of one vector from each tangent space on M (i.e. an assignment of one tangent vector at each point in M – this is a "section" of the tangent bundle). Let (U_{x_0}, ϕ_{x_0}) be a coordinate patch that contains x_0. We can form a basis for the tangent space $T_{x_0}(M)$ by using the velocity vectors associated with curves where all but one coordinate remains constant. A *geometric structure* on M is the assignment of a differentiable inner product to each tangent space of M. Denote this inner product by g. Let $e_i(x)$ be the basis of velocity vectors to coordinate lines at $x \in M$. (The $e_i(x)$ span $T_x(M)$.) g is differentiable if $g(e_i(x), e_j(x))$ is a differentiable function on $\mathbb{R}$ in the ordinary sense. The differentiable inner product is the *metric tensor* on M. We will develop the necessary differential geometry in the context of surfaces as we need it.

Now suppose we start with an even $2n$-dimensional topological manifold M, and build a coordinate atlas where the charts map open subsets of M homeomorphically to open subsets of $\mathbb{C}^n$. An *analytic atlas* or complex analytic atlas is one where all of the transition functions are holomorphic (analytic). Two analytic atlases are equivalent when their union remains an analytic atlas. An equivalence class of analytic atlases is called a *complex structure*, and a $2n$ real dimensional topological manifold with a complex structure is called a complex n-manifold. Two complex analytic manifolds are essentially equivalent when there is a biholomorphic map (the bijective map and its inverse are holomorphic) between them.

Finally, we may formally define the object of interest in this chapter. A *Riemann surface* is a connected 1-dimensional complex manifold. For the special case of one-dimensional complex manifolds, the complex structure is also called a conformal structure. If there is a biholomorphic map (a "holomorphic homeomorphism") between 2 Riemann surfaces, then the 2 surfaces are said to be conformally equivalent.

The string path integral requires us to sum over conformally inequivalent Riemann surfaces, which in turn requires a classification of Riemann surfaces to carry out this sum. We will take up this subject starting in the next section, where we will begin concentrating exclusively on surfaces. Before we do so, however, let us introduce another concept from topology called compactness.

Let $\{U_\alpha, \alpha \in A\}$ be an open cover of a topological space X. That is, $\{U_\alpha\}$ is a family of open sets whose union is X, $\cup_{\alpha\in A} U_\alpha = X$. X is *compact* if for *any* open cover of X, there is a finite number of the U_α's (a finite subfamily) that also cover X, $X = U_{\alpha_1} \cup U_{\alpha_2} \cup \cdots \cup U_{\alpha_m}$. Note that any open cover of X that involves only a finite number of open sets trivially satisfies the finite refinement requirement since the cover itself is finite. Therefore, the only covers that might lack a finite subfamily that also covers are those involving an infinite number of open subsets (i.e. the index set A has an infinite number of elements). Thus, any space with a finite number of points is compact. The real line is not compact because the open cover comprising the union of open intervals $(n - 5/8, n + 5/8)$ for every integer n covers $\mathbb{R}$, but if we remove just one of the intervals from the family, the remainder no longer cover $\mathbb{R}$. Any open interval of $\mathbb{R}$ is homeomorphic to $\mathbb{R}$ so it is also noncompact. Any closed, bounded interval in $\mathbb{R}$ is compact. In fact, the Heine-Borel theorem states that any subset of $\mathbb{R}^n$ is compact if it is closed and bounded. This intuitively characterizes any compact topological space.

21.2 RIEMANN SURFACES

Our goal in this section is to classify all Riemann surfaces. Let's begin with the topological structures.

Recall that 2 surfaces are topologically equivalent or homeomorphic if there is a bijective continuous map between them whose inverse is also continuous. Thus, to show that 2 surfaces are homeomorphic, we find a homeomorphism. This may not be the simplest or the most efficient way to go about this. If we want to show that 2 surfaces are not homeomorphic, then we must show that there are no homeomorphisms. Instead of trying to directly show that there are no homeomorphisms, we can try to use topological properties to show that they cannot be homeomorphic. For example, the closed interval, $[a, b]$, on the line cannot be homeomorphic to the line because the closed interval is compact and the real line is not. The same argument holds for the sphere and the plane. The real line cannot be homeomorphic to the plane, because if we delete a point from each, the real line breaks into 2 disconnected pieces, while the plane minus a point remains connected. We will use a similar argument to first show that the surfaces with differing genus (number of handles) cannot be homeomorphic.

Let's start with 2-sided surfaces that are compact and have no boundary. (We will not consider 1-sided surfaces since all Riemann surfaces are orientable.) Consider the sphere (genus 0) and the torus (genus 1). On the sphere, every simple closed curve γ separates the sphere into two disconnected parts. (See figure 21.6.) If we cut the sphere along γ, it falls into 2 parts. However, there are simple closed curves γ' that do not cut the torus into two parts. Therefore, the two cannot be homeomorphic. We cannot deform one into the other in a continuous way.

Next consider the torus and a surface with genus 2 as in figure 21.7. If we make any two cuts in the torus along two nonintersecting curves γ_1 and γ_2, the torus always falls into 2 parts. However, we can find 2 nonintersecting closed curves, γ_1' and γ_2', on the genus 2 surface that leave the surface connected. Thus, the torus and a genus 2 surface cannot be homeomorphic.

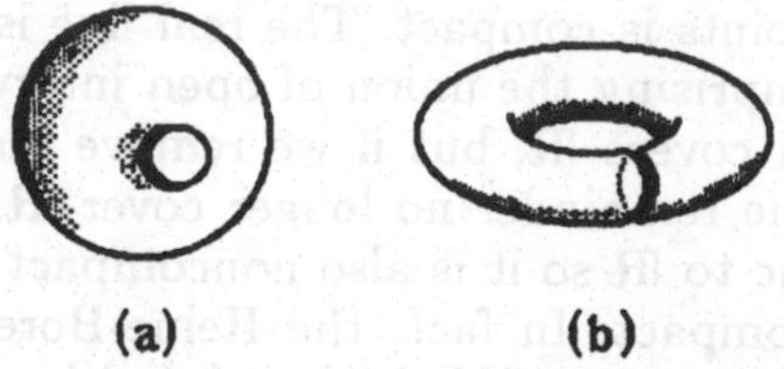

Figure 21.6 A demonstration that (a) the sphere (genus 0) is not homeomorphic to (b) the torus (genus 1). On the sphere, every simple closed curve separates the sphere into 2 disconnected pieces. However, there are simple closed curves that do not cut the torus into 2 parts.

One can easily see how to generalize this argument for any genus surface (any n closed nonintersecting curves on a genus $(n-1)$ surface always separate it). Compact surfaces with no boundary and different genus cannot be homeomorphic. If we now consider compact 2-sided surfaces with boundaries (which cannot be homeomorphic to surfaces without boundary because of the boundaries), we can see that we can use the same argument except now we take curves that start on and end on the boundary. In this way we can see that two surfaces of the same genus but with a different number of boundary components (i.e. holes) cannot be homeomorphic. The genus 0 case with 1 boundary component vs. 2 boundary components is illustrated in figure 21.8.

For the case of noncompact surfaces, we can *define* the genus of the surface to be the maximum number of noninteracting closed curves on the surface which do not divide the surface into disconnected pieces. Then, trivially, two surfaces with differing genus cannot be homeomorphic.

So far we have argued that surfaces with different genus cannot be homeomorphic. If one thinks of the surface as being flexible, then it is easy to imagine that a continuous deformation (no tearing or folding) of the surface is not going to change the maximum number of nonintersecting

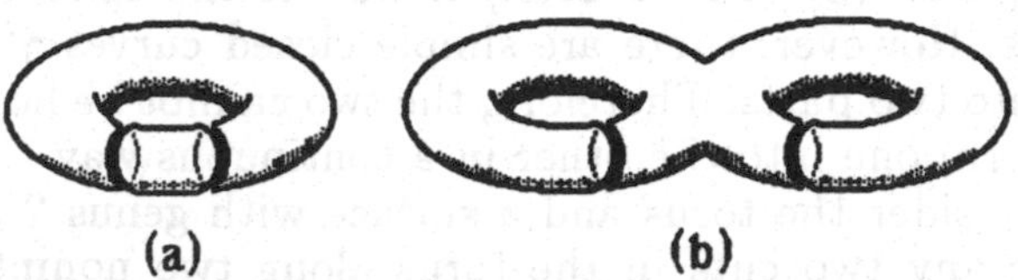

Figure 21.7 Same argument as in figure 21.6, except between (a) the torus (genus 1) and (b) a genus 2 surface. Any two nonintersecting closed curves separate the torus into two parts. This is not true for the genus 2 surface, hence the two cannot be homeomorphic.

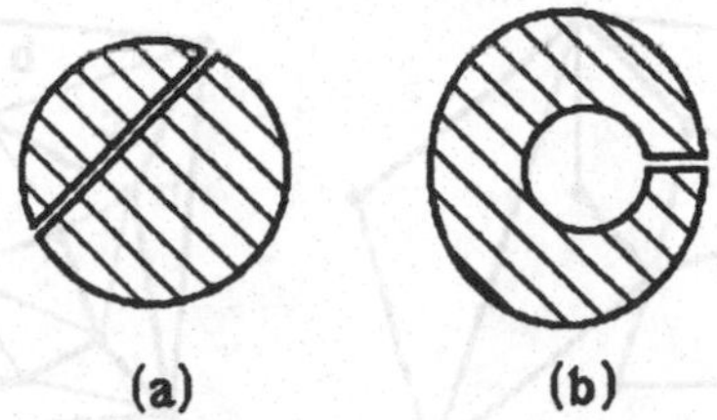

Figure 21.8 Demonstration that (a) a genus 0 surface with one boundary component is not homeomorphic with (b) a genus 0 surface with 2 boundary components (1 hole).

cuts that leave the surface in one piece. Therefore, this quantity is a topological invariant, and surfaces with the same genus and the same number of boundary components are homeomorphic.

A topological invariant that we will encounter in the string path integral is the *Euler-Poincaré characteristic*, χ. The origin of this invariant is Euler's formula for polyhedra. For a simple polyhedron (no holes, genus 0), it states that

$$V - E + F = 2, \tag{21.1}$$

where V is the number of vertices, E is the number of edges, and F is the number of faces.

We can prove this by the following argument: We start with any simple polyhedron and cut out one face. We have not removed any edges or vertices, just one face, thus $V - E + F$ has decreased by 1. Now flatten out the remaining polyhedron surface. An example of a polyhedron with a face missing that has been flattened is given in figure 21.9(a). The flattening does not change the number of vertices, edges, or faces. Next, we want to triangulate the flattened surface. If any face is not a triangle, then we draw straight lines between vertices of the face until it is broken into triangles. A triangulation of the surface in figure 21.9(a) is given in 21.9(b). (The triangulation is not unique.) Each time we connect two vertices to form a triangle, we are adding an edge and a face. Therefore, $V - E + F$ remains constant. Now we start to cut triangles off the triangulated flattened surface, starting from the outside, and only cutting along the edges. If it takes only one cut to remove a triangle (e.g. the triangle marked "a" in figure 21.9(b)), then we are removing 1 face, 1 vertex, and 2 edges. Thus the sum $V - E + F$ stays the same for the remaining connected triangles. If it takes two cuts to remove a triangle (e.g. as for the two triangles labeled "b" in figure 21.9(b)), then we are removing 1 edge, 1 face, but no vertices. Once again the sum $V - E + F$ has not changed. After the two triangles labeled "b" are removed, the triangle "c" requires only one cut, so the same argument as for the triangle labeled "a" can be applied. Eventually, as we cut and remove triangles, the only thing we will have left is just 1 triangle. The remaining triangle will have 3 edges,

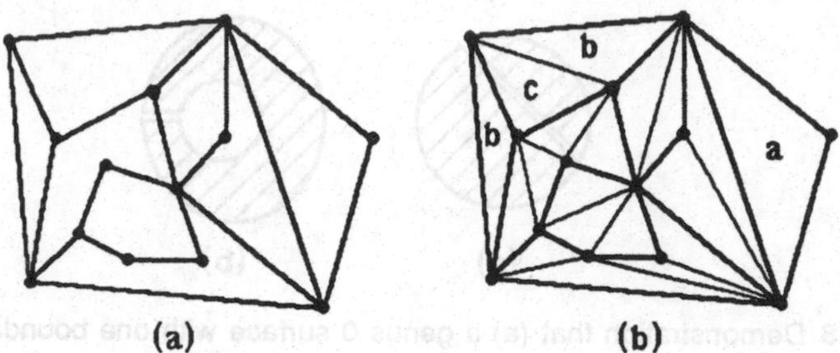

Figure 21.9 (a) A simple polyhedron that has been flattened after one face has been removed (here a 5-sided face). (b) Triangulation of the flattened surface in (a). When we remove the triangle labeled "a" by cutting along 1 edge, we are removing 1 face, 2 edges, and 1 vertex from the surface. When we remove the triangle labeled "b" by cutting along 2 edges, we are removing 1 face and 1 edge.

3 vertices, and 1 face, hence $V - E + F = 1$. Since this value did not change as we removed triangles, then the two flattened surfaces in figure 21.9 both have $V - E + F = 1$. Since we removed a face from the original polyhedron before flattening it, then the original polyhedron must have $V - E + F = 2$, which is eq.(21.1).

Any simple polyhedron is homeomorphic with the sphere. This can be seen by thinking of the polyhedron as a balloon and "inflating" it into a sphere. An example using a cube is given in figure 21.10. The polyhedron can be triangulated and thereby provide a triangulation of the sphere. Since the process of inflation does not change the number of vertices, edges, or faces of the triangulation, then the sum $V - E + F$ on the sphere will still be 2. The Euler-Poincaré characteristic of the sphere, $\chi(S^2) = 2$, is the sum $V - E + F$ of a triangulation of the sphere (or any polygonal decomposition of the sphere). Since inflation does change χ, it is a topological invariant.

$$\chi(\text{genus } 0, \text{no boundary}) = V - E + F = 2 \tag{21.2}$$

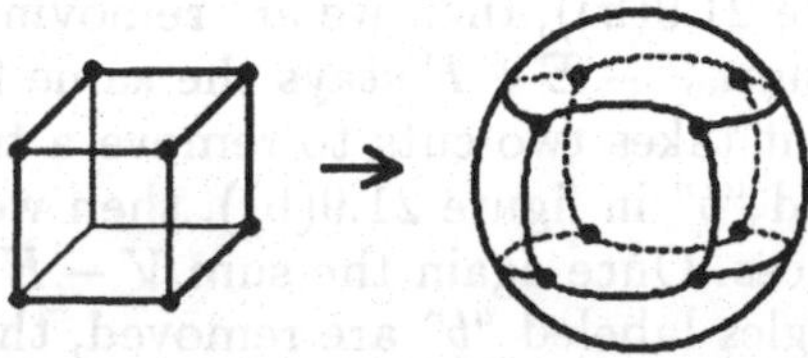

Figure 21.10 A cube can be continuously deformed into the sphere by inflating it like a balloon, therefore the cube and sphere are homeomorphic.

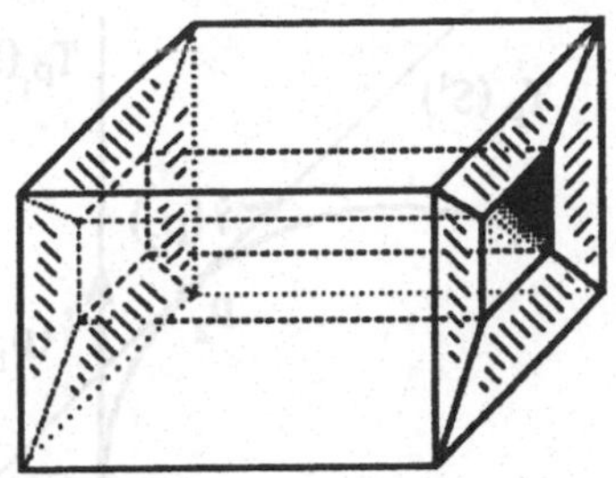

Figure 21.11 Non-simple polyhedron that can be inflated to give the torus. A count of vertices, edges, and faces yields $V - E + F = 0$.

Next, consider the non-simple polyhedron with 1 hole in figure 21.11. This can be inflated to give the torus. A quick count of vertices, edges, and faces gives $V - E + F = 0$. Is this true for any genus 1 polyhedron? Reconsider any simple genus 0 polyhedron, and triangulate it by drawing straight lines between vertices on each nontriangular face to make triangles. As we stated before, each new line adds one edge, but also one more face, so $V - E + F$ is still 2. Now cut out any *two* completely separate triangular faces (no common edges or vertices). $V - E + F$ is now 0 because we have decreased the number of faces by two without changing the number of vertices or edges. Next we draw lines internally connecting the three vertices of one missing face with the three on the other. This adds 3 edges and 3 new faces (which will be four-sided) and turns our polyhedron into genus 1. Since this operation adds the same number of edges and faces, the sum $V - E + F$ remains the same as before the internal lines were added, and we get $V - E + F = 0$ for any genus 1 polyhedron or surface.

To see what happens at genus 2, we take two genus 1 polyhedra that have been triangulated and cut one face out of each one. Next, we glue the two together, lining up the three vertices and edges on the missing faces. This results in a genus 2 polyhedron. In this process we eliminate 3 vertices, 3 edges, and the 2 faces we cut out. Thus, the net change in $V - E + F$ is -2. Since the genus 1 polyhedra start with $V - E + F = 0$, then the genus 2 polyhedron has $V - E + F = -2$. Now we repeat the process. Every time we glue on another genus 1 polyhedron to form higher-genus polyhedra, the net loss in $V - E + F$ is 2, hence the *Euler-Poincaré characteristic* for any polyhedron of genus g (and, by inflation, for any compact 2-sided genus g surface without boundary) is

$$\chi(S_g) = V - E + F = 2 - 2g. \tag{21.3}$$

For the case of a compact connected 2-sided genus 0 surface with a boundary, we can carry out the same argument by placing vertices in the interior

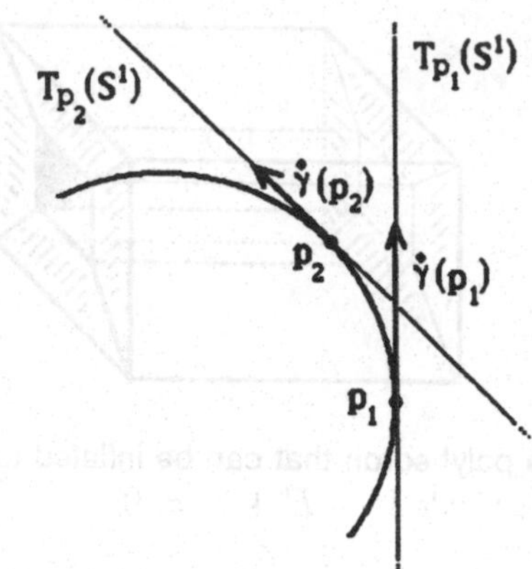

Figure 21.12 The tangent space to a point on the circle, S^1, is the real line. The velocity vectors of a curve in S^1 at two different points are in two different tangent spaces. From the Euclidean viewpoint of the imbedding, the difference in the two vectors at p_1 and p_2 is a vector perpendicular to $\dot{\gamma}(p_1)$, and therefore is in neither tangent space.

and on the boundary and triangulating. Higher-genus surfaces can be created by removing faces. The result is

$$\chi(S_g \text{ with boundary}) = V - E + F = 2 - 2g - b,$$

where b is the number of boundary components. This is left as an exercise.

For a compact surface, the Euler-Poincaré characteristic is also related to the total amount of curvature over the surface, a fact we will make explicit use of in the string path integral. Curvature is a geometric quantity associated with a geometric structure on a surface. Recall that a geometric structure is the assignment of an inner product or metric on the tangent space at each point of the surface. The metric may vary from point to point on the surface, the only requirement being that the coordinate representations of the components be differentiable.

Let $\gamma : \mathbb{R} \rightarrow S$ be a differentiable curve on a geometric surface S parametrized by $t \in [a, b]$. The velocity vector at the point $\gamma(t_0) = p \in S$ on the surface is in the tangent space of the surface at point p, $\dot{\gamma} \in T_p(S)$. Suppose we wanted to know how the velocity vector changes as we move along the curve (the acceleration). We immediately run into difficulty in that we cannot directly compare two velocity vectors at two different points on the curve, because the tangent vectors at two different places on the curve, say p_1 and p_2, are in two *different* tangent spaces, $T_{p_1}(S)$ and $T_{p_2}(S)$. In fact, the change in the velocity vector as we move along the curve is a vector that does not appear to belong to either tangent

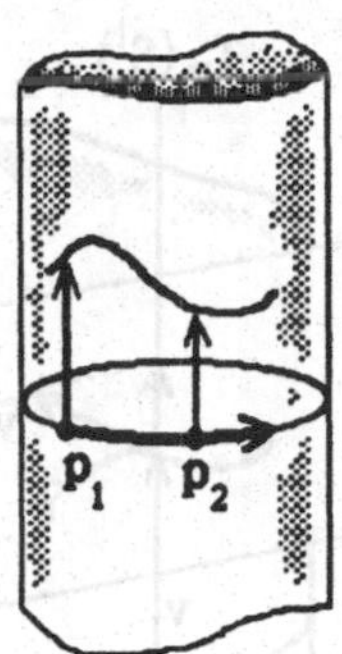

Figure 21.13 Tangent bundle, $TS^1 = S^1 \times \mathbb{R}$, for the circle. The curve γ in S^1 "lifts" to a curve $\tilde{\gamma}$ in the bundle. The arrow above p_1 is $\dot{\gamma}(p_1)$ and that above p_2 is $\dot{\gamma}(p_2)$.

space. This is illustrated for the 1-dimensional case in figure 21.12, where the space is a circle and the tangent space at the point $p \in S^1$ is the real line, $T_p(S^1) = \mathbb{R}$. Intuitively, the difference in the two vectors in figure 21.12 is perpendicular to $\dot{\gamma}(p_1)$. The point is that a change in a vector at $T_p(S^1)$ along the curve does not appear to be a vector (not an element in either tangent space). Well, as usual, comparison of objects in two different spaces is made with mappings, and the mappings that connect the various tangent spaces are called "connections."

For simplicity, let us continue with the 1-dimensional example. Even though the velocity vectors are in different tangent spaces, the velocity vector field along the curve is a curve in the tangent bundle, TS^1. The tangent bundle is the collection of all tangent spaces for the manifold, together with a projection mapping $\pi : TM \to M$, that projects from a point in the tangent bundle (a vector in some tangent space to the manifold) back into the manifold. Thus, if $\vec{v} \in T_p(M)$, then $\pi(\vec{v}) = p$. The n-manifold M is called the base space, and the tangent space at p is called a fiber, so the map π projects down the fiber back to the base space. The only other condition on π is that it must make the bundle look locally "trivial." This means that if U is a subset of M, then TM restricted to U is diffeomorphic to the product space $U \times \mathbb{R}^n$.

For our example, $M = S^1$, and each fiber is the real line. If we "turn" each tangent space on its end, then the tangent bundle for S^1 is the infinite cylinder, $S^1 \times \mathbb{R}$. See figure 21.13. The collection of velocity vectors along the curve γ in S^1 lifts to a curve $\tilde{\gamma}$ in TS^1. The change in the curve in TS^1 intuitively contains the information about how fast the velocity vectors are changing as we move along the curve in S^1. Thus, we want to consider vectors tangent to $\tilde{\gamma}$. However, these vectors are elements of the tangent spaces to the bundle and not of the bundle. This is a restatement of the

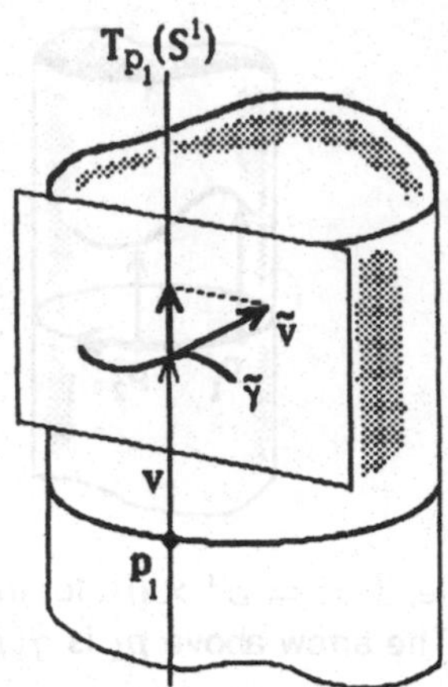

Figure 21.14 The fiber above point $p_1 \in S^1$ is the tangent space at p_1, $T_{p_1}(S^1)$. $\vec{v} \in T_{p_1}(S^1)$ points "vertically" along the fiber. The velocity vector, $\tilde{v}$, tangent to the lifted curve $\tilde{\gamma}$ in TS^1, is an element of the tangent space to the bundle TS^1. Any vertical projection of it along the fiber $T_{p_1}(S^1)$ can naturally be identified with a vector in $T_{p_1}(S^1)$. The "connection" is a vertical projection.

remark that changes in the velocity vectors do not appear to be vectors on S^1.

The vectors tangent to the bundle do have a component that can be identified with points in the bundle, and therefore will be vectors. If we consider the vector $\tilde{v}$ tangent to the lifted curve $\tilde{\gamma}$ in the bundle at a point that projects to p_1 in S^1 (figure 21.14), then the "vertical" component of this vector points along a single fiber. Since a single fiber is a tangent space to S^1, then it is a vector space, and we have a natural identification between a vertical vector tangent to the fiber and a point in the fiber. The mapping we mentioned earlier connecting tangent spaces at different points is equivalent to a rule for projecting vectors tangent to the bundle to vertical vectors (i.e. decomposing vectors tangent to the bundle into horizontal and vertical components). Thus, the *connection* is a "vertical" projection. There are many different ways of projecting a vector, hence the connection is certainly not unique. In a moment, we will specify which connection we want.

The point of going through all this business about connections is that we want to define covariant differentiation of a vector field on the base manifold such that the derivative of a vector is a vector. Since we want the connection to define a derivative, then we will want it to have the usual properties of a derivative. Let $\vec{u}$ and $\vec{w}$ be vectors tangent to a manifold M at point p. Let $\vec{v}$ and $\vec{q}$ be vector fields on M (i.e. a choice of one vector

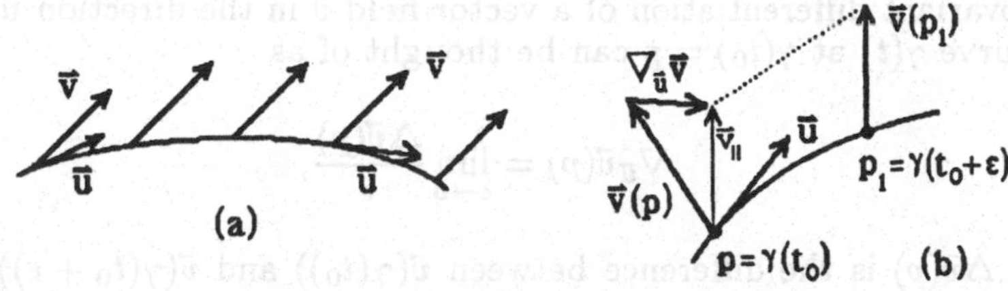

Figure 21.15 **(a) Parallel transport of a vector $\vec{v}$ along a curve whose velocity vector field is $\vec{u}$. Parallel transport implies that $\nabla_{\vec{u}}\vec{v} = 0$. (b) Definition of covariant differentiation of a vector field $\vec{v}$ along a curve γ. The vector at p_1, $\vec{v}(p_1)$, is parallel transported back to p along γ where the comparison is made with $\vec{v}(p)$.**

at each point on M—this is a section of TM). The *covariant derivative* of $\vec{v}$ in the direction of $\vec{u}$ is denoted as $\nabla_{\vec{u}}\vec{v}$. The connection is required to satisfy

$$\begin{aligned} \nabla_{a\vec{u}+b\vec{w}}\vec{v} &= a\nabla_{\vec{u}}\vec{v} + b\nabla_{\vec{w}}\vec{v} \\ \nabla_{\vec{u}}(a\vec{v} + b\vec{q}) &= a\nabla_{\vec{u}}\vec{v} + b\nabla_{\vec{u}}\vec{q}, \end{aligned} \tag{21.4}$$

and if f is a function on M,

$$\nabla_{\vec{u}}(f\vec{v}) = f\nabla_{\vec{u}}\vec{v} + (\partial_{\vec{u}}f)\vec{v}, \tag{21.5}$$

where $\partial_{\vec{u}}f$ is the usual directional derivative of a function f in the direction $\vec{u}$.

A coordinate patch on M provides us with coordinates on M. Choose a basis $\{e_i\}$ of vectors that span $T_p(M)$ at point p that are tangent to the coordinate lines. Then the covariant derivative on the various basis vectors is

$$\nabla_{e_i}e_j = \Gamma^k_{ij}e_k. \tag{21.6}$$

The Γ^k_{ij} are called the *connection coefficients* (or *Christoffel symbols* when the e_i are tangents to coordinate lines). If $\vec{u} = u^i e_i$, $\vec{v} = v^i e_i$, then, in terms of the connection coefficients,

$$\nabla_{\vec{u}}\vec{v} = u^i\partial_i v^j + u^i\Gamma^j_{ki}v^k. \tag{21.7}$$

A vector field $\vec{v}$ is said to be *parallel transported* along $\vec{u}$ if $\vec{v}$ is covariantly constant along $\vec{u}$, $\nabla_{\vec{u}}\vec{v} = 0$. If the $\vec{u}$ are velocity vectors of some curve γ, then $\nabla_{\vec{u}}\vec{v} = 0$ means that $\vec{v}$ is carried along the curve γ without changing its direction (figure 21.15(a)). A *geodesic* is a curve that parallel transports its own velocity vector (no acceleration). That is, if α is a geodesic, and $\vec{u}$ are the velocity vectors tangent to the curve, then $\nabla_{\vec{u}}\vec{u} = 0$.

Covariant differentiation of a vector field $\vec{v}$ in the direction $\vec{u}$ tangent to a curve $\gamma(t)$ at $\gamma(t_0) = p$ can be thought of as

$$\nabla_{\vec{u}}\vec{v}(p) = \lim_{\epsilon \to 0} \frac{\Delta\vec{v}(p)}{\epsilon},$$

where $\Delta\vec{v}(p)$ is the difference between $\vec{v}(\gamma(t_0))$ and $\vec{v}(\gamma(t_0+\epsilon))$ parallel transported back along the curve to $p = \gamma(t_0)$ (figure 21.15(b)). The comparison between vectors is easily made since the parallel transported vector is now in the same space as $\vec{v}(p)$. Thus, the connection connects neighboring tangent spaces via parallel transport.

The covariant derivative is not just restricted to comparison of contravariant vectors. We can consider the changes in other types of tensors as they are parallel transported. Let $\underline{w}$ be a *covariant vector* (dual to a contravariant vector). In component form, $\underline{w} = w_i\theta^i$, where θ^i are the dual basis vectors to the coordinate basis e_i. Covariant vectors are linear functionals of contravariant vectors which means that they map contravariant vectors to the real numbers. The dual basis satisfies $\theta^i(e_j) = \delta^i_j$, thus $\underline{w}(\vec{v}) = w_i v^i$. (Note that no metric is involved. The metric tensor is used to change covariant into contravariant tensors, etc.) In terms of the connection coefficients, the covariant derivative of a contravariant vector $\underline{w}$ along $\vec{u}$ is

$$\nabla_{\vec{u}}\underline{w} = u^i\partial_i w_j - u^i\Gamma^k_{ji}w_k. \tag{21.8}$$

For a second-rank tensor, $\mathbf{N}$, of mixed type (1,1) (1 contravariant index, and 1 covariant index),

$$\nabla_{\vec{u}}\mathbf{N} = u^i\partial_i N^j_k + u^i\Gamma^j_{ni}N^n_k - u^i\Gamma^n_{ki}N^j_n, \tag{21.9}$$

etc.

The metric tensor $\mathbf{g}$ is a symmetric second-rank tensor of type (0,2) (2 covariant indices), and changes contravariant types into covariant types, $v_i = g_{ij}v^j$. The inverse metric (type (2,0)—also just called the metric) does the reverse, $v^i = g^{ij}v_j$. The metric defines the scalar or dot product of two vectors. $\mathbf{g}(\vec{u},\vec{v}) = u^i g_{ij} v^j = u_i v^i = u^i v_i = \vec{u}\cdot\vec{v}$. The inverse metric does the same for dual vectors. As we mentioned previously, there are many possible connections. There is, however, a unique connection that is compatible with a given metric that satisfies $\Gamma^k_{ij} = \Gamma^k_{ji}$ (torsion free). A metric-compatible connection satisfies $\nabla_{\vec{u}}\mathbf{g} = 0$ for any $\vec{u}$. The metric parallel transported from point p_1 to p_2 agrees with the metric already at p_2 no matter what path is taken. (What a mess if it didn't.) In component form, the compatibility equation is

$$\nabla_i g_{jk} = \partial_i g_{jk} - \Gamma^n_{ji}g_{nk} - \Gamma^n_{ki}g_{jn} = 0. \tag{21.10}$$

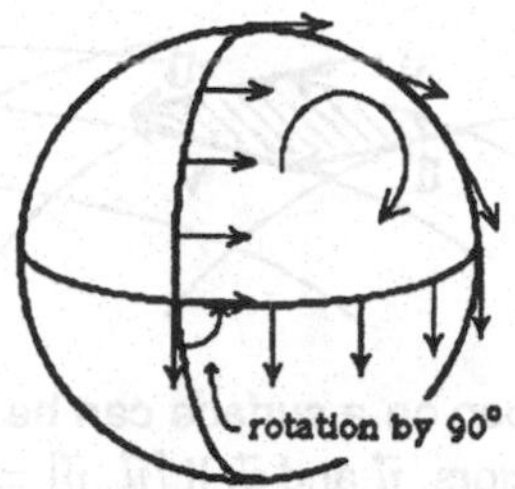

Figure 21.16 A vector rotates 90° when parallel transported around a closed right triangle on the sphere. This is a manifestation of the curvature of the sphere.

The torsion free connection compatible with the metric (also called the *Levi-Civitá connection*) is the solution to eq.(21.10),

$$\Gamma^k_{ij} = \frac{1}{2} g^{kn} \left(\partial_j g_{ni} + \partial_i g_{nj} - \partial_n g_{ij} \right). \tag{21.11}$$

One way that curvature manifests itself is when parallel transport of a vector from p_1 to p_2 depends on the path chosen to go from p_1 to p_2. Let γ_1 and γ_2 be two such curves. By first traversing γ_1 from p_1 to p_2, and then parallel transporting the result back to p_1 on γ_2^{-1}, we can equivalently say that curvature manifests itself when parallel transport of a vector around a closed loop does not reproduce the original vector. One standard example is illustrated in figure 21.16. A vector is parallel transported around a closed triangle on the 2-sphere. Starting at the equator, we move up towards the north pole along a line of constant longitude. At the north pole we turn 90° and head down towards the equator along another line of constant longitude. When we arrive at the equator we turn 90° again and head back along the equator to our starting point. Upon arrival, we find that the parallel transported vector has rotated 90° with respect to the original vector.

As we shrink the closed loop over which we parallel transport, we can get a better and better measure of the curvature near the starting point. For very tiny loops, we can approximate them by a parallelogram made up of two tangent vectors, $\vec{u}$ and $\vec{v}$, as in figure 21.17. For just a moment, intuitively define the "product," $\vec{u}\vec{v}$, as the point reached by moving along the vector $\vec{u}$, then along the vector $\vec{v}$. Assume that $[\vec{u}, \vec{v}] = 0$ so that the parallelogram closes (i.e. we reach the same point no matter if we travel along $\vec{u}$ first or $\vec{v}$ first). The ***curvature operator*** is the commutator of covariant derivatives along $\vec{u}$ and $\vec{v}$,

$$\mathcal{R}(\vec{u}, \vec{v}) = [\nabla_{\vec{u}}, \nabla_{\vec{v}}]. \tag{21.12}$$

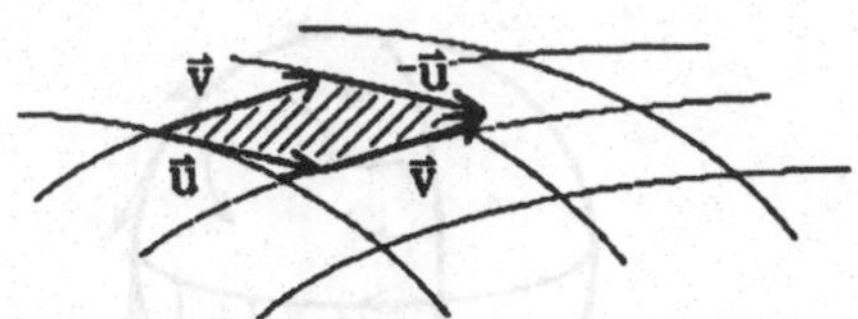

Figure 21.17 A tiny closed loop on a surface can be approximated by a parallelogram generated by two tangent vectors, $\vec{u}$ and $\vec{v}$. If $[\vec{u}, \vec{v}] = 0$, then we reach the same point no matter which direction we go around the parallelogram, hence the parallelogram closes.

Compare this with the discussion preceding eq.(6.16). If $[\vec{u}, \vec{v}] \neq 0$, subtract $\nabla_{[\vec{u},\vec{v}]}$ from the right-hand side of eq.(21.12), where $[\vec{u}, \vec{v}]$ is the vector pointing from $\vec{u}\vec{v}$ to $\vec{v}\vec{u}$ that closes the parallelogram. Note that $\mathcal{R}(\vec{u}, \vec{v}) = -\mathcal{R}(\vec{v}, \vec{u})$. If we parallel transport a vector $\vec{w}$ around this tiny parallelogram, then the change in $\vec{w}$ is $\mathcal{R}(\vec{u}, \vec{v})\vec{w} = \nabla_{\vec{u}}\nabla_{\vec{v}}\vec{w} - \nabla_{\vec{v}}\nabla_{\vec{u}}\vec{w}$. The result of this is a vector, hence to get a number out of this we must contract it on a covariant vector. Thus, the curvature tensor defined by the curvature operator is a fourth-rank tensor of type (1,3) called the *Riemann tensor*. In component form,

$$R^l_{kij} = \theta^l \mathcal{R}(e_i, e_j)e_k = \langle \mathcal{R}(e_i, e_j)e_k, e_l \rangle,$$

where $\langle , \rangle$ denotes the inner product. In terms of the connection coefficients,

$$R^l_{kij} = \partial_i \Gamma^l_{kj} - \partial_j \Gamma^l_{ki} + \Gamma^l_{ni}\Gamma^n_{kj} - \Gamma^l_{nj}\Gamma^n_{ki}. \tag{21.13}$$

The *Ricci tensor* is a particular contraction of the Riemann tensor, $R_{kj} = R^i_{kij}$. The *scalar curvature* is the trace of the Ricci tensor, $R = R^k_k$.

With some insight or tedious manipulation of indices, one can show that $R^l_{kij} = -R^l_{kji}$ (which follows from eq.(21.12)), $R^l_{kij} + R^l_{ijk} + R^l_{jki} = 0$ (because the connection coefficients are symmetric in the lower 2 indices), $R_{lkij} = -R_{klij}$ (true when the connection is compatible with the metric —note that we needed the metric to lower the first index), and $R_{lkij} = R_{ijlk}$ (which follows from the first 3 relations). Altogether this means that the Riemann tensor has $n^2(n^2-1)/12$ independent components on an n-manifold. Luckily, on a surface ($n = 2$), there is just one independent component. Curvature on a surface can be described by a single scalar function on the surface known as Gaussian curvature.

There are many equivalent ways to characterize the Gaussian curvature at a point on a geometric surface. Let us consider one, based on the local version of the *Gauss-Bonnet theorem*, that does not depend on how the surface is imbedded into a Euclidean space.

A geodesic triangle on a surface is a triangle whose sides connecting the 3 vertices are geodesics. The local form of the Gauss-Bonnet theorem

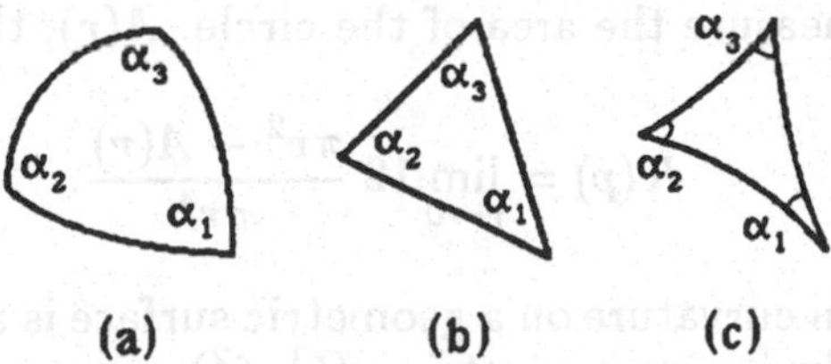

Figure 21.18 Three types of geodesic triangles. In (a), the sum of the interior angles is $> \pi$, so the integrated curvature inside the triangle is positive. In (b), the sum is π, so the total curvature vanishes. In (c), the sum is $< \pi$, so the integrated curvature is negative.

(this part was proven by Gauss) states that the sum of the 3 interior angles of the triangle minus π is the integral of the Gaussian curvature (traditionally denoted as K) over the interior of the triangle. If the interior angles are α_1, α_2, and α_3, then

$$(\alpha_1 + \alpha_2 + \alpha_3) - \pi = \int K\,dA. \tag{21.14}$$

For example, the geodesic triangle on the sphere of radius r illustrated in figure 21.16 has $\alpha_1 = \alpha_2 = \alpha_3 = \pi/2$. The area of the triangle is $(4\pi r^2)/8$, so the Gaussian curvature, which happens to be constant over the triangle, is $K = 1/r^2$ which is positive.

Three types of geodesic triangles are illustrated in figure 21.18. The first has an integrated curvature that is positive, the second is flat, and the third has negative integrated curvature. As we shrink the triangle down, we get a better estimate of the curvature at one point, p, in the triangle, Δ, hence

$$K(p) = \lim_{\Delta \to \text{point } p} \frac{(\alpha_1 + \alpha_2 + \alpha_3) - \pi}{\text{Area of } \Delta}. \tag{21.15}$$

The same kind of argument can be used for nongeodesic triangles, except that eq.(21.14) must include the curvature of the boundaries of the triangle (this is the part Bonnet proved).

An equivalent definition along the same lines involves geodesic circles. A geodesic circle is constructed around point p by traveling a distance r away from point p on a geodesic in every direction. We then compare the measurement of the circumference of the circle, $C(r)$, with $2\pi r$, the flat case result. If the curvature is positive, the circumference will be less than $2\pi r$. If the curvature is negative, it will be more. The curvature at point p is

$$K(p) = \lim_{r \to 0} \frac{2\pi r - C(r)}{\pi r^3}. \tag{21.16}$$

If, instead, we measure the area of the circle, $A(r)$, then

$$K(p) = \lim_{r\to 0} 12\,\frac{\pi r^2 - A(r)}{\pi r^4}. \tag{21.17}$$

The Gaussian curvature on a geometric surface is also 1/2 of the scalar curvature, R. Thus, given a metric $g(\xi^1, \xi^2)$ expressed in terms of local coordinates, ξ^1 and ξ^2, the Gaussian curvature, in terms of the Riemann tensor, is

$$K(p) = \frac{R_{1212}}{\det(g)}, \tag{21.18}$$

where, as usual, $\det(g) = g_{11}g_{22} - (g_{12})^2$. This computation, based on eqs.(21.11) and (21.13), may still be tedious. We cannot resist the opportunity to introduce differential forms to speed up the computation. This economy is a result of the antisymmetry of the Riemann tensor under the exchange of various indices.

A *1-form* is a linear functional on a vector field (takes vectors to real numbers), i.e. a covariant vector. If $\vec{v}$ is a vector, and $\underline{w}$ a 1-form, then $\underline{w}(\vec{v}) \in \mathbb{R}$. A *2-form* is a skew-symmetric or antisymmetric bilinear tensor of type (0,2) (2 covariant indices). If $\vec{u}$ and $\vec{v}$ are vectors, and $\mathbf{B}$ is a 2-form, then $\mathbf{B}(\vec{u}, \vec{v}) = -\mathbf{B}(\vec{v}, \vec{u}) \in \mathbb{R}$. A 3-form would be a completely antisymmetric tensor of type (0,3), but we need not consider any forms higher than a 2-form on a surface since the antisymmetry in 2-dimensions automatically makes them vanish. By convention, a *0-form* is just an ordinary function.

To make higher-rank tensors from lower-rank tensors, we can use the *tensor product*, $\otimes$. The equivalent product for forms is called the *wedge product*, $\wedge$. It is just an antisymmetric tensor product. If $\underline{w}$ and $\underline{r}$ are two 1-forms, then

$$\underline{w} \wedge \underline{r} = \underline{w} \otimes \underline{r} - \underline{r} \otimes \underline{w} \tag{21.19}$$

is a 2-form. Note that this means that $\underline{w} \wedge \underline{w} = 0$. To write this out in component form, let us introduce a basis of 1-forms, θ^i ($i = 1, 2$ on a surface). These basis 1-forms span the *cotangent space* at point p on the surface, the space of covariant vectors dual to the tangent space at p. In component form, $\underline{w} = w_i\theta^i$, $\underline{r} = r_i\theta^i$, and

$$\begin{aligned} \underline{w} \wedge \underline{r} &= w_i r_j \theta^i \otimes \theta^j - r_i w_j \theta^i \otimes \theta^j \\ &= \sum_{i<j} (w_i r_j - w_j r_i)\theta^i \wedge \theta^j. \end{aligned} \tag{21.20}$$

Note that $\theta^i \wedge \theta^j$ form a basis for 2-forms. The wedge of a 0-form with a 1-form or a 2-form is just ordinary scalar multiplication.

Another way to form 2-forms from 1-forms or 1-forms from 0-forms is with the *exterior derivative*, d. It is just an antisymmetric derivative. If $f(\xi^1, \xi^2)$ is a 0-form, then df is the ordinary gradient,

$$df = \frac{\partial f}{\partial \xi^i} d\xi^i = \partial_i f \, d\xi^i. \tag{21.21}$$

Note that the coordinates ξ^i can be thought of as functions, thus $d\xi^i$ is a 1-form and can be used to form a coordinate basis of 1-forms (i.e. we can take $\theta^i = d\xi^i$ in a coordinate basis). If $\underline{w} = w_i(\xi^1, \xi^2) d\xi^i$, then

$$\begin{aligned} d\underline{w} &= \frac{\partial w_i}{\partial \xi^j} d\xi^j \wedge d\xi^i \\ &= \sum_{i>j} (\partial_j w_i - \partial_i w_j) d\xi^j \wedge d\xi^i \end{aligned} \tag{21.22}$$

which is just the curl. Note that

$$\begin{aligned} d^2 f = d(df) &= \frac{\partial^2}{\partial \xi^j \partial \xi^i} d\xi^j \wedge d\xi^i \\ &= \sum_{i>j} \left(\frac{\partial^2 f}{\partial \xi^j \partial \xi^i} - \frac{\partial^f}{\partial \xi^i \partial \xi^j} \right) d\xi^j \wedge d\xi^i \\ &= 0 \end{aligned} \tag{21.23}$$

by equality of mixed partials. In a space that supports 3-forms, we could use eq.(21.22) to show that $d^2 \underline{w} = 0$ for the same reason. $d^2 = 0$ holds in general.

Now let us return to geometry on a surface. A ***moving frame field*** on an orientable surface S is a choice of an orthonormal vector field $\{e_1, e_2\}$ for the tangent space, $T_p(S)$, at each point $p \in S$. Let $\{\theta^1, \theta^2\}$ be the dual basis for the cotangent space at each point. $\theta^i(e_j) = \delta^i_j$. The ***connection 1-forms***, ω^i_j, are defined by the ***first structural equation***,

$$d\theta = -\omega \wedge \theta, \tag{21.24}$$

or, $d\theta^i = \omega^i_j \wedge \theta^j$, i.e.

$$\begin{aligned} d\theta^1 &= -\omega^1_2 \wedge \theta^2 \\ d\theta^2 &= -\omega^2_1 \wedge \theta^1. \end{aligned} \tag{21.25}$$

Expand the connection 1-forms in the dual basis,

$$\omega^i_j = \Gamma^i_{jk} \theta^k. \tag{21.26}$$

The coefficients of the connection 1-form are the Christoffel symbols or connection coefficients, eq.(21.11). Substituting eq.(21.26) into eq.(21.25) implies that $\omega_1^2 = -\omega_2^1$.

The *second structural equation* defines the *curvature 2-forms*, Ω^i_j, via

$$d\omega = -\omega \wedge \omega + \Omega, \tag{21.27}$$

or

$$d\omega^i_j = -\omega^i_k \wedge \omega^k_j + \Omega^i_j. \tag{21.28}$$

The coefficients of the curvature 2-form in the $\theta^i \wedge \theta^j$ basis are the components of the Riemann tensor,

$$\Omega^i_j = \frac{1}{2} R^i_{jkl} \theta^k \wedge \theta^l. \tag{21.29}$$

In 2 dimensions, $\omega_2^1 \wedge \omega_1^2 = 0$, hence $d\omega_2^1 = \Omega_2^1$, which in turn is related to R_{1212}. In fact, in a frame field, $\det(g) = 1$, so by eq.(21.18), the Gaussian curvature at p, $K(p)$, is just the $\theta^1 \wedge \theta^2$ component of Ω_2^1.

$$d\omega_2^1 = K(p)\theta^1 \wedge \theta^2. \tag{21.30}$$

As an example, let us do the sphere of radius r again. The usual metric on the sphere is (ϑ measured from north pole, ϕ azimuthal)

$$(ds)^2 = r^2 \sin^2(\vartheta)\,(d\phi)^2 + r^2(d\vartheta)^2 \tag{21.31}$$

Thus,

$$\theta^1 = r\sin(\vartheta)\,d\phi, \qquad \theta^2 = r\,d\vartheta. \tag{21.32}$$

The first structural equation gives

$$\begin{aligned} d\theta^1 &= r\cos\vartheta\; d\vartheta \wedge d\phi \\ &= -\cos\vartheta\; d\phi \wedge (r d\vartheta) \\ d\theta^2 &= 0 \end{aligned} \tag{21.33}$$

Thus, $\omega_2^1 = \cos(\vartheta)\,d\phi$. The second structural equation yields

$$\begin{aligned} d\omega_2^1 &= -\sin(\vartheta)\; d\vartheta \wedge d\phi \\ &= \frac{1}{r^2}(r\,\sin(\vartheta)\,d\phi) \wedge (r\,d\vartheta), \end{aligned} \tag{21.34}$$

hence $K = 1/r^2$ everywhere on the sphere.

Finally, let us return to the topological connection with the geometry of the surface. The *global Gauss-Bonnet theorem* states that the total

Gaussian curvature of a compact orientable geometric surface, S, is equal to 2π times the Euler-Poincaré characteristic of the surface,

$$\int_S K\sqrt{g}\,d^2\xi = 2\pi\chi(S). \tag{21.35}$$

This result can be established from the local Gauss-Bonnet theorem by considering any (nongeodesic) triangulation of S and applying the local Gauss-Bonnet theorem to each triangle. The curvature associated with each nongeodesic edge will cancel when the local theorem is applied to the neighboring triangles that share the edge. Also, interior angles associated with all triangles sharing any one common vertex must add up to 2π. Therefore, the sum of all interior angles for all triangles must add up to 2π times the number of vertices, $2\pi V$. From eq.(21.14), we also see that there is a contribution of $-\pi$ for each triangle, the sum of which is $-\pi F$, where F is the number of faces. However, if there are F faces, then there are $3F/2$ edges (each triangle or face has three edges, but each edge is connected to 2 neighboring triangles, hence $3F = 2E$). Therefore, $-\pi F = -3\pi F + 2\pi F = -2\pi E + 2\pi F$. When we add this to the sum of interior angles, we get $2\pi\chi(S)$. The proof of the local Gauss-Bonnet theorem (as with all the proofs we have left out in this chapter) may be found in the references in the Bibliography.

One consequence of the Gauss-Bonnet theorem is that any compact orientable surface without boundary that has positive total Gaussian curvature must be diffeomorphic with the sphere (the only surface with positive χ). If the total Gaussian curvature vanishes, then the surface is genus 1, and so on.

Before we move on to the conformal classification of Riemann surfaces, let us introduce one more algebraic quantity that is a topological invariant. This is the fundamental group of a surface from homotopy theory. Before we begin, we want to define an arc-wise or path-wise connected topological space as a space where any two points in the space can be joined by an arc. An arc-wise connected space is connected, but the converse is not necessarily true. The classic example is the "topologist's sine curve," which consists of the points in the plane, (x, y), satisfying $y = \sin(1/x)$ for $0 \le x \le 1$ and the origin, $(0,0)$. With the induced topology from the plane, one can see that this space is connected. It is not arc-wise connected because there is no continuous path connecting the origin to any point (x, y) for $x > 0$. All surfaces under consideration are arc-wise connected since every manifold is.

A path, γ, on a surface S is a continuous mapping from a closed interval, $[a, b]$, of the real line to the surface, $\gamma : [a, b] \to S$. Two paths, γ_1, and γ_2, with the same initial and final endpoints ($\gamma_1(a) = \gamma_2(a)$, $\gamma_1(b) = \gamma_2(b)$), are equivalent or *homotopic* if one path may be continuously deformed into the other while holding the endpoints fixed.

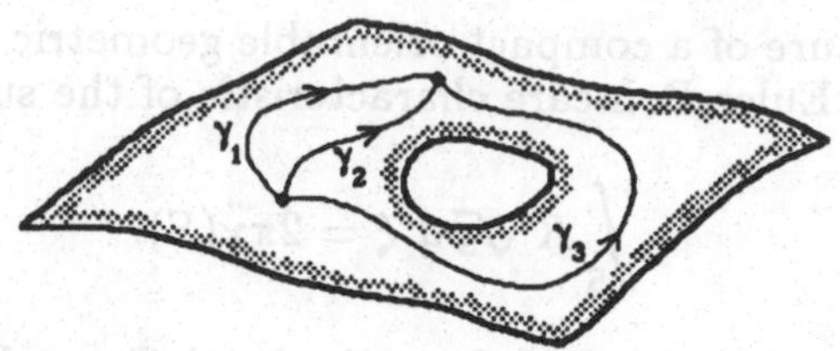

Figure 21.19 Paths γ_1 and γ_2 are homotopic, since each can be deformed continuously into the other, keeping the endpoints fixed. γ_2 and γ_3 are not homotopic, since we would have to break γ_3 to deform it into γ_2 due to the hole.

In other words, γ_1 and γ_2 are homotopic if there is a continuous map $f : [a,b] \times I \to S$ (I is the unit interval $[0,1]$), such that $f(s,0) = \gamma_1(s)$, $f(s,1) = \gamma_2(s)$, $f(a,t) = \gamma_1(a) = \gamma_2(a)$, and $f(b,t) = \gamma_1(b) = \gamma_2(b)$. For example, paths γ_1 and γ_2 in figure 21.19 are homotopic, while γ_2 and γ_3 are not because of the intervening hole. Note that γ_1 is also not homotopic with γ_3, and that homotopy of paths forms an equivalence relation. Implicit in the definition of this equivalence relation is the orientation of the path, i.e. the direction we move on the path.

Let α and β be two paths in S such that the final endpoint of α is the same as the initial endpoint of β. For convenience, let us always assume we have used homeomorphisms so that the domain of all paths is the unit interval. The requirement on α and β is $\alpha(1) = \beta(0)$. Let us define the product of two such paths, $\alpha \cdot \beta$ as a path that goes along α, then along β. That is, if $\gamma = \alpha \cdot \beta$, then

$$\gamma(t) = \begin{cases} \alpha(2t) & 0 \le t \le 1/2 \\ \beta(2t-1) & 1/2 \le t \le 1. \end{cases} \tag{21.36}$$

Notice that this product is compatible with the equivalence relation we just defined. If α_1 and α_2 are homotopic, $\alpha_1 \sim \alpha_2$, and β_1 and β_2 are also homotopic, and if $\alpha_1(1) = \alpha_2(1) = \beta_1(0) = \beta_2(0)$, then $\alpha_1 \cdot \beta_1$ is homotopic to $\alpha_2 \cdot \beta_2$.

We would like to use this multiplication to build a group out of equivalence classes of paths. Define the inverse of the path $\alpha(t)$ as $\bar{\alpha}(t) = \alpha^{-1}(t) = \alpha(1-t)$ (travel in the opposite direction). The paths α and α^{-1} are not in the same equivalence class because the orientation of α^{-1} is the opposite of α. The identity map, e_p is the constant curve, $\alpha(t) = p \in S$. If $\gamma(0) = p$, then $\gamma \cdot \gamma^{-1}$ is in the same equivalence class as e_p. If $\gamma(1) = q$, then $\gamma^{-1} \cdot \gamma$ is in the same equivalence class as e_q. The path product of equivalence classes, when defined, is also associative, i.e. $(\alpha \cdot \beta) \cdot \gamma$ is homotopic to $\alpha \cdot (\beta \cdot \gamma)$. Thus, the set of all equivalence classes of paths on S along with equivalence class multiplication, defined above, satisfies the axioms of a group *except* that the path product is not always defined between equivalence classes of paths. This arises only because the final endpoint of one equivalence class does not match the initial endpoint of another. However, if we consider the subset of equivalence classes consisting of closed paths starting and ending at point p on the surface, then the path product is always defined and we have a group. This group is called the *fundamental group* of S at p, denoted $\pi_1(S,p)$.

Consider two points on the surface, p_1 and p_2, and the fundamental groups based at the two points, $\pi_1(S,p_1)$ and $\pi_1(S,p_2)$. If S is arc-wise connected, then we can connect p_1 and p_2 with a path γ. If α is a closed path based at p_2, then $\gamma \cdot \alpha \cdot \gamma^{-1}$ is a closed path based at p_1. Thus, the path γ defines an isomorphism between $\pi_1(S,p_1)$ and $\pi_1(S,p_2)$. So, if S is arc-wise connected, all of the fundamental groups on S are isomorphic. In this case there is no need to always specify a base point, and the fundamental group is a property of the surface and not the particular point on the surface where the closed paths are based. In addition, if two topological spaces are homeomorphic, then the homeomorphism can be used to define an isomorphism between the fundamental groups on the two spaces. Thus, the fundamental group, and whatever algebraic properties it has, are topological invariants.

Let us consider a couple of examples. On the sphere, any closed loop is homotopic with the identity map at a point, hence the fundamental group is trivial and consists of one element, the equivalence class containing the constant closed curves. This is true for any simply connected space. One way this is often stated is that all loops are contractible to a point. We may think of contracting a loop by pulling in the curve from the final endpoint side as if pulling in the free end of a garden hose. If nothing prevents us from completely pulling in the hose (like a hole or handle in the surface or a tree in the yard), then the loop is contractible to a point. If all loops are so, then the space is simply connected.

The example surface in figure 21.19 is not simply connected because of the hole. The closed loop consisting of $\gamma_3 \cdot \gamma_2^{-1}$ cannot be contracted to a point without breaking the loop. The fundamental group on this surface is isomorphic to the set of integers, $\mathbb{Z}$, under addition, with the integers representing the number of times the loop is wrapped around the hole. The sign of the integer corresponds to the orientation of the loop as it wraps around the hole. The fundamental group of a circle is also isomorphic to the set of integers, $\mathbb{Z}$. The curve,

$$\gamma(t) = \begin{cases} e^{i\pi t} & 0 \le t \le \frac{1}{2} \\ e^{-i\pi t} & \frac{1}{2} \le t \le 1 \end{cases},$$

is contractible to the identity map, $\gamma(t) = 0$. However, the map $\gamma(t) = \exp(i2\pi t)$, which wraps once around the circle, is not.

If X and Y are arc-wise connected topological spaces with fundamental groups $\pi_1(X)$ and $\pi_1(Y)$, then the product space $X \times Y$ will have fundamental group $\pi_1(X) \times \pi_1(Y)$. Thus, the fundamental group for the torus, $T^1 = S^1 \times S^1$, is isomorphic to $\mathbb{Z} \times \mathbb{Z}$.

Now let us turn to the classification of Riemann surfaces. To begin, let's review some definitions from analysis on the complex plane, $\mathbb{C}$. The function $f : A \rightarrow \mathbb{C}$ is *holomorphic* or *analytic* in A if the first derivative of f, $f^{(1)}(z)$, exists for all points $p \in A$. (Some mathematicians distinguish between "holomorphic" and "analytic," reserving "analytic" for maps between plane domains of $\mathbb{C}$ and "holomorphic" for maps between Riemann surfaces. Others adopt the opposite convention. We will average the two and just use the two terms interchangeably.)

The existence of a derivative in the complex plane is far more restrictive than on the real line. The first derivative is defined, as usual, as

$$f^{(1)}(z) = \lim_{\Delta z \rightarrow 0} \frac{f(z + \Delta z) - f(z)}{\Delta z}, \tag{21.38}$$

but the limit must be independent of the path taken from $z + \Delta z$ to z. If $z = x + iy$ and $f(z) = u(x, y) + iv(x, y)$, then eq.(21.38) becomes

$$f^{(1)}(z) = \lim_{\Delta x \rightarrow 0} \frac{\Delta u + i\Delta v}{\Delta x} = \frac{\partial u}{\partial x} + i\frac{\partial v}{\partial x} \tag{21.39}$$

when $\Delta z = \Delta x$ (approach z along a line of constant y), and

$$f^{(1)}(z) = -i\frac{\partial u}{\partial y} + \frac{\partial v}{\partial y} \tag{21.40}$$

when $\Delta z = i\Delta y$. In order for the derivative to exist, eq.(21.39) must agree with eq.(21.40), hence

$$\frac{\partial u}{\partial x} = \frac{\partial v}{\partial y}, \quad \text{and} \quad \frac{\partial v}{\partial x} = -\frac{\partial u}{\partial y}. \tag{21.41}$$

Eqs.(21.41) are known as the *Cauchy-Riemann equations*, and are conditions necessary for the existence of the complex derivative. They are sufficient when the first partials of u and v exist and are continuous. Note that if $\bar{z} = x - iy$, the complex conjugate of z, then in the formal sense, eqs.(21.41) hold if and only if

$$\frac{\partial f}{\partial \bar{z}} = \frac{\partial f}{\partial x} + i\frac{\partial f}{\partial y} = 0. \tag{21.42}$$

A function that is analytic in $A \subset \mathbb{C}$ is *univalent* if it is also one-to-one (injective). A function that is analytic on the entire complex plane

$\mathbb{C}$ is called *entire*. By Liouville's theorem, a function that is entire and bounded is a constant.

The condition of analyticity is sufficiently strong so as to guarantee that the derivatives of an analytic function are also analytic. This means that all higher-order derivatives of an analytic function exist and are continuous. The higher-order derivatives may be defined through Cauchy's integral formula,

$$f^{(n)}(z_0) = \frac{n!}{2\pi i} \oint_C \frac{f(z)}{(z - z_0)^{n+1}} dz, \tag{21.43}$$

where C is any closed contour in A and z_0 is any interior point of the curve. This also implies that there exists a power series expansion of f about z_0 in A.

For a nonconstant function f on A, f has zeroes of order n at $z_0 \in A$ if $f(z_0)$ vanishes along with the first $n-1$ derivatives, $f^{(1)}(z_0) = f^{(2)}(z_0) = \cdots = f^{(n-1)}(z_0) = 0$, $f^{(n)}(z_0) \neq 0$. If f is analytic in a neighborhood of z_0, but is not analytic at z_0, then z_0 is called an isolated singularity of f. If there is a function g that is analytic in the neighborhood of z_0 and at z_0, and is equal to f for $z \neq z_0$, then the singularity at z_0 is called *removable*. If $f(z_0) = \infty$ at the isolated singularity, then f is said to have a *pole* at z_0. The pole is of order n, if $(z - z_0)^n f(z)$ is finite as $z \to z_0$. A pole of order 1 is a removable singularity if $(z - z_0)f(z) = 0$ as $z \to z_0$. If an isolated singularity is neither a removable singularity, nor a pole of finite order, then it is an *essential singularity*. If the function f is analytic in $A \subset \mathbb{C}$ except for poles, then it is said to be *meromorphic* in A. The quotient of two analytic functions, $f(z)/g(z)$ is meromorphic provided that $g(z)$ is not identically zero.

A holomorphic or analytic map $f : A \to \mathbb{C}$ is *conformal* at z_0 if $f^{(1)}(z_0) \neq 0$. A conformal map preserves angles between intersecting smooth curves and maps tiny "circles" to "circles." This arises because the Jacobian of the mapping $(f(z) = u(x, y) + iv(x, y))$, after application of the Cauchy-Riemann equations, is

$$J = \begin{pmatrix} \partial u/\partial x & \partial u/\partial x \\ \partial u/\partial y & \partial v/\partial y \end{pmatrix} = \begin{pmatrix} \partial u/\partial x & \partial v/\partial x \\ -\partial v/\partial x & \partial u/\partial x \end{pmatrix}, \tag{21.44}$$

which is $|f^{(1)}(z)|^2 = (\partial u/\partial x)^2 + (\partial v/\partial x)^2$ times an orthogonal matrix of unit determinant. The orthogonal matrix represents a solid 2-dimensional rotation, thus preserving angles, while $|f^{(1)}(z)|$ is the local magnification at the point z. A (nonholomorphic) map that maps circles to ellipses is called quasi-conformal. Note that eq.(21.44) also implies that all Riemann surfaces are orientable.

Clearly a conformal map is holomorphic. But how about the other way around? If $f : A \to \mathbb{C}$ is holomorphic in A, then f will be conformal at $z_0 \in A$ if and only if f is univalent (one-to-one) in a neighborhood of z_0 contained in A. Thus, if f is analytic and univalent in A, it is conformal. However, ***in addition***, if f is analytic and univalent, then $f(A)$ is open when A is open, and $f^{-1} : f(A) \to A$ is *also* analytic.

We can use the complex structures on Riemann surfaces to transfer the definitions and results above to maps on and between Riemann surfaces. For example, a map between two surfaces, $f : S_1 \to S_2$ is analytic if its coordinate expression, $\phi_2 \circ f \circ \phi_1^{-1} : \mathbb{C} \to \mathbb{C}$, is analytic, for all combinations of coordinate patches in the analytic atlases of the two surfaces. Similarly, $f : S_1 \to S_2$ is conformal if its coordinate expressions are, etc. Applying the results of the preceeding paragraph, if $f : S_1 \to S_2$ is an analytic bijection, then f, which must be locally univalent, is also conformal. In addition, f^{-1} must also be analytic and conformal. Thus, two surfaces are conformally equivalent if (and only if) there is an analytic bijection (i.e. holomorphic homeomorphism) between them. When two surfaces are conformally equivalent, they have the same analytic properties and cannot be distinguished in terms of their complex structures. This is why a complex structure on a Riemann surface is often also called a conformal structure. Interestingly, though, the same topological surface with two incompatible complex structures (i.e. two different Riemann surfaces) may still be conformally equivalent.

A conformal homeomorphism $f : S \to S$ of a surface into itself is called an ***automorphism***. That is, an automorphism of S is a univalent holomorphic map of the surface into itself. Automorphisms play a central role in the classification of Riemann surfaces.

Because of their importance in classifying all Riemann surfaces, we want to examine three particular Riemann surfaces in more detail. These surfaces are: the complex plane ($\mathbb{C}$), the open unit disc ($\mathcal{D}$) (or upper half-plane, $\mathcal{U}$), and the extended complex plane or Riemann sphere (Σ or $\mathbb{C}_\infty$). All three are simply connected and conformally inequivalent. This is obvious for Σ, since it is compact while $\mathbb{C}$ and $\mathcal{D}$ are not. However, $\mathbb{C}$ and $\mathcal{D}$ are homeomorphic. They are not conformally equivalent, because any function $f : \mathbb{C} \to \mathcal{D}$ that is holomorphic everywhere must be bounded and therefore a constant by Liouville's theorem. Such a function cannot also be a homeomorphism.

Let's start with the Riemann sphere. The Riemann sphere is a 1-point compactification of the complex plane, also called the "extended complex plane," $\mathbb{C} \cup \{\infty\}$, denoted as Σ or $\mathbb{C}_\infty$. The identification of the extended complex plane with the unit sphere, S^2, is easily made using stereographic projection, as illustrated in figure 21.20. The unit sphere centered about the origin is identified bijectively with points in the (x, y) plane (i.e. $\mathbb{C}$) by drawing straight lines from the north pole $N = (0, 0, 1)$ down through the (x, y) plane. The point where this line intersects the sphere, p is mapped

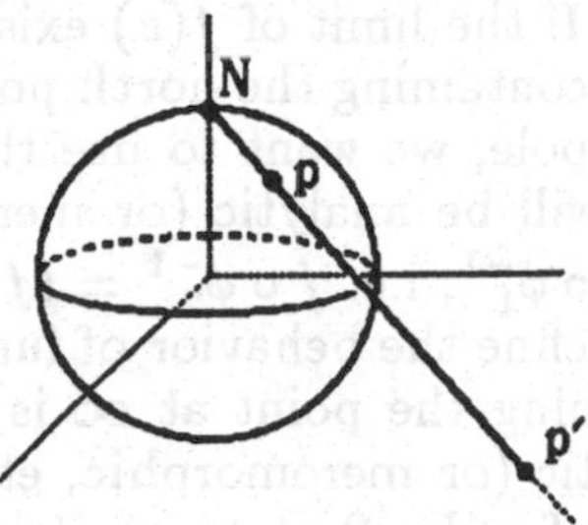

Figure 21.20 Stereographic projection of the unit sphere onto the plane. A line is drawn from the north pole $N = (0,0,1)$ down through the $z = 0$ plane. The point of intersection of the line with the sphere, p, is identified with the point of intersection, p', of the line with the (x, y) plane.

to the point of intersection of the plane, p'. The equator is identically mapped to the unit circle, the south pole with the origin, and the north pole with the point at infinity, denoted ∞.

We can use the projection mapping to build an analytic atlas on Σ consisting of two coordinate patches. (Every atlas on Σ will contain at least 2 patches.) The first patch, (ϕ_1, A_1), covers the entire sphere except for the north pole, while the second patch, (ϕ_2, A_2), covers the entire sphere except for the south pole. The chart on the first patch, ϕ_1, is just stereographic projection described in the previous paragraph. It is a bijective map of A_1 onto $\mathbb{C}$, $\phi_1 : A_1 \to \mathbb{C}$. The inverse of the projection, $\phi_1^{(-1)}$, places the coordinates of $\mathbb{C}$ back onto the sphere. The bijective chart on the second patch, $\phi_2 : A_2 \to \mathbb{C}$, is just stereographic projection from the south pole.

Another useful map is the inversion, $I : \mathbb{C}_\infty \to \mathbb{C}_\infty$, defined by $z \mapsto 1/z$. This takes the unit circle or equator into itself. Everything inside the circle is mapped to the outside, and vice versa. The origin is mapped to the point at ∞ and vice versa, so the net effect of an inversion is the rotation of the sphere 180° about the x-axis. The second chart in the atlas given above is related to the first by composing the first chart with the inversion, $\phi_2 = \phi_1 \circ I$.

We can use the atlas to transfer properties (e.g. analyticity of maps) from the plane to the extended plane or sphere. We can then use the inversion map to define the behavior of functions at ∞. First consider functions f that map a subset $B \subset \Sigma$ into the plane $\mathbb{C}$, $f : B \to \mathbb{C}$. Suppose that B does not contain the north pole. Since ϕ_1 is analytic, we can say that f is analytic in B if $f \circ \phi_1^{-1} : \mathbb{C} \to \mathbb{C}$ is analytic at every point in $f(B) \subset \mathbb{C}$, f is meromorphic in B if $f \circ \phi_1^{-1}$ is meromorphic, etc. Suppose now that f is defined in a neighborhood of the north pole but not at the north pole. If the limit of $f(z)$ exists for $z \to \infty$, then we can define f in a subset B containing the north pole by using this limit. Since B contains the north pole, we want to use the second coordinate patch in the atlas. Then f will be analytic (or meromorphic, etc.), if $f \circ \phi_2^{-1}$ is. Note that $\phi_2^{-1} = I \circ \phi_1^{-1}$, i.e. $f \circ \phi_2^{-1} = (f \circ I) \circ \phi_1^{-1}$, so we are using the inversion map to define the behavior of functions at ∞ by saying that a map f on B containing the point at ∞ is analytic (or meromorphic, etc.), if $f \circ I$ is analytic (or meromorphic, etc.) at $z = 0$. For example, $f(z) = z^2$ has a pole of order 2 at ∞ so it is meromorphic there, while $f(z) = 1/(1+z)$ is holomorphic at ∞ with a zero of order 1.

Now consider what happens if we extend the range of f to include the point at ∞, $f : B \to \Sigma$. Now that ∞ is in the range, poles are no longer singularities and meromorphic functions become analytic. In fact, we use this to *define* a meromorphic function, g, on any Riemann surface, S, as a function that is analytic on Σ. That is, g is meromorphic on S if $g : S \to \Sigma$ is analytic. Thus, there is no distinction between meromorphic and analytic or holomorphic for maps of the sphere into itself, $f : \Sigma \to \Sigma$. In this case, one can show that all meromorphic or analytic functions $f : \Sigma \to \Sigma$ are rational functions, i.e. functions that are the quotient of two polynomials, $p(z)/q(z)$, where $q(z)$ is not identically zero.

Recall that an automorphism of a surface is a univalent holomorphic map (a conformal homeomorphism) of a surface into itself. For the case of the sphere, a univalent holomorphic function is the same as a meromorphic bijection, and all holomorphic functions of the sphere into itself are rational, $p(z)/q(z)$. Obviously, we can assume that $p(z)$ and $q(z)$ do not have any common factors since we can just cancel them. The degree of rational or analytic functions on Σ is the maximum of the degrees of the polynomials, $p(z)$ and $q(z)$. In order that an analytic function on Σ be univalent (one-to-one or injective), its degree must be 1. Thus, an automorphism of Σ must be of the form,

$$T(z) = \frac{az+b}{cz+d} \qquad a, b, c, d \in \mathbb{C}, \tag{21.45}$$

where $ad - bc \neq 0$ (i.e. no common factors). The collection of all automorphisms on Σ is denoted as $Aut(\Sigma)$.

Transformations of the form eq.(21.45) are called Möbius or linear fractional transformations. Note that if $\alpha \in \mathbb{C}$, then αa, αb, αc, αd define the same transformation provided that $\alpha \neq 0$.

One can show that the composition of two Möbius transformations is again a Möbius transformation. The composition is associative. The identity transformation has $a = d \neq 0$ with $b = c = 0$ or $T(z) = z$, and the inverse of eq.(21.45) is

$$T^{-1}(z) = \frac{dz-b}{-cz+a}. \tag{21.46}$$

Therefore, the Möbius transformations form a group. A matrix representation of this group can be defined by associating the matrix

$$M(T) = \begin{pmatrix} a & b \\ c & d \end{pmatrix} \tag{21.47}$$

with the transformation (21.45). Composition corresponds to matrix multiplication, hence the mapping of matrices of the type eq.(21.47) to Möbius transformations, eq.(21.45), is a group homomorphism. The group of matrices of the form eq.(21.47) with nonvanishing determinant is the general linear group, $GL(2, \mathbb{C})$.

The general linear group of 2×2 matrices is not isomorphic to the group of Möbius transformations, $Aut(\Sigma)$, because $\alpha M(T)$, $\alpha \in \mathbb{C}$, $\alpha \neq 0$, defines the same T. If $M : GL(2, \mathbb{C}) \to Aut(\Sigma)$ is the group homomorphism, then the kernel of M, $\Lambda = \ker(M)$, is the set of matrices that are mapped to the identity element of $Aut(\Sigma)$, the transform $T(z) = z$. These matrices are of the form $a = d = \lambda \neq 0$, and $b = c = 0$, or $\Lambda = \lambda \mathbf{1}$ for all nonvanishing $\lambda \in \mathbb{C}$, where $\mathbf{1}$ is the 2×2 unit matrix. The group of Möbius transformations will be the quotient group, $Aut(\Sigma) = GL(2, \mathbb{C})/\Lambda$ which is known as the projective general linear group, $PGL(2, \mathbb{C})$. The special linear group, $SL(2, \mathbb{C})$, is the group of all 2×2 matrices that are unimodular (determinant = 1). Since λM defines the same Möbius transformation as $M \in GL(2, \mathbb{C})$, then we can take $\lambda^2 = 1/\det M$ so that λM is an element of $SL(2, \mathbb{C})$. If we define the projective special linear group, $PSL(2, \mathbb{C})$, to be the subset of $PGL(2, \mathbb{C})$ with unit determinant, $PSL(2, \mathbb{C}) = SL(2, \mathbb{C})/\Lambda \cap SL(2, \mathbb{C})$, then we also have that $Aut(\Sigma)$ is isomorphic to $PSL(2, \mathbb{C})$ (i.e. $PSL(2, \mathbb{C})$ is isomorphic to $PGL(2, \mathbb{C})$).

Anti-automorphisms of Σ are defined by

$$T(z) = \frac{a\bar{z} + b}{c\bar{z} + d}, \tag{21.48}$$

where $a, b, c, d \in \mathbb{C}$ and $ad - bc \neq 0$ (which we can take as $= 1$). $\bar{z}$ is the complex conjugate of z. The anti-automorphisms also form a group. Automorphisms preserve the orientation of Σ while anti-automorphisms reverse it.

$PGL(2, \mathbb{C})$ is generated by the following 4 Möbius transforms:

$$\begin{aligned} T^1(z) &= e^{i\theta} z \\ T^2(z) &= rz \qquad r \in \mathbb{R} \text{ and } r > 0 \\ T^3(z) &= z + t \qquad t \in \mathbb{C} \\ T^4(z) &= 1/z = I(z). \end{aligned} \tag{21.49}$$

T^1 represents a rotation of the sphere by angle θ about the axis that goes through the north and south pole. T^4, as we have already discussed, is a rotation of Σ by π about the x-axis, T^2 leaves $z = 0$ and $z = \infty$ fixed and expands or contracts the plane by r, and T^3 is a translation that leaves ∞ fixed. We can construct any element of $PGL(2, \mathbb{C})$ by starting with T^1 and applying T^2 with $r\exp(i\theta) = -c^2$. Next we apply T^3 with $t = -cd$. The result so far is $T^3 \circ T^2 \circ T^1 = -c^2 z - cd$. Now apply T^4, then T^3 again with $t = a/c$. The result is

$$T^3(t = \frac{a}{c}) \circ T^4 \circ T^3(t = -cd) \circ T^2 \circ T^1 = \frac{a}{c} - \frac{1}{c(cz + d)} = \frac{az + (ad - 1)/c}{cz + d}, \tag{21.50}$$

which is eq.(21.45) when $ad - bc = 1$.

A circle on the sphere is the intersection of any non-tangent plane with the unit sphere. Using stereographic projection, this defines a circle on Σ. Note that any circle on the unit sphere containing the north pole will project to a straight line on Σ, hence circles on Σ include straight lines. Interestingly, one can show that each of the generators maps circles to circles, therefore all Möbius transformations (elements of $PGL(2, \mathbb{C})$) map circles to circles.

The ***stabilizer*** of the circle is the subgroup of $PGL(2, \mathbb{C})$ that maps a circle onto itself. The Möbius transformation,

$$T(z) = \frac{(z - z_1)(z_2 - z_3)}{(z_1 - z_2)(z_3 - z)}, \tag{21.51}$$

maps z_1 to 0, z_2 to 1, and z_3 to ∞. The circle containing these three points is the x-axis and the point at ∞, $\mathbb{R}_\infty = \mathbb{R} \cup \{\infty\}$. Since any three distinct points determine a plane, then these three points also determine a circle, hence $T(z)$ above maps this circle to the circle $\mathbb{R}_\infty$. If we knew the stabilizer of $\mathbb{R}_\infty$, then we would know the stabilizer for any circle by using $T(z)$ above, the stabilizer of $\mathbb{R}_\infty$, and T^{-1} of eq.(21.51) above. Since $\mathbb{R}_\infty$ is completely real, any mapping of $\mathbb{R}_\infty$ into itself of the form eq.(21.49) must involve real coefficients. If z_1, z_2, z_3 are in $\mathbb{R}_\infty$ in eq.(21.51), then when $z \in \mathbb{R}_\infty$, so will be $T(z)$. Let $W(z)$ be equal to $T(z)$ of eq.(21.51) with $w_1, w_2, w_3 \in \mathbb{R}_\infty$ replacing z_1, z_2, z_3. Then $W^{-1} \circ T(z)$ takes z_1, z_2, z_3 to w_1, w_2, w_3 and has real coefficients, hence the stabilizer of $\mathbb{R}_\infty$ is the collection of Möbius transformations with real coefficients, $PGL(2, \mathbb{R})$.

For a circle C in Σ, the interior is a disc D. The Möbius transform $T(z)$ that takes C into $\mathbb{R}_\infty$ must map the interior D to either the upper half-plane $\mathcal{U}$ ($Im(z) > 0$), or the lower half-plane $\mathcal{L}$ ($Im(z) < 0$). Inversion, $T^4(z)$, flips $\mathcal{U}$ into $\mathcal{L}$ and vice versa, hence we can always map a disc into the upper half-plane. The stabilizer of $\mathcal{U}$ must be similar to that for $\mathbb{R}_\infty$, because any mapping of $\mathcal{U}$ into itself must map $\mathbb{R}_\infty$ (its boundary as a circle) to itself. Therefore, the stabilizer of $\mathcal{U}$ must be a subgroup of $PGL(2, \mathbb{R})$, the stabilizer of $\mathbb{R}_\infty$. By an examination of an arbitrary element of $PGL(2, \mathbb{R})$, we can see that if $ad - bc < 0$ $(a, b, c, d \in \mathbb{R})$, then $\mathcal{U}$ is taken into $\mathcal{L}$. Thus, we must have $ad - bc > 0$ in order that $\mathcal{U}$ go into $\mathcal{U}$. By projection, we can always take $ad - bc = 1$, so the stabilizer of $\mathcal{U}$ is $PSL(2, \mathbb{R})$. The upper half-plane plays a central role. We will return to it in a moment.

As we have already noted, the automorphisms are univalent, holomorphic maps, and therefore are conformal homeomorphisms. This can be seen from the explicit form, eq.(21.45). What is not obvious, is that every conformal map of Σ into itself is an automorphism. By writing out the stereographic projection in terms of coordinates, one can see that it too is conformal. We can use it to project the usual metric on the sphere onto the extended plane. The projection, being bijective, will be an isometry (preserve distances), because, by projecting the metric on the plane, the distance between points p' and q' on the plane will be the same as between points p and q on the sphere (as illustrated in figure 21.21). The usual metric on the unit sphere is

$$(ds)^2 = g_{ij}\, d\xi^i \otimes d\xi^j = \sin^2(\vartheta)(d\phi)^2 + (d\vartheta)^2.$$

Let (x, y) be the coordinates in the plane. By writing out the explicit form of the projection in coordinates, we find that the projected metric on the plane is

$$\begin{aligned} (ds)^2 &= \frac{4}{(1 + (x^2 + y^2))^2}\left((dx)^2 + (dy)^2\right) \\ &= \frac{4}{(1 + |z|^2)^2}|dz|^2, \end{aligned} \tag{21.52}$$

where $z = x + iy$.The details of the computation are left as an exercise. $(ds)^2 = |dz|^2 = (dx)^2 + (dy)^2$ is the flat Euclidean metric on the plane. Eq.(21.52) is of the form

$$(ds)^2 = \lambda(z)|dz|^2, \tag{21.53}$$

and is called a *conformally flat* metric. Coordinates that put the metric on a surface into this form are called *isothermal*.

A conformally flat metric is conformal with the flat Euclidean metric in that the angle between two vectors is the same with both metrics. If $\tilde{v}$ and $\tilde{w}$ are vectors in a tangent space with a conformally flat metric, then the angle, ψ, between them is given by

$$\cos(\psi) = \frac{\tilde{v} \cdot \tilde{w}}{||\tilde{v}||\,||\tilde{w}||} = \frac{\tilde{v} \cdot \tilde{w}}{\sqrt{\tilde{v} \cdot \tilde{v}}\sqrt{\tilde{w} \cdot \tilde{w}}}.$$

The conformal factor in the metric, $\lambda(z)$, cancels out on the right-hand side, and we get the same result as with the flat metric.

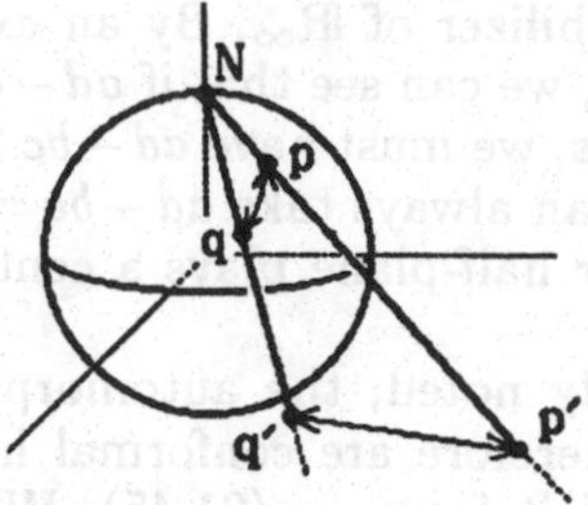

Figure 21.21 Stereographic projection of the usual metric on the unit sphere (which is the induced metric from 3-dimensional Euclidean space where the unit sphere sits) to the extended plane. By construction, the projection is an isometry, hence the distance between p' and q' in the plane will be the great circle distance between p and q on the sphere.

Even though the extended plane with conformally flat metric, eq.(21.52), may look flat, it certainly is not. For illustration, let us quickly compute the Gaussian curvature of the metric, eq.(21.52), using forms. The orthonormal basis 1-forms of a frame field are

$$\theta^1 = \frac{2}{1+(x^2+y^2)}dx, \qquad \theta^2 = \frac{2}{1+(x^2+y^2)}dy.$$

Application of the first structural equation, eq.(21.24), gives

$$\begin{aligned} d\theta^1 &= -\frac{4y}{(1+(x^2+y^2))^2}\,dy\wedge dx = \frac{2y}{1+(x^2+y^2)}dx\wedge\theta^2 \\ d\theta^2 &= -\frac{4x}{(1+(x^2+y^2))^2}\,dx\wedge dy = \frac{2y}{1+(x^2+y^2)}dy\wedge\theta^1, \end{aligned} \tag{21.54}$$

and so implies that

$$\omega^1_2 = -\frac{2y}{1+(x^2+y^2)}\,dx + \frac{2x}{1+(x^2+y^2)}\,dy. \tag{21.55}$$

Application of the second structural equation, eq.(21.30), to the connection 1-form ω^1_2 in eq.(21.55) gives

$$d\omega^1_2 = \theta^1\wedge\theta^2, \tag{21.56}$$

hence the extended plane with this metric has constant positive unit Gaussian curvature, just as with the original sphere.

Isometries of the unit sphere into itself are solid rotations of the sphere. By stereographic projection, rotations of $\mathbb{C}_\infty$ or Σ are isometries. Since the rotations are conformal, the group of rotations on the sphere or the extended complex plane forms a subgroup of the group of Möbius transformations or automorphisms on Σ, i.e. a subgroup of $PGL(2,\mathbb{C})$ or $PSL(2,\mathbb{C})$. Rotations of the sphere leave two points fixed, the points of intersection of the sphere with the axis of rotation. Projecting these two points onto the extended plane and requiring a Möbius transformation to fix only these points implies that a rotation of Σ is

$$T(z) = \frac{az+b}{-\bar{b}z+\bar{a}} \qquad a\bar{a}+b\bar{b}=1, \tag{21.57}$$

where $\bar{a}$ is the complex conjugate of a. It is easy to verify that the matrix of coefficients of eq.(21.57) is unitary, and that its determinant is 1, hence the rotations of Σ are the group $PSU(2,\mathbb{C})$. $PSU(2,\mathbb{C})$ is the quotient group, $SU(2,\mathbb{C})/I_2$, where I_2 is the group containing the 2×2 unit matrix, and -1 times the unit matrix. The group of rotations of the sphere is isomorphic to $SO(3,\mathbb{R})$, thus $PSU(2,\mathbb{C})$ is isomorphic to $SO(3,\mathbb{R})$.

Let us move on to the complex plane, $\mathbb{C}$. An atlas can be built with one patch, $\mathbb{C}$ with the identity map. What about its automorphisms, $Aut(\mathbb{C})$? Any automorphism is holomorphic, and since the range, $f:\mathbb{C}\to\mathbb{C}$, does not contain ∞, f cannot have any singularities. Therefore, f can be written as a power series with infinite radius of convergence, $f = a_0 + a_1 z + a_2 z^2 + \ldots$. f must also be univalent, so f cannot be a polynomial with degree greater than 1 (for example, if $f=(z-a)(z-b)$, then $f(a)=f(b)=0$, which is not univalent). Thus, the automorphisms of $\mathbb{C}$ must be of the form,

$$f(z) = az+b = T^3\circ T^2\circ T^1, \qquad a,b\in\mathbb{C},\ a\neq 0. \tag{21.58}$$

The last of the three canonical surfaces, the open unit disc, $\mathcal{D} = \{w\in\mathbb{C}\,||w|^2<1\}$ is conformally equivalent to the upper half-plane, $\mathcal{U} = \{z\in\mathbb{C}\,|Im(z)>0\}$. To explicitly demonstrate this, consider the Möbius transformation, $M:\mathcal{D}\to\mathcal{U}$,

$$M(w) = \frac{w+i}{iw+1}. \tag{21.59}$$

This maps the disc to the upper half-plane, because the imaginary part of $M(w)$ is >0 when $|w|^2<1$. In terms of the sphere and extended complex plane Σ, M is a rotation of the sphere by $\pi/2$ about the x-axis and therefore takes $\mathbb{R}_\infty$ to the unit circle. One can easily check that M is conformal.

In discussing the stabilizer of the circle on the sphere, we also arrived at the stabilizer of the upper half-plane. The stabilizer of $\mathcal{U}$ is a subset of

Möbius transformations that map $\mathcal{U}$ to itself, and is therefore the group of automorphisms of $\mathcal{U}$. Thus, $Aut(\mathcal{U}) = PSL(2, \mathbb{R})$. Using M and its inverse,

$$M^{-1}(z) = \frac{z - i}{-iz + 1}, \tag{21.60}$$

which maps $\mathcal{U}$ to $\mathcal{D}$, $M^{-1} \circ T \circ M : \mathcal{D} \to \mathcal{D}$ will be an automorphism of $\mathcal{D}$ when $T \in PSL(2, \mathbb{R})$,

$$T(z) = \frac{az + b}{cz + d} \qquad a, b, c, d \in \mathbb{R}, \quad ad - bc = 1. \tag{21.61}$$

By matrix multiplication we have

$$M^{-1}TM = \begin{pmatrix} 1 & -i \\ -i & 1 \end{pmatrix} \begin{pmatrix} a & b \\ c & d \end{pmatrix} \begin{pmatrix} 1 & i \\ i & 1 \end{pmatrix} = \begin{pmatrix} u & v \\ \bar{v} & \bar{u} \end{pmatrix}, \tag{21.62}$$

where $u = (a + d) + i(b - c)$, and $v = (b + c) + i(a - d)$. Therefore, automorphisms of $\mathcal{D}$ are of the form

$$T(w) = \frac{uw + v}{\bar{v}w + \bar{u}}, \qquad |u|^2 - |v|^2 = 1. \tag{21.63}$$

Compare this with eq.(21.57) which represents rotations of the sphere. The difference between eqs.(21.63) and (21.57) suggests that a different geometry is involved. Rotations of the sphere are isometries on the sphere (positive curvature). Transformations of the form eq.(21.63) represent isometries of hyperbolic geometry (negative curvature). The connection between hyperbolic geometry and $\mathcal{D}$ or $\mathcal{U}$ was discovered by Poincaré. He observed that the metric on the upper half-plane,

$$(ds)^2 = \frac{|dz|^2}{y^2}, \tag{21.64}$$

where $z = x + iy$, is invariant under the automorphisms of $\mathcal{U}$, $PSL(2, \mathbb{R})$. We leave this demonstration as an exercise. The upper half-plane equipped with this metric is called the Poincaré half-plane. Note that the metric is conformally flat.

To compute the curvature associated with this metric, we take $\theta^1 = 1/y\ dx$ and $\theta^2 = 1/y\ dy$ as basis 1-forms. Then $d\theta^1 = 1/y\ dx \wedge \theta^2$ and $d\theta^2 = 0$, so $\omega^1_2 = -1/y\ dx$. The curvature is read off from $d\omega^1_2 = 1/y^2\, dy \wedge dx = -\theta^1 \wedge \theta^2$, hence the Gaussian curvature of the Poincaré half-plane is a constant, -1.

We can use M^{-1}, eq.(21.60), to move the metric from the Poincaré half-plane to the disc, $\mathcal{D}$. The result is

$$(ds)^2 = \frac{4}{(1-|w|^2)^2}\,|dw|^2, \tag{21.65}$$

which is conformally flat. The disc $\mathcal{D}$ with this metric is sometimes called the hyperbolic plane. Even though the disc does not remind us of a plane, the distance between the center of the disc and the unit circle at the edge is infinite with this metric. Compare the form of eq.(21.65) with the projected metric on the extended plane, eq.(21.52).

The importance of the three canonical surfaces is given by the generalized Riemann Mapping theorem which is part of the Uniformization theorem. The original Riemann Mapping theorem states that any open subset A of the plane $\mathbb{C}$ that is not the entire plane, $A \neq \mathbb{C}$, is conformally equivalent to the disc, $\mathcal{D}$. This conformally classifies all simply connected open subsets of $\mathbb{C}$. The generalized Riemann Mapping theorem classifies all simply connected Riemann surfaces. It states that every simply connected Riemann surface is conformally equivalent to either $\mathbb{C}$, $\mathcal{D}$ (or $\mathcal{U}$), or Σ. Since Σ is the only compact surface of the three, this means that every simply connected compact Riemann surface is conformally equivalent to the Riemann sphere. There is only one analytic type of compact simply connected genus 0 surface. For noncompact simply connected surfaces, if the geometry is hyperbolic, then it is conformally equivalent to the disc $\mathcal{D}$. If it is parabolic, then it is conformal with the plane $\mathbb{C}$.

The remaining part of the Uniformization theorem classifies *all* remaining Riemann surfaces (i.e. non-simply connected) in that it states that the universal covering surface of any Riemann surface is one of the three canonical surfaces, $\mathbb{C}$, $\mathcal{D}$, or Σ. In addition, the cover group is a subgroup of Möbius transformations.

A ***smooth covering surface*** of a surface S is a surface $\tilde{S}$ and a continuous map $f : \tilde{S} \to S$ that is a local homeomorphism. This means that if $\tilde{U} \subset \tilde{S}$ is an open neighborhood of a point $\tilde{p} \in \tilde{S}$, then f restricted to $\tilde{U}$ is a homeomorphism. $f : \tilde{S} \to S$ is continuous and onto (surjective), but is not globally one-to-one (injective), only locally. f is called a ***covering map*** or a ***projection***. If $f(\tilde{p}) = p \in S$, then $\tilde{p}$ is said to lie over p. The collection of points of $\tilde{S}$ that lie over p (i.e. $f^{-1}(p)$) is called the fiber above p. A simple 1-dimensional example is illustrated in figure 21.22. Here the real line $\mathbb{R}$ covers S^1 when the covering map is $f(t) = (\cos(2\pi t), \sin(2\pi t))$. The fiber over $p \in S^1$ is the set of integers, $\mathbb{Z}$. Another example is the

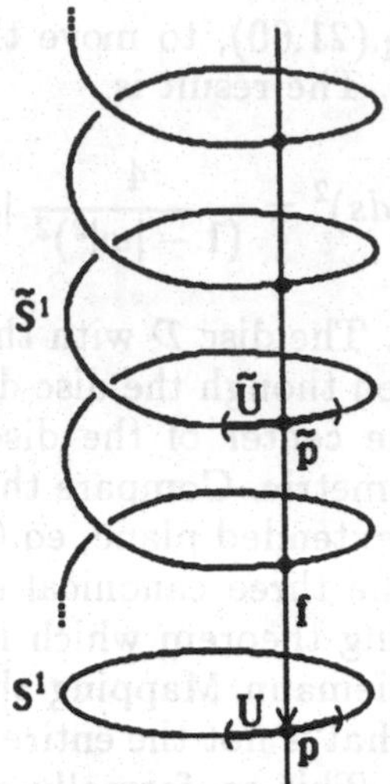

Figure 21.22 The real line as a covering surface, $\tilde{S}^1$, of the circle, S^1. The projection, $f : \tilde{S}^1 \rightarrow S^1$, is a local homeomorphism, which means that the open neighborhood, $\tilde{U}$, of $\tilde{p} \in \tilde{S}^1$ is homeomorphic under f to the open neighborhood, U, of $p \in S^1$. f is not a global homeomorphism, since it is not one-to-one. When $f(t) = (\cos(2\pi t), \sin(2\pi t))$, the fiber above p, $f^{-1}(p)$, is the set of integers $\mathbb{Z}$.

Riemann surface of the function $\log(z)$, which covers the Riemann sphere Σ minus its north and south pole.

A *covering transformation*, g, of a covering surface $(\tilde{S}, f)$ is a homeomorphism $g : \tilde{S} \rightarrow \tilde{S}$ that maps each fiber, $f^{-1}(p)$, to itself. The collection of covering transformations forms a group, G, under composition, the *covering group*. The covering group of a connected covering surface $\tilde{S}$ acts *discontinuously* on $\tilde{S}$. This means that every $\tilde{p}$ of $\tilde{S}$ has a neighborhood $\tilde{U}$, such that $g(\tilde{U}) \cap \tilde{U} = \emptyset$ for every $g \in G$ except the identity. This means that any $g \in G$ that is not the identity can have no fixed points. If, in addition, for any two points, $\tilde{p}_1$ and $\tilde{p}_2$, in $\tilde{S}$ on the same fiber $(f(\tilde{p}_1) = f(\tilde{p}_2))$, there is a $g \in G$ such that $\tilde{p}_1 = g(\tilde{p}_2)$, then the covering group acts *transitively* on the fiber, and the covering surface is called *regular*.

If $\tilde{S}$ is a regular covering surface of S with covering group G, then S is homeomorphic with the quotient surface $\tilde{S}/G$ ($\tilde{S}$ modded by G). One can begin to see how the classification of Riemann surfaces can be accomplished by noting that the covering surface $\tilde{S}$ of a Riemann surface S has a unique complex structure (the "lifted" complex structure) such that the projection f is analytic. If $\tilde{S}$ is connected, then $\tilde{S}$ is also a Riemann surface. When $\tilde{S}$ is given this complex structure, the elements of the covering group are automorphisms of $\tilde{S}$. Thus, the covering group is a subgroup of $Aut(\tilde{S})$. In addition, if the covering surface $\tilde{S}$ of Riemann surface S with the lifted complex structure is regular, then $\tilde{S}/G$ is conformally equivalent to S.

If we had a classification of covering surfaces $\tilde{S}$ of all Riemann surfaces, then we would have a conformal classification of all Riemann surfaces. We do have this classification because every Riemann surface has a regular covering surface that is simply connected. The simply connected regular covering surfaces are called ***universal covering surfaces*** because any two universal covering surfaces of a surface are homeomorphic, and because the universal covering surfaces cover all of the covering surfaces of S. That is, the universal covering surface of S is a covering surface of all covering surfaces of S. Once the universal covering surface is given the lifted complex structure, then we know that since it is simply connected, it must be conformally equivalent to $\mathbb{C}$, Σ, or $\mathcal{D}$ (equivalently $\mathcal{U}$). Collecting these results together, we finally arrive at the ***Uniformization theorem*** (Poincaré-Klein-Koebe). This theorem states that if S is any Riemann surface, $\tilde{S}$ its universal covering surface ($\mathbb{C}$, Σ, or $\mathcal{U}$), and G the covering group, then

1. *S is conformally equivalent to $\tilde{S}/G$.*
2. *G is a Möbius group (a subgroup of $Aut(\tilde{S})$) which acts discontinuously on $\tilde{S}$ (and, therefore, the non-identity elements of G have no fixed points).*
3. The covering group G is isomorphic to the fundamental group of S, $\pi_1(S)$.

In summary, let us restate the results of this theorem. Universal covering surfaces are simply connected, and therefore all universal covering surfaces are conformally equivalent to $\mathbb{C}$, Σ, or $\mathcal{U}$ (equivalently $\mathcal{D}$). The universal covering surfaces are regular, hence the covering group G acts discontinuously on the universal covering surface, $\tilde{S}$. Thus, $g \in G$ has no fixed points except for the identity element of G. In addition, since the universal covering surfaces are regular, then S is homeomorphic to $\tilde{S}/G$. The homeomorphism can be used to transfer the complex structure from S to $\tilde{S}/G$ by lifting it to $\tilde{S}$, hence S is conformal to $\tilde{S}/G$, G is a subgroup of $Aut(\tilde{S})$, and the projection f is holomorphic.

If $\tilde{S} = \Sigma$, then $Aut(\tilde{S}) = PSL(2, \mathbb{C})$. There are *no* non-identity elements of $PSL(2, \mathbb{C})$ that are free of fixed points, hence G contains just one element, the identity map. This means that Σ/G is conformally equivalent to Σ, so there is just 1 conformal structure on Σ or any compact genus 0 Riemann surface (without boundary).

If $\tilde{S} = \mathbb{C}$, then the automorphisms are given by eq.(21.58). If $a \neq 1$, then $b/(1-a)$ is a fixed point, hence $g \in G$ means that $g(z) = z + b$. In order that G act discontinuously, there must be an open neighborhood U of $z \in \mathbb{C}$ that is disjoint with $g(U)$ ($g(U) \cap U = \emptyset$) for *every* $g \in G$ except the identity. This means that G must be a discrete additive subgroup of $\mathbb{C}$ in order to prevent a non-vanishing $|b|$ from being arbitrarily small. There are only 3 distinct subgroups of $\mathbb{C}$, one isomorphic to the identity $\{0\}$, one to the integers $\mathbb{Z}$, and one to the lattice $\mathbb{Z} \times \mathbb{Z}$. In the first case G is just the identity element and $\mathbb{C}/G = \mathbb{C}$. When G is a lattice, $\mathbb{C}/G$ is a torus. When G is isomorphic to the integers, $\mathbb{C}/G$ is conformally equivalent to the "punctured" plane, the plane minus the origin, $\mathbb{C} - \{0\}$.

If the Riemann surface S is not conformally equivalent to the sphere Σ, the plane $\mathbb{C}$, a torus, or the punctured plane, then S is conformally equivalent to $\mathcal{U}/G$. Since $Aut(\mathcal{U}) = PSL(2, \mathbb{R})$, G is a discontinuous subgroup of $PSL(2, \mathbb{R})$. The discrete subgroups of $PSL(2, \mathbb{R})$ are called ***Fuchsian groups.*** Fuchsian groups do not necessarily act discontinuously on $\mathcal{U}$, but almost. Instead, they act *"properly discontinuously."* This means that there is a neighborhood $\tilde{U}$ around each point $\tilde{p} \in \tilde{S}$, such that $\tilde{U}$ is disjoint with its image under g, $g(\tilde{U}) \cap \tilde{U} = \emptyset$, for all but a *finite* number of non-identity elements $g \in G$. Note that G acting properly discontinuously on $\tilde{S}$ with no fixed points is equivalent to G acting discontinuously on $\tilde{S}$. If we relax the restriction of no fixed points in the Uniformization theorem, then every Riemann surface can be represented by $\mathcal{U}/G$.

Now that we have a classification of all Riemann surfaces up to a conformal equivalence, let us turn our attention to spaces of conformal equivalence classes of compact Riemann surfaces, the moduli spaces, and their covering spaces, the Teichmüller spaces. Application of the Faddeev-Popov procedure to the string path integral based on Polyakov's bosonic string action, eq.(19.22), results in a reduction of the string path integral in the critical dimension to integration over the moduli spaces. In the next section, we will first examine the moduli space for the torus (genus 1), where we will use the Fuchsian group, $PSL(2, \mathbb{Z})$, called the *modular group*, to parametrize the set of all conformal equivalence classes of tori.

21.3 MODULI SPACES

The set of all conformal equivalence classes of compact genus g Riemann surfaces is called the moduli space of genus g, $\mathcal{M}_{g,0}$, the topic of this section. A covering space of $\mathcal{M}_{g,0}$ is T_g, called a Teichmüller space. The points of the moduli space are the numerical parameters that describe the properties of Riemann surfaces. There are several equivalent ways to represent the moduli spaces, based on the uniformization (the set of numerical parameters) used. So far we have mainly described uniformization by Fuchsian groups. Two Fuchsian groups, G_1 and G_2 (without fixed points), will define conformally equivalent surfaces if and only if they are conjugate. G_1 and G_2 are two *conjugate subgroups* of a group G, if there is a $g \in G$ such that $gG_1g^{-1} = G_2$. Moduli spaces can then be described as conjugacy classes of fixed-point free Fuchsian groups. As we are about to see, the structure of this result carries over to genus 1 surfaces, where the discrete subgroups arise from $Aut(\mathbb{C})$. We will initially use this representation to determine the moduli space for compact genus 1 surfaces (with no boundaries or punctures).

We also observed that each of the universal covering surfaces possesses a canonical conformal metric of constant curvature. Via the Uniformization theorem, this determines a canonical conformal metric on any given Riemann surface. This means that this metric defines a complex structure, as we stated at the beginning of this chapter. Moduli spaces can then be viewed as spaces of diffeomorphically inequivalent conformal metrics of constant curvature on a surface of fixed topological type. This viewpoint has been used extensively in string theory (it matches the representation of the symmetries of the Polyakov string action and Faddeev-Popov reduction), so we will examine it in some detail.

Another way to describe moduli spaces makes use of their canonical complex structure. This representation is based on the period matrix of a canonical basis of holomorphic 1-forms on compact Riemann surfaces. The period matrix of a Riemann surface determines the complex structure of the surface up to a biholomorphic or conformal equivalence. As this viewpoint is also useful in string theory, we will discuss it, too.

We may also arrive at a moduli space by first constructing the Teichmüller space and projecting. Teichmüller spaces are equivalence classes of Riemann surfaces under quasi-conformal maps. A representation is also obtained from Beltrami differentials (the non-conformal part of the metric on the surface). There are, of course, many other representations.

The moduli space for compact genus 0 Riemann surfaces is trivial since it is just 1 point. There are no fixed-point free subgroups of $Aut(\Sigma) = PSL(2, \mathbb{C})$, hence all compact genus 0 surfaces without boundary are conformally equivalent. Therefore, let us immediately move on to genus 1, the first nontrivial moduli space.

Recall from the Uniformization theorem that all tori can be represented by $\mathbb{C}/\Omega$, where Ω, a discrete subgroup of $Aut(\mathbb{C})$, is a lattice,

$$\Omega(\omega_1, \omega_2) = \{m\omega_1 + n\omega_2 \big| m, n \in \mathbb{Z}\}, \qquad Im\left(\frac{\omega_1}{\omega_2}\right) \neq 0. \qquad (21.66)$$

The two points in the plane, $\{\omega_1, \omega_2\}$, generate the lattice and are called a *basis* for Ω. We can define a fundamental region anchored at the origin as the points on and in a parallelogram defined by $m, n = 0, 1$ as illustrated in figure 21.23(a). The basis for a lattice is not unique. $\Omega(\omega_1, \omega_1+\omega_2)$ defines the same lattice as $\Omega(\omega_1, \omega_2)$ (figure 21.23(b)). When do two different bases define the same lattice? Consider two bases, $\{\omega_1, \omega_2\}$ and $\{q_1, q_2\}$, where $Im(\omega_1/\omega_2) \neq 0$ and $Im(q_1/q_2) \neq 0$. If the two bases define the same lattice, then q_1 must be a lattice point of $\Omega(\omega_1, \omega_2)$, $q_1 = c\omega_1 + d\omega_2$, $c, d \in \mathbb{Z}$, and q_2 must also be a lattice point, $q_2 = a\omega_1 + b\omega_2$, $a, b \in \mathbb{Z}$. That is,

$$\begin{pmatrix} q_2 \\ q_1 \end{pmatrix} = B \begin{pmatrix} \omega_2 \\ \omega_1 \end{pmatrix} = \begin{pmatrix} a & b \\ c & d \end{pmatrix} \begin{pmatrix} \omega_2 \\ \omega_1 \end{pmatrix} \qquad a, b, c, d \in \mathbb{Z}. \tag{21.67}$$

Similarly, ω_1 and ω_2 must be lattice points of $\Omega(q_1, q_2)$, thus

$$\begin{pmatrix} \omega_2 \\ \omega_1 \end{pmatrix} = B^{-1} \begin{pmatrix} q_2 \\ q_1 \end{pmatrix} = \begin{pmatrix} a' & b' \\ c' & d' \end{pmatrix} \begin{pmatrix} q_2 \\ q_1 \end{pmatrix} \qquad a', b', c', d' \in \mathbb{Z}. \tag{21.68}$$

Since B^{-1} is the inverse of B, then $a' = d/\det(B)$, $b' = -b/\det(B)$, etc. The only way that a', b', c', d' will be integers when a, b, c, d are integers is when $\det(B) = \pm 1$. Thus, two bases define the same lattice when the bases are related by a transformation of the form eq.(21.67) with determinant $= \pm 1$.

Next, let us consider when two lattices Ω_1 and Ω_2 produce conformally equivalent tori. Assuming that $\mathbb{C}/\Omega_1$ and $\mathbb{C}/\Omega_2$ are conformally equivalent, then there is a univalent holomorphic map $f : \mathbb{C}/\Omega_1 \rightarrow \mathbb{C}/\Omega_2$. By the Uniformization theorem, we know that there are two holomorphic projections, $\pi_1 : \mathbb{C} \rightarrow \mathbb{C}/\Omega_1$ and $\pi_2 : \mathbb{C} \rightarrow \mathbb{C}/\Omega_2$. A *lift* of f is a map $\tilde{f} : \mathbb{C} \rightarrow \mathbb{C}$ that satisfies $f \circ \pi_1 = \pi_2 \circ \tilde{f} : \mathbb{C} \rightarrow \mathbb{C}/\Omega_2$. In order that f be conformal, the lift, $\tilde{f}$, must be an automorphism of $\mathbb{C}$. Therefore, $\mathbb{C}/\Omega_1$ and $\mathbb{C}/\Omega_2$ will be conformally equivalent tori if there is an automorphism of $\mathbb{C}$ that takes one lattice into the other. Automorphisms of $\mathbb{C}$ are of the form eq.(21.58), $\tilde{f} = az + b$, $a, b \in \mathbb{C}$. This means that Ω_1 and Ω_2 will produce conformally equivalent tori if a translation of one lattice point of Ω_1 to one of Ω_2, followed by a rotation and uniform expansion or contraction of Ω_1 about this point, causes Ω_1 to fall on top of Ω_2. In other words, $\mathbb{C}/\Omega_1$ and $\mathbb{C}/\Omega_2$ are conformally equivalent if Ω_1 is geometrically similar to Ω_2. If Ω_1 is geometrically similar to Ω_2, then for any basis $\{\omega_1, \omega_2\}$ of Ω_1, $\{a\omega_1, a\omega_2\}$ will be a basis for Ω_2 for some nonvanishing $a \in \mathbb{C}$.

We can use the result above to conformally relate any torus generated by a lattice $\Omega(\omega_1, \omega_2)$ with basis $\{\omega_1, \omega_2\}$ to that of a torus generated by $\Omega(\tau)$ with basis $\{1, \tau\}$, $Im(\tau) > 0$ (figure 21.23(c)). If $Im(\omega_2/\omega_1) > 0$, then one simply chooses $a = 1/\omega_1$ so that $\tau = \omega_2/\omega_1$. If $Im(\omega_1/\omega_2) > 0$, then one chooses $a = 1/\omega_2$. Since $\Omega(\omega_1, \omega_2) = \Omega(\omega_2, \omega_1)$, we will assume that we have already ordered the basis so that $Im(\omega_2/\omega_1) > 0$. $\tau = \omega_2/\omega_1$ is called the *modulus* of the basis. In this standard basis, the point τ in

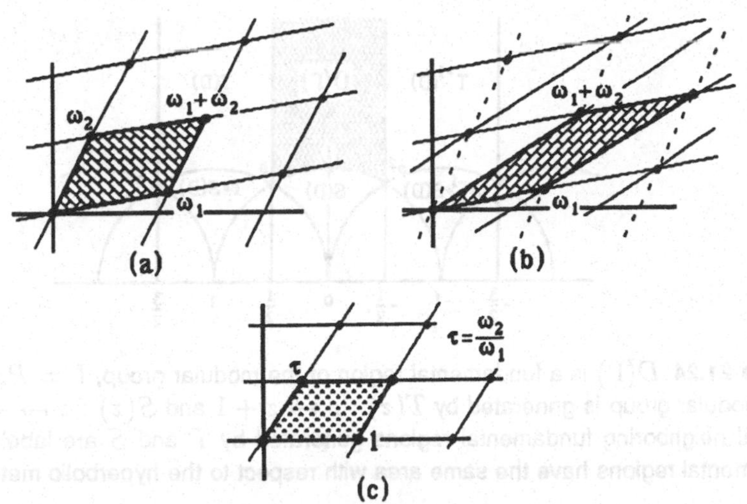

Figure 21.23 (a) A lattice $\Omega(\omega_1, \omega_2)$ generated by the basis $\{\omega_1, \omega_2\}$. The shaded parallelogram is a fundamental region of Ω, and represents a single copy of the torus when the edges are identified under the action of $\Omega : z \mapsto z + m\omega_1 + n\omega_2$. (b) Lattice generated by the basis $\{\omega_1, \omega_1 + \omega_2\}$. This lattice is identical to the lattice in (a). Note that the shaded parallelogram has the same area as in (a). (c) Lattice generated by the standard basis $\{1, \tau\}$. This lattice is geometrically similar to (a) and (b), and therefore, the torus generated by (c) is conformally equivalent to the tori generated by (a) and (b). The modulus of the basis of (c) is identical to (a), and related to (b) by an element of the modular group, $PSL(2, \mathbb{Z})$.

the upper half-plane generates a lattice and every lattice with the same modulus is geometrically similar to a lattice generated in this fashion.

The upper half-plane is not the moduli space for the torus (genus 1) because two different moduli, τ and τ', can still generate the same lattice or geometrically similar lattices. In fact, every lattice generates an infinite number of moduli, one for each basis. Recall that two bases, $\{\omega_1, \omega_2\}$ and $\{q_1, q_2\}$ generate the same lattice when they are related by eq.(21.67). Eq.(21.67) also relates the two moduli of the bases, $\tau = \omega_2/\omega_1$ and $\tau' = q_2/q_1$,

$$\tau' = \frac{a\tau + b}{c\tau + d} \qquad a, b, c, d \in \mathbb{Z}, \tag{21.69}$$

except now we must have $ad - bc = 1$ in order that $Im(\tau') > 0$. All transformations of this type form the group $\Gamma = PSL(2, \mathbb{Z})$, the *modular group*. The two bases, $\{1, \tau\}$ and $\{1, \tau'\}$, generate similar lattices (and therefore conformal tori) when their moduli satisfy eq.(21.69). Therefore, the moduli space of the unpunctured torus is the upper half-plane modded by the modular group, $\mathcal{M}_{1,0} = \mathcal{U}/\Gamma$. $\mathcal{U}/\Gamma$ is the set of all conformal

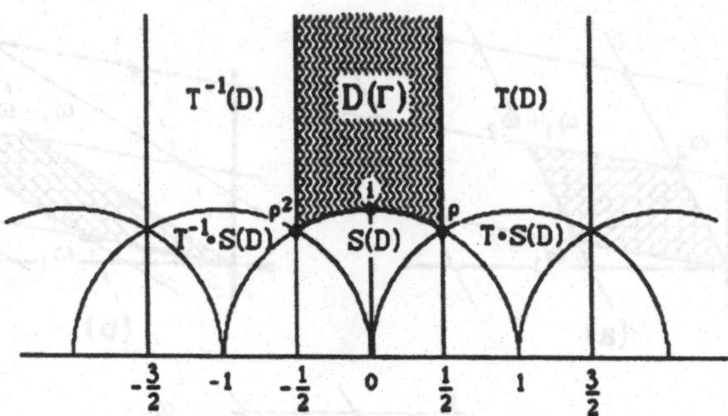

Figure 21.24 $D(\Gamma)$ is a fundamental region of the modular group, $\Gamma = PSL(2,\mathbb{Z})$. The modular group is generated by $T(z) : z \mapsto z+1$ and $S(z) : z \mapsto -1/z$, and several neighboring fundamental regions generated by T and S are labeled. All the fundamental regions have the same area with respect to the hyperbolic metric.

structures that can be imposed on a compact genus 1 surface. Γ is a discrete subgroup of $PSL(2,\mathbb{R})$, hence it is Fuchsian. By the Uniformization theorem, $\mathcal{M}_{1,0}$ is itself a Riemann surface. $\mathcal{U}$, which is a covering space of $\mathcal{M}_{1,0}$, is the Teichmüller space for the torus.

Any lattice Ω divides $\mathbb{C}$ up into parallelograms that share boundaries but not interior points. The *orbit* of a point $z \in \mathbb{C}$ is the image of z under the lattice mapping $z \mapsto z + m\omega_1 + n\omega_2$, $m, n \in \mathbb{Z}$. Points that are directly opposite each other on the boundaries of the parallelograms fall on the same orbit, which is why the parallelogram can be identified with the torus by glueing together opposite boundaries. The vertices of the parallelogram all fall on the same orbit. A *fundamental region* for the lattice group $\Omega(\omega_1,\omega_2)$ is a subset of $\mathbb{C}$ that does not contain two points on the same orbit except possibly on its boundary. A fundamental region is required to contain at least one point from each orbit. It is easy to see that any parallelogram generated by 4 lattice points as vertices with sides of length $|\omega_1|$ and $|\omega_2|$ is a fundamental region. Of course, it is not necessary that the vertices be lattice sites. Any four points in the plane on the same orbit separated by ω_1 and ω_2 will generate a parallelogram that is a fundamental region. Nor are fundamental regions restricted to parallelograms. The *Dirichlet region* for a lattice Ω,

$$D(\Omega) = \{z \in \mathbb{C} \big| |z| \leq |z-\omega|, \omega \in \Omega\}, \tag{21.70}$$

employs the metric, and usually is not a parallelogram. It is centered at zero.

The upper half-plane contains infinitely many copies of $\mathcal{M}_{1,0}$, just as the plane $\mathbb{C}$ contains infinitely many copies of the torus, $\mathbb{C}/\Omega$. As with the lattice, we can define a fundamental region of the modular group Γ on $\mathcal{U}$ in exactly the same way as the lattice. To begin, we note that $T(z) : z \mapsto z + 1$ is an element of Γ. In order to avoid having two points in the interior of the fundamental region on the same orbit, the vertical width of the fundamental region must be ≤ 1. Following the example with the lattice, and to avoid the possibility of missing an orbit, points on the boundary can lie in the same orbit, hence we take the vertical width to be 1. Centering the region about the y-axis (by symmetry of the Poincaré metric), this means that we have $|\text{Re } z| \leq 1/2$. Next, we note that $S(z) : z \mapsto -1/z$ is also an element of Γ. This inverts about the unit half-circle, so the half-circle must be the other boundary. Hence,

$$D(\Gamma) = \left\{ z \in \mathcal{U} \middle| |\text{Re } z| \leq \frac{1}{2}, |z| \geq 1 \right\}, \tag{21.71}$$

which is illustrated in figure 21.24. Using the hyperbolic metric on $\mathcal{U}$, eq.(21.64), one can show that $D(\Gamma)$ is a Dirichlet region centered about ρi, $\rho > 1$. Other fundamental regions can be obtained by following orbits of points in $D(\Gamma)$ under Γ. Some examples are shown in figure 21.24. The connected fundamental regions of a particular lattice all have the same area. The same holds true for Γ when the hyperbolic metric is used.

Before we move on to the moduli spaces for genus ≥ 2 and the metric deformation viewpoint, let us briefly mention the algebraic approach to genus 1 surfaces. This follows from the fact that $w = \sqrt{p(z)}$, where $p(z)$ is any cubic polynomial with 3 distinct roots, is a genus 1 Riemann surface. A cubic polynomial with 3 distinct roots can be written as $p(z) = (z - z_1)(z - z_2)(z - z_3)$. The roots of the polynomial, z_1, z_2, and z_3, define the *discriminant* of p,

$$\Delta_p = 16(z_1 - z_2)^2(z_2 - z_3)^2(z_3 - z_1)^2. \tag{21.72}$$

Since the roots are distinct, $\Delta_p \neq 0$. Using an automorphism of $\mathbb{C}$ ($z \mapsto az + b$), any cubic polynomial can be put into *Weierstrass normal form*,

$$p(z) = 4z^3 - g_2 z - g_3, \qquad g_2, g_3 \in \mathbb{C}, \tag{21.73}$$

in which case,

$$\Delta_p = g_2^3 - 27g_3^2. \tag{21.74}$$

Given a lattice Ω on $\mathbb{C}$, define

$$\begin{aligned} g_2(\Omega) &= 60 \sum_{q \in \Omega}{}' \frac{1}{q^4} \\ g_3(\Omega) &= 140 \sum_{q \in \Omega}{}' \frac{1}{q^6}, \end{aligned} \tag{21.75}$$

where the prime on the sum means sum over the lattice sites, $q = m\omega_1 + n\omega_2$, $m, n \in \mathbb{Z}$, except the origin, $m = n = 0$. Plugging eq.(21.75) into eq.(21.74), one can show that $\Delta_p \neq 0$. Therefore, we can safely define the *modular function*, $J(\Omega)$, as

$$J(\Omega) = \frac{g_2^3(\Omega)}{\Delta_p(\Omega)}. \tag{21.76}$$

Similar lattices determine the same value of J, since if Ω' is similar to $\Omega(\omega_1, \omega_2)$, then there is a basis for Ω' of the form $\{a\omega_1, a\omega_2\}$, $a \in \mathbb{C}$, and $a \neq 0$. This means that $q' \in \Omega'$ can be written as aq, and that $\Delta_p(\Omega') = a^{-12}\Delta_p(\Omega)$, $g_2(\Omega') = a^{-4}g_2(\Omega)$, so $J(\Omega') = J(\Omega)$. Since similar lattices yield identical values for J, we can express J in terms of the lattice basis $\{1, \tau\}$. J then becomes a function of the modulus, τ.

$$\begin{aligned} g_2(\tau) &= 60 \sum_{m,n\in\mathbb{Z}} {}'(m+n\tau)^{-4} \\ g_3(\tau) &= 140 \sum_{m,n\in\mathbb{Z}} {}'(m+n\tau)^{-6}, \end{aligned} \tag{21.77}$$

etc. Recall that two moduli, τ and τ', generate similar lattices if $\tau' = T(\tau)$, where $T \in \Gamma$, the modular group. Hence, $J(T(\tau)) = J(\tau)$. The set of all moduli is the upper half-plane, $\mathcal{U}$. g_2, g_3, Δ_p, and J all are analytic functions on $\mathcal{U}$ into $\mathbb{C}$. $\Delta_p(\tau) = \Delta(\tau)$ may also be expressed as the infinite product,

$$\Delta(\tau) = (2\pi)^{12}\, e^{2i\pi\tau} \prod_{n=1}^{\infty} \left(1 - e^{2i\pi n\tau}\right)^{24}. \tag{21.78}$$

$\Delta(\tau)$ makes an explicit appearance in this form in the genus 1 contribution in the string path integral in the critical dimension.

We have touched upon the parametrization of conformal classes of compact Riemann surfaces via metrics of constant curvature in our discussion of the three universal covering surfaces in the previous section. Let us take up this topic again by observing that isothermal coordinates define a natural conformal or complex structure. To see this, recall from the discussion around eq.(21.53) that the metric becomes conformally flat,

$$(ds)^2 = \lambda(z)|dz|^2, \tag{21.53}$$

in an isothermic coordinate patch. Suppose the surface S is covered with isothermal coordinate patches, and let (U, z) and (V, w) be neighboring patches, with local expressions for the metric given by eq.(21.53) for U

and $(ds)^2 = \lambda_1(w)|dw|^2$ for V. In the overlap region, $U \cap V$, the transition function $\psi_{UV} : U \to V$, maps $z \mapsto w$. Since there is only one metric, then

$$\lambda^{1/2}(z)|dz| = \lambda_1^{1/2}(w)|dw|, \tag{21.79}$$

in the region of overlap, $U \cap V$. Eq.(21.79) implies that $|w'(z)| \neq 0$, and therefore, the transition function ψ_{UV} is holomorphic and conformal, so this atlas defines a complex structure. Thus, a metric defines a complex structure provided that isothermal coordinates can be found.

An arbitrary metric on a surface can always be put into the form

$$(ds)^2 = \lambda(z)|dz + \mu d\bar{z}|^2, \tag{21.80}$$

where $|\lambda| > 0$, and $|\mu| < 1$, in order that the metric be positive definite. To relate the metric eq.(21.80) to a complex structure, we need to find a diffeomorphism $w = f(z)$ that changes the coordinates z in eq.(21.80) into isothermal coordinates. Such a diffeomorphism f must satisfy the *Beltrami equation*,

$$\partial_{\bar{z}} f = \mu \, \partial_z f, \tag{21.81}$$

(where $\partial_z = (\partial_x - i\partial_y)/2$ and $\partial_{\bar{z}} = (\partial_x + i\partial_y)/2$), so that the isothermal metric, $\lambda_1(w)|dw|^2$, will be conformally equivalent to eq.(21.80). To see this, note that if $w = f(z)$ satisfies (21.81), then

$$\begin{aligned} |dw|^2 &= |\partial_z w \, dz + \partial_{\bar{z}} w \, d\bar{z}|^2 \\ &= |\partial_z w|^2 \, |dz + \mu \, d\bar{z}|^2 . \end{aligned} \tag{21.82}$$

Gauss was the first to show that local solutions (solutions in the neighborhood of a point) of the Beltrami equation exist when the components of the metric are real-analytic. Local existence was proven in a more general setting by Korn and Lichtenstein.

In summary, any metric defines a conformal or complex structure on S because one can always find isothermal coordinates, and the isothermal coordinates naturally define a complex structure. The connection between the original metric and the resulting complex structure is expressed through the dependence of the Beltrami coefficient, μ, on the components of the metric. Another common way of expressing the same thing is by defining the tensor,

$$J_b^a = \sqrt{g} \, \epsilon_{bc} \, g^{ca}, \tag{21.83}$$

where $g = \det(g_{ab})$, and $\epsilon_{12} = -\epsilon_{21} = 1$, $\epsilon_{11} = \epsilon_{22} = 0$. J_b^a is called an "*almost complex structure*" because it satisfies $J_b^a J_a^b = J^2 = -1$ and therefore defines "i" (which we get from the orthogonality defined by the metric). J_b^a makes the tangent space, upon which the metric is defined, into a complex vector space. J_b^a is "almost" a complex structure because, while it makes the tangent spaces complex, this complex structure cannot necessarily be transferred from the tangent spaces to the manifold below. When the manifold is a surface (i.e. 2-dimensional), the almost complex structure, J^a_b, always does define a complex structure on the surface because, as we have discussed, the Beltrami equation,

$$J^a_b\,\partial_a w = i\partial_b w, \tag{21.84}$$

possesses solutions locally. The solutions, w, are isothermal coordinates.

Denote the space of all Riemannian metrics on compact surfaces of a fixed topological type without boundary (genus g, no punctures) as $\mathcal{G}_{g,0}$. Since every metric determines a complex structure, the space of all conformal equivalence classes, the moduli space $\mathcal{M}_{g,0}$, must be represented as a subspace of $\mathcal{G}_{g,0}$. Of course, not every metric defines a distinct complex structure, since we already argued that we can always find isothermal coordinates. In addition, the angle between two vectors with metric

$$(ds)^2 = \lambda_1(z)|dz + \mu_1 d\bar{z}|^2 \tag{21.85}$$

will be the same as with eq.(21.80) provided that $\mu = \mu_1$. This follows from the argument given for eq.(21.53). Any change in the metric that amounts to only a change in the conformal factor, λ, will determine conformally equivalent surfaces. Such conformal rescalings of the metric on the surface S (not arising from coordinate reparametrizations), are called Weyl transformations. The collection of all such transformations, which forms a group under composition, is denoted as *Conf*(S) or *Weyl*(S). Since the conformal rescaling won't change the conformal structure, then $\mathcal{M}_{g,0}$ must be represented by a subspace of $\mathcal{G}_{g,0}/\,Weyl(S)$.

Conformal rescalings can also arise from coordinate reparametrizations or diffeomorphisms of the surface into itself. That is, the diffeomorphism $w = f(z)$ that maps the surface to itself will be conformal if the metric, eq.(21.80), becomes

$$(ds)^2 = \lambda_2(z)|dw + \mu\, d\bar{w}|^2 \tag{21.86}$$

with the same μ. However, not all diffeomorphisms of the surface into itself amount to just a conformal rescaling. The intrinsic geometry of the surface as determined by the metric tensor is a coordinate-free notion. The metric tensor itself is a coordinate-free object (defined independently of coordinates), thus *any* coordinate reparametrization or diffeomorphism of the surface to itself does not change the intrinsic metric, and thereby the complex structure. The collection of all diffeomorphisms of a surface to itself, which forms a group under composition, is denoted as *Diff*(S).

Other changes in the metric produce conformally inequivalent surfaces, thus the moduli space, $\mathcal{M}_{g,0}$ can be represented as

$$\mathcal{M}_{g,0} \sim \frac{\mathcal{G}_{g,0}}{Weyl(S) \times Diff(S)}. \tag{21.87}$$

Not every diffeomorphism is homotopic to the identity map, $z \mapsto z$, on a surface. The subgroup of *Diff*(S) that is homotopic to the identity is called *Diff*$_0(S)$. The quotient of the two is the *Mapping Class group*,

$$MCG = \frac{Diff(S)}{Diff_0(S)}. \tag{21.88}$$

The Teichmüller space can be represented as $\mathcal{G}_{g,0}/(Diff_0(S) \times Weyl(S))$, thus $\mathcal{M}_{g,0} = T_{g,0}/MCG$.

From the definition of the complex structure, J^a_b, eq.(21.83), it is easy to see that J^a_b is invariant under Weyl transformations. Since *Diff*(S) transformations do not change the intrinsic metric, and, therefore, J^a_b, then the moduli space $\mathcal{M}_{g,0}$ can be represented by the collection of all complex structures J^a_b on compact genus g surfaces divided by *Diff*(S).

$$\mathcal{M}_{g,0} \sim \frac{\{J^a_b \text{ on } S,\ J^a_b J^b_c = -\delta^a_c\}}{Diff(S)} \tag{21.89}$$

As an alternative to eq.(21.87), we can take one metric (i.e. one point in $\mathcal{G}_{g,0}$) on each orbit of *Weyl*(S) in $\mathcal{G}_{g,0}$, divide it by *Diff*(S), and obtain another representation of $\mathcal{M}_{g,0}$. In other words, we can use the gauge freedom provided by the Weyl symmetry to fully fix the conformal factor $\lambda(z)$ in any metric, eq.(21.80). One such choice of Weyl gauge is the requirement that the metric possess constant curvature: $+1$ for genus 0, 0 for genus 1, and -1 for genus ≥ 2. To fully fix the conformal factor for genus 1, we must also require the surface to have unit area.

Let $g \in \mathcal{G}_{g,0}$ be any arbitrary metric with Gaussian curvature K_g. Suppose $\widehat{g} \in \mathcal{G}_{g,0}$ is conformal with g but has constant curvature $K_{\widehat{g}}$ $(= \pm 1$ or 0). Since $\widehat{g}$ is conformal with g, then there is a Weyl transformation relating the two metrics,

$$\widehat{g}_{ab} = \delta\lambda\, g_{ab}. \tag{21.90}$$

Since $|\lambda| > 0$ in eq.(21.80), it is traditional to write $\lambda = \exp(\phi)$. Then, $\delta\lambda = \exp(\delta\phi) \equiv \exp(\sigma)$. By computing the curvature on both sides of eq.(21.90), we can see that σ must satisfy

$$K_{\widehat{g}} = e^{\sigma}(K_g - \Delta_g\, \sigma), \tag{21.91}$$

where Δ_g is the Laplacian of the g_{ab} metric. Eq.(21.91) is an inhomogeneous Liouville equation which is known to possess unique solutions. Thus, the requirement that $K_{\widehat{g}}$ be constant fully fixes the Weyl symmetry. If $\mathcal{G}^{\text{const}}_{g,0}$ is the set of metrics $\widehat{g}$ on S with constant normalized curvature, then

$$T_{g,0} \sim \frac{\mathcal{G}^{\text{const}}_{g,0}}{Diff_0(S)}, \quad \mathcal{M}_{g,0} \sim \frac{\mathcal{G}^{\text{const}}_{g,0}}{Diff(S)}. \tag{21.92}$$

Another way to validate eq.(21.92) is with the Uniformization theorem. Recall that each of the universal covering spaces, Σ, $\mathbb{C}$, and $\mathcal{U}$, came equipped with a constant curvature metric (eq.(21.52), $|dz|^2$, and eq.(21.64)). We also have a holomorphic projection h from the covering surfaces $\widetilde{S} = (\Sigma, \mathbb{C}, \text{ or } \mathcal{U})$ to the Riemann surface $S = \widetilde{S}/G$, where G is the cover group. Not every metric on the covering surface $\widetilde{S}$ can be projected to the base surface S. Any metric on $\widetilde{S}$ that is not invariant along the orbits of G won't be single-valued upon projection. Thus, the metrics on $\widetilde{S}$ that are invariant under the cover group G can be projected to $S = \widetilde{S}/G$. On the other hand, every metric on S can be lifted or "pulled back" to a metric on $\widetilde{S}$. The flat metric $|dz|^2$ is certainly invariant on the orbits generated by any lattice, so every torus will admit a flat metric. The Poincaré metric, eq.(21.64), on the upper half-plane can also be projected safely, since all elements of $Aut(\mathcal{U})$ are isometries of this metric. Thus, all Riemann surfaces possess metrics of constant curvature. The question now is whether all metrics on any Riemann surface are conformally or diffeomorphically equivalent to a constant-curvature metric.

Let us consider the most interesting case where $\widetilde{S} = \mathcal{U}$ (i.e. genus $\geq$ 2). If g is a metric on S, then the lifted or pulled-back metric, g^*, is automatically invariant under the cover group, G. There is only one conformal class on $\mathcal{U}$, hence all metrics on $\mathcal{U}$ are conformal under a diffeomorphism to the Poincaré metric, eq.(21.64), on $\mathcal{U}$. The reason for this is that the diffeomorphism, $f : (\mathcal{U}, g^*) \to (\mathcal{U}, g_{poincare})$, that pulls back the Poincaré metric to a metric conformal with g^* must satisfy the Beltrami equation (μ determined by g^*). We already know that solutions always exist (on the one patch that covers $\mathcal{U}$). Thus, we start with a metric on $S = \mathcal{U}/G$, and pull it back to a metric on $\mathcal{U}$ under the action of the projection, h. We then use a diffeomorphism, f, on $\mathcal{U}$, to go from $(\mathcal{U}, g^*)$ to $(\mathcal{U}, g_{poincare})$, then project again. Since f is mapping a metric invariant on orbits of G to another invariant on the same orbits, the diffeomorphism itself projects to a diffeomorphism on $S = \mathcal{U}/G$ to itself. Thus, all metrics on S are conformal under a diffeomorphism to a constant-curvature metric, and again we arrive at eq.(21.92).

Briefly returning to the torus for illustration, consider a torus generated by the lattice $\Omega(1,\tau)$ (figure 21.23(c)). An isothermal metric is

$$f(x,y)\left((dx)^2+(dy)^2\right). \tag{21.93}$$

On the plane $\mathbb{C}$, the function f must only be positive (so we write $f(x,y) = \exp(\phi(x,y))$). However, to be a metric on the torus, f or ϕ must be invariant under Ω,

$$\phi(x+m+n\tau_1, y+n\tau_2) = \phi(x,y), \tag{21.94}$$

where $m, n \in \mathbb{Z}$, and $\tau = \tau_1 + i\tau_2$. As it stands, the only dependence on the modular parameter, τ, is in the conformal factor, $\exp(\phi)$. We can explicitly introduce a dependence on the modular parameter in the nonconformal part of the metric with the change of coordinates,

$$\begin{aligned} x &= \sigma_1 + \tau_1\sigma_2 \\ y &= \tau_2\sigma_2, \end{aligned} \tag{21.95}$$

which maps the unit square in the (σ_1, σ_2) plane to the parallelogram in figure 21.23(c). In (σ_1, σ_2) coordinates, the pull-back of the metric, eq.(21.93), is

$$\exp(\phi(\sigma_1,\sigma_2))\,|d\sigma_1 + \tau\, d\sigma_2|^2, \tag{21.96}$$

and eq.(21.94) translates to the requirement,

$$\phi(\sigma_1+m, \sigma_2+n) = \phi(\sigma_1,\sigma_2), \qquad m,n \in \mathbb{Z}. \tag{21.97}$$

The Beltrami coefficient associated with eq.(21.96) is

$$\mu = \frac{1+i\tau}{1-i\tau}, \tag{21.98}$$

and the solution to the Beltrami equation takes us back to the (x, y) isothermal coordinates.

The conformal factor in eq.(21.96) can be uniquely fixed by requiring the metric to have constant zero curvature and by requiring the unit square to have unit area. $\phi(\sigma_1, \sigma_2) =$ *constant* will make the metric flat. The unit (σ_1, σ_2) square will have unit area if $\sqrt{g} = 1$. Therefore, the element of $\mathcal{G}_{1,0}^{\text{const}}$ associated with eq.(21.96) is

$$(ds)^2 = g_{ab}d\sigma^a\, d\sigma^b = \frac{1}{\tau_2}\,|d\sigma_1 + \tau\, d\sigma_2|^2. \tag{21.99}$$

Any metric on a torus can be brought by a diffeomorphism into the form eq.(21.99). Consider an arbitrary metric on the plane $\mathbb{C}$ with periodicity,

$$g_{ab}(\sigma_1 + m\omega_1 + nq_1, \sigma_2 + m\omega_2 + nq_2) = g_{ab}(\sigma_1, \sigma_2), \tag{21.100}$$

for $m, n \in \mathbb{Z}$. Such a metric is invariant along the orbits of the lattice $\Omega(\omega, q)$, where $\omega = \omega_1 + i\omega_2$, and $q = q_1 + iq_2$ (assume $Im(q/\omega) > 0$), and therefore, is a metric on the torus $\mathbb{C}/\Omega(\omega, q)$. This metric will determine some Beltrami coefficient, and the solution to the Beltrami equation provides us with a diffeomorphism that transforms eq.(21.100) into an isothermal metric, $\exp(\phi(\sigma_1, \sigma_2))((d\sigma_1)^2 + (d\sigma_2)^2)$, where $\phi(\sigma_1 + m\omega_1 + nq_1, \sigma_2 + m\omega_2 + nq_2) = \phi(\sigma_1, \sigma_2)$. The diffeomorphism,

$$\begin{aligned} \sigma_1 &\rightarrow \frac{\omega_1}{|\omega|^2}\sigma_1 + \frac{\omega_2}{|\omega|^2}\sigma_2 \\ \sigma_2 &\rightarrow -\frac{\omega_2}{|\omega|^2}\sigma_1 + \frac{\omega_1}{|\omega|^2}\sigma_2, \end{aligned} \tag{21.101}$$

maps the lattice $\Omega(\omega, q)$ to $\Omega(1, \tau)$, where $\tau = q/\omega$ (figure 21.23(a), with $\omega = \omega_1$, $q = \omega_2$, to figure 21.23(c)). This transformation preserves the isothermal nature of the metric, and so we arrive at eq.(21.93) and eq.(21.94). Eq.(21.95) transforms this metric to eq.(21.96), and from here we arrive back at eq.(21.99).

The metric, eq.(21.99), is parametrized by the complex number τ with $Im(\tau) > 0$. Since an arbitrary metric can be brought to the form eq.(21.99) by an element of $Diff_0$, then eq.(21.92) indicates that the Teichmüller space of the torus is the upper half-plane, $\mathcal{U}$.

In order to find the moduli space, $\mathcal{M}_{1,0}$, by this approach, we must determine the mapping class group, MCG, for the torus. Diffeomorphisms that preserve the form of the metric, eq.(21.96), are linear in σ_1 and σ_2,

$$\sigma_1 \rightarrow a\sigma_1 + b\sigma_2, \qquad \sigma_2 \rightarrow c\sigma_1 + d\sigma_2. \tag{21.102}$$

Diffeomorphisms that also preserve the periodicity (i.e. map vertices of the unit square to vertices of the unit lattice, $\Omega(1,1)$) must have $a, b, c, d \in \mathbb{Z}$. In order that the orientation be preserved, and that a bijective inverse exist that also preserves the periodicity, we must also require that $ad - bc = 1$. Transformations of this type are the elements of the modular group, $\Gamma = PSL(2, \mathbb{Z})$. Diffeomorphisms of this type, however, are not homotopic to the identity map (except, of course, for the identity element of Γ). An illustration of this is given in figure 21.25 for the case, $a = 2$, $b = 1$, $c = 1$, $d = 1$. The unit square in figure 21.25(a) is mapped to the parallelogram in figure 21.25(b). In both cases, points in the (σ_1, σ_2) plane on the same orbit of $\Omega(1, 1)$ are equivalent. The dashed line in the

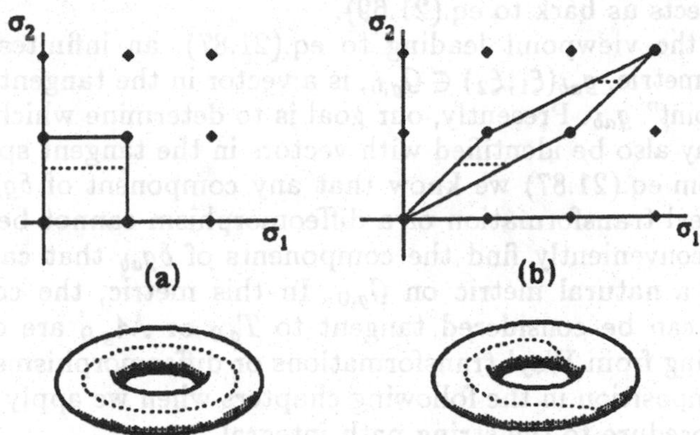

Figure 21.25 The unit square in (a) is mapped to the parallelogram in (b) by a modular transformation of the form eq.(21.102) with $a = 2, b = c = d = 1$. The dashed line in (a) and (b) is the same constant-σ_2 line. When the edges of the parallelogram in (b) are glued together to form the torus, the dashed line wraps once around the torus and cannot be continuously deformed to the dashed line in (a).

square in 21.25(a) is a line of constant-σ_2. The same line as it appears on the torus is illustrated below the square. The same constant-σ_2 line is given in (b). When the edges of the parallelogram are glued together to form the same torus, this line wraps once around the torus. There is no way to continuously deform the dashed line on the torus in (b) into the dashed line on the torus in (a) without breaking it. Therefore, the diffeomorphism that maps (a) to (b) is not homotopic with the identity map.

Any diffeomorphism of the torus to itself can be written as an element of Γ composed with an element of Diff_0, hence

$$MCG = \frac{\mathit{Diff}}{\mathit{Diff}_0} = \Gamma = PSL(2, \mathbb{Z}), \tag{21.103}$$

and using eq.(21.92), we once again arrive at

$$\mathcal{M}_{1,0} = \frac{\mathcal{U}}{\Gamma}. \tag{21.104}$$

Under a diffeomorphism $\in \Gamma$, the metric, eq.(21.99), transforms to

$$(ds)^2 = \frac{1}{\tau_2'} \left| d\sigma_1 + \tau' d\sigma_2 \right|^2, \tag{21.105}$$

where

$$\tau' = \frac{a\tau - b}{d - c\tau}. \tag{21.106}$$

This connects us back to eq.(21.69).

From the viewpoint leading to eq.(21.87), an infinitesimal change, δg_{ab}, in a metric, $g_{ab}(\xi_1, \xi_2) \in \mathcal{G}_{g,0}$, is a vector in the tangent space of $\mathcal{G}_{g,0}$ at the "point" g_{ab}. Presently, our goal is to determine which components of δg_{ab} may also be identified with vectors in the tangent space of $T_{g,0}$ or $\mathcal{M}_{g,0}$. From eq.(21.87) we know that any component of δg_{ab} that arises from a Weyl transformation or a diffeomorphism cannot be identified as such. To conveniently find the components of δg_{ab} that can be, we will introduce a natural metric on $\mathcal{G}_{g,0}$. In this metric, the components of δg_{ab} that can be considered tangent to $T_{g,0}$ or $\mathcal{M}_{g,0}$ are orthogonal to those arising from Weyl transformations or diffeomorphisms. We will use this decomposition in the following chapters when we apply the Faddeev-Popov procedure to the string path integral.

Let $\delta g^{(1)}_{ab}$ and $\delta g^{(2)}_{ab}$ be two infinitesimal variations in the metric g_{ab} (i.e. two vectors in the tangent space of $\mathcal{G}_{g,0}$ at g_{ab}). The most general metric on $\mathcal{G}_{g,0}$ that does not involve coordinate derivatives is

$$\langle \delta g^{(1)}, \delta g^{(2)} \rangle = \int d^2\xi \sqrt{g} \left(g^{ac} g^{bd} + c g^{ab} g^{cd} \right) \delta g^{(1)}_{ab} \, \delta g^{(2)}_{cd} , \tag{21.107}$$

where c is an arbitrary constant. It will turn out that c never enters into any physical answers, so with hindsight we will set it to zero. Components of a general δg_{ab} that can be considered in the tangent space of $T_{g,0}$ or $\mathcal{M}_{g,0}$ are orthogonal with respect to eq.(21.107) with the gauge variations arising from Weyl transformations and diffeomorphisms. To identify these, we first need to determine the effects of Weyl transformations and diffeomorphisms on g_{ab}.

If we write the conformal factor in the usual fashion, $\lambda(\xi^1, \xi^2) = \exp(\phi(\xi^1, \xi^2))$, then the effect of an infinitesimal Weyl transformation is simply

$$\delta g_{ab} = \delta\phi \, g_{ab} . \tag{21.108}$$

In order to write down the effect of an infinitesimal diffeomorphism, we need to digress for a moment.

Recall that a vector field on a surface S is a choice of one tangent vector at each point of S. Since each point of S has its own tangent space, this amounts to a selection of one vector from each tangent space. If the choice of vector varies smoothly from point to point, then we can draw curves on the surface S such that the vector at point $p \in S$ is tangent to the curve passing through p. Such curves are called *integral curves* of the vector field. As long as the vector field never vanishes at any point on S, the different curves will never intersect. There will always be just one curve through each point, hence the set of integral curves covers the surface S. Such a set of curves is called a *congruence*.

A congruence provides a natural mapping of the surface S into itself. If the vector field is C^∞ (i.e. the integral curves, $(\xi^1(t), \xi^2(t))$, are infinitely differentiable with respect to t), then the resulting congruence is a diffeomorphism (homotopic to the identity) of the surface into itself. An infinitesimal diffeomorphism can, therefore, be parametrized by a C^∞ vector field, $\vec{v}(\xi^1, \xi^2)$.

Suppose that (ξ^1, ξ^2) are nonsingular coordinates on S. Then the lines of constant ξ^1 form a congruence as do the lines of constant ξ^2. Let $e_1(\xi^1, \xi^2)$ denote the unit tangent vector to the line of constant ξ^2 at the point (ξ^1, ξ^2), and e_2 denote the unit tangent to the line of constant ξ^1. These vectors form a basis for the tangent space at (ξ^1, ξ^2), so if $\vec{v}$ is the vector of the vector field at (ξ^1, ξ^2), then we may write $\vec{v} = v^a e_a$.

Let $f(\xi^1, \xi^2)$ be a scalar function on S. We can define the operation of the vector field, $\vec{v}(\xi^1, \xi^2)$ on f, denoted $\vec{v}[f]$, as the change in f as we move infinitesimally along the integral curves of $\vec{v}$. This change in f is simply the directional derivative of f in the direction of the vector $\vec{v}$ tangent to the integral curve, hence

$$\vec{v}[f] = v^a e_a[f] = v^a \frac{\partial f}{\partial \xi^a}. \tag{21.109}$$

This motivates the extremely useful notation, $e_a = \partial / \partial \xi^a$. In general, if the integral curve is parametrized by t, then the tangent vector can be written as $\partial / \partial t$, and $v^a = \partial \xi^a / \partial t$.

Suppose that the integral lines of a congruence are parametrized by t. Given a function f on S, we can define a new function, $\widetilde{f}_{\Delta t}$, by carrying f along the congruence a parameter distance Δt, $\widetilde{f}_{\Delta t}(\xi^1(t+\Delta t), \xi^2(t+\Delta t)) = f(\xi^1(t), \xi^2(t))$. If $\widetilde{f}_{\Delta t} = f$ for all Δt, then f is said to be *Lie dragged*. If f is Lie dragged, then it is constant along the integral curves of the congruence and $\partial f / \partial t = 0$.

We clearly can also do the same thing to a vector field. Let $\vec{w}$ be a second smooth vector field on S. We can define a new vector field, $\vec{w}_{\Delta t}$, by carrying the vector field along the congruence generated by the vector field $\vec{v}$ by Δt. If the vector field is invariant for all Δt, then it is Lie dragged. If a vector field is Lie dragged, then the integral curves of $\vec{w}$ are identical to the integral curves of $\vec{w}_{\Delta t}$, i.e. the congruence generated by $\vec{w}$ is the same as $\vec{w}_{\Delta t}$ for any Δt.

We can define a derivative along a congruence in an obvious fashion. It is called the *Lie derivative*. Let the congruence be generated by the vector field $\vec{v}$ and the integral curves be parametrized by t. For a function, we Lie drag the function at point $t + \Delta t$ back to t along the integral curve and compare it to the value of the function at this point. The Lie derivative,

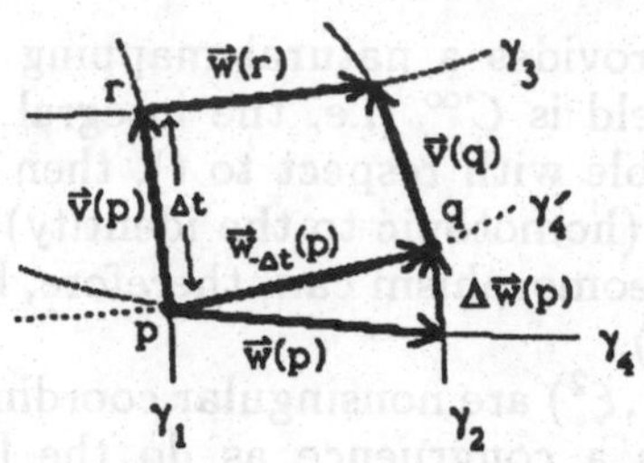

Figure 21.26 γ_1 **and** γ_2 **are two integral curves of the vector field** $\vec{v}$**.** γ_3 **and** γ_4 **are integral curves of** $\vec{w}$**.** γ_4' **is an integral curve of the Lie dragged vector field** $\vec{w}_{-\Delta t}$**. The failure of the vector fields** $\vec{v}$ **and** $\vec{w}$ **to close,** $[\vec{v}, \vec{w}]$**, is the vector** $\Delta\vec{w}(p) = \vec{w}_{-\Delta t}(p) - \vec{w}(p)$**, which in the limit is the Lie derivative,** $\pounds_{\vec{v}}\vec{w}$**.**

$\pounds_{\vec{v}} f$, is the difference between the two values, divided by Δt, in the limit that $\Delta t \to 0$.

$$\pounds_{\vec{v}} f = \lim_{\Delta t \to 0} \frac{f(t+\Delta t) - f(t)}{\Delta t} = \frac{df}{dt} = v^a \frac{\partial f}{\partial \xi^a} = \vec{v}[f] \qquad (21.110)$$

This is just the directional derivative of f again. f is Lie dragged if $\pounds_{\vec{v}} f = 0$.

The Lie derivative of a vector field is defined in a similar manner. Let $\vec{w}$ be another vector field on S and let the integral curves of $\vec{w}$ be parametrized by q. Let γ_1 and γ_2 be two integral curves of the vector field $\vec{v}$, as illustrated in figure 21.26. γ_3 and γ_4 are integral curves of the vector field $\vec{w}$. The vector $\vec{w}(r)$ at $r = \gamma_1(t+\Delta t)$ is Lie dragged back to $p = \gamma_1(t)$ and is labeled $\vec{w}_{-\Delta t}(p)$. The Lie derivative of $\vec{w}$ along $\vec{v}$ is

$$\pounds_{\vec{v}}\vec{w} = \lim_{\Delta t \to 0} \frac{\vec{w}_{-\Delta t}(p) - \vec{w}(p)}{\Delta t}. \qquad (21.111)$$

The entire integral curve γ_3, Lie dragged back Δt along $\vec{v}$, forms the dashed curve γ_4', which is an integral curve of the Lie dragged vector field $\vec{w}_{-\Delta t}$. The four vectors, $\vec{v}(p)$, $\vec{w}(r)$, $\vec{v}(q)$, and $\vec{w}_{-\Delta t}$, form a closed quadrilateral, hence $[\vec{v}, \vec{w}_{\text{dragged}}] = 0$. The failure of the vector fields, $\vec{v}$ and $\vec{w}$, to close is $\vec{w}_{-\Delta t}(p) - \vec{w}(p) = [\vec{v}, \vec{w}]$. In the limit this is $\pounds_{\vec{v}}\vec{w}$, hence

$$\pounds_{\vec{v}}\vec{w} = [\vec{v}, \vec{w}] = \frac{\partial}{\partial t}\frac{\partial}{\partial q} - \frac{\partial}{\partial q}\frac{\partial}{\partial t} = v^a \frac{\partial w^b}{\partial \xi^a}\frac{\partial}{\partial \xi^b} - w^a \frac{\partial v^b}{\partial \xi^a}\frac{\partial}{\partial \xi^b}. \qquad (21.112)$$

Note that $\pounds_{\vec{v}}\vec{w} = -\pounds_{\vec{w}}\vec{v}$, and that eqs.(21.110) and (21.112) imply that the Lie derivative satisfies the Leibniz rule,

$$\pounds_{\vec{v}}(f\vec{w}) = (\pounds_{\vec{v}} f)\,\vec{w} + f\,\pounds_{\vec{v}}\vec{w}. \qquad (21.113)$$

The Leibniz rule along with the Lie derivative for scalars and (contravariant) vectors can be used to define the Lie derivative of covariant vectors (1-forms) and higher-rank tensors. For example, if $\underline{u}$ is a covariant vector, then $\underline{u}(\vec{w})$ is a scalar. Applying the Leibniz rule, we have

$$\mathcal{L}_{\vec{v}}(\underline{u}(\vec{w})) = (\mathcal{L}_{\vec{v}}\underline{u})(\vec{w}) + \underline{u}(\mathcal{L}_{\vec{v}}\vec{w}). \tag{21.114}$$

We can solve eq.(21.114) for $\mathcal{L}_{\vec{v}}\underline{u}$ by using eq.(21.110) on the left-hand side, and eq.(21.112) on the right. Similarly, for a second-rank covariant tensor g_{ab}, we have

$$\mathcal{L}_{\vec{v}}(g(\vec{u},\vec{w})) = (\mathcal{L}_{\vec{v}}g)(\vec{u},\vec{w}) + g(\mathcal{L}_{\vec{v}}\vec{u},\vec{w}) + g(\vec{u},\mathcal{L}_{\vec{v}}\vec{w}). \tag{21.115}$$

Once again, we can apply eqs.(21.110) and (21.112) to solve for $\mathcal{L}_{\vec{v}}g$.

The point of this discussion is that the change in the metric, g_{ab}, due to an infinitesimal diffeomorphism, is the Lie derivative of g_{ab} along the vector field $\vec{v}$ that generates the diffeomorphism,

$$\delta g_{ab} = \mathcal{L}_{\vec{v}}g. \tag{21.116}$$

When we solve for $\mathcal{L}_{\vec{v}}g$ in eq.(21.115), and apply eq.(21.10), we find

$$\delta g_{ab} = \nabla_a v_b + \nabla_b v_a, \tag{21.117}$$

where ∇_a is the covariant derivative compatible with g_{ab}. This computation is left as an exercise.

Eq.(21.117) together with eq.(21.108) represents the change in the metric due to Weyl transformations and diffeomorphisms. It is not difficult to see, using the metric, eq.(21.107), on $\mathcal{G}_{g,0}$, that δg_{ab} arising from eq.(21.117) is not orthogonal to eq.(21.108). It will turn out to be convenient to orthogonally decompose the combined effects of Weyl transformations and diffeomorphisms. This is easily done by removing the trace from eq.(21.117), and combining it with eq.(21.108). Thus, the combined effects of Weyl transformations and diffeomorphisms is written as

$$\begin{aligned}\delta g_{ab} &= \delta\phi\, g_{ab} + \nabla_a v_b + \nabla_b v_a \\ &= (\delta\phi + \nabla_c v^c) g_{ab} + (P_1\vec{v})_{ab},\end{aligned} \tag{21.118}$$

where

$$(P_1\vec{v})_{ab} = \nabla_a v_b + \nabla_b v_a - g_{ab}\nabla_c v^c. \tag{21.119}$$

The operator P_1 takes vectors into symmetric traceless second-rank tensors. The first component in eq.(21.118) that is proportional to g_{ab} is orthogonal to the second, traceless, component.

From eq.(21.87) we can see that the components of δg_{ab} that can be identified with elements of a tangent space of $T_{g,0}$ or $\mathcal{M}_{g,0}$ will be orthogonal to components in eq.(21.118) above. Let us denote these components as $\delta g^{\perp}$. From eq.(21.118) we can see that a Weyl transformation can always eliminate the trace, hence $\delta g^{\perp}$ will be traceless. Since $\delta g^{\perp}$ is orthogonal to eq.(21.119), then

$$\langle \delta g^{\perp}, P_1 \vec{v} \rangle = 0 = \langle P_1^{\dagger} \delta g^{\perp}, \vec{v} \rangle. \tag{21.120}$$

Using the explicit form for P_1, eq.(21.119), and for the metric $\langle\, ,\, \rangle$, eq. (21.107), and integrating by parts, we find

$$(P_1^{\dagger}\, \delta g)_a = -2\nabla^b\, \delta g_{ab}. \tag{21.121}$$

Since $\vec{v}$ in eq.(21.120) is an arbitrary smooth vector field, then $P_1^{\dagger}\delta g^{\perp} = 0$, that is, $\delta g^{\perp} \in Ker\ P_1^{\dagger}$. In fact, $Ker\ P_1^{\dagger}$ represents the (co)tangent space of $\mathcal{T}_{g,0}$ and $\mathcal{M}_{g,0}$. By eq.(21.92), this implies that any $\delta g^{\perp} \in Ker\ P_1^{\dagger}$ cannot change the curvature. It is straightforward to verify that this is so.

The metric, eq.(21.107), may also be used on $\mathcal{G}_{g,0}^{\text{const}}$. This metric is invariant under all diffeomorphisms (i.e. diffeomorphisms are isometries). Since this metric is invariant on the orbits of *Diff* (S), then we can also use it on $\mathcal{T}_{g,0}$ or $\mathcal{M}_{g,0}$. When pulled back to $\mathcal{T}_{g,0}$ or $\mathcal{M}_{g,0}$, this metric is known as the Weil-Petersson metric.

From our previous discussion we know that the dimension of the Teichmüller and moduli spaces for compact genus 0 surfaces is 0 and 1 (complex) for the torus. Since the tangent space and the space it is tangent to have the same dimension, then we can find the dimension of $\mathcal{T}_{g,0}$ and $\mathcal{M}_{g,0}$ by knowing the dimension of the space $Ker\ P_1^{\dagger}$. In order to determine this, we appeal to the result of one version of the Riemann-Roch theorem, the details of which are beyond the present discussion. This result is

$$\dim(Ker\ P_1) - \dim(Ker\ P_1^{\dagger}) = \frac{3}{2}\chi(S) = 3 - 3g, \tag{21.122}$$

where $\chi(S) = 2 - 2g$ is the Euler-Poincaré characteristic (for compact surfaces without boundary). A vector field $\vec{u}$ that generates a diffeomorphism that leaves the metric invariant, $\delta g_{ab} = \pounds_{\vec{u}} g = 0$, is known as a *Killing vector field.* A vector field that leaves the metric invariant up to a conformal transformation is a *conformal Killing vector field.* An element $\vec{u}$ of $Ker\ P_1$ satisfies $P_1\vec{u} = 0$, and therefore, by eq.(21.118), is a conformal Killing vector. For genus 0 and 1, we can apply the results we have already discussed and work backwards to find that (complex) $\dim(Ker\ P_1) = 3, 1$ respectively. For genus 2 and greater, there are no conformal Killing vectors, hence $\dim(Ker\ P_1^{\dagger}) = 3g - 3$ complex or $6g - 6$ real dimensions. Thus,

$$\dim(\mathcal{T}_{g,0}) = \dim(\mathcal{M}_{g,0}) = 3g - 3. \tag{21.123}$$

Since each puncture requires one more complex dimension to describe the location of the puncture on the surface, then $\dim(\mathcal{M}_{g,n}) = 3g - 3 + n$.

Before we move on to the actual computation of the string path integral in the next chapter, let us briefly describe the use of the periods of Abelian differentials as moduli. This generalizes our previous discussion on $\mathcal{M}_{1,0}$ for the torus, and appears to be the natural setting for bosonic string theory in the critical dimension. In order to define the period matrix of a Riemann surface, we must first introduce homology groups, another concept from algebraic topology.

Consider again a triangulation of a Riemann surface, S_g, of genus g. Let the points $p_1, p_2, \ldots$ of S_g denote the locations of the vertices of the triangulation. Rename the vertices as *0-simplices.* The edges are called *1-simplices* and the triangular faces are called *2-simplices.* Denote the edge connecting the vertices p_i, p_j as $\langle p_i p_j\rangle$, and the triangle or 2-simplex formed by vertices p_i, p_j, and p_k as $\langle p_i p_j p_k\rangle$. Since the surface is oriented, we can orient the n-simplices and associate a direction with each edge, and a sense of traversal around each triangle. Thus, $\langle p_j p_i\rangle = -\langle p_i p_j\rangle$, and $\langle p_k p_j p_i\rangle = - < p_i p_j p_k\rangle$.

An *n-chain* is a linear combination of positively oriented n-simplices with integer coefficients. The collection of n-chains, C_n, $n = 0, 1, 2$ forms an Abelian group under addition.

We can define a *boundary map* $\delta_n : C_n \to C_{n-1}$ which takes n-chains to $(n-1)$-chains where the image of δ_n on an n-chain is its boundary. A vertex has no boundary, hence $\delta_0\langle p_i\rangle = 0$. The boundary of an oriented edge is the two endpoints, $\delta_1\langle p_i p_j\rangle = \langle p_j\rangle - \langle p_i\rangle$, and the boundary of a triangular face is the three oriented edges, $\delta_2\langle p_i p_j p_k\rangle = \langle p_i p_j\rangle + \langle p_j p_k\rangle - \langle p_i p_k\rangle$. Since an n-chain is just a linear combination of n-simplices, the action of δ_n on an n-simplex given above defines the action of δ_n on any n-chain. The boundary of a boundary is zero, hence $\delta_{n-1}\delta_n = 0$. Since C_n for $n > 2$ is just the identity element, and since $\delta_0\langle p_i\rangle = 0$, to show that $\delta_{n-1}\delta_n = 0$, we need only check that $\delta_1\delta_2\langle p_i p_j p_k\rangle = 0$. This is straightforward to do.

An n-chain, $c \in C_n$, is *closed* if it has no boundary, $\delta_n(c) = 0$. An n-chain that is closed is also called an *n-cycle.* For example, closed 1-chains form one or more loops. The collection of closed n-chains is called $Z_n = Ker\ \delta_n$. The image of δ_{n+1}, i.e. $\delta_{n+1}(C_{n+1})$, is the collection of n-chains that are boundaries of $(n+1)$-chains, and is denoted B_n. Since the boundary of a boundary is zero, then $\delta_n(B_n) = \delta_n\delta_{n+1}(C_{n+1}) = 0$, and all elements of B_n are closed. This makes B_n a subset of Z_n, $B_n \subseteq Z_n$. The quotient, $H_n = Z_n/B_n$, is a group called the nth *simplicial homology group.*

At present we are interested in H_1. It can be shown to be the abelianization of the fundamental group, $\pi_1(S)$. The three edges that form the boundary of a triangle form a closed 1-chain or 1-cycle. It is, however, the boundary of a triangle, a 2-chain, hence it is homologous to the identity element in H_1. A set of edges that forms a closed loop on a surface is also

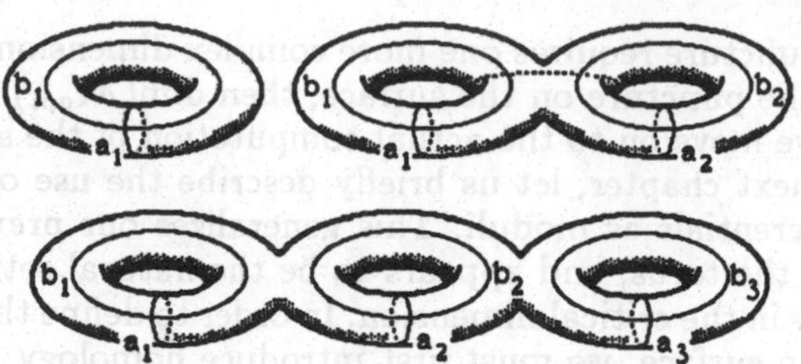

Figure 21.27 Canonical homology bases for genus 1, 2, and 3. The dashed 1-cycle on the genus 2 surface is the cycle $a_1^{-1}a_2$ (see exercise 17).

a 1-cycle. If the bounded region is simply connected, then the triangles inside form a 2-chain and this 1-cycle is also homologous to the identity element in H_1. However, if the bounded region contained a hole, the 1-cycle would no longer be the boundary of a 2-chain (unless the boundary around the hole was contained in the 1-cycle). This 1-cycle represents an element of a equivalence class that is nontrivial in H_1. Similarly, the loop a_1 on the torus in figure 21.27 is a 1-cycle, but is not the boundary of a 2-chain. (The 2-chain consisting of all of the 2-simplices that form the particular triangulation of the torus has no boundary since the torus is without boundary.) The loop b_1 on the torus in figure 21.27 is also a closed 1-chain, but not a boundary. Loops a_1 and b_1 do not belong to the same equivalence class because no linear combination of the two (with nonzero integer coefficients) is an element of B_1. However, every element of Z_1 is homologous with a 1-cycle of the form ma_1+nb_1, $m, n \in \mathbb{Z}$, hence the loops a_1 and b_1 provide a basis for H_1 on the torus. This means that $H_1(T)$ is isomorphic to $\mathbb{Z}\oplus\mathbb{Z}$. Bases for genus 2 and 3 are also illustrated in figure 21.27. In general, for a compact surface, S_g, of genus g without boundary, $H_1(S_g)$ is isomorphic to $\mathbb{Z}^{2g}$.

Associated with the homology is an alternating bilinear form (antisymmetric) called the *intersection product*. The intersection product of two cycles, a and b, denoted as $a \star b$, counts the number of times cycle a intersects cycle b. The product is antisymmetric, $a \star b = -b \star a$, so that a cycle a is considered not to intersect itself, $a \star a = 0$. The intersection product is used to define the *intersection matrix* of a basis of the first homology group, H_1. For the bases described above and illustrated in figure 21.27, we have $a_i \star a_j = 0$, $b_i \star b_j = 0$, and $a_i \star b_j = -b_i \star a_j = \delta_{ij}$ for $i, j = 1, \ldots g$. Hence, the intersection matrix for this basis is

$$J = \begin{pmatrix} 0 & I \\ -I & 0 \end{pmatrix}, \tag{21.124}$$

where I is the $g\times g$ unit matrix. Bases in which the intersection matrix takes this form are called ***canonical or standard homology bases.***

A canonical homology basis is not unique. Another basis, (a'_i, b'_i) for $i = 1, \ldots, g$, defined by

$$a'_i = D_{ij}a_j + C_{ij}b_j$$
$$b'_i = B_{ij}a_j + A_{ij}b_j,$$

where A, B, C, D are $g\times g$ matrices with integer entries, will also be canonical if the intersection matrix is invariant. $2g\times 2g$ matrices, M, with integer elements that preserve the symplectic form J form the symplectic group, $Sp(2g, \mathbb{Z})$. $M \in Sp(2g, \mathbb{Z})$ implies that $M^T JM = J$.

In local coordinates z on a Riemann surface, a differential form can be written as

$$\omega = f(z)dz + g(z)d\bar{z}. \tag{21.125}$$

ω is called a ***meromorphic differential*** if $g(z) = 0$ and $f(z)$ is a meromorphic function. Under a change of coordinates $z \to z'$, we must have

$$\omega = f(z)dz = \tilde{f}(z')dz' \quad \text{or} \quad \tilde{f}(z') = f(z)\frac{\partial z}{\partial z'}. \tag{21.126}$$

Meromorphic differentials are also called ***Abelian differentials.*** If $f(z)$ is holomorphic, then ω is called a ***holomorphic differential.*** Note that a differential can be integrated along a 1-chain or 1-cycle.

On a compact Riemann surface, S_g, of genus g, the space of holomorphic differentials, $\mathcal{H}^1(S_g)$, is a vector space of dimension g (another result of the Riemann-Roch theorem). Let $\{\omega_1, \ldots, \omega_g\}$ be a set of holomorphic differentials that form a basis of $\mathcal{H}^1(S_g)$. It is possible to choose a unique basis on $\mathcal{H}^1(S_g)$ that satisfies

$$\int_{a_i} \omega_j = \delta_{ij}, \tag{21.127}$$

i.e. the integral of the ω_j basis differentials over the a_i 1-cycles in the homology basis of $H^1(S_g)$ is δ_{ij}. The ***canonical $g\times g$ period matrix*** $\Pi = \pi_{ij}$ is defined as

$$\pi_{ij} = \int_{b_i} \omega_j. \tag{21.128}$$

One can use the so-called Riemann bilinear relations to show that $\pi_{ij} = \pi_{ji}$ and $Im(\Pi) > 0$. The set of all complex $g\times g$ symmetric matrices that have positive definite imaginary part is called the ***Siegel upper half-space***, $\mathcal{H}_g$.

Given a compact Riemann surface, S_g of genus g, and a canonical homology basis, we get a unique basis of holomorphic differentials, and a unique period matrix, Π. As we have discussed, the canonical homology basis is not unique. We can transform to another canonical basis using an element of $Sp(2g, \mathbb{Z})$. Such a change of homology basis will also transform the period matrix. If

$$M = \begin{pmatrix} A & B \\ C & D \end{pmatrix} \tag{21.129}$$

is an element of $Sp(2g, \mathbb{Z})$, then the period matrix transforms to

$$\Pi' = \frac{A\Pi + B}{C\Pi + D}. \tag{21.130}$$

This clearly generalizes eq.(21.69).

In fact, for the torus, the differential dz forms a canonical basis of holomorphic differentials for the canonical homology basis given in figure 21.27. The 1×1 period matrix is

$$\tau = \int_{b_1} dz, \tag{21.131}$$

where $Im(\tau) > 0$. The Siegel upper half-space for $g = 1$ is the upper half-plane, $\mathcal{U}$. One can also easily check that the modular group, $\Gamma = PSL(2, \mathbb{Z})$ preserves the intersection matrix,

$$J = \begin{pmatrix} 0 & 1 \\ -1 & 0 \end{pmatrix}. \tag{21.132}$$

Therefore, $PSL(2, \mathbb{Z}) = Sp(2, \mathbb{Z})$. For this reason, the group $Sp(2g, \mathbb{Z})$ is also called the modular group or *Siegel modular group*.

At genus 1, the modular group is equal to the mapping class group, MCG. However, at genus 2 and greater, the modular group, $Sp(2g, \mathbb{Z})$, is a subgroup of the MCG. The quotient, $MCG/Sp(2g, \mathbb{Z}) = T$, is called a *Torelli group*.

The importance of the canonical period matrix is that its elements can serve as complex analytic moduli. In fact, the period matrix defines a map (the period map) from the Teichmüller space, $T_{g,0}$, into the Siegel upper half-space, $\mathcal{H}_g$. By Torelli's theorem, two conformally inequivalent Riemann surfaces cannot have the same period matrix. However, two period matrices related by a $Sp(2g, \mathbb{Z})$ transformation describe the same Riemann surface. In addition, not all period matrices (i.e. elements of $\mathcal{H}_g$) describe Riemann surfaces. Which do is called the Schottky problem (which was solved in 1984). One can see how this problem arises by comparing the dimension of $T_{g,0}$ or $\mathcal{M}_{g,0}$ $(= 3g - 3)$ with $\mathcal{H}_g$ $(= g(g + 1)/2)$

for $g \geq 2$. The two dimensions agree at $g = 2$ and 3 (3, 6 respectively), but the trouble begins at $g = 4$, where the dimension of $\mathcal{H}_g$ rapidly starts to grow larger than $T_{g,0}$ or $\mathcal{M}_{g,0}$.

The moduli space, $\mathcal{M}_{g,0}$ can be embedded into the quotient space, $\mathcal{H}_g/Sp(2g,\mathbb{Z})$. This embedding is accomplished using the Jacobian map. The details may be found in the references in the Bibliography.

21.4 EXERCISES

1. Show by triangulation that the Euler-Poincaré characteristic for a genus g surface with b boundary components is

$$\chi(S_g \text{ with boundary}) = V - E + F = 2 - 2g - b.$$

2. Solve eq.(21.10) for Γ^i_{jk}.
3. Write eq.(21.12) in terms of components of the Riemann tensor, R^i_{jkl}.
4. Apply eqs.(21.16) and (21.17) to the sphere of radius r.
5. What is the fundamental group of a compact genus 2 surface without boundary?
6. Use eq.(21.44) to show that Riemann surfaces are orientable.
7. Find coordinate expressions for stereographic projection from the north pole of a unit sphere onto the plane. Verify that it is conformal.
8. What are the transition functions for the atlas on the Riemann sphere consisting of two coordinate patches that is described in the text on page 567?
9. Show that the Möbius transformations of the form eq.(21.57) are isometries of the metric, eq.(21.52). Which transformations of $Aut(\mathbb{C})$ are isometries on $\mathbb{C}$ with the flat metric?
10. Show that the metric, eq.(21.64), is invariant under $Aut(\mathcal{U})$.
11. Pull back the metric, eq.(21.64), from the Poincaré half-plane, $\mathcal{U}$, to the unit disc, $\mathcal{D}$, to arrive at eq.(21.65). Compute the Gaussian curvature of the metric, eq.(21.65).
12. Compute the Gaussian curvature of the metric,

$$(ds)^2 = \lambda(x,y)\left((dx)^2 + (dy)^2\right).$$

13. Draw the Dirichlet region for the lattice $\Omega(\omega_1, \omega_2)$.
14. Show that the transformations,

$$T(z) = z + 1, \qquad S(z) = -\frac{1}{z},$$

generate the modular group, $\Gamma = PSL(2,\mathbb{Z})$.

15. Show that under a diffeomorphism in $PSL(2,\mathbb{Z})$ the metric, eq. (21.99), pulls back to eqs.(21.105) and (21.106).
16. Starting from eq.(21.115), use eqs.(21.110), (21.112), and (21.10) to derive eq.(21.117).
17. Dehn Twists: Another viewpoint of the global diffeomorphisms (not homotopic to the identity) that connects the torus in figure 21.25(a) to the torus in 21.25(b) is the following: We can arrive at 21.25(b) from 21.25(a) by cutting 21.25(a) along the a_1 1-cycle of the homology basis for the torus in figure 21.27, twisting one side of the cut by 2π, and glueing the torus back together again. Such a cut, twist, and reglue is called a *Dehn Twist.* Let D_{a_1} denote a Dehn Twist about the a_1 basis 1-cycle of figure 21.27. The Dehn Twist leaves the a_1-cycle invariant, $D_{a_1}(a_1) = a_1$, but changes the b_1-cycle into the sum of an a_1-cycle and a b_1-cycle, $D_{a_1}(b_1) = a_1 + b_1$. If we represent the 2 basis cycles as a little vector,

$$\begin{pmatrix} a_1 \\ b_1 \end{pmatrix},$$

find matrix representations for Dehn Twists about the a_1-cycle, D_{a_1}, and about the b_1-cycle, D_{b_1}. Note that both matrices are elements of the modular group, $\Gamma = PSL(2,\mathbb{Z}) = Sp(2,\mathbb{Z})$. Show that these two Dehn Twists generate Γ. For genus ≥ 2, the set of all Dehn Twists is isomorphic to the mapping class group. At genus 2, the Dehn Twists around a canonical homology basis, D_{a_1}, D_{a_2}, D_{b_1}, D_{b_2}, generate the modular group $Sp(4,\mathbb{Z})$. To generate the mapping class group for genus 2, we must also include a Dehn Twist about a cycle that connects the neighboring handles. In figure 21.27, the dashed 1-cycle, $a_1^{-1}a_2$ in the genus 2 surface represents the additional Dehn Twist, $D_{a_1^{-1}a_2}$. This Dehn Twist generates the Torelli group at genus 2. Find matrix representations of the 5 generators above and verify that the first 4 are elements of $Sp(4,\mathbb{Z})$ while the last is not.

CHAPTER **22**

Polyakov's Integral

The Partition Function for Genus 0

The path integral approach to string theory was revitalized by A.M. Polyakov in 1981, using a sum-over-surfaces viewpoint that maintains manifest Lorentz, reparametrization, and conformal invariance. The details of this classic calculation are the topic of this chapter.

In Polyakov's approach, the metric on the world-sheet, g_{ab}, as well as the mapping of the world-sheet into space-time, $x^\mu(\xi^1, \xi^2)$, are treated as independent variables to be quantized, and therefore are integrated over in the "path" integral.

$$Z = \mathcal{N} \int \mathcal{D}g_{ab}\, \mathcal{D}x^\mu \;\exp(-S[g_{ab}, x^\mu]) \tag{22.1}$$

The classical action, $S[g_{ab}, x^\mu]$, in which g_{ab} must appear explicitly, has the advantage of manifest reparametrization and conformal invariance. In addition, this formulation allows the role of the quantized metric in a noncritical dimension ($D \leq 26$) to appear explicitly through the conformal anomaly.

The action used in eq.(22.1) is the Polyakov action, eq.(19.22). The integral over all metrics and embeddings amounts to a sum over random surfaces. However, in order to apply the results from the previous chapter, we need to switch from Minkowski to Euclidean space. This amounts to replacing the time-like coordinates on the world-sheet, τ or ξ^0, and in space-time, x^0, by $i\tau = \xi^2$ and $ix^0 = x^D$, respectively. Such an operation is commonly called a "Wick rotation." After the calculation is done, we may return to Minkowski space by analytic continuation.

The point of doing the Wick rotation is that the metric, g_{ab}, becomes positive definite. The action, eq.(19.22), becomes

$$S = \frac{1}{2} \int d^2\xi \, \sqrt{g} \, g^{ab} \partial_a x^\mu \partial_b x^\mu . \tag{22.2}$$

We shall concentrate solely on closed, oriented bosonic string theory. We shall also restrict the sum over surfaces, eq.(22.1), to compact genus 0 surfaces without boundaries or punctures. The higher-genus contributions to eq.(22.1) will be considered in the next chapter. Since we are summing only over compact surfaces with no boundaries, and there are no sources present in eq.(22.1), by direct analogy with statistical physics, eq.(22.1) is referred to as a partition function (see also, for example, $Z[0]$ in eq.(14.6)).

The action S, eq.(22.2), in eq.(22.1) is reparametrization and Weyl invariant. In order to make eq.(22.1) well defined, we must first apply the Faddeev-Popov procedure. This we do in the next section. This will insert a functional determinant into eq.(22.1), the Faddeev-Popov determinant. The integration over maps to space-time, $x^\mu(\xi^1, \xi^2)$, also leads to a functional determinant, since the action is quadratic in x^μ. The general computation of functional determinants is discussed in the second section, and the methods are then applied in the third section to the determinants that appear in the computation of eq.(22.1). The result is the conformal anomaly. Only in 26 space-time dimensions will the conformal dependence of the Faddeev-Popov determinant exactly cancel that of the x^μ determinant and leave a quantized theory that is conformally invariant.

22.1 FADDEEV-POPOV PROCEDURE

The action in the sum over surfaces in eq.(22.1) is reparametrization and conformally invariant. This means that, in the absence of anomalies, conformally equivalent surfaces contribute identically to the sum. In order for this integral to make sense, we need to sum only over conformally inequivalent surfaces. We know from the previous chapter that the moduli space for compact genus 0 surfaces without boundaries or punctures is just a single point. All compact genus 0 surfaces without punctures or boundaries are conformally equivalent. Therefore, when we restrict eq.(22.1) to genus 0 surfaces, the sum of conformally inequivalent surfaces consists of 1 term or state. Hence, in the absence of anomalies, once we apply the Faddeev-Popov procedure to separate out the gauge volumes associated with reparametrizations and Weyl transformations, and absorb them into the normalization, we are free to adjust the normalization of the one term so that, at the end of the calculation, $Z = 1$. We will find that $Z = 1$ is possible only in the critical dimension, $D = 26$. For $D < 26$, the conformal anomaly spoils the classical conformal invariance and it is not possible to separate out the gauge volume associated with the Weyl transformations.

The integral over all metrics in Z, eq.(22.1), is an integral over $\mathcal{G}_{0,0}$, the set of admissible metrics on compact genus 0 surfaces with no boundaries or punctures. Before we apply the Faddeev-Popov procedure, we must make sense of $\mathcal{D}g_{ab}$. Following Polyakov, we can accomplish this by introducing a metric, eq.(21.107) – call it G – on $\mathcal{G}_{0,0}$.

$$G(\delta g^{(1)}, \delta g^{(2)}) = \langle \delta g^{(1)}, \delta g^{(2)} \rangle = \int d^2\xi \, \sqrt{g} \, g^{ac} g^{bd} \, \delta g^{(1)}_{ab} \, \delta g^{(2)}_{cd} \qquad (22.3)$$

The measure on $\mathcal{G}_{0,0}$ that is invariant under coordinate transformations involves $\sqrt{\det G}$. We can represent all surfaces using a fixed region in the ξ-plane, and all integrals $\int d^2\xi$ are over this fixed region.

A metric $g_{ab} \in \mathcal{G}_{0,0}$ has three independent components which we can take as g_{11}, g_{12}, and g_{22}. Let us first determine $\mathcal{D}g_{ab}$ in terms of "coordinates" g_{11}, g_{12}, and g_{22} using eq.(22.3). An infinitesimal deformation of the metric, δg_{ab}, is a vector tangent to $\mathcal{G}_{0,0}$ at the "point" g_{ab}. Using the coordinates above, we can define three tangent vectors,

$$\delta g^{(a)} = \delta g_{11} \begin{pmatrix} 1 & 0 \\ 0 & 0 \end{pmatrix}, \quad \delta g^{(b)} = \frac{\delta g_{12}}{\sqrt{2}} \begin{pmatrix} 0 & 1 \\ 1 & 0 \end{pmatrix}, \quad \delta g^{(c)} = \delta g_{22} \begin{pmatrix} 0 & 0 \\ 0 & 1 \end{pmatrix}. \qquad (22.4)$$

These three span the tangent space at g_{ab}, and it is easy to verify using eq.(22.3) that the three tangent vectors are mutually orthogonal. Thus, for $\delta g = \delta g^{(a)} + \delta g^{(b)} + \delta g^{(c)}$, we have

$$||\delta g||^2 = G(\delta g, \delta g) = \int d^2\xi \, \sqrt{g} \left((\delta g_{11})^2 + (\delta g_{12})^2 + (\delta g_{22})^2 \right). \qquad (22.5)$$

From eq.(22.5) we can read off that $\sqrt{\det G} = 1$, and

$$\mathcal{D}g_{ab} = \prod_{g \in \mathcal{G}_{0,0}} dg_{11} \, dg_{12} \, dg_{22}, \qquad (22.6)$$

which is no surprise.

To apply the Faddeev-Popov procedure to separate out the gauge volumes associated with diffeomorphisms and Weyl transformations, we first pick a gauge, i.e. a particular form the metric g_{ab} can be put into via diffeomorphisms and conformal transformations. Let's call it $\widehat{g}_{ab}$. Next, we consider an arbitrary metric $g_{ab} \in \mathcal{G}_{0,0}$ and change variables from

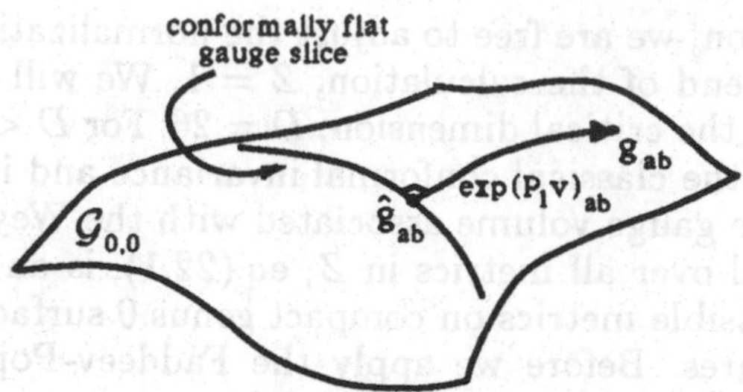

Figure 22.1 The conformally flat gauge slice in $\mathcal{G}_{0,0}$ is the collection of conformally flat metrics. Diffeomorphically inequivalent metrics not on the slice are on orbits generated by $(P_1\vec{v})$. The orbits intersect the slice at right angles.

the coordinates representing g_{ab} to the parameters associated with the diffeomorphisms and conformal transformations that take us from $\widehat{g}_{ab}$ to g_{ab}. The Jacobian of this transformation is the desired Faddeev-Popov determinant.

Let us choose the "conformally flat" gauge where the metric $\widehat{g}_{ab}$ is isothermal,

$$\begin{aligned}\widehat{g}_{ab} &= \exp(\phi(\xi^1,\xi^2))\left((d\xi^1)^2 + (d\xi^2)^2\right) \\ &= \exp(\phi(z))\,|dz|^2,\end{aligned} \tag{22.7}$$

where $z = \xi^1 + i\xi^2$. We know from the previous chapter that this choice can always be made. The conformally flat gauge slice in $\mathcal{G}_{0,0}$ is the collection of conformally flat metrics (see figure 22.1)

Weyl transformations preserve the isothermal form of $\widehat{g}_{ab}$, hence they slide along the gauge slice. The infinitesimal Weyl transformation,

$$\delta g_{ab} = \delta\phi\ g_{ab}, \tag{22.8}$$

at $g_{ab} = \widehat{g}_{ab}$ is a vector tangent to the gauge slice in $\mathcal{G}_{0,0}$. Diffeomorphisms (homotopic to the identity) slide along and move off the gauge slice. Since all diffeomorphisms on compact genus 0 surfaces without boundaries are homotopic to the identity (i.e. $KerP_1^\dagger = 0$), then every point in $\mathcal{G}_{0,0}$ is reachable from $\widehat{g}_{ab}$ by a combination of conformal transformations and diffeomorphisms. The change of coordinates we want to make is from g_{11}, g_{12}, and g_{33}, to the parameters of the gauge slice, ϕ, and the parameters of the diffeomorphisms.

Recall from the previous chapter that the variation in a metric due to an infinitesimal diffeomorphism is

$$\delta g_{ab} = \mathcal{L}_{\vec{v}}g = \nabla_a v_b + \nabla_b v_a, \tag{22.9}$$

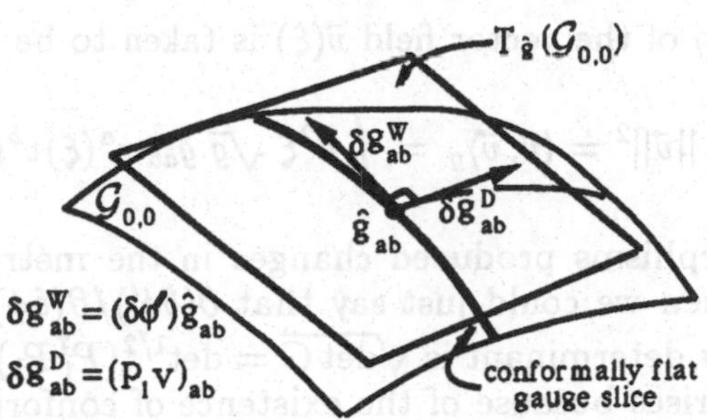

Figure 22.2 δg^W_{ab} is a vector in the tangent space $T_{\hat{g}}(\mathcal{G}_{0,0})$ at $\hat{g}_{ab}$ and is tangent to the gauge slice. δg^D_{ab} is a vector tangent to the orbit generated by $(P_1\vec{v})$ and is orthogonal to δg^W_{ab}.

where the vector field v^a on the surface generates the diffeomorphism. In the metric G, eq.(22.3), the vector, eq.(22.9), is not orthogonal to the vector, eq.(22.8), tangent to the slice. However, we can orthogonally decompose eq.(22.9) by separating the trace out of eq.(22.9).

$$\begin{aligned}\delta g_{ab} &= (\nabla_c v^c)g_{ab} + \delta g^D_{ab} \\ \delta g^D_{ab} &= (P_1\vec{v})_{ab} = \nabla_a v_b + \nabla_b v_a - (\nabla_c v^c)g_{ab}\end{aligned} \tag{22.10}$$

The first term in eq.(22.10), $(\nabla_c v^c)g_{ab}$, is tangent to the slice at $\hat{g}_{ab}$. Integral curves of δg^D_{ab} form gauge orbits in $\mathcal{G}_{0,0}$ that intersect the gauge slice at right angles, as illustrated in figure 22.1.

The combined effects of infinitesimal Weyl transformations and diffeomorphisms are

$$\delta g_{ab} = \delta g^W_{ab} + \delta g^D_{ab}, \tag{22.11}$$

where

$$\delta g^W_{ab} = (\delta\phi + \nabla^c v_c)g_{ab} = (\delta\phi')g_{ab}. \tag{22.12}$$

δg^W_{ab} and δg^D_{ab} are orthogonal and span the tangent space to $\mathcal{G}_{0,0}$ at $\hat{g}_{ab}$ (figure 22.2) as did $\delta g^{(a)}$, $\delta g^{(b)}$, and $\delta g^{(c)}$ of eq.(22.4). In order to determine the Jacobian of the coordinate transformation, we rewrite $||\delta g||^2$ as given by eq.(22.3) in terms of $\delta\phi$ and v^a (as per the discussion leading to eqs.(15.24) and (15.25)). Plugging eqs.(22.11), (22.12), and (22.10) into eq.(22.3) and integrating by parts, we have

$$\begin{aligned}||\delta g||^2 = G(\delta g, \delta g) &= \int d^2\xi\,\sqrt{g}\,\left((\delta\phi')^2 + v^a(P_1^\dagger P_1\vec{v})_a\right) \\ &= ||\delta\phi'||^2 + ||P_1\vec{v}||^2,\end{aligned} \tag{22.13}$$

where the norm of the vector field $\vec{v}(\xi)$ is taken to be

$$||\vec{v}||^2 = \langle \vec{v}, \vec{v} \rangle_g = \int d^2\xi \, \sqrt{g} \, g_{ab} \, v^a(\xi) v^b(\xi). \qquad (22.14)$$

If all diffeomorphisms produced changes in the metric $\widehat{g}_{ab}$ that took it off the slice, then we could just say that $\partial(\delta\phi')/\partial(\delta\phi) = 1$ and that the Faddeev-Popov determinant is $\sqrt{\det G} = \det^{1/2}(P_1^\dagger P_1)$. However, a small complication arises because of the existence of conformal Killing vectors for genus 0 surfaces. Recall from the previous chapter that these vector fields – call them $\bar{v}^a$ – are elements of $Ker(P_1)$, i.e. they satisfy $P_1 \bar{v}^a = 0$. Therefore, diffeomorphisms generated by conformal Killing vector fields move $\widehat{g}_{ab}$ along the slice but not off it and have exactly the same effect as a Weyl transformation. If we naively switch coordinates from g_{11}, g_{12}, and g_{22}, to ϕ and v^a, then we will be counting the conformally flat gauge slice twice. In other words, this transformation is not 1-1 and $\det^{1/2}(P_1^\dagger P_1)$ vanishes due to the existence of zero modes (i.e. vanishing eigenvalues).

The cure for this, of course, is to switch to ϕ and $\vec{v}'$, where $\vec{v}'$ is in the range of P_1. The Jacobian will include a factor to take care of this. We compute it in the same fashion as eq.(15.24) and eq.(15.25). Namely, we have

$$\begin{aligned} 1 &= \int \mathcal{D}g_{ab} \exp(-||\delta g||^2/2) \\ &= J \int \mathcal{D}\phi(\xi) \mathcal{D}v^{a\prime}(\xi) \exp\Big(-(||\delta\phi||^2 + ||P_1 \vec{v}'||^2)/2\Big), \end{aligned}$$

so,

$$J = \left(\det{}'(P_1^\dagger P_1)\right)^{1/2}, \qquad (22.15)$$

up to a constant factor. The prime on the determinant above means that the zero modes have been deleted from the determinant.

Under this coordinate change, eq.(22.1) becomes

$$Z = \mathcal{N} \int \mathcal{D}\phi(\xi) \mathcal{D}v^{a\prime}(\xi) \mathcal{D}x^\mu \left(\det{}'(P_1^\dagger P_1)\right)^{1/2} \exp(-S[g, x^\mu]). \quad (22.16)$$

$\int \mathcal{D}v^{a\prime}(\xi)$ is defined using the metric, eq.(22.14), in the same way. It is the volume, $V^{\perp}_{Diff_0}$, the gauge volume of the group of diffeomorphisms generated by $\vec{v}'$. The next step would be to extract $V^{\perp}_{Diff_0}$ from the integral and absorb it into the normalization. The trouble is, we really want to extract V_{Diff_0} from the integral and not just $V^{\perp}_{Diff_0}$, since V_{Diff_0} is gauge invariant.

Let $V^{\text{ckv}}_{Diff_0}$ be the gauge volume of the diffeomorphisms generated by the conformal Killing vectors. Since the two types of diffeomorphisms are orthogonal to each other, $V_{Diff_0} = V^{\perp}_{Diff_0} \times V^{\rm ckv}_{Diff_0}$. From eq.(22.14) we can explicitly see that $||\vec{v}||^2$ is independent of the conformal factor of the metric, and therefore, so is $\mathcal{D}v^a(\xi)$. Since the integrand is independent of v^a (the action is reparametrization invariant), there is nothing to keep us from pulling V_{Diff_0} out and absorbing it into $\mathcal{N}$. So what we must do is express $V^{\perp}_{Diff_0}$ in terms of V_{Diff_0}.

Let $\psi_i^{(0)}(\xi)$, $i = 1, \ldots, \dim Ker(P_1)$ (= 6 for genus 0) be a basis for $Ker(P_1)$. Then $\vec{v}$ is the linear combination, $\vec{v}(\xi) = \alpha_i \psi_i^{(0)}(\xi)$. Returning to eq.(22.14) we have

$$\begin{aligned} 1 &= \int \mathcal{D}v^a(\xi)\, \exp(-||\vec{v}||^2/2) \\ &= \int \mathcal{D}\alpha_i\, \exp\left(-\frac{1}{2}\,\alpha_i\alpha_j \int d^2\xi\, \sqrt{g}\, \psi_i^{(0)a}\, g_{ab}\psi_j^{(0)b}\right) \\ &\qquad \int \mathcal{D}v^{a\prime}(\xi)\, \exp(-||\vec{v}'||^2/2), \end{aligned}$$

hence,

$$V^{\perp}_{Diff_0} = V_{Diff_0} \left(\det \langle \psi_i^{(0)}, \psi_j^{(0)}\rangle_g\right)^{-1/2}, \tag{22.17}$$

up to a constant factor.

Substituting eq.(22.17) into eq.(22.16) and extracting V_{Diff_0} we get

$$Z = \mathcal{N}' \int \mathcal{D}\phi(\xi)\, \mathcal{D}x^\mu \left(\frac{\det'(P_1^\dagger P_1)}{\det\langle \psi_i^{(0)}, \psi_j^{(0)}\rangle_{\hat{g}}}\right)^{1/2} \exp(-S[\hat{g}, x^\mu]), \tag{22.18}$$

and the integral over $\mathcal{G}_{0,0}$ now is only over the gauge slice.

For the purpose of illustration, let us rederive eq.(22.16) using the traditional approach embodied in eq.(15.19). Specifically, we insert $1 = \Delta_{FP}\Delta_{FP}^{-1}$ into Z, eq.(22.1), where

$$\Delta_{FP}^{-1} = \int \mathcal{D}\phi(\xi)\, \mathcal{D}v^{a\prime}(\xi)\, \delta[g_{ab} - e^\phi \delta_{ab}], \tag{22.19}$$

and

$$\Delta_{FP} = \det \left|\frac{\partial(\delta g_{ab})}{\partial(\phi, v^{a\prime})}\right|_{g=\hat{g}}, \tag{22.20}$$

which is the Faddeev-Popov determinant. Z becomes

$$Z = \mathcal{N} \int \mathcal{D}\phi\, \mathcal{D}v^{a\prime}(\xi)\, \mathcal{D}x^\mu\, \mathcal{D}g_{ab}\, \delta(g_{ab} - e^\phi\, \delta_{ab})\, \Delta_{FP}\, \exp(-S[g, x^\mu]). \tag{22.21}$$

The integral $\int \mathcal{D}g_{ab}$ is now easy to do and sets $g = \widehat{g}$ in the action. As we have already discussed, general variations in the metric can be orthogonally decomposed into a piece that is symmetric and traceless – call it h_{ab} – and a conformal type change,

$$\delta g_{ab} = h_{ab} + \delta\lambda\, g_{ab}.$$

Δ_{FP} is the Jacobian of the transformation,

$$h_{ab} = (P_1 \vec{v}')_{ab}$$
$$\delta\lambda = \delta\phi + \nabla^c v'_c,$$

which is straightforward to compute.

$$\Delta_{FP} = \det \left| \frac{\partial(\lambda, h_{ab})}{\partial(\phi, v^{a\prime})} \right| = \det{}' \begin{vmatrix} 1 & \text{stuff} \\ 0 & P_1 \end{vmatrix} = \det{}'(P_1) = (\det{}'(P_1^\dagger P_1))^{1/2}, \tag{22.22}$$

and again we arrive at eq.(22.16). (The technical term "stuff" was introduced in this context by O. Alvarez.)

Starting from eq.(22.18) we next do the x^μ integral. Since the action is quadratic in x^μ, this is easy to do by applying eq.(9.48) D times (one for each x^μ, $\mu = 1, \ldots, D$). The result is another determinant, but again we must deal with a zero mode. The zero mode in this case is the constant vector, $x_0^\mu(\xi) = \text{constant}$.

The measure $\mathcal{D}x^\mu$ is defined by the metric

$$||\delta x^\mu||^2 = \int d^2\xi\, \sqrt{g}\, \delta x^\mu(\xi)\, \delta x^\mu(\xi) \tag{22.23}$$

in the usual way by requiring

$$1 = \int \mathcal{D}x^\mu \, \exp(-||\delta x^\mu||^2/2). \tag{22.24}$$

Note that the metric is reparametrization invariant but not Weyl invariant. This is the origin of the conformal anomaly for the embedding field x^μ. When δx_0^μ is independent of ξ, we have

$$||\delta x_0^\mu||^2 = \delta x_0^\mu\, \delta x_0^\mu \left(\int d^2\xi\, \sqrt{g} \right).$$

Thus,

$$\mathcal{D}x_0^\mu = \left(\frac{\int d^2\xi\, \sqrt{g}}{2\pi} \right)^{D/2} dx_0^1 \cdots dx_0^D.$$

Let $x^\mu = x_0^\mu + x'^\mu$. Then

$$\begin{aligned}
&\int \mathcal{D}x^\mu \, \exp(-S[\widehat{g}, x^\mu]) \\
&= \int \mathcal{D}x_0^\mu \int \mathcal{D}x'^\mu \, \exp\left(-\int d^2\xi \, \sqrt{\widehat{g}} \, \widehat{g}_{ab} \, \partial_a x'^\mu \partial_b x'^\mu\right) \\
&= \int \mathcal{D}x_0^\mu \int \mathcal{D}x'^\mu \, \exp\left(\int d^2\xi \, \sqrt{\widehat{g}} \, \frac{1}{\sqrt{\widehat{g}}} \, x'^\mu \left(\partial_a \sqrt{\widehat{g}} \, \widehat{g}^{ab} \partial_b\right) x'^\mu\right) \qquad (22.25) \\
&= \mathcal{V} \left(\frac{\int d^2\xi \, \sqrt{\widehat{g}}}{2\pi}\right)^{D/2} \left(\det{}'(\Delta_{\widehat{g}})\right)^{-D/2},
\end{aligned}$$

where $\mathcal{V} = \int dx_0^1 \cdots dx_0^D$ is the volume of space-time, and where

$$\Delta_g = -\frac{1}{\sqrt{g}} \partial_a \sqrt{g} \, g^{ab} \partial_b \qquad (22.26)$$

is the Laplace-Beltrami operator. Note that Δ_g is not Weyl invariant due to the factor of $1/\sqrt{g}$ in front of eq.(22.26). This factor arose in the third line of eq.(22.25), due to the need to associate a factor of $\sqrt{g}$ with $d^2\xi$, as required by the metric for δx^μ, eq.(22.23). Since Δ_g is not Weyl invariant, $\det'(\Delta_g)$ will depend on the conformal factor, $\phi(\xi)$, and this leads to the conformal anomaly for x^μ. The same comment applies to eqs.(22.3), (22.13), (22.14), and (22.15), of course. Again, the prime on the determinant means that the zero mode is deleted in the determinant. Altogether, we have

$$Z = \mathcal{N} \int \mathcal{D}\phi(\xi) \left(\frac{\det{}'(P_1^\dagger P_1)}{\det\langle\psi_i^{(0)}, \psi_j^{(0)}\rangle_{\widehat{g}}}\right)^{1/2} \left(\frac{\det{}'(\Delta_{\widehat{g}})}{\int d^2\xi \sqrt{\widehat{g}}}\right)^{-D/2}, \qquad (22.27)$$

where we have absorbed $\mathcal{V}$ and other constant factors into $\mathcal{N}$.

22.2 EVALUATION OF FUNCTIONAL DETERMINANTS

The Faddeev-Popov procedure and x^μ integral have reduced the calculation of Z to a computation of 2 functional determinants. In this section we will describe methods for evaluating functional determinants and apply these methods to the determinants we encountered in earlier chapters. In the next section we will complete the computation of Z by evaluating the determinants in eq.(22.27).

Consider an operator A with eigenfunctions $\psi_n(x)$,

$$A\psi_n(x) = \lambda_n \psi_n(x). \tag{22.28}$$

Assume that the eigenfunctions are orthonormal and complete,

$$\int dx\, \psi_n^*(x)\psi_m(x) = \delta_{nm}, \qquad \sum_n \psi_n^*(x)\psi_n(x') = \delta(x - x'). \tag{22.29}$$

The determinant of A is the product of its eigenvalues,

$$\det(A) = \prod_n \lambda_n. \tag{22.30}$$

Most methods for computing eq.(22.30) involve manipulating the various Green's functions associated with A.

For example, consider the Green's function, $G_\lambda(x, x')$, satisfying $(A - \lambda)G_\lambda(x, x') = \delta(x - x')$. In terms of the eigenfunctions,

$$G_\lambda(x, x') = \sum_n \frac{\psi_n(x)\psi_n^*(x')}{\lambda_n - \lambda}. \tag{22.31}$$

If we integrate the trace of G_λ with respect to λ we get

$$\int d\lambda \int dx\, G_\lambda(x, x) = -\sum_n \ln(\lambda_n - \lambda) = -\ln\left(\prod_n (\lambda_n - \lambda)\right). \tag{22.32}$$

Therefore,

$$\det(A) = \exp\left(-\left(\int d\lambda \int dx\, G_\lambda(x, x)\right)\bigg|_{\lambda=0}\right), \tag{22.33}$$

up to an overall factor that arises from the constant of integration in eq.(22.32). Often we need only know $\det(A)$ to this level, since the overall factor is absorbed in the normalization.

Let us apply eq.(22.33) to compute $\det(A)$ for

$$A = -(\partial_x^2 + \omega^2), \tag{22.34}$$

which is a determinant we encountered in chapter 12 (eq.(12.53) with $m = 2\pi i$). For boundary conditions, take $\psi_n(x) = \psi_n(x_0) = 0$. We have already computed the Green's function for A when $\lambda = 0$ in chapter 12 (eq.(12.70) with $m = 0$). When $\lambda \neq 0$, we simply replace ω by $\widetilde{\omega} = \sqrt{\omega^2 + \lambda}$. The trace of G_λ is

$$\begin{aligned}\int_{x_0}^{x} d\bar{x}\, G_\lambda(\bar{x}, \bar{x}) &= \frac{1}{\widetilde{\omega}\,\sin(\widetilde{\omega}X)} \int_{x_0}^{x} d\bar{x}\, \sin(\widetilde{\omega}(x - \bar{x}))\, \sin(\widetilde{\omega}(\bar{x} - x_0)) \\ &= \frac{1}{2}\left(\frac{1}{\widetilde{\omega}^2} - \frac{X}{\widetilde{\omega}}\frac{\cos(\widetilde{\omega}X)}{\sin(\widetilde{\omega}X)}\right),\end{aligned} \tag{22.35}$$

where $X = x - x_0$. The integral of eq.(22.35) over λ is also elementary,

$$\int d\lambda \int d\bar{x}\, G_\lambda(\bar{x}, \bar{x}) = -\ln\left(\frac{\sin(\widetilde{\omega}X)}{\widetilde{\omega}}\right). \tag{22.36}$$

Therefore,

$$\det(A) = \frac{\sin(\omega X)}{\omega}, \tag{22.37}$$

which is identical to eq.(12.60) when $m = 2\pi i$. Note that the $\omega = 0$ limit is also easily computed.

Another Green's function used in the computation of $\det(A)$ is the heat kernel, $G(x, x', t)$, which satisfies

$$\frac{\partial}{\partial t} G = -A\,G \tag{22.38}$$

and the boundary condition, $G(x, x', 0) = \delta(x - x')$. Here t is just an auxiliary variable. The eigenfunction expansion of G is

$$G(x, x', t) = \sum_n e^{-\lambda_n t}\psi_n(x)\psi_n^*(x'). \tag{22.39}$$

Again, we take the trace of G,

$$Y(t) = \int dx\, G(x, x, t) = \sum_n e^{-\lambda_n t}. \tag{22.40}$$

One way to extract $\det(A)$ is to multiply the trace by $\exp(-\lambda t)$, integrate with respect to t, then with respect to λ.

$$\int d\lambda \int_0^\infty dt\, Y(t)\, e^{-\lambda t} = \sum_n \int d\lambda \frac{1}{\lambda_n + \lambda} = \sum_n \ln(\lambda_n + \lambda) \tag{22.41}$$

Then,

$$\det(A) = \exp\left(\int d\lambda \int_0^\infty dt\, Y(t) e^{-\lambda t}\bigg|_{\lambda=0}\right) \tag{22.42}$$

up to an overall factor related to the integration constant from the integral over λ. If we exchange the order of integration, eq.(22.42) becomes

$$\det(A) = \exp\left(-\int_0^\infty dt\, t^{-1}\, Y(t)\right). \tag{22.43}$$

The advantage of using the heat kernel, eq.(22.39), for Laplacian-type operators with continuous eigenvalues is that the sum over eigenvalues in eq.(22.39) becomes a Gaussian which is quite doable. As an example, let's consider $A = -\partial_x^2 + \omega^2$ for $x \in \mathbb{R}$. The eigenvalues of A are all positive making eq.(22.39) well defined. In fact,

$$G(x, x', t) = \frac{1}{\sqrt{4\pi t}} \exp\left(-\frac{(x-x')^2}{4t}\right) \exp\left(-\omega^2 t\right). \tag{22.44}$$

From eq.(22.43) we have

$$\begin{aligned} \ln \det(A) &= -\frac{1}{\sqrt{4\pi}} \int_0^\infty dt\, t^{-3/2}\, e^{-\omega^2 t} \int dx \\ &= -\frac{\omega X}{\sqrt{4\pi}} \int_0^\infty dy\, y^{-3/2}\, e^{-y}, \end{aligned} \tag{22.45}$$

where $X = \int dx$. If we integrate eq.(22.45) by parts, we find

$$\ln \det(A) = -\frac{\omega X}{\sqrt{4\pi}} \left(y^{-1/2}\, e^{-y} \Big|_{y=0}^{\infty} - 2\sqrt{\pi} \right). \tag{22.46}$$

Note that the determinant is regularized by replacing the lower limit of the t integral in eq.(22.43) with ϵ. Throwing away the divergent factor we have

$$\det(A) = \exp(\omega X). \tag{22.47}$$

We can disguise the fact that we are discarding the infinite term by noticing that the integral in eq.(22.45) is a representation of the Γ-function,

$$\ln \det(A) = -\frac{\omega X}{\sqrt{4\pi}}\, \Gamma\left(-\frac{1}{2}\right), \tag{22.48}$$

where the Γ-function is defined in this region by analytical continuation.

If we compute $\det(-\partial_x^2 + \omega^2)$ by eq.(22.33), we find

$$\det(A) = \frac{\sinh(\omega X)}{\omega} \xrightarrow{X \to \infty} \frac{1}{2\omega} \exp(\omega X), \tag{22.49}$$

which agrees with eq.(22.47) up to an overall constant factor.

Another way to use the heat kernel to define $\det(A)$ is with the generalized ζ-function,

$$\zeta(s) = \sum_n \frac{1}{\lambda_n^s}. \tag{22.50}$$

In terms of the ζ-function,

$$\det(A) = \exp\left(-\frac{d\zeta(s)}{ds}\bigg|_{s=0}\right). \tag{22.51}$$

The trace of the heat kernel, $Y(t)$, is related to the ζ-function by a Mellin transformation,

$$\zeta(s) = \frac{1}{\Gamma(s)}\int_0^\infty dt\; t^{s-1}\,Y(t). \tag{22.52}$$

The determinant computed via $\zeta(s)$ is identical to that by eq.(22.43) when they are regularized by replacing the lower limit by ϵ.

Returning to our example, $A = -\partial_x^2 + \omega^2$, we find

$$\zeta(s) = \frac{1}{\sqrt{4\pi}}\frac{\Gamma(s-1/2)}{\Gamma(s)}(\omega^2)^{1/2-s}X. \tag{22.53}$$

Using the fact that

$$\frac{1}{\Gamma(s)} = s + \mathcal{O}(s^2), \tag{22.54}$$

we have

$$\zeta'(0) = \frac{1}{\sqrt{4\pi}}\Gamma\left(-\frac{1}{2}\right)\omega X, \tag{22.55}$$

and again we arrive at eq.(22.47).

As another example, let us compute the determinant we encountered in the computation of the effective potential in chapter 18. In this case, $A = \partial_\mu\partial^\mu + m_0^2 + \lambda_0\phi_c^2/2$, where ϕ_c is a constant. In order to obtain positive definite eigenvalues, we perform a Wick rotation and compute $\det(A)$ in Euclidean space. Then, $A = -\partial^2 + m_0^2 + \lambda_0\phi_c^2/2$. The heat kernel in 4 dimensions is

$$G(\vec{x},\vec{x}',t) = \frac{1}{16\pi^2t^2}\exp\left(-\frac{1}{4t}|\vec{x}-\vec{x}'|^2\right)\exp\left(-\left(m_0^2 + \frac{\lambda_0\phi_c^2}{2}\right)t\right). \tag{22.56}$$

The computation of $\zeta(s)$ is straightforward,

$$\zeta(s) = \frac{1}{16\pi^2}\left(m_0^2 + \frac{\lambda_0\phi_c^2}{2}\right)^{2-s}\frac{\Gamma(s-2)}{\Gamma(s)}\int d^4x. \tag{22.57}$$

Note how the divergence of $\Gamma(s-2)$ at $s=0$ cancels that of $\Gamma(0)$, making $\zeta(0)$ regular,

$$\frac{\Gamma(s-2)}{\Gamma(s)} = \frac{1}{(s-2)(s-1)}. \tag{22.58}$$

Applying eq.(22.51), we find

$$\ln\ \det(A) = \frac{1}{32\pi^2}\left(m_0^2 + \frac{\lambda_0\phi_c^2}{2}\right)^2\left(-\frac{3}{2} + \ln\left(m_0^2 + \frac{\lambda_0\phi_c}{2}\right)\right)\int d^4x. \tag{22.59}$$

According to eq.(18.83), the effective potential at 1-loop, $V^{(1)}[\phi_c]$, is

$$V^{(1)}[\phi_c]\int d^4x = \frac{1}{2}\ln\ \det\left(-\partial^2 + m_0^2 + \lambda_0\phi_c^2/2\right) - \frac{1}{2}\ln\ \det\left(-\partial^2 + m_0^2\right). \tag{22.60}$$

To arrive at eq.(18.64), we apply eq.(18.56) to $V[\phi_c] = V_{tree}[\phi_c] + V^{(1)}[\phi_c]$, eqs.(18.34), (22.59), and (22.60), and use the result to replace m_0 by m and λ_0 by λ. This is left as an exercise.

Now let us turn our attention back to the string integral, eq.(22.27). We are interested in finding the ϕ dependence of the determinants. We will find that there are divergent and finite ϕ-dependent pieces. The divergent piece may be cancelled by adding a counterterm to the Polyakov action, eq.(22.2). There is no local reparametrization invariant counterterm to cancel the finite piece and we are led to the conformal anomaly. The only escape from the anomaly is in $D = 26$, where the finite ϕ-dependent pieces from the 2 determinants in eq.(22.27) cancel each other.

22.3 CONFORMAL ANOMALY

In eq.(22.27), we have two determinants to compute. Let's start with $\det'(\Delta_{\widehat{g}})$, where

$$\Delta_{\widehat{g}} = -e^{-\phi}(\partial_1^2 + \partial_2^2) \tag{22.61}$$

is the Laplace-Beltrami operator for $\widehat{g}_{ab} = \exp(\phi)\delta_{ab}$ and where $\partial_i = \partial/\partial\xi^i$. In order to apply either eq.(22.43) or eq.(22.51), we must first determine the heat kernel for the operator $\Delta_{\widehat{g}}$. Considering that $\phi(\xi)$ is an arbitrary function of ξ^1 and ξ^2, this is no simple task. Fortunately, as we shall see, to find the divergent parts we need only to compute the heat kernel perturbatively for small times. In addition, the fact that the functional derivative of $\Delta_{\widehat{g}}$ with respect to the conformal factor ϕ is proportional to $\Delta_{\widehat{g}}$,

$$\frac{\delta\Delta_{\widehat{g}}}{\delta\phi} \propto \Delta_{\widehat{g}}, \tag{22.62}$$

will allow us to find the ϕ-dependent divergent and finite parts of $\det'(\Delta_{\hat{g}})$ as the solution to a simple functional differential equation.

In order to see that we only need to know the heat kernel for $\Delta_{\hat{g}}$ perturbatively, begin by writing the operator A as $-\partial^2 + V$, where $\partial^2 = \partial_1^2 + \partial_2^2$. To simplify things for the moment, let us work in flat space and take $\sqrt{g} = 1$. When $V = 0$, the heat kernel, $G^{(0)}(\vec{x}, \vec{x}', t)$, in 2 dimensions is

$$G^{(0)}(\vec{x}, \vec{x}', t) = \frac{1}{4\pi t} \exp\left(-\frac{|\vec{x} - \vec{x}'|^2}{4t}\right) \tag{22.63}$$

and satisfies $G^{(0)}(\vec{x}, \vec{x}', 0) = \delta^2(x - x')$. The corresponding trace is

$$Y(t) = \frac{1}{4\pi t} \int d^2x, \tag{22.64}$$

and the determinant, via eq.(22.43) with a regulating cutoff ϵ, is

$$\ln\, \det(-\partial^2) = -\int_\epsilon^\infty dt\, t^{-1}\, Y(t) = -\frac{1}{4\pi\epsilon} \int d^2x. \tag{22.65}$$

Let G denote the heat kernel of A, i.e. G satisfies eq.(22.38). G also is the solution of the integral equation,

$$G(\vec{x}, \vec{x}', t)$$
$$= G^{(0)}(\vec{x}, \vec{x}', t) - \int_0^t dt_1 \int d^2x_1 G^{(0)}(\vec{x}, \vec{x}_1, t - t_1) V(\vec{x}_1) G(\vec{x}_1, \vec{x}', t_1). \tag{22.66}$$

In order to see that eq.(22.66) is equivalent to eq.(22.38), apply $\partial_t - \partial^2$ to both sides of eq.(22.64). Also compare the form of eq.(22.64) with eq.(7.82) and eq.(12.90).

The integral equation (22.66) can be solved perturbatively in a Neumann series by successive approximation. If we write

$$G(\vec{x}, \vec{x}', t) = G^{(0)}(\vec{x}, \vec{x}', t) + G^{(1)}(\vec{x}, \vec{x}', t) + G^{(2)}(\vec{x}, \vec{x}', t) + \ldots,$$

then

$$G^{(1)}(\vec{x}, \vec{x}', t) = -\int_0^t dt_1 \int d^2x_1 G^{(0)}(\vec{x}, \vec{x}', t - t_1) V(\vec{x}_1) G^{(0)}(\vec{x}_1, \vec{x}', t_1)$$
$$G^{(2)}(\vec{x}, \vec{x}', t) = \int_0^\infty dt_2 \int d^2x_2 \int_0^\infty dt_1 \int d^2x_1 G^{(0)}(\vec{x}, \vec{x}_2, t - t_2) V(\vec{x}_2)$$
$$G^{(0)}(\vec{x}_2, \vec{x}_1, t_2 - t_1) V(\vec{x}_1) G^{(0)}(\vec{x}_1, \vec{x}', t_1), \tag{22.67}$$

etc. The potential V is independent of t. If V is a constant, then by eq.(22.44) we can see that $G^{(n)}/G^{(0)} \propto t^n$. Taylor expand when V is not constant. The integrals $\int d^2x_i$ are just moments of Gaussians, and again each term $G^{(n)}/G^{(0)}$ will contribute positive powers of t. Therefore, we can write

$$Y(t) = \left(\frac{1}{4\pi t} + \sum_{n=0}^{\infty} B_n t^n\right) \int d^2x. \tag{22.68}$$

Since the eigenvalues of A are positive, we can ignore the upper limit on the integral over t, and we have

$$\ln\ \det(A) = \left(-\frac{1}{4\pi\epsilon} + B_0 \ln(\epsilon) + \mathcal{O}(\epsilon)\right) \int d^2x. \tag{22.69}$$

The regulator is removed by taking $\epsilon \to 0$. Therefore, to find the divergent parts of $\det(A)$, we need only determine B_0, i.e. we need to find

$$\lim_{t\to 0} Y(t). \tag{22.70}$$

Fortunately, B_0 is completely determined by the first order term, $G^{(1)}$.

For the case at hand, eq.(22.70) is also sufficient to determine the ϕ-dependent part of the determinant. This arises because of eq.(22.62). Starting from eq.(22.43), now with $\sqrt{g} \neq 1$, we have

$$\begin{aligned}\delta(\ln\ \det(A)) &= -\int_{\epsilon}^{\infty} dt\, t^{-1}\, \delta Y(t) \\ &= -\int_{\epsilon}^{\infty} dt\, t^{-1}\, \delta\left(\int d^2x\, \sqrt{g}\, G(\vec{x}, \vec{x}, t)\right).\end{aligned} \tag{22.71}$$

The idea is to find the variation in $Y(t)$ when we change the operator A by varying ϕ. If the change in the operator A is δA, then by eq.(22.38) we have

$$\frac{\partial}{\partial t}(G + \delta G) = -(A + \delta A)(G + \delta G),$$

or,

$$\left(\frac{\partial}{\partial t} + A_{\vec{x}}\right) \delta G(\vec{x}, \vec{x}', t) = -(\delta A_{\vec{x}}) G(\vec{x}, \vec{x}', t), \tag{22.72}$$

where $G(\vec{x}, \vec{x}', 0) = (1/\sqrt{g'})\delta^2(x - x')$ and $\sqrt{g'} = \sqrt{g(\vec{x}')}$, and therefore, $\delta(\sqrt{g'}G(\vec{x}, \vec{x}', 0)) = 0$. The inverse of $\partial_t + A_{\vec{x}}$ is G; therefore

$$\begin{aligned}&\sqrt{g'}\, \delta G = \delta\left(\sqrt{g'} G(\vec{x}, \vec{x}', t)\right) \\ &= -\int_0^t dt_1 \int d^2x_1 \sqrt{g_1} \sqrt{g'}\, G(\vec{x}, \vec{x}_1, t - t_1)(\delta A_{\vec{x}_1}) G(\vec{x}_1, \vec{x}', t_1),\end{aligned} \tag{22.73}$$

and

$$\delta Y(t) = -\int_0^t dt_1 \int d^2x_1\sqrt{g_1}\int d^2x\sqrt{g}G(\vec{x},\vec{x}_1,t-t_1)\,\delta A_{\vec{x}_1}\,G(\vec{x}_1,\vec{x},t_1)$$
$$= -\int_0^t dt_1 \int d^2x_1\sqrt{g_1}\int d^2x\sqrt{g}\,G(\vec{x},\vec{x}_1,t_1)\,\delta A_{\vec{x}_1}\,G(\vec{x}_1,\vec{x},t-t_1), \tag{22.74}$$

where we shifted integration variable t_1 to $t-t_1$ in the last step of eq.(22.74). If we think of G as a propagator of a particle, then $G(\vec{x},\vec{x}_1,t_1)$ propagates a particle from $(\vec{x}_1,0)$ to $(\vec{x},t_1)$, and $G(\vec{x}_1,\vec{x},t-t_1)$ from $(\vec{x},t_1)$ to $(\vec{x}_1,t)$. If we sum over the intermediate points $\vec{x}$, we must just get propagation from $(\vec{x}_1,0)$ to $(\vec{x}_1,t)$, hence

$$\int d^2x\,\sqrt{g}\,G(\vec{x},\vec{x}_1,t_1)\,\delta A_{\vec{x}_1}\,G(\vec{x}_1,\vec{x},t-t_1) = \delta A_{\vec{x}_1}\,G(\vec{x}_1,\vec{x}_1,t), \tag{22.75}$$

and

$$\delta Y(t) = -t\int d^2x_1\,\sqrt{g_1}\;\delta A_{\vec{x}_1}\;G(\vec{x}_1,\vec{x}_1,t), \tag{22.76}$$

where $\delta A_{\vec{x}_1}$ only operates on the first argument of G in both eq.(22.75) and eq.(22.76).

Now we make use of eq.(22.62) and assume that $\delta A_{\vec{x}_1} = \beta\,\delta\phi(\vec{x}_1)\,A_{\vec{x}_1}$ and see what happens. We can make use of eq.(22.38) to arrive at

$$\delta Y(t) = -t\beta\int d^2x_1\,\sqrt{g_1}\;\delta\phi(\vec{x}_1)\,A_{\vec{x}_1}G(\vec{x}_1,\vec{x}_1,t)$$
$$= \beta t\,\frac{d}{dt}\int d^2x_1\,\sqrt{g_1}\;\delta\phi(\vec{x}_1)\,G(\vec{x}_1,\vec{x}_1,t), \tag{22.77}$$

and, therefore, by eq.(22.71),

$$\delta(\ln\,\det(A)) = \lim_{\epsilon\to 0}\beta\int d^2x\,\sqrt{g}\,\delta\phi(\vec{x})\,G(\vec{x},\vec{x},\epsilon). \tag{22.78}$$

Note that this is exactly the same calculation we needed for eq.(22.69) (i.e. $G(\vec{x},\vec{x},\epsilon)$ when $\epsilon\to 0^+$).

Of course, what we really need to compute in eq.(22.27) is $\det'(A)$, the determinant without the zero modes. So let's assume that $\dim\,Ker(A)\neq 0$, and see what modifications to eq.(22.69) and eq.(22.77) this requires.

Starting from eq.(22.43) we have

$$\ln\,\det'(A) = -\lim_{\epsilon\to 0}\int_\epsilon^\infty dt\,t^{-1}\,Y'(t),$$

where $Y'(t)$ is the trace of the heat kernel for A minus the zero modes. Let ψ_n be any eigenfunction of A (including the zero modes) and $\psi_i^{(0)}$ represent a zero mode, $\psi_i^{(0)} \in Ker(A)$. The heat kernel for A, minus the zero modes is

$$G'(x,x',t) = \sum_n e^{-\lambda_n t}\psi_n(x)\,\psi_n^*(x') - \sum_{i=1}^{dim\,Ker(A)} \psi_i^{(0)}(x)\psi_i^{(0)*}(x'). \quad (22.79)$$

The second term is independent of time (since $\lambda_i = 0$) and is just the projection operator of any function that A operates on to $Ker(A)$. Since we are assuming all of the eigenfunctions are orthonormal, then

$$Y'(t) = \int dx\,\sqrt{g}\,G'(x,x,t) = Y(t) - \dim Ker(A), \quad (22.80)$$

and eq.(22.69) becomes

$$\ln\,\det{}'(A) = \left(-\frac{1}{4\pi\epsilon} + (B_0 - \dim Ker(A))\ln(\epsilon) + \mathcal{O}(\epsilon)\right)\int d^2x. \quad (22.81)$$

Of more interest is the modification to eq.(22.78). From eq.(22.80) we see that the difference between $Y'(t)$ and $Y(t)$ is just an integer, hence $\delta Y'(t) = \delta Y(t)$ and eq.(22.77) remains valid. The difference now is that $Y(t)$ contains contributions from zero modes and no longer vanishes as $t \to \infty$. Thus, eq.(22.78) becomes

$$\begin{aligned}&\delta(\ln\,\det{}'(A))\\ &= \lim_{\epsilon\to 0}\beta\int d^2x\,\sqrt{g}\,\delta\phi(\vec{x})\,G(\vec{x},\vec{x},\epsilon) - \lim_{t\to\infty}\beta\int d^2x\,\sqrt{g}\,\delta\phi(\vec{x})\,G(\vec{x},\vec{x},t).\end{aligned} \quad (22.82)$$

By examining the eigenfunction expansion of G, one can see that the only parts that survive as $t \to \infty$ are the zero modes, because they are not damped by $\exp(-\lambda_n t)$. Thus, the second term in eq.(22.82) can be written as

$$\lim_{t\to\infty}\beta\int d^2x\,\sqrt{g}\,\delta\phi(\vec{x})\,G(\vec{x},\vec{x},t) = \beta\int d^2x\,\sqrt{g}\,\delta\phi(\vec{x})\sum_i \psi_i^{(0)}(x)\psi_i^{(0)*}(x). \quad (22.83)$$

The right-hand side of eq.(22.83) looks like the variation of a determinant that involves only zero modes. Comparing it with eq.(22.14) and eq.(22.17), we can see that $\int d^2x\sqrt{g}$ represents the metric on $Ker(A)$ where the basis is orthonormal. Thus,

$$\beta\int d^2x\sqrt{g}\,\delta\phi(\vec{x})\sum_i \psi_i^{(0)}(x)\psi_i^{(0)*}(x)=\delta(-\ln\,\det\langle\psi_i^{(0)},\psi_i^{(0)}\rangle_g). \quad (22.84)$$

Switching to a non-orthonormal basis and plugging this back into eq.(22.82), we find that eq.(22.78) becomes

$$\delta\ln\left(\frac{\det'(A)}{\det\langle\psi_i^{(0)},\psi_j^{(0)}\rangle_g}\right)=\lim_{\epsilon\to 0}\beta\int d^2x\sqrt{g}\,\delta\phi(\vec{x})\,G(\vec{x},\vec{x},\epsilon). \quad (22.85)$$

With the two results, eqs.(22.81) and (22.85), in hand, we are ready to compute $\det'(\Delta_{\hat{g}})$. From eq.(22.61) we can read off that

$$V(\xi)=-(e^{-\phi(\xi)}-1)\partial^2.$$

To lowest order in $\phi(\xi)$, this is just $V(\xi)=\phi(\xi)\partial^2$. Taylor expanding $\phi(\xi)$ about $\xi^a=\xi_0^a$, we have

$$\phi(\xi)=\phi(\xi_0)+\partial_a\phi(\xi_0)\,(\xi-\xi_0)^a+\frac{1}{2}\partial_a\partial_b\phi(\xi_0)\,(\xi-\xi_0)^a(\xi-\xi_0)^b+\ldots \quad (22.86)$$

To simplify the calculation we switch to Riemann-normal coordinates centered at ξ_0. In Riemann-normal coordinates, the metric at ξ_0 is flat, $g_{ab}(\xi_0)=\delta_{ab}$, and its first derivative also vanishes, $\partial_c g_{ab}(\xi_0)=0$. There is no way to eliminate the tidal forces at a point, so the second derivative cannot vanish. On the gauge slice, these two conditions translate into $\phi(\xi_0)=0$ and $\partial_a\phi(\xi_0)=0$, which eliminates the first two terms of eq.(22.86). The order t^0 contribution to the heat kernel in this coordinate system for $\Delta_{\hat{g}}$ at $\xi=\xi'=\xi_0$ is

$$-\frac{1}{32\pi^2}\partial_a\partial_b\phi(\xi_0)\int_0^t dt_1\,\frac{1}{t-t_1}\frac{1}{t_1}$$

$$\int d^2\xi_1\,(\xi-\xi_0)^a(\xi-\xi_0)^b\exp\left(-\frac{|\vec{\xi}_0-\vec{\xi}_1|^2}{4(t-t_1)}\right)\partial^2\exp\left(-\frac{|\vec{\xi}_1-\vec{\xi}_0|^2}{4t_1}\right)$$

$$=-\frac{1}{24\pi}(\partial_1^2+\partial_2^2)\phi(\xi_0). \quad (22.87)$$

The scalar curvature, $R(\xi_0)$, (which is -2 times the Gaussian curvature, K, defined in the previous chapter) for a conformally flat metric (exercise 12 of the previous chapter) is

$$R=e^{-\phi}\partial^2\phi. \quad (22.88)$$

In Riemann-normal coordinates, $R(\xi_0)$ is simply $\partial^2\phi(\xi_0)$, hence eq.(22.87) implies that

$$G(\xi_0,\xi_0,t) = \frac{1}{4\pi t} - \frac{1}{24\pi}R(\xi_0) + \mathcal{O}(t) \tag{22.89}$$

in the original coordinate system.

Eq.(22.89) is sufficient knowledge of the heat kernel of $\Delta_{\hat{g}}$ to compute the relevant parts of $\det'(\Delta_{\hat{g}})$. Recall from the discussion surrounding eq.(22.25) that $\Delta_{\hat{g}}$ has one zero mode (x_0^μ for fixed μ), and therefore, by eq.(22.23), $\det\langle x_0^\mu, x_0^\mu\rangle = \int d^2\xi\,\sqrt{\hat{g}}$ up to a constant factor. Placing this and eq.(22.89) into the functional differential equation (22.85) we find ($\beta = -1$),

$$\delta\,\ln\left(\frac{\det'(\Delta_{\hat{g}})}{\int d^2\xi\,\sqrt{\hat{g}}}\right) = -\lim_{\epsilon\to 0}\int d^2\xi\,\sqrt{\hat{g}}\,\delta\phi(\xi)\left(\frac{1}{4\pi\epsilon} - \frac{1}{24\pi}R(\xi)\right). \tag{22.90}$$

This equation is easily integrated to give

$$\ln\left(\frac{\det'(\Delta_{\hat{g}})}{\int d^2\xi\,\sqrt{\hat{g}}}\right) = -\lim_{\epsilon\to 0}\int d^2\xi\left(\frac{e^\phi}{4\pi\epsilon} + \frac{1}{48\pi}\partial_a\phi\partial_a\phi\right)$$
$$+\ \phi\text{–independent terms}, \tag{22.91}$$

where we have integrated by parts in the second term of (22.91). The divergence will have to be renormalized, the result of which is to replace it by an arbitrary renormalized coupling μ^2. We will return to this in a bit. Since μ^2 is arbitrary, we may define it in such a way that we can factor $1/(24\pi)$ out of the integral. The result is

$$\ln\left(\frac{\det'(\Delta_{\hat{g}})}{\int d^2\xi\,\sqrt{\hat{g}}}\right) = -\frac{1}{24\pi}\int d^2\xi\left(\frac{1}{2}(\partial_a\phi)^2 + \mu^2 e^\phi\right)$$
$$+\ \phi\text{–independent terms.} \tag{22.92}$$

Eq.(22.92) does not quite tell the whole story since it is missing a ϕ-independent divergent piece. This piece is represented by B_0 in eq.(22.69) or eq.(22.81). From eq.(22.81) and eq.(22.89) we find that the divergent parts of $\det'(\Delta_{\hat{g}})$ are

$$\ln\,\det'(\Delta_{\hat{g}}) = -\lim_{\epsilon\to 0}\int d^2\xi\,\sqrt{\hat{g}}\left(\frac{1}{4\pi\epsilon} + \frac{1}{24\pi}R(\xi)\ln(\epsilon)\right)$$
$$-6\ln(\epsilon) + \text{ finite terms.} \tag{22.93}$$

Comparison with eq.(22.92) indicates that a ϕ-independent divergent piece that we missed in eq.(22.92) is

$$-\lambda \int d^2\xi \sqrt{\widehat{g}}\, R, \tag{22.94}$$

where again we have introduced a renormalized coupling λ to absorb the divergence. At first sight, eq.(22.94) appears to be ϕ-dependent, even though it is not in eq.(22.92). However, recall from the previous chapter that by eq.(21.35), the global Gauss-Bonnet theorem, eq.(22.94) is just a topological number.

$$-\lambda \int d^2\xi \sqrt{g} R = \lambda 4\pi\chi, \tag{22.95}$$

where χ is the Euler-Poincaré characteristic of the surface (= 2 here since the sum is restricted to genus 0 surfaces). Since it is just a number and independent of ϕ, we can throw it into the normalization.

$\widehat{g}$ in eq.(22.92) is a metric somewhere along the conformally flat gauge slice. Each metric on the slice represents a choice of gauge. There is one metric on the slice that is independent of ϕ, namely $\widehat{g}_{0\,ab} = \delta_{ab}$ (i.e. $\phi = 0$). We can neatly write the ϕ-independent pieces in eq.(22.92) in terms of $\widehat{g}_0$ merely by setting $\phi = 0$ in eq.(22.92). The final result for $\det'(\Delta_{\widehat{g}})$ can be written as

$$\ln\left(\frac{\det'(\Delta_{\widehat{g}})}{\int d^2\xi \sqrt{\widehat{g}}}\right) = \ln\left(\frac{\det'(\Delta_{\widehat{g}_0})}{\int d^2\xi \sqrt{\widehat{g}_0}}\right) - \frac{1}{24\pi}\int d^2\xi \left(\frac{1}{2}(\partial_a\phi)^2 + \mu^2(e^\phi - 1)\right). \tag{22.96}$$

Moving on to the calculation of the Faddeev-Popov determinant, $\det'^{1/2}(P_1^\dagger P_1)$, we first need to find expressions for P_1 and $P_1^\dagger$ that are convenient for computation. This can be accomplished by changing to complex coordinates $z = \xi^1 + i\xi^2$, $\bar{z} = \xi^1 - i\xi^2$. By comparing the metric written in both bases, $(ds)^2 = \exp(\phi(\xi))((d\xi^1)^2 + (d\xi^2)^2) = \exp(\phi)(dz\, d\bar{z} + d\bar{z}\, dz)/2$, we find that

$$\widehat{g}_{zz} = \widehat{g}_{\bar{z}\bar{z}} = 0, \qquad \widehat{g}_{z\bar{z}} = \widehat{g}_{\bar{z}z} = \frac{1}{2}\exp(\phi). \tag{22.97}$$

As usual,

$$\partial_z = \partial = \frac{1}{2}(\partial_1 - i\partial_2),\ \partial_{\bar{z}} = \bar{\partial} = \frac{1}{2}(\partial_1 + i\partial_2),$$

so $\partial\bar{\partial} = \partial^2/4$.

A vector, v^a, written in complex index notation is

$$v^z = v^1 + iv^2, \qquad v^{\bar{z}} = \overline{v^z} = v^1 - iv^2.$$

A symmetric, traceless second-rank tensor, T^{ab}, has 2 independent real components and, therefore, can be represented as

$$T^{zz} = T^{11} + iT^{12}, \quad T^{\bar{z}\bar{z}} = T^{11} - iT^{12}. \tag{22.98}$$

P_1 will diagonalize if we switch basis from (v^1, v^2) to $(v^z, v^{\bar{z}})$. Actually, the proper thing to do is to work entirely with z indices, and no $\bar{z}$ indices. Thus, the basis we will switch to is (v^z, v_z), where $v_z = \widehat{g}_{z\bar{z}} v^{\bar{z}}$. $P_1^\dagger$ will diagonalize in a (T^{zz}, T_{zz}) basis.

Starting from eq.(21.119) or eq.(22.10),

$$(P_1\vec{v})_{ab} = \nabla_a v_b + \nabla_b v_a - (\nabla_c v^c)\widehat{g}_{ab},$$

and explicitly computing the connection coefficients, Γ^k_{ij}, for the metric $\exp(\phi)\delta_{ab}$, it follows that

$$(P_1\vec{v})^{zz} = \widehat{g}^{z\bar{z}}\partial_{\bar{z}} v^z = 2e^{-\phi}\bar{\partial} v^z. \tag{22.99}$$

Since

$$(P_1\vec{v})_{zz} = (\widehat{g}_{z\bar{z}})^2 (P_1\vec{v})^{\bar{z}\bar{z}} = (\widehat{g}_{z\bar{z}})^2 \overline{(P_1\vec{v})^{zz}},$$

then it follows from eq.(22.99) that

$$(P_1\vec{v})_{zz} = (\widehat{g}_{z\bar{z}})^2 \widehat{g}^{z\bar{z}} \partial_z v^{\bar{z}} = \widehat{g}_{z\bar{z}} \partial_z \widehat{g}^{z\bar{z}} v_z.$$

Therefore, P_1 acting on (v^z, v_z) is

$$P_1 = \begin{pmatrix} 2e^{-\phi}\bar{\partial} & 0 \\ 0 & e^{\phi}\partial e^{-\phi} \end{pmatrix}. \tag{22.100}$$

The computation leading to eq.(22.99) is left as an exercise.

Recall that P_1 maps vectors to symmetric, traceless second-rank tensors and that $P_1^\dagger$ maps in the other direction. Applying eq.(21.121) to a symmetric, traceless second-rank tensor, T^{ab}, we find

$$(P_1^\dagger T)^z = -4\left(\partial + 2(\partial\phi)\right) T^{zz},$$

or equivalently,

$$(P_1^\dagger T)^z = -4(\widehat{g}^{z\bar{z}})^2 \partial_z (\widehat{g}_{z\bar{z}})^2 T^{zz}. \tag{22.101}$$

It follows from eq.(22.101) by complex conjugation that

$$(P_1^\dagger T)_z = -4\,\widehat{g}^{z\bar{z}} \partial_{\bar{z}} T_{zz},$$

so $P_1^\dagger$ acting on (T^{zz}, T_{zz}) is

$$P_1^\dagger = \begin{pmatrix} -4\,e^{-2\phi}\partial e^{2\phi} & 0 \\ 0 & -8\,e^{-\phi}\bar{\partial} \end{pmatrix}. \tag{22.102}$$

Eq.(22.101) may, of course, be derived directly from the metrics, eqs.(22.3) and (22.14), using eq.(22.99). In complex index notation, these metrics are

$$\begin{aligned} ||T||^2 &= 8\int dz\,d\bar{z}\,\sqrt{g}\,(g_{z\bar{z}})^2 T^{\bar{z}\bar{z}}T^{zz} \\ ||\vec{v}||^2 &= 2\int dz\,d\bar{z}\,\sqrt{g}\,(g_{z\bar{z}})\,v^{\bar{z}}v^{z}. \end{aligned} \tag{22.103}$$

The numerical coefficients in front above make eq.(22.103) exactly agree with eq.(22.14) and eq.(22.3) applied to symmetric, traceless tensors. Plugging in $(P_1\vec{v})^{zz}$, eq.(22.99), for T^{zz} above, integrating by parts to bring $\bar{\partial}$ onto $T^{\bar{z}\bar{z}}$, and comparing with $||\vec{v}||^2$ above yields eq.(22.101). This is left as an exercise.

Eqs.(22.100) and (22.102) combine together to yield

$$P_1^\dagger P_1 = \begin{pmatrix} -8e^{-2\phi}\partial e^{\phi}\bar{\partial} & 0 \\ 0 & -8e^{-\phi}\bar{\partial}e^{\phi}\partial e^{-\phi} \end{pmatrix} \equiv 2\begin{pmatrix} \Delta_{-1}^{(-)} & 0 \\ 0 & \Delta_1^{(+)} \end{pmatrix}. \tag{22.104}$$

The factor of 2 appears on the right-hand side above so that the Laplacians $\Delta_{-1}^{(-)}$ and $\Delta_1^{(+)}$ are defined to agree with the ordinary flat-space Laplacian in the limit $\phi \to 0$. The notation for the Laplacians is explained at the end of the second section of chapter 26. Since the bottom diagonal components of P_1 and $P_1^\dagger$ are obtained from the top essentially by complex conjugation, then

$$\left(\det{}'(P_1^\dagger P_1)\right)^{1/2} \propto \left(\det{}'(\Delta_1^{(+)})\det{}'(\Delta_{-1}^{(-)})\right)^{1/2} = \det{}'(\Delta_{-1}^{(-)}). \tag{22.105}$$

The factor $\det{}'(2)$ arising from eq.(22.104) can be disposed of in the normalization by renormalizing μ^2. See exercise 2. For notational convenience, let D_1 and $D_1^\dagger$ be proportional to the upper diagonal components of P_1 and $P_1^\dagger$ respectively, so that $\Delta_{-1}^{(-)} = D_1^\dagger D_1$.

In this basis we can see that $D_1^\dagger D_1$ is similar to $\Delta_{\hat{g}}$, so we can attempt to apply the same methods employed for $\Delta_{\hat{g}}$ to compute the determinant. However, we immediately run into a difficulty in that

$$\frac{\delta(D_1^\dagger D_1)}{\delta\phi} = -2\,\delta\phi\,D_1^\dagger D_1 - 4\,e^{-2\phi}\partial\,\delta\phi\,e^{\phi}\bar{\partial}. \tag{22.106}$$

Thus, the simplification that led to eq.(22.85) no longer will happen since the variation of $D_1^\dagger D_1$ is not strictly proportional to $D_1^\dagger D_1$. Amazingly, though, we can find a simple functional differential equation for $\det{}'(D_1^\dagger D_1)$ that is similar to eq.(22.85).

Let $G(z, z', t)$ be the heat kernel for $D_1^\dagger D_1$. Rewrite the second term in eq.(22.106) in terms of D_1 and $D_1^\dagger$.

$$4e^{-2\phi}\partial\,\delta\phi\, e^{\phi}\bar\partial = 4e^{-2\phi}\partial e^{2\phi}\,\delta\phi\, e^{-\phi}\bar\partial = -D_1^\dagger\delta\phi D_1$$

Assume, just for a moment, that there are no zero modes. If we put eq.(22.106) into eq.(22.76) and back into eq.(22.71), we arrive at

$$\begin{aligned}
&\delta(\ln\ \det(D_1^\dagger D_1)) \\
&\quad = \lim_{\epsilon\to 0}\int_\epsilon^\infty dt\int d^2z\,\sqrt{\hat g}\left(-2\delta\phi(z)D_1^\dagger D_1 + D_1^\dagger\,\delta\phi(z)\,D_1\right)G(z,z,t) \\
&\quad = -2\lim_{\epsilon\to 0}\int d^2z\,\sqrt{\hat g}\,\delta\phi(z)\,G(z,z,\epsilon) \\
&\qquad + \int_\epsilon^\infty dt\int d^2z\,\sqrt{\hat g}D_1^\dagger\delta\phi(z)D_1G(z,z,t). \qquad (22.107)
\end{aligned}$$

The integral $\int d^2z\,\sqrt{\hat g}$ represents taking the infinite-dimensional trace of $D_1^\dagger\,\delta\phi\,D_1G$. We can use the cyclic property of a trace to rewrite the second term in eq.(22.107).

$$\begin{aligned}
&\int_\epsilon^\infty dt\int d^2z\,\sqrt{\hat g}D_1^\dagger\delta\phi(z)D_1G(z,z,t) \\
&\qquad\qquad = \int_\epsilon^\infty dt\int d^2z\,\sqrt{\hat g}\,\delta\phi(z)\,D_1G(z,z,t)D_1^\dagger
\end{aligned}$$

It turns out that the last term can be written as an operator acting on its heat kernel. Let $H(z, z', t)$ be the heat kernel for the operator $D_1D_1^\dagger$. Then,

$$\begin{aligned}
GD_1^\dagger &= \exp(-tD_1^\dagger D_1)\,D_1^\dagger \\
&= \left(1 - tD_1^\dagger D_1 + \frac{1}{2}t^2D_1^\dagger D_1D_1^\dagger D_1 + \cdots\right)D_1^\dagger \\
&= D_1^\dagger\left(1 - tD_1D_1^\dagger + \frac{1}{2}t^2D_1D_1^\dagger D_1D_1^\dagger + \cdots\right) \\
&= D_1^\dagger\,\exp(-tD_1D_1^\dagger),
\end{aligned}$$

so

$$G\,D_1^\dagger = D_1^\dagger\,H, \qquad (22.108)$$

and eq.(22.107) becomes

$$\delta(\ln\ \det(D_1^\dagger D_1)) = \lim_{\epsilon\to 0}\int d^2z\,\sqrt{\hat g}\,\delta\phi(z)\,(-2G(z,z,\epsilon) + H(z,z,\epsilon)). \qquad (22.109)$$

Now let's account for the zero modes. At genus 0 we have dim $Ker(P_1^\dagger)$ = dim $Ker(D_1^\dagger) = 0$, hence H has no zero modes and vanishes as $t \to \infty$. G does contain zero modes, so we apply eq.(22.85). The zero modes of D_1 are spanned by $\psi_i^{(0)z}$, thus the zero mode determinant for eq.(22.109) is $\det\langle\psi_i^{(0)z}, \psi_j^{(0)z}\rangle_{\widehat{g}}$. By the same argument leading to eq.(22.105), we have

$$\det{}^{1/2}\langle\psi_i^{(0)}, \psi_j^{(0)}\rangle_{\widehat{g}} = \det\langle\psi_i^{(0)z}, \psi_j^{(0)z}\rangle_{\widehat{g}}.$$

Thus,

$$\delta \ln\left(\frac{\det'(D_1^\dagger D_1)}{\det\langle\psi_i^{(0)z}, \psi_j^{(0)z}\rangle_{\widehat{g}}}\right) = \lim_{\epsilon\to 0}\int d^2z\,\sqrt{\widehat{g}}\,\delta\phi(z)\left(-2G(z,z,\epsilon) + H(z,z,\epsilon)\right). \quad (22.110)$$

Eq.(22.110) reduces the computation of the determinant to calculating the asymptotic behavior of the heat kernels of $D_1^\dagger D_1$ and

$$D_1 D_1^\dagger = -4\exp(-\phi)\bar\partial\exp(-2\phi)\partial\exp(2\phi)$$

at small times to which we now turn.

Since

$$D_1^\dagger D_1 = -4e^{-2\phi}\partial e^{\phi}\bar\partial = -e^{-\phi}\partial^2 - 4\,e^{-\phi}(\partial\phi)\bar\partial, \quad (22.111)$$

we can use the result, eq.(22.89), for the first term. The contribution of the second term to $G(z,z,t)$ is given by

$$4\int_0^t dt_1 \int d^2z_1\, G^{(0)}(z,z_1,t-t_1)\partial\phi(z_1)\bar\partial G^{(0)}(z_1,z,t_1), \quad (22.112)$$

where the $\exp(-\phi(z_1))$ from V cancels the $\sqrt{g_1}$ from the integral over z_1. Integrating by parts yields

$$\begin{aligned} G^{(0)}\partial\phi(z_1)\bar\partial G^{(0)} &= -\bar\partial\left(G^{(0)}\partial\phi(z_1)\right)G^{(0)} \\ &= -\bar\partial G^{(0)}\partial\phi(z_1)G^{(0)} - G^{(0)}(\bar\partial\partial\phi(z_1))G^{(0)}, \end{aligned} \quad (22.113)$$

so

$$G^{(0)}\partial\phi(z_1)\bar\partial G^{(0)} = -\frac{1}{2}G^{(0)}(\bar\partial\partial\phi(z_1))G^{(0)}, \quad (22.114)$$

where we changed integration variables $t_1 \to t - t_1$ to match the first term on the right-hand side of eq.(22.113) with the left-hand side. Switching to Riemann-normal coordinates, the integrals in eq.(22.112) become easy to do, with the result

$$-\frac{1}{2}\partial^2\phi(z)\, t\, G^{(0)}(z,z,t), \tag{22.115}$$

and, therefore,

$$G(z,z,t) = \frac{1}{4\pi t} - \frac{1}{24\pi}R(z) - \frac{1}{8\pi}R(z) + \mathcal{O}(t). \tag{22.116}$$

Moving on to $D_1 D_1^\dagger$ and H, we have

$$D_1 D_1^\dagger = -e^{-\phi}\partial^2 - 2e^{-\phi}\partial^2\phi, \tag{22.117}$$

and again we can use eq.(22.89). In fact, we can also use eq.(22.115). Integrating by parts the contribution of the second term in eq.(22.117) to $H(z,z,t)$, we have

$$2\,G^{(0)}\partial^2\phi G^{(0)} = -8\,\bar{\partial}G^{(0)}\partial\phi G^{(0)},$$

which, after a shift $t_1 \rightarrow t - t_1$, is just what we finished computing. Hence,

$$H(z,z,t) = \frac{1}{4\pi t} - \frac{1}{24\pi}R(z) + \frac{1}{4\pi}R(z) + \mathcal{O}(t). \tag{22.118}$$

Plugging eq.(22.116) and eq.(22.118) back into eq.(22.110) we find that the famous factor of 13 has finally appeared:

$$\delta\,\ln\left(\frac{\det'(D_1^\dagger D_1)}{\det\langle\psi_i^{(0)z},\psi_j^{(0)z}\rangle_{\hat{g}}}\right) = \lim_{\epsilon\rightarrow 0}\int d^2z\,\sqrt{\hat{g}}\,\delta\phi(z)\left(-\frac{1}{4\pi\epsilon} + \frac{13}{24\pi}R(z)\right). \tag{22.119}$$

Eq.(22.119) is trivial to integrate. Again, we renormalize the divergent piece. The result is

$$\ln\left(\frac{\det'(D_1^\dagger D_1)}{\det\langle\psi_i^{(0)z},\psi_j^{(0)z}\rangle_{\hat{g}}}\right) = \frac{1}{2}\ln\left(\frac{\det'(P_1^\dagger P_1)}{\det\langle\psi_i^{(0)},\psi_j^{(0)}\rangle_{\hat{g}}}\right)$$
$$= -\frac{13}{24\pi}\int d^2\xi\left(\frac{1}{2}(\partial_a\phi)^2 + \mu^2 e^\phi\right) + \phi\text{–independent terms.} \tag{22.120}$$

As with the calculation of $\det'(\Delta_{\hat{g}})$, we are missing a ϕ-independent divergent piece which we can find by plugging eq.(22.116) into eq.(22.43). The resulting $\ln(\epsilon)$ term is again proportional to the Euler characteristic.

As in the case of $\det'(\Delta_{\widehat{g}})$, we can rewrite the ϕ-independent part of eq.(22.120) in terms of $\widehat{g}_0$ with the result

$$\frac{1}{2}\ln\left(\frac{\det'(P_1^\dagger P_1)}{\det\langle\psi_i^{(0)},\psi_j^{(0)}\rangle_{\widehat{g}}}\right)$$
$$=\frac{1}{2}\ln\left(\frac{\det'(\widetilde{P}_1^\dagger \widetilde{P}_1)}{\det\langle\widetilde{\psi}_i^{(0)},\widetilde{\psi}_j^{(0)}\rangle_{\widehat{g}_0}}\right)-\frac{13}{24\pi}\int d^2\xi\left(\frac{1}{2}(\partial_a\phi)^2+\mu^2(e^\phi-1)\right), \tag{22.121}$$

where $\widetilde{P}_1^\dagger$ and $\widetilde{P}_1$ are computed with respect to $\widehat{g}_0=\delta_{ab}$.

Finally, we can use eq.(22.96) and eq.(22.121) to complete the calculation of Z, eq.(22.27), with the result

$$Z=\mathcal{N}\int\mathcal{D}\phi(\xi)\,\exp\left(-\frac{26-D}{48\pi}\int d^2\xi\left(\frac{1}{2}(\partial_a\phi)^2+\mu^2(e^\phi-1)\right)\right). \tag{22.122}$$

Any and all ϕ-independent determinants and factors have been tossed into the normalization. Only when $D=26$ does the conformal anomaly cancel. In this case the conformal gauge volume, $\int\mathcal{D}\phi(\xi)$, can be absorbed into the normalization leaving $Z=1$.

In the process of arriving at eq.(22.122) we performed a couple of renormalizations. To formally carry this out, we should add 2 counterterms to the original Polyakov action, eq.(22.2). Thus, the action we should start with in eq.(22.1) is

$$S[g_{ab},x]=\frac{1}{2}\int d^2\xi\,\sqrt{g}\,g^{ab}\partial_a x^\mu\partial_b x_\mu+\mu_0^2\int d^2\xi\,\sqrt{g}+\lambda_0\int d^2\xi\,\sqrt{g}\,R, \tag{22.123}$$

where μ_0^2 and λ_0 are the bare or cutoff-dependent couplings. These two terms are the only two possible counterterms that are invariant (consistent with the symmetries) and whose couplings are dimensionless or have negative (length) dimension. The renormalization of the surface area term required to arrive at eq.(22.122) is accomplished by defining μ_0 as

$$\mu_0^2=\mu^2\left(\frac{26-D}{48\pi}\right)+\frac{D-2}{8\pi\epsilon}. \tag{22.124}$$

In a noncritical dimension $D<26$, eq.(22.122) can be interpreted as the path integral for a 2-dimensional scalar quantum field theory, where the scalar field is $\phi(\xi)$. The field equation for ϕ associated with the action in eq.(22.122) is the Liouville equation, thus eq.(22.122) is the partition function or sourceless generating functional for the quantum Liouville theory. Classically, the Liouville theory is completely integrable. If the integrability survives quantization, then it may be possible to compute eq.(22.122) exactly. We will return to this topic in chapter 25. Next, let us move on to the higher-genus contributions to the partition function.

22.4 EXERCISES

1. Verify eq.(22.13).
2. Show that
$$\ln\,\det(\beta A) = -\int_{\beta\epsilon}^{\infty} dt\; t^{-1}\, Y(t),$$
and argue that the effect of changing β is a finite renormalization. Also show that
$$\det(\beta A) = (\beta)^{\zeta(0)}\,\det(A).$$
Compare the two results.
3. Verify eq.(22.57).
4. Show that eq.(22.59) and (22.60) lead to the same result computed earlier, eq.(18.64).
5. Recompute eq.(22.65) using the ζ-function.
6. Show that eq.(22.66) is equivalent to eq.(22.38).
7. Derive eq.(22.76) using the eigenfunction expansion of the heat kernel G.
8. Verify eq.(22.87).
9. Using the results from the previous chapter, compute the connection coefficients, Γ^k_{ij}, for the metric $\exp(\phi(\xi))((d\xi^1)^2 + (d\xi^2)^2)$. Use the connection coefficients to verify eqs.(22.99) through (22.104).
10. Using eq.(22.98), verify that eq.(22.103) is identical to eq.(22.3) and eq.(22.14). Derive eq.(22.101) by defining the adjoint of P_1 via the metrics, eq.(22.103).
11. Show that $P_1^\dagger P_1$ and $P_1 P_1^\dagger$ have the same spectrum.

CHAPTER 23

Higher-Genus Integrals

In the previous chapter, we evaluated the genus 0 contribution to the closed bosonic string partition function. We followed the approach of Polyakov, where the partition function is viewed as a sum over random surfaces. Our task in this chapter is to discuss the contribution of higher-genus surfaces to the sum.

The starting point is the sourceless string "path" integral,

$$Z = \mathcal{N} \int \mathcal{D}g_{ab}\, \mathcal{D}x^\mu \, \exp\left(- \int d^2\xi \, \sqrt{g} \, (\tfrac{1}{2} g^{ab} \partial_a x^\mu \partial_b x_\mu \, + \, \mu_0^2 \, + \, \lambda_0 R) \right), \tag{23.1}$$

where we have included the two counterterms encountered in the genus 0 calculation. By the global Gauss-Bonnet theorem, eq.(21.35), the last counterterm is a topological invariant and can be pulled out of the integral.

$$Z = \sum_{h=0}^{\infty} \mathcal{N}' \exp\Big(4\pi\lambda_0(2-2h)\Big)$$
$$\int_{\text{fixed } h} \mathcal{D}g_{ab}\, \mathcal{D}x^\mu \, \exp\left(- \int d^2\xi \, \sqrt{g} \, (\tfrac{1}{2} g^{ab} \partial_a x^\mu \partial_b x_\mu \, + \, \mu_0^2) \right) \tag{23.2}$$

The sum over genus h represents a sum over topologies, and amounts to a perturbative expansion. We intend to concentrate on the critical dimension. In this case the normalization $\mathcal{N}$ implicitly contains $1/(V_{Conf}V_{Diff_0})$ which will be cancelled after application of the Faddeev-Popov procedure.

We will begin by applying the Faddeev-Popov procedure to eq.(23.2) to reduce each term, in the critical dimension, to an integral over the moduli space, $\mathcal{M}_{h,0}$. D. Friedan was the first to call attention to the role of moduli in eq.(23.1), and much of the formalism for the reduction was worked out by O. Alvarez. A complete general formalism for $h \geq 2$ was first done by Moore and Nelson, then by D'Hoker and Phong. After applying the procedure, we will then examine the genus 1 term in more detail. The evaluation of this term from the sum-over-surfaces viewpoint was first carried out by J. Polchinski. The genus 1 modular integrand can easily be seen to be a function of the modular parameter τ (which generalizes to the period matrix at higher genus) times the square of a holomorphic function. The fact that this structure holds for arbitrary genus is called holomorphic factorization, and is a result due to Belavin and Knizhnik. We shall discuss this result in the final section.

23.1 REDUCTION TO MODULI

We will carry out the Faddeev-Popov procedure assuming that the genus h of the surfaces we are summing over is greater than or equal to 2. In the next section, this result will be modified for the genus 1 case. The principal difference between $h \geq 2$ and the $h = 0$ calculation in the previous chapter is that now dim $Ker\ P_1^{\dagger}$ is no longer zero (but, on the other hand, dim $Ker\ P_1$ is now zero). This means that there is more than one complex structure that can be placed on the surface. Since a conformal class of metrics determines a complex structure, then, unlike the genus 0 case, the admissible metrics on a surface are not all related to each other by diffeomorphisms or Weyl transformations. Additional degrees of freedom are needed to label the metrics by their conformal class. These parameters are called moduli or Teichmüller parameters.

Let $\mathcal{G}_{h,0}$ be the set of all admissible metrics on surfaces of genus h. In order to define a measure $\mathcal{D}g_{ab}$ on $\mathcal{G}_{h,0}$ we again use the metric, eq.(21.107) or eq.(22.3), and require that

$$1 = \int \mathcal{D}g_{ab} \exp(-\langle \delta g, \delta g \rangle /2). \tag{23.3}$$

An infinitesimal variation of the metric, δg_{ab}, is a vector tangent to $\mathcal{G}_{h,0}$ at the point g_{ab}. The change in the metric due to an infinitesimal Weyl

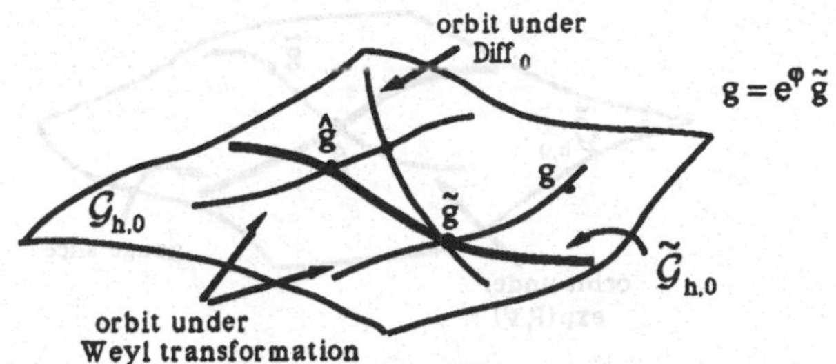

Figure 23.1 $\mathcal{G}_{h,0}$ **is the collection of all admissible metrics on a compact surface of genus** h. $\widehat{g}$ **is a point on the gauge slice.** $\widetilde{g}$ **is related to** $\widehat{g}$ **by a variation that is symmetric and traceless. The collection of all metrics related to the gauge slice in this way is** $\widetilde{\mathcal{G}}_{h,0}$. g **and** $\widetilde{g}$ **are conformally related and fall on the same Weyl orbit.**

transformation and diffeomorphism is given by eq.(21.118) or eq.(22.10). If the diffeomorphism is generated by a vector field $v^a(\xi)$, then

$$\delta g_{ab} = (\delta\phi + \nabla_c v^c)g_{ab} + (P_1\vec{v})_{ab}$$
$$(P_1\vec{v})_{ab} = \nabla_a v_b + \nabla_b v_a - (\nabla_c v^c)g_{ab}.$$

P_1 generates variations that are symmetric and traceless. Deformations $\delta g^{\perp}$ that are elements of $Ker\ P_1^{\dagger}$, $P_1^{\dagger}\delta g^{\perp} = 0$, are orthogonal to those generated by diffeomorphisms and conformal transformations above.

$$\langle \delta g^{\perp}, P_1\vec{v}\rangle = \langle P_1^{\dagger}\delta g^{\perp}, \vec{v}\rangle = 0 \tag{23.4}$$

The metric for the vector field $\vec{v}(\xi)$ used in eq.(23.4) is given by eq.(22.14). Metrics g_{ab} and $g_{ab} + \delta g_{ab}^{\perp}$ belong in different conformal classes and are not related to each other by a Weyl transformation or diffeomorphism. This does exhaust the possibilities so all deformations of a metric can be written as a linear combination of $\delta\phi\, g_{ab}$, $(P_1\vec{v})_{ab}$, and $\delta g_{ab}^{\perp}$.

Let t_i be the moduli or Teichmüller parameters used to label the conformal classes of metrics. The number of real parameters we need is dim $Ker\ P_1^{\dagger} = 6h-6$ for $h \geq 2$. The change of variables we want to perform in eq.(23.2) is from g_{ab} to ϕ, v^a, and t_i. In order to apply the Faddeev-Popov procedure, we must first make a choice of gauge. An acceptable choice of gauge is any slice of $\mathcal{G}_{h,0}$ that is transversal to the orbits of Weyl transformations and $Diff_0$, and intersects each orbit only once. A choice is illustrated in figures 23.1 and 23.2. Let $\widehat{g}(t_i)$ denote a metric on the gauge slice. Let $\widetilde{g}$ be related to $\widehat{g}(t_i)$ by a transformation generated by $(P_1\vec{v})$. This means that $\widetilde{g}$ and $\widehat{g}$ fall on the same orbit or integral curve generated by $(P_1\vec{v})$ as illustrated in figure 23.2. Denote the collection of all metrics

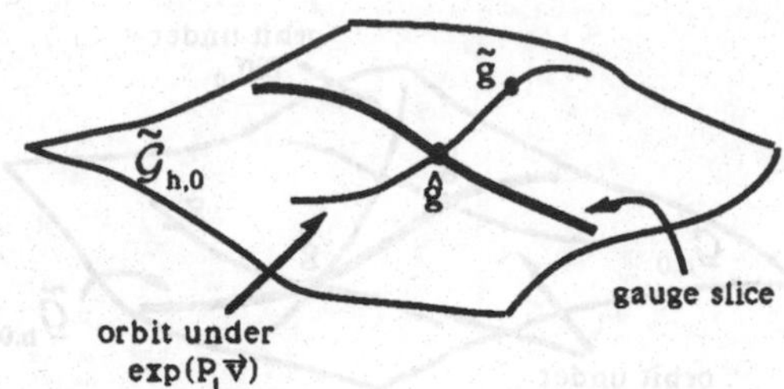

Figure 23.2 $\hat{g}$ is a point on the gauge slice. $\tilde{g}$ is related to $\hat{g}$ by a transformation generated by $(P_1\vec{v})$. $\tilde{\mathcal{G}}_{h,0}$ is the collection of all metrics related to the gauge slice by such symmetric traceless transformations.

related to the gauge slice in this way as $\tilde{\mathcal{G}}_{h,0}$. The remaining metrics in $\mathcal{G}_{h,0}$ fall on orbits of Weyl transformations through $\tilde{\mathcal{G}}_{h,0}$ (figure 23.1), i.e. $g_{ab} = \exp(\phi)\tilde{g}_{ab}$. If the change in a metric in $\tilde{\mathcal{G}}_{h,0}$ is traceless, then the new metric will remain in $\tilde{\mathcal{G}}_{h,0}$. Infinitesimal changes in $\tilde{g}_{ab} \in \tilde{\mathcal{G}}_{h,0}$ that are traceless are tangent to $\tilde{\mathcal{G}}_{h,0}$ at $\tilde{g}_{ab}$. Thus, the orbits of Diff_0 in general will not lie in $\tilde{\mathcal{G}}_{h,0}$ (except when $\tilde{\mathcal{G}}_{h,0} = \mathcal{G}^{\text{const}}_{h,0}$). Of course, any metric on an orbit of Diff_0 through $\tilde{g}$ can be brought back to $\tilde{\mathcal{G}}_{h,0}$ by a compensating Weyl transformation.

Recall from the discussion in chapter 21 that the Uniformization theorem tells us that there is a unique metric $\tilde{g}$ that is conformal to any $g_{ab} \in \mathcal{G}_{h,0}$ which has constant unit negative curvature, $R_{\tilde{g}} = -1$. We denoted the collection of such metrics in chapter 21 as $\mathcal{G}^{\text{const}}_{h,0}$. One common and convenient way of fixing the gauge is to choose the gauge slice to lie in $\mathcal{G}^{\text{const}}_{h,0}$, i.e. take $\tilde{\mathcal{G}}_{h,0} = \mathcal{G}^{\text{const}}_{h,0}$. Since a diffeomorphism does not change the curvature, then the orbits of Diff_0 lie entirely in $\mathcal{G}^{\text{const}}_{h,0}$. That is, $\tilde{\mathcal{G}}_{h,0}$ in this case is chosen so as to contain the entire orbits of Diff_0 that intersect the slice.

A general infinitesimal variation of a metric $g(t_i) \in \mathcal{G}_{h,0}$ is

$$\begin{aligned}\delta g_{ab} &= \delta\phi\, g_{ab} + \nabla_a v_b + \nabla_b v_a + \delta t_i \frac{\partial}{\partial t_i} g_{ab} \\ &= \left(\delta\phi + \nabla_c v^c + \frac{1}{2}\left(g^{cd}\,\delta t_i \frac{\partial}{\partial t_i} g_{cd}\right)\right) g_{ab} + (P_1\vec{v})_{ab} + T^i_{ab}\,\delta t_i,\end{aligned} \tag{23.5}$$

where the second line above results from breaking the variation up into trace and traceless pieces. The first term involving the traces is a vector tangent to Weyl orbits in $\mathcal{G}_{h,0}$ at g_{ab}. The two traceless pieces are vectors

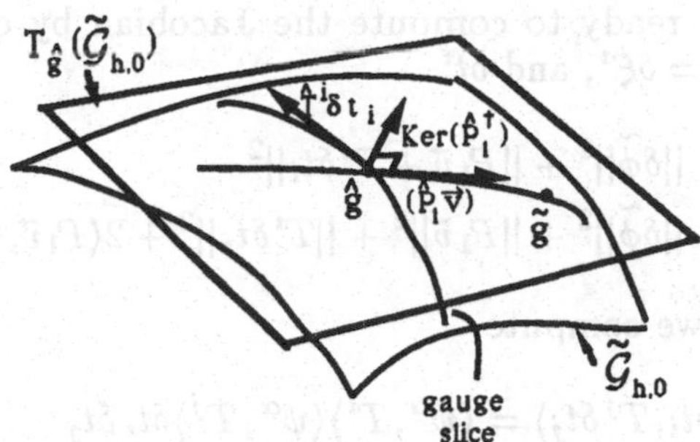

Figure 23.3 $T_{\hat{g}}(\widetilde{\mathcal{G}}_{h,0})$ is the tangent space to $\widetilde{\mathcal{G}}_{h,0}$ at $\widehat{g}_{ab}$. $(\widehat{P}_1\vec{v})$ are vectors tangent to the projected orbits of $Diff_0$ in $\widetilde{\mathcal{G}}_{h,0}$. $\widehat{T}^i\,\delta t_i$ are tangents to the gauge slice. Tangent vectors in $Ker(\widehat{P}_1^\dagger)$ are orthogonal to $(\widehat{P}_1\vec{v})$ and point in directions that change the class of complex structure associated with $\widehat{g}_{ab}$. Vectors tangent to Weyl orbits are elements of $T_{\hat{g}}(\mathcal{G}_{h,0})$ but not $T_{\hat{g}}(\widetilde{\mathcal{G}}_{h,0})$.

orthogonal to the tangents of Weyl orbits. $\delta t_i\,\partial_{t_i} g_{ab}$ are tangents to the coordinate lines of the Teichmüller parameters. The traceless part,

$$T^i_{ab}\,\delta t_i = \left(\frac{\partial}{\partial t_i} g_{ab} - \frac{1}{2}\left(g^{cd}\frac{\partial}{\partial t_i} g_{cd}\right) g_{ab}\right)\delta t_i, \tag{23.6}$$

is the projection of these tangent vectors to directions orthogonal to Weyl orbits. Let $\widehat{T}^i\,\delta t_i$ denote eq.(23.6) evaluated on the slice at $\widehat{g}_{ab}$. When the variation of the metric is made at $\widehat{g}_{ab}$ on the slice, then the $\widehat{T}^i\,\delta t_i$, which are vectors in $T_{\widehat{g}}(\mathcal{G}_{h,0})$ and $T_{\widehat{g}}(\widetilde{\mathcal{G}}_{h,0})$, are tangent to the slice. The tangent space, $T_{\widehat{g}}(\widetilde{\mathcal{G}}_{h,0})$ at $\widehat{g}_{ab}$ is illustrated in figure 23.3.

Let ψ^α_{ab}, $\alpha = 1, \ldots,$ dim $Ker(P_1^\dagger) = 6h - 6$ be an orthonormal basis for $Ker(P_1^\dagger)$. The ψ^α_{ab} are tangent vectors in $T_g(\mathcal{G}_{h,0})$ and are orthogonal to $(P_1\vec{v})_{ab}$ by the metric G, eq.(21.107) or eq.(22.3), on $T_g(\mathcal{G}_{h,0})$. Since the orbits generated by $(P_1\vec{v})$ do not in general intersect the Teichmüller coordinate lines at right angles, then the vectors T^i_{ab} tangent to the coordinate lines are not orthogonal to either $(P_1\vec{v})$ or ψ^α_{ab}. Using the metric, eq.(22.3), decompose T^i_{ab} as a linear combination of ψ^α_{ab} and $(P_1\vec{v})_{ab}$,

$$T^i_{ab}\,\delta t_i = \langle\psi^\alpha, T^i\rangle\psi^\alpha_{ab}\,\delta t_i + \frac{\langle P_1\vec{v}, T^i\rangle}{||P_1\vec{v}||^2}(P_1\vec{v})_{ab}\,\delta t_i. \tag{23.7}$$

We are now ready to compute the Jacobian by expressing $||\delta g||^2$ in terms of $\delta\phi$, $v^a = \delta\xi^a$, and δt^i.

$$\begin{aligned}||\delta g||^2 &= ||\delta\widetilde{\phi}||^2 + ||P_1\vec{v} + T^i\delta t_i||^2 \\ &= ||\delta\widetilde{\phi}||^2 + ||P_1\vec{v}||^2 + ||T^i\delta t_i||^2 + 2\langle P_1\vec{v}, T^i\delta t_i\rangle\end{aligned} \tag{23.8}$$

From eq.(23.7) we compute

$$\begin{aligned}||T^i\delta t_i||^2 = \langle T^i\delta t_i, T^j\delta t_j\rangle &= \langle\psi^\alpha, T^i\rangle\langle\psi^\alpha, T^j\rangle\delta t_i\,\delta t_j \\ &\quad + \frac{\langle P_1\vec{v}, T^i\rangle\langle P_1\vec{v}, T^j\rangle}{\langle P_1\vec{v}, P_1\vec{v}\rangle}\delta t_i\,\delta t_j.\end{aligned} \tag{23.9}$$

Thus,

$$||\delta g||^2 = ||\delta\widetilde{\phi}||^2 + ||P_1\widetilde{v}||^2 + \langle\psi^\alpha, T^i\rangle\langle\psi^\alpha, T^j\rangle\delta t_i\,\delta t_j, \tag{23.10}$$

where

$$\delta\widetilde{\phi} = \delta\phi + \nabla_c v^c + \frac{1}{2}\left(\widehat{g}^{cd}\delta t_i\frac{\partial}{\partial t_i}\widehat{g}_{cd}\right), \tag{23.11}$$

and

$$\widetilde{v} = \left(1 + \frac{\langle P_1\vec{v}, T^i\delta t_i\rangle}{\langle P_1\vec{v}, P_1\vec{v}\rangle}\right)\vec{v}. \tag{23.12}$$

Since $\partial(\delta\widetilde{\phi})/\partial(\delta\phi) = 1$ and $\widetilde{v}$ is proportional to $\vec{v}$, then the Jacobian $(\sqrt{\det G})$ is

$$\det{}^{1/2}(P_1^\dagger P_1)\det\langle\psi^\alpha, T^j\rangle. \tag{23.13}$$

Recall that the ψ^α basis of $Ker(P_1^\dagger)$ is assumed to be orthonormal. For an arbitrary basis, eq.(23.13) becomes

$$\det{}^{1/2}(P_1^\dagger P_1)\,\frac{\det\langle\psi^i, T^j\rangle}{\det^{1/2}\langle\psi^i, \psi^j\rangle}. \tag{23.14}$$

Before we insert this result back into eq.(23.2), let's also do the x^μ integral. This integral is not any different than the one we computed in the previous chapter, hence we may just use eq.(22.25). Eqs.(23.14) and (22.25) reduce the genus h term in eq.(23.2) to

$$\begin{aligned}\mathcal{VN}\int_{\text{fixed }h}\mathcal{D}\phi\,\mathcal{D}v^a(\xi)\,dt^1\cdots dt^{6h-6}\left(\frac{2\pi}{\int d^2\xi\sqrt{g}}\det{}'(\Delta_g)\right)^{-D/2} \\ \det{}^{1/2}(P_1^\dagger P_1)\,\frac{\det\langle\psi^i, T^j\rangle}{\det^{1/2}\langle\psi^i, \psi^j\rangle},\end{aligned} \tag{23.16}$$

where $\mathcal{V}$ is the volume of space-time. At this point it is apparent that nothing depends on v^a since it is all reparametrization invariant. We can pull $V_{Diff_0} = \int \mathcal{D}v^a(\xi)$ out of the integral. Unlike the genus 0 case, not all diffeomorphisms are connected to the identity. The gauge volume we want to absorb into the normalization is V_{Diff} and not just V_{Diff_0}. We will return to this in a moment, but for the time being we will carry V_{Diff_0} along explicitly.

Next, we would like to extract $\int \mathcal{D}\phi(\xi)$ from the integral. A quick examination of

$$\langle \psi^i, T^j \rangle = \int d^2\xi \sqrt{g}\, g^{ab} g^{cd} \psi^i_{ac} T^j_{bd} \tag{23.17}$$

indicates that it is independent of ϕ, and therefore so is $\det\langle \psi^i, T^j \rangle$. However, as in the genus 0 case, the remaining determinants still depend upon ϕ and again we meet the conformal anomaly. Fortunately, the anomaly is a local object and therefore has the same form for all genus h. This means that we can use eq.(22.90),

$$\delta \ln\left(\frac{\det'(\Delta_g)}{\int d^2\xi \sqrt{g}}\right) = -\lim_{\epsilon\to 0} \int d^2\xi \sqrt{g}\, \delta\phi(\xi)\Big(\frac{1}{4\pi\epsilon} - \frac{1}{24\pi} R(\xi)\Big). \tag{23.18}$$

If $\widehat{R}(\xi)$ is the scalar curvature of the metric $\widehat{g}$ on the gauge slice that is conformally related to the metric g by $g_{ab} = \exp(\phi)\widehat{g}_{ab}$, then

$$R(\xi) = \exp(-\phi)(\widehat{R} + \widehat{g}^{ab}\partial_a\partial_b\phi). \tag{23.19}$$

Inserting eq.(23.19) into eq.(23.18) and integrating, we have

$$\ln\left(\frac{\det'(\Delta_g)}{\int d^2\xi \sqrt{g}}\right) = \ln\left(\frac{\det'(\Delta_{\widehat{g}})}{\int d^2\xi \sqrt{\widehat{g}}}\right) - \frac{1}{24\pi} S_L, \tag{23.20}$$

where

$$S_L = \int d^2\xi \sqrt{\widehat{g}} \left(\frac{1}{2}(\partial_a\phi)^2 + \mu^2(e^\phi - 1) + \widehat{R}\phi\right), \tag{23.21}$$

is the Liouville action.

The computation of $\det(P_1^\dagger P_1)$ is similar to that in the previous chapter. The difference here is that dim $Ker(P_1) = 0$ so that we do not need to worry about zero modes in $\det(P_1^\dagger P_1)$. However, in evaluating this determinant in the previous chapter, we found that the heat kernel of the operator $P_1 P_1^\dagger$ entered the computation (eq.(22.109)). Since dim $Ker(P_1^\dagger) \neq 0$, there are $P_1^\dagger$ zero modes present in this heat kernel which survive the limit

$t \to \infty$. The contribution of this heat kernel in the limit $t \to \infty$ is precisely the determinant of the $P_1^\dagger$ zero modes, hence

$$\delta \ln\left(\frac{\det(P_1^\dagger P_1)}{\det\langle\psi^i,\psi^j\rangle}\right) = \lim_{\epsilon\to 0}\int d^2z\,\sqrt{g}\,\delta\phi(\xi)\Big(-\frac{1}{2\pi\epsilon}+\frac{13}{12\pi}R(z)\Big),$$

and

$$\ln\left(\frac{\det(P_1^\dagger P_1)}{\det\langle\psi^i,\psi^j\rangle}\right) = \ln\left(\frac{\det(\widehat{P}_1^\dagger \widehat{P}_1)}{\det\langle\widehat{\psi}^i,\widehat{\psi}^j\rangle}\right) - \frac{26}{24\pi}S_L, \tag{23.22}$$

where the hats on the operators and basis zero modes means that they are evaluated on the gauge slice at the point $\widehat{g}_{ab}$ in $\mathcal{G}_{h,0}$. Combining eqs.(23.22), (23.20), and (23.16), the genus h contribution to Z is

$$\mathcal{V}\mathcal{N}V_{Diff_0}\int_{\text{fixed}\,h}\mathcal{D}\phi\, d^{6h-6}t\,\det\langle\widehat{\psi}^i,\widehat{T}^j\rangle\left(\frac{\det'(\Delta_{\widehat{g}})}{\int d^2\xi\,\sqrt{\widehat{g}}}\right)^{-D/2}$$
$$\left(\frac{\det(\widehat{P}_1^\dagger \widehat{P}_1)}{\det\langle\widehat{\psi}^i,\widehat{\psi}^j\rangle}\right)^{1/2}\exp\left(-\frac{26-D}{48\pi}S_L\right). \tag{23.23}$$

In the critical dimension, $D = 26$, the ϕ dependence disappears from the integrand and $V_{Conf} = \int \mathcal{D}\phi$ may be absorbed into the normalization. The remaining finite-dimensional integral with respect to the t^i is an integral over $\mathcal{G}_{h,0}/(Diff_0 \times Weyl) = \mathcal{T}_{h,0}$, the Teichmüller space for genus h. As we mentioned earlier, the integrand is invariant under the full diffeomorphism group *Diff* and not just *Diff*$_0$. This means that by integrating over all of Teichmüller space we are again over-counting. This residual gauge freedom is fixed by integrating only over the moduli space, $\mathcal{M}_{h,0}$ $(= \mathcal{G}_{h,0}/(Diff \times Weyl))$. Thus, the genus h contribution to Z for $h \geq 2$ reduces to

$$\int_{\mathcal{M}_{h,0}} d^{6h-6}t\,\det\langle\widehat{\psi}^i,\widehat{T}^j\rangle\left(\frac{2\pi}{\int d^2\xi\,\sqrt{\widehat{g}}}\det'(\Delta_{\widehat{g}})\right)^{-13}\left(\frac{\det(\widehat{P}_1^\dagger \widehat{P}_1)}{\det\langle\widehat{\psi}^i,\widehat{\psi}^j\rangle}\right)^{1/2}, \tag{23.24}$$

which is an integral over the finite-dimensional moduli space $\mathcal{M}_{h,0}$. All factors that are independent of the coordinates on $\mathcal{M}_{h,0}$ have been absorbed into the normalization.

When the gauge choice is $\widetilde{\mathcal{G}}_{h,0} = \mathcal{G}_{h,0}^{\text{const}}$, then $\widehat{g}$ is a constant-curvature metric with $\widehat{R} = -1$. The metric on $\mathcal{G}_{h,0}$, eq.(21.107), induces a metric on the gauge slice and on the moduli space $\mathcal{M}_{h,0} = \mathcal{G}_{h,0}^{\text{const}}/Diff$, called the

Weil-Petersson metric. With this gauge choice, eq.(23.24) can be written as

$$\int_{\mathcal{M}_{h,0}} d(\text{Weil}-\text{Petersson})\,\det{}^{1/2}(\widehat{P}_1^\dagger \widehat{P}_1)\left(\frac{2\pi}{\int d^2\xi\,\sqrt{\widehat{g}}}\det{}'(\Delta_{\widehat{g}})\right)^{-13}. \tag{23.25}$$

In the next section we will make the minor modification to eq.(23.24) for the genus 1 case that is necessary because dim $Ker(P_1) = 2$ when $h = 1$. We will also evaluate the genus 1 determinants.

23.2 GENUS 1

The situation at genus 1 is described in some detail in chapter 21. For genus 1, dim $Ker(P_1^\dagger) = 2$ (real). The Teichmüller space, $T_{1,0}$, is the upper half-plane, $\mathcal{U}$, and the modular group, which is also the mapping class group, is $\Gamma = PSL(2,\mathbb{Z}) = Sp(2,\mathbb{Z})$. The moduli space is $\mathcal{M}_{1,0} = \mathcal{U}/\Gamma$, which can be identified with $D(\Gamma)/\mathbb{Z}_2$, where $D(\Gamma)$ is the fundamental region of the modular group, eq.(21.71).

However, we also have dim $Ker(P_1) = 2$, so we must modify eq.(23.24) to account for the zero modes. Since we already had to account for conformal Killing vectors in the previous chapter for the genus 0 computation, we can just make use of the result embodied in eq.(22.15) and (22.17). Let χ^α be an arbitrary basis for $Ker(P_1)$. To account for the zero modes, we replace

$$\det{}^{1/2}(\widehat{P}_1^\dagger \widehat{P}_1) \longrightarrow \left(\frac{\det{}'(\widehat{P}_1^\dagger \widehat{P}_1)}{\det\langle\widehat{\chi}^\alpha, \widehat{\chi}^\beta\rangle}\right)^{1/2} \tag{23.26}$$

in eq.(23.24). The metric for vector fields employed in $\langle\chi^\alpha,\chi^\beta\rangle$ above is eq.(22.14). Thus, the contribution to Z from genus 1 is

$$\int_{\mathcal{M}_{1,0}} d^2t\,\frac{\det\langle\widehat{\psi}^i,\widehat{T}^j\rangle}{\det^{1/2}\langle\widehat{\psi}^i,\widehat{\psi}^j\rangle}\left(\frac{\det{}'(\widehat{P}_1^\dagger \widehat{P}_1)}{\det\langle\widehat{\chi}^\alpha, \widehat{\chi}^\beta\rangle}\right)^{1/2}\left(\frac{2\pi}{\int d^2\xi\,\sqrt{\widehat{g}}}\det{}'(\Delta_{\widehat{g}})\right)^{-13} \tag{23.27}$$

The indices i, j, α, and β all run from 1 to 2.

In order to proceed with an explicit computation, we must make a choice of gauge, i.e. a choice of $\widehat{g}$. Following Polchinski, we will take

$$\widehat{g}_{ab} = |d\xi^1 + \tau\, d\xi^2|^2, \tag{23.28}$$

which is on the same Weyl orbit as eq.(21.96). The (ξ^1, ξ^2) coordinate region is fixed to be the unit square. The modular parameters are $\tau = \tau_1 + i\tau_2$. From eq.(23.28) we have $\sqrt{\hat{g}} = \tau_2$, hence

$$\int d^2\xi \sqrt{\hat{g}} = \tau_2. \tag{23.29}$$

A straightforward computation yields

$$\widehat{T}^1_{ab} = \begin{pmatrix} -\tau_1 & 1-\tau_1 \\ 1-\tau_1 & \tau_1 \end{pmatrix}, \quad \widehat{T}^2_{ab} = \begin{pmatrix} -\tau_2 & -\tau_2 \\ -\tau_2 & \tau_2 \end{pmatrix}. \tag{23.30}$$

The metric $\hat{g}_{ab}$ is flat and since its components are independent of ξ^1 and ξ^2, so are the components of $\widehat{T}^i_{ab}$. Since the metric is flat and the components of T^i_{ab} are constants, then $(P_1^\dagger \widehat{T}^i)_b = -2\partial^a \widehat{T}^i_{ab} = 0$, hence $\widehat{T}^i_{ab} \in Ker(P_1^\dagger)$ and can serve as a basis for $Ker(P_1^\dagger)$ (i.e. $\widehat{\psi}^i = \widehat{T}^i$). This simplifies the computation of the determinants in eq.(23.27) to

$$\frac{\det\langle \widehat{\psi}^i, \widehat{T}^j\rangle}{\det^{1/2}\langle \widehat{\psi}^i, \widehat{\psi}^j\rangle} \longrightarrow \det{}^{1/2}\langle \widehat{T}^i, \widehat{T}^j\rangle, \tag{23.31}$$

which also follows directly from eq.(23.8). The metric $\langle,\rangle$ employed in eq.(23.31) is eq.(22.3). The inverse metric $\hat{g}^{ab}$ is

$$\hat{g}^{ab} = \frac{1}{\tau_2^2}\begin{pmatrix} |\tau|^2 & -\tau_1 \\ -\tau_1 & 1 \end{pmatrix}, \tag{23.32}$$

and it follows that

$$\det\langle \widehat{T}^i, \widehat{T}^j\rangle = \left(\int d^2\xi \sqrt{\hat{g}}\right)^2 \frac{4}{\tau_2^4} = \frac{4}{\tau_2^2}. \tag{23.33}$$

Next, let us move on to the Killing vector determinant. It is clear that the vector field $v^a =$ constant, which is just a translation, leaves the metric invariant, $\mathcal{L}_{\vec{v}}\hat{g} = 0$. Thus, the vectors $\chi^1 = (1,0)$ and $\chi^2 = (0,1)$ span $Ker(P_1)$. Using the metric for vector fields, eq.(22.14), it follows that

$$\langle \chi^i, \chi^j\rangle = \int d^2\xi \sqrt{\hat{g}}\, \hat{g}_{ij}, \tag{23.34}$$

hence

$$\det\langle \chi^i, \chi^j\rangle = \left(\int d^2\xi \sqrt{\hat{g}}\right)^2 \det \hat{g}_{ij} = \tau_2^4. \tag{23.35}$$

For the differential determinants, $\Delta_{\hat{g}}$ is $-\hat{g}^{ab}\partial_a\partial_b$. Since the metric $\hat{g}$ is flat, covariant derivatives are just ordinary derivatives and $(P_1^\dagger P_1)_{ab}v^b = 2\Delta_{\hat{g}}v_a = 2\delta_{ab}\,\Delta_{\hat{g}}v^b$, thus

$$\det{}'(P_1^\dagger P_1) = \left(\det{}'(2\Delta_{\hat{g}})\right)^2. \tag{23.36}$$

It follows from the properties of determinants that $\det'(2\Delta_{\hat{g}}) = \det'(2)\,\det'(\Delta_{\hat{g}})$. The prime on $\det'(2)$ means that one "zero mode" has been extracted (i.e. the zero mode of $\Delta_{\hat{g}}$). Since all "eigenvalues" in the product $\det(2)$ are the same, then we must have

$$\alpha \det{}'(\alpha) = \det(\alpha) \tag{23.37}$$

for any constant α, and

$$\det{}'(2\Delta_{\hat{g}}) = \frac{1}{2}\det(2)\det{}'(\Delta_{\hat{g}}). \tag{23.38}$$

The determinant of a constant can be computed in a formal sense using the various definitions of a determinant in the previous chapter. The result is a divergence proportional to the area of the surface, $\int d^2\xi\,\sqrt{g}$. It can therefore be eliminated by adjusting μ_0^2 in the counterterm in eq.(23.1).

Returning eqs.(23.33), (23.35), (23.36), and (23.38) to eq.(23.27) and eliminating the constant determinant by renormalization, we have

$$Z_{\text{torus}} = \int_{\mathcal{M}_{1,0}} d^2\tau\,\frac{\tau_2^{10}}{(2\pi)^{13}}\det{}'(\Delta_{\hat{g}})^{12}. \tag{23.39}$$

All that remains is the evaluation of $\det'(\Delta_{\hat{g}})$ to which we now turn.

The eigenfunctions, $\psi_{n_1 n_2}(\xi^1,\xi^2)$, of $\Delta_{\hat{g}}$, which must be periodic on the unit square are $\exp(2\pi i n_1\xi^1)\exp(2\pi i n_2\xi^2)$, with eigenvalues

$$\lambda_{n_1 n_2} = \frac{4\pi^2}{\tau_2^2}\left(|\tau|^2 n_1^2 - 2\tau_1 n_1 n_2 + n_2^2\right) = \frac{4\pi^2}{\tau_2^2}\left(n_2 - n_1\tau\right)\left(n_2 - n_1\bar{\tau}\right). \tag{23.40}$$

Thus,

$$\begin{aligned}\ln\,\det{}'(\Delta_{\hat{g}}) &= \ln\prod_{n_1,n_2}{}'\frac{4\pi^2}{\tau_2^2}\left(n_2 - n_1\tau\right)\left(n_2 - n_1\bar{\tau}\right)\\ &= \sum_{n_1,n_2}{}' - \ln(\tau_2^2) + \ln(2\pi(n_2 - n_1\tau)) + \ln(2\pi(n_2 - n_1\bar{\tau})),\end{aligned} \tag{23.41}$$

where the prime means that $n_1 = n_2 = 0$ is eliminated from the product or sum. The first term represents the det′ of a constant. We apply eq.(23.37) and eliminate $\det(1/\tau_2^2)$ by renormalization.

To evaluate the second term in eq.(23.41) we will represent the determinant via eq.(22.33). Hence, we want to compute

$$-\lim_{\lambda\to 0}\sum_{n_1,n_2}{}' \int d\lambda \,\frac{1}{2\pi(n_2 - n_1\tau) - \lambda} = -\lim_{\lambda\to 0}\sum_{n_1,n_2}{}' \int d\lambda \,\frac{1}{(n_2 - n_1\tau) - \lambda}. \tag{23.42}$$

The sum over n_2 can be converted to an integral by thinking of the sum as a sum of residues and employing the Sommerfeld-Watson transformation. This transformation is based on the result of the following contour integral,

$$\int_\gamma dz\,\frac{f(z)}{\sin(\pi z)} = 2\pi i \sum_{n_2} (-)^{n_2}\,\frac{1}{\pi}\,f(n_2), \tag{23.43}$$

where the contour γ surrounds the real axis. Applying eq.(23.43) to eq. (23.42), we have

$$-\lim_{\lambda\to 0}\sum_{n_1}{}' \int d\lambda \int_\gamma \frac{dz}{2i}\,\frac{e^{i\pi z}}{\sin(\pi z)}\,\frac{1}{z - (n_1\tau + \lambda)}. \tag{23.44}$$

Now we want to deform the contour and break it into 2 closed pieces. The first passes on top of the real axis and closes in the upper half-plane. Since $\tau_2 > 0$, the contour encloses one pole at $z_0 = n_1\tau + \lambda$. The other piece passes below the real axis and closes in the lower half-plane. We use $\exp(-i\pi z)$ in the integrand so that it converges. Both new contours are oppositely oriented from the original contour, so we pick up a minus sign. Since the lower contour does not enclose any poles, it does not contribute to the integral, eq.(23.44). Therefore, the integral over z in eq.(23.44) is $-2\pi i\ \mathrm{Res}(z_0)$ and eq.(23.44) becomes

$$\begin{aligned}\pi \lim_{\lambda\to 0}\sum_{n_1}{}' \int d\lambda\,\frac{e^{i\pi(n_1\tau+\lambda)}}{\sin(\pi(n_1\tau + \lambda))} &\\ = \pi \lim_{\lambda\to 0}\sum_{n_1}{}' \int d\lambda\,\cot(\pi(n_1\tau + \lambda)) + i & \qquad (23.45)\\ = \sum_{n_1}{}' \ln\Big(\sin(\pi n_1\tau)\Big). &\end{aligned}$$

The same procedure can be applied to the third term in eq.(23.41) except that this time only the lower contour contributes. All together, eq.(23.41) becomes

$$\begin{aligned}\ln\,\det{}'(\Delta_{\hat{g}}) &= \ln(\tau_2^2) + \sum_{n_1}{}' \ln\,\sin(\pi n_1 \tau) + \ln\,\sin(\pi n_1 \bar{\tau})\\ &= \ln(\tau_2^2) + 2\sum_{n_1=1}^{\infty} \ln\,\sin(\pi n_1 \tau) + \ln\,\sin(\pi n_1 \bar{\tau})\\ &= \ln(\tau_2^2) + 2\sum_{n_1=1}^{\infty} \ln\left|1 - e^{2i\pi n_1 \tau}\right|^2 + 2\pi n_1 \tau_2.\end{aligned} \tag{23.46}$$

In the last step, another constant piece is eliminated by renormalization. The last term in eq.(23.46) contains $\sum n_1$. We regularized and evaluated this sum in chapter 20 when we were considering the zero point energies of the modes of the string (see eqs.(20.101) and (20.111)). The origin of the third term above is identical.The result is

$$\sum_{n_1=1}^{\infty} n_1 = -\frac{1}{12}. \tag{23.47}$$

This completes the evaluation of the determinant with the result

$$\det{}'(\Delta_{\hat{g}}) = \tau_2^2\, e^{-\pi\tau_2/3} \left|\prod_{n=1}^{\infty} 1 - e^{2i\pi n\tau}\right|^4. \tag{23.48}$$

Plugging this back into eq.(23.39) we finally arrive at

$$\begin{aligned}Z_{\text{torus}} &= \int_{\mathcal{M}_{1,0}} \frac{d^2\tau}{2\pi\tau_2^2}\, e^{4\pi\tau_2}\,(2\pi\tau_2)^{-12} \left|\prod_{n=1}^{\infty} 1 - e^{2i\pi n\tau}\right|^{-48}\\ &= \frac{1}{2}\int_{D(\Gamma)} \frac{d^2\tau}{2\pi\tau_2^2}\, e^{4\pi\tau_2}\,(2\pi\tau_2)^{-12} \left|\prod_{n=1}^{\infty} 1 - e^{2i\pi n\tau}\right|^{-48}.\end{aligned} \tag{23.49}$$

$D(\Gamma)$ in the second line of eq.(23.49) is the fundamental region of the modular group $PSL(2,\mathbb{Z})$, eq.(21.71). Integrating over $D(\Gamma)$ is equivalent to integrating twice over $\mathcal{M}_{1,0}$. The factor of 1/2 in the second line accounts for this. In other words, integrating only over $\mathcal{M}_{1,0}$ fully fixes the gauge. Integrating over $D(\Gamma)$ does not. The residual gauge freedom not fixed is the diffeomorphism, $\xi^1 \to -\xi^1$, $\xi^2 \to -\xi^2$. The factor of 1/2 removes the over-counting.

23.3 HOLOMORPHIC FACTORIZATION

The viewpoint we have discussed up to now could be labeled as the real analytic approach, because we have written everything in terms of real coordinates on the moduli space. The result is an expression of the partition function for each particular genus in terms of a set of determinants. As we have seen, determinants of operators are not trivial objects to calculate. Even so, these determinants can be expressed in terms of the Selberg ζ-function. The asymptotics and divergences can then be explored. When $\mathcal{G}^{\text{const}}_{h,0}$ is used as the gauge slice, and when Fenchel-Nielsen coordinates are used as real coordinates on the Teichmüller space, then the explicit expression for the Weil-Petersson measure is particularly simple. The determinants can also be expressed in terms of these coordinates and their asymptotics explicitly analyzed. The drawback with this approach, though, is that the real coordinates are appropriate for the Teichmüller space and not really for the moduli space because the action of the Mapping Class group in this coordinate system is complicated. As a result, the simple complex-analytic structure of the critical, closed, oriented bosonic string, discovered by A. Belavin and V. Knizhnik, is completely obscured. When the natural complex structure on the moduli space $\mathcal{M}_{h,0}$ is employed to provide complex analytic coordinates on $\mathcal{M}_{h,0}$, the simple complex analytic structure becomes accessible. Belavin and Knizhnik were able to show that the higher-genus string measure can be written as the absolute square of a holomorphic function divided by the 13th power of the determinant of the imaginary part of the period matrix. Since the integrand essentially splits into a holomorphic and an antiholomorphic part, this result is called holomorphic factorization. Intuitively it says that in the critical dimension, the left-moving modes and right-moving modes of the quantum closed string act independently and make separate contributions to the partition function. In addition, the holomorphic function contains no zeroes on $\mathcal{M}_{h,0}$, and a second order pole on the 1-point compactification of $\mathcal{M}_{h,0}$. These two properties are enough to determine this function uniquely up to a constant factor. This means that the string integrand can be determined, and, in principle, completely written down without having to explicitly evaluate the functional determinants.

The genus 1 result exhibits almost the same properties. Let us recast the genus 1 result, eq.(23.49), into a form to illustrate this. As complex coordinates we can take the moduli τ. Therefore we write $d^2\tau$ as $(i/2)\, d\tau\, d\overline{\tau}$, or more properly as $(i/2)\, d\tau \wedge d\overline{\tau}$. Recall from the discussion in chapter 21 that the period matrix is the $1{\times}1$ matrix τ, hence $\det(\operatorname{Im}\Pi_{ij}) = \tau_2$. The remaining terms in eq.(23.49) can be expressed as the discriminant, $\Delta(\tau)$, eq.(21.78). Hence,

$$Z_{\text{torus}} = \int_{\mathcal{M}_{1,0}} \frac{i}{2} \frac{d\tau \wedge d\overline{\tau}}{2\pi} \, (2\pi)^{12} \, (\operatorname{Im}\tau)^{-14} \, |\Delta(\tau)|^{-2}, \tag{23.50}$$

where

$$\Delta(\tau) = (2\pi)^{12}\, e^{2i\pi\tau} \prod_{n=1}^{\infty} (1 - e^{2i\pi n\tau})^{24}, \tag{23.51}$$

or

$$Z_{\text{torus}} = \int_{\mathcal{M}_{1,0}} \frac{i}{2} \frac{dy \wedge d\overline{y}}{2\pi} \frac{1}{|y|^2} (2\pi)^{10} (\ln|y|)^{-14} |F(y)|^2, \tag{23.52}$$

where $y = \exp(2\pi i \tau)$, and

$$F(y) = \left(y \prod_{n=1}^{\infty} (1 - y^n)^{24} \right)^{-1}. \tag{23.53}$$

Up to a power of the logarithm, the measure in eq.(23.52) is the absolute square of a holomorphic function. In addition, this holomorphic function does not vanish anywhere on $\mathcal{M}_{1,0}$. Note, however, that the holomorphic function does develop a second order pole ($|y|^{-1}|F(y)| \to |y|^{-2}$ as $y \to 0$) on the boundary of $\mathcal{M}_{1,0}$ at $y = 0$. y vanishes as $\tau \to i\infty$, the "point at infinity" on the boundary of $\mathcal{M}_{1,0}$. This pole will lead to a divergence. If we trace back where this factor comes from, we are led to the sum of the string mode oscillator zero-point energies, i.e. the tachyon.

The generalization of eq.(23.52) given by Belavin and Knizhnik is

$$Z_h = \int_{\mathcal{M}_{h,0}} \left(\frac{i}{2}\right)^{3h-3} dy_1 \wedge \cdots \wedge dy_{3h-3} \wedge d\overline{y}_1 \wedge \cdots \wedge d\overline{y}_{3h-3} \exp\left(W(y_i, \overline{y}_i)\right), \tag{23.54}$$

where

$$\exp\left(W(y_i, \overline{y}_i)\right) = |F(y_1, \ldots, y_{3h-3})|^2 \left(\det \operatorname{Im} \Pi\right)^{-13}, \tag{23.55}$$

and Π is the period matrix (see eq.(21.127)). Again, $F(y)$ is a nonzero holomorphic function on $\mathcal{M}_{h,0}$, and has second order poles in $\overline{\mathcal{M}}_{h,0}$ (the 1-point compactification of $\mathcal{M}_{h,0}$). In other words, $F(y_i)$ is meromorphic on $\overline{\mathcal{M}}_{h,0}$ with a second order pole at $\overline{\mathcal{M}}_{h,0}/\mathcal{M}_{h,0}$.

In order to discuss how they arrived at eqs.(23.54) and (23.55), we must first describe how the complex coordinates y_i on $\mathcal{M}_{h,0}$ are chosen. Recall from the discussion in chapter 21 that isothermal coordinates define a natural complex structure on a Riemann surface S. In an isothermal coordinate patch, the metric expressed in these coordinates is conformally flat,

$$(ds)^2 = \widehat{\rho}|dz|^2 = 2\widehat{g}_{z\overline{z}}\, dz d\overline{z}. \tag{23.56}$$

An infinitesimal variation of this metric is

$$\delta\rho'\,dzd\overline{z} + h_{zz}\,dz\,dz + h_{\overline{z}\overline{z}}\,d\overline{z}\,d\overline{z}, \tag{23.57}$$

where $\overline{h}_{zz} = h_{\overline{z}\overline{z}}$. As we have already discussed, these variations arise from Weyl transformations, diffeomorphisms, and changes in the complex structure. An infinitesimal diffeomorphism generated by the vector field v^z $(= v^1 + iv^2)$ induces a change in the coordinates by

$$\begin{aligned} z &\longrightarrow z + v^z(z,\overline{z}) \\ \overline{z} &\longrightarrow \overline{z} + v^{\overline{z}}(z,\overline{z}). \end{aligned} \tag{23.58}$$

Since

$$\begin{aligned} dz &\longrightarrow dz \,+\, \partial_z v^z dz \,+\, \partial_{\overline{z}} v^z d\overline{z} \\ d\overline{z} &\longrightarrow d\overline{z} + \partial_z v^{\overline{z}} dz \,+\, \partial_{\overline{z}} v^{\overline{z}} d\overline{z}, \end{aligned} \tag{23.59}$$

then the contribution to h_{zz} and $h_{\overline{z}\overline{z}}$ from the diffeomorphism is

$$h_{zz} = \widehat{\rho}\,\partial_z v_z, \quad h_{\overline{z}\overline{z}} = \widehat{\rho}\partial_{\overline{z}} v_{\overline{z}}, \tag{23.60}$$

where $v_z = g_{z\overline{z}}v^{\overline{z}}$. Variations f_{zz} and $f_{\overline{z}\overline{z}}$ that arise in eq.(23.57) from changes in the complex structure are orthogonal to eq.(23.60). Using the metric on the tangent space of $\mathcal{G}_{h,0}$ at $\widehat{g}$, eq.(22.3) or eq.(22.103), we see that orthogonality implies that

$$\partial_{\overline{z}} f_{zz} = 0, \quad \partial_z f_{\overline{z}\overline{z}} = 0. \tag{23.61}$$

Tensors of the form $f_{zz}(dz)^2$ (type (2,0) with 2 lower z indices and no lower $\overline{z}$ indices) satisfying eq.(23.61) are called holomorphic quadratic differentials. Of course, this discussion is no different than that surrounding eq.(21.120). Hence, $Ker(P_1^\dagger)$ is the space of holomorphic quadratic differentials.

An alternative but equivalent approach to the parametrization of variations of the metric and complex structure is with Beltrami coefficients and differentials. A general metric can always be put into the form eq.(21.80),

$$(ds)^2 = \widetilde{\rho}|dz + \mu_{\overline{z}}^z\, d\overline{z}|^2. \tag{23.62}$$

If eq.(23.62) represents an infinitesimal change of the metric, eq.(23.56), then the Beltrami coefficients parametrize the variations due to changes in the conformal structure along with diffeomorphisms. From eq.(23.59) it is easy to see that changes due to the latter are

$$\mu_{\overline{z}}^z = \partial_{\overline{z}} v^z. \tag{23.63}$$

Any other changes deform the complex structure, hence the tangent space of the moduli space $\mathcal{M}_{h,0}$ consists of the collection of Beltrami differentials modulo the range of $\partial_{\bar{z}}$ on vectors, i.e. excluding eq.(23.63).

The collection of Beltrami differentials is dual to the space of quadratic differentials because the Beltrami differentials are linear functionals on the quadratic differentials. The linear map from the quadratic differentials to the reals is given by

$$\mu_{\bar{z}}^{z}[f_{zz}] = \int dz\, d\bar{z}\, \mu_{\bar{z}}^{z}\, f_{zz}. \tag{23.64}$$

Note that the $\mu_{\bar{z}}^{z}$ and f_{zz} are made for each other in that the combination, $\mu_{\bar{z}}^{z} f_{zz}$ is a type (1,1) tensor that can be combined with $dz\, d\bar{z}$ and immediately integrated without further reference to a metric on the surface. Also note that the Beltrami differentials arising from diffeomorphisms, eq.(23.63), annihilate the quadratic differentials via eq.(23.64) (integrate by parts).

By the way, it is easy to see from eq.(23.63) that the z and $\bar{z}$ indices on $\mu_{\bar{z}}^{z}$ occur in a natural way. The z index appears as the upper index since we are operating on a vector. The $\bar{z}$ index appears on the bottom since it occurs in conjugation with a derivative with respect to a coordinate. On the other hand, to arrive at eq.(23.60) where both indices are on the bottom, we need to use the metric. The Beltrami differentials modulo the type $\partial_{\bar{z}} v^{z}$ form the tangent space of $\mathcal{M}_{h,0}$, while, technically speaking, the quadratic differentials, $Ker(P_1^{\dagger})$, form the cotangent space. The distinction is essential only when there is no metric present to raise or lower indices. Since a metric is present, we will continue to call them both tangent vectors.

Let f_i, $i = 1, \ldots, 3h-3$, be a basis of holomorphic quadratic differentials for $Ker(P_1^{\dagger})$. Define a dual basis, $\mu^j = (\mu_{\bar{z}}^{z})^j$, $j = 1, \ldots, 3h-3$, for the space of Beltrami differentials via the requirement

$$\int d^2z\, \mu^j f_i = \delta_i^j. \tag{23.65}$$

Now let $\mu_{\bar{z}}^{z}$ be an infinitesimal deformation of the complex structure related to the isothermal coordinates z and metric in eq.(23.56). This Beltrami differential can be written as a linear combination of the basis μ^j,

$$\mu_{\bar{z}}^{z}(y) = y_j \mu^j. \tag{23.66}$$

y_j is a complex $(3h-3)$-tuple, and an element of $\mathbb{C}^{3h-3}$. Although beyond the scope of this discussion, it also serves to define the complex structure on $\mathcal{M}_{h,0}$, and serves as complex coordinates for $\mathcal{M}_{h,0}$ near eq.(23.56). These are the coordinates on $\mathcal{M}_{h,0}$ that appear in eq.(23.54). When we solve the Beltrami equation, $\partial_{\bar{z}} w = \mu_{\bar{z}}^{z}(y)\partial_z w$, we find the coordinate transformation, $w(z)$, that takes the metric associated with the complex structure $\mu_{\bar{z}}^{z}$, eq.(23.62), into its isothermal form proportional to $|dw|^2$.

To describe how eqs.(23.54) and (23.55) come about, we must carry out the Faddeev-Popov reduction with the new coordinates, y_i, in mind. We start by choosing a metric, $\widehat{g}$, and a set of isothermal coordinates for it, hence

$$\widehat{g} = \widehat{\rho}|dz|^2. \tag{23.67}$$

The y_i coordinate system on $\mathcal{M}_{h,0}$ will be set up such that the conformal class represented by eq.(23.67) will be at the origin, $y_i = 0$. From eqs.(23.57) and (23.60), a general infinitesimal variation of the metric $\widehat{g}$ is

$$\begin{aligned} \delta\widehat{g} &= \widehat{\rho}\,\delta\widehat{\phi}'\,dzd\overline{z} + \widehat{\rho}(\overline{\partial}v^z(d\overline{z})^2 + \partial v^{\bar{z}}(dz)^2) + (f_{zz}(dz)^2 + f_{\bar{z}\bar{z}}(d\overline{z})^2) \\ \delta\widehat{\phi}' &= \widehat{\rho}\,\delta\widehat{\phi} + \partial v^z + \overline{\partial}v^{\bar{z}}. \end{aligned} \tag{23.68}$$

The first two terms are just eqs.(22.10)-(22.12) rewritten in complex index notation. $f_{zz}(dz)^2$ is a holomorphic quadratic differential satisfying eq.(23.61) and represents a change in the complex structure. All three terms are mutually orthogonal in the metric eq.(22.3) on $\mathcal{G}_{h,0}$. Rewriting the metric for vector fields, eq.(22.14), in complex notation,

$$||\vec{v}||^2 = 2\int d^2z\,\widehat{\rho}^2\,v^z v^{\bar{z}}, \tag{23.69}$$

and applying eq.(22.103), we find, after integrating by parts, that the Faddeev-Popov determinant is

$$\det{}'(-4\widehat{\rho}^{-2}\partial\widehat{\rho}\overline{\partial}) = \det{}'(\Delta_{-1}^{(-)}), \tag{23.70}$$

which, of course, agrees with the earlier result, eq.(22.104).

Now let's examine the contribution of the third term to $\langle\delta g, \delta g\rangle$. Choosing a basis $(f_i)_{zz}$ for the holomorphic quadratic differentials, we have

$$f_{zz} = C^i (f_i)_{zz}. \tag{23.71}$$

Plugging this into the metric, eq.(22.3), the contribution of the third term in eq.(23.68) to $\langle\delta g, \delta g\rangle$ is

$$\overline{C^i}\,(N_2)_{ij}\,C^j, \tag{23.72}$$

where

$$(N_2)_{ij} = 8\int d^2z\,\widehat{\rho}^{-1}\,(f_i)_{\bar{z}\bar{z}}\,(f_j)_{zz}. \tag{23.73}$$

The y_i coordinates, however, are associated with the Beltrami differentials, $\mu_{\bar{z}}^{z}(y)$. The infinitesimal Beltrami coefficient associated with f_{zz} is

$$\delta\mu_{\bar{z}}^{z} = g^{z\bar{z}} f_{\bar{z}\bar{z}}. \tag{23.74}$$

Using eq.(23.66), the left-hand side of eq.(23.74) can be written as

$$\delta\mu_{\bar{z}}^{z} = \delta y_j (\mu^j)_{\bar{z}}^{z}, \tag{23.75}$$

where μ^j is the dual basis to f_i and therefore must satisfy eq.(23.65). Plugging eq.(23.75) and eq.(23.71) back into eq.(23.74) we have

$$\delta y_j (\mu^j)_{\bar{z}}^{z} = g^{z\bar{z}}\, \overline{C}^i (f_i)_{\bar{z}\bar{z}}. \tag{23.76}$$

Multiply both sides by $(f_k)_{zz}$, integrate over z, and apply eq.(23.65). The result is

$$\delta y_j \int d^2z\, (\mu^j)_{\bar{z}}^{z} (f_k)_{zz} = \delta y_k = \overline{C}^i \int d^2z\, g^{z\bar{z}} (f_i)_{\bar{z}\bar{z}} (f_k)_{zz} = \overline{C}^i (N_2)_{ik}. \tag{23.77}$$

Thus,

$$\overline{C}^i = (N_2^{-1})^{ik}\, \delta y_k. \tag{23.78}$$

Therefore, from eq.(23.72), we find that the contribution to $\langle \delta g, \delta g \rangle$ in terms of the y_i is

$$\delta\overline{y}_i (N_2^{-1})^{ij} \delta y_j. \tag{23.79}$$

Application of the Faddeev-Popov procedure yields

$$\mathcal{D}g_{ab} = \mathcal{D}\phi\, \mathcal{D}v^z \left(\frac{i}{2}\right)^{3h-3} dy_1 \wedge \cdots \wedge d\overline{y}_{3h-3} \frac{\det'(\Delta_{-1}^{(-)})}{\det(N_2)}. \tag{23.80}$$

Upon doing the x^μ integral, the contribution to eq.(23.2) for fixed genus $h \geq 2$ is

$$Z_h = \int_{\mathcal{M}_{h,0}} \left(\frac{i}{2}\right)^{3h-3} dy_1 \wedge \cdots \wedge d\overline{y}_{3h-3} \frac{\det'(\Delta_{-1}^{(-)})}{\det(N_2)} \left(\frac{2\pi}{\int d^2\xi \sqrt{\hat{g}}} \det'(\Delta_{\hat{g}}) \right)^{-13}. \tag{23.81}$$

To compare this with eq.(23.24), note that in the first section we used a real basis, ψ^i, for $Ker(P_1^\dagger)$. In terms of the complex basis,

$$\psi_{ab}^{i}\, d\xi^a \otimes d\xi^b = (f_i)_{zz}(dz)^2 + (f_i)_{\bar{z}\bar{z}}(d\overline{z})^2, \tag{23.82}$$

hence

$$\det(N_2) = \det{}^{1/2} \langle \psi^i, \psi^j \rangle. \tag{23.83}$$

So far, the only simplification that appears to have resulted from this approach versus the real analytic approach in the first section is that explicit mention of the gauge slice has disappeared. However, we have now expressed the integrand in complex coordinates on $\mathcal{M}_{h,0}$, where the complex analytic properties of the integrand can be examined.

In particular, we can now determine when and how the integrand,

$$\exp(W(y_i,\overline{y}_i)) \equiv \frac{\det'(\Delta_{-1}^{(-)})}{\det(N_2)}\left(\frac{2\pi}{\int d^2\xi\sqrt{\widehat{g}}}\det'(\Delta_{\widehat{g}})\right)^{-13}, \tag{23.84}$$

is related to the square of a holomorphic function. To do this, consider variations of the metric that result in changes in the moduli coordinates $y_i, \overline{y}_i$. In other words, consider infinitesimal variations of the metric that result purely in a change of complex structure,

$$\begin{aligned} \delta\widehat{g} &= \widehat{g}_{z\overline{z}}\left(\mu_{\overline{z}}^{z}(d\overline{z})^2 + \mu_{z}^{\overline{z}}(dz)^2\right) \\ \mu_{\overline{z}}^{z} &= (\mu^j)_{\overline{z}}^{z}\,\delta y_j . \end{aligned} \tag{23.85}$$

For notational simplicity, denote $\mu = \mu_{\overline{z}}^{z}$, $\overline{\mu} = \mu_{z}^{\overline{z}}$. If eq.(23.84) is the square of a holomorphic function in y_i, then we must have

$$\delta_\mu\,\delta_{\overline{\mu}}\,W = 0. \tag{23.86}$$

By a lengthy calculation, but one similar to $\delta/\delta\phi\,\det'(\Delta_{-1}^{(-)})$ done in the previous chapter, one can show that

$$\delta_\mu\,\delta_{\overline{\mu}}\ \ln\frac{\det'(\Delta_{-1}^{(-)})}{\det(N_2)} = -\frac{13}{24\pi}\int d^2z\,\sqrt{\widehat{g}}\,\widehat{g}^{z\overline{z}}(\overline{\partial}\mu_{zz}\partial\mu_{\overline{z}\overline{z}} + \mu_{zz}\mu_{\overline{z}\overline{z}}\overline{\partial}\partial\widehat{\phi}), \tag{23.87}$$

where $\widehat{\rho} = \exp(\widehat{\phi})$, and $\mu_{zz} = g_{z\overline{z}}\mu_{z}^{\overline{z}}$. In the same way one can also show that

$$\delta_\mu\,\delta_{\overline{\mu}}\ \ln\left(\frac{\det'(\Delta_{\widehat{g}})}{\frac{1}{2\pi}\int d^2z\,\sqrt{\widehat{g}}}\right) = -\frac{1}{24\pi}\int d^2z\,\sqrt{\widehat{g}}\widehat{g}^{z\overline{z}}(\overline{\partial}\mu_{zz}\partial\mu_{\overline{z}\overline{z}} + \mu_{zz}\mu_{\overline{z}\overline{z}}\widehat{\partial\overline{\partial}\phi}) + \delta_\mu\delta_{\overline{\mu}}\ln\,\det(N_1). \tag{23.88}$$

The extra factor arising on the right-hand side of eq.(23.88) is another zero-mode determinant. Recall from the computation of $\delta/\delta\phi\,\det'(P_1^\dagger P_1)$ in the previous chapter that the determinant of the zero modes of $P_1^\dagger$ as well as P_1 enter the calculation. $\det(N_1)$ arises in an analogous fashion to the determinant of the zero modes of $P_1^\dagger$. The zero modes in $\det(N_1)$ are the holomorphic Abelian differentials, $\omega_z\,dz$, satisfying $\partial_{\overline{z}}\omega = 0$.

Combining eqs.(23.87) and (23.88), we find that the analytic anomaly cancels in the critical dimension, up to this determinant of zero modes, hence

$$\delta_\mu \delta_{\bar{\mu}} W = -13\, \delta_\mu \delta_{\bar{\mu}} (\ln \det(N_1)), \tag{23.89}$$

and we can conclude that

$$\exp(W(y_i, \overline{y}_i)) = |F(y_i)|^2 (\det(N_1))^{-13}. \tag{23.90}$$

In arriving at eq.(23.88), Belavin and Knizhnik assumed that the basis of holomorphic Abelian differentials, ω_z^j, $j = 1, \ldots, h$, have a holomorphic dependence on the coordinates y_i (i.e. $\delta_{\bar{\mu}} \omega_z^j = 0$). One such basis was described in chapter 21, where the ω_z^j differentials are chosen to satisfy

$$\int_{a_i} dz\, \omega_z^j = \delta_{ij}$$

on the a-cycles of a canonical homology basis. In this case, $\det(N_1) = \det(\mathrm{Im}\ \Pi)$, where Π is the period matrix. With this identification in eq.(23.90), we arrive at eq.(23.55). The fact that $F(y_i)$ has no zeroes or poles in $\mathcal{M}_{h,0}$ follows from the fact that it arises from determinants. Any zeroes or poles of $F(y_i)$ in $\mathcal{M}_{h,0}$ translate back into additional zero modes of $\Delta_{-1}^{(-)}$ or $\Delta_{\widehat{g}}$, which do not exist. On can show, just as in the genus 1 case, that $F(y_i)$ does develop a second order pole on the boundary of $\mathcal{M}_{h,0}$ at the point $\overline{\mathcal{M}}_{h,0}/\mathcal{M}_{h,0}$, i.e. the point at infinity of the 1-point compactification of $\mathcal{M}_{h,0}$. Again, this produces a divergence and arises from the presence of the tachyon.

The holomorphy of $F(y_i)$ together with its lack of zeroes or poles in $\mathcal{M}_{h,0}$ determines $F(y_i)$ uniquely up to a constant. For reasons we discussed at the end of chapter 21, in the case $h = 2$ or 3, we can use the independent entries of the period matrix as the coordinates y_i. Given the properties of the integrand discussed above, we can write down explicit formulae for F in terms of generalized ϑ-functions. The details may be found in the references in the Bibliography.

There exists an elegant interpretation of F as a global holomorphic nonvanishing section of the tensor products of 2 bundles over $\mathcal{M}_{h,0}$. $\det(N_1)$ can be identified with the natural metric on the dual of one of these bundles. In this approach, solving string theory becomes an algebraic geometric exercise (as opposed to computing spectral invariants). The details, which require developing a case of bundle fibrosis, are in the references in the Bibliography.

23.4 EXERCISES

1. **Derive eq.(23.14) starting with a non-orthonormal basis of $Ker(P_1^\dagger)$.**
2. **Derive eq.(23.19).**
3. **Verify eqs.(23.30), (23.33), (23.35), and (23.36). Also, derive eq. (23.37) directly from eq.(23.8).**
4. **Show that the determinant of a constant is proportional to the area of the surface and determine the size of the divergences. Is there a finite piece?**
5. **Relate the coefficients g_2 and g_3 of the Weierstrass normal form, eqs.(21.73) and (21.77), to determinants of operators.**
6. **Derive eq.(23.70) from eqs.(23.68), (23.69), and (22.103).**

CHAPTER **24**

Scattering Amplitudes

Up to this point we have discussed the free open and closed bosonic strings in the operator and Schrödinger representations and the interacting closed, oriented bosonic string partition function in the path integral representation. Applications of interacting strings to particle physics require the computation of string scattering amplitudes. We turn to this now, where we will continue to concentrate on the closed, oriented bosonic string theory in the critical dimension and stick to the sum-over-surfaces path integral representation developed in the previous two chapters.

24.1 VERTEX OPERATORS

Consider the 4-string amplitude. A perturbative expansion of this amplitude is represented diagrammatically in figure 24.1 as a sum over surfaces of all topologies with 4 fixed boundary components. The initial and final string states are encoded on the boundaries.

In the critical dimension we know to sum only over conformally inequivalent surfaces. Now, however, in addition to the handles, there are moduli associated with the boundaries. For the classic example, consider the cylinder of radius a and height b as in figure 24.2(a). If we unroll the cylinder, we have a rectangle of length b and width $2\pi a$ with 2 sides

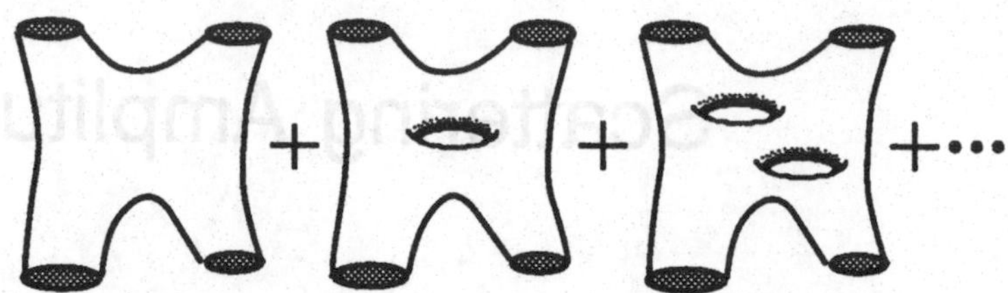

Figure 24.1 Diagrammatic expansion of the 4-string amplitude which amounts to a sum over surfaces of all topologies with 4 fixed boundary components.

identified. This is similar to the genus 1 case except that the second pair of sides remains independent. As in the genus 1 case, a second cylinder of radius a_1 and height b_1 will be conformally equivalent to the first cylinder if the two rectangles are geometrically similar. This occurs if $b_1/a_1 = b/a$.

Take the metric on the cylinder to be flat,

$$(ds)^2 = (d\zeta)^2 + a^2(d\varphi)^2, \tag{24.1}$$

where $-b/2 \le \zeta \le b/2$ and $0 \le \varphi \le 2\pi$. The transformation,

$$\begin{aligned} \zeta &\mapsto a \ln(r) \\ \varphi &\mapsto \varphi, \end{aligned} \tag{24.2}$$

under which the metric becomes conformally flat,

$$(ds)^2 = \frac{a^2}{r^2}\left((dr)^2 + r^2(d\varphi)^2\right), \tag{24.3}$$

demonstrates that cylinders are conformal to annuli (figure 24.2(b)). Let r_1 be the inner radius of the annulus, and r_2 be the outer radius. From eq.(24.2) we have

$$\frac{r_2}{r_1} = \exp\left(\frac{b}{a}\right), \tag{24.4}$$

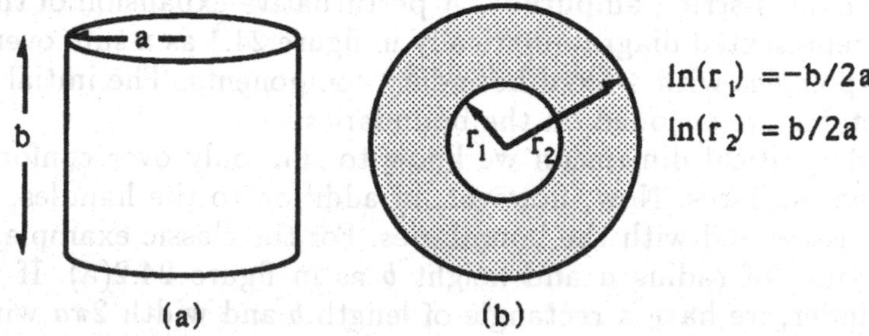

Figure 24.2 The cylinder (a) is conformal to the annulus (b) as demonstrated by the transformation (24.2). The modular parameter associated with the boundaries is the ratio of the outer to inner radius of the annulus.

hence two annuli are conformal if their ratios of outer to inner radii are identical. This ratio serves as the modular parameter associated with the boundaries of the cylinder or annulus. Since the outer radius is bigger than the inner radius, the moduli space for the cylinder is the open real interval $(1, \infty)$.

To carry out the sum represented by figure 24.1 in its full glory with all of the boundary moduli, etc., is too complicated to do. Instead, we follow the lead of the asymptotic state analysis for ordinary point particle theories. Far from the region of interaction, the strings will act free. We know the spectrum of states of the free string from chapter 20. Thus, we evaluate scattering amplitudes with external string states on their physical mass shells. Since each pure on shell string state is supposed to represent a type of particle, the connection with particle physics is naturally established. From these amplitudes we can compute particle S-matrix elements and cross sections. At the low energies at which we carry out experiments, we in fact need only compute scattering amplitudes of the massless string states (assuming the theory is tachyon-free). Of course, there are many issues that must be dealt with before string theory can be applied to any real experiment, e.g. the dimension of space-time, just to mention one.

If the external string states are asymptotic, then the tubes in figure 24.1 extend off to the distant past and distant future of space-time. The evaluation of these sums of surfaces now simplifies. To get an idea of how, return to the cylinder of figure 24.2(a). The ends extend to the distant future and past, hence $-\infty < \zeta < \infty$. The transformation, eq.(24.2), maps the incoming string to the origin of the plane and the outgoing string to the point of infinity ($r_1 \to 0, r_2 \to \infty$). Using stereographic projection, we can map this extended plane to the Riemann sphere. The incoming point ends up at the south pole and the outgoing point at the north pole. If we conformally transform the metric in eq.(24.3) by taking

$$\frac{a^2}{r^2} \longrightarrow \frac{4}{(1+r^2)^2}, \tag{24.5}$$

then by examining eq.(21.52) we can see that the metric projected onto the sphere by stereographic projection will be the usual metric.

The result of the transformation is that the infinitely long cylinder is conformally mapped to the compact sphere. The external string states appear as points at the north and south poles as illustrated in figure 24.3(a). The data describing the external states must be encoded at the points. One possibility is to treat the surface in figure 24.3(a) as a sphere with 2 punctures at the poles. At each puncture we associate a wave functional to describe the external string state. The genus h contribution to the sum of surfaces for this "2-point" amplitude will reduce to an integral over the moduli space of surfaces of genus h with 2 punctures, $\mathcal{M}_{h,2}$. Alternatively,

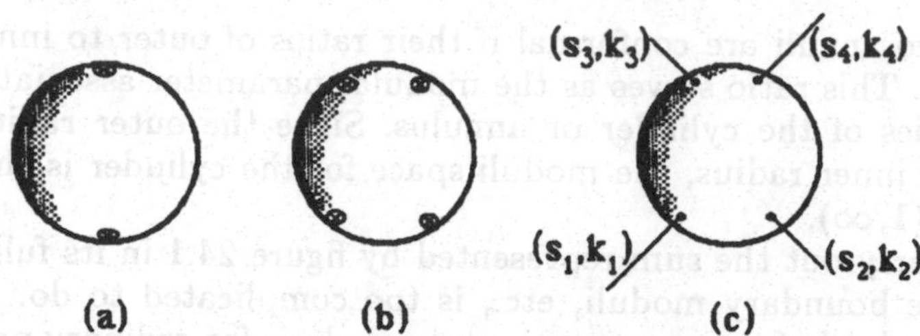

Figure 24.3 (a) The infinitely long cylinder is conformal to the sphere with two punctures. The punctures correspond to the boundaries of the cylinder. The punctures may be filled in and replaced with vertex operators. (b) The genus 0 contribution of figure 24.1, where the boundaries extend off to infinity, is conformally equivalent to a surface with four punctures or four vertex operators. (c) A line and the quantum numbers associated with the emission or absorption of a string represent the operation of the vertex operator at that point.

we can consider the punctures as filled in and associate an operator at each external point that describes the emission and absorption of the particular external asymptotic string state. Such operators are called vertex operators.

Let $V_s(\xi, k)$ represent the vertex operator for species s of string state (e.g. tachyon, graviton, dilaton, etc.) and momentum k^μ operating at the point ξ on the surface. In terms of the string path integral, the genus 0 2-point amplitude contribution of figure 24.3(a) can be written as

$$A_2^{(0)}(s_1, k_1; s_2, k_2) = \int_{\text{genus } 0} \mathcal{D}g_{ab}\, \mathcal{D}x^\mu$$

$$\int d^2\xi_1\, d^2\xi_2 \sqrt{g(\xi_1)}\sqrt{g(\xi_2)} V_{s_1}(\xi_1, k_1)\, V_{s_2}(\xi_2, k_2) \exp(-S[g, x]), \quad (24.6)$$

where S is the action appearing in eq.(23.1). Note that the total amplitude requires that we sum over all possible positions for emission and absorption, hence the appearance of $\int d^2\xi_i \sqrt{g(\xi_i)}$ in eq.(24.6) above. For notational convenience, the vertex operator $V_s(k)$ is defined as

$$V_s(k) = \int d^2\xi \sqrt{g}\, V_s(\xi, k). \quad (24.7)$$

It is possible to carry out a similar conformal transformation to map the genus 0 contribution of figure 24.1 (with long tubes) to the sphere with 4 vertex operators as in figure 24.3(b). Often, a line and the quantum numbers s_i, k_i are associated with each emission or absorption to represent the operation of $V_s(\xi, k)$ at that point. In fact, the conformal factor of the metric can always be chosen so that any number of long tubes can be mapped to points on compact surfaces without boundaries. Thus, the genus h contribution to the n-point amplitude is

$$A_n^{(h)}(s_1, k_1; \cdots; s_n, k_n) = \kappa_0^n \int_{\text{genus } h} \mathcal{D}g_{ab}\, \mathcal{D}x^\mu$$

$$\int d^2\xi_1 \cdots d^2\xi_n \sqrt{g(\xi_1)} \cdots \sqrt{g(\xi_n)}\, V(s_1, k_1) \cdots V(s_n, k_n) \exp(-S[g, x]). \tag{24.8}$$

κ_0 is the coupling constant associated with each string interaction. The computation of asymptotic scattering amplitudes on shell becomes the evaluation of expectation values of products of vertex operators where the sum over surfaces is identical to those in the computation of the partition function.

For the closed bosonic string, let us write down the first few vertex operators. In general we will have

$$V_s(k) = \int d^2\xi \sqrt{g}\, W_s[x^\mu]\, e^{ik\cdot x}. \tag{24.9}$$

$W_s[x^\mu]$ is the wave functional for the particular state. It must be spacetime Lorentz invariant to make the entire operator invariant. The simplest invariant wave functionals will be polynomials in x^μ:

$$\begin{aligned} W_T &= 1 \\ W_G &= \epsilon_{\mu\nu}\, g^{ab} \partial_a x^\mu \partial_b x^\nu \\ W_D &= g^{ab} \partial_a x^\mu \partial_b x_\mu, \end{aligned} \tag{24.10}$$

for the tachyon (T), graviton (G), and dilaton (D). The tachyon and dilaton are Lorentz scalars. W_D is chosen to be orthogonal to W_T. $\epsilon_{\mu\nu}$ is the polarization tensor for the graviton. The momentum wave functional, $\exp(ik\cdot x)$, included in eq.(24.9) above is necessary since the external state is an eigenstate of momentum and so that the vertex operator transforms correctly under a global translation, $x^\mu \to x^\mu + a^\mu$, a^μ = constant. Finally, k^μ is the total momentum of the state. Recall from the operator representation results of chapter 20 that $k^2 = 8\pi$ (tension $T = 1$) for the tachyon, and $k^2 = 0$ for the graviton and dilaton.

When used in the operator representation, the vertex operators will require normal ordering. As an application of the results of this section and the previous two chapters, let us compute the tree-level and 1-loop tachyon scattering amplitudes.

24.2 TACHYON TREE AMPLITUDES

The scattering amplitude of n on shell tachyons at tree level is

$$A_n^{(0)}(k_1, \ldots, k_n) = \kappa_0^n \mathcal{N} \int_{\text{genus } 0} \mathcal{D}g_{ab}\, \mathcal{D}x^\mu\, V_T(k_1) \cdots V_T(k_n)\, \exp(-S[g, x]), \tag{24.11}$$

where

$$V_T(k_j) = \int d^2\xi_j \sqrt{g(\xi_j)}\, e^{ik_j \cdot x(\xi_j)}. \tag{24.12}$$

Applying the Faddeev-Popov procedure as in chapter 22 we have

$$A_n^{(0)}(k_1, \cdots, k_n) =$$

$$\mathcal{N}\kappa_0^n \int \mathcal{D}v^a\, \mathcal{D}\phi\, \mathcal{D}x^\mu \left(\frac{\det'(P_1^\dagger P_1)}{\det(\psi_i, \psi_j)_{\hat{g}}}\right)^{\frac{1}{2}} V_T(k_1) \cdots V_T(k_n)\, \exp(-S[\hat{g}, x]), \tag{24.13}$$

where $\hat{g}_{ab} = \exp(\phi)\,\delta_{ab}$ is conformally flat and where ψ_i span $Ker(P_1)$. The x^μ integral is now easy to do by completing the squares in the exponential. However, recall from chapter 22 that the zero modes must be treated separately. Let $G'(\xi, \xi')$ be the Green's function for $\Delta_{\hat{g}}$ with zero modes removed. In terms of an eigenfunction expansion, symbolically we have

$$G'(\xi, \xi') = \sum_n{}' \frac{\varphi_n(\xi)\varphi_n^*(\xi')}{\lambda_n}, \tag{24.14}$$

where the prime on the sum means that the zero modes have been left out. $G'(\xi, \xi')$ satisfies

$$\Delta_{\hat{g}} G'(\xi, \xi') = \tilde{\delta}^2(\xi - \xi'), \tag{24.15}$$

where $\tilde{\delta}$ is the δ-function with the zero modes removed,

$$\begin{aligned} \tilde{\delta}^2(\xi - \xi') &= \sum_n{}' \varphi_n(\xi)\varphi_n^*(\xi') \\ &= \sum_n \varphi_n(\xi)\varphi_n^*(\xi') - \varphi_0(\xi)\varphi_0^*(\xi') \\ &= \delta^2(\xi - \xi') - \varphi_0(\xi)\varphi_0^*(\xi'). \end{aligned} \tag{24.16}$$

Recall that the zero modes are constants, $\varphi_0(\xi) = c$. Orthonormality requires

$$\int d^2\xi \sqrt{\hat{g}}\, \varphi_0(\xi)\varphi_0^*(\xi) = c^2 \int d^2\xi \sqrt{\hat{g}} = 1, \tag{24.17}$$

hence $G'(\xi,\xi')$ satisfies

$$\Delta_{\widehat{g}} G'(\xi,\xi') = \delta^2(\xi-\xi') - \frac{1}{\int d^2\xi\,\sqrt{\widehat{g}}}. \tag{24.18}$$

In terms of $G'(\xi,\xi')$, the x^μ integral is

$$\int \mathcal{D}x^\mu \,\exp\left(-\frac{1}{2}\int d^2\xi\,\sqrt{\widehat{g}}\,\widehat{g}^{ab}\partial_a x^\mu \partial_b x^\mu + i\sum_{j=1}^{n} k_j^\mu x^\mu(\xi_j)\right)$$

$$= (2\pi)^D\,\delta^D(\textstyle\sum_j k_j)\left(\dfrac{2\pi}{\int d^2\xi\,\sqrt{\widehat{g}}}\det{}'(\Delta_{\widehat{g}})\right)^{-\frac{D}{2}} \exp\left(-\dfrac{1}{2}\displaystyle\sum_{i,j=1}^{n} k_i^\mu k_j^\mu G'(\xi_i,\xi_j)\right). \tag{24.19}$$

The factor $(2\pi/\int d^2\xi\,\sqrt{\widehat{g}})^{-D/2}$ arises from the measure $\mathcal{D}x^\mu$ for the zero modes (see eq.(22.25)), while the factor $(2\pi)^D\,\delta^D(\sum_j k_j)$ is the result of the zero-mode integral in eq.(24.19). With eq.(24.19), the expression for the amplitude, eq.(24.13), becomes

$$A_n^{(0)}(k_1,\ldots,k_n) = \mathcal{N}\kappa_0^n (2\pi)^D \delta^D(\textstyle\sum_j k_j)\displaystyle\int \mathcal{D}\phi(\xi)\int d^2\xi_1 \cdots d^2\xi_n\,\sqrt{\widehat{g}_1}\cdots\sqrt{\widehat{g}_n}$$

$$\left(\frac{\det{}'(P_1^\dagger P_1)}{\det\langle\psi_i,\psi_j\rangle_{\widehat{g}}}\right)^{\frac{1}{2}}\left(\frac{2\pi}{\int d^2\xi\,\sqrt{\widehat{g}}}\det{}'(\Delta_{\widehat{g}})\right)^{-\frac{D}{2}}\exp\left(-\frac{1}{2}\sum_{i,j=1}^{n} k_i^\mu k_j^\mu G'(\xi_i,\xi_j)\right) \tag{24.20}$$

where the volume of *Diff*$_0$ has been absorbed into the normalization. We would now like to switch to the critical dimension to eliminate the ϕ dependence from the integrand by applying eqs.(22.96) and (22.121). However, at first sight it looks as though we have picked up additional ϕ dependence through $G'(\xi_i,\xi_j)$ and the n factors $\int d^2\xi\,\sqrt{\widehat{g}}$.

To examine this issue, let us look at the explicit expression for G'.

$$G'(\xi_i,\xi_j,\phi) = G(\xi_i,\xi_j,\phi) + f(\xi_i,\phi) + f(\xi_j,\phi), \tag{24.21}$$

where

$$G(\xi_i,\xi_j,\phi) = -\frac{1}{2\pi}\ln(|\xi_i-\xi_j|) \tag{24.22}$$

is the ordinary Green's function for $\Delta_{\widehat{g}}$ without the zero mode removed and where

$$f(\xi_i,\phi) = \frac{1}{2\pi\int d^2\xi\,\sqrt{\widehat{g}}}\int d^2\xi\,\ln(|\xi_i-\xi|)\,e^{\phi(\xi)}. \tag{24.23}$$

When we plug eq.(24.21) into the exponential of eq.(24.20), we find

$$\begin{aligned}\frac{1}{2}\sum_{i,j=1}^{n} k_i^\mu k_j^\mu G'(\xi_i,\xi_j) &= \frac{1}{2}\sum_{i,j=1}^{n} G(\xi_i,\xi_j,\phi)k_i^\mu k_j^\mu + 2\sum_{i=1}^{n} k_i^\mu f(\xi_i,\phi)\sum_{j=1}^{n} k_j^\mu \\ &= \frac{1}{2}\sum_{i=1}^{n} G(\xi_i,\xi_i,\phi)k_i^2 + \sum_{i<j} G(\xi_i,\xi_j,\phi)k_i^\mu k_j^\mu .\end{aligned} \tag{24.24}$$

The term containing $f(\xi_i,\phi)$ is eliminated by the presence of the $\delta(\sum_j k_j)$ in eq.(24.20). Next we note from eq.(24.22) that $G(\xi_i,\xi_j,\phi)$ is independent of ϕ when $\xi_i \neq \xi_j$. Any ϕ dependence must reside in $G(\xi_i,\xi_i,\phi)$. From eq.(24.22) we easily see that G diverges as $\xi_j \to \xi_i$ and therefore must be regularized in a covariant manner. Following Polyakov, we find that

$$G(\xi,\xi,\phi) \longrightarrow -\frac{1}{4\pi}\ln(\varepsilon) + \frac{1}{4\pi}\phi(\xi), \tag{24.25}$$

where ε is the short distance invariant (length)^2 cutoff, by the following heuristic argument: We want the regulator to be covariant, therefore we set the minimum invariant (length)^2 to $(\Delta s)^2 = \epsilon$. However, $(\Delta s)^2 = (\Delta\xi)^2 \exp(\phi(\xi))$, hence $(\Delta\xi)^2 = \exp(-\phi(\xi))\,\varepsilon$. Plugging this back into eq.(24.22) we arrive at eq.(24.25).

Using eq.(24.25) and eq.(24.24) along with $k_i^2 = 8\pi$ in the exponential in eq.(24.20), we find

$$\begin{aligned}&\exp\left(-\frac{1}{2}\sum_{i,j=1}^{n} k_i^\mu k_j^\mu G'(\xi_i,\xi_j)\right) \\ &\quad = \exp\left(\sum_{i=1}^{n}\ln(\varepsilon)\right)\exp\left(-\sum_{i=1}^{n}\phi(\xi_i)\right)\exp\left(-\sum_{i<j} G(\xi_i,\xi_j,\phi=0)k_i^\mu k_j^\mu\right).\end{aligned} \tag{24.26}$$

Notice that the ϕ dependence in eq.(24.26) just cancels the $\sqrt{g_i}$ factors from $\int d^2\xi_i\sqrt{g_i}$ in eq.(24.20). Therefore, the integrand will be ϕ-independent in the critical dimension. Also note that the cutoff dependence can be absorbed into the coupling, $\kappa = \kappa_0\varepsilon$. Applying eqs.(22.96), (22.121), (24.22), and (24.26) to eq.(24.20), the amplitude $A_n^{(0)}$ in the critical dimension becomes

$$\begin{aligned}A_n^{(0)}(k_1,\ldots,k_n) &= \mathcal{N}\kappa^n(2\pi)^{26}\delta^{26}(\textstyle\sum_j k_j) \\ &\left(\frac{1}{\det(\widetilde{\psi}_i,\widetilde{\psi}_j)_{\widehat{g}_0}}\right)^{1/2}\int d^2\xi_1\cdots d^2\xi_n\prod_{i<j}|\xi_i-\xi_j|^{k_i\cdot k_j/2\pi}.\end{aligned} \tag{24.27}$$

The determinants $\det'(\widetilde{P}_1^\dagger \widetilde{P}_1)^{1/2}$ and $\det'(\Delta_{\widehat{g}_0})^{-13}$, evaluated at $\widehat{g} = \widehat{g}_0 = \delta_{ab}$ along with $\int \mathcal{D}\phi$, have been absorbed into the normalization. All are independent of the remaining factors. The determinants are ultimately absorbed by normalizing the amplitude by the partition function.

However, note that we let the determinant of basis vectors of $Ker(P_1)$ remain. Recall that the elements of $Ker(P_1)$ are conformal Killing vectors. These vector fields generate diffeomorphisms that leave the metric form invariant up to a conformal transformation. Diffeomorphisms of this type are therefore conformal homeomorphisms, hence the elements of $Ker(P_1)$ generate via exponentiation the group $Aut(\Sigma)$, the group of automorphisms on Σ. From the discussion in chapter 21 we know that $Aut(\Sigma) = PSL(2, \mathbb{C})$. Thus, the squareroot of the determinant appearing in eq.(24.27) represents the volume of this group (see eq.(22.17)).

The reason for keeping this determinant in view is that the integral $\int d^2\xi_1 \cdots d^2\xi_n$ in eq.(24.27) diverges. The divergence arises because the integrand is invariant under $Aut(\Sigma)$. The addition of the vertex operators has led to the need for additional gauge-fixing. To eliminate this divergence we must carry out the Faddeev-Popov procedure on this residual gauge freedom. Once the volume of $Aut(\Sigma)$ is separated out, it will cancel the remaining determinant in eq.(24.27).

From the discussion in chapter 21 we know that the group $Aut(\Sigma) = PSL(2, \mathbb{C})$ is the group of Möbius transformations,

$$z' = T(z) = \frac{az + b}{cz + d}, \quad a, b, c, d \in \mathbb{C}, \tag{24.28}$$

where $ad - bc = 1$. Clearly there are 3 complex or 6 real degrees of freedom. Gauge-fixing the integrand in eq.(24.27) amounts to fixing three of the locations, ξ_i, of the vertex operators on the genus 0 surface.

From eq.(24.28) we find that

$$dz' = dT(z) = \frac{1}{(cz + d)^2} dz, \tag{24.29}$$

and

$$z_i' - z_j' = T(z_i) - T(z_j) = \frac{z_i - z_j}{(cz_i + d)(cz_j + d)}. \tag{24.30}$$

If we write $z_j = \xi_j^1 + i\xi_j^2$, then one can explicitly verify, using eqs.(24.29) and (24.30), that the integrand in eq.(24.27) is $PSL(2, \mathbb{C})$ invariant. This is left as an exercise. However, from eqs.(24.29) and (24.30) it is also easy to see that

$$\frac{d^2 z_i \, d^2 z_j \, d^2 z_k}{|z_i - z_j|^2 \, |z_j - z_k|^2 \, |z_k - z_i|^2} \tag{24.31}$$

is also invariant under $PSL(2,\mathbb{C})$. Since eq.(24.31) involves the same number of complex degrees of freedom as $PSL(2,\mathbb{C})$, then the integral of eq.(24.31) is proportional to the volume of $PSL(2,\mathbb{C})$.

Let us use eq.(24.31) to carry out the Faddeev-Popov procedure on eq.(24.27). Choose a "gauge" by fixing z_1, z_2, and z_3. Multiplying and dividing the integrand by $|z_1-z_2|^2|z_2-z_3|^2|z_3-z_1|^2$, we have

$$\int d^2\xi_1\cdots d^2\xi_n \prod_{i<j}|\xi_i-\xi_j|^{k_i\cdot k_j/2\pi}=\int d^2z_1\cdots d^2z_n\prod_{i<j}((z_i-z_j)(\overline{z}_i-\overline{z}_j))^{k_i\cdot k_j/4\pi}$$

$$=\int\frac{d^2z_1\,d^2z_2\,d^2z_3}{|z_1-z_2|^2|z_2-z_3|^2|z_3-z_1|^2}$$

$$\int d^2z_4\cdots d^2z_n\,|z_1-z_2|^2|z_2-z_3|^2|z_3-z_1|^2\prod_{i<j}|z_i-z_j|^{k_i\cdot k_j/2\pi}. \tag{24.32}$$

The integral over z_1, z_2, and z_3 in the last line of eq.(24.32) is invariant under $PSL(2,\mathbb{C})$. Since the original integral in eq.(24.27) is also invariant, then the integral over $z_4,\ldots,z_n$ in the last line of eq.(24.32) must also be invariant. This can be explicitly verified using eqs.(24.29) and (24.30). Thus the integral over z_1, z_2, and z_3 is $Vol(PSL(2,\mathbb{C}))$ and cancels the determinant in eq.(24.27). The result is

$$A_n^{(0)}(k_1,\ldots,k_n)=\mathcal{N}\kappa^n(2\pi)^{26}\delta^{26}(\textstyle\sum_j k_j)|z_1-z_2|^2|z_2-z_3|^2|z_3-z_1|^2$$
$$\int d^2z_4\cdots d^2z_n\prod_{i<j}|z_i-z_j|^{k_i\cdot k_j/2\pi}, \tag{24.33}$$

where z_1, z_2, and z_3 are fixed.

The customary choice of gauge is to take $z_1=0$, $z_2=1$, and $z_3=\infty$. In this case,

$$|z_1-z_2|^2|z_2-z_3|^2|z_3-z_1|^2\longrightarrow|z_3|^4, \tag{24.34}$$

while using momentum conservation, $\sum_j k_j=0$, we have

$$\prod_{j\neq 3}|z_3-z_j|^{k_3\cdot k_j/2\pi}\longrightarrow|z_3|^{(\sum_j k_j-k_3)/2\pi}=|z_3|^{-k_3^2/2\pi}=|z_3|^{-4} \tag{24.35}$$

which cancels eq.(24.34). Thus, the final result for the scattering amplitude ($n\geq 3$) is

$$A_n^{(0)}(k_1,\ldots,k_n) = \mathcal{N}\kappa^n(2\pi)^{26}\,\delta^{26}(\textstyle\sum_j k_j)$$

$$\int d^2z_4\cdots d^2z_n \prod_{j=4}^{n} |z_j|^{k_1\cdot k_j/2\pi}\,|1-z_j|^{k_2\cdot k_j/2\pi} \prod_{4\le i<j}^{n} |z_i - z_j|^{k_i\cdot k_j/2\pi}. \quad (24.36)$$

For $n = 3$ this is simply

$$A_3^{(0)}(k_1,k_2,k_3) = \mathcal{N}\kappa^3(2\pi)^{26}\delta^{26}(k_1+k_2+k_3). \quad (24.37)$$

Relabeling $k_4 \to k_1$, $k_3 \to k_4$, $k_2 \to k_3$, $k_1 \to k_2$, and using

$$|z|^{-p} = \frac{1}{\Gamma(p/2)}\int_0^\infty dt\, t^{p/2}\exp(-t|z|^2), \quad (24.38)$$

the 4-point Virasoro-Shapiro amplitude can be written as

$$A_4^{(0)}(k_1,\ldots,k_4) = \mathcal{N}\kappa^4(2\pi)^{26}\delta^{26}(\textstyle\sum_j k_j)\displaystyle\frac{\Gamma(-1-s/8\pi)\Gamma(-1-t/8\pi)\Gamma(-1-u/8\pi)}{\Gamma(2+s/8\pi)\Gamma(2+t/8\pi)\Gamma(2+u/8\pi)}, \quad (24.39)$$

where

$$\begin{aligned} s &= -(k_1+k_2)^2 \\ t &= -(k_2+k_3)^2 \\ u &= -(k_1+k_3)^2 \end{aligned} \quad (24.40)$$

are the Mandelstam variables.

From the properties of the Γ-function, we can see that the amplitude eq.(24.39) displays a simple pole when any of the arguments of the Γ-functions in the numerator are negative integers ≤ 0. The denominator guarantees that any poles remain simple, as they must, when two of the arguments in the numerator are simultaneously ≤ 0. Even though in this situation we expect to pick up a double pole from the numerator, the denominator also picks up a pole, leading to simple pole behavior over all.

Often the arguments of the Γ-functions in eq.(24.39) are expressed in terms of Regge trajectories,

$$\alpha(q) = 2 + \frac{q}{4\pi}, \quad (24.41)$$

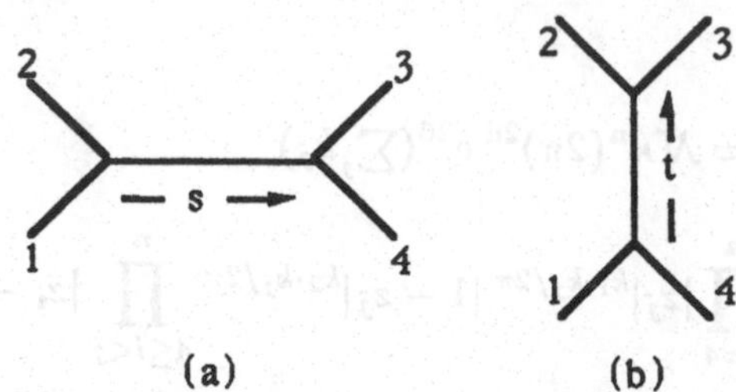

Figure 24.4 Tree-level contributions to a 4-point amplitude of a point particle theory where the exchange occurs (a) in the s channel and (b) in the t channel. The total amplitude is the sum of the two. Ordinarily, the two contributions are not equal and the theory is not dual.

for $q = s$, t, or u. The poles in eq.(24.39) then arise equally spaced along linear Regge trajectories.

The amplitude eq.(24.39) is "manifestly dual" in that it is invariant under the interchange of s and t (and u for that matter). In addition, the Γ-functions, being factorial functions, can be factorized such that the amplitude can be written as a sum of s-channel poles with t-dependent factorized residues or as a sum of t-channel poles with s-dependent residues. Ordinary point particle theories are not dual. Two tree-level diagrams for the 4-point amplitude of a point particle theory with cubic interaction are given in figure 24.4. The exchange occurs in the s channel in figure 24.4(a), and in the t channel in figure 24.4(b). Ordinarily, the two amplitudes from the two diagrams do not equal each other and to get the total amplitude we need to sum the two. String theory can lead to dual amplitudes like eq.(24.39) because the string theory equivalent of diagram 24.4(a) is identical to the string theory equivalent of diagram 24.4(b). There is just 1 tree-level string diagram. For closed strings, this diagram is the genus 0 surface of figure 24.1.

24.3 1-LOOP TACHYON AMPLITUDES

The scattering amplitude for n on shell tachyons at genus 1, according to eq.(24.8), is

$$A^{(1)}(k_1, \ldots, k_n) = \mathcal{N}\kappa_0^n \int_{h=1} \mathcal{D}g_{ab}\mathcal{D}x^\mu$$

$$\int d^2\xi_1 \cdots d^2\xi_n \sqrt{g(\xi_1)} \cdots \sqrt{g(\xi_n)}\, \exp(ik_j^\mu x^\mu(\xi_j)) \exp(-S[g,x]). \quad (24.42)$$

As we did in chapter 23, we will take the coordinate region in the (ξ^1, ξ^2) plane to be the unit square and choose the "gauge" (eq.(23.28)),

$$\widehat{g}_{ab} = |d\xi^1 + \tau d\xi^2|^2. \tag{24.43}$$

Applying the Faddeev-Popov procedure as in chapter 23, we have

$$A_n^{(1)}(k_1, \cdots, k_n) = \kappa_0^n \mathcal{N} \int \mathcal{D}v^a \, \mathcal{D}x^\mu \, \mathcal{D}\phi \, d^2\tau \left(\frac{\det'(\widetilde{P}_1^\dagger \widetilde{P}_1)}{\det\langle \widetilde{\chi}^i, \widetilde{\chi}^j \rangle_{\widetilde{g}}} \right)^{\frac{1}{2}} \frac{\det\langle \widetilde{\psi}^i, \widetilde{T}^j \rangle_{\widetilde{g}}}{\det^{1/2}\langle \widetilde{\psi}^i, \widetilde{\psi}^j \rangle_{\widetilde{g}}}$$

$$\int d^2\xi_1 \cdots d^2\xi_n \sqrt{\widetilde{g}(\xi_1)} \cdots \sqrt{\widetilde{g}(\xi_n)} \exp(ik_j^\mu x^\mu(\xi_j)) \exp(-S[\widetilde{g}, x]), \tag{24.44}$$

where $\widetilde{g} = \exp(\phi)\widehat{g}$ is a conformal rescaling of $\widehat{g}$, eq.(24.43). The conformal Killing vector basis $\widetilde{\chi}^i$ span $Ker(\widetilde{P}_1)$ at $\widetilde{g} \in \mathcal{G}_{1,0}$ while $\widetilde{\psi}^i$ span $Ker(\widetilde{P}_1^\dagger)$. Upon absorbing $\int \mathcal{D}v^a$, we next do the x^μ integral with the result

$$A_n^{(1)}(k_1, \cdots, k_n) = \kappa_0^n \mathcal{N} (2\pi)^D \, \delta^D(\textstyle\sum_j k_j)$$

$$\int \mathcal{D}\phi \, d^2\tau \left(\frac{\det'(\widetilde{P}_1^\dagger \widetilde{P}_1)}{\det\langle \widetilde{\chi}^i, \widetilde{\chi}^j \rangle_{\widetilde{g}}} \right)^{\frac{1}{2}} \frac{\det\langle \widetilde{\psi}^i, \widetilde{T}^j \rangle_{\widetilde{g}}}{\det^{1/2}\langle \widetilde{\psi}^i, \widetilde{\psi}^j \rangle_{\widetilde{g}}} \left(\frac{2\pi}{\int d^2\xi \sqrt{\widetilde{g}}} \det'(\Delta_{\widetilde{g}}) \right)^{-\frac{D}{2}}$$

$$\int d^2\xi_1 \cdots d^2\xi_n \sqrt{\widetilde{g}(\xi_1)} \cdots \sqrt{\widetilde{g}(\xi_n)} \exp\left(-\frac{1}{2} k_i^\mu k_j^\mu G'(\xi_i, \xi_j, \tau, \phi) \right). \tag{24.45}$$

$G'(\xi_i, \xi_j, \tau, \phi)$, the Green's function for $\Delta_{\widetilde{g}}$ on the torus minus the zero modes, still satisfies eq.(24.18). The ϕ dependence of G' arises from the need to regularize the short-distance behavior, a local phenomenon and therefore insensitive to topology. Thus, the ϕ dependence will cancel that from $\sqrt{\widetilde{g}(\xi_1)} \cdots \sqrt{\widetilde{g}(\xi_n)}$. Applying eqs.(23.20), (23.22), and (23.31), we find

$$A_n^{(1)}(k_1, \cdots, k_n) = \kappa_0^n \mathcal{N} (2\pi)^D \, \delta^D(\textstyle\sum_j k_j) \displaystyle\int \mathcal{D}\phi \, d^2\tau \left(\frac{\det'(\widehat{P}_1^\dagger \widehat{P}_1)}{\det\langle \widehat{\chi}^i, \widehat{\chi}^j \rangle_{\widehat{g}}} \right)^{\frac{1}{2}}$$

$$\det^{1/2}\langle \widehat{T}^i, \widehat{T}^j \rangle_{\widehat{g}} \left(\frac{2\pi}{\int d^2\xi \sqrt{\widehat{g}}} \det'(\Delta_{\widehat{g}}) \right)^{-D/2} \exp\left(-\frac{26 - D}{48\pi} S_L \right)$$

$$\int d^2\xi_1 \cdots d^2\xi_n \sqrt{\widehat{g}(\xi_1)} \cdots \sqrt{\widehat{g}(\xi_n)} \exp\left(-\frac{1}{2} k_i^\mu k_j^\mu G'(\xi_i, \xi_j, \tau, \phi = 0) \right). \tag{24.46}$$

In the critical dimension, the ϕ dependence disappears and $\int \mathcal{D}\phi$ falls into the normalization. We evaluated the determinants in the previous chapter (eqs.(23.33), (23.35), and (23.48)). Substitution of these results into eq.(24.46) produces

$$A_n^{(1)}(k_1, \cdots, k_n) = \kappa_0^n \mathcal{N} (2\pi)^{26}\, \delta^{26}(\textstyle\sum_j k_j)$$

$$\int_{\mathcal{M}_{1,0}} d^2\tau \, \frac{1}{2\pi\tau_2^2}\, e^{4\pi\tau_2}\, (2\pi\tau_2)^{-12}\, \tau_2^n \left| \prod_{n=1}^{\infty} 1 - e^{2i\pi n\tau} \right|^{-48}$$

$$\int d^2\xi_1 \cdots d^2\xi_n \, \exp\Big(-\frac{1}{2} k_i^\mu k_j^\mu G'(\xi_i, \xi_j, \tau) \Big). \qquad (24.47)$$

The Green's function for the ordinary Laplacian ($(ds)^2 = dz\, d\overline{z}$) on the torus represented by a parallelogram of sides 1 and τ is

$$G(z, z'; \tau) = -\frac{1}{2\pi} \ln \left| \frac{\vartheta_1(z - z'|\tau)}{\vartheta_1'(0|\tau)} \right| + \frac{1}{2\tau_2} (\mathrm{Im}(z - z'))^2. \qquad (24.48)$$

ϑ_1 is one of the four Jacobi ϑ-functions,

$$\vartheta_1(\nu|\tau)$$
$$= 2 \prod_{m=1}^{\infty} (1 - e^{4mi\pi\tau})\, e^{i\pi\tau/4} \sin(\pi\nu) \prod_{n=1}^{\infty} (1 - 2e^{2i\pi n\tau} \cos(2\pi\nu) + e^{4i\pi n\tau}),$$
$$(24.49)$$

from which it follows that

$$\vartheta_1'(0|\tau) = \frac{\partial}{\partial\nu} \vartheta_1(\nu|\tau) \Big|_{\nu=0} = 2\pi (\eta(\tau))^3$$
$$\eta(\tau) = e^{i\pi\tau/12} \prod_{n=1}^{\infty} (1 - e^{2\pi i n\tau}). \qquad (24.50)$$

The requisite periodicity of G follows from

$$\vartheta_1(\nu + 1|\tau) = -\vartheta_1(\nu|\tau)$$
$$\vartheta_1(\nu + \tau|\tau) = e^{-i\pi\tau - 2\pi i\nu}\, \vartheta_1(\nu|\tau). \qquad (24.51)$$

$G(z, z'; \tau)$, eq.(24.48), is not the Green's function in eq.(24.47), in that the torus is represented in eq.(24.47) by the unit (ξ^1, ξ^2) square with metric eq.(24.43) instead of the $(1,\tau)$ parallelogram with $(ds)^2 = dz\,d\overline{z}$. The two are related by the conformal transformation,

$$\begin{aligned} z &= \xi^1 + \tau\xi^2 \\ \overline{z} &= \xi^1 + \overline{\tau}\xi^2. \end{aligned} \tag{24.52}$$

To apply eq.(24.48) to eq.(24.47) we can either transform G from $z \to \xi$ and insert into eq.(24.47) or transform $\xi_i \to z_i$ in eq.(24.47) and insert $G(z, z'; \tau)$, eq.(24.48), directly. In the latter case, the factor of τ_2^n in eq.(24.47) is canceled by the n Jacobians arising from $d^2\xi_i \to J\,d^2 z_i$. G diverges as $z \to z'$ and is regularized in the same way as the genus 0 case. Since $(ds)^2 = (\Delta z)^2 = \epsilon$ is the invariant cutoff, eq.(24.45) still holds and $\kappa = \kappa_0\epsilon$ as before. The final result is

$$A_n^{(1)}(k_1, \cdots, k_n) = (2\pi)^{26}\,\delta^{26}(\textstyle\sum_j k_j)\mathcal{N}\kappa^n$$

$$\frac{1}{2}\int_{D(\Gamma)} d^2\tau\,\frac{1}{2\pi\tau_2^2}\,e^{4\pi\tau_2}\,(2\pi\tau_2)^{-12}\left|\prod_{n=1}^{\infty} 1 - e^{2i\pi n\tau}\right|^{-48}$$

$$\int d^2 z_1 \cdots d^2 z_n \prod_{i<j} F(z_i, z_j)^{k_i\cdot k_j/2\pi}, \tag{24.53}$$

where

$$F(z_i, z_j) = \exp\left(-\frac{\pi(\mathrm{Im}(z_i - z_j))^2}{\tau_2}\right)\left|\frac{\vartheta_1(z_i - z_j|\tau)}{\vartheta_1'(0|\tau)}\right|, \tag{24.54}$$

and where $D(\Gamma)$ is the fundamental region of the modular group and is given by eq.(21.71). Note that the presence of the tachyon causes this amplitude to diverge for the same reason as given in the previous chapter in the discussion following eq.(23.53).

24.4 EXERCISES

1. Use eqs.(24.29) and (24.30) to verify that eqs.(24.27) and (24.31) are $PSL(2, \mathbb{C})$ invariant.
2. Use eq.(24.38) to write the 4-string amplitude, $A_4^{(0)}(k_1, \cdots, k_4)$, from eq.(24.36) to (24.39).
3. Use eq.(24.49) to verify eqs.(24.50) and (24.51).

4. Compute the Green's function, $G'(\xi_i, \xi_j; \tau)$, on the unit (ξ^1, ξ^2) square for the Laplacian $\Delta_{\widehat{g}}$ where $\widehat{g}$ is given by eq.(24.43). Insert the result into eq.(24.47) and compare with eqs.(24.53) and (24.54).
5. Compute the n-point tree-level amplitude for $n-2$ tachyons and 2 dilatons.
6. Verify that $G(z, z'; \tau)$ is the Green's function of the ordinary Laplacian on the torus represented by a parallelogram formed by the points 1 and τ.

CHAPTER **25**

Noncritical Dimensions

One of the most notable features of string theory is the quantum appearance of critical embedding space-time dimensions. In the critical dimension the world-sheet gravity decouples from the matter or embedding fields. However, many potential applications of strings (or sums over surfaces) require the more ordinary space-time dimensions of 3 or 4, which are noncritical. In this case, the additional modes from the world-sheet metric must also be quantized; hence this subject is also referred to as 2-dimensional quantum gravity. For the bosonic string actions and boundary conditions we have discussed so far, this means that we must quantize the Liouville action. This is the topic of the first section of this chapter.

As we discussed in chapter 19, one application of noncritical strings is to rewrite pure Yang-Mills as a quantum theory of Wilson loops. Alternatively, the leading order of the $1/N$ nonperturbative expansion of QCD appears to be a string theory. If this leading term can be solved exactly, then it may be possible to compute the higher-order corrections adequately enough to make real predictions with QCD. However, it is likely that Liouville strings are not the kind of strings that describe QCD. One alternative action can be built by including the extrinsic curvature of the world-sheet. Whether these rigid strings can exactly, or at least phenomenologically, describe QCD is still a completely open question. The application of strings to QCD is the topic of the second section of this chapter.

Before we move on, let us mention that the evidence points to the fact that the critical 3-dimensional Ising model is equivalent to some kind of fermionic string (i.e. superstring). It is known that critical superstrings compactified to 3 dimensions cannot be the right kind of fermionic string for the 3-dimensional Ising model. In the next chapter we will consider the genus 0 supersurface contribution to the partition function for closed oriented RNS superstrings. The superconformal anomaly present at $D = 3$ leads to the super Liouville model (see eq.(26.165)). This fermionic string is a candidate for the 3-dimensional Ising model, but the exact Ising-string equivalence remains an open question.

25.1 2-DIMENSIONAL QUANTUM GRAVITY

2-dimensional quantum gravity is a very active area of research at present. There have been many contributions to this topic over the past decade. To keep within scope and length constraints, let us concentrate on discussing the quantization of the Liouville model in the conformal gauge. We will follow the work of David, Distler, and Kawai, and the presentation by Polchinksi. Their work was motivated by the results obtained by Polyakov, Zamolodchikov, and Knizhnik in the light-cone gauge.

To begin, let us first note how the correlation functions of the quantized Liouville model enter into scattering amplitudes in noncritical dimensions. For concreteness, consider again the tachyon scattering amplitude at genus 0, eq.(24.20). In a noncritical dimension, the $\phi(\xi)$ dependence of the two determinants does not cancel and instead leaves $\exp(-\gamma^{-2} S_L[g,\phi])$, where $\gamma^{-2} = (26 - D)/48\pi$, and where $S_L[g,\phi]$ is the Liouville action, eq.(23.21).

$$S_L[g,\phi] = \int d^2\xi \sqrt{g}\left(\frac{1}{2} g^{ab}\partial_a\phi\partial_b\phi + R\phi + \mu^2 e^{\phi}\right) \tag{25.1}$$

There is additional dependence on $\phi(\xi)$ in eq.(24.20) from the factors $\sqrt{g_1}\cdots\sqrt{g_n}$, and from the cutoff of the Green's function $G(\xi_i,\xi_i)$ at coincident points (see eq.(24.25)). In a noncritical dimension these no longer cancel. All together, the ϕ dependence of the amplitude amounts to the path integral representation of the Liouville correlations of the exponential of $\phi(\xi_i)$, evaluated at the points on the surface where the vertex operators are inserted. To proceed, we must solve the quantum Liouville model.

This solution has proved to be notoriously difficult to find, and still no satisfactory solution exists for physically reasonable space-time dimensions. Despite the fact that the classical Liouville model is completely integrable, the quantization itself is nontrivial. To appreciate why this is so, let's start at the beginning with the expression for the genus h contribution to the partition function, eq.(23.2).

$$\begin{aligned} Z_h &= \mathcal{N}\int \mathcal{D}g_{ab}\,\mathcal{D}x^\mu\,\exp\Big(-S_m[g,x] - \int d^2\xi\,\sqrt{g}\,\mu_0^2\Big) \\ S_m[g,x] &= \frac{1}{2}\int d^2\xi\,\sqrt{g}\Big(g^{ab}\partial_a x^\mu \partial_b x_\mu\Big) \end{aligned} \tag{25.2}$$

$\mathcal{D}x^\mu$ is defined via eq.(22.23), and $\mathcal{D}g_{ab}$ by eq.(22.3). Pick a gauge slice $\widehat{g}_{ab}$ and let $g_{ab} = \exp(\phi)\,\widehat{g}_{ab}$. To carry out the Faddeev-Popov procedure, we change variables from g_{ab} to $v^a(\xi)$ and $\phi(\xi)$, where the vector fields $v^a(\xi)$ parametrize infinitesimal diffeomorphisms. The change of variables introduces the Jacobian $\Delta_{\rm FP}(g)$; hence eq.(25.2) becomes

$$Z_h = \mathcal{N}\int d^m t\,(\mathcal{D}\phi)_g\,(\mathcal{D}x^\mu)_g\,\Delta_{\rm FP}(g)\,\exp\Big(-S_m[g,x] - \int d^2\xi\,\sqrt{g}\,\mu_0^2\Big). \tag{25.3}$$

$d^m t$ is symbolic for the integration measure on the moduli space. The volume of the diffeomorphism group has been absorbed in the normalization. Since we are not going to work in the critical dimension, we do not expect to extract the volume of the group of Weyl transformations from the integral. The subscript on the functional measures means that they are defined with respect to the surface metric g_{ab}.

Let's examine the ϕ dependence of the measures and integrand. The integration measure on the moduli spaces can be set up to be independent of ϕ. The action for the matter or embedding fields, S_m, is Weyl invariant, but the "cosmological" term proportional to μ_0^2 is not. This type of term acquires an anomalous dimension which we will return to in a moment. From the results of chapter 22, specifically eq.(22.96) or eq.(23.20), we also know that

$$\begin{aligned} &\mathcal{N}\int (\mathcal{D}x^\mu)_g\,\exp(-S_m[g,x]) \\ &\qquad = \mathcal{N}'\left(\frac{\det'(\Delta_g)}{\int d^2\xi\,\sqrt{g}}\right)^{-D/2} \\ &\qquad = \mathcal{N}'\left(\frac{\det'(\Delta_P\widehat{g})}{\int d^2\xi\,\sqrt{\widehat{g}}}\right)^{-D/2}\exp\Big(\frac{D}{48\pi}\,S_L[\widehat{g},\phi]\Big) \\ &\qquad = \mathcal{N}\exp\Big(\frac{D}{48\pi}\,S_L[\widehat{g},\phi]\Big)\int(\mathcal{D}x^\mu)_{\widehat{g}}\,\exp(-S_m[\widehat{g},x]). \end{aligned} \tag{25.4}$$

Hence

$$(\mathcal{D}x^\mu)_g = \exp\Big(\frac{D}{48\pi}\,S_L[\widehat{g},\phi]\Big)\,(\mathcal{D}x^\mu)_{\widehat{g}}. \tag{25.5}$$

Furthermore, eq.(22.121) or eq.(23.22) implies that

$$\Delta_{\mathrm{FP}}(g) = \Delta_{\mathrm{FP}}(\widehat{g}) \exp\left(-\frac{26}{48\pi} S_L[\widehat{g}, \phi]\right). \tag{25.6}$$

Finally, to examine $(\mathcal{D}\phi)_g$, we define it by taking

$$||\delta\phi||^2 = \int d^2\xi \sqrt{g}\, \delta\phi(\xi)\, \delta\phi(\xi), \tag{25.7}$$

and herein lies the trouble. Eq.(25.7) is not Weyl invariant and depends on the conformal factor itself. In addition, the theory must be regulated in a covariant manner. As we discussed in chapter 24, this means that the regulator itself depends on ϕ (see the discussion leading to eq.(24.25)). This makes $(\mathcal{D}\phi)_g$ very complicated and unknown.

To proceed, a conjecture is made on the form of $(\mathcal{D}\phi)_g$. In particular, it is assumed that

$$\int (\mathcal{D}\phi)_g \exp\left(-\frac{26-D}{48\pi} S_L[g,\phi]\right) = \int (\mathcal{D}\phi)_{\widehat{g}} \exp\left(-\int d^2\xi \sqrt{\widehat{g}}\left(\alpha \widehat{g}^{ab}\partial_a\phi\partial_b\phi + \beta\widehat{R}\phi + \mu'^2 e^{\zeta\phi}\right)\right), \tag{25.8}$$

where α, β, ζ, and μ' are constants that we must determine. Built into eq.(25.8) is the assumption that the regularized measure can be written as the exponential of a local action and that the local action is renormalizable. In other words, all counterterms are local and satisfy general coordinate invariance.

Combining eqs.(25.5), (25.6), and (25.8) into Z_h, eq.(25.3), we now have

$$\begin{aligned} Z_h &= \mathcal{N} \int d^m t\, (\mathcal{D}\phi)_g\, \Delta_{\mathrm{FP}}(g) \int (\mathcal{D}x^\mu)_g \exp\left(-S_m[g,x] - \int d^2\xi \sqrt{g}\, \mu_0^2\right) \\ &= \mathcal{N} \int d^m t\, \Delta_{\mathrm{FP}}(\widehat{g}) \int (\mathcal{D}\phi)_g \exp\left(-\frac{26-D}{48\pi} S_L[g,\phi]\right) \\ &\qquad \int (\mathcal{D}x^\mu)_{\widehat{g}} \exp\left(-S_m[\widehat{g},x]\right) \\ &= \mathcal{N} \int d^m t\, \Delta_{\mathrm{FP}}(\widehat{g}) \\ &\qquad \int (\mathcal{D}\phi)_{\widehat{g}} \exp\left(-\int d^2\xi \sqrt{\widehat{g}}\left(\alpha\widehat{g}^{ab}\partial_a\phi\partial_b\phi + \beta\widehat{R}\phi + \mu'^2 e^{\zeta\phi}\right)\right) \\ &\qquad \int (\mathcal{D}x^\mu)_{\widehat{g}} \exp\left(-S_m[\widehat{g},x]\right). \end{aligned} \tag{25.9}$$

In the first line of eq.(25.9) note that Z_h depends only on g_{ab} and not on $\widehat{g}_{ab}$ and ϕ separately. That is, $\widehat{g}_{ab}$ and ϕ always appear together in Z_h in the form $g = \widehat{g}e^{\phi}$. The same must be true in the last line of eq.(25.9). Therefore, the transformation

$$\begin{aligned} \widehat{g}_{ab} &\longrightarrow \widehat{g}_{ab}\, e^{\sigma(\xi)} \\ \phi(\xi) &\longrightarrow \phi(\xi) - \sigma(\xi) \end{aligned} \tag{25.10}$$

will leave Z_h unchanged and independent of $\sigma(\xi)$. We may determine the unknown coefficients in eq.(25.8) by transforming the last line of eq.(25.9) with eq.(25.10) and then requiring

$$\frac{\delta}{\delta\sigma} Z_h[\widehat{g} \to \widehat{g}e^{\sigma}, \phi \to \phi - \sigma] = 0. \tag{25.11}$$

Upon transforming Z_h, which pieces separately depend on σ? From eq.(25.5) and eq.(25.6) we have

$$\begin{aligned} (\mathcal{D}x^{\mu})_{\widehat{g}e^{\sigma}} &= \exp\left(\frac{D}{48\pi} S_L[\widehat{g}, \sigma]\right) (\mathcal{D}x^{\mu})_{\widehat{g}} \\ \Delta_{\mathrm{FP}}(\widehat{g}e^{\sigma}) &= \exp\left(-\frac{26}{48\pi} S_L[\widehat{g}, \sigma]\right) \Delta_{\mathrm{FP}}(\widehat{g}). \end{aligned} \tag{25.12}$$

Since σ is a conformal transformation independent from ϕ, then $(\mathcal{D}\phi)_{\widehat{g}}$ transforms in the same way as a single component ($D = 1$) of $(\mathcal{D}x^{\mu})_{\widehat{g}}$; hence

$$(\mathcal{D}\phi)_{\widehat{g}e^{\sigma}} = \exp\left(\frac{1}{48\pi} S_L[\widehat{g}, \sigma]\right) (\mathcal{D}\phi)_{\widehat{g}}. \tag{25.13}$$

In addition to the explicit σ dependence that comes from transforming the generalized Liouville action in the last line of eq.(25.9), the $\exp(\zeta\phi)$ term is renormalized and possesses an anomalous dimension. The σ dependence enters via a covariant regulator *à la* eq.(24.25). To calculate this we examine the propagator, $\langle e^{\zeta\phi(\xi')} e^{\zeta\phi(0)} \rangle$. This computation is nearly identical to that leading to eq.(24.24). By completing the squares, we find

$$\begin{aligned} \langle e^{\zeta\phi(\xi')} e^{\zeta\phi(0)} \rangle &= \int \mathcal{D}\phi\; e^{\zeta\phi(\xi')} e^{\zeta\phi(0)} \exp\left(-\int d^2\xi\, \sqrt{\widehat{g}}\, \alpha\, \widehat{g}^{ab} \partial_a\phi\partial_b\phi\right) \\ &= \exp\left(\frac{1}{4}\frac{\zeta^2}{\alpha}\, G(\xi', 0; \sigma)\right), \end{aligned} \tag{25.14}$$

where $G(\xi', 0; \sigma)$ is the Green's function for Δ_g. As $\xi' \to 0$, we use eq.(24.25) and arrive at

$$\langle e^{\zeta\phi(\xi')} e^{\zeta\phi(0)} \rangle = \exp\left(-\frac{\zeta^2}{16\pi\alpha} \ln(\epsilon) + \frac{\zeta^2}{16\pi\alpha}\sigma \right). \tag{25.15}$$

The first term (which in eq.(25.15) is for genus 0) can be absorbed by redefining $(\mu')^2$. The second term, $\exp(\zeta^2\sigma(\xi)/16\pi\alpha)$, which is the same for each genus, is the dependence we are looking for. Thus, under the transform eq.(25.10),

$$\int d^2\xi \, \sqrt{\hat{g}} \, (\mu')^2 \, e^{\zeta\phi(\xi)} \longrightarrow \int d^2\xi \, \sqrt{\hat{g}} \, (\mu')^2 \, e^{-(\zeta^2/16\pi\alpha)\,\sigma} e^{\zeta(\phi-\sigma)} e^{\sigma}. \tag{25.16}$$

Collecting the results, eqs.(25.12), (25.13), and (25.16), together, the computation in eq.(25.11) yields

$$\begin{aligned} 0 = {} & \left(\frac{D-25}{48\pi} - \beta \right) R - (2\alpha - \beta)\left(\Delta_{\hat{g}}\sigma - \Delta_{\hat{g}}\phi \right)\sqrt{\hat{g}} \\ & - (\mu')^2 \left(1 - \zeta - \frac{\zeta^2}{16\pi\alpha} \right) \sqrt{\hat{g}} \, \exp(\phi) \, \exp\Big((1 - \zeta - \zeta^2/16\pi\alpha)\sigma \Big). \end{aligned} \tag{25.17}$$

In this computation we have made use of eq.(23.19), and of

$$\frac{\delta}{\delta\sigma} S_L[g, \sigma] = R, \tag{25.18}$$

which follows from the results in chapter 22 or 23 (e.g. eq.(23.18)). In arriving at eq.(25.18), we have implicitly absorbed the remaining terms into the coefficients of eq.(25.8).

From eq.(25.17), we have

$$\begin{aligned} \beta &= \frac{D-25}{48\pi} \\ 2\alpha &= \beta \\ \zeta &= \frac{1}{12}\left(25 - D \pm \sqrt{(D-25)(D-1)} \right). \end{aligned} \tag{25.19}$$

The limit $D \to -\infty$ corresponds to the (weak coupling) semi-classical limit of the Liouville model ($\gamma^2 = 48\pi/(26-D)$, the Liouville coupling, vanishes in this limit). In order to agree with results from saddle-point methods in this limit, we must choose the + sign in ζ in eq.(25.19) above. However, independent of the sign, note that ζ becomes complex when

$1 < D < 25$. Unfortunately, despite all of the effort to get to this point, the method fails in the range of physically interesting space-time dimensions. When $D > 25$, the kinetic term in the Liouville action develops the wrong sign indicating that ϕ has become a ghost field of negative norm and that trouble is brewing. See Polchinski's article in the Bibliography for this chapter, for interpreting this region.

As an application of the results in eq.(25.19), and to illustrate the problems encountered with extending the results to $1 < D < 25$, let us compute the critical exponent called the string susceptibility. To define this quantity start again at Z_h, eq.(25.2). The idea is to rewrite this expression as an integral over the area A of the surfaces in the sum.

$$Z_h = \int_0^\infty dA\, \Gamma_h(A)\, \exp(-\mu_0^2 A) \tag{25.20}$$

Since the area of the surface is simply

$$A = \int d^2\xi\, \sqrt{g}, \tag{25.21}$$

it follows from eq.(25.2) that

$$\Gamma_h(A) = \mathcal{N} \int \mathcal{D}g_{ab}\, \mathcal{D}x^\mu\, \delta\Big(\int d^2\xi\, \sqrt{g} - A\Big)\, \exp(-S_m[g,x]). \tag{25.22}$$

The asymptotic behavior of Γ_h for large A is expected to be

$$\Gamma_h(A) \sim \exp(\mu^2 A)\, A^{\gamma(h)-3}, \tag{22.23}$$

where μ^2 is cutoff-dependent. $\gamma(h)$ is the string susceptibility and is cutoff-independent.

To make use of the results in eq.(25.19) we express Z_h using the third line in eq.(25.9) and rewrite it in the form of eq.(25.20). There are no terms in the exponent of eq.(25.9) that are directly equal to A since $\zeta \neq 1$. We can remedy this by requiring $\int d^2\xi\, \sqrt{\hat{g}}\, \exp(\zeta\phi)$ to be the area A via the δ-function in eq.(25.22). Hence, using eq.(25.9) we have

$$\begin{aligned}\Gamma_h(A) = \mathcal{N} \int d^m t\, \Delta_{\mathrm{FP}}(\hat{g}) & \\ \int (\mathcal{D}\phi)_{\hat{g}}\, \exp\Big(-\int d^2\xi\, \sqrt{\hat{g}}\Big(\alpha\hat{g}^{ab}\partial_a\phi\partial_b\phi + \beta\hat{R}\phi\Big)\Big) & \\ \int (\mathcal{D}x^\mu)_{\hat{g}}\, \exp(-S_m[g,x])\; \delta\Big(\int d^2\xi\, \sqrt{\hat{g}}\, e^{\zeta\phi} - A\Big), &\end{aligned} \tag{25.24}$$

where α, β, and ζ are given by eq.(25.19).

Following David, Distler, and Kawai, we extract $\gamma(h)$ from eq.(25.24) by shifting ϕ by a constant:

$$\phi \longrightarrow \phi + \rho. \tag{25.25}$$

The measures and determinants are invariant under this shift as long as ρ is a constant. The only changes occur in the exponent and the δ-function in eq.(25.24).

$$\begin{aligned}\int d^2\xi\,\sqrt{\hat{g}}\,\beta\hat{R}\phi &\longrightarrow \int d^2\xi\,\sqrt{\hat{g}}\,\beta\hat{R}\phi + \beta\rho\int d^2\xi\,\sqrt{\hat{g}}\hat{R} \\ &= \int d^2\xi\,\sqrt{\hat{g}}\,\beta\hat{R}\phi - 8\pi\,\beta\rho(1-h)\end{aligned} \tag{25.26}$$

The last line follows from the definition of the Euler characteristic, eq. (22.95). The change in the δ-function follows from the property $\delta(ax) = a^{-1}\delta(x)$.

$$\delta\Big(\int d^2\xi\,\sqrt{\hat{g}}\,e^{\zeta\phi} - A\Big) \longrightarrow e^{-\zeta\rho}\,\delta\Big(\int d^2\xi\,\sqrt{\hat{g}}\,e^{\zeta\phi} - e^{-\zeta\rho}A\Big) \tag{25.27}$$

Eqs.(25.26) and (25.27) together imply that

$$\Gamma_h(A) \longrightarrow e^{8\pi\,\beta\rho(1-h)}e^{-\zeta\rho}\,\Gamma_h(e^{-\zeta\rho}A). \tag{25.28}$$

This result is compatible with eq.(25.23) only if we take

$$\gamma(h) = \frac{8\pi\,\beta}{\zeta}(1-h) + 2. \tag{25.29}$$

Using eq.(25.19), this is

$$\gamma(h) = \frac{1}{12}(1-h)\left(D - 25 - \sqrt{(D-25)(D-1)}\right) + 2. \tag{25.30}$$

Note that γ is complex when $1 < D < 25$, which formally means that the sum over surfaces would be ill-defined at the large A end in this region. In fact, what happens is that at $D = 1$ we cross over from the weak coupling regime to the strong coupling regime. In such a region there is no guarantee that an expansion in topology makes any sense at all.

The formulation we have discussed is perturbative with respect to the topology of the surfaces we are summing over, in that we calculate the result for each genus and then sum. A possible nonperturbative formulation is found by discretizing the surfaces and taking scaling limits of the resulting random-matrix models. Results from this approach are rapidly expanding. However, these models suffer the same restriction in space-time dimension ($1 < D < 25$ does not make sense and is excluded). Thus, the noncritical Liouville string in physically reasonable space-time dimensions remains unsolved.

25.2 QCD AND RIGID STRINGS

Many of the attempts to connect QCD to strings involve the Wilson loop. If γ is a closed curve in space-time, then the Wilson loop is the trace of the path-ordered exponential of the vector potential, $A^\mu = igA^\mu_a\lambda^a/2$, integrated along the curve γ.

$$W[\gamma] = \mathrm{tr}\,\mathrm{P}\exp\left(\oint_\gamma A^\mu(x)\,dx_\mu\right) \tag{25.31}$$

The vacuum expectation of $W[\gamma]$ is related to the heavy quark (static) potential. For example, consider quickly separating a quark-antiquark pair over a distance R, holding them apart for a time t, and then quickly putting them back together again. The world-lines of the quark pair will form a closed rectangular loop. If H is the Hamiltonian for the system, then $\exp(-Ht)$ is the time-evolution operator and the amplitude for the process is $\langle\exp(-Ht)\rangle$. The path integral representation of this amplitude is

$$\langle\exp(-Ht)\rangle = \mathcal{N}\int \mathcal{D}A^\mu\,\exp\left(-S[A] + \int d^4x\,A^\mu(x)\,J_\mu(x)\right), \tag{25.32}$$

where $S[A]$ is the gauge theory action and the external massive quark current is J^μ. Of course, there is some work involved in making sense of the integral in eq.(25.32) (gauge-fixing, Faddeev-Popov, etc.). The current has support only on the world-lines of the quark pair, hence

$$\begin{aligned}\langle\exp(-Ht)\rangle &= \mathcal{N}\int \mathcal{D}A^\mu\,\exp\left(-S[A] + \oint_\gamma A^\mu(x)\,dx_\mu\right)\\ &= \langle W[\gamma]\rangle.\end{aligned} \tag{25.33}$$

As t gets large, only the lowest energy state contributes to the expectation, thus

$$\langle\exp(-Ht)\rangle \longrightarrow e^{-E_0 t}\,\langle 0\,|\,0\rangle. \tag{25.34}$$

The energy E_0 of the (nondynamical) quark pair relative to the vacuum is the heavy quark static potential, $V(R)$. Therefore,

$$V(R) = -\frac{1}{t}\ln\langle W[\gamma]\rangle \tag{25.35}$$

as t gets large.

A gauge invariant criterion for quark confinement was developed by K. Wilson and is based on the Wilson loop. It states that the theory will confine quarks if and only if the expectation of the Wilson loop for large loops grows as $\exp(-\alpha A)$, where α is a constant and A is the minimal area spanning the loop. In the above example, the minimal area of the loop is simply Rt. Thus, if the quarks are confined, the static potential is linear, $V(R) \propto R$.

One method that has been used to explore the QCD-string connection is to think of $W[\gamma]$ as the creation operator of a closed string. $\langle W[\gamma]\rangle$ can then be thought of as a wave functional for the string. The connection is made by attempting to find a string-like functional Schrödinger equation for $\langle W[\gamma]\rangle$ from QCD. For example, let the vector potential A^μ be fixed and let the curve γ be parametrized by σ (i.e. the embedded curve γ is given by $x^\mu(\sigma)$). The first functional derivative of $\langle W[\gamma]\rangle$ with respect to the curve $x^\mu(\sigma)$ can be shown to equal

$$\frac{\delta\langle W\rangle}{\delta x^\mu(\sigma)} = \mathrm{tr}\,\mathrm{P}\left(x'_\mu(\sigma)F_{\mu\nu}[x]\,\exp\left(\oint_\gamma A_\beta\,dx^\beta\right)\right). \tag{25.36}$$

$F_{\mu\nu}$ is the usual gauge field strength and x'_μ is the derivative of x_μ with respect to σ. The second functional derivative is

$$\begin{aligned}\frac{\delta^2\langle W\rangle}{\delta x^\mu(\sigma)^2} = {} & \mathrm{tr}\,\mathrm{P}\left(D_\alpha F_{\alpha\nu}[x]x'_\nu(\sigma)\,\exp\left(\oint_\gamma A_\beta\,dx^\beta\right)\right)\delta(0) \\ & + \mathrm{tr}\,\mathrm{P}\left(\left(F_{\alpha\nu}[x]x'_\nu(\sigma)\right)^2\exp\left(\oint_\gamma A_\beta\,dx^\beta\right)\right).\end{aligned} \tag{25.37}$$

D_α is the usual gauge covariant derivative. $D_\alpha F_{\alpha\nu} = 0$ is a classical field equation for Yang-Mills (no sources). Therefore, to connect with QCD we drop the first term which leaves

$$\frac{\delta^2\langle W\rangle}{\delta x^\mu(\sigma)^2} = \mathrm{tr}\,\mathrm{P}\left(\left(F_{\alpha\nu}[x]x'_\nu(\sigma)\right)^2\exp\left(\oint_\gamma A_\beta\,dx^\beta\right)\right). \tag{25.38}$$

This equation has a string-like flavor. Compare eq.(25.38) to the functional Schrödinger equation for the bosonic string in the light-cone gauge for $E = 0$, eq.(20.80). The correspondence between eqs.(25.38) and (20.80) would become more exact if we could somehow replace $(F_{\alpha\nu})^2$ in eq.(25.38) by its expectation and pull it out of the path-ordering. In any case, the use of eq.(25.38) is complicated by the fact that it must be renormalized due to the presence of the second functional derivative. In addition, at the same time the solutions must be reparametrization invariant. They cannot depend on how we choose σ to parametrize the curve.

Different functional equations for $\langle W\rangle$ can be derived using different definitions of derivatives with respect to a loop, etc. Also, Polyakov has shown that the lattice version of the equation for $\langle W\rangle$ can be interpreted as a chiral field on loop space.

Instead of trying to develop string-like functional differential equations for $\langle W \rangle$ from QCD, we can attempt to find functional integral representations of $\langle W \rangle$ based on string theories. In this case, the expectation of the Wilson loop $\langle W[\gamma] \rangle$ is written as the sum over all surfaces whose boundary is the curve γ, where each surface in the sum is weighted by the exponential of some string action. This particular form for $\langle W[\gamma] \rangle$ is motivated by the lattice strong coupling expansion. In the lattice version, the expectation of a Wilson loop is a sum over all surfaces formed on the lattice that span the loop. It is also motivated by the fact that strings give linear potentials.

As an example, consider

$$\langle W[\gamma] \rangle = \int \mathcal{D}x^\mu \, \exp\left(-T \int_S d^2\xi \, \sqrt{g} \right), \tag{25.39}$$

where g_{ab} is the induced metric on the surface S (eq.(19.19)),

$$g_{ab} = \frac{\partial x^\mu}{\partial \xi^a} \frac{\partial x_\mu}{\partial \xi^b}, \tag{25.40}$$

and where the boundary of the surface S is γ. Since the string action in eq.(25.39) is the Nambu-Goto action, we can call this the Nambu-Goto model.

From the explicit expression adopted for $\langle W[\gamma] \rangle$, we can, in principle, compute the static potential $V(R)$ by choosing the appropriate γ (e.g. the rectangle with sides R and t). The model in eq.(25.39) belongs to a universality class of string actions that produces a Coulomb potential in addition to the desired linear potential.

$$V(R) = TR - \frac{\pi(D-2)}{24R} + \ldots \tag{25.41}$$

Since QCD is asymptotically free, we expect that the short distance potential is Coulombic. Perturbation theory verifies this. However, the Coulomb contribution to eq.(25.41) is a long-distance effect. The term is universal for this class of string actions, since it depends only on the dimension of space-time. It is not known for sure whether QCD does possess a long-distance Coulomb potential. In any case, the existence of such universality classes provides one criterion for selecting the correct class of string actions to put into eq.(25.39).

The static potential for the Nambu-Goto model can be computed exactly. However, it turns out to be easier in the operator representation. In this case we quantize x^μ in exactly the same way as we did in chapter 20, only this time we fix the ends of the string for all time at a distance R apart. Since the strings in chapter 20 had free ends, the mode expansion in this case is slightly different from eq.(20.17). The static potential is the vacuum energy of the string. This turns out to be

$$V(R) = \left(T^2 R^2 - \frac{\pi(D-2)T}{12} \right)^{1/2}. \tag{25.42}$$

This result was first found by Arvis. Eq.(25.41) follows by expanding $V(R)$. The term with the space-time dependence arises from the zero-point fluctuations of the quantum mode oscillators, which is the same as in the free-ends case.

Just as in the free-ends case studied in chapter 20, this string contains a tachyon and is only consistent in the critical dimension $D = 26$. Since QCD does not contain tachyons, this of course means that eq.(25.39) cannot be the correct representation for $\langle W[\gamma] \rangle$. $V(R)$ in eq.(25.42) doesn't even make sense as $R \to 0$. Note, however, that eq.(25.39) is not in line with Polyakov's approach, since the intrinsic metric on the surface is not considered dynamical and is not integrated over (no 2-dimensional quantum gravity). Eq.(25.39) is not renormalizable, while Polyakov's approach is. If we modify eq.(25.39) to include the metric, then we will find an effective action that contains the conformal anomaly. There will also be terms that are associated with the fixed boundary of the surfaces in the sum. These terms have been computed by O. Alvarez, and by Durhuus, Nielsen, Olesen, and Petersen, but we are once again faced with quantizing the Liouville model in physically reasonable space-time dimensions. Thus, the results from this model remain essentially unknown.

If we do not insist on exactly reproducing QCD with models such as eq.(25.39), but only desire an effective theory valid over a limited range of distances, then the Nambu-Goto model may still be useful. As pointed out by Olesen, the inconsistencies (i.e. critical dimension) in this model, as well as the tachyon, asymptotically disappear at large distances. If QCD does indeed possess a long-distance Coulombic term, then the Nambu-Goto model is a viable candidate for an effective theory description of QCD at long distances.

Another link between QCD and strings arises in 't Hooft's $1/N$ expansion of QCD. The motivation for this expansion comes from the fact that the number of colors, N, is the only free dimensionless parameter in the theory. The leading order of this expansion is some kind of free string theory that has yet to be identified. Let us briefly describe some of the string-like aspects of the leading order.

The appropriate Lagrangian density is

$$\mathcal{L} = \frac{N}{g^2} \operatorname{tr} \left(-\frac{1}{4} F^{\mu\nu} F_{\mu\nu} + \overline{\psi}(i\partial_\mu + A_\mu)\gamma^\mu \psi - m\overline{\psi}\psi \right). \tag{25.43}$$

Figure 25.1 (a) Representation of the quark propagator. (b) Double-line representation of the gluon propagator.

The quark fields are in the fundamental representation of $U(N)$. (If we only care about the leading order, then there is no difference between $SU(N)$ and $U(N)$.) This means that they carry one group matrix element index, hence $\overline{\psi}\psi = \overline{\psi}^n \psi_n$, etc. The vector potential is in the adjoint representation, so it is a matrix $A_{\mu m}{}^n$ (i.e. $A_{\mu m}{}^n = A_\mu^a (T^a)_m{}^n$, where T_a are the appropriate generators of the gauge group). Similarly, the field strength is also a matrix, $F_{\mu\nu m}{}^n$. The trace in eq.(25.43) is taken on these indices.

From eq.(25.43) we can read off that the quark propagator is

$$\langle 0 | T(\psi^m(x)\overline{\psi}_n(y)) | 0 \rangle = \frac{1}{N} \delta^m_n \, S_F(x-y), \tag{25.44}$$

where $S_F(x-y)$ is the usual Dirac propagator as in chapter 4. For the gluons we have

$$\langle 0 | T(A_{\mu n}{}^m(x) A_{\nu q}{}^p(y)) | 0 \rangle = \frac{1}{N} \delta^m_q \, \delta^p_n \, D_{\mu\nu}(x-y), \tag{25.45}$$

where $D_{\mu\nu}(x-y)$ is the usual free photon propagator as in chapter 5. The diagrammatic representation convenient for eq.(25.44) is given in figure 25.1(a). The usual diagrammatic representation of eq.(25.45) is given in figure 15.5(a). However, since we are explicitly tracking matrix elements of an individual generator, then the natural representation of the propagator invented by 't Hooft is a double line as in figure 25.1(b). With this representation, any diagram, such as figure 15.6(a), breaks into a set of continuous single lines (figure 25.2).

Now consider vacuum-to-vacuum diagrams with no external legs (i.e. contributions to the partition function). Without external legs, any diagram must be composed of a set of closed loops (e.g. figure 25.3). We can identify each diagram with an oriented surface by gluing together neighboring lines. For example, figure 25.3(a) becomes a disk, i.e., a sphere with one boundary component, while figure 25.3(b) becomes a torus.

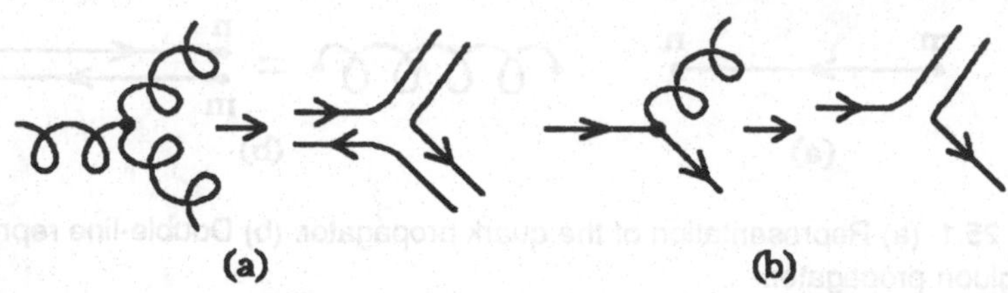

(a) (b)

Figure 25.2 (a) Double-line equivalent of the 3-gluon vertex. (b) Double-line equivalent of the quark-gluon vertex.

Now let's count the power of N that is associated with each vacuum diagram. Each diagram will be composed of V vertices, E edges, and F faces. With each vertex, eq.(25.43) tells us to associate one factor of N. Each edge is a propagator, so by eqs.(25.44) and (25.45) we associate a factor of N^{-1}. Each face represents a loop and there are N different matrix indices that can travel around the loop so we associate a factor of N. Altogether, we find each vacuum diagram multiplied by a factor

$$N^{V-E+F} = N^{\chi}, \tag{25.46}$$

where $\chi = 2 - 2h - b$ is the usual Euler characteristic for a surface with h handles and b boundary components.

Eq.(25.46) tells us that the leading order is N^2, which is the set of genus 0 surfaces with no boundaries. Since quark loops produce boundaries, then the leading order is entirely made up of gluons. Also, we can remove a face from the surface and flatten it out as we did in chapter 21. When we do this, no edges will cross one another, so the leading order is entirely composed of planar diagrams (no gluon propagators cross over another propagator). Figure 25.3(a) is planar while 25.3(b) is not. The next order is $h = 0$, $b = 1$, and therefore involves 1 quark loop. The boundary represents the face already removed; hence the flattened-out diagram consists of a single quark loop on the boundary and is planar. The solution to leading order consists of summing all leading-order graphs, i.e. it is some kind of free string theory (free because the surfaces are all genus 0). The problem is that we do not know which string theory (i.e. we don't know how to define the sum and weight the terms without calculating each diagram separately). One way to go about this is to return to the Wilson loop,

$$W[\gamma] = \frac{1}{N}\text{tr P}\left(\exp\left(\oint_{\gamma} A_{\mu}\,dx^{\mu}\right)\right). \tag{25.47}$$

In the large N limit, $\langle W[\gamma]\rangle$ becomes a classical field since $\langle W[\gamma_1]W[\gamma_2]\rangle \rightarrow \langle W[\gamma_1]\rangle\langle W[\gamma_2]\rangle$. In this limit, $\langle W[\gamma]\rangle$ satisfies a string-like functional dif-

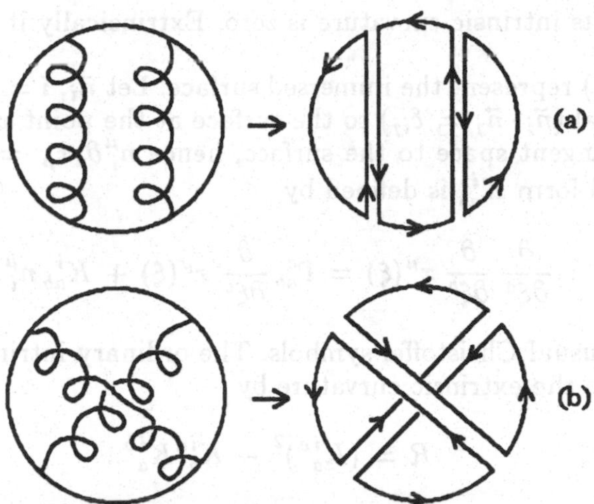

Figure 25.3 (a) Double-line diagram can be identified as a disk or genus 0 surface with one boundary component. This planar diagram is of order N^1. (b) Double-line diagram can be identified as a torus or surface of genus 1 with no boundaries. This nonplanar diagram is of order N^0.

ferential equation, but this equation requires renormalization, and no solutions to it have been found yet.

The free string theory that is equivalent to the leading order must be consistent in 4 space-time dimensions. One might expect the leading candidate to be the strongly coupled Liouville model. However, the evidence (from discretized models, etc.) indicates that the Liouville model at $D = 4$ is not the correct string theory. Since we want a free string theory, the surfaces in the sum should be dominated by smooth surfaces with no surface tension and without self-intersections. It is thought that the dominant surfaces in the Liouville sum are highly crumpled, with a finite surface tension.

One way to suppress the crumpling is to make the surfaces pay with energy for crumpling up. To do this we must relate the energy to the way they are immersed. For this purpose, Polyakov has introduced the extrinsic curvature into the string action. Such strings are called rigid strings.

The extrinsic curvature of a surface is a measure of how fast the unit normal to the surface changes as we move about the surface. Since the unit normal cannot be in any tangent space to the surface, then the surface must be immersed in a larger space in order to talk about its extrinsic curvature. The standard example is the cylinder with its usual embedding in 3-space. Its intrinsic curvature is zero. Extrinsically it certainly is not flat.

Let $x^\mu(\xi)$ represent the immersed surface. Let $\vec{n}_i$, $i = 1, \ldots, D-2$, be a unit normal ($\vec{n}_i \cdot \vec{n}_j = \delta_{ij}$) to the surface at the point x^μ. $\partial x^\mu / \partial \xi^a$ will be in the tangent space to the surface, hence $n_i^\mu \partial_a x_\mu = 0$. The second fundamental form K^i_{ab} is defined by

$$\frac{\partial}{\partial \xi^a} \frac{\partial}{\partial \xi^b} x^\mu(\xi) = \Gamma^c_{ab} \frac{\partial}{\partial \xi^c} x^\mu(\xi) + K^i_{ab} n^\mu_i. \tag{25.48}$$

Γ^c_{ab} are the usual Christoffel symbols. The ordinary intrinsic curvature is connected to the extrinsic curvature by

$$R = (K^{i\,a}_{a})^2 - K^{ia}_{b} K^{ib}_{a}. \tag{25.49}$$

The generalization of the ordinary string action to the one for rigid strings is

$$S = \mu_0^2 \int d^2\xi \, \sqrt{g} + \frac{1}{2} \kappa \int d^2\xi \; K^{ia}_{b} K^{ib}_{a}, \tag{25.50}$$

where κ is the bending rigidity modulus. The ordinary term involving the surface tension μ_0^2 is just the Nambu-Goto action. To make the second term look something like an ordinary field theory action, we can rewrite it as

$$\frac{1}{2} \kappa \int d^2\xi \; K^{ia}_{b} K^{ib}_{a} = \frac{1}{2} \kappa \int d^2\xi \, \sqrt{g} \, g^{ab} \partial_a t_{\mu\nu} \, \partial_b t^{\mu\nu}, \tag{25.51}$$

where

$$t_{\mu\nu} = \frac{\varepsilon^{ab}}{\sqrt{g}} \partial_a x_\mu \partial_b x_\nu. \tag{25.52}$$

The idea is to use rigidity to allow for the possibility of lowering the surface tension to zero. How might this come about? There is a topological invariant of an immersed surface that is special to $D = 4$. It is

$$\nu = \frac{1}{4} \int d^2\xi \, \sqrt{g} \, g^{ab} \varepsilon^{\mu\nu\lambda\rho} \, \partial_a t_{\mu\nu} \partial_b t_{\lambda\rho}, \tag{25.53}$$

and it counts the number of times the immersed surface self-intersects. This term can be added to the action with a coupling usually denoted as ϑ. (It is similar to the vacuum angle in Yang-Mills.) Symbolically, the partition function becomes

$$Z = \sum_{\text{surfaces}} e^{iS + i\vartheta\nu}. \tag{25.54}$$

When $\vartheta = \pi$, this becomes

$$Z = \sum_{\text{surfaces}} e^{iS} (-1)^{\nu}. \tag{25.55}$$

Crumpled surfaces with a large number of self-intersections are suppressed by this additional factor, since "neighboring" highly crumpled surfaces in the sum with slightly different ν will cancel each other's contribution to the sum. The properties of quantum rigid strings in physical space-time dimensions have yet to be worked out.

String theory was originally developed as dual resonance models to explain hadronic physics. It was almost abandoned after the invention of QCD and the discovery of asymptotic freedom. QCD has resisted solution due to the strong coupling at long distances or low energies. It may just be that string theory will ultimately come to the rescue by offering an alternative but tractable method of solution.

CHAPTER **26**

Introduction to Superstrings

26.1 THE RNS ACTION

The purpose of this chapter is to briefly introduce what has become known as the "old" superstring, the spinning string, the dual pion model, or the Ramond-Neveu-Schwarz model. The original model began in 1971 as a set of equations of motion introduced by P. Ramond to incorporate space-time fermions into a dual model structure. What became the bosonic sector was introduced separately by A. Neveu and J. Schwarz at about the same time. The synthesis of the two sectors was initiated by Neveu and Schwarz shortly thereafter. Supersymmetry was discovered through this model in 1971 by J.-L. Gervais and B. Sakita. At that time, only a covariant gauge-fixed action was known. An action with manifest local world-sheet supersymmetry and reparametrization invariance was written down by L. Brink, P. Di Vecchia, and P. Howe, and independently by S. Deser and B. Zumino, but not until 1976. This action is our starting point. We will rewrite the Euclidean version of this action in terms of an action on supersurfaces. We will then describe the evaluation of the contribution of genus 0 supersurfaces to the partition function for oriented closed RNS superstrings.

In order to correctly incorporate fermionic degrees of freedom into the string, we must couple the fermions to the world-sheet in a reparametrization invariant fashion. General coordinate invariance is associated with gravity, hence we need to couple fermions to world-sheet gravity. Let us first discuss the problem of coupling fermions to gravity in an n-dimensional space-time, then return to the special case of the string.

Under a general space-time coordinate transformation, the bosonic fields will transform under a representation of $GL(n, \mathbb{R})$. The difficulty in adding fermions in a reparametrization invariant manner is that there are no spinorial representations of $GL(n, \mathbb{R})$. In the flat space case this is not an issue, because we only require Lorentz invariance and there are spinorial representations of the Lorentz group.

For example, under a change of coordinates $x^\mu \to x'^\mu \equiv x^{\mu'}$, a vector field $v^\mu(x)$ is transformed to $v'^\mu(x')$ via

$$v'^\mu(x') = v^{\mu'} = \frac{\partial x^{\mu'}}{\partial x^\mu} v^\mu = T^{\mu'}{}_\mu \, v^\mu. \tag{26.1}$$

$T^{\mu'}{}_\mu$ falls into the fundamental representation of $GL(n, \mathbb{R})$ and therefore is an $n \times n$ nonsingular matrix with real entries. The bilinear covariant, $\overline{\psi}(x)\gamma^\mu\psi(x)$, should also transform as a space-time vector, hence

$$\overline{\psi}(x')\gamma^{\mu'}\psi(x') = T^{\mu'}{}_\mu \overline{\psi}(x)\gamma^\mu\psi(x). \tag{26.2}$$

Ideally, the spinors would transform under a spinorial representation of $GL(n, \mathbb{R})$ and the transformation of γ^μ would be constructed so that the overall transformation of $\overline{\psi}\gamma^\mu\psi$ would build $T^{\mu'}{}_\mu$ and fall into the correct representation of $GL(n, \mathbb{R})$. Without spinorial representations, we cannot accomplish this in a straightforward manner.

In flat space-time there is no problem. Let us use Latin indices starting from the middle of the alphabet to represent flat space quantities. In flat space, eq.(26.1) becomes

$$v^{m'} = \Lambda^{m'}{}_m \, v^m, \tag{26.3}$$

where $\Lambda^{m'}{}_m$ is an ordinary Lorentz transformation (or its Euclidean equivalent, an $SO(n)$ transformation). Eq.(26.2) becomes

$$\overline{\psi}(x')\gamma^{m'}\psi(x') = \Lambda^{m'}{}_m \overline{\psi}(x)\gamma^m\psi(x). \tag{26.4}$$

Let $S(\Lambda)$ be a spinorial representation of the Lorentz group. Then we take

$$\begin{aligned} \psi(x') &= S\,\psi(x) \\ \overline{\psi}(x') &= \overline{\psi}(x)\,S^{-1}. \end{aligned} \tag{26.5}$$

In order to fulfill eq.(26.4) we must have

$$S^{-1}\gamma^{m'}S = \Lambda^{m'}{}_{m}\gamma^{m}. \tag{26.6}$$

It is possible to find such S and γ^m that satisfy eqs.(26.4)-(26.6), since eq.(26.6) leaves the γ^m algebra, $\{\gamma^m, \gamma^n\} = 2\eta^{mn}$, invariant:

$$\begin{aligned} S^{-1}\{\gamma^{m'}, \gamma^{n'}\}S &= \{S^{-1}\gamma^{m'}S, S^{-1}\gamma^{n'}S\} \\ &= \Lambda^{m'}{}_{m}\Lambda^{n'}{}_{n}\{\gamma^m, \gamma^n\} \\ &= 2\eta^{m'n'}. \end{aligned} \tag{26.7}$$

Without spinorial representations of $GL(n, \mathbb{R})$, we do not have a straightforward way to replace $\Lambda^{m'}{}_{m}$ by $T^{\mu'}{}_{\mu}$, and η^{mn} in eq.(26.7) by $g^{\mu\nu}$. The way out of this difficulty is to use the fact that any space-time is locally flat and borrow the results above. The cost will be the need to introduce another gauge field.

In chapter 21 we discussed the concept of a moving frame field to simplify the computations of the connection coefficients and curvature using exterior calculus. A moving frame field is a choice of an orthonormal basis of tangent vectors, $e^{\mu}_{m}(x)$, for each tangent space $T_x(M)$ of space-time. For n-dimensional space-time, the tangent space at $x \in M$ is a n-dimensional vector space and every basis will contain n vectors. The index m keeps track of which vector in the basis we are using while μ is the ordinary space-time index, i.e. μ labels the components of the mth vector. The orthonormal basis field $e^{\mu}_{m}(x)$ is called a zweibein in 2 dimensions, vierbein in 4 dimensions, and a vielbein in n dimensions.

Orthonormality implies

$$g_{\mu\nu}\, e^{\mu}_{m} e^{\nu}_{n} = \eta_{mn} \tag{26.8}$$

with $\eta_{mn} \to \delta_{mn}$ in the Euclidean version. Applying η^{mn} to both sides of eq.(26.8) we get

$$g_{\mu\nu}(e^{\mu}_{m} e^{\nu}_{n}\eta^{mn}) = \eta^{mn}\eta_{mn} = \delta^{m}_{m} = \delta^{\mu}_{\mu} \tag{26.9}$$

which implies that

$$e^{\mu}_{m}(x)e^{\nu}_{n}(x)\,\eta^{mn} = g^{\mu\nu}(x). \tag{26.10}$$

Associated with each moving vector frame is an orthonormal basis $e^{m}_{\mu}(x)$ of the cotangent space $T^*_x(M)$. These covariant vectors are dual to $e^{\mu}_{m}(x)$, since it follows from eq.(26.8) or eq.(26.10) that

$$e^{m}_{\mu} e^{m}_{\nu} = g^{\mu}_{\nu} = \delta^{\mu}_{\nu}; \quad e^{m}_{\mu} e^{\mu}_{n} = \eta^{m}_{n} = \delta^{m}_{n}. \tag{26.11}$$

The moving frame field is certainly not unique. Any Lorentz transformation (or rotation in the Euclidean version) will produce another orthonormal basis and preserve eq.(26.8).

$$g_{\mu\nu}e^{\mu}_{m'}e^{\nu}_{n'} = g_{\mu\nu}\Lambda^{m}{}_{m'}\Lambda^{n}{}_{n'}\, e^{\mu}_{m}e^{\nu}_{n} = \Lambda^{m}{}_{m'}\Lambda^{n}{}_{n'}\eta_{mn} = \eta_{m'n'} \tag{26.12}$$

Since the m index on e^{μ}_{m} transforms via a Lorentz transformation,

$$e^{\mu}_{m'} = \Lambda^{m}{}_{m'}e^{\mu}_{m}, \tag{26.13}$$

it is also called the Lorentz index or flat space index. Note that this transformation occurs in the space tangent to one point $x \in M$. Under a change of space-time coordinates, the μ index of e^{μ}_{m} still transforms as a space-time vector,

$$e^{\mu'}_{m} = T^{\mu'}{}_{\mu}e^{\mu}_{m}. \tag{26.14}$$

The fact that the two indices of e^{μ}_{m} transform in two different ways, eqs.(26.13) and (26.14), allows us to satisfy eq.(26.2) using eqs.(26.4)-(26.6). To do this define

$$\gamma^{\mu}(x) = e^{\mu}_{m}(x)\,\gamma^{m}. \tag{26.15}$$

γ^{m} are the usual "flat space" Dirac γ matrices satisfying $\{\gamma^{m}, \gamma^{n}\} = 2\eta^{mn}$. Using eqs.(26.10) and (26.15) we find that

$$\{\gamma^{\mu}(x), \gamma^{\nu}(x)\} = 2g^{\mu\nu}(x). \tag{26.16}$$

Plugging eq.(26.15) into $\bar{\psi}\gamma^{\mu}\psi$ and using eqs.(26.4), (26.13), and (26.14), we find that we do indeed satisfy eq.(26.2).

$$\begin{aligned}\bar{\psi}(x')\gamma^{\mu'}(x')\psi(x') &= \bar{\psi}(x')\,e^{\mu'}_{m'}(x')\,\gamma^{m'}\,\psi(x') \\ &= T^{\mu'}{}_{\mu}\,\Lambda^{m'}{}_{m}\,\bar{\psi}(x)e^{\mu}_{m'}\gamma^{m}\psi(x) \\ &= T^{\mu'}{}_{\mu}\,\bar{\psi}\gamma^{\mu}\psi\end{aligned} \tag{26.17}$$

Therefore, to couple fermions in a reparametrization invariant manner, we must use γ^{μ} as defined by eq.(26.15).

However, we are not finished yet. Clearly we are free to perform a different Lorentz transformation on each tangent space and still get the same result. To avoid having the final result depend on the choice of moving frame, we must gauge this local Lorentz transformation. This means we must add another gauge field to accomplish this. This gauge field is called the spin connection.

One way to see that a connection is needed is to examine how the derivative of ψ transforms under $\psi \to S(\Lambda(x))\psi$, eq.(26.5).

$$\begin{aligned} \partial_\mu S(\Lambda)\psi &= S(\Lambda)\partial_\mu\psi + (\partial_\mu S(\Lambda))\psi \\ &= S(\Lambda)(\partial_\mu + S^{-1}\partial_\mu S)\psi \end{aligned} \tag{26.18}$$

The ordinary derivative is not covariant, since $\partial_\mu\psi$ does not transform in the same way as ψ. The problem is the $S^{-1}\partial_\mu S$ term. As in the case for Yang-Mills (see the discussion in chapter 6 starting at eq.(6.12)), we define a covariant derivative D_μ by adding a correction piece involving the spin connection 1-form, ω_μ.

$$D_\mu\psi = (\partial_\mu + \omega_\mu)\,\psi \tag{26.19}$$

$D_\mu\psi$ will transform in the same way as ψ by eliminating the $S^{-1}\partial_\mu S$ piece, provided that ω_μ transforms as

$$\omega_\mu \to S\omega_\mu S^{-1} - (\partial_\mu S)S^{-1}. \tag{26.20}$$

Just as in the Yang-Mills system, the connection 1-form is expanded in terms of the gauge group generators, in this case the spinorial representation of the Lorentz group, $\sigma_{ab} = [\gamma_a, \gamma_b]/2$.

$$\omega_\mu = \frac{1}{4}\omega_\mu^{ab}\sigma_{ab} \tag{26.21}$$

The introduction of moving frames by itself does not change the geometry of the space, and so one can expect that the spin connection is related to the Levi-Cività connection and the Christoffel coefficients discussed in chapter 21. When the spin connection is torsion-free, the spin connection coefficients are related to the Christoffel symbols via

$$\omega_{\mu n}{}^{m} = \Gamma^m_{np}\, e^p_\mu. \tag{26.22}$$

The covariant derivative based on ω_μ can be thought of as correcting the Lorentz indices,

$$D_\mu v^m = \partial_\mu v^m + \omega_{\mu n}{}^{m} v^n, \tag{26.23}$$

just as the Christoffel symbols correct the ordinary space-time indices,

$$D_\mu v^\nu = \partial_\mu v^\nu + \Gamma^\nu_{\mu\alpha} v^\alpha. \tag{26.24}$$

The $\Gamma^\alpha_{\beta\gamma}$'s are compatible with the metric in that $D_\sigma(g_{\mu\nu}) = 0$. The choice eq.(26.22) implies that the frame field is also covariantly constant,

$$D_\mu e^m_\nu = 0 = \partial_\mu e^m_\nu - \Gamma^\alpha_{\mu\nu} e^m_\alpha + \omega_{\mu n}{}^{m} e^n_\nu, \tag{26.25}$$

which is certainly compatible with eq.(26.10).

Eq.(26.25) may be inverted to express the spin connection coefficients in terms of the frame field. The result has already been given in chapter 21 in the form of the first structural equation, eq.(21.24). In the language of forms and exterior calculus, where $e^n = e^n_\mu \, dx^\mu$ and $d = dx^\mu \, \partial_\mu$,

$$de^m = -\omega^m_n \wedge e^n. \tag{26.26}$$

Comparison of eq.(21.26) with eq.(26.22) shows that the connection 1-forms introduced in chapter 21 starting at eq.(21.24) are in fact the spin connection coefficients. (Note that the indices displayed in eqs.(21.24)-(21.34) are Lorentz indices. e^m_μ are the components of the 1-form θ^m of chapter 21. We have switched notation here to stick with the convention that reserves θ for superspace coordinates.)

The second structural equation,

$$\begin{aligned} d\omega^m_n &= -\omega^m_p \wedge \omega^p_n + \Omega^m_n \\ \Omega^m_n &= \frac{1}{2} R^m{}_{npq} \, e^p \wedge e^q, \end{aligned} \tag{26.27}$$

defines the curvature 2-form, Ω^m_n, and Riemann tensor in terms of the e^m_μ and its derivatives.

The curvature 2-form transforms covariantly,

$$\Omega \rightarrow \Lambda^{-1} \Omega^m_n \Lambda. \tag{26.28}$$

Compare this with the color field strength tensor of Yang-Mills, eq.(6.17). The curvature 2-form may also be defined in terms of the commutator of covariant derivatives:

$$(\Omega_{\mu\nu})^m_n = [D_\mu, D_\nu]^m_n. \tag{26.29}$$

So, to add fermions in a reparametrization invariant manner, we use γ^μ as defined by eq.(26.15), and the covariant derivative, eq.(26.19). Imposing additional symmetries, such as supersymmetry, may require using a non-minimal spin connection that has torsion. The generalization of the first structural equation involving torsion is

$$de^m = -\omega^m_n \wedge e^n + T^m, \tag{26.30}$$

and the generalization of eq.(26.29) is

$$[D_\mu, D_\nu] = T^\rho_{\mu\nu} D_\rho + \Omega_{\mu\nu}. \tag{26.31}$$

T^m is the torsion 2-form,

$$T^m = \frac{1}{2} T^m_{np} \, e^n \wedge e^p,$$

or

$$(T_{\mu\nu})^m = \sum_{n<p} T^m_{np}\, e^n_\mu e^p_\nu = D_\mu e^m_\nu - D_\nu e^m_\mu. \tag{26.32}$$

It is clear from eq.(26.32) that a spin connection with torsion can no longer support covariantly constant frame fields, and that the covariant derivative using a spin connection with torsion is different from the covariant derivative defined using the Levi-Cività connection.

Now let us return to the string. We will soon run out of types of labels for indices so let's adopt the following convention for now: Lowercase Latin letters a through l will denote world-sheet coordinate indices, while lowercase Latin letters m through y will be used for flat space or Lorentzian indices. Greek letters starting from the middle of the alphabet will be used for D-dimensional space-time indices, and Greek letters at the beginning of the alphabet for spinorial indices, e.g. the components of a spinor of the matrix elements of a γ matrix.

We will first write down the action for Minkowski signature space-time, then switch to Euclidean space. In flat 2-dimensional Minkowski space, the two γ matrices we will use are

$$\gamma^0 = \begin{pmatrix} 0 & -i \\ i & 0 \end{pmatrix}, \quad \gamma^1 = \begin{pmatrix} 0 & i \\ i & 0 \end{pmatrix}, \tag{26.33}$$

which satisfy $\{\gamma^m, \gamma^n\} = 2\eta^{mn}$. "$\gamma^5$" $= \gamma^0\gamma^1$ and $\sigma_{01} = \gamma_5$. In 2 dimensions there is only one independent spin connection coefficient, so the spin covariant derivative, eq.(26.19), can be expressed as

$$D_a\psi = \partial_a\psi + \frac{1}{2}\omega_a\gamma_5\psi. \tag{26.34}$$

The Dirac operator, $i\partial\!\!\!/$, is completely real, hence we are dealing with Majorana spinors with 2 real independent components.

The complete RNS action is ($T = 1$)

$$S_{\rm NRS}[g,\psi,\chi,x] = -\frac{1}{2}\int d\sigma\, d\tau\, \sqrt{-g}(g^{ab}\partial_a x^\mu \partial_b x_\mu - i\overline{\psi}^\mu\gamma^a\partial_a\psi_\mu)$$
$$-\frac{1}{2}\int d\sigma\, d\tau\, \sqrt{-g}\,\overline{\chi}_a\gamma^b\gamma^a\psi^\mu\partial_b x_\mu - \frac{1}{8}\int d\sigma\, d\tau\, \sqrt{-g}\,\overline{\psi}_\mu\psi^\mu\overline{\chi}_a\gamma^b\gamma^a\chi_b, \tag{26.35}$$

where $\overline{\psi}^\mu = (\psi^\mu)^T\gamma^0$. ψ^μ is a collection of D 2-component Majorana spinors (spinor index suppressed). The collection transforms like a vector in D-dimensional space-time even though the individual components of the spinor are Grassmann variables. ψ^μ is the world-sheet supersymmetric partner of x^μ. χ_b is a world-sheet spin 3/2 gravitino with 1 world-sheet vector index, and one (suppressed) world-sheet spinor index. It is the supersymmetric partner of the metric g_{ab}. The curved space γ matrices are $\gamma^a = e^a_m \gamma^m$, where e^a_m is a world-sheet zweibein defining a moving frame,

$$g^{ab} = e^a_m e^b_m. \tag{26.36}$$

$\sqrt{-g} = \sqrt{|\det g_{ab}|}$ may also be written as $e = \det(e^m_a)$. The use of the frame field insures that eq.(26.35) is reparametrization invariant. However, at first sight it would appear that local Lorentz invariance is not maintained because of the absence of the spin covariant derivative, eq.(26.34), in the kinetic part of the action. Lorentz invariance is maintained, and the reason that $\overline{\psi} i \not\partial \psi$ appears instead of $\overline{\psi} i \not D \psi$ is that for Majorana spinors in 2 dimensions we have

$$\overline{\psi}\gamma^m \gamma_n \gamma_p \psi = 0, \tag{26.37}$$

so that

$$\omega^{np}_m \overline{\psi}\gamma_m \sigma_{np}\psi = 0, \tag{26.38}$$

hence $\overline{\psi} i \not D \psi = \overline{\psi} i \not\partial \psi$. The spin connection, however, will show up in a moment. Finally, note that there is no kinetic term for the gravitino. This is another property of 2 dimensions. Such a term would look like $\overline{\chi}_a \Gamma^{abc} D_b \chi_c$, where Γ^{abc} must be antisymmetric. There is no support for any third-rank antisymmetric tensor in 2 dimensions. The kinetic term for the metric or e^m_a is a topological invariant in 2 dimensions (the Euler characteristic). Since we want the gravitino to be the supersymmetric partner of the metric, it is expected that it too is classically not dynamical.

The local world-sheet supersymmetry is parametrized by an infinitesimal arbitrary Majorana spinor, $\alpha(\sigma, \tau)$. The action eq.(26.35) is invariant on shell up to a total derivative under the local supersymmetry transformation:

$$\begin{aligned} &\delta x^\mu = \overline{\alpha}\psi^\mu, \quad \delta\psi^\mu = -i\gamma^a\,\alpha(\partial_a x^\mu - \frac{1}{2}\overline{\psi}^\mu \chi_a) \\ &\delta\chi_a = 2D_a\alpha, \quad \delta e^m_a = -i\overline{\alpha}\gamma^m\chi_a. \end{aligned} \tag{26.39}$$

The covariant derivative D_a is given by eq.(26.34), except that a nonminimal choice for the spin connection must be made in order to accomplish the symmetry.

$$\omega_a = \omega^{(0)}_a + \frac{i}{2}\overline{\chi}_a \gamma_5 \gamma^b \chi_b \tag{26.40}$$

$\omega^{(0)}_a$ is the minimal spin connection defined via eq.(26.25). The torsion generated by eq.(26.40) (via eq.(26.32)) is

$$T^m_{ab} = \frac{i}{2}\overline{\chi}_a \gamma^m \chi_b. \tag{26.41}$$

In addition to local supersymmetry, the action eq.(26.35) is invariant under the Weyl transformation,

$$\begin{aligned} x^\mu &\rightarrow x^\mu, \quad e_a^m \rightarrow \Lambda e_a^m \\ \psi^\mu &\rightarrow \Lambda^{-1/2}\psi^\mu, \quad \chi_a \rightarrow \Lambda^{1/2}\chi_a, \end{aligned} \tag{26.42}$$

where $\Lambda = \Lambda(\sigma, \tau)$ is an arbitrary function. The action is also invariant under

$$\chi_a \rightarrow \chi_a + \gamma_a \eta, \tag{26.43}$$

where $\eta(\sigma, \tau)$ is an arbitrary Majorana spinor. This symmetry results from the identity

$$\gamma^m \gamma_n \gamma_m = 0, \tag{26.44}$$

which is special to 2 dimensions. The symmetry arising from eqs.(26.42) and (26.43) together means that the action describes a superconformal theory.

In summary, the RNS action possesses world-sheet reparametrization invariance, local Lorentz invariance, local world-sheet supersymmetry, and super Weyl invariance. The action contains no kinetic terms for e_a^m or χ_a, so classically they act like Lagrange multipliers. Their equations of motion generate the conservation of stress-energy and supercurrent.

$$\begin{aligned} -\frac{2}{e}\frac{\delta S_{\text{RNS}}}{\delta e_a^m} &= T_m^a = 0 \\ -\frac{1}{e}\frac{\delta S_{\text{RNS}}}{\delta \chi^a} &= J_a = \frac{1}{2}\gamma^b \gamma_a \psi^\mu \partial_b x_\mu = 0 \end{aligned} \tag{26.45}$$

The coefficients of the Fourier mode expansions of these constraints satisfy the super-Virasoro algebra that is the supersymmetric extension of eq.(20.132).

Now let us turn our attention to the Euclidean version of the RNS action. Rotate time to $\xi^2 = i\tau$. To maintain $\gamma^2\, \partial/\partial\xi^2 = \gamma^0\, \partial/\partial\tau$, we take $\gamma^2 \equiv i\gamma^0$. Using eq.(26.33), this implies that $\gamma^m = i\sigma^m$, and that

$$\{\gamma^m, \gamma^n\} = -2\,\delta^{mn}, \tag{26.46}$$

as required. Here we take $\gamma^5 = i\gamma^1\gamma^2 = \sigma^3$.

There is no exact Euclidean translation of a Majorana spinor. To see this, examine the Dirac equation in flat space, $i\partial\!\!\!/\psi = 0$, using the γ matrices defined above.

$$i\partial\!\!\!/\psi = -2\begin{pmatrix} 0 & \partial_z \\ \partial_{\bar{z}} & 0 \end{pmatrix}\begin{pmatrix} \psi_u \\ \psi_d \end{pmatrix} = 0 \tag{26.47}$$

Once again we are using $z = \xi^1 + i\xi^2$ and $\partial_z = (\partial_1 - i\partial_2)/2$. From eq.(26.47) we can see that the upper and lower components of the spinor satisfy conjugate equations, so the complex conjugate of any lower component can serve as the upper component and vice versa. Thus, $\psi_u^* = \psi_d$. Since the components are complex, there are still 2 independent degrees of freedom.

The Euclidean RNS action is

$$S^{\mathrm{E}}_{\mathrm{RNS}} = \frac{1}{2}\int d^2\xi\,\sqrt{g}\Big(g^{ab}\partial_a x^\mu\,\partial_b x_\mu - i\psi_\mu \partial\!\!\!/ \psi^\mu\Big)$$
$$-\frac{1}{2}\int d^2\xi\,\sqrt{g}\left(\chi_a\gamma^b\gamma^a\psi_\mu\right)\left(\partial_b x^\mu - \frac{1}{4}(\chi_b\psi^\mu)\right). \tag{26.48}$$

Notably absent are Hermitian conjugate spinors involving γ^0. Instead, the spinorial products above can be easily expressed using spinorial component indices. Recall that our convention uses Greek letters at the beginning of the alphabet to denote spinorial indices. If ψ and χ are 2 spinors, then

$$\psi\chi = \psi^\alpha\,\chi_\alpha. \tag{26.49}$$

For reasons that will become apparent when we switch to supersurfaces, we will denote the 2 values of the spinor index α as θ and $\overline{\theta}$. Hence,

$$\psi\chi = \psi^\theta\chi_\theta + \psi^{\overline{\theta}}\chi_{\overline{\theta}}. \tag{26.50}$$

The spinor metric, $g_{\alpha\beta}$, in the $\theta, \overline{\theta}$ basis that we use to raise and lower spinorial indices, is

$$g_{\alpha\beta} = \begin{pmatrix} 0 & 1 \\ -1 & 0 \end{pmatrix} = \delta_{\alpha\beta}. \tag{26.51}$$

Since the components of the spinors are Grassmann variables, we must keep track of their order. Eq.(26.49) defines the ordering used with the metric eq.(26.51). The positive ordering requires the left-hand spinor to have its index up, while the right-hand spinor must have its index down. Thus

$$\psi_\alpha = \psi^\beta\,\delta_{\beta\alpha} = -\delta_{\alpha\beta}\psi^\beta, \tag{26.52}$$

which implies that

$$\psi_{\overline{\theta}} = \psi^\theta \quad \text{and} \quad \psi_\theta = -\psi^{\overline{\theta}}, \tag{26.53}$$

and that

$$\psi\chi = \psi^\alpha\chi_\alpha = -\psi_\alpha\chi^\alpha. \tag{26.54}$$

With the ordering of contracted indices given by eq.(26.49) in mind, the spinor kinetic piece in eq.(26.48) is

$$\psi_\mu \partial\!\!\!/ \psi^\mu = e^a_m\,\psi^\alpha_\mu(\gamma^m)_\alpha{}^\beta\,\partial_a\psi^\mu_\beta. \tag{26.55}$$

Note that $g_{\alpha\beta} = \delta_{\alpha\beta} = \gamma^2 = i\gamma^0$, which explains the disappearance of $\overline{\psi}^\mu = (\psi^\mu)^\dagger\gamma^0$ from the action (26.48).

The action (26.48) simplifies significantly in the superconformal gauge. This is also a convenient gauge in which to introduce superspace and supersurfaces. We shall also use the superconformal gauge in the computation of the genus 0 superstring partition function.

In the superconformal gauge, we take

$$e_a^m = \exp(\phi(\xi)/2)\delta_a^m, \quad \chi^a = \gamma^a\chi, \tag{26.56}$$

where χ is a constant spinor. In this gauge, χ disappears from the gauge-fixed action because of relation eq.(26.44). ϕ disappears from the remaining two terms, leaving

$$S_{\text{sconf}} = \frac{1}{2}\int d^2\xi\left(\delta^{mn}\partial_m x^\mu\partial_n x_\mu - i\psi_\mu\gamma^m\partial_m\psi^\mu\right). \tag{26.57}$$

The remnant of local supersymmetry is global supersymmetry:

$$\delta x^\mu = \varepsilon^\alpha\psi^\mu_\alpha, \quad \delta\psi^\mu_\alpha = -i(\gamma^m)_\alpha{}^\beta\,\varepsilon_\beta\,\partial_m x^\mu, \tag{26.58}$$

where ε is a constant spinor.

In this gauge we may use complex index notation for the world-sheet and Lorentz indices. With $z = \xi^1 + i\xi^2$, $\partial_z = (\partial_1 - i\partial_2)/2$, the metric in the $z, \overline{z}$ basis is

$$g = \frac{1}{2}e^\phi\begin{pmatrix} 0 & 1 \\ 1 & 0 \end{pmatrix}. \tag{26.59}$$

For illustration, let's again examine the spinor kinetic piece of eq. (26.57). In the $z, \overline{z}$ basis, $\not\partial = \gamma^z\partial_z + \gamma^{\overline{z}}\partial_{\overline{z}}$. Following the rule implied by eq.(26.54), the spinorial indices are arranged as

$$\begin{aligned}\psi_\mu\not\partial\psi^\mu &= \psi_\mu(\gamma^z\partial_z + \gamma^{\overline{z}}\partial_{\overline{z}})\psi^\mu \\ &= \psi^\alpha_\mu(\gamma^z)_\alpha{}^\beta\,\partial_z\psi^\mu_\beta + \psi^\alpha_\mu(\gamma^{\overline{z}})_\alpha{}^\beta\,\partial_{\overline{z}}\psi^\mu_\beta.\end{aligned} \tag{26.60}$$

From our choice for the γ's, $\gamma^m = i\sigma^m$, we have

$$\gamma^z = \gamma^1 + i\gamma^2 = \begin{pmatrix} 0 & 2i \\ 0 & 0 \end{pmatrix}, \tag{26.61}$$

hence the only nonzero element is $(\gamma^z)_\theta{}^{\overline{\theta}} = 2i = -(\gamma^z)_{\theta\theta}$. It follows from eq.(26.61) that

$$\gamma^{\overline{z}} = \gamma^1 - i\gamma^2 = \begin{pmatrix} 0 & 0 \\ 2i & 0 \end{pmatrix}, \tag{26.62}$$

so that eq.(26.60) becomes

$$\frac{i}{2}\psi_\mu \partial\!\!\!/ \psi^\mu = \frac{i}{2}\left(\psi^\theta_\mu(\gamma^z)_\theta{}^{\overline{\theta}}\,\partial_z\psi^\mu_{\overline{\theta}} + \psi^{\overline{\theta}}_\mu(\gamma^{\overline{z}})_{\overline{\theta}}{}^{\theta}\,\partial_{\overline{z}}\psi^\mu_\theta\right) \\ = -\psi^\theta_\mu\,\partial_z\psi^\mu_{\overline{\theta}} - \psi^{\overline{\theta}}\,\partial_{\overline{z}}\psi^\mu_\theta. \tag{26.63}$$

When we compare this result with $(i/2)\,\psi_\mu \partial\!\!\!/ \psi^\mu$, using $\partial\!\!\!/ = \gamma^1\partial_1 + \gamma^2\partial_2$ and spinor components as in eq.(26.47), we find that $\psi_\theta = \psi_u$ and $\psi_{\overline{\theta}} = \psi_d$, so that $\psi_\theta = \overline{\psi}_{\overline{\theta}}$, as you would expect from the spinorial-index notation convention.

26.2 SUPERSURFACE ACTION

It is not particularly obvious ahead of time that the action eq.(26.57) is globally world-sheet supersymmetric and that the action eq.(26.48) is locally supersymmetric. The actions can be written on a supersurface in order to make the supersymmetry manifest.

A supersurface is parametrized by the usual two real coordinates, ξ^1 and ξ^2, and two real Grassmann variables, θ^1 and θ^2. The complex versions of these coordinates are $z = \xi^1 + i\xi^2$, $\overline{z} = \xi^1 - i\xi^2$, $\theta = \theta^1 + i\theta^2$, and $\overline{\theta} = \theta^1 - i\theta^2$. These are combined into one supercoordinate, $z^A = (z, \overline{z}, \theta, \overline{\theta})$. (All superspace indices will be denoted with uppercase Latin letters.) A scalar superfield, $B(z, \overline{z}, \theta, \overline{\theta})$, is a function on the supersurface. A flat supersurface will possess the metric

$$g_{MN} = \begin{pmatrix} \delta_{mn} & 0 \\ 0 & \delta_{\alpha\beta} \end{pmatrix}, \tag{26.64}$$

where δ_{mn} is given by eq.(26.59) with $\phi = 0$, and $\delta_{\alpha\beta}$ by eq.(26.51).

The point of introducing superspace is that supersymmetry becomes the geometric transform (i.e. a type of superdiffeomorphism),

$$\delta z = -i\varepsilon^\theta\,\gamma^z_{\theta\theta}\,\theta = -2\,\varepsilon^\theta\,\theta, \quad \delta\overline{z} = -i\varepsilon^{\overline{\theta}}\gamma^{\overline{z}}_{\overline{\theta}\overline{\theta}}\,\overline{\theta} = 2\,\varepsilon^{\overline{\theta}}\,\overline{\theta} \\ \delta\theta = \varepsilon^\theta, \quad \delta\overline{\theta} = \varepsilon^{\overline{\theta}}, \tag{26.65}$$

where ε is a Grassmann constant spinor. Under an infinitesimal supersymmetry transformation, eq.(26.65), the superfield B transforms as

$$\delta B = \delta z\,\partial_z B + \delta\overline{z}\,\partial_{\overline{z}} B + \delta\theta\,\partial_\theta B + \delta\overline{\theta}\,\partial_{\overline{\theta}} B \\ = \varepsilon^\theta Q_\theta B + \varepsilon^{\overline{\theta}} Q_{\overline{\theta}} B, \tag{26.66}$$

where

$$Q_\theta = \partial_\theta - i\theta\gamma^{z}_{\theta\theta}\partial_z = \partial_\theta - 2\theta\partial_z$$
$$Q_{\overline{\theta}} = \partial_{\overline{\theta}} - i\overline{\theta}\gamma^{\overline{z}}_{\overline{\theta}\overline{\theta}}\partial_{\overline{z}} = \partial_{\overline{\theta}} + 2\overline{\theta}\partial_{\overline{z}}. \tag{26.67}$$

Eq.(26.66) can be written as $\delta B = \varepsilon QB$. Supersymmetric quantities are easy to form out of superfields. For example, the quantity

$$I = \int d^2z\, d^2\theta\, B$$

is supersymmetric, because $\delta I = \int d^2z\, d^2\theta\ \varepsilon QB$ vanishes. In all that follows, the ordering and constants in $\int d^2\theta$ are defined so that

$$\int d^2\theta\, \overline{\theta}\, \theta = -1. \tag{26.68}$$

Interesting quantities of the form of I above must contain derivatives. To maintain the invariance, these derivatives must be supercovariant (i.e. anticommute with a supersymmetry transformation). Let us continue this discussion in flat superspace so as to rewrite the action eq.(26.57) as an integral on a supersurface. We will then generalize to curved supersurfaces to rewrite the action eq.(26.48) as an integral on a curved supersurface.

On a flat supersurface, the supercovariant derivative D_θ is easy to construct. One can explicitly verify that

$$D_\theta = \partial_\theta + 2\theta\partial_z, \quad D_{\overline{\theta}} = \partial_{\overline{\theta}} - 2\overline{\theta}\partial_{\overline{z}} \tag{26.69}$$

satisfies

$$\{D_\theta, Q_\theta\} = \{D_{\overline{\theta}}, Q_{\overline{\theta}}\} = \{D_\theta, Q_{\overline{\theta}}\} = 0, \tag{26.70}$$

which fits the bill. Therefore, the quantity

$$I = \int d^2z\, d^2\theta\, D_{\overline{\theta}}B\, D_\theta B \tag{26.71}$$

is supersymmetric.

To connect with the component formalism discussed in the previous section, the superfields are written as expansions in the anticommuting coordinates. Due to the Grassmann nature of these coordinates, the expansions are always finite in length. For example, we can expand the superfield B as

$$B(z,\overline{z},\theta,\overline{\theta}) = b(z,\overline{z}) + \theta\zeta_\theta(z,\overline{z}) + \overline{\theta}\zeta_{\overline{\theta}}(z,\overline{z}) + \overline{\theta}\theta f(z,\overline{z}). \tag{26.72}$$

This kind of expansion provides the motivation for using θ and $\overline{\theta}$ as complex spinorial indices. In component form, the behavior of B under a supersymmetry transformation is given by

$$\begin{aligned}\delta B &= \delta b + \theta\,\delta\zeta_\theta + \overline{\theta}\,\delta\zeta_{\overline{\theta}} + \overline{\theta}\theta\,\delta f;\\ \varepsilon QB &= \varepsilon(\zeta_\theta - \overline{\theta} f) - 2\varepsilon\theta(\partial_z b + \overline{\theta}\partial_z\zeta_{\overline{\theta}})\\ &\quad + \overline{\varepsilon}(\zeta_{\overline{\theta}} + \theta f) + 2\overline{\varepsilon}\overline{\theta}(\partial_{\overline{z}} b + \theta\partial_{\overline{z}}\zeta_\theta),\end{aligned} \tag{26.73}$$

from which we find ($\varepsilon = \varepsilon^\theta, \overline{\varepsilon} = \varepsilon^{\overline{\theta}}$)

$$\begin{aligned}\delta b &= \varepsilon\zeta_\theta + \overline{\varepsilon}\zeta_{\overline{\theta}}\\ \delta\zeta_\theta &= 2\varepsilon\partial_z b - \overline{\varepsilon} f\\ \delta f &= 2\varepsilon\partial_z\zeta_{\overline{\theta}} + 2\overline{\varepsilon}\partial_{\overline{z}}\zeta_\theta.\end{aligned} \tag{26.74}$$

Now let us return to the action (26.57). Define the collection of D superfields, X^μ ($\mu = 1,\ldots,D$), as

$$X^\mu = x^\mu + \theta\psi^\mu_\theta + \overline{\theta}\psi^\mu_{\overline{\theta}} + \overline{\theta}\theta f^\mu. \tag{26.75}$$

Then the action

$$S = \frac{1}{2}\int d^2z\, d^2\theta\, D_{\overline{\theta}}X^\mu\, D_\theta X_\mu \tag{26.76}$$

is equal to eq.(26.57) when $f^\mu = 0$. Since $f^\mu = 0$ is the trivial equation of motion resulting from eq.(26.76), the two actions are equivalent. The only difference is that the supersymmetry transformation resulting from eq.(26.67) and eq.(26.75) closes off shell or without the need to invoke the equations of motion. The details of the comparison of eq.(26.76) with eq.(26.57) are left as exercises.

The superspace formulation of the full locally supersymmetric RNS action, eq.(26.48), requires curved supersurfaces. Let's first discuss some aspects of the geometry of such supersurfaces, since we will need these in the Faddeev-Popov reduction of the superstring path integral.

The supersurface extension of the ordinary zweibein is the superzweibein E^M_A. Since the indices A and M each range over 4 values, the superzweibein consists of 16 superfield components. The A index transforms as an ordinary supersurface coordinate under the superdiffeomorphism group, while the M index is a flat space, tangent-space, or Lorentzian index. For an ordinary surface, the Euclidean Lorentz group acting on the tangent space is just $O(2) = U(1)$. Vectors v^m infinitesimally transform according to $\delta v^{m'} = \lambda\epsilon^{m'}_m v^m$, while spinors ψ^α infinitesimally transform as $\delta\psi^{\alpha'} = \lambda\psi^\alpha\frac{1}{2}(\gamma_5)_\alpha{}^{\alpha'}$ (which arises from $\sigma_{01} = \gamma_5$). Together these imply that a generator of a supersurface Euclidean Lorentz tangent space group transformation can be written as

$$L_M^N = L\,\mathcal{E}_M^N$$

$$\mathcal{E}_m^n = \epsilon_m^n, \quad \mathcal{E}_\alpha^\beta = \frac{1}{2}(\gamma_5)_\alpha{}^\beta, \quad \mathcal{E}_m^\beta = \mathcal{E}_\alpha^n = 0, \tag{26.77}$$

where L is a scalar superfield. The effect of an infinitesimal tangent space transformation on the superzweibein is

$$\delta E_A^N = E_A^M\, L_M^N. \tag{26.78}$$

In correspondence with the case of ordinary Riemann surfaces, the superzweibein E_A^M and superconnection 1-form, $W_N^M = dz^A\,(W_A)_N^M$ (along with constraints), specify the geometry. In view of eq.(26.77), we may write $(W_A)_N^M = W_A\,\mathcal{E}_N^M$. From the superconnection we define the supercovariant derivative, $\mathcal{D}_A = \partial_A + W_A$. The action on supercovariant and supercontravariant vectors is

$$\begin{aligned} \mathcal{D}_A V_N &= \partial_A V_N + (W_A)_N^M V_M \\ \mathcal{D}_A V^M &= \partial_A V^M - (-)^{an} V^N (W_A)_N^M. \end{aligned} \tag{26.79}$$

The symbol $(-)^{an}$ is equal to -1 if both a and n are odd or anticommuting indices and is 1 otherwise (e.g. $(-)^{\theta\theta} = -1$, $(-)^{z\theta} = 1$). To arrive at eq.(26.79) we start from the invariant expression involving differential forms. Let $\mathcal{D} = dz^A\,\mathcal{D}_A$. Then

$$\begin{aligned} \mathcal{D}V^M &= dV^M - V^N (W)_N^M \\ &= dz^A\,\partial_A V^M - V^N\,dz^A (W_A)_N^M \\ &= dz^A\Big(\partial_A V^M - (-)^{an} V^N (W_A)_N^M\Big). \end{aligned}$$

The super torsion and super curvature are given by the generalizations of the first and second structural equations, eqs.(26.30) and (26.31). The super first structural equation is

$$\mathcal{D}E^M = dE^M - E^N \wedge W_N^M = T^M \tag{26.80}$$

$$T^M = \frac{1}{2}\,E^N \wedge E^P\,T_{NP}^M,$$

where $E^M = dz^A\,E_A^M$. T^M is the super torsion 2-form. Expanding this expression in the basis $dz^A \wedge dz^B$ we find

$$\mathcal{D}_A E_B^M = \frac{1}{2}\,(-)^{(a+n)b+ab}\,E_A^N\,E_B^P T_{NP}^M = \frac{1}{2}\,(-)^{nb}\,E_A^N\,E_B^P T_{NP}^M. \tag{26.81}$$

$(-)^{(a+n)b}$ is equal to -1 when index b is odd and the combination of a and n is odd.

The inverse superzweibein E^A_M satisfies $E^A_M E^N_A = \delta^M_N$. Using the inverse superzweibein, we can express the second structural equation as

$$[\mathcal{D}_M, \mathcal{D}_N\} = T^P_{MN}\mathcal{D}_P + R_{MN}, \qquad (26.82)$$

where $\mathcal{D}_M = E^A_M \mathcal{D}_A$, etc. $[,\}$ is the standard notation for a graded commutator. $[,\}$ is a commutator except when M and N are spinor indices, in which case it is an anticommutator (i.e. $[\mathcal{D}_M, \mathcal{D}_N\} = \mathcal{D}_M\mathcal{D}_N - (-)^{mn}\mathcal{D}_N\mathcal{D}_M$).

To complete the specification of the supergeometry, we need to impose constraints on the superzweibein E^M_A. The superzweibein is composed of 16 superfields. The gauge-fixed action eq.(26.71) indicates that only 1 superfield is necessary to describe a conformally flat supersurface. Four superfields are necessary to describe superdiffeomorphic gauge degrees of freedom (2 for ordinary diffeomorphisms and 2 for local supersymmetry transformations) and 1 superfield is needed to parametrize tangent space Lorentz transformations (i.e. rotations). The remaining 10 superfield degrees of freedom of E^M_A are redundant and the constraints are there to remove them. The constraints must be imposed in a covariant manner. Following Howe, we place the following constraints on the torsion:

$$T^m_{\alpha\beta} = 2i(\gamma^m)_{\alpha\beta}, \quad T^\alpha_{\beta\gamma} = T^m_{np} = 0. \qquad (26.83)$$

These constraints are sufficient to express the superconnection coefficients in terms of the superzweibein and to express the curvature and the remaining torsion components in terms of a single superfield.

The supersurface version of the Euclidean RNS action (26.48) is

$$S^{\mathrm{E}}_{\mathrm{RNS}} = \frac{1}{4}\int d^2z\, d^2\theta\, E\, \mathcal{D}^\alpha_{(0)} X^\mu \mathcal{D}^{(0)}_\alpha X_\mu = \frac{1}{2}\int d^2z\, d^2\theta\, E\, \mathcal{D}^{(0)}_{\bar\theta} X^\mu \mathcal{D}^{(0)}_\theta X_\mu. \qquad (26.84)$$

X^μ is the same collection of superfields as in the action eq.(26.76). Since it is a collection of superfield scalars, $\mathcal{D}^{(0)}_\alpha X^\mu = E^A_\alpha \partial_A X^\mu$ (i.e. the notation (0) on the supercovariant derivative indicates that it is acting on a scalar or tensor of weight 0). The superzweibein must satisfy eq.(26.83) and $E = \mathrm{sdet}(E^M_A)$.

The superdeterminant, sdet, of a supermatrix S^M_N can be defined using the supertrace,

$$\mathrm{sdet}(S) = \exp(\mathrm{str}(\ln\, S)), \qquad (26.85)$$

where the supertrace is taken to be

$$\mathrm{str}(S) = S^m_m - S^\alpha_\alpha. \qquad (26.86)$$

One can show, using this definition, that $\mathrm{str}(S_1S_2) = \mathrm{str}(S_2S_1)$, as would be expected from an ordinary trace. In conjunction with eq.(26.85), this property preserves the determinant product relationship, $\mathrm{sdet}(S_1S_2) = \mathrm{sdet}(S_1)\,\mathrm{sdet}(S_2)$. Alternatively, we can define the superdeterminant as the result of a superspace Gaussian functional integral:

$$(\mathrm{sdet}(S))^{-1} = \int d^2z\, d^2z'\, d^2\theta\, d^2\theta' \exp\left(-z'^M S_M^N z_N\right). \tag{26.87}$$

The exponent in eq.(26.87) is

$$z'^M S_M{}^N z_N = z'^m S_m{}^n z_n + z'^m S_m{}^\beta \theta_\beta + \theta'^\alpha S_\alpha{}^n z_n + \theta'^\alpha S_\alpha^\beta \theta_\beta.$$

Completing the squares and shifting variables we find

$$\begin{aligned} z'^M S_M{}^N z_N &= z'^m S_m{}^n z_n + \theta'^\alpha\Big(s_\alpha{}^\beta - S_\alpha{}^n (S^{-1})_n{}^m S_m{}^\beta\Big)\theta_\beta \\ &= z'^m\Big(S_m{}^n - S_n{}^\beta (S^{-1})_\beta{}^\alpha S_\alpha{}^n\Big) z_n + \theta'^\alpha S_\alpha{}^\beta \theta_\beta. \end{aligned} \tag{26.88}$$

Hence

$$\begin{aligned} \mathrm{sdet}(S) &= \det(S_m{}^n)\det{}^{-1}\Big(S_\alpha{}^\beta - S_\alpha{}^n (S^{-1})_n{}^m S_m{}^\beta\Big) \\ &= \det\Big(S_m{}^n - S_n{}^\beta (S^{-1})_\beta{}^\alpha S_\alpha{}^n\Big)\det{}^{-1}(S_\alpha{}^\beta). \end{aligned} \tag{26.89}$$

The two definitions, eqs.(26.85) and (26.87), agree, of course.

To compare the action (26.84) with the component version, eq.(26.48), the superfields are written as expansions in the anticommuting coordinates as before. The expansion for X^μ is exactly the same as before, eq.(26.75). The expansion of the superzweibein starts as

$$\begin{aligned} E_a^m &= e_a^m - i\theta\gamma_{\theta\alpha}^m \chi_a^\alpha - i\overline{\theta}\gamma_{\overline{\theta}\alpha}^m \chi_a^\alpha + \dots \\ E_a^\alpha &= \frac{1}{2}\chi_a^\alpha + \dots \\ E_\beta^m &= -i\theta\gamma_{\theta\beta}^m - i\overline{\theta}\gamma_{\overline{\theta}\beta}^m \\ E_\beta^\alpha &= \delta_\beta^\alpha + \dots . \end{aligned} \tag{26.90}$$

Since we will not need these general expressions, we will not dwell on the details. For reference, the flat superzweibein corresponding to flat superspace is

$$\widehat{E}_a^m + \delta_a^m, \quad \widehat{E}_a^\alpha = 0, \quad \widehat{E}_\beta^m = -i\theta\gamma_{\theta\beta}^m - i\overline{\theta}\gamma_{\overline{\theta}\beta}^m, \quad \widehat{E}_\beta^\alpha = \delta_\beta^\alpha. \tag{26.91}$$

From eq.(26.91) it is easy to check, using the inverse superzweibein, that $\mathcal{D}_\theta = D_\theta$, eq.(26.69), and that the constraints eq.(26.83) are satisfied. Substitution of $\widehat{E}_M^A$ into the action eq.(26.84) reduces it to eq.(26.76).

Application of the constraints (26.83) reduces the number of independent superfields of the superzweibein E_A^M from 16 to 6. The remainder can be fixed by super reparametrizations (4 superfields), local Lorentz rotations (1 superfield), and what should amount to a superconformal transformation (1 superfield). Under a superdiffeomorphism $z^A \rightarrow z^{A'}$ (ordinary diffeomorphisms and local supersymmetry transformations), the superzweibein transforms as

$$E_A^M(z) = \frac{\partial z^{A'}}{\partial z^A} E_{A'}^M(z'). \tag{26.92}$$

The transformation under a infinitesimal local Lorentz rotation is given by eq.(26.77). The generalization of the super Weyl transformation to the superzweibein is not as straightforward, due to the fact that super Weyl transformations must preserve the constraints (26.83). The straightforward generalization from the purely bosonic case is to take

$$\widetilde{E}_A^m = \Lambda E_A^m, \quad \widetilde{E}_A^\alpha = \Lambda^{\frac{1}{2}} E_A^\alpha, \tag{26.93}$$

which implies that

$$\widetilde{E}_m^A = \Lambda^{-1} E_m^A, \quad \widetilde{E}_\alpha^A = \Lambda^{-\frac{1}{2}} E_\alpha^A, \tag{26.94}$$

and that

$$\begin{aligned} \widetilde{\mathcal{D}}_\alpha^{(0)} &= \widetilde{E}_\alpha^A \partial_A + \Lambda^{-\frac{1}{2}} E_\alpha^A \partial_A = \Lambda^{-\frac{1}{2}} \mathcal{D}_\alpha^{(0)} \\ \widetilde{\mathcal{D}}_m^{(0)} &= \Lambda^{-1} \mathcal{D}_m^{(0)}. \end{aligned} \tag{26.95}$$

$\mathcal{D}_\alpha^{(0)}$ satisfies the constraint (26.83),

$$\{\mathcal{D}_\alpha^{(0)}, \mathcal{D}_\beta^{(0)}\} = 2i\, \gamma_{\alpha\beta}^m \mathcal{D}_m^{(0)}, \tag{26.96}$$

but $\widetilde{\mathcal{D}}_\alpha^{(0)}$ as defined above does not.

$$\begin{aligned} &\{\widetilde{\mathcal{D}}_\alpha^{(0)}, \widetilde{\mathcal{D}}_\beta^{(0)}\} \\ &\quad = \Lambda^{-1}\{\mathcal{D}_\alpha^{(0)}, \mathcal{D}_\beta^{(0)}\} + \Lambda^{-\frac{1}{2}}(\mathcal{D}_\alpha^{(0)}\Lambda^{-\frac{1}{2}})\mathcal{D}_\beta^{(0)} + \Lambda^{-\frac{1}{2}}(\mathcal{D}_\beta^{(0)}\Lambda^{-\frac{1}{2}})\mathcal{D}_\alpha^{(0)} \\ &\quad \neq 2i\gamma_{\alpha\beta}^m \widetilde{\mathcal{D}}_m^{(0)} \end{aligned} \tag{26.97}$$

It will satisfy the constraint if we redefine $\widetilde{\mathcal{D}}_m^{(0)}$ to absorb the extra terms:

$$\widetilde{\mathcal{D}}_m^{(0)} = \Lambda^{-1}\mathcal{D}_m^{(0)} + \frac{i}{4}\Lambda^{-2}(\gamma_m)^{\alpha\beta}\mathcal{D}_\beta^{(0)}\Lambda\mathcal{D}_\alpha^{(0)}. \tag{26.98}$$

Eq.(26.98) implies that the correct generalization of the super Weyl transform on the inverse superzweibein is

$$\begin{aligned}\widetilde{E}_m^A &= \Lambda^{-1}E_m^A + \frac{i}{4}\Lambda^{-2}(\gamma_m)^{\alpha\beta}\mathcal{D}_\beta^{(0)}\Lambda E_\alpha^A \\ \widetilde{E}_\alpha^A &= \Lambda^{-\frac{1}{2}}E_\alpha^A,\end{aligned} \tag{26.99}$$

and, therefore, on the superzweibein is

$$\begin{aligned}\widetilde{E}_A^m &= \Lambda E_A^m \\ \widetilde{E}_A^\alpha &= \Lambda^{\frac{1}{2}}E_A^\alpha - \frac{i}{4}\Lambda^{-\frac{1}{2}}E_A^m(\gamma_m)^{\alpha\beta}\mathcal{D}_\beta^{(0)}\Lambda.\end{aligned} \tag{26.100}$$

By analogy with the case of an ordinary surface, the superconformally flat gauge is defined as when the superzweibein is superconformally equivalent to the flat space superzweibein. From eq.(26.100) we have that $\widetilde{E}_A^M$ is in the superconformally flat gauge if

$$\begin{aligned}\widetilde{E}_A^m &= \Lambda\widehat{E}_A^m \\ \widetilde{E}_A^\alpha &= \Lambda^{\frac{1}{2}}\widehat{E}_A^\alpha - \frac{i}{4}\Lambda^{-\frac{1}{2}}\widehat{E}_A^m(\Gamma_m)^{\alpha\beta}D_\beta\Lambda,\end{aligned} \tag{26.101}$$

where $\widehat{E}_A^m$ is the flat space superzweibein, eq.(26.91), and D_β is the flat space supercovariant derivative, eq.(26.69). In this gauge we have

$$\begin{aligned}\widetilde{\mathcal{D}}_z^{(0)} &= \widetilde{E}_z^A\partial_A = \Lambda^{-1}\partial_z - \frac{1}{4}\Lambda^{-2}(D_\theta)D_\theta \\ \widetilde{\mathcal{D}}_\theta^{(0)} &= \widetilde{E}_\theta^A\partial_A = \Lambda^{-\frac{1}{2}}D_\theta,\end{aligned} \tag{26.102}$$

and the connection coefficients are

$$W_A = \Lambda^{-1}\widehat{E}_A^m\varepsilon_m^n(D_n\Lambda) + \Lambda^{-1}\widehat{E}_A^\beta(\gamma_5)_\beta{}^\gamma(D_\gamma\Lambda). \tag{26.103}$$

In the next section we will discuss the evaluation of the genus 0 contribution to the superstring partition function. In correspondence with the bosonic string, we will carry out the Faddeev-Popov procedure about the superconformally flat gauge slice. In this process we will encounter the need for explicit forms of the covariant derivatives on tensor objects of various rank along with the corresponding Laplacians. Since these are the "super" generalizations of the objects we met in the bosonic case in chapter 22, let us begin by considering only ordinary indices associated with the commuting coordinates (ξ^1, ξ^2) or $(z, \bar{z})$.

The tangent space or Euclidean Lorentz transformation for a surface is just a simple rotation of the plane. A finite rotation by ϑ on a vector V^m results in $V^{m'} = V^m \Lambda_m{}^{m'}(\vartheta)$, or as usual,

$$\begin{pmatrix} V^{1'} & V^{2'} \end{pmatrix} = \begin{pmatrix} V^1 & V^2 \end{pmatrix} \begin{pmatrix} \cos(\vartheta) & \sin(\vartheta) \\ -\sin(\vartheta) & \cos(\vartheta) \end{pmatrix}. \tag{26.104}$$

$d\Lambda_m{}^{m'}(\vartheta)/d\vartheta$ at $\vartheta = 0$ is $\varepsilon_m{}^{m'}$, as stated in eq.(26.77) ($\varepsilon_{12} = 1$). Switching to complex index notation, we find from eq.(26.104) that

$$V^z \to e^{-i\vartheta} V^z, \quad V_z \to e^{i\vartheta} V_z. \tag{26.105}$$

Recall that $V_z = g_{z\bar{z}} v^{\bar{z}}$, hence all upper $\bar{z}$ indices are equivalent to lower z indices and all lower $\bar{z}$ indices are equivalent to upper z indices. Since

$$V^{z'} = \left(\frac{\partial z'}{\partial z}\right) V^z, \quad V_{z'} = \left(\frac{\partial z}{\partial z'}\right) V_z, \tag{26.106}$$

we learn from eq.(26.105) that

$$T^{z'z'} = \left(\frac{\partial z'}{\partial z}\right)\left(\frac{\partial z'}{\partial z}\right) T^{zz} = e^{-2i\vartheta} T^{zz}, \quad T_{zz} = e^{2i\vartheta} T_{zz}, \tag{26.107}$$

and that an object with p upper z indices and q lower z indices will transform as

$$T^{z'\cdots z'}_{z'\cdots z'} = \exp(i(q-p)\vartheta) T^{z\cdots z}_{z\cdots z}, \tag{26.108}$$

so that V^z and T^{zz}_z transform in the same way.

Let n be the number of lower z indices minus the number of upper z indices. All tensors with the same n transform in the same way, and we collect them together into a set denoted as T^n. The covariant derivative acting on T^n will be denoted $\mathcal{D}^{(n)}_A$. If $T \in T^n$, then

$$\mathcal{D}^{(n)}_A T = \partial_A T - n W_A T. \tag{26.109}$$

When the index A takes on the value z, covariant differentiation adds one lower z index and thereby increases n by 1. Thus, $\mathcal{D}^{(n)}_z T \in T^{n+1}$. Similarly, when the index A is $\bar{z}$, we are adding one lower $\bar{z}$ index, which is equivalent to adding an upper z index, so n decreases by 1 and $\mathcal{D}^{(n)}_{\bar{z}} T \in T^{n-1}$.

We can now form 2 Laplacians: $\Delta_n^{(+)} = \mathcal{D}_{\bar{z}}^{(n+1)}\mathcal{D}_z^{(n)}$, and $\Delta_n^{(-)} = \mathcal{D}_z^{(n-1)}\mathcal{D}_{\bar{z}}^{(n)}$. Neither, of course, changes the weight n of the tensor it is applied to. Returning to the expression for $P_1^\dagger P_1$ for the bosonic case, eq.(22.104), the upper component acts on $v^z \in \mathcal{T}^{-1}$. Since $\partial_{\bar{z}}$ appears to the right of ∂_z, the Laplacian involved is $\Delta_{-1}^{(-)}$. Similarly, the lower diagonal component of $P_1^\dagger P_1$ acts on $v_z \in \mathcal{T}^1$. ∂_z appears to the right of $\partial_{\bar{z}}$, hence the Laplacian is $\Delta_1^{(+)}$. Finally, we note that in the conformally flat gauge, the Laplace-Beltrami operator, Δ_g, eq.(22.26), is $\Delta_g = \Delta_0^{(+)} = \Delta_0^{(-)}$.

Now consider the super generalization of these results by adding the spinorial indices back in again. From the fact that the only nonzero element of the matrix γ^z is the $\theta\theta$ component, $\gamma^z_{\theta\theta} \neq 0$, we can conclude that 2 lower θ indices are worth 1 lower z index. Similarly, $\gamma^{\bar{z}}_{\bar{\theta}\bar{\theta}} \neq 0$ implies that 1 upper z index is worth 2 upper θ indices. This is, of course, consistent with the fact that it takes a rotation of 720° to return a spinor to its original state. Hence, under a tangent space rotation by angle ϑ the spinor ψ^α transforms as

$$\psi^\theta \to e^{-i\vartheta/2}\,\psi^\theta, \quad \psi_\theta \to e^{i\vartheta/2}\psi_\theta. \tag{26.110}$$

The γ matrices allow us to convert z indices into θ indices; hence we can write any tensor in terms of upper and lower θ indices only. Again we define n to be the difference between the number of lower θ indices and the number of upper indices divided by 2, and group objects with identical n into collections $\mathcal{T}^n$. For example, ψ^θ and $\chi_\theta^{\theta\theta}$ are both in $\mathcal{T}^{-1/2}$. The covariant derivative $\mathcal{D}_\theta^{(n)}$ adds one lower θ index and thereby increases n by 1/2, while $\mathcal{D}_{\bar{\theta}}^{(n)}$ adds one upper θ index and decreases n by 1/2. Again we can form 2 Laplacians for each n: $\Delta_n^{(-)} = \mathcal{D}_\theta^{(n-1/2)}\mathcal{D}_{\bar{\theta}}^{(n)}$, and $\Delta_n^{(+)} = \mathcal{D}_{\bar{\theta}}^{(n+1/2)}\mathcal{D}_\theta^{(n)}$.

To connect the superconformally flat gauge to the conformally flat gauge used in chapter 22, take $\Lambda = \exp(\Phi/2)$ in eqs.(26.101)-(26.103). We then have

$$\begin{aligned} \widetilde{\mathcal{D}}_z^{(0)} &= \exp(-\Phi/2)\left(\partial_z - \frac{1}{8}(D_\theta\Phi)D_\theta\right) \\ \widetilde{\mathcal{D}}_\theta^{(0)} &= \exp(-\Phi/4)\,D_\theta, \end{aligned} \tag{26.111}$$

and

$$W_\theta = \widetilde{E}_\theta^A W_A = \frac{1}{2}\exp(-\Phi/4)\,D_\theta\Phi, \quad W_{\bar{\theta}} = -\frac{1}{2}\exp(-\Phi/4)\,D_{\bar{\theta}}\Phi, \tag{26.112}$$

which implies that

$$\begin{aligned}
\widetilde{\mathcal{D}}_{\theta}^{(n)} &= \exp(-\Phi/4)\Big(D_\theta - \frac{n}{2} D_\theta \Phi\Big) \\
&\qquad\qquad = \exp\Big((n-1/2)\Phi/2\Big) D_\theta \exp(-n\Phi/2) \\
\widetilde{\mathcal{D}}_{\bar\theta}^{(n)} &= \exp(-\Phi/4)\Big(D_{\bar\theta} + \frac{n}{2} D_{\bar\theta} \Phi\Big) \\
&\qquad\qquad = \exp\Big(-(n+1/2)\Phi/2\Big) D_{\bar\theta} \exp(n\Phi/2).
\end{aligned} \tag{26.113}$$

In this gauge the Laplacians are

$$\begin{aligned}
\Delta_0 = \Delta_0^{(-)} &= -\Delta_0^{(+)} = \exp(-\Phi/2)\, D_\theta D_{\bar\theta}, \\
\Delta_{-1}^{(-)} &= \exp(-\Phi)\, D_\theta \exp(\Phi)\, D_{\bar\theta} \exp(-\Phi/2), \\
\Delta_1^{(+)} &= \exp(-\Phi)\, D_{\bar\theta} \exp(\Phi)\, D_\theta \exp(-\Phi/2),
\end{aligned} \tag{26.114}$$

and so on. The super Laplacians above are those that appear in the genus 0 contribution to the superstring partition function, to which we now turn.

26.3 SUM OVER GENUS 0 SUPERSURFACES

The sourceless RNS superstring path integral or partition function defined as

$$Z_{\mathrm{RNS}} = \mathcal{N} \int \mathcal{D}E_A^M\, \mathcal{D}X^\mu \exp\left(-S_{\mathrm{RNS}}^{\mathrm{E}}[E_A^M, X^\mu]\right), \tag{26.115}$$

where the action is given by eq.(26.84), can be thought of as a sum over random supersurfaces by analogy with the bosonic string partition function considered in chapters 22 and 23. To the action eq.(26.84) we can add a term consisting of the integral of the scalar supercurvature over the supersurface, since the dimension of the coupling of this counterterm is consistent with renormalizability. As in the purely bosonic case, this term is a topological invariant equal to the "super" Euler number, which just reduces to the ordinary Euler number. Thus the partition function can be evaluated as a perturbative series consisting of terms which sum over supersurfaces of fixed genus. We shall consider the genus 0 contribution to the partition function,

$$Z_0 = \mathcal{N} \int_{\text{genus } 0} \mathcal{D}E_A^M\, \mathcal{D}X^\mu \exp\left(-S_{\mathrm{RNS}}^{\mathrm{E}}[E_A^M, X^\mu]\right). \tag{26.116}$$

In correspondence with the bosonic string, the computation of the functional integral amounts to the evaluation of several superdeterminants.

The superdeterminants are not super Weyl invariant and their dependence on the superconformal degree of freedom of the superzweibein leads to the superconformal anomaly. The numerology of the determinants leads to a cancellation of the anomaly and the restoration of superconformal symmetry in the critical space-time dimension of $D = 10$.

The measures in eq.(26.116) are implicitly defined by Gaussian integration by the requirement,

$$\begin{aligned}\int \mathcal{D}X^\mu \exp\left(-\langle\delta X^\mu, \delta X^\mu\rangle/2\right) &= 1\\ \int \mathcal{D}(\delta E_A^M) \exp\left(-\langle\delta E_A^M, \delta E_A^M\rangle/2\right) &= 1,\end{aligned} \tag{26.117}$$

which, therefore, ultimately rests on how the metrics $\langle,\rangle$ are defined for δX^μ and δE_A^M. For δX^μ we take

$$\langle\delta X^\mu, \delta X^\mu\rangle = \int d^2z\, d^2\theta\, E\, \delta X^\mu \delta X_\mu, \tag{26.118}$$

where $E = \text{sdet}(E_A^M)$. Once again we must treat the zero mode, $\overline{X}^\mu =$ constant, separately. Let $X^\mu = \overline{X}^\mu + X'^\mu$. Noting that $\langle\delta\overline{X}^\mu, \delta\overline{X}^\mu\rangle \propto \int d^2z\, d^2\theta\, E$ for $\delta\overline{X}^\mu =$ constant, and then integrating the action eq.(26.84) by parts, we find

$$\begin{aligned}&\int \mathcal{D}X^\mu \exp\left(-S[X^\mu, E_A^M]\right)\\ &= \int \mathcal{D}\overline{X}^\mu \mathcal{D}X'^\mu \exp\left(-\frac{1}{2}\int d^2z\, d^2\theta\, E X^\mu \Delta_0 X_\mu\right)\\ &= \mathcal{V}\left(\frac{2\pi}{\int d^2z\, d^2\theta\, E}\right)^{-D/2} \left(\text{sdet}'(-\Delta_0)\right)^{-D/2},\end{aligned} \tag{26.119}$$

where $\Delta_0 = -E^{-1}E_\theta^A \partial_A(E\, E_\theta^B)\partial_B$ and $\mathcal{V}$ is the volume of D-dimensional space-time. The extra minus sign in the determinant in eq.(26.119) arises from the convention adopted in eq.(26.68). Thus, eq.(26.116) becomes

$$Z_0 = \mathcal{N}\int_{\text{genus } 0} \mathcal{D}E_A^M \left(\frac{2\pi}{\int d^2z\, d^2\theta\, E}\right)^{-D/2} \left(\text{sdet}'(-\Delta_0)\right)^{-D/2}, \tag{26.120}$$

where $\mathcal{V}$ has been absorbed into $\mathcal{N}$.

To complete the computation of Z_0 we must do two things. First we must define $\mathcal{D}E_A^M$, which as we have already mentioned, means that we must choose a metric for the vectors δE_A^M. Second, we must fix the gauge dependent components of E_A^M by choosing a gauge, then carry out the Faddeev-Popov procedure. This amounts to changing variables from the gauge dependent components of E_A^M to the parameters of the gauge transformations. The requisite Jacobian of the transformation or Faddeev-Popov determinant is found by re-expressing the norm of δE_A^M in terms of the infinitesimal parameters of the gauge transformation and then using eq.(26.117).

The superzweibein E_A^M consists of 16 superfields. Ten of these are eliminated by the torsion constraints. The choice of metric to specify $\langle \delta E_A^M, \delta E_A^M \rangle$ and ultimately define $\mathcal{D}E_A^M$ must reflect the fact that the superzweibein is torsion-constrained. One way to accomplish this is to build a metric for δE_A^M based only on the unconstrained components of E_A^M. The metric applies to tangent space tensors, so we want to construct a quantity that involves only tangent space or Lorentz indices. δE_A^M involves both supersurface or superworld-sheet indices and tangent space indices, hence the infinitesimal variation of the superzweibein we want to work with is $H_N^M = E_N^A \, \delta E_A^M$. After a little investigation, one finds that the independent quantities can be chosen as $H_\alpha{}^m$, $H_\alpha{}^\alpha$, and $(\gamma_5)_\alpha{}^\beta H_\beta{}^\alpha$. Hence, we take

$$\langle H_1, H_2 \rangle = \int d^2z\, d^2\theta\, E \left(H_1^{\alpha m} H_{2\alpha m} + c_1 (\gamma_5)_\alpha{}^\beta H_{1\beta}{}^\alpha (\gamma_5)_\rho{}^\sigma H_{2\sigma}{}^\rho + c_2\, H_{1\alpha}{}^\alpha H_{2\beta}{}^\beta \right), \tag{26.121}$$

where c_1 and c_2 are constants. In a manner analogous to the bosonic case, these constants do not enter in the final expression in a fundamental way, so they may be set to unity. This metric defines $\mathcal{D}E_A^M$ through eq.(26.117) and leads to the superconformal anomaly, since eq.(26.121) is not super Weyl invariant.

Next we carry out the Faddeev-Popov procedure. The remaining 6 superfields of E_A^M not fixed by the torsion constraints are accounted for by superdiffeomorphisms (ordinary diffeomorphisms and local supersymmetry transformations which together can be parametrized by 4 superfields), local Lorentz tangent space transformations (1 superfield), and super Weyl transformations (1 superfield). So let the infinitesimal superdiffeomorphisms be parametrized by the super vector field V^A, the infinitesimal local Lorentz transforms by δL, and the infinitesimal super Weyl transformations by $\delta\Phi$. The Faddeev-Popov reduction is carried out by changing variables from the remaining gauge dependent components of E_A^M to V_A, L, and Φ. The Jacobian is found via eq.(26.117) by reexpressing $\langle \delta E_A^M, \delta E_A^M \rangle$, defined by eq.(26.121), in terms of V_A, δL, and $\delta\Phi$.

The effect of an infinitesimal tangent group transformation on the superzweibein is given by eq.(26.78). From this it quickly follows that

$$H_N^M = E_N^A \, \delta E_A^M = \delta L \, \mathcal{E}_N^M . \tag{26.122}$$

The action of a super Weyl transformation on the superzweibein is given by eq.(26.100). With $\Lambda = \exp(\Phi/2)$, the infinitesimal version leads to

$$H_n^m = \frac{1}{2}\,\delta\Phi\,\delta_n^m, \quad H_\alpha^\beta = \frac{1}{4}\,\delta\Phi\,\delta_\alpha^\beta, \quad H_\alpha^m = 0, \quad H_n^\alpha = i\frac{1}{8}(\gamma_n)^{\alpha\beta}\mathcal{D}_\beta^{(0)}(\delta\Phi). \tag{26.123}$$

Finally, for a superdiffeomorphism generated by the vector field V^A, the infinitesimal change in coordinates is $z^A \to z^A + V^A$. From this it follows that

$$dz^A \to dz^A + dV^A = dz^A + dz^B \partial_B(V^A). \tag{26.124}$$

For $E^M = dz^A E_A^M$, eq.(26.124) implies that

$$\delta E^M = dz^B(\partial_B V^A)\, E_A^M + dz^A\, V^C \partial_C E_A^M , \tag{26.125}$$

or

$$\delta E_B^M = (\partial_B V^A) E_A^M + V^C \partial_C E_B^M . \tag{26.126}$$

The first term on the right-hand side of eq.(26.126) can be written as

$$(\partial_B V^A) E_A^M = \partial_B V^M - (-)^{ab} V^A \partial_B E_A^M , \tag{26.127}$$

where $V^M = V^A E_A^M$. The first structural equation, eq.(26.81), is used to evaluate the derivatives of the superzweibein. The result is

$$H_N^M = \mathcal{D}_N V^M + V^P W_P \mathcal{E}_N^M - V^P T_{NP}^M . \tag{26.128}$$

Using the results eqs.(26.122), (26.123), and eq.(26.128), we find

$$\begin{aligned} H_\alpha{}^\alpha &= \frac{1}{2}\delta\Phi + \mathcal{D}_\alpha V^\alpha \\ (\gamma_5)_\alpha{}^\beta H_\beta{}^\alpha &= H_\theta{}^\theta - H_{\bar\theta}{}^{\bar\theta} = \delta L + (\gamma_5)_\alpha{}^\beta \mathcal{D}_\beta V^\alpha + V^P W_P \\ H_\alpha{}^m &= \mathcal{D}_\alpha V^m - V^P T_{\alpha P}^m = \mathcal{D}_\alpha V^m - 2iV^\beta \gamma_{\alpha\beta}^m . \end{aligned} \tag{26.129}$$

Note that $H_\alpha{}^\alpha$ can be eliminated by the appropriate super Weyl transformation and $(\gamma_5 H)_\alpha{}^\alpha$ by the appropriate tangent space rotation. We can therefore write $H_\alpha{}^\alpha = \delta\Phi'$ and $(\gamma_5 H)_\alpha{}^\alpha = \delta L'$ by analogy with what we did in eq.(22.12). Plugging eq.(26.129) back into eq.(26.121) we find

$$||H||^2 = ||\delta\Phi'||^2 + ||\delta L'||^2 + \int d^2z\, d^2\theta\, E\, V^M \mathcal{P}_M{}^N V_N , \tag{26.130}$$

where

$$\begin{aligned} \mathcal{P}_m{}^n &= -\mathcal{D}^\alpha \mathcal{D}_\alpha \delta_m{}^n, \quad \mathcal{P}_\beta{}^n = -2i(\gamma^n)_\beta{}^\alpha \mathcal{D}_\alpha \\ \mathcal{P}_m{}^\gamma &= 2i\mathcal{D}^\alpha (\gamma_m)_\alpha{}^\gamma, \quad \mathcal{P}_\beta{}^\gamma = -4(\gamma^m)_\beta{}^\alpha (\gamma_m)_\alpha{}^\gamma. \end{aligned} \tag{26.131}$$

Eqs.(26.130) and (26.117) imply that the Faddeev-Popov determinant is $\mathrm{sdet}'^{1/2}(\mathcal{P})$. Since $\mathcal{P}_\beta{}^\gamma$ does not involve any derivatives, it is convenient to express $\mathrm{sdet}'(\mathcal{P})$ using the second line of eq.(26.89). In fact, since $\mathcal{P}_\beta{}^\gamma = 8\delta_\beta{}^\gamma$, then

$$\mathrm{sdet}'(\mathcal{P}) = \mathrm{const} \cdot \det{}' \left(\mathcal{P}_m{}^n - \frac{1}{8} \mathcal{P}_m{}^\beta \mathcal{P}_\beta{}^n \right). \tag{26.132}$$

From eq.(26.131) we find

$$\begin{aligned} \mathcal{P}_{mn} - \frac{1}{8}\mathcal{P}_m{}^\beta \mathcal{P}_{\beta n} &= -(\mathcal{D}^\alpha \mathcal{D}_\alpha)\delta_{mn} - \frac{1}{2}\mathcal{D}^\alpha (\gamma_m)_\alpha{}^\beta (\gamma_n)_\beta{}^\sigma \mathcal{D}_\sigma \\ &= \frac{1}{2}\mathcal{D}^\alpha (\gamma_n)_\alpha{}^\beta (\gamma_m)_\beta{}^\sigma \mathcal{D}_\sigma, \end{aligned} \tag{26.133}$$

where we have used the γ matrix algebra, eq.(26.46), to arrive at the second line. Thus, the Faddeev-Popov determinant is

$$\mathrm{const} \cdot \mathrm{sdet}'^{1/2}(-\mathcal{D}^\alpha (\gamma_n \gamma_m)_\alpha{}^\beta \mathcal{D}_\beta). \tag{26.134}$$

For $n = z$, $m = \bar{z}$, the only nonzero element of the γ-product is $(\gamma_z\gamma_{\bar{z}})_{\bar{\theta}}{}^{\bar{\theta}} = -1$, thus $\mathcal{D}^\alpha(\gamma_z\gamma_{\bar{z}})_\alpha{}^\beta = -\mathcal{D}^{\bar{\theta}}\mathcal{D}_{\bar{\theta}} = \mathcal{D}_\theta \mathcal{D}_{\bar{\theta}}$. To determine which Laplacian this is, we note that $\mathcal{D}_\beta$ in eq.(26.134) acts on V^n, i.e. V^z when $n = z$. This is a vector of weight -1 (1 up z index); thus $\mathcal{D}_{\bar{\theta}}$ in this case is $\mathcal{D}_{\bar{\theta}}^{(-1)}$. $\mathcal{D}_\theta$ acts on $\mathcal{D}_{\bar{\theta}}^{(-1)} V^z$ which is weight $-1/2$. This means that $\mathcal{D}_\theta \mathcal{D}_{\bar{\theta}}$ for $n = z$, $m = \bar{z}$ is $\Delta_{-1}^{(-)}$. By the same procedure we find that $-\mathcal{D}_{\bar{\theta}}\mathcal{D}_\theta$ for $n = \bar{z}$, $m = z$ is $\Delta_1^{(+)}$. Therefore, the Faddeev-Popov determinant, eq.(26.134), is

$$\mathrm{const} \cdot \mathrm{sdet}'^{1/2}(-\Delta_1^{(+)})\, \mathrm{sdet}'^{1/2}(-\Delta_{-1}^{(-)}). \tag{26.135}$$

Note that this is the supersymmetric extension of eq.(22.104). By analogy with the bosonic case, we can define

$$\begin{aligned} (\mathcal{P}_1 V)_{\alpha m} &\propto (\gamma_n \gamma_m)_\alpha{}^\beta \mathcal{D}_\beta V^n \\ (\mathcal{P}_1^\dagger H)_m &\propto (\gamma_n \gamma_m)_\alpha{}^\beta \mathcal{D}_\beta H^{\alpha n}, \end{aligned} \tag{26.136}$$

in which case the Faddeev-Popov Jacobian can be written as $\mathrm{sdet}'^{\frac{1}{2}}(\mathcal{P}_1^\dagger \mathcal{P}_1)$. Naively, the elements of $Ker(\mathcal{P}_1)$ are the zero modes removed from the evaluation of the determinant, and are called superconformal Killing vectors and spinors.

With the result eq.(26.135), the partition function for genus 0 supersurfaces, eq.(26.120), is

$$Z_0 = \mathcal{N} \int \mathcal{D}\Phi \left(\frac{\text{sdet}'(-\Delta_1^{(+)})\,\text{sdet}'(-\Delta_{-1}^{(-)})}{\text{sdet}\langle \Psi_M, \Psi_N \rangle} \right)^{\frac{1}{2}} \left(\frac{2\pi}{\int d^2z\, d^2\theta\, \widetilde{E}} \right)^{-\frac{D}{2}} \left(\text{sdet}'(-\Delta_0^{(-)})\right)^{-\frac{D}{2}}, \quad (26.137)$$

where the superzweibein is in the superconformal gauge, eq.(26.101), and where Ψ_M is a basis of the space of appropriate zero modes.

Now let us turn to the evaluation of the Φ dependence of the determinants above, which leads to the superconformal anomaly. Following the procedure used in the bosonic case in chapter 22, we try to define the regularized determinant as

$$\ln\,\text{sdet}'(-\Delta_n) = -\int_\epsilon^\infty \frac{dt}{t}\,\text{str}'\left(e^{-t(-\Delta_n)}\right). \quad (26.138)$$

The trouble with this definition is that the spectrum of Δ_n is not bounded from below, so eq.(26.138) is divergent in all the wrong spots. The problem occurs even in flat space and is illustrated by the appearance of single derivatives in

$$\begin{aligned} D_\theta D_{\bar\theta} X &= D_\theta D_{\bar\theta}\left(x + \theta\psi + \overline{\theta\psi} + \bar\theta\theta f\right) \\ &= 4\bar\theta\theta\partial_z\partial_{\bar z}x + 2\theta\partial_z\overline{\psi} + 2\bar\theta\partial_{\bar z}\psi + f. \end{aligned}$$

Note, however, that if we apply $D_\theta D_{\bar\theta}$ twice, we have

$$D_\theta D_{\bar\theta} D_\theta D_{\bar\theta} X = 4\partial_z\partial_{\bar z} X, \quad (26.139)$$

which is just the ordinary Laplacian and is no problem. Following Martinec, this suggests that we take

$$\ln\,\text{sdet}'(-\Delta_n) = \frac{1}{2}\ln\,\text{sdet}'(-(\Delta_n^2)) = -\frac{1}{2}\int_\epsilon^\infty \frac{dt}{t}\,\text{str}'\left(e^{-t(-\Delta_n^2)}\right). \quad (26.140)$$

Using the square of the Laplacians potentially introduces some problems with the way the determinants of norms of the zero modes are defined, but we will just gloss over such subtleties.

Next we consider the functional derivative of eq.(26.140) with respect to the superconformal factor. What counts is $\delta(\Delta_n^2)$. Let's start with $\Delta_n^{(+)2}$. In the superconformal gauge this is

$$\begin{aligned}\delta(\Delta_n^{(+)})^2 &= 2\Delta_n^{(+)}\,\delta\Delta_n^{(+)} = \delta\left(\mathcal{D}_{\bar\theta}^{(n+1/2)}\mathcal{D}_\theta^{(n)}\right)\\ &= 2\Delta_n^{(+)}\,\delta\left(\exp\left(-(n+1)\frac{\Phi}{2}\right)D_{\bar\theta}\exp(n\Phi)D_\theta\exp\left(-\frac{n}{2}\Phi\right)\right)\\ &= 2\Delta_n^{(+)}\left(-\frac{(n+1)}{2}\delta\Phi\,\Delta_n^{(+)} + n\mathcal{D}_{\bar\theta}^{(n+1/2)}\delta\Phi\,\mathcal{D}_\theta^{(n)} - \frac{n}{2}\Delta_n^{(+)}\,\delta\Phi\right).\end{aligned} \tag{26.141}$$

Due to the fact that this variation appears inside a trace in the definition of $\delta(\ln\ \mathrm{sdet}'(-\Delta_n^2))$, we can cyclically rearrange the order of the factors in each term. Of particular interest is the second term in the last line of eq.(26.141) which reads

$$\begin{aligned}\Delta_n^{(+)}\mathcal{D}_{\bar\theta}^{(n+1/2)}\delta\Phi\,\mathcal{D}_\theta^{(n)} &= \mathcal{D}_{\bar\theta}^{(n+1/2)}\mathcal{D}_\theta^{(n)}\mathcal{D}_{\bar\theta}^{(n+1/2)}\,\delta\Phi\,\mathcal{D}_\theta^{(n)}\\ &= \mathcal{D}_\theta^{(n)}\mathcal{D}_{\bar\theta}^{(n+1/2)}\mathcal{D}_\theta^{(n)}\mathcal{D}_{\bar\theta}^{(n+1/2)}\,\delta\Phi\\ &= (\Delta_{n+\frac{1}{2}}^{(-)})^2\delta\Phi.\end{aligned}$$

Hence, inside the trace eq.(26.141) can be written as

$$\delta(\Delta_n^{(+)})^2 = -(2n+1)(\Delta_n^{(+)})^2\delta\Phi + 2n\left(\Delta_{n+\frac{1}{2}}^{(-)}\right)^2\delta\Phi, \tag{26.142}$$

and it follows that

$$\begin{aligned}&\delta(\ln\ \mathrm{sdet}'(-\Delta_n^{(+)}))\\ &= \frac{1}{2}\int_\epsilon^\infty dt\ \mathrm{str}'\Bigg((2n+1)(\Delta_n^{(+)})^2\delta\Phi e^{-t(-\Delta_n^{(+)2})}\\ &\qquad\qquad - 2n\left(\Delta_{n+\frac{1}{2}}^{(-)}\right)^2\delta\Phi\, e^{-t(-\Delta_{n+\frac{1}{2}}^{(-)\,2})}\Bigg)\\ &= \frac{1}{2}\int_\epsilon^\infty dt\ \mathrm{str}'\left((2n+1)\delta\Phi\frac{\partial}{\partial t}\,e^{-t(-\Delta_n^{(+)2})} - 2n\delta\Phi\frac{\partial}{\partial t}\,e^{-t(-\Delta_{n+\frac{1}{2}}^{(-)\,2})}\right)\\ &= -\frac{1}{2}\mathrm{str}'\left((2n+1)\delta\Phi\ e^{-\epsilon(-\Delta_n^{(+)2})} - 2n\delta\Phi\ e^{-\epsilon(-\Delta_{n+\frac{1}{2}}^{(-)\,2})}\right),\end{aligned} \tag{26.143}$$

where we have made use of an argument similar to that leading to eq. (22.108). By a similar computation,

$$\delta(\ln \text{sdet}'(-\Delta_n^{(-)})) = \frac{1}{2}\text{str}'\left((2n-1)\delta\Phi e^{-\epsilon(-\Delta_n^{(-)2})} - 2n\delta\Phi e^{-\epsilon(-\Delta_{n-\frac{1}{2}}^{(+)}{}^2)}\right). \tag{26.144}$$

Eqs.(26.143) and (26.144) reduce the computation of the Φ dependence of the determinants to a calculation of the short-time behavior of the trace of the super heat kernels, $G_n^{(\pm)}(z,z';\theta,\theta';\epsilon) \equiv \exp(-\epsilon(-\Delta_n^{(\pm)2}))$. These heat kernels satisfy

$$\left(\frac{\partial}{\partial t} - \Delta_n^{(\pm)2}\right)G_n^{(\pm)}(z,z';\theta,\theta';t) = \delta^2(z-z')\,\delta^2(\theta-\theta')\,\delta(t). \tag{26.145}$$

One can readily verify that

$$\delta^2(\theta-\theta') = (\theta-\theta')(\overline{\theta}-\overline{\theta}'), \tag{26.146}$$

when the convention eq.(26.68) is used. Thus, when $\Delta_n^{(\pm)2}$ is just the ordinary flat space Laplacian, $4\partial_z\partial_{\overline{z}}$, the super heat kernel is

$$G^{(0)}(z,z';\theta,\theta';t) = \frac{1}{4\pi t}\exp\left(-\frac{|z-z'|^2}{4t}\right)(\theta-\theta')(\overline{\theta}-\overline{\theta}'). \tag{26.147}$$

We can always express $\Delta_n^{(\pm)2}$ as $4\partial_z\partial_{\overline{z}} + V_n^{(\pm)}$. Since we only need the short-time behavior of the heat kernels, we can solve for $G_n^{(\pm)}$ as a Neumann series, treating $V_n^{(\pm)}$ as a perturbation off $G^{(0)}$. As in the bosonic case, we will only need $G_n^{(\pm)}$ to first order in $V_n^{(\pm)}$,

$$G_n^{(\pm)} = G^{(0)} + G^{(0)}V_n^{(\pm)}G^{(0)} + \mathcal{O}(t). \tag{26.148}$$

Since we only need the trace of $G_n^{(\pm)}$, we can concentrate on the diagonal piece of $G_n^{(\pm)}$, i.e. $z=z'$, $\theta=\theta'$. Note that because of the nature of $\delta^2(\theta-\theta')$, the diagonal of $G^{(0)}(z,z;\theta,\theta;t)$ vanishes (unlike the bosonic case). Thus,

$$G_n^{(\pm)}(z,z;\theta,\theta;t) = \int_0^t dt'\int d^2z'\,d^2\theta'\,G^{(0)}(z,z';\theta,\theta';t-t')V_n^{(\pm)}G^{(0)}(z',z;\theta',\theta;t') + \mathcal{O}(t). \tag{26.149}$$

Each factor of $G^{(0)}$ contains $\delta^2(\theta-\theta')$. Since $\int d^2\theta'\,(\delta^2(\theta-\theta'))^m = 0$ for $m>1$, then the only surviving terms in eq.(26.149) will come from the pieces of $V_n^{(\pm)}(z,\theta)$ proportional to $\partial_\theta\partial_{\overline{\theta}}$.

As an explicit example, consider $V_0^{(-)}$. Working in the superconformal gauge and using eq.(26.114), a straightforward calculation yields

$$V_0^{(-)} = -(1-\exp(-\Phi))\, D_\theta D_{\bar\theta} D_\theta D_{\bar\theta} + \exp(-\Phi)\left[\frac{1}{4}(D_\theta\Phi)(D_{\bar\theta}\Phi) - \frac{1}{2}(D_\theta D_{\bar\theta}\Phi) - \frac{1}{2}(D_\theta\Phi)D_{\bar\theta} + \frac{1}{2}(D_{\bar\theta}\Phi)D_\theta\right] D_\theta D_{\bar\theta}. \quad (26.150)$$

The terms proportional to $\partial_\theta\partial_{\bar\theta}$ are

$$\exp(-\Phi)\left[\frac{1}{4}(D_\theta\Phi)(D_{\bar\theta}\Phi) - \frac{1}{2}(D_\theta D_{\bar\theta}\Phi)\right]\partial_\theta\partial_{\bar\theta}. \quad (26.151)$$

As in the bosonic case, we can choose the coordinate system such that the conformal factor Φ vanishes at (z', θ') along with $D_{\theta'}\Phi$ and $D_{\bar\theta'}\Phi$. Hence, in this coordinate system,

$$\int d^2\theta'\, \delta^2(\theta-\theta')\, V_0^{(-)}(z',\theta')\, \delta^2(\theta'-\theta) = \frac{1}{2}(D_\theta D_{\bar\theta}\Phi), \quad (26.152)$$

and

$$G^{(0)} V G^{(0)} = \frac{1}{2}(D_\theta D_{\bar\theta}\Phi)\int_0^t dt' \int d^2z'\, \frac{1}{4\pi(t-t')}\exp\left(-\frac{|z-z'|^2}{4(t-t')}\right) \times \frac{1}{4\pi t}\exp\left(-\frac{|z'-z|^2}{4t}\right) = \frac{1}{8\pi} D_\theta D_{\bar\theta}\Phi. \quad (26.153)$$

As in the bosonic case, when we return to the original coordinate system, the result (26.153) is proportional to the scalar curvature.

Returning to the case for general n, a similar but longer computation yields

$$\int d^2\theta'\, \delta^2(\theta-\theta')\, V_n^{(\pm)}\, \delta^2(\theta'-\theta) = \frac{1}{2}(1 \pm 2n)(D_\theta D_{\bar\theta}\Phi), \quad (26.154)$$

from which it follows that

$$G_n^{(\pm)}(z,z;\theta,\theta;t) = \frac{1}{8\pi}(1 \pm 2n)\, D_\theta D_{\bar\theta}\Phi + \mathcal{O}(t). \quad (26.155)$$

To finish the computation of the determinants, we must find the primed supertrace of eq.(26.155) above. The primed supertrace is the full supertrace less the zero modes (technically the zero modes of $-\Delta_n^{(\pm)2}$). In the superconformal gauge we have

$$\mathrm{str}\left(\delta\Phi\, e^{-\epsilon(-\Delta_n^{(\pm)2})}\right) = \frac{1}{8\pi}(1 \pm 2n)(-)^{2n}\int d^2z d^2\theta\, (D_\theta D_{\bar\theta}\Phi)\delta\Phi + \mathcal{O}(\epsilon), \tag{26.156}$$

where the extra minus sign from the factor $(-)^{2n}$ arises from the definition of the supertrace, eq.(26.86) (minus is used on the odd-odd part). Therefore,

$$\delta \ln\left(\frac{\mathrm{sdet}'(-\Delta_n^{(\pm)})}{\mathrm{sdet}\langle\Psi_M^{(\pm)},\Psi_N^{(\pm)}\rangle}\right) = -\frac{1}{16\pi}(-)^{2n}(1\pm 4n)\int d^2z\, d^2\theta\, (D_\theta D_{\bar\theta}\Phi)\,\delta\Phi, \tag{26.157}$$

where $\mathrm{sdet}\langle\Psi_M^{(\pm)},\Psi_N^{(\pm)}\rangle$ is symbolic for the proper zero-mode contribution. These equations are easy to integrate. The integration constant is the log of the ratio of superdeterminants involving Laplacians that are independent of Φ. Since we are working in the superconformal gauge, these are the flat superspace Laplacians. Hence,

$$\ln\left(\frac{\mathrm{sdet}'(-\Delta_n^{(\pm)})}{\mathrm{sdet}\langle\Psi_M^{(\pm)},\Psi_N^{(\pm)}\rangle}\right) - \ln\left(\frac{\mathrm{sdet}'(-\widehat{\Delta}_n^{(\pm)})}{\mathrm{sdet}\langle\widehat{\Psi}_M^{(\pm)},\widehat{\Psi}_N^{(\pm)}\rangle}\right) = \frac{1}{16\pi}(-)^{2n}(1\pm 4n)\int d^2z\, d^2\theta\, \frac{1}{2}\, D_\theta\Phi\, D_{\bar\theta}\Phi. \tag{26.158}$$

Plugging this result back into Z_0, eq.(26.137), yields the superconformal anomaly,

$$Z_0 = \mathcal{N}\int \mathcal{D}\Phi\, \exp\left(\frac{(10-D)}{32\pi}\int d^2z\, d^2\theta\, \frac{1}{2}\, D_\theta\Phi\, D_{\bar\theta}\Phi\right), \tag{26.159}$$

and it is apparent that the critical dimension is 10. As usual, we have put all the determinant factors independent of Φ into the normalization.

We could have enlarged the action eq.(26.84) we started from by adding the potential counterterm

$$2\mu_0\int d^2z\, d^2\theta\, E. \tag{26.160}$$

In this case, eq.(26.159) becomes

$$Z_0 = \mathcal{N}\int \mathcal{D}\Phi\, \exp\left(\frac{(10-D)}{32\pi}\int d^2z\, d^2\theta\, \frac{1}{2}\, D_\theta\Phi\, D_{\bar\theta}\Phi + 2\mu\, \exp(\Phi/2)\right), \tag{26.161}$$

a result first obtained by Polyakov. The difference here with the bosonic case is that μ_0 is not renormalized because of the factor $\delta^2(\theta - \theta')$ in the heat kernel. The contribution to the term above from the bosons is exactly canceled by the fermions. There is no particular reason here not to choose $\mu = \mu_0 = 0$ other than the term with nonvanishing μ is allowed by the required symmetries of Z_0, etc.

As we mentioned, another possible counterterm is essentially the Euler number of the supersurfaces. If we apply eq.(22.69) we can see that this term is renormalized, i.e. the superdeterminants do possess a $\log(\epsilon)$ divergence.

The effective action in eq.(26.161) is manifestly supersymmetric and is the supersymmetric extension of the Liouville action, eq.(22.122), since we functionally only replace ordinary derivatives by superderivatives and ordinary fields by superfields. Therefore, it is called the supersymmetric Liouville model. For comparison, let us rewrite this effective action in components. With

$$\Phi(z,\theta) = \phi(z)\,\theta\zeta_\theta(z) + \overline{\theta}\zeta_{\overline{\theta}}(z) + \overline{\theta}\theta b(z), \tag{26.162}$$

we find that

$$\begin{aligned}\int d^2z\, d^2\theta\, \frac{1}{2}\, D_\theta\Phi\, D_{\overline{\theta}}\Phi + 2\mu\, \exp(\Phi/2) \\ = -\int d^2z\; 2(\partial_z\phi)(\partial_{\overline{z}}\phi) + \zeta_{\overline{\theta}}\partial_z\zeta_{\overline{\theta}} - \zeta_\theta\partial_{\overline{z}}\zeta_\theta - \frac{1}{2}b^2 \\ - 2\mu\, e^{-\phi/2}\Big(-\frac{1}{2}b + \frac{1}{4}\zeta_{\overline{\theta}}\zeta_\theta\Big).\end{aligned} \tag{26.163}$$

Eliminating the field b with its equation of motion, eq.(26.163) reduces to

$$-\int d^2\xi\, \frac{1}{2}\,(\partial_a\phi)^2 - \frac{i}{2}\zeta\partial\!\!\!/\zeta + \frac{1}{2}\mu^2 e^{\phi} - \mu e^{\phi/2}\,\zeta\zeta, \tag{26.164}$$

so that eq.(26.161) becomes

$$\begin{aligned}Z_0 = \mathcal{N}\int \mathcal{D}\phi\,\mathcal{D}\zeta\, \exp\bigg(-\frac{(10-D)}{32\pi}\int d^2\xi\,\Big(\frac{1}{2}\,(\partial_a\phi)^2 \\ - \frac{i}{2}\zeta\partial\!\!\!/\zeta + \frac{1}{2}\mu^2 e^{\phi} - \mu e^{\phi/2}\,\zeta\zeta\Big)\bigg).\end{aligned} \tag{26.165}$$

Higher-genus contributions to the partition function can be computed in an analogous manner. These terms will produce the same superconformal anomaly as given above, since the anomaly is a high energy, short-distance local property and therefore independent of the genus. The main additional complication in evaluating the higher-genus contributions is the fact that dim $Ker(\mathcal{P}_1^\dagger)$ is no longer 0 as it is in the genus 0 case. As in the bosonic string, this means that there is more than one superconformal class of (torsion-constrained) superzweibein, or, equivalently, more than one supercomplex structure that can be placed on a supersurface of genus ≥ 1. The additional parameters needed to label the superconformal classes are called supermoduli, and in the critical dimension the higher-genus contributions to Z will reduce to integrals over finite-dimensional supermoduli spaces.

What cannot be apparent in the above results is that the physical spectrum is tachyon-free. If we had quantized in the critical dimension in the operator representation we would have found a space-time bosonic state that is a tachyon. However, there is a truncation of the spectrum, the GSO projection, that eliminates the tachyon and results in a space-time supersymmetric spectrum. At tree level there is no reason to enforce the GSO projection other than the desire to get rid of the tachyon, etc. However, at 1-loop and higher, application of the GSO projection is needed for consistency (in particular, for modular invariance and unitarity).

In the path integral sum-over-supersurfaces formalism the supermodular invariance is manifest. The issue of ordinary modular invariance of the measure and scattering amplitudes is full of subtleties. Part of this invariant summation can be interpreted as a sum over spin structures on each surface.

To define a spin structure, note that a vector carried around a closed loop on a surface can pick up a phase. A spinor picks up the square root of this phase. With the square root, a choice of sign must be made. If the loop is contractible, then the obvious choice must be made for continuity. If the loop is not contractible, then there is a real choice. In order for noncontractible loops to exist, the surface must have a nontrivial topology. In fact, there are 2^h independent (i.e. nonhomotopic) noncontractible loops on a surface of genus h (the homology basis provides one example). A choice of + or − for the square root on each independent noncontractible loop is called a spin structure. Since there are two choices for each loop, there are 2^{2h} spin structures possible on a compact surface of genus h. Diffeomorphisms and the mapping class group mix spin structures of the same parity; hence we must sum over all spin structures to maintain modular invariance. The GSO projection can be interpreted as the sum over spin structures. The non-space-time supersymmetric states are eliminated from contributing by this sum, and with these goes the tachyon. For genus 0 there is only 1 spin structure. The necessity of the GSO projection only shows up beyond tree level.

By analogy with the bosonic case, there exists a superholomorphic anomaly that is canceled for $D = 10$. The supermoduli space does possess a complex structure, and the superdeterminants and partition function do factorize superholomorphically. The necessary developments in super algebraic geometry to take this structure to the same distance as in the bosonic case are the topic of active research. The details of the issues, subtleties, and results can be found in the references (and references within the references) in the Bibliography.

26.4 EXERCISES

1. Verify eqs.(26.37), (26.38), and (26.44).
2. Show that the integral I, eq.(26.68), is supersymmetric. Verify eq. (26.70) and show that eq.(26.71) is supersymmetric.
3. Verify the γ matrix algebra, eq.(26.46), in the z, $\overline{z}$ index basis. Expand the supersurface action, eq.(26.76), in components using eq. (26.75), and do the $\theta, \overline{\theta}$ integral. Rewrite the action eq.(26.57) in z, $\overline{z}$, θ, $\overline{\theta}$ index component form and compare. Write δX^μ in component form where δX^μ is the change of X^μ under an infinitesimal supersymmetry transformation. Write the transformation eq.(26.58) in z, $\overline{z}$, θ, $\overline{\theta}$ index component form and compare.
4. Using the definition of the supertrace, eq.(26.86), show that
$$\mathrm{str}(S_1 \cdots S_m) = \mathrm{str}(S_m S_1 \cdots S_{m-1}).$$
5. Starting from eq.(26.87), derive eq.(26.90).
6. Using eq.(26.91), verify that the covariant derivative $\mathcal{D}_\theta$ is D_θ, eq. (26.69). In addition, verify that D_θ satisfies the torsion constraints, eq.(26.83).
7. Verify eqs.(26.122) and (26.123). Using the first structural equation (26.81), derive eq.(26.128) from eq.(26.126). Use the results to derive eq.(26.134).
8. Derive eq.(26.144).
9. Using eq.(26.146), verify that
$$\int d^2\theta' \, \delta^2(\theta - \theta') = 1$$
and that
$$\int d^2\theta' \, F(z, \theta') \, \delta^2(\theta - \theta') = F(z, \theta).$$
10. Derive eq.(26.165) starting from eq.(26.161).

Bibliography

GENERAL REFERENCES

The following references were used in the preparation of this text. Together they cover many related topics in greater depth (e.g. electroweak theory), and include many original references.

Quantum Field Theory

D.J. Amit, *Field Theory, The Renormalization Group and Critical Phenomena* (McGraw-Hill, New York, 1978).

J.D. Bjorken and S.D. Drell, *Relativistic Quantum Mechanics* and *Relativistic Quantum Fields* (McGraw-Hill, New York, 1965).

S. Coleman, *Aspects of Symmetry* (Cambridge University Press, Cambridge, 1985).

R.P. Feynman and A.R. Hibbs, *Quantum Mechanics and Path Integrals* (McGraw-Hill, New York, 1965).

C. Itzykson and J.-B. Zuber, *Quantum Field Theory* (McGraw-Hill, New York, 1980).

P. Ramond, *Field Theory: A Modern Primer* (Addison-Wesley, Redwood City, 1989).

Strings

E. D'Hoker and D.H. Phong, *Rev. Mod. Phys.* **60**, 917 (1988).

M.B. Green, J.H. Schwarz, and E. Witten, *Superstring Theory* (Cambridge University Press, Cambridge, 1987), vols. 2 and I2.

M. Kaku, *Introduction to Superstrings* (Springer-Verlag, New York, 1988).

A.M. Polyakov, *Gauge Fields and Strings* (Harwood, Chur, 1987).

J.H. Schwarz, ed., *Superstrings* (World Scientific, Singapore, 1985), vols. 1 and 2.

ADDITIONAL REFERENCES

Chapter 2

E. Lieb and W. Liniger, *Phys. Rev.* **130**, 1605 (1963).

Chapter 9

F.A. Berezin, *The Method of Second Quantization* (Academic Press, New York, 1966).

Chapter 11

S. Enguehard and B. Hatfield, "Rayleigh-Schrödinger Perturbation Theory with Constraints," AMPR-91-09, June, 1991.

B.F. Hatfield, *Phys. Lett.* **147 B**, 435 (1984).

R. Jackiw, *Rev. Mod. Phys.* **52**, 661 (1980). (A review on Yang-Mills that includes a discussion of the vacuum angle.)

Chapter 12

R. Courant and H. Robbins, *What Is Mathematics?* (Oxford University Press, London, 1977).

Chapter 18

T. Barnes and G.I. Ghandour, *Phys. Rev.* **D22**, 924 (1980).

S. Coleman and E. Weinberg, *Phys. Rev.* **D7**, 1888 (1973).

Chapter 19

J.H. Schwarz, *Phys. Reports* **89**, 223 (1982).

J.H. Schwarz, *Int. J. Mod. Phys.* **A2**, 593 (1987).

F. Theiss, unpublished notes, University of California, San Diego.

G. Veneziano, *Europhys. Lett.* **2**, 199 (1986).

Chapter 20

J. Scherk, *Rev. Mod. Phys.* **47**, 123 (1975).

Chapter 21

A.F. Beardon, *A Primer on Riemann Surfaces* (Cambridge University Press, Cambridge, 1984).

R.L. Bishop and S.I. Goldberg, *Tensor Analysis on Manifolds* (Dover Publications, New York, 1980).

R. Courant and H. Robbins, *What Is Mathematics?* (Oxford University Press, London, 1977).

H.M. Farkas and I. Kra, *Riemann Surfaces* (Springer-Verlag, New York, 1980).

G.A. Jones and D. Singerman, *Complex Functions* (Cambridge University Press, Cambridge, 1987).

O. Lehto, *Univalent Functions and Teichmüller Spaces* (Springer-Verlag, New York, 1987).

W.S. Massey, *Algebraic Topology: An Introduction* (Springer-Verlag, New York, 1967).

S. Nag, *The Complex Analytic Theory of Teichmüller Spaces* (J. Wiley & Sons, New York, 1988).

B. O'Neill, *Elementary Differential Geometry* (Academic Press, New York, 1966).

M. Schlichenmaier, *An Introduction to Riemann Surfaces, Algebraic Curves and Moduli Spaces* (Springer-Verlag, New York, 1989).

M. Spivak, *Differential Geometry*, 5 vols., (Publish or Perish, Boston, 1975).

Chapter 22

O. Alvarez, *Nucl. Phys.* **B 216**, 125 (1983). (Especially.)

S.W. Hawking, *Commun. Math. Phys.* **55**, 133 (1977).

A.M. Polyakov, *Phys. Lett.* **B 103**, 207 (1981). (Of course.)

Chapter 23

L. Alvarez-Gaumé and P. Nelson, in *Supersymmetry, Supergravity and Superstrings 86* (World Scientific, Singapore, 1986).

A.A. Beilinson and Yu.I. Manin, *Commun. Math. Phys.* **107**, 359 (1986).

A.A. Belavin and V.G. Knizhnik, *Phys. Lett.* **B 168**, 201 (1986).

A.A. Belavin and V.G. Knizhnik, *Sov. Phys. JETP* **64**, 214 (1986).

E. D'Hoker and D.H. Phong, *Nucl. Phys.* **B 269**, 205 (1986).

D. Friedan, in *Recent Advances in Field Theory and Statistical Mechanics*, J.-B. Zuber and R. Stora, eds. (North-Holland, Amsterdam, 1984).

G. Moore and P. Nelson, *Nucl. Phys.* **B 266**, 58 (1986).

P. Nelson, *Phys. Rep.* **149**, 304 (1987).

J. Polchinski, *Commun. Math. Phys.* **104**, 37 (1986).

Chapter 25

O. Alvarez, *Phys. Rev.* **D 24**, 440 (1981).

O. Alvarez, *Nucl. Phys.* **B 216**, 125 (1983).

J.F. Arvis, *Phys. Lett.* **B 127**, 106 (1983).

F. David, *Mod. Phys. Lett.* **A 3**, 1651 (1988).

F. David, *Phys. Rep.* **184**, 221 (1989).

F. David and E. Guitter, *Nucl. Phys.* **B 295 [FS21]**, 332 (1988).

J. Distler and H. Kawai, *Nucl. Phys.* **B 321**, 509 (1989).

B. Durhuus, H.B. Nielsen, P. Olesen, and J.L. Petersen, *Nucl. Phys.* **B 196**, 498 (1982).

J.-L. Gervais and A. Neveu, *Phys. Rep.* **67**, 151 (1980).

V.G. Knizhnik, A.M. Polyakov, and A.B. Zamolodchikov, *Mod. Phys. Lett.* **A 3**, 819 (1988).

J.B. Kogut, *Rev. Mod. Phys.* **55**, 775 (1983).

M. Lüscher, *Nucl. Phys.* **B 180 [FS2]**, 317 (1981).

M. Lüscher, K. Symanzik, and P. Weisz, *Nucl. Phys.* **B 173**, 365 (1980).

Y. Nambu, *Phys. Lett.* **B 80**, 372 (1979).

P. Olesen, *Phys. Lett.* **B 160**, 144 (1985).

J. Polchinski, *Nucl. Phys.* **B 324**, 123 (1989).

A.M. Polyakov, *Phys. Lett.* **B 82**, 247 (1979).

A.M. Polyakov, *Nucl. Phys.* **B 164**, 171 (1979).

A.M. Polyakov, *Nucl. Phys.* **B 268**, 406 (1986).

A.M. Polyakov, *Mod. Phys. Lett.* **A 2**, 899 (1987).

G. 't Hooft, *Nucl. Phys.* **B 72**, 461 (1974).

K. Wilson, *Phys. Rev.* **D 10**, 2445 (1974).

A.B. Zamolodchikov, *Phys. Lett.* **B 117**, 87 (1982).

Chapter 26

R. Arnowitt, P. Nath, and B. Zumino, *Phys. Lett.* **B 56**, 81 (1975).

L. Brink, P. Di Vecchia, and P. Howe, *Phys. Lett.* **B 65**, 471 (1976).

S. Deser and B. Zumino, *Phys. Lett.* **B 65**, 369 (1976).

S.J. Gates, Jr., M.T. Grisaru, M. Roček, and W. Siegel, *Superspace* (Addison-Wesley, Reading, 1983).

P.S. Howe, *J. Phys. A: Math. Gen.* **12**, 393 (1979).

E. Martinec, *Phys. Rev.* **D 28**, 2604 (1983).

A.M. Polyakov, *Phys. Lett.* **B 103**, 211 (1981).

Index

abelian differential, 599
almost complex structure, 585
analytic, 564
anharmonic oscillator path integral, 298–300
anomalous dimension, 467
anomalous magnetic moment, 417
anomaly,
    chiral, 420
    conformal, 629, 637
    definition, 418
    superconformal, 717, 718
anticommuting variables, 190
    δ-functional, definition, 192
    differentiation, 191
    functional derivative, 192
    integration, 191
asymptotic freedom, 469–470
atlas, 541
    analytic, 544
    C^{∞} atlas, 541
    orientable, 542
automorphism, 566
automorphism group, *Aut(S)*, 568
axial current,
    and chiral anomaly, 420
    conserved, non gauge invariant, 420
    nonabelian, 422
    nonconserved, 420

baryons, 5
Belavin-Knizhnik theorem, 645
Beltrami coefficients, 646–647
Beltrami differentials, 647
Beltrami equation, 585
β-function,
    fixed points, 467–469
    $\lambda\phi^4$, 464–465
    QED, 406
bijective (1-1 correspondence), 539
bilinear covariants, 68–69
Bjorken scaling, 469–470
bosonic string,
    classical action, 492, 604
    classical equations of motion, 495, 496, 505
    classical solution (open), 500
    classical stress-energy tensor, 494
    covariant constraints, 504
    light-cone gauge, 509
    light-cone gauge Hamiltonian, 512
    metric as Lagrange multiplier, 494
    partition function, 603
    partition function (genus 0), 611
    solution of constraints, 509–510
    stress-energy constraints, 506, 526
    world-sheet symmetries, 505
bosonic string in operator rep,
    anomalous Lorentz generator in light-cone gauge, 516
    ghost states, 532
    graviton state, 519
    implementation of stress-energy constraints, 526
    Lorentz generators, 515
    lowest lying states (closed), 519
    lowest lying states (open), 516
    mass shell condition, 531
    mode expansion, 507, 508, 510, 517
    normal ordered Hamiltonian, 513
    quantization in covariant gauge, 525–529
    quantization in light-cone gauge, 512
    square mass operator, 511, 513, 518, 531
bosonic string in path integral rep,
    Belavin-Knizhnik theorem, 645
    Faddeev-Popov (genus 0), 605–610
    Faddeev-Popov (genus >0), 632–636, 649
    holomorphic factorization, 644
    partition function, 603, 631, 638, 639, 641, 643, 671

partition function (genus 0), 611, 629
partition function (genus 1), 639, 641, 643, 644, 645
partition function (genus ≥2), 638, 645
perturbative expansion, 631
reduction to integral over moduli, 638
scattering amplitudes, 657
tachyon 1-loop amplitude, 667
tachyon tree amplitude, 663
bosonic string in Schrödinger rep,
mode expansion, 520–521
regularization, 523
Schrödinger equation, 519
square mass operator, 522
vacuum energy, 522–524
vacuum wave functional, 521–524

Callen-Symanzik equation, 470–471
Cauchy-Riemann equations, 564
chain, 597
closed, 597
charge, 5
charge conjugation, 98
Dirac fields, 100–102
electromagnetic field, 102
scalar fields, 99–100
charged scalar fields, 48–52
chiral anomaly, 420
and gauge invariance, 419
and PCAC, 421
evaluation, 423
lack at higher loops, 427
physical consequences, 420–423
Christoffel symbols, 553
classical gauge theory, 77–81
Yang-Mills, 96–97
compact, 544
complex index notation, 623–624
spinorial, 696–698, 706–707
complex index tensors,
classification, 707
transformation, 706
complex structure, 544
conformal, 565
conformal anomaly, 629, 637
conformal Killing vector fields, 596, 608
conformally flat gauge, 606
conformally flat metric, 571, 584
congruence, 592
conjugate subgroups, 579
connected, 539
connection, 552
Levi-Cività, 555
1-forms, 559
connection coefficients, 553
continuous, 539
coordinate patch, 541
cotangent space, 558, 689
Coulomb gauge propagator, 85–86, 314
relation to covariant propagator, 86, 164, 338
coupling fermions to gravity, 688–693
covariant derivative, 553, 691
super, 699
covariant vector, 554
cover, 541
covering group, 575
covering map, 575
covering transformation, 575
critical dimension, 482
cross sections, 152–154, 176, 177
curvature operator, 555
curvature 2-forms, 560

Dehn twists, 602
det$(-\partial^2)$, 617
det$(-\partial_x^2 - \omega^2)$, 289, 613
det$(-\partial_x^2 + \omega^2)$, 614, 615
det$(-\partial^2 + m_0^2 + \lambda\phi_c^2/2)$, 454–455, 616
det$'(\Delta_g)$,
conformal dependence, 622–623, 637
genus 1, 641–643
det$'(\Delta_{-1}^{(-)})$, 650
det$'(P_1^\dagger P_1)$,
conformal dependence, 629
genus 1, 641, 643
diffeomorphism, 543
super, 698
differentiable structure, 542
dimensional regularization,
applied to QED electron self-energy, 389
applied to QED vacuum polarization, 392
applied to QED vertex correction, 395
Dirac equation, 60–62
completeness, orthogonality of spinors, 64, 73, 218
Dirac current, 62
momentum spinors, 63–64, 218
physical vacuum, 65–66

plane wave solutions, 62–64
propagator, 72–74
Dirac γ matrices, 61
algebra, 689
chiral representation, 67
complex index notation, 697
curved space matrices, 690
γ_5, 419
Majorana representation, 66
trace identities, 75
2-d Euclidean, 695
Dirichlet region, 582
discrete symmetries, 98–106
discrete topology, 538
discriminant, 583
duality, 664

effective action,
as generating functional of proper functions, 436–439
defined (operator rep), 432, 443
defined (path integral rep), 443–444
defined (Schrödinger rep), 444
Legendre transform of vacuum energy, 457
Legendre transform of $W[J]$, 435
relation to other generating functionals, 435
static action, 456
effective potential,
defined, 440
$\lambda\phi^4$ (1-loop), 448–450, 616
relation to stable vacua, 440–441
effective potential for $\lambda\phi^4$ (1-loop),
evaluated (operator rep), 446–448
evaluated (path integral rep), 453–455, 616
evaluated (Schrödinger rep), 457–461
Gaussian approximation, 462
static potential, 456
when $m^2 = 0$ (1-loop), 451
electromagnetic field strength tensor, 78
dual field strength tensor, 78–79
electron-electron scattering, 175
from Schrödinger rep, 268–269
electron mass shift in QED,
from operator rep, 398
from Schrödinger rep, 264–267
electron self-energy diagram, 386
evaluated, 386–391
regularized (dimensional), 389
regularized (Pauli-Villars), 386

entire, 565
Euler-Poincaré characteristic, 547, 549, 550, 682
and Gauss-Bonnet theorem, 561
extended complex plane, 566
exterior derivative, 559
extrinsic curvature, 683–684
relation to intrinsic curvature, 684

Faddeev-Popov determinant,
bosonic string, 608-610, 671
conformal dependence, 672
ghost representation, 356
QED (Coulomb gauge), 352
QED (Lorentz gauge), 352
simple example, 345–348
superstring, 712
Yang-Mills, 355, 356, 359
Faddeev-Popov ghosts, 356
Faddeev-Popov metric method,
bosonic string, 605–608, 632–636, 649
simple example, 349
superstring, 710–712
Faddeev-Popov procedure,
applied to bosonic string, 605–610, 632–636, 649
applied to QED, 351
applied to string tachyon amplitude, 661–662
applied to superstring, 710–712
applied to Yang-Mills, 354–356
integrate over all slices, 348, 353
simple example, 347
fermionic single particle path integral, 319–321, 324–326
Feynman diagrams,
in quantum mechanics, 296–297
$\lambda\varphi^4$, 137–150
QED, 164–165
Yang-Mills, 357–358
Feynman rules,
$\lambda\varphi^4$, 141, 143
QED, 167–168
first structural equation, 559, 692
flat supersurface metric, 698
Fock space, 26
forces, 4
forms, 558
free particle path integral, 281–285
free photon field theory,
in path integral rep, 315–317
in Schrödinger rep, 211–212, 224

in the Coulomb gauge, 81–84, 224, 316
in the Lorentz gauge, 86–90
in the temporal gauge, 211
free scalar field theory,
in operator rep, 42–48
in path integral rep, 303–309
in Schrödinger rep, 200–201
free spinor field theory,
in operator rep, 69–72
in path integral rep, 318
in Schrödinger rep, 217–219
Fuchsian group, 578
functional, definition, 179
functional derivative, 180
fermionic, 192
functional determinants,
conformal dependence, 619
divergent parts, 618, 620
harmonic oscillator, 287–289
heat kernel techniques, 613–615
methods of evaluation, 612–615
role of zero modes, 620–621
Sommerfeld-Watson transformation, 642
zeta-function regularization, 615
functional differential equations, 182–186
power counting technique to solve, 184
functional directional derivative, 180–181
as regulator in string, 523
use in Coulomb gauge, 224
functional Fourier transform, 210
functional integrals,
δ-functional, 187, 192
Gaussian, bosonic, 188–189
Gaussian, fermionic, 193–195
Gaussian, with complex-valued functions, 195–197
Gaussian, with sources, 189–190
functional Schrödinger equation,
bosonic string, 519
free photon field theory, 211, 214
free scalar field theory, 201, 209
free spinor field theory, 217, 219
$\lambda\phi^4$, 460
Yang-Mills, 244, 252
fundamental group, 563
of the circle, 563
of the torus, 564
fundamental region, 582
of the modular group, 583
Furry's theorem, 168

gauge covariant derivative, 95, 691
gauge-fixed action,
QED, 353
superstring, 697
Yang-Mills, 356, 357
gauge orbit,
bosonic string, 633–634
simple example, 343
Yang-Mills, 355
gauge slice,
bosonic string, 606, 633–634
conformally flat, 606, 633–634
QED, 351, 352
simple example, 344
superconformal, 697
Yang-Mills, 355
gauge volume,
of *Diff*$_0$ (bosonic string), 608–609
QED, 352
simple example, 344
Yang-Mills, 355
Gauss-Bonnet theorem,
global, 561, 623
local, 557
Gaussian curvature, 556–558
Gaussian effective potential, 462
Gauss's law constraint,
free electromagnetic field, 212
QED, 255
relation to gauge invariance, 212, 245
use in solving Schrödinger eq., 249, 258
Yang-Mills, 245
genus, 536, 546
geodesic, 553
geometric structure, 543–544
ghost-free gauge, 359
ghost states in covariant bosonic string, 532
ghosts (negative normed states), 89–90,
no ghosts in the electrodynamic Lorentz gauge, 92
representation of Faddeev-Popov determinant, 356
Goldstone bosons, 443
Grassmann variables, 190
δ-functional, definition, 192
differentiation, 191
functional derivative, 192
in fermionic wave functionals, 219

integration, 191, 323
supersurface δ-function, 715
gravitino, 693
Green's function for Δ_g,
conformal dependence, 660
genus 0, 659–660
genus 1, 666
short-distance behavior, 660
GSO projection, 719
gyromagnetic ratio, 417

hadrons, 5
harmonic oscillator, 16–19
classical Green's function, 291, 311
Heisenberg picture, 16–18
path integral representation, 286–294
Schrödinger representation, 18–19
Hausdorff space, 539
heat kernel,
perturbative expansion, 618–619
role of zero modes, 620–621
use in evaluation of determinants, 613–615, 617–621
Higg's mechanism, 471
holomorphic, 564
holomorphic differential, 599
holomorphic factorization, 644
homeomorphism, 540
homology group, 597
basis, 599, 651
homotopic, 561
hyperbolic metric, 574

image, 539
indiscrete topology, 538
induced world-sheet metric, 492, 505
infrared divergences in QED, 410
injective (1-1), 539
in-out fields,
electromagnetic, 162
fermionic, 159
scalar, 113–114
instantaneous Coulomb energy, 152, 338
integral curves, 592
intersection matrix, 598
intersection product, 598
isothermal, 571, 584, 648, 645
coordinates and complex structure, 584–585

Jacobi ϑ-functions, 666

Killing vector field, 596
conformal, 596
Klein-Gordon,
current, 37
equation, 36
propagator, 53

Lamb shift, 408,417
Landau ghost, 428
Laplace-Beltrami operator, 611
leptons, 5
Levi-Cività connection, 555
Lie derivative, 593
metric tensor, 595, 606
scalar, 594
vector, 594
Lie dragged, 594
linear fractional transformation, 568
Liouville model, 629, 637, 670
quantization in conformal gauge, 672–674
supersymmetric extension, 717–718
loop expansion,
as expansion in powers of $\hbar$, 374
scalar field theory, 374
Lorentz algebra, 57
Lorentz gauge propagator, 91
Lorentz generators,
scalar field, 56
spinor field, 75–76
Lorentz group,
spinorial representations, 691

magnetic moment of electron, 416
Mandelstam variables, 663
manifold, 540
differentiable, 542
Riemannian, 543
Mapping Class group, 587, 600
torus (genus 1), 591
meromorphic, 565
meromorphic differential, 599
mesons, 5
metric,
on space of metrics, 592, 605
metric tensor, 544, 554
flat supersurface, 698
spinorial, 696
microcausality, 47, 72
Möbius transformation, 568
mode expansion,
bosonic string, 507, 508, 510, 517
free electromagnetic field, 83–84, 88

free scalar field, 43–45
free spinor field, 70, 318
functional rep (bosonic string), 520
functional rep (photon), 216
functional rep (scalar), 207
functional rep (spinor), 218–219
modular function, 584
modular group, 578
moduli space, 537, 578, 587, 588
dimension, 596
genus 1 (torus), 581
in bosonic string partition function, 638
of cylinder, 655
one-point compactification, 645
modulus of a basis, 580
momentum operator,
functional representation, 207
scalar field, 56
moving frame field, 559, 689
nonuniqueness, 690

Nambu-Goto action, 492
negative energy problem, 39
nonabelian phase factor, 480
transformation property, 480
Nonlinear Schrödinger model, 30–34
thermodynamic limit, 32
2-Body Sector, 34
nonlinear sigma model, 98
normal ordering, 45, 513, 529
Nöther's theorem, 22

on mass shell, 365
one-particle irreducible diagram,371
operator product expansion, 32, 34
operator-valued distribution, 45

parallel transport, 553
parity, 102
Dirac fields, 103–104
electromagnetic field, 104
scalar fields, 102–103
partition function,
bosonic string, 603, 631, 638, 639, 641, 645, 671
bosonic string (genus 0), 611, 629
bosonic string (genus 1), 639, 641, 643, 644, 645
bosonic string (genus ≥2), 638, 645
perturbative expansion, 631
reduction to integral over moduli, 638
superstring, 708, 717, 718
path integral in qm,
anharmonic oscillator integral, 298–300
constructive def, 271–276
forced harmonic oscillator, 293
free particle integral, 285
harmonic oscillator integral, 289
matrix elements, 280
perturbative expansion, 294–297
recovering diagrams, 296
recovering wave functions, 289, 299
symmetric operator ordering, 277
vacuum expectation values, 279
path-ordered exponential, 480
Pauli-Villars regularization, 382–383
applied to QED chiral anomaly, 426
applied to vacuum polarization, 385
PCAC hypothesis, 421
PCT theorem, 106
period matrix, 599, 645, 651
perturbation theory,
operator representation, 125–132
path integral rep, 294–297, 330, 339
Schrödinger rep, 227
perturbed wave functionals, $\lambda\varphi^4$
from Rayleigh-Schrödinger pert theory, 230–234
direct solution from Schröd. eq., 234–237, 458, 460
perturbed wave functionals, QED
direct solution from Schröd. eq., 257–260
perturbed wave functionals, Yang-Mills
direct solution from Schröd. eq., 247–251
Planck length, 477
Planck mass, 485
Poincaré half-plane, 574
Polyakov action, 492, 604, 629
proper diagrams, 371
proper functions, 373
proper vertex, 373
asymptotic behavior, 467
properly discontinuous, 578

QCD $1/N$ expansion, 680
connection to free strings, 683
double-line representation, 681
QED,
chiral anomaly, 420
Coulomb gauge quantization, 155–158, 336–338

counterterms, 402
Faddeev-Popov procedure, 351
gauge-fixed action, 353
generating functional, 336, 339
operator rep, 155
path integral rep, 336
regularized action, 383
renormalized action, 402
Schrödinger rep, 255
temporal gauge, 255
quantum field theory,
nonrelativistic, 23–27
relativistic, 38–42
quantum mechanics, 11–16
Heisenberg (operator) representation, 15
path integral representation, 271–276
Schrödinger representation, 14

Rayleigh-Schrödinger perturbation theory, 227
applied to $\lambda\varphi^4$, 228–234
failure in gauge theories, 246, 256
reduction formula,
electromagnetic, 162
fermionic, 161
scalar, 124
Regge trajectory, 663
regularization,
dimensional, 382
introduced, 362
momentum cutoff, 266, 450
Pauli-Villars, 382–383
regularized photon propagator, 384
relativistic point particle,
classical action, 487–489
renormalization,
basic idea, 362
of Coulomb's law, 408
of electric charge, 401
of electron mass, 266, 398
of $\lambda\phi^4$ via effective potential, 444, 449–450
of photon propagator, 399, 405
of QED electron propagator, 397, 409–412
of QED vertex, 400, 412
of surface tension in string partition function, 629
renormalization group, 463
renormalization group equations,
for asymptotic behavior of proper vertices, 466, 468
for interacting scalar fields, 465, 470
for massless ϕ^4, 464
use, 465–466
renormalization of electron mass,
from Schrödinger rep, 265–268
renormalization prescription,
for electron propagator, 409
for QED vertex correction, 412
for vacuum polarization, 405
introduced, 368
Ricci tensor, 556
Riemann Mapping theorem, 575
Riemann-Roch theorem, 596
Riemann sphere, 566
Riemann surface, 544
topological classification, 545–546
Riemann tensor, 556
rigid strings,
action, 684
partition function, 684
ϑ term, 684
running coupling constant, 407

S-matrix, 107, 118–120
operator representation, 129, 182
Schrödinger rep, 238
scalar curvature, 556
scalar field theory,
free, 42–48
$\lambda\varphi^4$, 111–112
$\lambda\varphi^4$, Schrödinger rep, 226
Schottky problem, 600
Schrödinger equation,
free photon field theory, 211, 214
free scalar field theory, 201, 209
free spinor field theory, 217, 219
$\lambda\phi^4$, renormalized, 460
Yang-Mills, 244, 252
$\mathrm{sdet}'(-\Delta_n^{(\pm)})$, 717
second fundamental form, 684
second quantization, 20–30
of the nonrelativistic Schrödinger equation, 20, 23–27
relation to first quantization, 27–28
second structural equation, 560, 692
Shapiro-Virasoro model, 519
Siegel modular group, 600
Siegel upper half-plane, 599
simplices, 597
simplicial homology group, 597
smooth covering surface, 575

Sommerfeld-Watson transformation, 642
spectral density, 116, 160
spin and statistics, 71–72
spin connection,
  covariant derivative, 691, 693
  first structural equation, 692
  relation to Christoffel symbols, 691
  second structural equation, 692
  with torsion, 692, 694
spin structure, 719
spinor field theory,
  free, 69–72
spinor metric, 696
spontaneous breakdown of symmetry, 441
  relation to Goldstone bosons, 443
stabilizer, 570
standard homology basis, 599, 651
static quark potential,
  from Nambu-Goto model, 680
  relation to Wilson loop, 677–678
  universal string Coulombic piece, 679
stress-energy tensor, 56
  bosonic string, 494
string scattering amplitudes, 657
  conformal dependence, 660, 670
  duality, 664
  tachyon 1-loop amplitude, 667
  tachyon tree amplitude, 663
  Virasoro-Shapiro amplitude, 663
string susceptibility, 675–676
string theories,
  $\hbar$ not fundamental, 486
  natural units, 477, 486
  spontaneous breakdown of general coordinate invariance, 485
  types, 482–484
structural equations, 559, 560
stuff, 610
superconformal anomaly, 717, 718
  component form, 718
  supersurface form, 717
superconformal gauge, 697
  in terms of superzweibein, 705, 707
superconnection, 701
  in superconformally flat gauge, 705
supercovariant derivative, 699, 701
  in superconformally flat gauge, 705
supercurrent, 695
supercurvature, 701
superdeterminant, 702–703
  regularized, 713
superdiffeomorphism, 698
superficial degree of divergence,
  self-interacting scalar field, 369
  QED, 378
superfield, 698
  component form, 700
  Grassmann expansion, 699
  transformation, 698
super first structural equation, 701
super heat kernel, 715, 716
  perturbative expansion, 715
super Laplacians, 707
  in superconformally flat gauge, 707–708
superstrings,
  Faddeev-Popov procedure, 710–712
  partition function, 708
  partition function (genus 0), 708, 709, 713, 717, 718
  RNS action, 693
  RNS action (Euclidean), 696
  RNS action (supersurface), 702
  RNS gauge-fixed action, 697
  RNS gauge-fixed action (supersurface), 700
super second structural equation, 702
supersurface, 698
  flat metric, 698
  supercoordinate, 698
  superdiffeomorphism, 698
  superfield, 698
supersymmetric Liouville action, 718
supersymmetry,
  generator, 698–699
  global world-sheet transformation, 697
  local world-sheet transformation, 694
  superdiffeomorphism, 698
supertorsion, 701
  constraints, 702
supertrace, 702
super Weyl transformation, 695
superzweibein, 700
  for flat superspace, 703
  for superconformally flat superspace, 705
  inverse, 702
  transformation under superdiffeomorphism, 704
  transformation under super Weyl, 705
surjective (onto), 539

tangent bundle, 543, 551
tangent space, 543
tangent space transformation,
 super generators, 701
Teichmüller space, 578, 588
 dimension, 596
 parameters, 633
 relation to moduli space, 578
 torus, 581–582
tensor product, 558
time-evolution operator, 302
 free Dirac field, 318
 free scalar field, 307, 308
 limit leading to $Z[J]$, 311–312
 path integral rep, 307
 single fermionic mode, 319–326
 with sources, 310
time-ordered products, 130
 generating functional for connected diagrams, $W[J]$, 332, 334, 339
 generating functional for interacting scalar field, 182, 330
 perturbative expansion, 132, 163, 227, 330, 339
time-ordering operator, 53
time reversal, 105
topological space, 538
topology, 538
Torelli group, 600
torsion, 692–694
transition functions, 541
transverse δ-function, 83
triangle diagram, 424

U-matrix, 125
$U(1)$ problem, 422–423
Uniformization theorem, 577
univalent, 564
universal covering surfaces, 577

vacuum polarization,
 and Coulomb's law, 380, 407
 evaluation, 392–394
 physical interpretation, 406
 regularized, 392
vacuum-to-vacuum amplitude, Z,
 bosonic string, 603, 631, 638, 639, 641, 643, 644, 645, 671
 free photon field, 315–317
 free scalar field, 182, 312
 free spinor field, 322, 327
 interacting scalar field, 445
 $\lambda\phi^4$ (1-loop), 455
 $\lambda\phi^4$, perturbative expansion, 330
 loop expansion, 374
 QED, 336
 superstring, 708, 717, 718
 Yang-Mills, 357
vacuum wave functional,
 bosonic string, 521, 522
 bosonic string, regularized, 523
 exact, unnormalizable for Yang-Mills, 252
 free electromagnetic field, 215–216
 free scalar field theory, 201–203, 210, 309
 free spinor field theory, 220, 322
 $\lambda\varphi^4$, 233–234, 458, 460
 Yang-Mills, 247, 251
vector field, 544
Veneziano model, 482
vertex correction,
 and anomalous magnetic moment, 412–417
 evaluated, 395–396
 regularized, 395
 renormalized, 400, 412
vertex operators, 656, 657
Virasoro algebra, 516, 529–530
Virasoro-Shapiro amplitude, 663

Ward-Takahashi identities, 173–174, 403–404, 420
wave function renormalization,
 introduced, 368
 QED, 402
wedge product, 558
Weierstrass normal form, 583
Weil-Petersson metric, 596
Wick rotation, 604
Wick's theorem, 136
Wilson loop, 480, 677, 682
 confinement criterion, 677
 equation of motion, 678
 functional integral approximations, 679
 relation to static quark potential, 677–678
winding number of a gauge transformation, 423

Yang-Mills,
 exact, unnormalizable solution, 252
 Faddeev-Popov procedure, 354–356
 Gauss's law constraint, 245
 $1/N$ expansion, 680–683

perturbed wave functional, 251
Schrödinger rep, 244
temporal gauge, 244
vacuum-to-vacuum amplitude, 357

zeta-function regularization, 615
zweibein, 689
completeness, 689
orthonormality, 689